Annual Review of Biophysics and Biomolecular Structure

Production Editor: CLEO X. RAY
Bibliographic Quality Control: MARY A. GLASS
Color Graphics Coordinator: EMÉ O. AKPABIO
Electronic Content Coordinator: SUZANNE K. MOSES
Subject Indexer: KYRA KITTS

Annual Review of Biophysics and Biomolecular Structure

VOLUME 32, 2003

ROBERT M. STROUD, *Editor*
University of California, San Francisco

WILMA K. OLSON, *Associate Editor*
Rutgers University

MICHAEL P. SHEETZ, *Associate Editor*
Columbia University

www.annualreviews.org science@annualreviews.org 650-493-4400

ANNUAL REVIEWS
4139 El Camino Way • P.O. Box 10139 • Palo Alto, California 94303-0139

ANNUAL REVIEWS
Palo Alto, California, USA

International Standard Serial Number: 1056-8700
International Standard Book Number: 0-8243-1832-3
Library of Congress Catalog Card Number: 79-188446

∞ The paper used in this publication meets the minimum requirements of American National Standards for Information Sciences—Permanence of Paper for Printed Library Materials, ANSI Z39.48-1992.

TYPESET BY TECHBOOKS, FAIRFAX, VA
PRINTED AND BOUND BY MALLOY INCORPORATED, ANN ARBOR, MI

PREFACE

Fifty years. DNA. The perfect 10.

Fifty years ago Rosalind Franklin's diligently acquired fiber diffraction from pulled fibers of DNA under different hydration conditions provided the data that was the last essential key to the determination of the three-dimensional structure of DNA. The resolved simplicity of her most beautiful X-ray pattern, the 34 Å repeat along the fiber, the 3.4 Å between major steps on the ladder, the twofold symmetry of A form DNA, and her measured density of the DNA crystals she had grown that proved that the monoclinic crystal of DNA had only two strands. The model of DNA, brilliantly conceived by Francis Crick and James Watson, took account of all the available evidence and hinged on Chargraff's chemical evidence for equivalent ratios of A to T and G to C bases and the X-ray diffraction of Rosalind Franklin, without need for any further experiment. And it first explained the basis for replication.

After fifty years Rosalind's best "B-form" DNA X-ray pattern is perhaps the most compelling image supporting the nature of the structure of DNA and what it implies for replication. J.D. Bernal in his obituary in *Nature* wrote, "As a scientist Miss Franklin was distinguished by extreme clarity and perfection in everything she undertook. Her photographs are among the most beautiful X-ray photographs of any substance ever taken." Rosalind died April 16, 1958, too early to appreciate the magnitude of the impact that the structure of DNA had on the world of biology.

As the structure of the α-helix and the double helix of DNA began to elaborate the structures of cellular machines and the message for replication, more complex structures and physical measurements lead to unraveling the networks and interactions between proteins, membranes, enzymes, transcription factors, and nucleic acids in the cell. This year's volume brings us close to this synthesis.

This volume begins with Hoofnagle, Resing, and Ahn, who show how mass spectrometry has revolutionized insight into protein-protein and protein-ligand interfaces and conformational changes during protein folding or denaturation. Subirana and Soler-López offer a powerful perspective on electrostatic interactions between ions and DNA, the very subject that Rosalind Franklin had paid so much attention to, arguing that the phosphate groups had to lie on the outside.

Peters considers the emerging optical single transporter recording, which can measure transport kinetics in membrane patches. How the dynamics of the molecule and its environment contribute to catalytic mechanisms is the subject of the review by Daniel, Dunn, Finney, and Smith. Radaev and Sun focus on the structure and function of natural killer cell surface receptors and describe their ligand complexes. Theobald, Mitton-Fry, and Wuttke focus on nucleic acid recognition by OB-fold proteins, a compact structural motif frequently used for nucleic acid recognition.

Site-specific recombination, as explored by Chen and Rice, is seen in the Flp, tyrosine-based family of site-specific recombinases as they excise and invert DNA segments.

"The Power and Prospects of Fluorescence Microscopies and Spectroscopies" are the subject for Michalet, Kapanidis, Laurence, Pinaud, Doose, Pflughoefft, and Weiss, who illuminate the renaissance of fluorescence microscopy techniques and applications, from live-animal multiphoton confocal microscopy to single-molecule fluorescence spectroscopy and imaging in living cells. Garavito and Mulichak take the case of cyclooxygenases that catalyze the committed step in prostaglandin synthesis, as major targets of nonsteroidal antiinflammatory drugs interact with COXs and show how differences in the structure of COX-2 result in enhanced selectivity toward COX-2 inhibitors. Chalikian in a beautifully thorough discussion discusses structural and thermodynamic characterizations of interactions aimed at solving the problems of protein folding and binding. He shows how volumetric techniques can be used to determine protein hydration and intraglobular packing, a relatively untapped way to tackle protein folding and binding.

Golbeck focuses on interactions within the Photosystem I reaction center, 15 years after the first report of crystals. Edidin discusses the determination of the characteristics of "lipid rafts" in membranes. Cartailler and Luecke analyze lipid-protein interactions in the bacteriorhodopsin purple membrane and the special roles of haloarchaeal lipids. They lead up to a discussion of the possibility of bacteriorhodopsin being an inward hydroxide pump versus an outward proton pump. Sixma and Smit describe their most remarkable discovery of the structure of the molluskan acetylcholine binding protein (AChBP), a secreted glial protein that provides a high-resolution model for the entire extracellular domain of pentameric ligand-gated ion channels. These include the nicotinic acetylcholine receptors, $GABA_A$, $5HT_3$ serotonin and glycine, which can be interpreted in the light of the 2.7 Å AChBP structure. The structural template provides critical details of the binding site and helps create models for toxin binding, mutational effects, and molecular gating.

In a scholarly history of the ideas and concepts that defined molecular recognition and docking, Brooijmans and Kuntz summarize the history of the key test of how well we actually understand molecular interactions. The structure of the first G protein–coupled receptor, rhodopsin, has had a revolutionary impact on predicted structures of other family members, as Filipek, Teller, Palczewski, and Stenkamp analyze. Proteome analysis by mass spectrometry is taken on by Ferguson and Smith, who review recent progress in the application of mass spectrometry–based techniques for the qualitative and quantitative analysis of global proteome samples. Warshel leads us into computer simulations of enzyme catalysis, which have ripened to the growth cone of knowledge of how enzymes function, with some dramatic recent contributions from experimental and theoretical enzymologists that include Tom Bruice and Steve Benkovic.

Another dramatic impact is made by the structure of the calcium pump, beautifully linked to its function by Stokes and Green. The mechanism by which active transport of cations by the large family of such ATP-dependent ion pumps is carried out is the reward. McConnell and Vrljic take on "condensed complexes" formed between cholesterol and some but not all phospholipids. The properties of defined cholesterol-phospholipid mixtures provide a conceptual foundation for several aspects of the biophysics of animal cell membranes.

I applaud the scholarship and insight, the care and attention that contributing authors, especially the Associate Editors and the Editorial Committee, bring to these critical Reviews. Especially crucial is Cleo X. Ray, as the production editor, coordinator, and link between authors, editors, and policy, for her diligence, good humor, and professional excellence. Our goal remains to coordinate reviews of mature fields at lasting landmark stages, written by primary contributors, stepping stones to the future. May each fifty-year interval bring more perfect 10s!

Robert M. Stroud
Editor

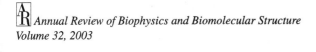

Annual Review of Biophysics and Biomolecular Structure
Volume 32, 2003

Contents

ERRATA

An online log of corrections to *Annual Review of Biophysics and Biomolecular Structure* chapters may be found at http://biophys.annualreviews.org/errata.shtml

Related Articles

From the ***Annual Review of Biochemistry***, Volume 72, 2003

Challenges in Enzyme Mechanism and Energetics, Daniel A. Kraut, Kate S. Carroll, and Daniel Herschlag

Covalent Trapping of Protein-DNA Complexes, Gregory L. Verdine and Derek P.G. Norman

Dynamics of Cell Surface Molecules During T Cell Recognition, Mark M. Davis, Michelle Krogsgaard, Johannes B. Huppa, Cenk Sumen, Marco A. Purbhoo, Darrell J. Irvine, Lawren C. Wu, and Lauren Ehrlich

Proteomics, Heng Zhu, Metin Bilgin, and Michael Snyder

Semisynthesis of Proteins by Expressed Protein Ligation, Tom W. Muir

Signals for Sorting of Transmembrane Proteins to Endosomes and Lysosomes, Juan S. Bonifacino and Linton M. Traub

The Dynamics of Chromosome Organization and Gene Regulation, David L. Spector

The Establishment, Inheritance, and Function of Silenced Chromatin in Saccharomyces cerevisiae, Laura N. Rusche, Ann L. Kirchmaier, and Jasper Rine

The Many Faces of Vitamin B_{12}: Catalysis by Cobalamin-Dependent Enzymes, Ruma Banerjee and Stephen W. Ragsdale

The Structural Basis of Large Ribosomal Subunit Function, Peter B. Moore and Thomas A. Steitz

TRK Receptors: Roles in Neuronal Signal Transduction, Eric J. Huang and Louis F. Reichardt

The Rotary Motor of Bacterial Flagella, Howard C. Berg

Function and Structure of Complex II of the Respiratory Chain, Gary Cecchini

Protein Disulfide Bond Formation in Prokaryotes, Hiroshi Kadokura, Federico Katzen, and Jon Beckwith

From the ***Annual Review of Cell and Developmental Biology***, Volume 18, 2002

A Cell Biological Perspective on Alzheimer's Disease, Wim Annaert and Bart De Strooper

Type III Protein Secretion in Yersinia *Species*, Kumaran S. Ramamurthi and Olaf Schneewind

Break Ins and Break Outs: Viral Interactions with the Cytoskeleton of Mammalian Cells, Gregory A. Smith and Lynn W. Enquist

Receptor Kinase Signaling in Plant Development, Philip Becraft

Chromosome-Microtubule Interactions During Mitosis, J. Richard McIntosh, Ekaterina L. Grishchuk, and Robert R. West

Cellular Control of Actin Nucleation, Matthew D. Welch and R. Dyche Mullins

Membrane Fusion in Eukaryotic Cells, Andreas Mayer

Actin Cytoskeleton Regulation in Neuronal Morphogenesis and Structural Plasticity, Liqun Luo

Striated Muscle Cytoarchitecture: An Intricate Web of Form and Function, Kathleen A. Clark, Abigail S. McElhinny, Mary C. Beckerle, and Carol C. Gregorio

From the ***Annual Review of Genetics***, Volume 36, 2002

DNA Topology-Mediated Control of Global Gene Expression in Escherichia coli, G. Wesley Hatfield and Craig J. Benham

Xist RNA and the Mechanism of X Chromosome Inactivation, Kathrin Plath, Susanna Mlynarczyk-Evans, Dmitri A. Nusinow, and Barbara Panning

Genetics of Influenza Viruses, David A. Steinhauer and John J. Skehel

Allosteric Cascade of Spliceosome Activation, David A. Brow

Genetics of Sensory Mechanotransduction, Glen G. Ernstrom and Martin Chalfie

The Genetics of RNA Silencing, Marcel Tijsterman, René F. Ketting, and Ronald H.A. Plasterk

Toward Maintaining the Genome: DNA Damage and Replication Checkpoints, Kara A. Nyberg, Rhett J. Michelson, Charles W. Putnam, and Ted A. Weinert

From the ***Annual Review of Pharamacology and Toxicology***, Volume 43, 2003

Induction of Drug-Metabolizing Enzymes: A Path to the Discovery of Multiple Cytochromes P450, Allan H. Conney

Protein Flexibility and Computer-Aided Drug Design, Chung F. Wong and J. Andrew McCammon

Signal Transduction–Directed Cancer Treatments, Edward A. Sausville, Yusri Elsayed, Manish Monga, and George Kim

Monoamine Transporters: From Genes to Behavior, Raul R. Gainetdinov and Marc G. Caron

Trafficking of NMDA Receptors, Robert J. Wenthold, Kate Prybylowski, Steve Standley, Nathalie Sans, and Ronald S. Petralia

Telomere Inhibition and Telomere Disruption as Processes for Drug Targeting, Evonne M. Rezler, David J. Bearss, and Laurence H. Hurley

Pharmacology and Physiology of Human Adrenergic Receptor Polymorphisms, Kersten M. Small, Dennis W. McGraw, and Stephen B. Liggett

Gene Therapy with Viral Vectors, Neeltje A. Kootstra and Inder M. Verma

K+ Channel Structure-Activity Relationships and Mechanisms of Drug-Induced QT Prolongation, Colleen E. Clancy, Junko Kurokawa, Michihiro Tateyama, Xander H.T. Wehrens, and Robert S. Kass

Alzheimer's Disease: Molecular Understanding Predicts Amyloid-Based Therapeutics, Dennis J. Selkoe and Dale Schenk

The Role of Drug Transporters at the Blood-Brain Barrier, A.G. de Boer, I.C.J. van der Sandt, and P.J. Gaillard

From the ***Annual Review of Physical Chemistry***, Volume 53, 2002

Physical Chemistry of Nucleic Acids, Ignacio Tinoco, Jr

Scanning Tunneling Microscopy Studies of the One-Dimensional Electronic Properties of Single-Walled Carbon Nanotubes, Min Ouyang, Jin-Lin Huang, and Charles M. Lieber

Chemical Shifts in Amino Acids, Peptides, and Proteins: From Quantum Chemistry to Drug Design, Eric Oldfield

Molecular Theory of Hydrophobic Effects: "She is Too Mean to Have Her Name Repeated," Lawrence R. Pratt

From the ***Annual Review of Physiology***, Volume 65, 2003

Permeation and Selectivity in Calcium Channels, William A. Sather and Edwin W. McCleskey

Processive and Nonprocessive Models of Kinesin Movement, Sharyn A. Endow and Douglas S. Barker

Application of Microarray Technology in Environmental and Comparative Physiology, Andrew Y. Gracey and Andrew R. Cossins

Nuclear Receptors and the Control of Metabolism, Gordon A. Francis, Elisabeth Fayard, Frédéric Picard, and Johan Auwerx

Insulin Receptor Knockout Mice, Tadahiro Kitamura, C. Ronald Kahn, and Domenico Accili

The Gastric Biology of Helicobacter pylori, George Sachs, David L. Weeks, Klaus Melchers, and David R. Scott

In Vivo NMR Studies of the Glutamate Neurotransmitter Flux and Neuroenergetics: Implications for Brain Function, Douglas L. Rothman, Kevin L. Behar, Fahmeed Hyder, and Robert G. Shulman

Hyperpolarization-Activated Cation Currents: From Molecules to Physiological Function, Richard B. Robinson and Steven A. Siegelbaum

Mammalian Urea Transporters, Jeff M. Sands

Getting Ready for the Decade of the Lipids, Donald W. Hilgemann

Regulation of TRP Channels via Lipid Second Messengers, Roger C. Hardie

Phosphoinositide Regulation of the Actin Cytoskeleton, Helen L. Yin and Paul A. Janmey

Structure and Mechanism of Na, K-ATPase: Functional Sites and Their Interactions, Peter L. Jorgensen, Kjell O. Håkansson, and Steven J. Karlish

G Protein–Coupled Receptor Rhodopsin: A Prospectus, Slawomir Filipek, Ronald E. Stenkamp, David C. Teller, and Krzysztof Palczewski

ANNUAL REVIEWS is a nonprofit scientific publisher established to promote the advancement of the sciences. Beginning in 1932 with the *Annual Review of Biochemistry*, the Company has pursued as its principal function the publication of high-quality, reasonably priced *Annual Review* volumes. The volumes are organized by Editors and Editorial Committees who invite qualified authors to contribute critical articles reviewing significant developments within each major discipline. The Editor-in-Chief invites those interested in serving as future Editorial Committee members to communicate directly with him. Annual Reviews is administered by a Board of Directors, whose members serve without compensation.

Annual Reviews of

Anthropology
Astronomy and Astrophysics
Biochemistry
Biomedical Engineering
Biophysics and Biomolecular Structure
Cell and Developmental Biology
Earth and Planetary Sciences
Ecology and Systematics
Entomology
Environment and Resources
Fluid Mechanics
Genetics
Genomics and Human Genetics
Immunology
Materials Research
Medicine
Microbiology
Neuroscience
Nuclear and Particle Science
Nutrition
Pharmacology and Toxicology
Physical Chemistry
Physiology
Phytopathology
Plant Biology
Political Science
Psychology
Public Health
Sociology

SPECIAL PUBLICATIONS
Excitement and Fascination of Science, Vols. 1, 2, 3, and 4

Annu. Rev. Biophys. Biomol. Struct. 2003. 32:1–25
doi: 10.1146/annurev.biophys.32.110601.142417

First published online as a Review in Advance on February 18, 2003

Protein Analysis by Hydrogen Exchange Mass Spectrometry

Andrew N. Hoofnagle,[1] Katheryn A. Resing,[1] and Natalie G. Ahn[1,2]

[1]Department of Chemistry and Biochemistry and [2]Howard Hughes Medical Institute, University of Colorado, Boulder, Colorado 80309; email: Andrew.Hoofnagle@uchsc.edu; Katheryn.Resing@colorado.edu; Natalie.Ahn@colorado.edu

Key Words protein dynamics, folding, allostery, electrospray ionization, matrix-assisted laser desorption ionization

■ **Abstract** Mass spectrometry has provided a powerful method for monitoring hydrogen exchange of protein backbone amides with deuterium from solvent. In comparison to popular NMR approaches, mass spectrometry has the advantages of higher sensitivity, wider coverage of sequence, and the ability to analyze larger proteins. Proteolytic fragmentation of proteins following the exchange reaction provides moderate structural resolution, in some cases enabling measurements from single amides. The technique has provided new insight into protein-protein and protein-ligand interfaces, as well as conformational changes during protein folding or denaturation. In addition, recent studies illustrate the utility of hydrogen exchange mass spectrometry toward detecting protein motions relevant to allostery, covalent modifications, and enzyme function.

CONTENTS

1056-8700/03/0609-0001$14.00

INTRODUCTION

Hydrogen exchange at protein backbone amides was first analyzed by scintillation counting, infrared and ultraviolet spectroscopies, neutron diffraction, and nuclear magnetic resonance (NMR) spectroscopy. Recent advances in mass spectrometry (MS) allow increased sensitivity and the ability to analyze larger proteins and protein complexes than currently possible with NMR, although generally at the cost of reduced structural resolution. This review surveys the theory of hydrogen exchange, the methods for hydrogen exchange mass spectrometry (HX-MS), and the application to various biophysical problems, including folding and conformational changes. Special attention is paid to new insights into protein dynamics provided by HX-MS.

THEORY

In short peptides, amide hydrogen exchange involves proton abstraction described by a chemical exchange rate (k_{ch}) for a second-order reaction that depends on an "intrinsic" rate of exchange for that hydrogen (k_{int}) as well as the concentration of available catalyst, including OH^-, H_3O^+, water, and acidic or basic solutes ($k_{ch} = k_{int}$ [catalyst]) (Figure 1*a*). The chemical exchange rate is minimal near pH_{read} 2.5. Below this pH, exchange occurs via proton addition, catalyzed by D_3O^+. Above this pH, exchange occurs by proton abstraction predominantly catalyzed by OH^-. Because chemical exchange rates of amide deuterium and tritium are slower than hydrogen, with little solvent isotope effect, proton abstraction is rate limiting in reactions above pH_{read} 2.5 (9, 16). Importantly, the chemical exchange rate is also influenced by the amino acid sequence surrounding an amide hydrogen in two ways. First, the intrinsic rate of exchange for that hydrogen (k_{int}) depends on local inductive effects of adjacent side chains that alter the pKa of the hydrogen atom. Second, the local concentration of available catalyst can be altered by the presence of adjacent reactive side chain groups. The inductive, catalytic, and steric effects of adjacent residues on amide chemical hydrogen exchange rates in peptides have been elegantly quantified by Bai et al. (8), whose analysis allows rapid calculation of chemical exchange rates of peptide amide hydrogen atoms based on sequence.

Although chemical exchange occurs rapidly for amide hydrogens in peptides at neutral pH ($k_{ch} \sim 10^1 - 10^3$ sec^{-1}), observed exchange of backbone amide hydrogens in proteins can occur much more slowly, with half lives ranging from milliseconds to years. The exchange rate of a given proton depends on two factors. The most important factor is the degree of solvent protection and hydrogen bonding within the protein. In general, hydrogen exchange rates are slower when protons are removed from the solvent-protein interface and when more stable hydrogen bond contacts are made with surrounding residues in the secondary and tertiary structure (62).

Because higher-order structure has such a profound role, hydrogen exchange is markedly affected by protein flexibility and mobility (9, 27, 38). Even protons that a

$$A\text{-}H + OD^- \xrightarrow{k_{int}[OD^-]} A^- + H\text{-}OD$$

$$A^- + D_3O^+ \longrightarrow D_2O + A\text{-}D$$

b

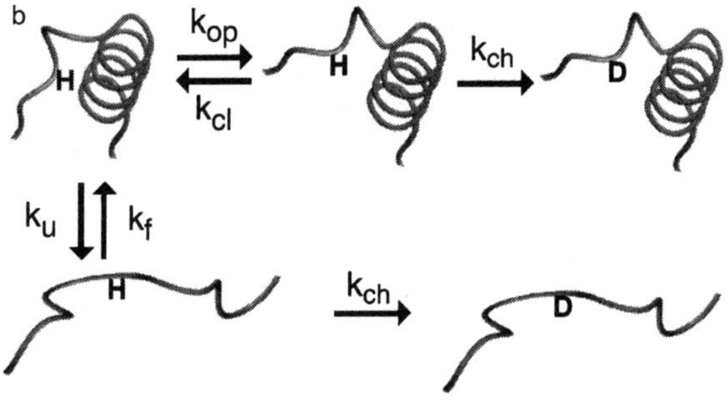

c

$$k_{obs} = \frac{k_{op} \cdot k_{ch}}{k_{cl} + k_{ch} + k_{op}} \approx \frac{k_{op} \cdot k_{ch}}{k_{cl} + k_{ch}}$$

$$\text{(EX1)} \quad k_{cl} << k_{ch} \; ; \qquad k_{obs} = k_{op}$$

$$\text{(EX2)} \quad k_{cl} >> k_{ch} \; ; \quad k_{obs} = \frac{k_{op}}{k_{cl}} \cdot k_{ch}$$

Figure 1 Mechanism of hydrogen exchange. (*a*) Amide exchange at neutral pH involves base catalyzed proton abstraction and acid catalyzed transfer of deuterium from solvent. Measurable isotope effects on the amide hydrogen and a lack of a solvent isotope effect indicate that proton abstraction is rate limiting. (*b*) Hydrogen exchange of a buried amide is facilitated by different mechanisms, involving small-amplitude fluctuations (*upper pathway*) on one extreme, and complete unfolding (*lower pathway*) on the other. The observed rate of exchange (k_{obs}) for small-amplitude fluctuations is a function of the rate of structural opening (k_{op}), the rate of structural closing (k_{cl}), and the chemical rate of exchange ($k_{ch} = k_{int}$ [catalyst]), where catalyst is OH^- or buffer. In native proteins, the rate of opening is assumed to be much slower than the rate of closing, which results in a simplified rate expression (upper equation, *far right*). The observed rates of small-amplitude fluctuations lie on a continuum described by EX1 and EX2 conditions, as described in the text.

are highly buried or hydrogen bonded can exchange through fluctuations in the molecule that allow transient solvent penetration. The amplitude of these fluctuations can be small enough to involve the breaking of a single hydrogen bond or large enough to involve complete unfolding of the protein (4, 15, 69). Many hydrogens exchange by mechanisms involving small-amplitude fluctuations and may include low-energy explorations of conformational space as well as higher-energy local unfolding events. These can be described by equilibria between solvent-exposed versus solvent-protected states, governed by rate constants for opening and closing (k_{op}, k_{cl}) (Figure 1*b*, upper pathway). On the other hand, hydrogens buried in the middle of large stable protein domains exchange by mechanisms involving complete unfolding (Figure 1*b*, lower pathway), which are enhanced in the presence of heat or denaturants (7, 26).

The small-amplitude fluctuations that convert solvent-protected hydrogens to solvent-exposed (Figure 1*b*, upper pathway) are assumed to be completely reversible (6, 15, 27, 38). They represent a continuum of hydrogen exchange mechanisms: At one end of the continuum (termed the EX1 regime), chemical exchange occurs quickly after conversion to the solvent-exposed form, and the observed rate (k_{obs}) can be described by the rate of structural opening ($k_{obs} = k_{op}$) (15, 27) (Figure 1*c*). These motions can be described as local unfolding events and occur on timescales of milliseconds to days. In general, local unfolding involves many residues of the protein and leads to simultaneous solvent exposure of many amides (4, 28, 69). At the other end (termed EX2), reconversion of the solvent-exposed form back to the protected form occurs much faster than the rate of chemical exchange. These motions can be described as native state fluctuations or protein breathing motions, assumed to occur on timescales of microseconds to milliseconds. In this extreme, k_{obs} depends on the equilibrium of protected and exposed forms and on the chemical exchange rate ($k_{obs} = k_{op}/k_{cl} \cdot k_{ch}$) (15, 27) (Figure 1*c*). Thus, hydrogen exchange measurements reveal information about folding as well as internal motions of the folded state.

Because the chemical exchange rate is proportional to hydroxide ion concentration (Figure 1*a*), the pH dependence of observed hydrogen exchange rates reveals where protein motions reside on the continuum between EX1 and EX2. In the EX1 regime, k_{obs} is independent of chemical exchange, and in most cases complete pH independence will be observed provided that protein structure and the opening and closing rates are not affected by pH (Figure 1*b*,*c*). On the other hand, k_{obs} is strongly pH dependent in the EX2 regime because the observed rate is directly proportional to the chemical exchange rate (Figure 1*b*,*c*). For native state proteins, experimental evidence based on the pH dependence of k_{obs} confirms the predominance of small-amplitude protein fluctuations in the mechanism of exchange (9, 15, 27, 38). An important advantage of MS as an analytical tool for the measurement of hydrogen exchange is that EX1 and EX2 motions can be distinguished by examining the distribution of mass spectral peaks.

In native state hydrogen exchange experiments in the EX2 regime, the degree of protection of individual hydrogens can be quantified. As described by Bai et al. (6), the ratio of the chemical exchange rate to the observed exchange rate provides a measure of the equilibrium constant describing the distribution of open versus closed states in solution ($k_{ch}/k_{obs} \approx k_{cl}/k_{op} = 1/K_{op}$). This ratio is termed the "protection factor" (P) and is proportional to ΔG_{op}, the thermodynamic barrier over which a protein structure must cross to enable solvent exposure and subsequent hydrogen exchange (6, 7, 15). Log P typically ranges from 2 to

9 in native state proteins, suggesting motions with free energy barriers of 2–12 kcal mol^{-1}. This interpretation is an approximation that assumes that constant chemical exchange rates (k_{ch}) for hydrogens are determined solely by the primary structure and the concentration of available catalyst. However, local side chains in the three-dimensional microenvironment of the hydrogen may alter the chemical exchange rate to values that cannot be quantified easily using model compounds or peptides (47, 79, 80). Nevertheless, the approximation indicates that free energies of protein motions that lead to hydrogen exchange are consistent with low-energy fluctuations, hydrogen bond disruptions, and local unfolding events.

Hydrogen exchange studies of native state proteins are used to explore conformational properties of folded proteins. For example, hydrogen exchange rates are commonly used in NMR determinations of protein structure, where very slowly exchanging amide hydrogens are assumed to be hydrogen bonded within regions of secondary structure. Such information can be included as constraints in simulated annealing protocols for structure calculations. Another interpretation of native state exchange experiments is that the slowest exchanging amides, which typically form a core near the center of the molecule, constitute a folding core and potential nucleation site for secondary structure formation on the protein folding pathway (47, 79, 80). This has been supported by Φ analysis experiments, which characterize effects of mutations on in vitro folding rates (43). In general, the slowest exchanging amide hydrogens are observed on residues with the highest Φ values, which are equated with residues that nucleate first to form the folding core. This suggests that native state exchange measurements in some cases may lend insight into folding pathways.

In contrast to studies with native proteins, hydrogen exchange measurements have been used to more directly examine protein folding and unfolding mechanisms, utilizing equilibrium and pre-equilibrium approaches. We mention this broad area only briefly, as detailed reviews can be found elsewhere (15, 26, 30). Three key hydrogen exchange strategies have been used in folding studies. First, the Baldwin laboratory labeled denatured ribonuclease A with tritium and then measured exchange-out of radiolabel during folding. The results showed significant protection of backbone amides from exchange prior to complete folding of ribonuclease A, as monitored by tyrosine fluorescence measurements (67), demonstrating the presence of at least one partially structured intermediate in the folding reaction of ribonuclease A. Second, an NMR technique reported simultaneously by the Baldwin and Englander laboratories involved complete deuteration of unfolded protein at low pH, initiation of folding followed by pulse labeling with H_2O at increased pH, and quenching by decreasing pH. The results provided unequivocal support for the existence of intermediate states in the folding reactions of ribonuclease A (71) and cytochrome *c* (64). Third, a method developed by the Englander laboratory (7) subsequently identified cooperative secondary structural elements during unfolding of cytochrome *c* by measuring hydrogen exchange rates titrated over a range of low concentrations of denaturant. This important finding established firm support for a stepwise folding pathway, in contrast to a folding energy landscape with shallow minima.

HISTORICAL PERSPECTIVE

Measurements of the incorporation of deuterium and tritium into protein molecules have been performed for more than 40 years. Lenormant & Blout were among the first to report the process of amide hydrogen exchange in their attempts to assign the 1550 cm^{-1} band in infrared spectra of protein solutions, demonstrating decreased absorbance at this wavelength upon deuterium incorporation at high pH (46). This band has been subsequently assigned to N-H bond bending. Since then, hydrogen exchange has been measured by a variety of methods. Englander (25) carried out pioneering studies using tritium and scintillation counting to measure isotope exchange into full-length proteins. Ultraviolet spectroscopy and neutron diffraction have also been used to determine exchange rates (10, 23, 54). The application and interpretation of these techniques, as well as early experiments with one-dimensional NMR, are reviewed by Barksdale & Rosenberg (9). Because deuterium is an NMR-inactive isotope, reduced areas under NMR proton absorbance peaks are used to monitor exchange of protons for deuterons at individual amides. However, the disadvantages of NMR approaches are that the experiments require large amounts of protein and that assignment of spectral peaks can be arduous. NMR techniques to analyze hydrogen exchange rates were refined by Englander (26, 29, 30), who made substantial contributions to the fields of hydrogen exchange and protein folding and reviewed these approaches.

In 1990, Chowdhury et al. (14) demonstrated the use of electrospray ionization mass spectrometry (ESI-MS) to probe the conformational distribution of cytochrome *c* in solution (14). Because MS measures mass/charge, a protein ESI-MS spectrum consists of a number of differently charged ions. The distribution of this protein charge state envelope reflects the exposure of ionizable side chains to solvent, in that greater numbers of basic side chains protonated in solution will widen the charge state distribution in the gas phase. Thus, it was shown that unfolded cytochrome *c* at low pH has a wider charge state distribution than native cytochrome *c* at higher pH. Less than a year later, Katta & Chait (42) published the first use of MS to analyze hydrogen exchange rates. Using ESI-MS, they quantified the incorporation of deuterium into bovine ubiquitin, noting that some amides remained nonexchanging even after days under denaturing conditions.

With the exception of multidimensional NMR pulse techniques, methods for measuring hydrogen exchange were limited to low resolution until Rosa & Richards (65) reported protein fragmentation by proteolysis for measuring tritium incorporation into localized regions of ribonuclease S. The approach involved complete exchange of all polar protons with tritium, followed by measurement of tritium loss through back-exchange to water at different pH values. At varying times, the reactions were quenched at pH 2.8 and the proteins digested with pepsin, generating at least six different peptides that were separated by reversed-phase high-performance liquid chromatography (HPLC). Exchange rates for different peptides were then quantified as the loss of tritium over time. Also presented in their report was the first discussion using this medium resolution technique to identify sites of protein-peptide interactions. Later, Englander et al. (24) improved the original protocol by decreasing the temperature of the HPLC separation of peptic peptides, which minimized further back-exchange of tritium radiolabel after quenching.

Zhang & Smith (84) first reported success using protein fragment separation in HX-MS, employing fast atom-bombardment (FAB) for ionization of peptides from cytochrome *c*. The FAB-MS approach suffered from the drawback that the number of peptides recovered from a hydrogen exchange experiment was only 59%, much lower than needed for complete coverage of backbone amides. Johnson & Walsh (40) first employed ESI-MS with HPLC fragment separation in hydrogen exchange analyses. This improved amide coverage to 89% and demonstrated regions of horse skeletal muscle myoglobin, which were stabilized by heme binding in vitro (40). The short amount of time it took to perform this experiment, involving hours for data collection, combined with the superior coverage and micromolar quantities of protein needed for the entire experiment, made hydrogen exchange ESI-MS (HX-ESI-MS) a rapid, easy technique for analyzing protein structure and dynamics. Although the back-exchange in this experiment reached ~50%, further improvements in the MS protocol have reduced the back-exchange to 10%–20% (36, 61), in agreement with empirical measurements on model peptides (8).

Improvements in ionization methods have been accompanied by improvements in the resolution of mass spectrometers. The development of orthogonal time-of-flight mass analyzers has permitted HX-ESI-MS studies with mass resolution <5 ppm, a significant increase over the 200 ppm resolution of quadrupole and ion trap instruments (13). Sample data is presented in Figure 2. Fourier transform ion cyclotron resonance (FT-ICR) mass spectrometers provide even higher resolution of ~1 ppm (68). These instruments enable detection of lower abundance peptides in digests, providing greater coverage and peptide overlap of sequences. FT-ICR mass spectrometers also allow gas phase fragmentation of full-length proteins as they enter the orifice of the instrument, replacing fragmentation by pepsin digestion (1, 11, 31, 41). This technique, alternatively called capillary-skimmer dissociation or nozzle-skimmer dissociation, is nearly identical in theory (although more efficient in practice) to source fragmentation by non-FT-ICR ESI-MS, where low-energy collisions with gas molecules excite protein molecules to break down into smaller fragments. Protein fragmentation in the absence of the proteolysis step has the potential advantage of eliminating back-exchange from HX-MS experiments.

A mass spectral technique recently developed in the Komives laboratory to analyze hydrogen exchange is matrix-assisted laser desorption ionization (MALDI) MS (50). HX-MALDI-MS has drawbacks of high back-exchange (30%–40%) and lower coverage than ESI-MS due to less efficient ionization and overlap of peaks in complex spectra. Nevertheless, the method removes the HPLC step, which greatly speeds the rate of data collection. The first experiments using this technique identified binding sites for protein kinase inhibitor (PKI) peptide and ATP in cAMP-dependent protein kinase (PKA), and the site for thrombomodulin

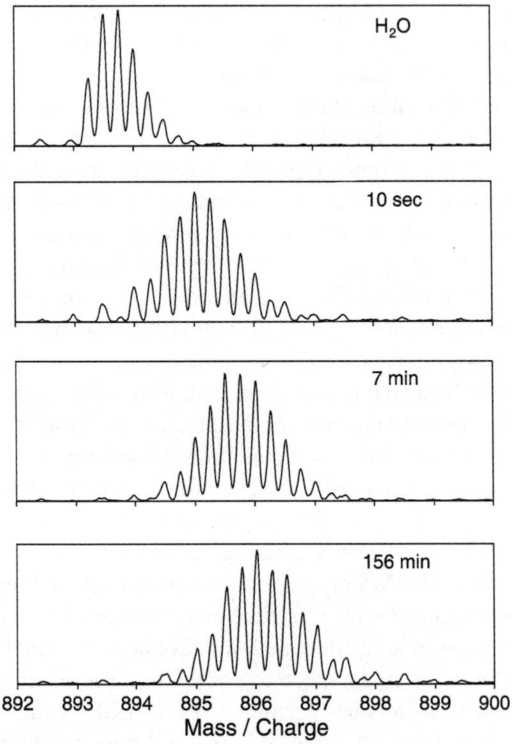

Figure 2 Example of hydrogen exchange mass spectrometric data. The mass spectrum of a MH_4^{+4} peptide ion (undeuterated monoisotopic mass = 3568.99 Da) recovered from pepsin digestion of p38α MAPK is shown. Incubation times of protein with D_2O are indicated. Isotopic resolution is achieved using a quadrupole orthogonal time-of-flight mass spectrometer and electrospray ionization. Weighted average masses are calculated by dividing the sum of the products of the mass and intensity of each isotopic peak by the total intensity for the ion.

binding in thrombin (48, 49). In both cases, the results agreed with X-ray structural studies of co-crystallized complexes. The simplicity of data acquisition and strict agreement of the hydrogen exchange data with the available structural data make pepsin fragmentation HX-MALDI-MS an easy method for identifying putative ligand binding sites and protein-protein interfaces at medium resolution.

TECHNIQUES

In this section, we describe detailed methods for hydrogen exchange analysis by ESI-MS. Two labeling strategies are commonly used in this experiment. The general scheme for exchange-in reactions involves dissolving lyophilized protein or diluting concentrated protein solutions into D_2O and measuring increased mass

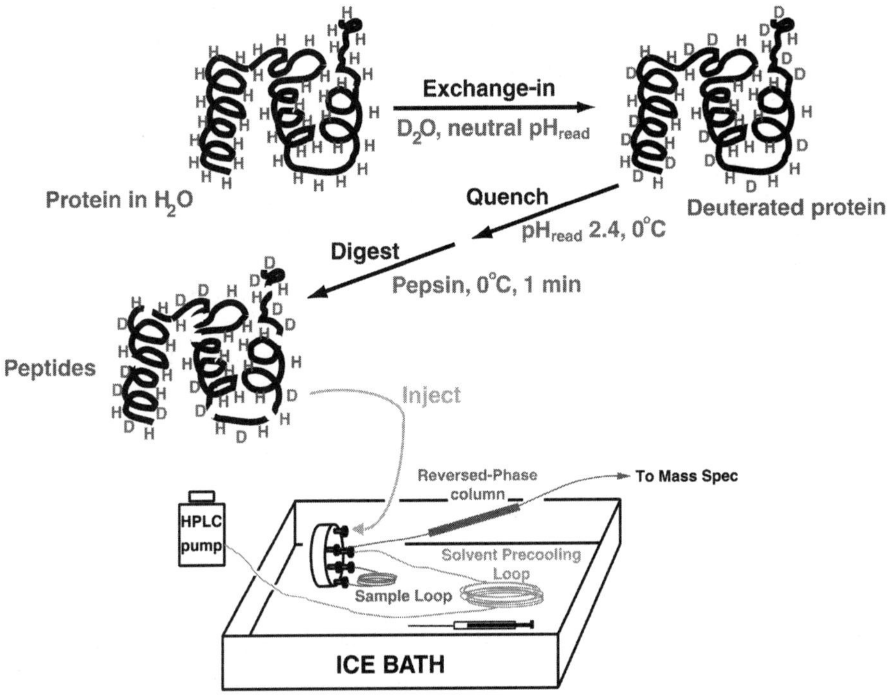

Figure 3 Apparatus for HX-MS by fragment separation. Proteins are incubated with D_2O, quenched in acid and lower temperature, and digested with pepsin. Peptides are separated with reversed-phase HPLC prior to analysis by ESI-MS. Parts of the HPLC, including the injection syringe, solvent precooling loop, sample loop, injector, and capillary column, are all immersed in ice to minimize back-exchange.

to follow deuterium incorporation (Figures 2 and 3). Alternatively, exchange-out reactions can be performed by fully deuterating the sample and pulse labeling with H_2O, measuring the exchange of protein-bound deuterium for solvent hydrogen. In binding studies performed by MALDI-MS described above, a brief exchange-in period was followed by varying times of exchange-out (48–50). This had the advantage of reducing line broadening due to deuteration of slowly exchanging amides and focusing on the rapidly exchanging amides, which typically occur on the surface of the protein and are most relevant to binding interfaces. Exchange-in and exchange-out methods are similar, and exchange-in protocols are easily adapted to measure exchange-out reactions.

Data Collection

For simplicity, we discuss our protocols used to monitor exchange-in reactions by HPLC-MS (61, 62) (Figure 3). The reactions are initiated by the addition of nine volumes of D_2O to a concentrated protein sample in buffer ($pH_{read} \sim 7.4$).

After varying times of exchange-in, the reaction is quenched by rapidly lowering the temperature (to 0°C) and decreasing pH_{read} to 2.4 by addition of 1:1 (v/v) citrate/succinate buffer. The protein is digested with an acid-stable protease; in most cases, pepsin is added to a final level of 1:1 (w/w) protein:pepsin.

Pepsin is a nonspecific protease that cleaves preferentially at hydrophobic residues and has some degree of secondary and tertiary structural specificity (62). Thus, a partially unfolded protein will show a different pepsin cleavage pattern than the same protein in its native state (78). Due to this unusual specificity, pepsin generates many peptides with overlapping sequence, which is of great advantage in increasing coverage of amide hydrogens in the experiment. When used with ESI-MS, we have achieved 95%–100% coverage of amide hydrogens for 42-kDa proteins. Overlapping peptides also increase the structural resolution. However, the use of pepsin in the HX-MS experiment requires peptide sequencing in order to identify the cleavage sites. Digests are analyzed by tandem mass spectrometry (MS/MS), and peptides are identified by de novo sequencing in combination with accurate peptide mass measurements. With current instrument and analysis software, it is possible to identify the constituent peptides of large proteins in a few hours. Pepsin can be used in solution or immobilized on a solid support. The latter was suggested by Rosa & Richards (65) in their original fragment separation paper and was demonstrated for HX-ESI-MS by the Smith laboratory (77).

In order to minimize sample handling (which impacts back-exchange as well as artifactual in-exchange), the sample is loaded into the sample loop immediately after addition of pepsin. Peptides are then injected onto a reversed-phase HPLC column in 0.05% trifluoroacetic acid (TFA) for 4 min, during which time further proteolysis occurs. Most commonly, C_{18} or C_4 reversed-phase resins are used to separate peptides, which are eluted using a gradient of acetonitrile. We use fused silica capillary columns, which can be prepared in-house at low cost (60). The entire apparatus, which includes the injector, sample loop, and column, is tightly packed in ice (Figure 3). In addition, the solvent passes through a precooling loop as it exits the pump and before entering the injector. After the binding step, the column is washed to remove salts, before diverting flow to the mass spectrometer in order to minimize problems with ionization and contamination of the orifice.

Because side chain deuterons back-exchange for hydrogen during the wash step, only deuterons at the backbone amides are retained. The maximum number of observable amides equals the number of residues in the peptide minus one, less the number of proline residues. Back-exchange of the amide hydrogen at the second amide position can also be accelerated (78). In our experience, this is not universally observed in every peptide, most likely because our protocol has been optimized to minimize back-exchange.

Following washing, peptides are eluted using a gradient of acetonitrile in 0.05% TFA. The gradient used depends on the size of the protein and complexity of the digest, and in general should be steep enough to elute the peptides quickly in order to minimize back-exchange, yet shallow enough to adequately separate the peptides. Use of a step gradient of acetonitrile loaded into the sample loop of the injector, together with isocratic solvent delivery of 0.05% TFA/H_2O, reduces the dead time between the end of the wash step and the beginning of the gradient and also permits the incorporation of the solvent precooling loop (Figure 3). Generally, elution of the last peptide can be achieved within 20 min after quenching.

Rigorous attempts to automate the LC-MS analysis of hydrogen exchange reactions have also proven successful. Using an autosampler equipped with a dry ice cooling system, Woods & Hamuro (78) have automated the analysis of quenched, frozen reactions by HPLC-ESI-MS using a quadrupole ion trap mass spectrometer, which allows data collection at a rate of 20–50 hydrogen exchange reactions per day. Ghaemmaghami et al. (32) have also proposed methods for autosampler MALDI target plate spotting for analyzing protein stability by hydrogen exchange at rates of hundreds per day in order to find potential protein specimens for structural genomics.

The number of time points required in the HX-MS experiment depends on the application. For example, in the binding and stability studies described below, which involve large changes in solvent protection, three to five short time points (<30 min) are sufficient to draw qualitative conclusions. However, in order to quantify exchange rates for estimation of protection factors, measurements are optimal with 20–30 data points across a wide range of time. Manual sample manipulation limits the shortest time for hydrogen exchange to a few seconds. Alternative mixing approaches using rapid quench apparatuses have been reported, which extend the exchange time to the millisecond regime (20). Thus, it is currently possible to achieve millisecond to hour resolution in observed hydrogen exchange rate measurements.

Data Reduction

Two important sources of error during the LC-MS analysis of hydrogen exchange experiments must be corrected during data reduction. First, after quenching and during digestion, deuterium in the reaction mixture continues to exchange with peptide hydrogens, leading to artifactual in-exchange. Although this occurs at a slow rate due to the low pH and temperature of the mixture, it may lead to 3%–10% elevated amide deuterium incorporation. A simple control experiment is to quench the reaction at pH_{read} 2.4 prior to the addition of D_2O, followed by pepsin digestion and LC-MS analysis. The amount of artifactual in-exchange can then be corrected for by the following equations (62):

$$M_{t,corr(IE)} = (M_{t,wa} - LM_{\infty,90})/(1 - L) \qquad 1.$$

and

$$L = (M_o - M_{calc})/(M_{\infty,90} - M_{calc}), \qquad 2.$$

where $M_{t,corr(IE)}$ is the artifactual in-exchange-corrected peptide mass at time t, $M_{t,wa}$ is the observed weighted average mass at time t, M_o is the observed mass for the peptide in the control experiment, M_{calc} is the theoretical average mass of the peptide, and $M_{\infty,90}$ is the theoretical mass of the peptide at 90% total exchange [for a complete discussion of the derivation, see (62)].

A second source of error is back-exchange, where deuterons incorporated into peptides exchange-out with hydrogen from water. Deuterons may also back-exchange with water vapor during mass spectral analysis by ESI-MS or with matrix protons by MALDI-MS. Following an estimation of the back-exchange, the data may be corrected by the following equation:

$$M_{t,corr(BE)} = M_{t,corr(IE)} + [BE \cdot (M_{t,corr(IE)} - M_{calc})], \qquad 3.$$

where $M_{t,corr(BE)}$ is the artifactual in-exchange- and back-exchange-corrected mass of the peptide at time t, BE is the fractional back-exchange, $M_{t,corr(IE)}$ is the mass at time t, corrected for in-exchange, and M_{calc} is the theoretical average mass of the peptide (84).

Three methods can be used to estimate the back-exchange. First, the back-exchange following quenching can be directly measured by protein deuteration followed by digestion and mass analysis (61, 84). In cases where this method is not sufficient to achieve complete deuteration, an alternative method is to generate the peptic peptides in water and pool them after purification by reversed-phase HPLC. After lyophilization, the peptides are resuspended in D_2O and incubated for 90 min at 90°C, which enables full deuteration, and then analyzed by LC-MS after quenching. The fractional back-exchange can then be calculated for each peptide:

$$BE = (M_{\infty,90} - M_{BE})/(M_{\infty,90} - M_{calc}), \qquad 4.$$

where $M_{\infty,90}$ is the theoretical mass of the peptide with 90% of the backbone amide hydrogens exchanged for deuterium (for experiments performed in 90% D_2O), M_{BE} is the observed mass of the peptide using the back-exchange experiment, and M_{calc} is the theoretical average mass of the peptide.

A second method estimates back-exchange by measuring reduction in peptide mass with varying gradients of peptide elution (62). In our experimental configuration, the back-exchange increases approximately 1% for each minute the peptide is retained on the column, leading to the following empirical equation:

$$BE = L + [(\text{peptide elution time from HPLC in min} + 6\ \text{min}) \cdot 0.01\ \text{min}^{-1}], \qquad 5.$$

where L is the fraction of artificial in-exchange from Equation 2. The third method estimates back-exchange using the measurements of chemical exchange rates from peptides free in solution:

$$BE_{amide} = 1 - [k_{calc} \cdot (\text{elution time in min})], \qquad 6.$$

where BE_{amide} is the fractional back-exchange estimate for each backbone amide hydrogen, and k_{calc} is the rate of exchange calculated for the amide hydrogen (8, 36). For each peptide the fractional back-exchange may be calculated by:

$$BE = \Sigma(BE_{amide})/(M_{\infty} - M_{calc}), \qquad 7.$$

where Σ (BE_{amide}) is the sum of BE_{amide} for every amide hydrogen in the peptide, from Equation 6, M_{∞} is the theoretical average mass of the peptide with every backbone amide hydrogen exchanged for deuterium, and M_{calc} is the theoretical average mass of the peptide.

Following data correction for artifactual in-exchange and back-exchange, the time courses can be modeled by a sum of exponentials, in which each amide hydrogen exchanges with a deuteron at a given rate. In theory, each amide backbone hydrogen would be represented by a separate exponential term:

$$Y = N_{th} - e^{-k_1 t} - e^{-k_2 t} - e^{-k_3 t} - e^{-k_4 t} \ldots - e^{-k_w t}, \qquad 8.$$

where Y is the mass of the peptide, w is the total number of amide hydrogens (minus prolines and the amino terminal amide hydrogen), k_1, k_2, k_3, k_4, and k_w are the rates of exchange for the amide hydrogen (units: time-unit^{-1}), t is time, and N_{th} is the theoretical mass of the peptide when 100% of the amide hydrogens are in-exchanged (if the exchange reaction were carried out in 100% D_2O). However, in practice amide hydrogen exchange rates are averaged into fast, intermediate, and slow rates, and time courses can be fit to between one and three exponential terms, as shown in Figure 4.

STRUCTURAL RESOLUTION

Ideally, single amide resolution is needed to calculate protection factors in estimating free energies of exchange. This is limited with HX-MS owing to the medium resolution obtainable through proteolysis. In contrast, hydrogen exchange rates can be assigned to single amides by multidimensional NMR spectroscopy. However, whereas comprehensive coverage of all amides is usually not possible using NMR, high sequence coverage can usually be achieved by MS. In addition, the process of assigning NMR peaks can be lengthy, the concentrations of protein needed are typically much higher than physiological, and the upper mass limit for proteins is relatively low. For this reason, improved strategies are needed to achieve single amide resolution by MS and are a focus of current developments in this field.

Increased Resolution Achieved with Overlapping Peptides

An effective strategy uses peptides with overlapping sequences to deconvolute exchange rates. Because pepsin is a nonspecific protease, the degree of overlap can be significant, reaching 3 to 4 overlapping peptides for a given region of the protein molecule. Least squares fitting of time courses provide estimates of the number of amides with fast, intermediate, or slow rates. Amides in peptides can be modeled to these rates and compared with overlapping peptides to increase resolution. As an example, a collection of peptides used in a study of MAPK demonstrates the ability of overlapping sequences to improve the resolution of measurements to within 3 to 4 amino acids (Figure 5). The assigned rates were consistent with hydrogen bonding patterns in regional secondary structures identified by X-ray crystallography

a

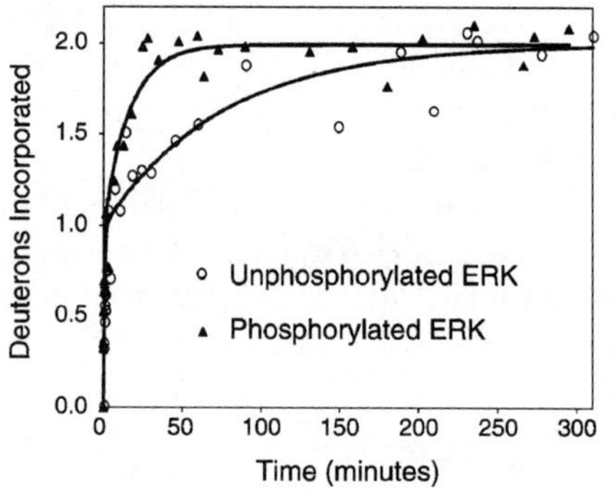

b

$$Y = N - Ae^{-k_1t} - Be^{-k_2t} - Ce^{-k_3t}$$

Parameter	Unphos	Phos
A	1.0	1.0
B	1.0	1.0
k_1	1.17 (0.25)	1.35 (0.27)
k_2	0.013 (0.003)	0.061 (0.014)

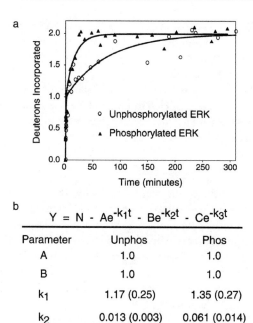

Figure 4 Data reduction. (*a*) Weighted average masses are calculated from isotopic peaks and plotted versus time of incubation with D_2O, as in this peptide derived from phosphorylated (▲) and unphosphorylated (o) forms of ERK2. (*b*) Data are corrected for D_2O dilution, artifactual in-exchange, and back-exchange, then fit by nonlinear least squares to a sum of exponentials. Fitted parameters include N, the mass of the peptide at equilibrium, and A, B, and C, the number of amides exchanging, respectively, at apparent rates k_1, k_2, and k_3. In this example, the data could be modeled to two amides exchanging at 1.17 and 0.013 min^{-1}, the slower of which increased to 0.061 min^{-1} following ERK2 phosphorylation.

(62, 82). Peptide sequence overlap can also be enhanced by performing pepsin digestion under slightly denaturing conditions (78), although with this approach it is important that data correction (particularly artifactual in-exchange) be performed appropriately under each experimental condition. Alternatively, overlap can be increased using multiple acid stable proteases with differing specificities (78). Combining these strategies could theoretically generate a collection of peptides that would achieve single amino acid resolution.

Tandem Mass Spectrometry

A second approach to increasing the resolution of HX-MS is to fragment peptide ions into daughter ions by collisions in the gas phase (MS/MS). By comparing weighted average masses of successive daughter ions, deuteration of individual amides should be revealed when mass increments are 1 Da greater than the residue

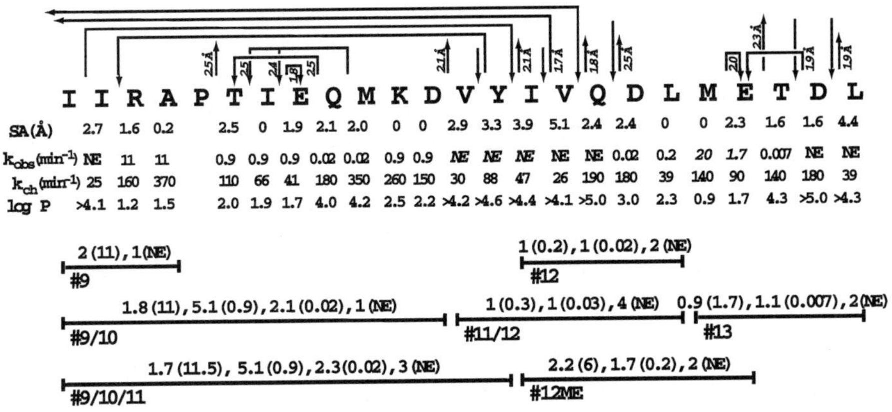

Figure 5 Increased resolution using overlapping peptic peptides. Partial digestion resulted in seven overlapping peptides in the β5-αD region of ERK2. At the top, *arrows pointing up* indicate amide hydrogen bonds and bond lengths (Å) to various acceptors at residue side chains or carbonyl oxygens are indicated. *Arrows pointing down* indicate hydrogen bonding to carbonyls adjacent to the amides. At the bottom, observed peptides are shown by *horizontal bars*, indicating stoichiometries and rates (min^{-1}) determined by nonlinear least squares. In the middle, distance to the surface of the protein and chemical exchange rate (k_{ch}) are indicated for each amide hydrogen (8). Assignment of an approximate k_{obs} to each amide is based on analysis of peptides 9, 9/10, 9/10/11, 11/12, 12, 12ME, and 13, where rates in italics indicate tentative assignments. The estimated protection factor ($P = k_{obs}/k_{ch}$) for nonexchanging (NE) amides was calculated from $k_{obs} < 0.002\ min^{-1}$.

mass. Most attempts to sequence peptides after hydrogen exchange have employed collision-induced dissociation (CID) for peptide fragmentation. The first group to report success was Anderegg et al. (3), who examined several model peptides by ESI-MS/MS and reported that it was possible to determine the extent of hydrogen exchange at individual amide hydrogens. In contrast, Johnson & Walsh (39) demonstrated shortly thereafter that model peptides in the gas phase showed complete scrambling of amide hydrogens due to migration to other positions in the peptide.

More recent reports using digested proteins (19, 44) have observed localization of deuterium incorporation at specific amide hydrogens, with no extensive scrambling. In addition, an HX-ESI-MS/MS study of a model α-helical peptide incorporated into lipid qualitatively demonstrated greater deuteration at the ends of the peptide than in the lipid-protected center, although they were unable to discern whether proton migration occurred during their experiment. Studies examining full-length proteins by HX-FT-ICR-MS/MS reported that no scrambling took place during capillary-skimmer dissociation (i.e., eliminating the proteolytic step) (41).

It is important to note that in each of the aforementioned examples, nonintegral deuterium incorporation was seen at single amide positions when quantified.

Although this does not prove that scrambling took place, it could in fact be attributed to partial scrambling. Other studies examining organic model compounds have reported significant hydrogen scrambling in the gas phase, in some cases relocating from polar groups to aliphatic carbon centers (59). For the most part, these molecules are smaller than the peptides examined in the studies described above, nevertheless the conclusions may be instructive for peptides. A recent review of proton migration in aromatic compounds highlights the necessity for further experimentation and cautious interpretation (45).

In summary, the seemingly contradictory results in this area demonstrate the need for further experimentation. Rigorous experiments examining the time-dependent deuteration of individual amides with MS/MS over a range of peptide chemistries are needed to clarify the mechanisms and extent of scrambling under different experimental conditions. Furthermore, it must be remembered that at any time point an amide may only be partially exchanged and that amides will back-exchange at different rates in the same peptide because of primary and secondary structure effects. While quantitative interpretation of hydrogen exchange rates by these methods may be suspect, qualitative interpretations most likely yield valid information.

APPLICATIONS

Folding and Stability

Mass spectrometry has the particular advantage of resolving folding states, which are revealed by bimodal populations of proteins or peptides resulting from varying extents of deuteration. These reflect different populations exchanging through different mechanisms, where native forms exchange through equilibrium (EX2) mechanisms and denatured or partly unfolded forms exchange through unfolding (EX1) mechanisms. The latter can be favored by increasing denaturant concentration or by varying temperature or pH. In contrast, these populations average together and are not distinguishable when analyzed by NMR.

With its higher sensitivity and higher-upper mass limit, HX-MS has been used to great advantage in examining pathways of protein folding. For example, large portions of aldolase and cellular retinoic acid-binding protein I (CRABP-I) unfold cooperatively and exchange before refolding can take place, helping to define the structure of folding intermediates (18, 31). Transient cooperative fluctuations leading to localized unfolding have also been observed in certain genetic mutants of human lysozyme and transthyretin under conditions that favor amyloid fibril formation. Such results explain the predisposition of mutants to aggregate in patients with amyloid diseases (12, 56). Rapid quenched-flow techniques have also been incorporated to examine the earliest stabilization of secondary structure in refolding experiments (51, 70).

One of HX-MS's greatest successes is to examine folding rates in the context of the chaperones that assist folding in vivo. Several proteins have been studied in this context and have revealed the mechanisms of chaperone assistance in the folding pathway for different substrates. For example, in the case of α-lactalbumin, the protein bound to the chaperone GroEL resembled the molten globule state (63), whereas human dihydrofolate reductase maintained a stable core of secondary structure during successive rounds of folding attempts by GroEL (34). This indicates that chaperones do not completely unfold proteins upon binding, but rather permit stable secondary structures in folding pathways to persist.

These methods have led to useful approaches for screening protein integrity. A clever technique developed by Ghaemmaghami and colleagues (32, 58) uses HX-MS and MALDI-MS to rapidly determine the relative stability of protein variants generated by recombinant expression. By titrating denaturant, stability can be measured from the midpoint at which hydrogen exchange becomes dominated by cooperative opening events, seen in these studies as a sharp increase in the total deuterium incorporation into full-length protein with increasing denaturant. This assay can even be performed on unpurified proteins expressed in *Escherichia coli*, as demonstrated in two studies (33, 66).

Practical applications include rapid screening for proteins suitable for structural analyses, which currently represents a significant bottleneck in genomic research initiatives. By using high-throughput techniques, these analyses could facilitate identification of expression systems among those made en masse that yield stable, folded proteins most amenable to NMR and X-ray crystallography. In addition, an ESI-MS approach has been used to characterize the ability of small molecules to inhibit transthyretin amyloid fibril formation in vitro, a promising strategy in drug development for amyloid diseases (52).

Ligand Binding, Aggregation, and Protein-Protein Interactions

A straightforward use of hydrogen exchange is to probe sites for molecular interaction by analyzing regions of solvent protection upon binding. In many cases, interfaces can be revealed by marked reductions in exchange rate caused by steric exclusion of solvent. Hydrogen exchange mass spectrometry has revealed interfaces for homomultimerization in aldolase and extracellular signal-regulated protein kinase-2 (ERK2) (36, 83). Binding interfaces for heteromeric protein-protein interactions have also been identified, illustrated by the examples of thrombomodulin-thrombin (49) and IGF1 binding protein-IGF1 (21). The technique has also proven useful for epitope mapping of monoclonal antibodies (5, 81). Although it has not been achieved to date, an important future application for HX-MS will be to map binding interfaces in large heteromultimeric protein complexes. Such an approach would provide rapid information to characterize the "interactome" and to facilitate the design of mutants and small molecules that modulate function.

Hydrogen exchange mass spectrometry with fragment separation has been used to probe ligand interactions with proteins. Heme binding to myoglobin was among the first published experiments using this method to examine the effects of small ligand binding to a protein molecule (40). Other examples include phosphotyrosine peptide binding to SH2 domains (22), calcium binding to recoverin, calmodulin, and troponin C (55, 57, 74), and enzyme-active site interactions with small molecule inhibitors (35, 72, 73, 75). In principle, high-throughput techniques that have been described for HX-MS can be used to rapidly map protein binding sites for small molecule inhibitors (78).

Dynamics

In addition to mapping protein interfaces that become sterically protected following ligand binding, changes in protein dynamics accompanying ligand binding have been revealed by HX-MS. In many examples where hydrogen exchange measurements of ligand binding to proteins can be compared to NMR or crystallographic structures at high resolution, additional effects of ligand binding on exchange are observed, often at long distances from the site of interaction. For example, whereas heme binding to myoglobin led to decreased exchange rates that were attributable to steric exclusion of solvent based on structural evidence, binding also led to decreased exchange rates in regions of the protein located distal to the heme interface (40). This suggested that heme binding induced and stabilized secondary structure in these distal regions, which were postulated from NMR and X-ray studies to have little or no secondary structure in the apomyoglobin. Thus, when proteins bind small molecules or proteins, allosteric effects may lead to changes in the internal motions within local regions, which can be revealed in the hydrogen exchange experiment. Such motions may alter the opening/closing equilibrium that leads to exchange and therefore modulate observed exchange rates. For this reason, conclusions about the location of interaction sites that are based on regions of solvent protection must be treated cautiously.

On the other hand, such results reveal valuable information about allosteric contributions to protein motions that are still poorly understood. In studies probing the solvent-excluded thrombomodulin binding interface in thrombin, additional changes in solvent protection were observed upon binding in a surface loop proximal to the active site and distal to the binding interface (48). Lack of structural differences in this region in X-ray studies suggests that such changes may reflect allosteric effects that lead to reduced flexibility of the surface loop upon thrombomodulin binding. Allostery is also invoked in studies of cAMP binding to the type I regulatory subunit of cAMP-dependent protein kinase (2). In addition to solvent protection within the cAMP binding pocket, increased exchange rates were observed in a helical subdomain located at long distances from the binding site and implicated in catalytic subunit interactions. The results suggest that cyclic nucleotide binding promotes dissociation of regulatory and catalytic subunits by altering conformation or dynamics in the binding interface.

These studies highlight the potential of HX-MS for probing dynamic and conformational changes relevant to enzyme catalysis. This has been addressed in hypoxanthine-guanine phosphoribosyltransferase (HGPRT), where various protein-ligand complexes have been used to model enzyme intermediates (76). Global hydrogen exchange progressively decreased in unbound enzyme, enzyme complexed with nucleotide (binary), nucleotide and substrate (ternary equilibrium complex), or transition state inhibitor. Significant protection was observed in the catalytic loop, nucleotide phosphate binding loop, and subunit interface, which was particularly pronounced in the transition state inhibitor complex. The results suggest stronger subunit interactions and reduced mobility of catalytic site loops in this model of the catalytic intermediate. Similar approaches examining other enzyme complexes with inhibitors, substrates, and cofactors revealed further evidence for loop and domain movements during catalysis (35, 72, 73, 75).

Internal protein motions have also been monitored by HX-MS in response to enzyme activation. For example, by comparing wild-type, inactive mitogen-activated protein kinase kinase-1 (MKK1) to three constitutively active mutant forms, marked increases in hydrogen exchange rates were observed in the N-terminal ATP binding lobe of the molecule (61). Although a crystallographic structure has not been reported for this kinase, overlapping peptides localized the site of increased exchange to single amides. This suggests that this region is more flexible in the active form, assuming that structural changes would affect more than one amide. Importantly, such changes were correlated with the degree of enzyme activity, and recent evidence showed that this region is the site of binding for a specific noncompetitive inhibitor of MKK1 (17).

Hydrogen exchange mass spectrometry has also been used to document motional effects following activation by phosphorylation in ERK2 (36). Phosphorylation led to altered exchange rates observed in regions far from the site of covalent modification. Comparison with corresponding X-ray structures ascribed these effects to long-distance changes in conformational mobility or flexibility upon enzyme activation. In each case, these were regions of the molecule expected either to interact with substrate or ATP, or to undergo movements during catalysis. This study represents an important step in using HX-MS to understand the effects of phosphorylation on protein dynamics.

Effects of phosphorylation have also been studied using HX-MALDI-MS in CheB, a regulator of chemotaxis in bacteria (37). Histidine phosphorylation of CheB increases its methyltransferase activity toward the CheA receptor, and attenuates signaling along the chemotaxis pathway. Phosphorylation caused increased hydrogen exchange in regions flanking the contact interface for regulatory and catalytic domains in CheB, which forms the active site. This disproved a previous model explaining activation by detachment of the two domains and disruption of the interface. The results instead suggested that activation by phosphorylation involves increased motions within regions surrounding the active site.

Together these findings indicate that ligand binding and covalent modifications communicate motional information over long distances in enzymes. Thus, HX-MS provides insight into solution behavior of proteins in a manner complimentary to the information from high-resolution structural evidence. When structural changes can be assumed to not occur, hydrogen exchange behavior can reveal perturbations in protein motions due to allostery and covalent modification.

Although the work in these systems has been an exciting advance in documenting the intramolecular motions in macromolecules, the theoretical description of these motions remains to be determined. First, it is important to understand the timescales, amplitude, and directionality of the motions sampled by hydrogen exchange. These cannot be identified from equilibrium measurements of hydrogen exchange, and despite attempts, it has not been possible so far to correlate hydrogen exchange rates with parameters from model free analyses of relaxation rates in NMR spectroscopy. These motions likely span timescales ranging from microseconds to milliseconds, the same magnitude on which enzyme catalysis takes place. In cases where rate-limiting steps involve a transient conformational change, it is possible that hydrogen exchange can sample regions of flexibility and correlated motions needed for dynamic events during catalysis.

CONCLUSIONS

From its inception, hydrogen exchange has complimented every aspect of biophysical analysis. The method provides a measure of the dynamic nature of proteins in solution, a complement to X-ray crystallography. It has the ability to distinguish different conformers in solution that are averaged in an NMR experiment. The approach also provides important insights into modes of peptide and small ligand binding in solution at physiological concentrations of protein. The sensitivity of the technique, the ease with which data is acquired and analyzed, and the quality of the data now permit hydrogen exchange experiments to be performed on large proteins and heteromultimeric systems with great success. New methods promise higher throughput, sensitivity, and mass accuracy. All these technological milestones make HX-MS an attractive method for gaining new insight into the behavior of macromolecular systems.

LITERATURE CITED

1. Akashi S, Naito Y, Takio K. 1999. Observation of hydrogen-deuterium exchange of ubiquitin by direct analysis of electrospray capillary-skimmer dissociation with Fourier transform ion cyclotron resonance mass spectrometry. *Anal. Chem.* 71:4974–80
2. Anand GS, Hughes CA, Jones JM, Taylor SS, Komives EA. 2002. Amide H/^{2}H exchange reveals communication between the cAMP- and catalytic subunit-binding sites in protein kinase A. *J. Mol. Biol.* 323:377–86
3. Anderegg RJ, Wagner DS, Stevenson CL, Borchardt RT. 1994. The mass-spectrometry of helical unfolding in peptides. *J. Am. Soc. Mass Spectrom.* 5:425–33
4. Arrington CB, Robertson AD. 2000. Correlated motions in native proteins from MS analysis of NH exchange: evidence for a

manifold of unfolding reactions in ovomucoid third domain. *J. Mol. Biol.* 300:221–32

5. Baerga-Ortiz A, Hughes CA, Mandell JG, Komives EA. 2002. Epitope mapping of a monoclonal antibody against human thrombin by R/D-exchange mass spectrometry reveals selection of a diverse sequence in a highly conserved protein. *Protein Sci.* 11:1300–8
6. Bai Y, Englander JJ, Mayne L, Milne JS, Englander SW. 1995. Thermodynamic parameters from hydrogen exchange measurements. *Methods Enzymol.* 259:344–56
7. Bai Y, Sosnick TR, Mayne L, Englander SW. 1995. Protein folding intermediates: native-state hydrogen exchange. *Science* 269:192–97
8. Bai YW, Milne JS, Mayne L, Englander SW. 1993. Primary structure effects on peptide group hydrogen-exchange. *Proteins* 17:75–86
9. Barksdale AD, Rosenberg A. 1982. Acquisition and interpretation of hydrogen exchange data from peptides, polymers, and proteins. *Methods Biochem. Anal.* 28:1–113
10. Bentley GA, Delepierre M, Dobson CM, Wedin RE, Mason SA, Poulsen FM. 1983. Exchange of individual hydrogens for a protein in a crystal and in solution. *J. Mol. Biol.* 170:243–47
11. Buijs J, Hakansson K, Hagman C, Hakansson P, Oscarsson S. 2000. A new method for the accurate determination of the isotopic state of single amide hydrogens within peptides using Fourier transform ion cyclotron resonance mass spectrometry. *Rapid Commun. Mass Spectrom.* 14:1751–56
12. Canet D, Last AM, Tito P, Sunde M, Spencer A, et al. 2002. Local cooperativity in the unfolding of an amyloidogenic variant of human lysozyme. *Nat. Struct. Biol.* 9:308–15
13. Chernushevich IV, Loboda AV, Thomson BA. 2001. An introduction to quadrupole-time-of-flight mass spectrometry. *J. Mass Spectrom.* 36:849–65
14. Chowdhury SK, Katta V, Chait BT. 1990. Probing conformational changes in proteins by mass-spectrometry. *J. Am. Chem. Soc.* 112:9012–13
15. Clarke J, Itzhaki LS. 1998. Hydrogen exchange and protein folding. *Curr. Opin. Struct. Biol.* 8:112–18
16. Connelly GP, Bai YW, Jeng MF, Englander SW. 1993. Isotope effects in peptide group hydrogen-exchange. *Proteins* 17:87–92
17. Delaney AM, Printen JA, Chen H, Fauman EB, Dudley DT. 2002. Identification of a novel MAPKK activation domain recognized by the inhibitor PD184352. *Mol. Cell. Biol.* 22:7593–602
18. Deng YH, Smith DL. 1998. Identification of unfolding domains in large proteins by their unfolding rates. *Biochemistry* 37: 6256–62
19. Deng YZ, Pan H, Smith DL. 1999. Selective isotope labeling demonstrates that hydrogen exchange at individual peptide amide linkages can be determined by collision-induced dissociation mass spectrometry. *J. Am. Chem. Soc.* 121:1966–67
20. Dharmasiri K, Smith DL. 1996. Mass spectrometric determination of isotopic exchange rates of amide hydrogens located on the surfaces of proteins. *Anal. Chem.* 68:2340–44
21. Ehring H. 1999. Hydrogen exchange electrospray ionization mass spectrometry studies of structural features of proteins and protein/protein interactions. *Anal. Biochem.* 267:252–59
22. Engen JR, Gmeiner WH, Smithgall TE, Smith DL. 1999. Hydrogen exchange shows peptide binding stabilizes motions in Hck SH2. *Biochemistry* 38:8926–35
23. Englander JJ, Calhoun DB, Englander SW. 1979. Measurement and calibration of peptide group hydrogen-deuterium exchange by ultraviolet spectrophotometry. *Anal. Biochem.* 92:517–24
24. Englander JJ, Rogero JR, Englander SW. 1985. Protein hydrogen-exchange studied

by the fragment separation method. *Anal. Biochem.* 147:234–44
25. Englander SW. 1963. A hydrogen exchange method using tritium and Sephadex: its application to ribonuclease. *Biochemistry* 2:798–807
26. Englander SW. 2000. Protein folding intermediates and pathways studied by hydrogen exchange. *Annu. Rev. Biophys. Biomol. Struct.* 29:213–38
27. Englander SW, Downer NW, Teitelba H. 1972. Hydrogen-exchange. *Annu. Rev. Biochem.* 41:903–24
28. Englander SW, Kallenbach NR. 1984. Hydrogen-exchange and structural dynamics of proteins and nucleic-acids. *Q. Rev. Biophys.* 16:521–655
29. Englander SW, Mayne L. 1992. protein folding studied using hydrogen-exchange labeling and 2-dimensional NMR. *Annu. Rev. Biophys. Biomol. Struct.* 21:243–65
30. Englander SW, Sosnick TR, Englander JJ, Mayne L. 1996. Mechanisms and uses of hydrogen exchange. *Curr. Opin. Struct. Biol.* 6:18–23
31. Eyles SJ, Speir JP, Kruppa GH, Gierasch LM, Kaltashov IA. 2000. Protein conformational stability probed by Fourier transform ion cyclotron resonance mass spectrometry. *J. Am. Chem. Soc.* 122:495–500
32. Ghaemmaghami S, Fitzgerald MC, Oas TG. 2000. A quantitative, high-throughput screen for protein stability. *Proc. Natl. Acad. Sci. USA* 97:8296–301
33. Ghaemmaghami S, Oas TG. 2001. Quantitative protein stability measurement in vivo. *Nat. Struct. Biol.* 8:879–82
34. Gross M, Robinson CV, Mayhew M, Hartl FU, Radford SE. 1996. Significant hydro gen exchange protection in GroEL-bound DHFR is maintained during iterative rounds of substrate cycling. *Protein Sci.* 5:2506–13
35. Halgand F, Dumas R, Biou V, Andrieu JP, Thomazeau K, et al. 1999. Characterization of the conformational changes of acetohydroxy acid isomeroreductase induced by the binding of Mg2+ ions, NADPH, and a competitive inhibitor. *Biochemistry* 38:6025–34
36. Hoofnagle AN, Resing KA, Goldsmith EJ, Ahn NG. 2001. Changes in protein conformational mobility upon activation of extracellular regulated protein kinase-2 as detected by hydrogen exchange. *Proc. Natl. Acad. Sci. USA* 98:956–61
37. Hughes CA, Mandell JG, Anand GS, Stock AM, Komives EA. 2001. Phosphorylation causes subtle changes in solvent accessibility at the interdomain interface of methylesterase CheB. *J. Mol. Biol.* 307: 967–76
38. Hvidt A, Johansen G, Linderstrøm-Lang K. 1960. Deuterium and ^{18}O exchange. In *Laboratory Manual of Analytical Techniques in Protein Chemistry*, ed. P Alexander, JR Block, pp. 101–30. New York: Pergamon
39. Johnson RS, Krylov D, Walsh KA. 1995. Proton mobility within electrosprayed peptide ions. *J. Mass Spectrom.* 30:386–87
40. Johnson RS, Walsh KA. 1994. Mass-spectrometric measurement of protein amide hydrogen-exchange rates of apo-myoglobin and holo-myoglobin. *Protein Sci.* 3: 2411–18
41. Kaltashov IA, Eyles SJ. 2002. Crossing the phase boundary to study protein dynamics and function: combination of amide hydrogen exchange in solution and ion fragmentation in the gas phase. *J. Mass Spectrom.* 37:557–65
42. Katta V, Chait BT. 1991. Conformational-changes in proteins probed by hydrogen-exchange electrospray-ionization mass-spectrometry. *Rapid Commun. Mass Spectrom.* 5:214–17
43. Kim KS, Fuchs JA, Woodward CK. 1993. Hydrogen exchange identifies native-state motional domains important in protein folding. *Biochemistry* 32:9600–8
44. Kim MY, Maier CS, Reed DJ, Deinzer ML. 2001. Site-specific amide hydrogen/deuterium exchange in *E. coli* thioredoxins measured by electrospray ionization mass

spectrometry. *J. Am. Chem. Soc.* 123:9860–66
45. Kuck D. 2002. Half a century of scrambling in organic ions: complete, incomplete, progressive and composite atom interchange. *Int. J. Mass Spectrom.* 213:101–44
46. Lenormant H, Blout ER. 1953. Origin of the absorption band at 1,550 cm^{-1} in proteins. *Nature* 172:770–71
47. Li R, Woodward C. 1999. The hydrogen exchange core and protein folding. *Protein Sci.* 8:1571–90
48. Mandell JG, Baerga-Ortiz A, Akashi S, Takio K, Komives EA. 2001. Solvent accessibility of the thrombin-thrombomodulin interface. *J. Mol. Biol.* 306:575–89
49. Mandell JG, Falick AM, Komives EA. 1998. Identification of protein-protein interfaces by decreased amide proton solvent accessibility. *Proc. Natl. Acad. Sci. USA* 95:14705–10
50. Mandell JG, Falick AM, Komives EA. 1998. Measurement of amide hydrogen exchange by MALDI-TOF mass spectrometry. *Anal. Chem.* 70:3987–95
51. Matagne A, Jamin M, Chung EW, Robinson CV, Radford SE, Dobson CM. 2000. Thermal unfolding of an intermediate is associated with non-Arrhenius kinetics in the folding of hen lysozyme. *J. Mol. Biol.* 297:193–210
52. McCammon MG, Scott DJ, Keetch CA, Greene LH, Purkey HE, et al. 2002. Screening transthyretin amyloid fibril inhibitors. Characterization of novel multiprotein, multiligand complexes by mass spectrometry. *Structure* 10:851–63
53. Deleted in proof
54. Nakanishi M, Nakamura H, Hirakawa AY, Tsuboi M, Nagamura T, Saijo Y. 1978. Measurement of hydrogen-exchange at tryptophan residues of a protein by stopped-flow and ultraviolet spectroscopy. *J. Am. Chem. Soc.* 100:272–76
55. Nemirovskiy O, Giblin DE, Gross ML. 1999. Electrospray ionization mass spectrometry and hydrogen/deuterium exchange for probing the interaction of calmodulin with calcium. *J. Am. Soc. Mass Spectrom.* 10:711–18
56. Nettleton EJ, Sunde M, Lai Z, Kelly JW, Dobson CM, Robinson CV. 1998. Protein subunit interactions and structural integrity of amyloidogenic transthyretins: evidence from electrospray mass spectrometry. *J. Mol. Biol.* 281:553–64
57. Neubert TA, Walsh KA, Hurley JB, Johnson RS. 1997. Monitoring calcium-induced conformational changes in recoverin by electrospray mass spectrometry. *Protein Sci.* 6:843–50
58. Powell KD, Fitzgerald MC. 2001. Measurements of protein stability by H/D exchange and matrix-assisted laser desorption ionization mass spectrometry using picomoles of material. *Anal. Chem.* 73: 3300–4
59. Reed DR, Kass SR. 2001. Hydrogen-deuterium exchange at non-labile sites: a new reaction facet with broad implications for structural and dynamic determinations. *J. Am. Soc. Mass Spectrom.* 12:1163–68
60. Resing KA, Ahn NG. 1997. Protein phosphorylation analysis by electrospray ionization-mass spectrometry. *Methods Enzymol.* 283:29–44
61. Resing KA, Ahn NG. 1998. Deuterium exchange mass spectrometry as a probe of protein kinase activation. Analysis of wild-type and constitutively active mutants of MAP kinase kinase-1. *Biochemistry* 37:463–75
62. Resing KA, Hoofnagle AN, Ahn NG. 1999. Modeling deuterium exchange behavior of ERK2 using pepsin mapping to probe secondary structure. *J. Am. Soc. Mass Spectrom.* 10:685–702
63. Robinson CV, Gross M, Eyles SJ, Ewbank JJ, Mayhew M, et al. 1994. Conformation of GroEL-bound alpha-lactalbumin probed by mass spectrometry. *Nature* 372:646–51
64. Roder H, Elove GA, Englander SW. 1988. Structural characterization of folding intermediates in cytochrome-*c* by H-exchange

labeling and proton NMR. *Nature* 335: 700–4

65. Rosa JJ, Richards FM. 1979. Experimental procedure for increasing the structural resolution of chemical hydrogen-exchange measurements on proteins—application to ribonuclease S-peptide. *J. Mol. Biol.* 133: 399–416
66. Rosenbaum DM, Roy S, Hecht MH. 1999. Screening combinatorial libraries of de novo proteins by hydrogen-deuterium exchange and electrospray mass spectrometry. *J. Am. Chem. Soc.* 121:9509–13
67. Schmid FX, Baldwin RL. 1979. Detection of an early intermediate in the folding of ribonuclease A by protection of amide protons against exchange. *J. Mol. Biol.* 135:199–215
68. Shen Y, Tolic N, Zhao R, Pasa-Tolic L, Li L, et al. 2001. High-throughput proteomics using high-efficiency multiple-capillary liquid chromatography with online high-performance ESI FTICR mass spectrometry. *Anal. Chem.* 73:3011–21
69. Swint-Kruse L, Robertson AD. 1996. Temperature and pH dependences of hydrogen exchange and global stability for ovomucoid third domain. *Biochemistry* 35:171–80
70. Tsui V, Garcia C, Cavagnero S, Siuzdak G, Dyson HJ, Wright PE. 1999. Quench-flow experiments combined with mass spectrometry show apomyoglobin folds through and obligatory intermediate. *Protein Sci.* 8:45–49
71. Udgaonkar JB, Baldwin RL. 1988. NMR evidence for an early framework intermediate on the folding pathway of ribonuclease A. *Nature* 335:694–99
72. Wang F, Blanchard JS, Tang XJ. 1997. Hydrogen exchange/electrospray ionization mass spectrometry studies of substrate and inhibitor binding and conformational changes of *Escherichia coli* dihydrodipicolinate reductase. *Biochemistry* 36:3755–59
73. Wang F, Li W, Emmett MR, Hendrickson CL, Marshall AG, et al. 1998. Conformational and dynamic changes of *Yersinia* protein tyrosine phosphatase induced by ligand binding and active site mutation and revealed by H/D exchange and electrospray ionization Fourier transform ion cyclotron resonance mass spectrometry. *Biochemistry* 37:15289–99
74. Wang F, Li W, Emmett MR, Marshall AG, Corson D, Sykes BD. 1999. Fourier transform ion cyclotron resonance mass spectrometric detection of small Ca(2+)-induced conformational changes in the regulatory domain of human cardiac troponin C. *J. Am. Soc. Mass Spectrom.* 10:703–10
75. Wang F, Scapin G, Blanchard JS, Angeletti RH. 1998. Substrate binding and conformational changes of *Clostridium glutamicum* diaminopimelate dehydrogenase revealed by hydrogen/deuterium exchange and electrospray mass spectrometry. *Protein Sci.* 7:293–99
76. Wang F, Shi W, Nieves E, Angeletti RH, Schramm VL, Grubmeyer C. 2001. A transition-state analogue reduces protein dynamics in hypoxanthine-guanine phosphoribosyltransferase. *Biochemistry* 40: 8043–54
77. Wang L, Pan H, Smith DL. 2002. Hydrogen exchange-mass spectrometry: optimization of digestion conditions. *Mol. Cell. Proteomics* 1:132–38
78. Woods VL, Hamuro Y. 2001. High resolution, high-throughput amide deuterium exchange-mass spectrometry (DXMS) determination of protein binding site structure and dynamics: utility in pharmaceutical design. *J. Cell. Biochem.* 37(Suppl.):89–98
79. Woodward C. 1993. Is the slow exchange core the protein folding core? *Trends Biochem. Sci.* 18:359–60
80. Woodward C, Simon I, Tuchsen E. 1982. Hydrogen exchange and the dynamic structure of proteins. *Mol. Cell. Biochem.* 48: 135–60
81. Yamada N, Suzuki E, Hirayama K. 2002. Identification of the interface of a large

protein-protein complex using H/D exchange and Fourier transform ion cyclotron resonance mass spectrometry. *Rapid Commun. Mass Spectrom.* 16:293–99

82. Zhang F, Strand A, Robbins D, Cobb MH, Goldsmith EJ. 1994. Atomic structure of the MAP kinase ERK2 at 2.3 Å resolution. *Nature* 367:704–11
83. Zhang ZQ, Post CB, Smith DL. 1996. Amide hydrogen exchange determined by mass spectrometry: application to rabbit muscle aldolase. *Biochemistry* 35:779–91
84. Zhang ZQ, Smith DL. 1993. Determination of amide hydrogen-exchange by mass-spectrometry—a new tool for protein-structure elucidation. *Protein Sci.* 2:522–31

Annu. Rev. Biophys. Biomol. Struct. 2003. 32:27–45
doi: 10.1146/annurev.biophys.32.110601.141726

First published online as a Review in Advance on February 14, 2003

CATIONS AS HYDROGEN BOND DONORS: A View of Electrostatic Interactions in DNA

Juan A. Subirana[1] and Montserrat Soler-López[2]
[1]*Departament d'Enginyeria Química, Universitat Politècnica de Catalunya, 08028 Barcelona, Spain; email: juan.a.subirana@upc.es*
[2]*European Molecular Biology Laboratory, Grenoble Outstation, 38042 Grenoble Cedex 9, France; email: soler@embl-grenoble.fr*

Key Words magnesium, X-ray diffraction, crystal structure, solvation, amines

■ **Abstract** Cations are bound to nucleic acids in a solvated state. High-resolution X-ray diffraction studies of oligonucleotides provide a detailed view of Mg^{2+}, and occasionally other ions bound to DNA. In a survey of several such structures, certain general observations emerge. First, cations bind preferentially to the guanine base in the major groove or to phosphate group oxygen atoms. Second, cations interact with DNA most frequently via water molecules in their primary solvation shell, direct ion-DNA contacts being only rarely observed. Thus, the solvated ions should be viewed as hydrogen bond donors in addition to point charges. Finally, ion interaction sites are readily exchangeable: The same site may be occupied by any ion, including spermine, as well as by a water molecule.

CONTENTS

INTRODUCTION

Main Objectives

DNA is a polyelectrolyte. Although this is an apparently obvious statement, most standard biochemistry textbooks ignore this fact: DNA is presented as a double helix with no ions around it. The cations are usually inorganic and hydrated, but also positively charged amino acid side chains of proteins and polyamines may neutralize DNA charges.

Most theoretical studies describe the electric field of DNA as resulting from more or less fixed charges on the DNA phosphates. Counterions are usually described as point charges with their hydration not always taken into account. However, recent high-resolution studies obtained by X-ray crystallography have demonstrated two facts:

1. Phosphate groups are often disordered and may occupy alternative positions in the DNA duplex.
2. Divalent solvated cations have been precisely located in many cases.

In this review our attention is focused on these two features, comparing and analyzing the high-resolution data on ions associated with DNA. This analysis yields an improved understanding of DNA-cation interactions. We concentrate on B-form DNA, with some comparisons to other forms of DNA. We do not cover specific interactions of ions with nonstandard forms of DNA (quadruplexes, rings).

When cations are described in DNA structures, emphasis is usually given to their charge. In this review we focus on another feature of ions: their solvation sphere. Due to solvation, ions become powerful hydrogen bond donors. We think greater attention should be given to this fact.

Related Reviews

First, we should note the overview of oligonucleotide structures published by Dickerson (13), which is a most valuable general reference. In another review, Cheatham & Young (7) discuss the present status of molecular dynamics simulation of nucleic acids. The use of NMR methods in the study of ion-DNA interactions has been thoroughly covered by Halle & Denisov (20). Reviews are also available (30, 43, 76) that discuss in detail the influence of ions on DNA curvature. Recent work on this topic includes molecular dynamics simulations (22, 42), NMR (32), and electrophoresis studies (65).

Excellent reviews (18, 19, 23–26) on the geometry of metal-ligand interactions in proteins and related compounds have also been recently published. It should be pointed out that the main features of ions found in DNA are essentially identical to those found in proteins, although in the latter a much broader range of interactions is available.

Significance of High Resolution in Crystallographic Studies

High resolution (better than 1.3 Å) in oligonucleotide crystals may only be achieved using synchrotron sources and cryogenic temperatures (usually 100–120 K), which is presently the routine methodology. The phosphodiester backbone, ions, and solvent have a significant degree of freedom due to the large amount of solvent in crystals, so that most ions and water molecules are not visible in the electron density maps, even when high resolutions are achieved. Regions with a clear single conformation coexist with regions of alternative conformations and even with completely disordered solvent channels, where ions are also present. Attempts to model the latter are currently in progress (45); however, in this review we focus on those ions that are clearly localized in a single or a few positions when associated with DNA.

Analysis of high-resolution structures should be carried out carefully. In some of the structures we have looked at, it is obvious that the authors have not refined the ion-water system adequately. For example, in some cases the solvated ion structure has been constrained to a preconceived theoretical model, with fixed geometry and ion-water distances. An additional problem stems from the fact that sometimes, especially in short oligonucleotides, high resolution is achieved by carrying out crystallization in a solvent that induces strong dehydration.

Therefore, we should be aware that increasing resolution through freezing and dehydrating the crystals might lead to unusual structures (9). We should also be aware that many of the ions located in high-resolution structures have a fixed position due to their participation in intermolecular interactions with a precise geometry, which would not be found in free DNA.

Distribution of Charges on DNA: Phosphate Groups

Net charges on DNA are just one negative charge per phosphate group. However, the real charge distribution cannot be described in such a straightforward way because the single negative charge is distributed unequally over the four oxygen atoms of each phosphate. Furthermore, the phosphodiester backbone is rather mobile, with a variable solvation (55). High-resolution crystal data show that the phosphate groups may either have a fixed conformation, with low B factors, or a mobile, multiple conformation. Examples are shown in Figure 1. In general, where a fixed conformation is present, the particular phosphate group is involved in crystal interactions and is stabilized by fixed phosphate-water/ion-phosphate

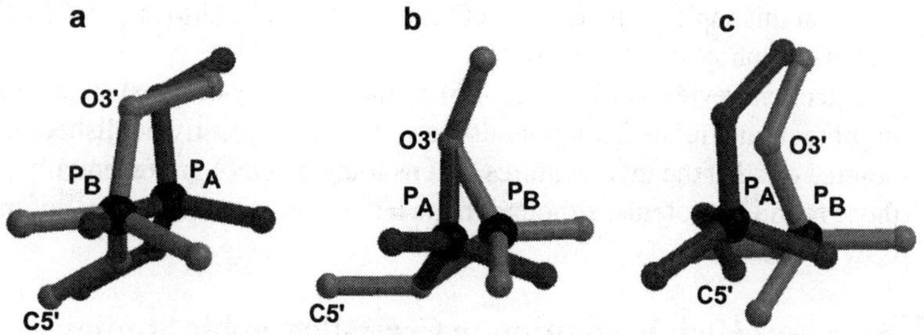

Figure 1 The double conformation of the phosphodiester backbone in different oligonucleotide structures: (*a*) Ca^{2+} salt of B-form DNA [BD0023, 0.74 Å resolution (26)]. (*b*) Mg^{2+} salt of B-form DNA (BD0037, 0.89 Å resolution). (*c*) Mg^{2+} salt of A-form DNA (AD0007, 0.84 Å resolution). Code names in parentheses refer to ID# in the Nucleic Acid Database (4).

interactions, which define the crystal lattice and restrict the position of the atoms involved. When the phosphate groups have either multiple conformations or high B factors, they face disordered solvent channels where ions and water molecules do not show well-defined positions in the electron density map. Such a range of phosphate conformations has been well documented in most high-resolution structures presently available. Hence, the phosphate negative charges are clearly not point charges found at a specific position on the DNA, and they should be considered as a broad electronegative cloud.

Distribution of Charges on DNA: Base Pairs

Although the Watson-Crick base pairs do not have any overall formal charge, they have electron donor and acceptor regions, which are essential for base-recognition processes, as already reported by Seeman et al. (56). For cation interactions, the A · T base pairs have two electronegative groups in their minor groove side, the N3 of adenine and the O2 of thymine. The G · C base pairs have an amino group between the equivalent atoms, which would interfere with any interaction with cations. Nevertheless, the minor groove in general has a strong electronegative potential (38). The situation on the major groove side is the opposite: The most favorable sites for cation interactions are the N7, O6 electron donor atoms of guanine. Thus DNA-cation interactions may occur not only at the phosphate groups but also on the minor groove in A · T-rich regions and on the major groove in G · C-rich regions. In practice, steric aspects also play an important role.

Crystal Structure of Mono- and Dinucleotides

The structure of small molecules can be determined with high accuracy. The structural data on crystals of mono- and dinucleotides might provide accurate information on oligonucleotide-ion interactions. In fact, the classical review of Swaminathan & Sundaralingam (67) offers a general view of the structure of solvated ions associated with oligonucleotides. Additional data on alkaline and alkaline earth ions associated with adenosine phosphate are also available (49). Such data provide accurate dimensions of the primary solvation shell of ions and their favored interactions. On the other hand, they do not give an adequate view of ions in a highly hydrated environment because the crystal structure usually contains a small number of water molecules. Under these conditions ions often show a variety of atoms in their primary solvation shell that substitute for water [for example see (10)]. Thus the structure of ions in low-molecular crystals cannot be extrapolated to the structure of ions around a fully hydrated DNA duplex.

Structure of Solvated Cations

When dissolved in water, most cations are surrounded by water molecules with a variable degree of order that depends on the nature of the individual cation. An excellent general description of ions in solution can be found in the book by Richens (53). At least three layers of water and other molecules can be distinguished around the central ion. Some ions have a rather rigid and stable primary solvation shell, which may be either tetrahedral (Li^+) or octahedral (Mg^{2+}, Co^{2+}, Ni^{2+}). A particular case is Zn^{2+}, which can have tetrahedral and octahedral solvation (18). Unfortunately there is only a single low-resolution structure (63) of an oligonucleotide crystal in the presence of Zn^{2+}.

Calcium and monovalent cations, with the exception of Li^+, have a much more variable first solvation shell. The number and distance of associated water molecules and other atoms in the primary solvation shell are highly variable (26) and they increase in parallel with the size of the ion. At the same time the ion-water interactions become more variable.

An important common feature to all cations is that the interaction with water molecules in the first solvation shell is a charge-dipole interaction. As a result, cations that are fully solvated with water molecules become extraordinary hydrogen bond donors. Because in this sense all cations are similar, it is easy to understand that in cases where the geometry (for example, groove width) is not a limiting factor, cation interaction sites on DNA might be occupied by different molecules. Thus a site of interaction may be simply substituted by a water molecule, explaining why cations are not always localized in crystal structures.

In the following pages we describe in greater detail the interaction of Mg^{2+} with DNA. Because this ion has a rigid first solvation shell and a comparatively high charge density, it is easier to visualize in crystal structures. Therefore it provides a useful model for the interaction of DNA with other cations, which in general have a less well-ordered primary solvation shell and weaker hydrogen bonds. Nevertheless, other ions are expected to interact with DNA in a way similar to solvated Mg^{2+}.

SOLVATED MAGNESIUM IONS AS HYDROGEN BOND DONORS

Magnesium in Aqueous Solutions

The primary solvation shell of Mg^{2+} is formed by six water molecules that coordinate the cation with octahedral geometry, as shown in Figure 2 (see color insert). The nature of the charge-dipole interactions is such that water molecules are oriented with hydrogen atoms pointing outward. These 12 hydrogens are predicted to lie in the equatorial planes of the octahedron, as shown in Figure 2. Neutron diffraction studies (72) have verified the octahedral geometry of the solvated ion, yielding water-Mg^{2+} distances in excellent agreement with those predicted by theory. However, certain hydrogen atoms were located out of the equatorial planes, depending on their hydrogen bond partners. In any event, the structures shown in Figure 2 clearly illustrate that water molecules in the primary solvation shell will not usually act as hydrogen bond acceptors. In this review we try to determine whether solvated Mg^{2+} in the presence of DNA behaves as shown in Figure 2:

- How far is the octahedral shape distorted?
- Are other atomic groups, such as oxygens/nitrogens from either phosphate, sugar, or bases, found in place of water molecules?
- Are the six water molecules in the primary solvation shell always hydrogen bond donors as expected or may they also accept hydrogen bonds?

The Primary Solvation Shell

In order to analyze the primary solvation shell of Mg^{2+} in oligonucleotide complexes, a few well-defined solvated ions are discussed first, mainly from B-form DNA, which are given in Table 1. We have selected the highest possible resolution data and have chosen Mg^{2+} ions with low B factors. Some of them are displayed in Figures 3 and 4 (see color inserts). We should note that Mg^{2+} ions in RNA show similar features to those described below [for example, in (34)], but high resolution is seldom attained.

Inspection of Table 1 shows that the dimensions of the solvated Mg^{2+} ion are constant, with an average Mg^{2+}-O distance to water molecules of 207.5 pm, practically identical to that found in theoretical studies (208 pm) and by neutron diffraction (207 pm). There does not appear to be any appreciable influence of temperature on this value. The O-Mg-O angles in the octahedron are also close to 90°, but no detailed statistics on them are presented.

The standard deviation of the Mg-O distances has to be interpreted with care. Most standard deviations are close to the average value of 4 pm, which appears to be a reasonable value. We have not included in Table 1 some Mg^{2+} ions with low standard deviations ($\sigma = 1$ pm or less), which appear to have been refined by strongly restraining the Mg-O distance. On the other hand, a high value of σ is detected when the ions are less well defined in the electron density map. It may also indicate strong asymmetries in the atmosphere of the Mg^{2+} ion. For example,

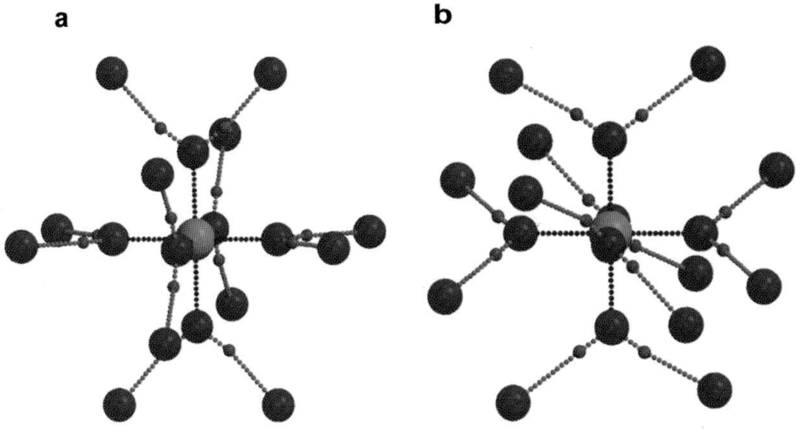

Figure 2 (*Left*) Predicted structure of a solvated Mg^{2+} ion according to Pavlov et al. (50). (*Right*) Experimental neutron diffraction structure of a solvated Mg^{2+} ion in a maleate salt (72). The structure of the octahedron formed by the six water molecules is practically the same in both cases, but the orientation of hydrogens differs significantly. Such orientation varies according to different authors (41), as discussed in the text.

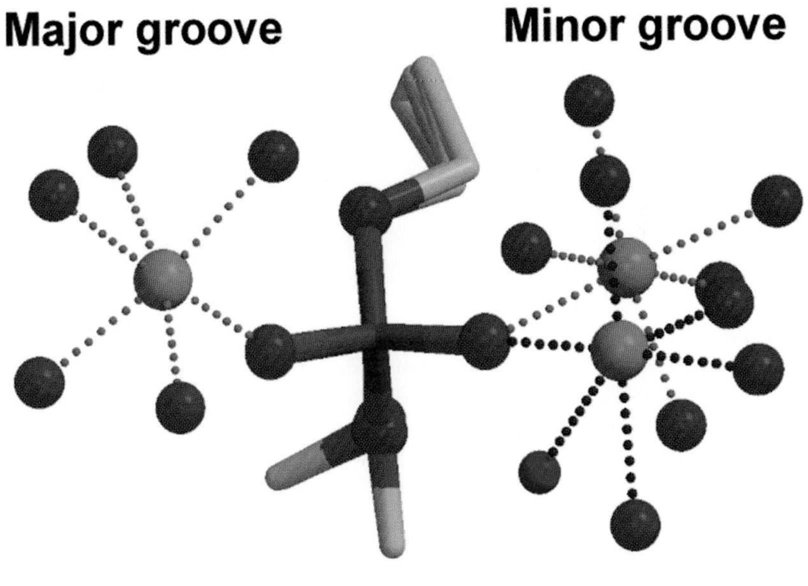

Figure 3 Direct Mg^{2+} interaction with phosphate oxygens. Three different structures have been superimposed on a fixed phosphate group. The cation may be found in both the major and minor groove sides. The phosphate may be in either the BI or BII conformation. No specific orientation is detected in such interaction.

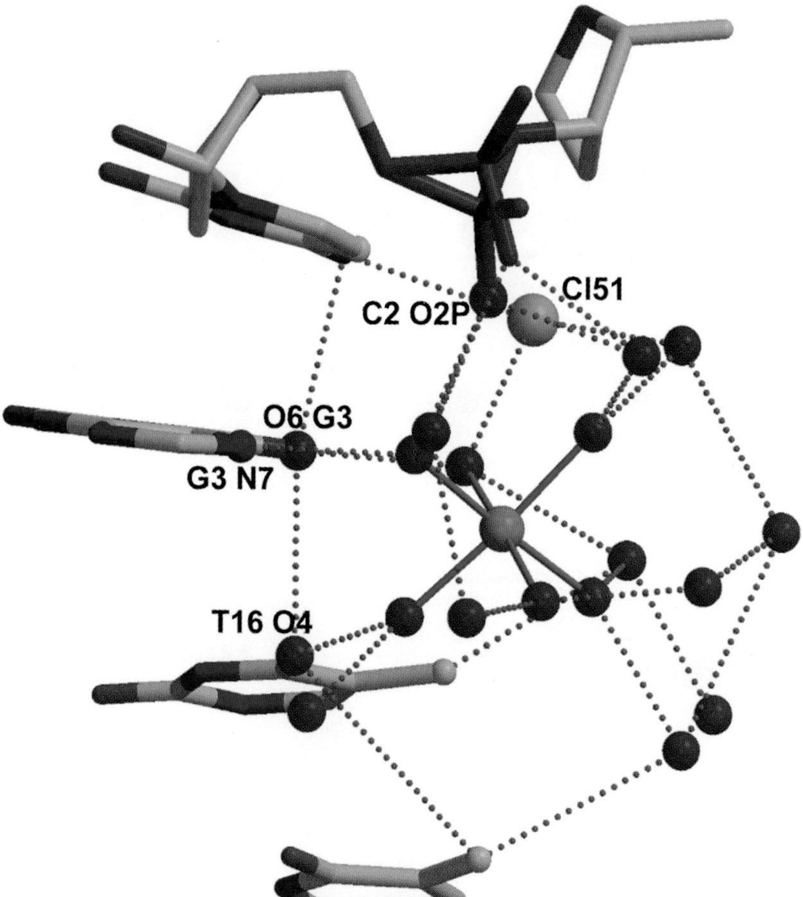

Figure 4 The secondary solvation shell of Mg19 in BD0037. Some of the hydrogen bonds and van der Waals interactions are shown as dotted lines. Waters in the primary solvation shell are connected by blue virtual bonds with the central Mg^{2+} ion. The secondary solvation shell forms an irregular cage around the primary solvation shell. Interactions between both shells are shown in greater detail in Figure 5.

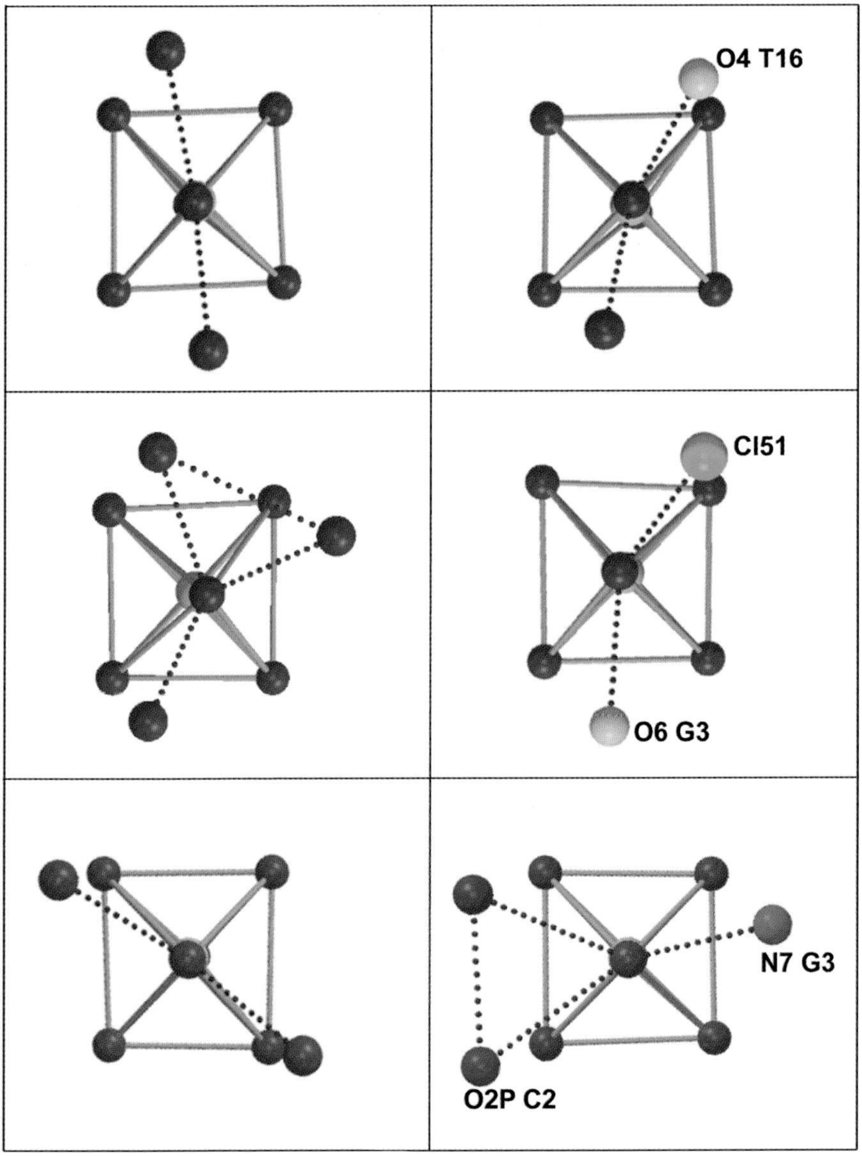

Figure 5 Views of Mg19 in BD0037 from each of its vertices. Water molecules are shown in red, one phosphate oxygen in purple, guanine N7 in blue, thymine and guanine oxygens in yellow, and chloride in green. Only about half of the atoms in the secondary solvation layer lie on an equatorial plane of the octahedron. Dotted lines indicate hydrogen bonds. Two of the waters of the primary layer show bifurcated hydrogen bonds. Water molecules in the primary hydration shell are connected by gray lines. They actually correspond to mainly repulsive oxygen-oxygen interactions.

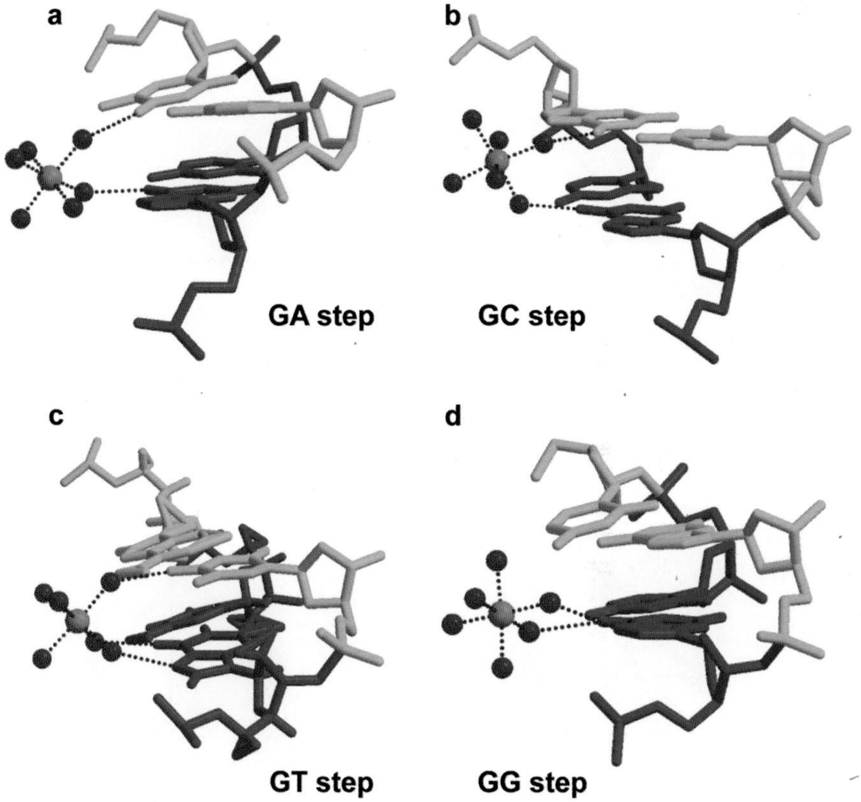

Figure 6 Binding of Mg to different guanine containing steps of B-form DNA. Guanine is always in the front side of the lower base pair. The ions have been taken from the following structures: (*a*) BD0037, (*b*) BD0007, (*c*) BD0033, and (*d*) BD0033.

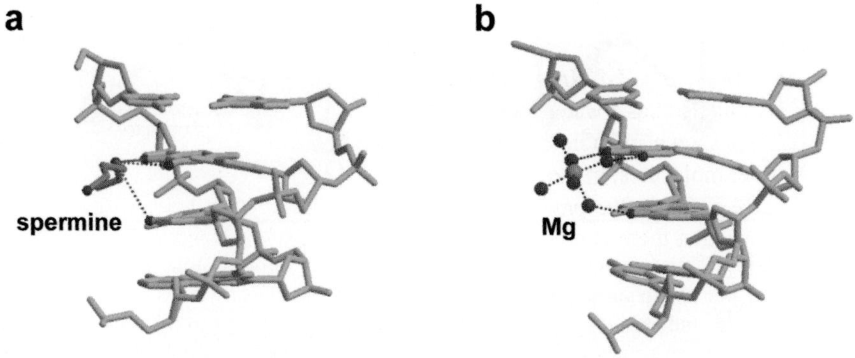

Figure 7 Comparison of spermine and Mg binding in a high-resolution structure of the classical Dickerson-Drew dodecamer (58). Both counterions are located in similar positions at opposite ends of the major groove of the duplex. Only the ends of the duplex with a sequence CGCG are shown in the figure.

TABLE 1 Dimensions of selected hexasolvated Mg^{2+} ions[a]

Structure	Atom No.	T(K)	Resolution (Å)	$\langle dH_2O \rangle$ (pm)	σ (pm)	d_{O^-} (pm)	d_N (pm)	Sequence	Reference
Theory				208	0				50
CIRVAA		300	.77	207	0.5				72
BDJ019	203	273	1.4	208	4.7			CCAACGTTGG	52
BDL084		136	1.4	207	2.2			CGCGAATTCGCG	57
BD0006	21	120	1.15	205	3.6			GGCCAATTGG	74
BD0006	22	120	1.15	212	9.9			GGCCAATTGG	74
BD0007	25	100	1.1	207	1.7			CGCGAATTCGCG	69
BD0007	32	100	1.1	206	4.6	207		CGCGAATTCGCG	69
BD0024		124	1.20	205	3.5	197		CCGAATGAGG	16
BD0030	25	120	.95	207	2.9			CGCGAATTCGCG	15
BD0030	32	120	.95	206	5.7			CGCGAATTCGCG	15
BD0033	111	100	.98	209	2.9			CCAACGTTGG	8
BD0033	114	100	.98	208	2.0			CCAACGTTGG	8
BD0037	19	120	.89	206	1.5			GCGAATTCG	62
BD0037	26	120	.89	206	3.7	201	233	GCGAATTCG	62
BD0037	38	120	.89	207	5.3	201		GCGAATTCG	62
BD0041		175	1.2	211	1.6			CGCGAATTCGCG	59
DD0003	26	293	1.6	211	3.7	207		CGCGAATTCGCG	9
AD0007	301	120	.84	207	2.6			GCGTATACGC	15
DDF027	14	—	1.0	208	12.3		234	CGCGCG	48
Average				207.5	4.1	203	233		

[a]The structures are described by their code in either the Cambridge Structural Database [CIRVAA (3)] or the Nucleic Acid Database (4). The atom number is that found in the .pdb file. Dimensions are given in picometers: ⟨dH2O⟩ is the average Mg-O distance to water molecules, σ is its standard deviation, d_{O^-} is the Mg-O distance to phosphate oxygens, and d_N is the Mg-N distance to the N7 atom of a guanine base.

one Mg^{2+} in a Z-DNA structure (DDF027) is not only associated with a guanine N7 but also has a water molecule shared with another Mg^{2+}-solvated ion.

As shown in Table 1, occasionally a phosphate oxygen substitutes one of the water molecules. In those cases the Mg-O distance decreases slightly; the average value found is 203 pm. Some of these ions are superimposed in Figure 3, which shows that there is no preferred location for Mg^{2+} ions when directly associated with DNA phosphates. In no case have sugar oxygens been found in the primary solvation shell of Mg^{2+}.

In exceptional cases there is a direct interaction of Mg^{2+} ions with the N7 atom of guanine. There are two reasons for such behavior. First, Mg^{2+} has no intrinsic affinity for complexation with nitrogen atoms. Second, such an interaction is unlikely in either A- or B-form DNA because the octahedral solvation shell would be strongly altered upon interaction (17). Thus the only two cases in which this interaction has been found are Z DNA (DDF027) and a terminal guanine base that forms a triplex (BD0037, Mg26) and has an unusual conformation. The Mg-N distance is significantly longer than the Mg-O distance, as is clearly apparent in Table 1. We have found no direct interactions of Mg^{2+} with the O6 atom of guanine in oligonucleotide crystals, although it has been described in RNA, where direct interactions with Mg^{2+} appear to occur more frequently than in DNA (34). This is due in part to the structural complexity of RNA, which has internal sites favorable for such interactions.

We should also note that detailed theoretical studies have been carried out (46, 51, 64) to determine the precise electronic properties of solvated Mg^{2+} ions when interacting with the DNA bases.

The Secondary Solvation Shell

The results presented in the previous section and in Table 1 show that the primary solvation shell is well organized as a regular octahedron with rigid dimensions. Only in a few cases do other atoms substitute water. We can now look at the secondary solvation shell. An example is given in Figure 4. A polyhedral network of hydrogen-bonded water molecules and other atoms, some of which are at Van der Waals distances, encloses the octahedral primary solvation shell.

In order to compare the structure of the secondary solvation shell with that expected from theory, a detailed view is given in Figure 5 (see color insert). Theory predicts (Figure 2) that all atoms in this shell will be hydrogen bond acceptors and should lie close to the equatorial planes of the octahedron (50). It is immediately clear from Figure 5 that about half of the hydrogen bond receptor atoms do not lie on the equatorial planes of the octahedron, as already found in the neutron diffraction structure (72) presented in Figure 2. An unexpected finding was that some of the waters in the primary solvation shell were hydrogen bonded to more than two atoms of the secondary solvation shell, apparently forming bifurcated hydrogen bonds. Inspection of the other ions given in Table 1 provides a similar view: a moderate deviation from the octahedral equatorial planes of the atoms in the secondary solvation shell and evidence for common bifurcated hydrogen bonds. We have found no clear evidence for any hydrogen bond being donated to water oxygens in the primary solvation shell.

Another feature of the secondary solvation shell is the occasional presence of hydrogen bonds among the water molecules in that shell. The model shown in Figure 2 (50) predicts that the latter waters do not form hydrogen bonds among themselves; rather they interact with water molecules in the tertiary solvation shell. Markham et al. (41) have analyzed other models in which every water in the secondary solvation shell is hydrogen bonded to one molecule of the primary solvation shell and two molecules in the secondary solvation shell. In the ions we have analyzed (Table 1) an intermediate situation is observed, with most water molecules in the secondary solvation shell interacting with only one water molecule in that shell. A few water molecules show either none or two hydrogen bonds as predicted by the different models (41).

With regard to the various pentamer groups involving the Mg^{2+} ion analyzed by Marhkam et al. (41), pentamers with two waters of the primary shell and two waters of the secondary shell are common, whereas those involving four water molecules of the secondary shell and one from the first shell are either rare or absent. In any case the models they analyzed (41) cannot be strictly compared with the structures found in DNA crystals because in DNA other atoms are found in the secondary solvation shell, as shown in Figures 4 and 5.

Therefore, the view we obtain by inspecting Table 1 and Figures 2, 3, 4, and 5 is that aqueous Mg^{2+} with 12 hydrogens in the primary solvation shell turns out to be a multiple hydrogen bond donor. Only in a few cases (Table 1) do we find a direct Mg-oxygen phosphate interaction, and even in those cases the Mg-O distance does not differ significantly from that found in Mg-water interactions. This is not unexpected because the phosphate charge is delocalized, as discussed in the Introduction (Figure 1).

Mg INTERACTION SITES

Interaction with Phosphates: A Crystallographic Glue

Practically all solvated Mg ions interact with one or more phosphate oxygens, either across the grooves or with different molecules in the crystal. Occasionally a phosphate oxygen, as shown in Figure 3 and Table 1, occupies one of the sites of the first solvation shell of Mg^{2+}. Interaction with phosphates in one duplex and bases in another is also common. Thus the solvated Mg ions stabilize the crystal lattice (8). This role as crystallographic glue is not specific; ions occupy different positions in different crystal structures.

Furthermore, Mg ions may be substituted by other ions, with or without changes in crystallographic packing, while the duplex structure does not change appreciably. A striking case is d(CGCGCG), which has been crystallized in isomorphous unit cells in the Z form with variable Mg-to-spermine ratios, as recently reviewed by Egli (14). In other cases it appears that counterions occupy many alternative sites in disordered regions of the crystal, such that the precise location cannot be determined. An example (33) is the highly symmetric structure of the barium salt of the classical d(CGCGAATTCGCG) duplex, in which no barium ions are detected in spite of their large electron density. In no crystal structure of oligonucleotides in the B form have all the expected counterions been found because some of them are in regions of disordered solvent. In other cases they are detected with partial occupancy.

Interaction with the Bases: Site Binding

As discussed in the Introduction, the bases have several electronegative sites, which in the case of A · T pairs are mainly at the minor groove side. However, the minor groove is often too narrow to accommodate octahedral solvated ions. Thus solvated divalent ions rarely penetrate inside the minor groove. Rather, they occupy a more external position and close the lips of the minor groove (59). Some exceptions are noted, either when the minor groove is wider than normal (40) or at the terminal steps of the duplexes (40, 61).

In the major groove side guanines offer an excellent anchoring point for solvated Mg ions. However, a single G · C pair does not appear to be sufficient to stabilize bound solvated Mg ions, and usually some water molecules are bound to bases in the next step. Examples of all the GN base steps in B-form DNA are presented in Figure 6 (see color insert). Thus it appears that there is no sequence specificity in binding; only guanine is required and the interaction is further stabilized by hydrogen bonding to a neighbor base pair. A similar situation prevails in RNA and DNA-RNA hybrids (54). In Figure 6 we have illustrated binding to the next step in the sequence, but binding to the previous step has also been found (8, 62). Another case is Mg in some A-form duplexes, where the major groove defines a central channel in the structure in which ions may be accommodated.

Influence of Experimental Conditions

The solvated ions discussed up to now only present minor variations from perfect octahedral geometry. However, in practice, solvated Mg ions may appear considerably distorted, even in high-resolution structures. A few examples are summarized in Table 2. The two structures BD0037 and UDI030 are isomorphous and correspond to the same nonamer d(GCGAATTCG) crystallized under different conditions. In the room temperature structure, only two Mg ions were detected, corresponding to Mg19 and Mg26 in BD0037. Surprisingly, they show larger dimensions and a much higher variability than those identified at low temperature, as indicated by the σ value. Evidently some form of disorder is present. It cannot be

TABLE 2 Influence of crystal order on apparent Mg features[a]

Structure	T(K)	Atom No.	Resolution (Å)	B factor[b]	⟨dH$_2$O⟩ (pm)	σ (pm)	Ref.
BD0037	120	19	.89	6.5	206	1.5	62
UDI030	291	19	2.05	35.8	227	16.8	73
BD0037	120	26	.89	7.6	206	3.7	62
UDI030	291	20	2.05	29.3	251	19.4	73
BD0037	120	31	.89	9.5	206	4.7	62
BD0037	120	38	.89	10.5	207	5.3	62
BD0037	120	44	.89	29.5	211	10.6	62
BD0037	120	49	.89	22.3	219	28.3	62
BD0007	100	25	1.1	12.8	207	1.7	69
BD0007	100	32	1.1	18.0	206	4.6	69
BD0007	100	38	1.1	21.6	208	7.8	69
BD0007	100	45[c]	1.1	25.4	215	6.1	69
BD0007	100	51[c]	1.1	23.9	224	24.1	69

[b]Displacement parameter of the Mg^{2+} atoms (Å^2), which may be related to the mean-square amplitude (u^2) of atomic vibration: $B = 8\pi^2 u^2$.

[c]Occupancy 50%.

due to an expansion of the hydration shell with temperature because the theoretical calculations show that the influence of temperature is negligible. Furthermore, as shown in Table 1, the experimentally determined octahedral dimensions in high-resolution structures at room temperature do coincide with those found at cryogenic temperatures. Disorder cannot be due to alternative positions of perfect octahedra either: The average of two or more displaced octahedra would give broader peaks for the atoms, but unchanged dimensions. Indeed, in some cases (8, 35) it is possible to model partially ordered ions as the sum of two or more ordered ions with partial occupancy. Therefore, the only possible explanation is that the Mg sites are only partially occupied and may be substituted either by other ions or by water molecules. The increase in dimensions and variability of the room-temperature (UDI030) Mg solvation when compared with the cryogenic structure (BD0037) of the same sequence should be attributed to partial displacement of the ion by either water or other ions present in the crystallization medium. Mg20 in UDI030, for example, has an average Mg-O distance of 251 pm, about half way between the standard Mg-O distance (207 pm) and a water hydrogen bond (280–300 pm), and it might indicate that in that structure this site is occupied about 50% of the time by Mg and the other 50% by a water molecule.

Tsunoda et al. (71) have obtained a direct demonstration of the possible substitution of a solvated Mg^{2+} ion by a network of water molecules. The authors compared two isomorphous structures of a dodecamer obtained in the presence and absence of Mg^{2+}. Water molecules in one case substituted the Mg^{2+} site in the other.

The variable degree of order of the ions within the same structure can be explained along similar lines. From the examples given in Table 2, it is clear that the apparent dimensions of solvated Mg ions increase as their B factors increase. Similarly, high B factors imply that the site is only partially occupied by Mg, and hence other ions and water may partially occupy its place.

OTHER CATIONS

Divalent Cations

Because most oligonucleotide crystals have been obtained from solutions containing Mg^{2+} and spermine, relatively little is known about the detailed structure of other divalent cations bound to DNA. Nevertheless, the few available structures indicate that such cations behave similarly to Mg^{2+}. We briefly consider four examples in particular. For additional views of the interaction of solvated ions with DNA, follow the Supplemental Material link on the Annual Reviews homepage at http://www.annualreviews.org/.

Ca^{2+} At least two high-resolution structures of a DNA-bound calcium ion have been reported (35, 44). Compared to that of Mg^{2+}, the primary solvation shell at Ca^{2+} is poorly defined (53), with a variable number of water molecules (six to eight) and a wider range of ion-oxygen distances (usually in the 230–290 pm range). As a result, it is difficult to define precisely the geometry of the solvated ion, as discussed in detail by Kielkopf et al. (35). Indeed, the ions are often modeled as two closely spaced ions with only partial occupancies. Nevertheless, the following three general observations can be made:

- The Ca^{2+} ions usually interact with DNA bases and phosphates through water molecules of the primary shell. In a few cases direct Ca-O (P) and Ca-base contacts are found. The latter always involve a terminal guanine at either the N7 or O6 atom.
- As with Mg^{2+}, the solvated Ca^{2+} ions act as net hydrogen bond donors; no single case of hydrogen bond acceptance in the primary shell has been detected.
- The solvated ions also act as a glue between different duplexes in the crystal.

Minasov et al. (44) have provided a detailed comparison of Mg^{2+} and Ca^{2+} associated with the same sequence in different space groups. They have shown how distortions of the duplex in the crystal are associated with ion binding at certain positions.

Ni^{2+}, Co^{2+} Most of the structures analyzed in the presence of these ions are complex and involve drugs associated with DNA in the crystal. Only a few of them contain standard DNA duplexes. These ions have an octahedral solvation shell with dimensions practically identical to Mg^{2+} (50, 53), although the interactions with DNA are dominated by the strong affinity of the transition metal ion for the N7 atom of guanine. As shown by Gao et al. (17), such interaction cannot occur within a standard A- or B-form DNA duplex, but rather can only occur in terminal or extra-helical guanine bases. This results in a unique crystal lattice, determined by guanine-N7-ion-N7-guanine bridges (1, 78). In some cases (1, 2) additional solvated ions have been detected, and they stabilize interactions among duplexes similar to those found in Mg^{2+} crystals.

Mn^{2+} This ion is considered to be quite similar to Mg^{2+} (5, 53) and it may be used alternatively in some enzymatic reactions (18). However, the ion-water distance is somewhat greater (about 230 pm). Unfortunately, there is no crystallographic structure available of oligonucleotides in the presence of this ion. There is only an NMR study (29) showing that this ion may enter the minor groove of A-track DNA. On the other hand, nucleosome core crystals have been obtained in its presence. In the Oak Ridge structure (27), 17 Mn^{2+} ions have been tentatively localized. Most of them interact directly with the N7 atom of guanines.

Monovalent Cations

Most monovalent cations have a solvation layer with a variable intrinsic number of water molecules, as well as variable positions and ion-water distances (26, 53). As a result, no Na^+ has been unambiguously located in either A- or B-form DNA structures. An additional complication is the difficulty in distinguishing Na^+ ions from water molecules as they exhibit equal electron densities. A method has been presented to distinguish Na^+ from water (47). It can be simplified since a solvated Na^+ ion should normally have at least five neighbors (in general water molecules), three of which should be at a distance shorter than 260 pm. The latter requirements are incompatible with water structure. The Na^+ ions that have been localized in oligonucleotide crystals do obey these rules. They have been analyzed by Tereshko et al. [see Table S1 in (70)]. Practically all of them belong to structures that are not standard and have special sites where the ions are located. One of the reasons for the clear localization of Na^+ in such structures is the low content of water in the crystal.

An extensive attempt to localize monovalent cations has been recently reported (70) in several related A-DNA decamers. Three different binding sites were localized in the crystal. These sites may be occupied by different monovalent cations with a variable occupancy and small geometrical differences. It is interesting to note that the average distance of the central ion to the atoms and water molecules in the first solvation layer tend to be higher than expected. The latter observation indicates that water molecules may occasionally displace the ions, as it is the case for Mg^{2+} ions as discussed above.

In the case of B-form DNA a considerable amount of work (21, 28, 58, 68–70, 77) has been carried out in order to determine up to what extent naked monovalent ions could penetrate the minor groove of DNA, which is usually occupied by water molecules, forming a sequence-dependent water spine (58, 61, 68). It seems that ion coordination in the minor groove is an isolated event that does not significantly alter the geometry of DNA. We should note that, compared to the octahedral geometry of the primary solvation shell of cations, the tetrahedral geometry of water molecules is more adequate for forming a hydration spine in the minor groove of B-form DNA. Nevertheless, some water molecules may be partially (10%–30%) displaced by monovalent cations (28). K^+ and Rb^+ might displace water molecules more easily than Na^+ and Cs^+ (68), although NMR experiments (12) indicate that all alkaline cations bind to the minor groove with similarly weak affinity. Ammonium is a special case because it enters in the minor groove (31) and induces a conformational change in the AATT sequence (60).

Amines and Basic Amino Acids

Amines and basic amino acids usually contribute to neutralize DNA charges. However, they are essentially different from inorganic ions because they have no clear solvation shell. In high-resolution crystals of low-molecular-weight peptides, amino groups interact directly with negatively charged counterions. Additional water molecules associated with either amino or guanidinium groups are seldom found [(6) and references within]. A similar situation is observed in proteins (39).

Basic amino acid side chains in proteins may interact with DNA in two ways: sequence recognition and charge neutralization. For specific recognition, the O6 and N7 atoms of guanine are favorite points of interaction for arginine, in a way similar to the interaction of solvated Mg^{2+} ions. Other types of interaction can also occur, as reviewed in detail by Luscombe et al. (39).

For charge neutralization, the relative location of DNA and charged amino acids is much more variable. The side chains approach phosphates either directly or through water molecules in any spatial orientation (27).

A particular case is spermine. It is often used as a coprecipitant to obtain oligonucleotide crystals. It has been observed in crystals of A and Z DNA. In the former case it may occupy different regions on the duplex (75). In Z DNA, it has been clearly localized, as reviewed by Egli (14). In B-DNA crystals it has only been partially detected in a few cases. An example is shown in Figure 7 (see color insert). Spermine may occupy regions similar to Mg^{2+}, although in general it cannot be easily detected in crystallographic studies because it may occupy many alternative positions in any crystal structure. Detailed molecular dynamics simulations (36, 37) confirm the versatility of spermine molecules, which may form "invisible" bridges among different duplexes in a crystal. It does not have any significant influence on the native structure of B DNA (11).

STATIC AND DYNAMIC VIEWS OF IONIC INTERACTIONS

The results presented in Table 1 and in Figures 4 and 5 show a static view of Mg^{2+} ions. Comparison with a lower-resolution structure (Table 2) allows a more dynamic view of ionic interactions: The solvated ions tend to be on the major groove side hydrogen bonded to G · C-rich regions and on the minor groove side facing A · T-rich regions. Different ions in the solvent may replace each other on these sites in a dynamic way (71). Substitution of ions has been also directly proven by Tereshko et al. (70). Spermine and Mg^{2+} may replace each other on the same sequence (Figure 7).

Direct cation-DNA contacts are not common, as shown in Table 1, with the noted exception of transition metals. They have been observed with monovalent cations such as Rb^{+} (14, 44) and Tl^{+} (28), which have a large ionic radius and an ill-defined first solvation shell (53). In the latter cases, owing to the loose geometric requirements, it is easier to replace a water molecule by a DNA atom. These sites appear to be in dynamic exchange with water molecules, occasionally part of a solvated ion.

Accurate molecular dynamics simulations, taking into account the detailed hydrogen bond geometry and energy of ion solvation, should obtain a dynamic view of ion-DNA interactions. Such studies are not currently available.

CONCLUSIONS

1. Mg^{2+} has a rigid octahedral primary solvation shell. Other ions have a more variable distribution of ligands.
2. Solvated ions are net hydrogen bond donors. Hexahydrated Mg^{2+}, for example, usually donates 12 or more hydrogen bonds. Values larger than 12 are due to bifurcation of some of the hydrogen bonds.

3. Ion-DNA interactions are usually mediated by water. Only in some cases may a phosphate oxygen substitute one of the water molecules in the primary solvation shell. More rarely the N7 and O6 atoms of guanine may be part of the primary solvation shell. On the other hand, transition metals have a strong tendency to interact directly with the N7 atom of guanine.
4. The most common interaction of solvated cations with the DNA bases takes place at guanines on the major groove. It usually involves the next base pair in the sequence. No sequence specificity appears to be present, only guanine is required.
5. The same site may be occupied by different ions in the solvent. Water molecules may also replace ions. As a result of such replacements, many ion sites show a substantial disorder (Table 2) and are often invisible in the electron density map. These effects are increasingly noticeable at lower resolutions. Diffraction data with limited resolution (~1.6 Å and upward) are clearly unable to determine metal-donor distances with precision, as is found in proteins (25).
6. Solvated cations are essential for the stability of oligonucleotide crystals. They form phosphate-phosphate and phosphate-base bridges among different duplexes, acting as a glue that stabilizes the crystal lattice.
7. Crystals obtained with a low water content, as is commonly found in mono- and dinucleotides, show an increasing number of direct ion–nucleic acid contacts. Such interactions will normally be absent in fully hydrated DNA molecules.

ACKNOWLEDGMENTS

We are thankful to Dr. B. Schneider for help in searching the NDB to find cations and to Drs. I. Fita, J. Luque, and C. Petosa for helpful suggestions. This work has been supported in part by PM98-0135 grant from the Ministerio de Ciencia y Tecnología and 2001SGR00250 grant from the Generalitat de Catalunya.

LITERATURE CITED

1. Abrescia NGA, Malinina L, Fernández LG, Huynh-Dinh T, Neidle S, Subirana JA. 1999. Structure of the oligonucleotide d(CGTATATACG) as a site-specific complex with nickel ions. *Nucleic Acids Res.* 27:1593–99
2. Abrescia NGA, Malinina L, Subirana JA. 1999. Stacking interaction of guanine with netropsin in the minor groove of d(CGTATATACG)$_2$. *J. Mol. Biol.* 294:657–66
3. Allen FH, Kennard O. 1993. 3D search using the Cambridge Structural Database. *Chem. Des. Autom. News* 8:31–37
4. Berman HM, Olson WK, Beveridge DL, Westbrook J, Gelbin A, et al. 1992. The

nucleic acid database: a comprehensive relational database of three-dimensional structures of nucleic acids. *Biophys. J.* 63: 751–59

5. Bock CW, Katz AK, Markham GD, Glusker JP. 1999. Manganese as a replacement for magnesium and zinc: functional comparison of the divalent ions. *J. Am. Chem. Soc.* 121:7360–72
6. Boqué L, Verdaguer N, Urpí L, Fita I, Subirana JA. 1989. Crystal and molecular structure of L-lysyl-L-valine hydrochloride—a new lysine conformation. *Int. J. Pept. Res.* 33:157–61
7. Cheatham TE III, Young MA. 2001. Molecular dynamics simulation of nucleic acids: successes, limitations, and promise. *Biopolymers* 56:232–56
8. Chiu TK, Dickerson RE. 2000. 1 Å crystal structures of B-DNA reveal sequence-specific binding and groove-specific binding of DNA by magnesium and calcium. *J. Mol. Biol.* 301:915–45
9. Clark GR, Squire CJ, Baker LJ, Martin RF, White J. 2000. Intermolecular interactions and water structure in a condensed phase B-DNA crystal. *Nucleic Acids Res.* 28:1259–65
10. Coll M, Solans X, Font-Altaba M, Subirana JA. 1987. Crystal and molecular structure of the sodium salt of the dinucleotide duplex d(CpG). *J. Biomol. Struct. Dyn.* 4:797–811
11. Deng H, Bloomfield VA, Benevides JM, Thomas GJ Jr. 2000. Structural basis of polyamine-DNA recognition: spermidine and spermine interactions with genomic B-DNAs of different GC content probed by Raman spectroscopy. *Nucleic Acids Res.* 28:3379–85
12. Denisov VP, Halle B. 2000. Sequence-specific binding of counterion to B-DNA. *Proc. Natl. Acad. Sci. USA* 97:629–33
13. Dickerson RE. 2001. Nucleic acids. In *International Tables for Crystallography. Vol. F*, ed. MG Rossman, E Arnold, pp. 588–622. Dordrecht: Kluwer
14. Egli M. 2002. DNA-cation interactions: Quo vadis? *Chem. Biol.* 9:277–86
15. Egli M, Tereshko V, Teplova M, Minasov G, Joachimiak A, et al. 1999. X-ray crystallographic analysis of the hydration of A- and B-form DNA at atomic resolution. *Biopolymers* 48:234–52
16. Gao YG, Robinson H, Sanishvili R, Joachimiak A, Wang AH-J. 1999. Structure and recognition of sheared tandem G·A base pairs associated with human centromere DNA sequence at atomic resolution. *Biochemistry* 38:16452–60
17. Gao YG, Sriram M, Wang AH-J. 1993. Crystallographic studies of metal ion-DNA interactions: different binding modes of cobalt(II), copper(II) and barium(II) to N^7 of guanines in Z-DNA and a drug-DNA complex. *Nucleic Acids Res.* 21:4093–101
18. Glusker JP. 1999. The binding of ions to proteins. In *Protein: A Comprehensive Treatise. Vol. II*, ed. G. Allen, pp. 99–152. Amsterdam: Elsevier
19. Glusker JP, Katz AK, Bock CW. 1999. Metal ions in biological systems. *Rigaku J.* 16:8–16
20. Halle B, Denisov VP. 1999. Water and monovalent ions in the minor groove of B-DNA oligonucleotides as seen by NMR. *Biopolymers* 48:210–33
21. Hamelberg D, McFail-Isom L, Williams LD, Wilson WD. 2000. Flexible structure of DNA: ion dependence of minor-groove structure and dynamics. *J. Am. Chem Soc.* 122:10513–20
22. Hamelberg D, Williams LD, Wilson WD. 2001. Influence of the dynamic positions of cations on the structure of the DNA minor groove: sequence-dependent effects. *J. Am. Chem. Soc.* 123:7745–55
23. Harding MM. 1999. The geometry of metal-ligand interactions relevant to proteins. *Acta Crystallogr. D* 55:1432–43
24. Harding MM. 2000. The geometry of metal-ligand interactions relevant to proteins. II. Angles at the metal atom, additional weak metal-donor interactions. *Acta Crystallogr. D* 56:857–67

25. Harding MM. 2001. Geometry of metal-ligand interactions in proteins. *Acta Crystallogr. D* 57:401–11
26. Harding MM. 2002. Metal-ligand geometry relevant to proteins and in proteins: sodium and potassium. *Acta Crystallogr. D* 58:872–74
27. Harp JM, Hanson BL, Timm DE, Bunick GJ. 2000. Asymmetries in the nucleosome core particle at 2.5 Å resolution. *Acta Crystallogr. D* 56:1513–34
28. Howerton SB, Sines CC, Vanderveer D, Williams LD. 2001. Locating monovalent cations in the grooves of B-DNA. *Biochemistry* 40:10023–31
29. Hud NV, Feigon J. 1997. Localization of divalent metal ions in the minor groove of DNA A-tracts. *J. Am. Chem. Soc.* 119:5756–57
30. Hud NV, Polak M. 2001. DNA-cation interactions: the major and minor grooves are flexible ionophores. *Curr. Opin. Struct. Biol.* 11:293–301
31. Hud NV, Sklenár V, Feigon J. 1999. Localization of ammonium ions in the minor groove of DNA duplexes in solution and the origin of DNA A-tract bending. *J. Mol. Biol.* 286:651–60
32. Jerkovic B, Bolton PH. 2001. Magnesium increases the curvature of duplex DNA that contains dA tracts. *Biochemistry* 40:9406–11
33. Johansson E, Parkinson G, Neidle S. 2000. A new crystal form for the dodecamer C-G-C-G-A-A-T-T-C-G-C-G: symmetry effects on sequence-dependent DNA structure. *J. Mol. Biol.* 300:551–61
34. Juneau K, Podell E, Harrington DJ, Cech TR. 2001. Structural basis of the enhanced stability of a mutant ribozyme domain and a detailed view of RNA-solvent interactions. *Structure* 9:221–31
35. Kielkopf CL, Ding S, Kuhn P, Rees DC. 2000. Conformational flexibility of B-DNA at 0.74 Å resolution: d(CCAGTACTGG)$_2$. *J. Mol. Biol.* 296:787–801
36. Korolev N, Lyubartsev AP, Laaksonen A, Nordenskiöld L. 2002. On the competition between water, sodium ions, and spermine in binding to DNA: a molecular dynamics computer simulation study. *Biophys. J.* 82:2860–75
37. Korolev N, Lyubartsev AP, Nordenskiöld L, Laaksonen A. 2001. Spermine: an "invisible" component in the crystals of B-DNA. A grand canonical Monte Carlo and molecular dynamics simulation study. *J. Mol. Biol.* 308:907–17
38. Lavery R, Pullman B. 1981. Molecular electrostatic potential on the surface envelopes of macromolecules: B-DNA. *Int. J. Quantum Chem.* 20:259–72
39. Luscombe NM, Laskowski RA, Thornton JM. 2001. Amino acid-base interactions: a three-dimensional analysis of protein-DNA interactions at an atomic level. *Nucleic Acids Res.* 29:2860–74
40. Mack DR, Chiu TK, Dickerson RE. 2001. Intrinsic bending and deformability at the T-A step of CCTTTAAAGG: a comparative analysis of T-A and A-T steps within A-tracts. *J. Mol. Biol.* 312:1037–49
41. Markham GD, Glusker JP, Bock CW. 2002. The arrangement of first- and second-sphere water molecules in divalent magnesium complexes: results from molecular orbital and density functional theory and from structural crystallography. *J. Phys. Chem. B* 106:5118–34
42. McConnell KJ, Beveridge DL. 2001. Molecular dynamics simulations of B′-DNA: sequence effects on A-tract-induced bending and flexibility. *J. Mol. Biol.* 314:23–40
43. McFail-Isom L, Sines CC, Williams LD. 1999. DNA structure: cations in charge? *Curr. Opin. Struct. Biol.* 9:298–304
44. Minasov G, Tereshko V, Egli M. 1999. Atomic-resolution crystal structures of B-DNA reveal specific influences of divalent metal ions on conformation and packing. *J. Mol. Biol.* 291:83–99
45. Misra VK, Draper DE. 2001. A thermodynamic framework for Mg^{2+} binding to

RNA. *Proc. Natl. Acad. Sci. USA* 98: 12456–61
46. Muñoz J, Sponer J, Hobza P, Orozco M, Luque FJ. 2001. Interactions of hydrated Mg^{2+} cation with bases, base pairs, and nucleotides. Electron topology, natural bond orbital, electrostatic, and vibrational study. *J. Phys. Chem. B* 105:6051–60
47. Nayal M, Di Cera E. 1996. Valence screening of water in protein crystals reveals potential Na^+ binding sites. *J. Mol. Biol.* 256:228–34
48. Ohishi H, Kunisawa S, Van der Marel G, Van Boom JH, Rich A, et al. 1991. Interaction between the left-handed Z-DNA and polyamine. The crystal structure of the $d(CG)_3$ and N-(2-aminoethyl)-1, 4-diamino-butane complex. *FEBS Lett.* 284:238–44
49. Padiyar SG, Seshadri TP. 1998. Metal-nucleotide interactions: crystal structures of alkali (Li^+, Na^+, K^+) and alkaline earth (Ca^{2+}, Mg^{2+}) metal complexes of adenosine 2′-monophosphate. *J. Biomol. Struct. Dyn.* 15:803–21
50. Pavlov M, Siegbahn PEM, Sandström M. 1998. Hydration of beryllium, magnesium, calcium, and zinc ions using density functional theory. *J. Phys. Chem. A* 102:219–28
51. Petrov AS, Lamm G, Pack GR. 2002. Water-mediated magnesium-guanine interactions. *J. Phys. Chem. B* 106:3294–300
52. Privé GG, Yanagi K, Dickerson RE. 1991. Structure of the B-DNA decamer C-C-A-A-C-G-T-T-G-G and comparison with the isomorphous decamers C-C-A-A-G-A-T-T-G-G and C-C-A-G-G-C-C-T-G-G. *J. Mol. Biol.* 217:177–99
53. Richens DT. 1997. *The Chemistry of Aqua Ions*. Chichester: Wiley. 567 pp.
54. Robinson H, Gao YG, Sanishvili R, Joachimiak A, Wang AH-J. 2000. Hexahydrated magnesium ions bind in the deep major groove and at the outer mouth of A-form nucleic acid duplexes. *Nucleic Acids Res.* 28:1760–66
55. Schneider B, Patel K, Berman HM. 1998. Hydration of the phosphate group in double-helical DNA. *Biophys. J.* 75:2422–34
56. Seeman NC, Rosenberg JM, Rich A. 1976. Sequence-specific recognition of double helical nucleic acid by proteins. *Proc. Natl. Acad. Sci. USA* 73:804–8
57. Shui X, McFail-Isom L, Hu GG, Williams LD. 1998. The B-DNA dodecamer at high resolution reveals a spine of water on sodium. *Biochemistry* 37:8341–55
58. Shui X, Sines CC, McFail-Isom L, VanDerveer D, Williams LD. 1998. Structure of the potassium form of CGCGAATTCGCG: DNA deformation by electrostatic collapse around inorganic cations. *Biochemistry* 37:16877–87
59. Sines CC, Mc-Fail-Isom L, Howerton SB, VanDerveer D, Williams LD. 2000. Cations mediate B-DNA conformational heterogeneity. *J. Am. Chem. Soc.* 122:11048–56
60. Snoussi K, Leroy JL. 2001. Ammonium induces a conformational change in the Dickerson dodecamer. *J. Biomol. Struct. Dyn.* 6:961 (Abstr.)
61. Soler-López M, Malinina L, Liu J, Huynh-Din T, Subirana JA. 1999. Water and ions in a high resolution structure of B-DNA. *J. Biol. Chem.* 274:23683–86
62. Soler-López M, Malinina L, Subirana JA. 2000. Solvent organization in an oligonucleotide crystal. *J. Biol. Chem.* 275:23034–44
63. Soler-López M, Malinina L, Tereshko V, Zarytova V, Subirana JA. 2002. Interaction of zinc ions with d(CGCAATTGCG) in a 2.9 Å resolution X-ray structure. *J. Biol. Inorg. Chem.* 7:533–38
64. Sponer J, Leszczynski J, Hobza P. 2002. Electronic properties, hydrogen bonding, stacking, and cation binding of DNA and RNA bases. *Biopolymers* 61:3–31
65. Stellwagen NC, Magnusdottir S, Gelfi C, Righetti PG. 2001. Preferential counterion binding to A-tract DNA oligomers. *J. Mol. Biol.* 305:1025–33
66. Deleted in proof

67. Swaminathan V, Sundaralingam M. 1979. The crystal structures of metal complexes of nucleic acids and their constituents. *CRC Crit. Rev. Biochem.* 6:245–336
68. Tereshko V, Minasov G, Egli M. 1999. A "hydrat-ion" spine in a B-DNA minor groove. *J. Am. Chem. Soc.* 121:3590–95
69. Tereshko V, Minasov G, Egli M. 1999. The Dickerson-Drew B-DNA dodecamer revisited at atomic resolution. *J. Am. Chem. Soc.* 121:470–71
70. Tereshko V, Wilds CJ, Minasov G, Prakash TP, Maier MA, et al. 2001. Detection of alkali metal ions in DNA crystals using state-of-the-art X-ray diffraction experiments. *Nucleic Acids Res.* 29:1208–15
71. Tsunoda M, Karino N, Ueno Y, Matsuda A, Takenaka A. 2001. Crystallization and preliminary X-ray analysis of a DNA dodecamer containing 2′-deoxy-5-formyl-uridine: What is the role of magnesium cation in crystallization of Dickerson-type DNA dodecamers? *Acta Crystallogr. D* 57:345–48
72. Vanhouteghem F, Lenstra ATH, Schweiss P. 1987. Magnesium bis(hydrogen maleate) hexahydrate, $[Mg(C_4H_3O_4)_2]$. $6H_2O$, studied by elastic neutron diffraction and ab initio calculation. *Acta Crystallogr. B* 43:523–28
73. Vlieghe D, Turkenburg JP, Van Meervelt L. 1999. B-DNA at atomic resolution reveals extended hydration patterns. *Acta Crystallogr. D* 55:1495–502
74. Vlieghe D, Van Meervelt L, Dautant A, Gallois B, Précigoux G, et al. 1996. Formation of (C · G) *G triplets in a B-DNA duplex with overhanging bases. *Acta Crystallogr. D* 52:766–75
75. Wahl MV, Sundaralingam M. 1997. Crystal structures of A-DNA duplexes. *Biopolymers* 44:15–63
76. Williams LD, Maher LJ III. 2000. Electrostatic mechanisms of DNA deformation. *Annu. Rev. Biophys. Biomol. Struct.* 29:497–521
77. Woods KK, McFail-Isom L, Sines CC, Howerton SB, Stephens RK, Williams LD. 2000. Monovalent cations sequester within the A-tract minor groove of $[d(CGCGAATTCGCG)]_2$. *J. Am. Chem. Soc.* 122:1546–47
78. Yang XL, Robinson H, Gao YG, Wang AH-J. 2000. Binding of a macrocyclic bisacridine and ametantrone to CGTACG involves similar unusual intercalation platforms. *Biochemistry* 39:10950–57

Annu. Rev. Biophys. Biomol. Struct. 2003. 32:47–67
doi: 10.1146/annurev.biophys.32.110601.142429

First published online as a Review in Advance on February 6, 2003

OPTICAL SINGLE TRANSPORTER RECORDING: Transport Kinetics in Microarrays of Membrane Patches

Reiner Peters
Institut für Medizinische Physik und Biophysik, Universität Münster, 48149 Münster, Germany: email: petersr@uni-muenster.de

Key Words confocal microscopy, membrane transport kinetics, membrane transporters, nuclear transport, photobleaching

■ **Abstract** Optical single transporter recording (OSTR) is an emerging technique for the fluorescence microscopic measurement of transport kinetics in membrane patches. Membranes are attached to transparent microarrays of cylindrical test compartments (TCs) ~0.1–100 μm in diameter and ~10–100 μm in depth. Transport across membrane patches that may contain single transporters or transporter populations is recorded by confocal microscopy. By these means transport of proteins through single nuclear pore complexes has been recorded at rates of <1 translocation/s. In addition to the high sensitivity in terms of measurable transport rates OSTR features unprecedented spatial selectivity and parallel processing. This article reviews the conceptual basis of OSTR and its realization. Applications to nuclear transport are summarized. The further development of OSTR is discussed and its extension to a diversity of transporters, including translocases and ATP-binding cassette (ABC) pumps, projected.

CONTENTS

INTRODUCTION

The eukaryotic cell, a microcosm of membrane-enclosed nano-compartments, maintains an abundance of membrane-spanning proteins mediating the selective transfer of all sorts of molecules. Here, all such proteins are subsumed as transporters discriminating between channels, carriers, and pumps (for a proposed classification system, see Reference 59). Channels are characterized by large transport rates (10^7–10^8 molecules s^{-1} at a concentration difference of ~150 mM) and an approximately linear dependence of the transport rate on concentration over a large concentration range. Carriers have much smaller transport rates (10^1–10^4 molecules s^{-1}) and show saturation in the micro- to millimolar range. Pumps, which are translocating substrates against electrochemical potential differences, have still smaller transport rates (10^0–10^3 molecules s^{-1}). The area density of transporters varies also in wide limits. The sodium potassium pump occurs at 500–5000 copies per μm^2 in nerve cell membranes (93), but less than 1 copy per μm^2 in erythrocyte membranes (4). The density of the nuclear pore complex (NPC), a large transporter spanning the nuclear envelope, varies between 3 copies μm^{-2} in avian erythrocytes and 70 copies μm^{-2} amphibian oocytes (49).

Emergence of Intracellular Transporters

Currently, a stream of new transporters is emerging from two sources. First, genomic analysis, already available for several pro- and eukaryotic organisms including *Homo sapiens*, is shedding new light on the abundance and subcellular distribution of transporters (2, 54, 59). Among the ~5600 protein-encoding genes of *Saccharomyces cerevisiae* at least 360, or 6%, encode transporters (90). Of these 360 putative transporters the function of only 140 has been established. Notably, the fraction of intracellular transporters is strongly increasing with increasing numbers of identified transporters. Major groups of intracellular transporters are translocases and ATP-binding cassette (ABC) transporters. The translocases (1a), having 12 members so far, mediate the transfer of proteins across or into organelle membranes. All of them, such as the translocon of the endoplasmic reticulum and the translocases of mitochondria and chloroplasts, are of fundamental physiological relevance. ABC transporters (13, 27, 33), comprising 48 known members, are ATP-driven pumps translocating a wide range of organic molecules and are implicated in a number of human diseases or pathological conditions including cystic fibrosis, adrenoleukodystrophy, and multidrug resistance. A second, virtually inexhaustible source of new transports is site-specific mutation, a powerful approach for elucidating the relations between structure and function. Alone in the case of bacteriorhodopsin, a relatively small light-driven proton pump (25), several hundreds of site-specific mutants have been generated for that purpose.

The Patch Clamp Paradigm

Traditionally, transporters are characterized with regard to substrate specificity, transport rate, and energy requirement by incubating intact cells or membrane vesicles with radiolabeled substrates. After different times aliquots are taken, cells or vesicles are separated from solution, and the uptake of substrate into cells/vesicles is determined. Many variants of the basic scheme have been devised (29). All these are sensitive, but the separation of cells/vesicles from the solution has remained a critical step. Alternatively, a number of optical methods involving light scattering, fluorescence spectroscopy, or confocal fluorescence microscopy (10, 94, 95) have been developed to follow transport continuously without separation. However, such methods are restricted to selected cases. Electrical methods offer excellent possibilities to measure the transport of charged substrates. Thus the hallmark among transport methods is still the patch clamp method (23, 53, 80), which revolutionized the study of ion transport by single-channel recording. The analysis of single channels, or single transporters in general, permits one to directly measure processes averaged out in nonsynchronized transporter populations, to obtain more precise kinetic data, to eliminate some macroscopic transport parameters, and, in heterogeneous populations, to resolve subfractions. However, based on electrical recording and a sensitivity limit of approximately three 10^5 monovalent ions s^{-1} at 2-kHz bandwidth (71), the single-transporter analysis by the patch clamp method is essentially restricted to ion channels. Electrophysiological studies of translocases incorporated into planar lipid bilayers are different. There, the leakage of ions through single translocases (26, 85) is recorded, proving the existence of an aqueous protein-conducting channel in translocases and providing important clues for the gating of these channels. The translocation of peptide substrates, however, could not be measured so far.

Optical Single Transporter Recording

In search of a technique that could open new avenues to membrane transport studies, we developed optical single transporter recording (OSTR). As a prologue we initiated in 1974 (68) the development (3, 15, 28, 64) of a now widely employed photobleaching method. Conceived for measuring translational diffusion in cell membranes, the method was extended in 1983 (61) to membrane transport (45, 62, 63, 67). The first chapter of OSTR was opened in 1990 with studies of single whole cells containing isolated transmembrane pores. The conceptual basis of OSTR was developed and the flux through single pores was measured (69, 81). Such measurements were also performed on several cells in parallel (87) using scanning microphotolysis (40, 66, 96). The second chapter brought the extension of OSTR to physiological receptor densities. The idea was to attach regular membranes to arrays of small test compartments (TCs). However, when the idea was conceived, suitable nanostructuring techniques for the generation of micro- and nano-wells had not yet been developed. Therefore, bonding commercially available track-etched membrane filters to solid substrates generated TC arrays (88). By these means OSTR became independent of photobleaching, which, however, continued to provide useful experimental options. In 1999 the potential of OSTR was demonstrated by applications to nuclear transport (31, 32). Nevertheless, more than a dozen technical improvements were required to develop a robust and easy-to-use transport assay (65, 83, 84).

This article discusses the current state and perspectives of OSTR. First, the conceptual basis of OSTR is considered and shown to be simple and perspicuous. In contrast, the realization of the concept, as summarized in the second part, was not trivial at all. Third, applications of OSTR to nuclear transport are reviewed. Nuclear transport has been studied previously only in intact and permeabilized (1) cells. OSTR now provides opportunities for experimentally testing previous ideas and hypothesis. Finally, the perspectives of OSTR are assessed, in particular its generalization and automatization.

CONCEPTUAL BASIS OF OSTR

Principle

The principle of OSTR is illustrated in Figure 1. In microchambers, intact cells, cell organelles, or isolated membranes are attached to transparent arrays of small TCs. After attachment, a transport solution containing at least two different substrates is added to the chamber. The transport substrate, exhibiting, for instance, a green fluorescence, is translocated by the transporters under study through the membrane patches covering the TCs. The control substrate, exhibiting, for instance, a red fluorescence, is similar both in size and properties to the transport substrate but is not translocated. The arrival of substrates in TCs is recorded by confocal fluorescence microscopy. Thus, TCs that are not covered by membranes, or in which the seal between membranes and TC array is insufficient, can be identified by an "immediate" increase of both green and red fluorescence. In contrast, TCs that are tightly sealed by membranes are characterized by a time-dependent increase of green fluorescence and simultaneous exclusion of red fluorescence. Tightly sealed TCs are used to measure transport kinetics by repetitive scanning.

In the described procedure, the import of substrates into the TCs is measured. For measuring export from the TCs, the OSTR chamber is loaded first with a transport

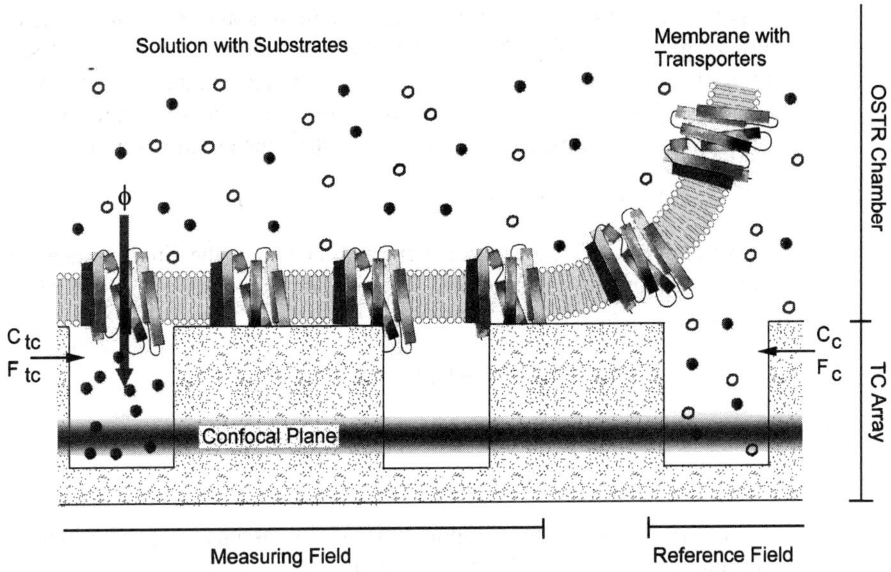

Figure 1 Principle of OSTR. In an OSTR chamber a cellular membrane is attached to a microarray of small test compartments (TCs). A transport substrate (filled circles) and a control substrate (open circles) are added to the OSTR chamber. The time-dependent appearance of the substrates in the TCs is recorded by confocal scans. In the measuring field the membrane seals the TCs and only the transport substrate appears in the TCs. In the reference field the TCs are not sealed, and both transport and control substrates appear in the TCs. C_{tc}, F_{tc}, C_c, and F_c are concentration and fluorescence in TCs and chamber, respectively. Figure is not drawn to scale.

solution. Then, membranes are attached and transport solution in the chamber is replaced by plain buffer. Alternative procedures for both import and export measurements are provided by application of photobleaching or photoactivation.

The scheme of Figure 1 shows a situation in which the TCs have a cross section approximately matching the average membrane area per transporter. With such "small" patches, subpopulations of patches can be discriminated carrying 0, 1, 2, . . . transporters. If the patches contain more than five transporters on average, the different classes of patches cannot be discriminated, a situation we refer to as "large patch" (see below for more details).

Transport Kinetics

The OSTR specimen can be modeled as a two-compartment system, which consists of one compartment (TC) of finite and the other (chamber) of "infinite" volume. Membrane transport in such systems has been treated extensively in the literature. Here, only aspects particular to OSTR are considered. It is assumed that transport across the membrane separating the two compartments is rate limiting so that substrates are always uniformly distributed in both TC and chamber. The condition for this assumption is specified below. Furthermore, it is assumed without much restriction of general validity, that C_{tc}, the substrate concentration in the TCs, is zero at $t = 0$ while C_c, the substrate concentration in the chamber, is constant.

For channel-type transporters the transport rate ϕ (translocated molecules m/time) is, to a first approximation, proportional to the concentration difference:

$$\phi(t) = dm/dt = -nP_1\Delta C(t) = -nP_1(C_{tc}(t) - C_c), \qquad 1.$$

where n is the number of transporters per patch and P_1 is the single-transporter permeability coefficient (volume/time). The time development of C_{tc} in a TC of volume V_{tc} is:

$$C_{tc}(t) = C_c\,(1 - e^{-nk_1 t}) \qquad 2a.$$

with transport rate constant

$$k_1 = 1/\tau_1 = P_1/V_{tc}, \qquad 2b.$$

where τ_1 is the characteristic transport time. If fluorescence is proportional to concentration, so that $C_{ct} = (F_{tc}/F_c)\,C_c$, then

$$F_{tc}(t) = F_c(1 - e^{-nk_1 t}). \qquad 3.$$

Carrier-type transporters show saturation kinetics according to

$$\phi(t) = -n\phi_{\max_1}\left(\frac{C_{tc}}{C_{tc} + K_m} - \frac{C_c}{C_c + K_m}\right), \qquad 4.$$

where ϕ_{max1} is the maximum transport rate of the single transporter and K_m is the Michaelis-Menten constant. The time development of C_c in a TC of volume V_{tc} covered with a patch containing n transporters is:

$$\frac{C_c + K_m}{K_m}C_{tc} + \frac{(C_c + K_m)^2}{K_m}\ln\frac{C_c - C_{tc}}{C_c} = -\frac{n\phi_{\max 1}}{V_{tc}}t. \qquad 5.$$

For $C_c \ll K_m$ Equation 5 transforms into Equation 2 with $k_1 = \phi_{max1}/K_m$.

Single-Transporter Analysis

Whether a particular membrane patch contains one, two, or several transporters cannot be concluded from the transport kinetics of that patch alone. Therefore, a statistical analysis is required to identify the number of transporters per patch (69). In reality the rate constant will not have a single value but rather a distribution of values around a mean value k_n. Assuming that the distribution g_n of k_n can be approximated by a normal distribution, then

$$g_n = \frac{1}{\sqrt{2\pi}\sigma_n}e^{-(k-k_n)^2/2\sigma_n^2}, \qquad 6.$$

where σ_n is the standard deviation of g_n with

$$\sigma_n = \sqrt{n}\sigma_1. \qquad 7.$$

Importantly, the standard deviation σ_n increases with $\sqrt{n}$, i.e., the more transporters in a patch the broader the rate constant distribution and the smaller its peak value. Another consideration concerns the distribution of transporters among patches. The mean number λ of transporters per patch is

$$\lambda = \rho A_{tc}, \qquad 8.$$

where ρ is the area density of transporters in the membrane and A_{tc} is the TC cross section. If the transporters are distributed at random in the membrane and λ is small (≤ 5), the fraction $f(n)$ of patches with $n = 0, 1, 2\ldots$ transporters is given by the Poisson distribution:

$$f(n) = \frac{\lambda^n}{n!}e^{-\lambda}. \qquad 9.$$

Together, the variability of k and the distribution of transporters among patches determine the frequency distribution $h(k)$ of the rate constant k in a large population of membrane patches:

$$h(k) = \sum f(n)g_n. \qquad 10.$$

By plotting the experimentally determined frequency distribution of k versus k and fitting it by Equation 10, the rate constant associated with patches containing a single transporter can be unequivocally determined (examples are given in Figure 6).

If λ is >5 the Poisson distribution gradually transforms into a normal distribution. Also, the standard deviations of the rate constants become large. Together, these parameters make it impossible to resolve patches with different numbers of transporters. On these grounds we discriminate "small" patches (≤ 5 transporters, multiple k-peaks) from "large" patches (>5 transporters, single k-peak) even if the dimensions of large patches may still be in the nanometer range. The rate constant measured with large patches is

$$\bar{k} = \frac{\rho P_1}{L_{tc}}, \qquad 11.$$

where L_{tc} is the length of the TCs. $\bar{k}$ is independent of the TC cross section. If transport is measured in the same membrane (constant transporter density) using small and large patches but leaving the TC length constant, then

$$\frac{\bar{k}}{k_1} = \rho A_{small}, \qquad 12.$$

where A_{small} is the cross section of the small patches.

The Test Compartment as an Unstirred Layer

Above, the assumption was made that the substrate concentration is uniform in both chamber and TC. While mixing by convection appears practical in the chamber, the TC is a perfect "unstirred layer." Therefore, the above equations are valid only if the time necessary for distributing the substrate within the TC by diffusion is substantially smaller than the transport time. For a cylindrical TC of cross section A_{tc} and length L_{tc} filled with substrate by diffusion from the entrance, the concentration at the bottom reaches ~90% of the concentration at the entrance after time (9)

$$t_{dif} = \frac{L_{tc}^2}{D}, \qquad 13.$$

where D is the diffusion coefficient of the substrate. On the other hand, for a TC covered with a membrane containing channel-type transporters at an area density ρ the characteristic transport time is

$$\tau = \frac{V_{tc}}{nP_s} = \frac{L_{tc}}{\rho P_s}. \qquad 14.$$

The condition $t_{dif} < \tau$ is therefore fulfilled if:

$$P_s < \frac{D}{\rho L_{tc}}. \qquad 15.$$

If the condition cannot be met it is still possible, within certain limits, to correct the measured rates for diffusion. This, however, involves a deconvolution of transport and diffusion processes.

EXPERIMENTAL REALIZATION OF THE CONCEPTUAL BASIS

Sequence of Events During an OSTR Experiment

The sequence of events during an OSTR experiment is sketched in Figure 2 (see color insert), using the recent procedure for oocyte nuclei (83, 84) as an example. The central experimental device is the OSTR chamber, a small cylindrical cavity with a TC array at the bottom. At first, the OSTR chamber is filled with a solution mimicking the intracellular milieu of oocytes. An isolated nucleus is deposited in the chamber and attached to the TC array by slight pressure. Transport can be started at this stage. Alternatively, the nucleus can be made permeable for macromolecules by perforation with a fine needle or fully dissected to generate an isolated and cleaned nuclear envelope. A transport substrate and a control substrate are added to the chamber. In a confocal microscope the nucleus or nuclear envelope is first localized at low magnification. Then, an area of the TC array covered by the nuclear envelope (measuring field) is imaged at higher magnification, and the appearance of transport substrate in and the exclusion of control substrate from TCs are recorded simultaneously by repetitive scanning. When a sufficient amount of data has been collected a region of the TC array not covered by the nucleus (reference field) is shifted into the scanned field and TCs are imaged. A dedicated computer program for the time-dependent fluorescence of each TC evaluates the image stack obtained in this manner. Finally, the experimental data are fitted by appropriate expressions to derive transport parameters.

OSTR Chambers and TC Arrays

A disposable OSTR chamber (Figure 3*A*, see color insert) may be created (31) by drilling a hole into the bottom of a plastic culture dish and covering the hole with a TC array. In addition many of the permanent chambers described in the literature for live cell observation and electrophysiology (20) should be useful in OSTR. TC arrays (Figure 3*B*) can be created by bonding small pieces of track-etched membrane filters to cover slips (83, 88). Recently, however, TC arrays that are custom-made by laser drilling or microlithography (Figure 3*B*, *left side*) have been employed (N. Kiskin & R. Peters, unpublished data), putting OSTR on new grounds (see Perspectives and Concluding Remarks, below).

In cellular membranes, both the area density of transporters and the transport rate vary over wide ranges. It is a decisive advantage of OSTR that the experimental conditions can be adjusted to these parameters independently. Thus by selecting an appropriate TC diameter, the number of transporters per patch can be chosen. The transport time depends on the TC length (Equation 14).

A comparison of patch sizes employed in the electrophysiological patch clamp method with those used in OSTR is instructive. The electrophysiological patch pipette usually has a tip diameter of approximately 1 μm, but the area of membrane patches, which bulge into the pipette, amounts to 2 to 25 μm^2 (80). In OSTR the attachment of membranes to TC arrays creates essentially planar patches with diameters between 0.016 and 100 μm and cross-sectional areas between 0.02 and 8000 μm^2.

Transport and Control Substrates

OSTR is based on fluorescence detection. This condition can be met by employing either fluorescent transport and control substrates (32, 69, 88) or substrate-specific fluorescent indicators (87). Fluorescent indicators were initially developed for Ca^{++} [reviewed in (22)] and are now available for a variety of inorganic cations and anions. Organic molecules indicators have been developed (89) that make use of fluorescence resonance energy transfer (FRET), a strategy that could be used to construct indicators for a diversity of substrates.

Attachment of Membranes to TC Arrays

Initially, we employed polylysine coating and blotting (31, 32, 88) to attach erythrocytes or oocyte nuclei to polycarbonate track-etched filters. We then found

(83, 84), however, that oocyte nuclei would attach spontaneously to polycarbonate filters if prepared and kept in a mock intracellular medium buffered to micromolar concentrations of free Ca^{++}. This observation permitted us to develop the procedure indicated in Figure 2 and made nuclear transport measurements easy to perform and reliable. We anticipate that the methods for the attachment of membranes to TC arrays can be further perfected. Patch clamp experiments have shown that virtually any cellular membrane can be tightly attached to the tip of glass pipettes. The main condition for the generation of "giga seals" is the cleanliness of the pipette tip (23). Meanwhile, TC arrays made from glass have become available and are currently tested in OSTR experiments (R. Peters, unpublished data).

Initiation of Transport by Mixing, Photobleaching, or Photoactivation

When an OSTR specimen has been set up, transport can be initiated by the addition of substrate to the chamber. Then, the specimen is mounted on the stage of a confocal microscope and several adjustments are made. This causes a delay of 1 to 2 min that frequently can be tolerated (Figure 4, see color insert). Mounting the specimen on the microscope stage and adjusting the microscope before adding transport substrate substantially reduce the delay. Chambers with perfusion systems (20) would further speed up the process.

A different timescale becomes accessible by photobleaching (31, 69, 84, 87) (Figure 5, see color insert). Transport substrate is added to the OSTR chamber and allowed to equilibrate between chamber and TCs. After a first scan individual TCs are selected and depleted from fluorescence by photobleaching using scanning microphotolysis (40, 96). The recovery of fluorescence by transport of fresh substrate from the chamber through the membrane patch is recorded by repetitive scanning. Sufficient depletion of fluorescence in a TC usually requires 0.1 s, which sets the time resolution of such a measurement.

An alternative method for the initiation of transport is photoactivation (87). A number of inorganic ions, fluorophores, nucleotides, neurotransmitters, and other biologically relevant ligands are available as caged compounds. These are photolabile covalent complexes from which the active substance can be released by short irradiation in the near UV. Suitable flash devices, which can be installed directly on confocal microscopes to photolyse caged compounds in the scanned field, are commercially available. The time required for releasing sufficient amounts of caged compounds is in the millisecond range.

Recording and Evaluation of Transport Kinetics

Recording of substrate concentration in TCs requires optical sectioning. Confocal laser scanning microscopy can achieve this. Optical sections can be also generated by other methods such as two-photon microscopy (35, 39), which has yet to be tested in OSTR. Stacks of confocal scans (Figures 4 and 5) are evaluated by

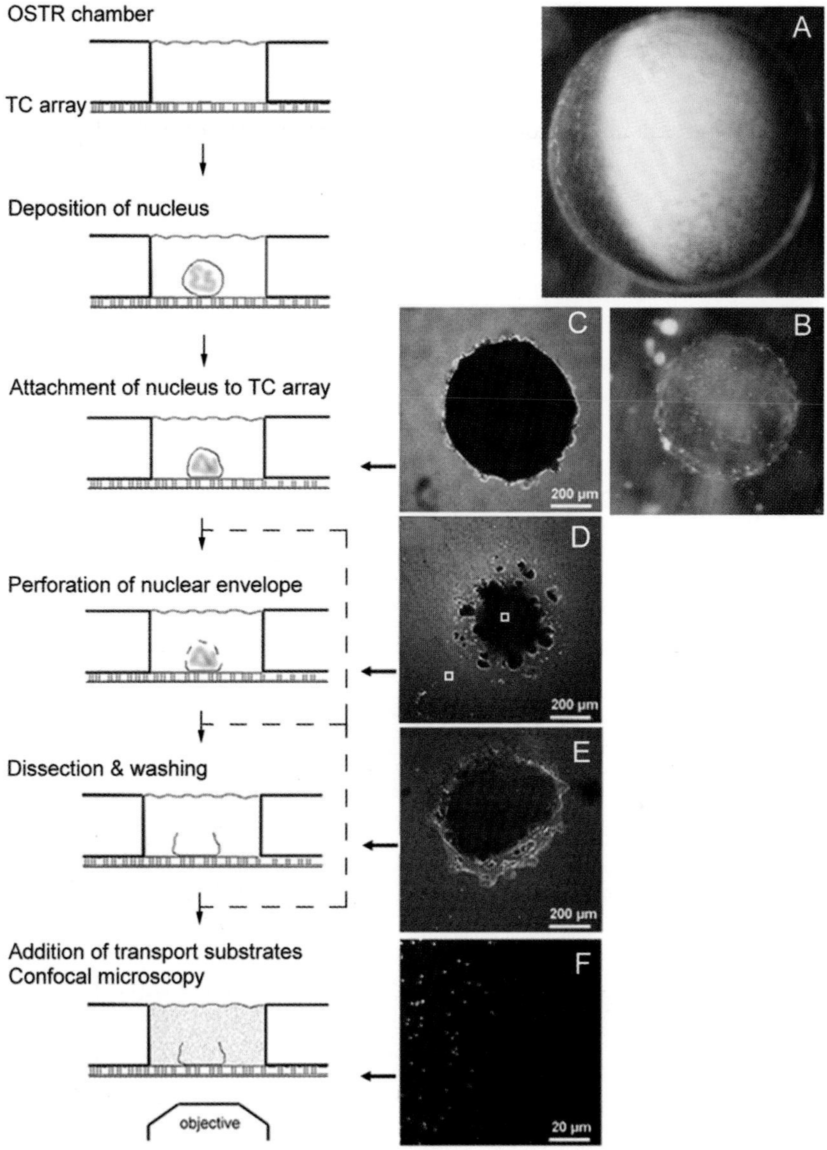

Figure 2 OSTR specimens for nuclear transport measurements. *Left:* Sequence of events. For details, see text. *Right:* Microscopic appearance. (*A*) Intact *Xenopus* oocyte. (*B*) Manually isolated nucleus. (*C*) Largest circumference of an intact TC-attached nucleus submersed in a control substrate, a 70-kD dextran. (*D*) Area of contact between nucleus and TC array. White squares indicate the typical positions and sizes of measuring and reference fields. (*E*) Isolated TC-attached nuclear envelope. (*F*) TC array at the boundary of the nuclear envelope-covered area. TC diameter = 0.78 μM, TC depth = 10 μm. *A*, *B*: Through-light microscopy. *C–F:* Confocal scans. Modified from Reference 83.

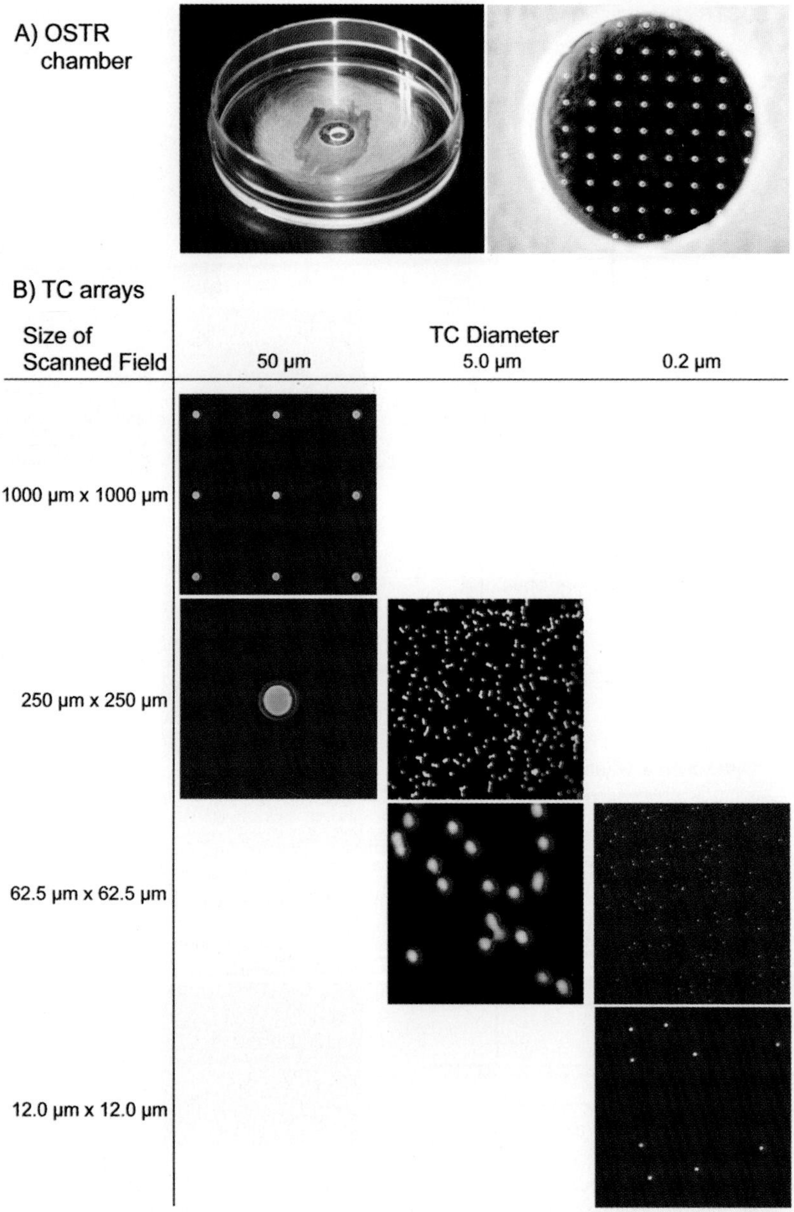

Figure 3 OSTR chamber and TC arrays. (*A*) Prototypic OSTR chamber made from a plastic culture dish by drilling a 3-mm-diameter hole into the bottom and covering the hole with a TC array. (*B*) Selection of TC arrays. *Left*: An array made by laser drilling. TC diameter = 50 μm, TC depth = 50 μm, pitch = 400 μm. *Middle*: Track-etched membrane filter. TC diameter = 5 μm, TC depth = 10 μm. *Right*: Track-etched membrane filter. TC diameter = 0.18 μm, TC depth = 10 μm.

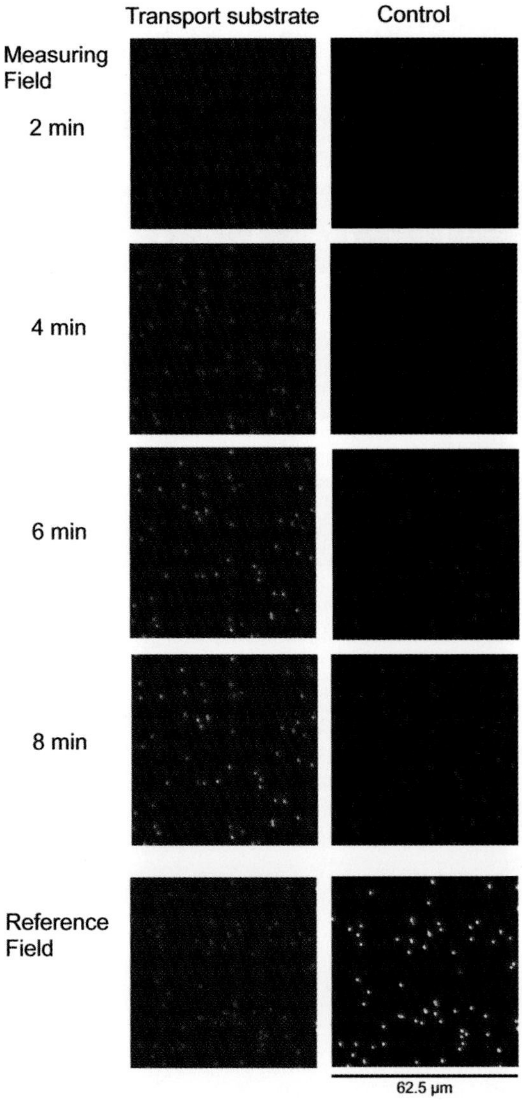

Figure 4 Example of an OSTR measurement. In an OSTR chamber an oocyte nucleus was attached to a TC array and made permeable for macromolecules by perforation. A transport solution was added containing a transport substrate (GG-NES), an impermeable control substrate (Texas-red-labeled 70-kD dextran), and the trinucleotide GTP. After the addition of the transport solution, TCs were imaged at the indicated times in a measuring field, i.e., an area covered by the nucleus. Subsequently, TCs were imaged in an area not covered by the nuclear envelope (reference field). By comparison it can be recognized that in the measuring field the transport substrate was exported from the perforated nucleus and accumulated in the TCs, whereas the control substrate was excluded from TCs. TC diameter = 0.72 μm, TC length = 10 μm. Modified from Reference 83.

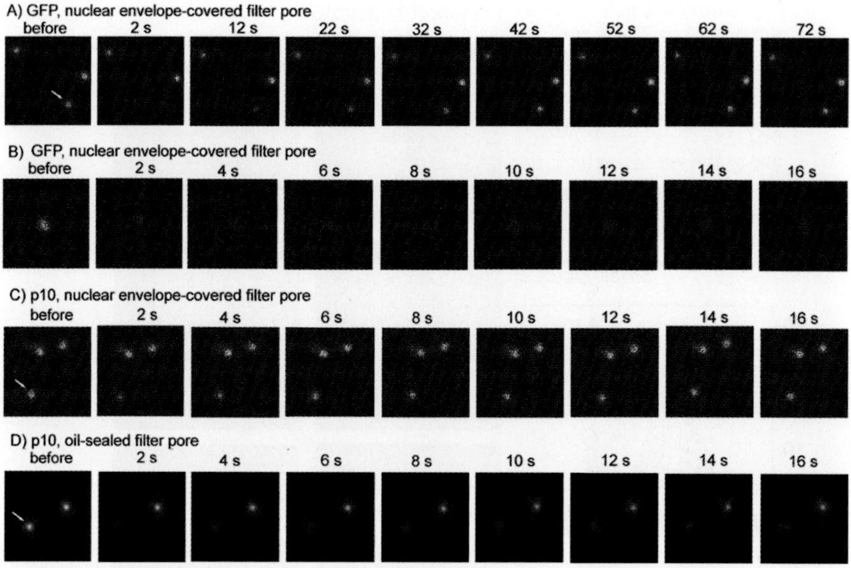

Figure 5 Use of photobleaching for initiation of transport. Isolated nuclear envelopes were attached to TC arrays (0.72-μm diameter, 10-μm depth) and incubated with transport solutions containing GFP or Alexa488-labeled NTF2. After equilibration filter pores were imaged at high magnification and individual filter pores (arrows) were photobleached for 1 s. Fluorescence recovery, owing to the export of fresh substrate, was monitored by a time series of confocal scans. Times after photobleaching are indicated. (*A, B*) Recovery time of GFP, measured on two timescales, was ~55 s. (*C*) The recovery time of NTF2 was 1 to 2 s. (*D*) Experiments with filter pores filled with the NTF2 transport solution but sealed on both ends by oil showed that NTF2 was effectively photobleached. Modified from Reference 84.

determining the time-dependent fluorescence of individual TCs. For this purpose, we use a computer program that aligns the image stack, localizes TCs, sets regions of interest (ROIs) over the TCs, determines the local background around individual TCs, corrects TC fluorescence for background, and fits the time-dependent intensity values of individual TCs by the appropriate expression. Finally, histograms are generated in which the frequency of a transport parameter is plotted and fitted by Equation 10 (Figure 6).

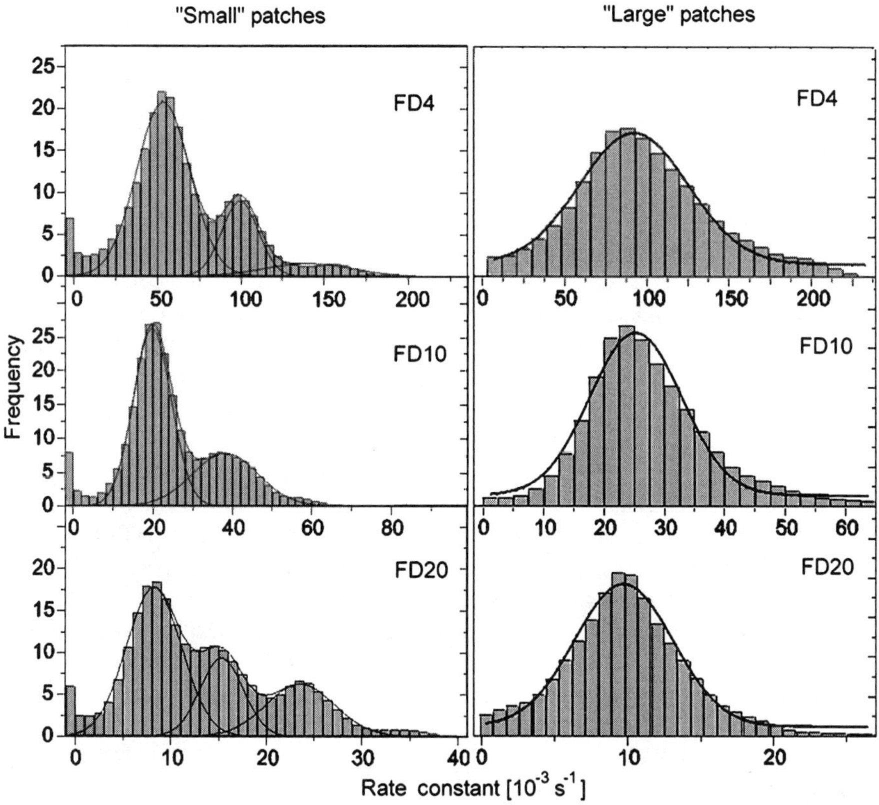

Figure 6 Single-transporter analysis by OSTR. Transport through the NPC of a homologous series of inert molecules (fluorescein-labeled dextrans) of 4-, 10-, and 20-kD mass (FD4, FD10, FD20, respectively) was measured by OSTR. Frequency of rate constants in populations of about 250 rate constants each was plotted. *Left*: "Small" patches had a diameter of 0.18 μm and contained 1.25 NPCs on average. The experimental rate constants (*bars*) split up into subpopulations and were fitted by a sum of weighted Gaussians (*lines*) according to Equation 10. *Right*: "Large" patches had a diameter of 1.8 μm and contained 125 NPCs on average. In the frequency diagram a single peak was observed and fitted by a single Gaussian. The mean value was about 1.25 times larger than the first peak in small patch measurements, as expected (Equation 12). Modified from Reference 31.

APPLICATIONS OF OSTR TO NUCLEAR TRANSPORT

The exchange of matter between nucleus and cytoplasm is mediated by the NPC (12, 48, 92, 98), a large protein complex spanning the nuclear envelope. Previously, nuclear transport has been studied in intact and permeabilized (1) cells, while attempts to employ simpler and better controllable systems such as isolated nuclei (91) or nuclear envelope vesicles (24) were not generally accepted. Studies with intact and permeabilized cells have shown that the NPC supports two basic types of transport. Molecules that neither contain nuclear transport signals nor interact specifically with NPC proteins (nucleoporins) and therefore can be considered as "inert" permeate the NPC by restricted diffusion. Their transport rates depend inversely on molecular size with an exclusion limit at about 9 nm (63). Molecules exceeding the exclusion limit are translocated through the NPC if they contain specific signals for import or export (nuclear localization signals, NLS; nuclear export signals, NES).

Signal-dependent transport involves additional components: nuclear transport receptors, collectively referred to as karyopherins (12); the small Ras-like protein Ran (86), which occurs in a GTP- and a GDP-bound form; and cofactors of Ran (8), which catalyze the exchange of GTP for GDP on Ran or activate the GTPase activity of Ran. In the case of nuclear import, first a complex of an NLS protein and an import receptor is formed. The import complex binds to the NPC and is translocated from the cytoplasmic to the nuclear side. In the nucleus, RanGTP binds to the import complex inducing the release of the NLS protein. The importin-RanGTP complex returns to the cytoplasmic side and dissociates after hydrolysis of Ran-bound GTP. In the case of export, a complex is formed in the nucleus consisting of an NES protein, an export receptor such as CRM1, and RanGTP. The export complex binds to and is translocated through the NPC. In the cytoplasm Ran-bound GTP is hydrolyzed, triggering the dissociation of the import complex. Thus, Ran functions as a molecular switch (21) making selective nuclear transport vectorial and permitting transport against concentration gradients. Energy is apparently not required for translocation through the NPC but only for the recharging of Ran with GTP (18, 34, 52, 76, 77).

Passive Permeability of the Nuclear Pore Complex

The passive permeability of the NPC was analyzed by OSTR (31) using both small and large TCs. Small TCs had a diameter of 0.18 μm (0.025 μm^2 cross section) so that membrane patches contained 1.25 NPCs on average (50 NPCs/μm^2). Large patches were 1.8 μm in diameter containing ~125 NPCs. Transport rates were measured for fluorescently labeled dextrans of 4-, 10-, and 17.2-kD molecular mass. For small patches the ensemble of transport rates split up into subpopulations for each of the substrates (Figure 6, left side). The mean rate constants of the subpopulations were integer multiples of a basic value and the magnitude of the fraction amounted to 2%–3%, 53%–70%, 21%–35%, and 1%–24%, in reasonable agreement with the distribution of NPCs among patches as predicted from electron micrographs. The fraction of patches with a single NPC could thus be identified and the single-transporter permeability coefficient P_1 could be determined. P_1 depended inversely on the molecular size of the transport substrate. By applying a frequently used hydrodynamic approach for restricted diffusion through cylindrical pores (57, 62), the equivalent diameter and length of the diffusion channel were derived as 10.7 and 45 nm, respectively. For large patches the frequency diagram of the rate constant showed a single peak (Figure 6, right side). The mean value was ~1.25 times larger than P_1, in agreement with expectation (Equation 12).

Facilitated Transport of Nuclear Transport Receptors

A major fraction of nucleoporins contains repeated sequences (FG repeats) composed of short hydrophobic stretches of the amino acid residues FXFG, GLFG, or FG, and hydrophilic linkers of variable length. Transport receptors interact with FG repeats (12, 73), although different receptors seem to bind to different subsets of FG repeats. Therefore, a direct connection between FG repeat binding and translocation through the NPC has long been suspected (6, 74). Recently, import kinetics of two FG repeat binding proteins—transportin, the receptor for proteins containing a NLS referred to as M9, and NTF2, the receptor of RanGDP—were measured in digitonin permeabilized cells (75). It was found that a passive process rapidly imports both molecules. At a concentration difference of 1 μM, the translocation rates were determined to be about 65 molecules/NPC/s for transportin and 250 molecules/NPC/s for NTF2. Significantly, bovine serum albumin (BSA), an inert molecule substantially smaller than transportin but above the exclusion limit of the diffusion channel, was not translocated at all, whereas green fluorescent protein (GFP), an inert molecule of virtually the same dimensions as NTF2, was translocated at only 2 molecules/NPC/s. The translocation of GFP and NTF2 through the *Xenopus* oocyte NPC was recently measured by OSTR (84). Employing TCs of 0.72-μm diameter, "large" membrane patches were generated containing ~15 NPCs without rim contact. To achieve an adequate time resolution transport was initiated by photobleaching (Figure 5). For GFP, the transport rate was 57 molecules/NPC/s at 15-μM concentration difference corresponding to 3.3 molecules/NPC/s upon linear interpolation to 1-μM concentration difference. The rate of NTF2 was 270, 907, and 1497 molecules/NPC/s at concentration differences of 1.5, 7.4, and 14.8 μM, respectively. Thus, the measurements confirmed that transport receptors, which interact with FG repeats, are translocated through the NPC much faster than inert molecules of the same size.

Signal-Dependent Nuclear Transport

Signal-dependent nuclear transport was studied by OSTR using the recombinant proteins GG-NLS and GG-NES (32, 83). Both proteins contain the moieties of glutathione transferase and GFP and either the NLS of the SV40 T antigen or the NES of the protein kinase A inhibitor.

First (32), the directionality of protein transport was analyzed. GG-NLS or GG-NES was applied to the cytoplasmic or nuclear side of the NPC. Likewise, predominantly cytoplasmic extracts prepared from whole *Xenopus* eggs or nuclear extracts prepared from isolated *Xenopus* oocyte nuclei were added either to the cytoplasmic or nuclear side. Transport was only observed when proteins and extracts were on their "correct" side, i.e., the NLS protein and the predominantly cytosolic extract on the cytosolic side of the NPC and the NES protein and the nuclear extract on the nuclear side of the NPC. Thus, in contrast to experiments with permeabilized cells (51) in which the import of an NES protein could be induced by artificially establishing a gradient of RanGTP between cytoplasm and nuclear interior, no reversal of the transport direction could be detected in oocyte nuclear envelope patches. It is conceivable that the oocyte NPC has an intrinsic asymmetry, perhaps related to docking and release sites for import and export complexes, that prevents an inversion of the transport direction.

More recently (83), the export of GG-NES was studied by OSTR in detail. In perforated *Xenopus* oocyte nuclei the export of GG-NES could be induced by the addition of RanGTP alone. Export continued against a concentration difference, was NES dependent, and was inhibited by leptomycin B, an inhibitor of CRM1, and GTPγS, a nonhydrolyzable GTP analog. Addition of RanBP3, a potential cofactor of CRM1-mediated export (17), showed only an inhibitory effect at high concentrations. Together the results show that perforated nuclei, although permeable for macromolecules, retain transport cofactors in sufficient amounts. Perforated nuclei therefore may prove useful for studying the complex export processes of ribonucleoprotein particles and ribosomal subunits that involve, in addition to translocation through the NPC, chromatin-dependent steps.

When filter-attached nuclei were dissected and residual nuclear envelopes purified by washing, export of GG-NES was virtually abolished in the presence of GTP alone. However, an export complex preformed from equimolar amounts of GG-NES, the export receptor CRM1, and RanGTP was exported at ~8 molecules/NPC/s (2-μM concentration difference). From a comparison of these transport rates with those measured previously by us or others (32, 55, 75), we concluded that the reconstitution of export was complete despite that accessory factors such as RanGAP or RanBP1 had not been added. Unexpectedly, export was strongly reduced when the export complexes contained GTPγS or the GTP form of RanG19V/Q69L, a GTPase-deficient Ran mutant. Previous experiments with intact (77) and permeabilized (18) cells had suggested that hydrolysis of GTP on Ran was required neither for the translocation of the export complex through the NPC center nor for its release from the cytoplasmic side of the NPC. The differences between our results and those mentioned previously may be technical and therefore have to be further analyzed. If, however, GTP hydrolysis should be required for efficient export, it would be necessary to modify recent models of translocation through the NPC (48, 75, 79) that are entirely based on diffusional mechanisms.

PERSPECTIVES

The current state of OSTR is apparent from nuclear transport studies. Thus, transport rates as small as one translocation/s/transporter could be easily detected and analyzed. The smallest membrane patches studied so far had a diameter of 180 nm. The time resolution was ~1 min when transport was initiated by mixing and ~1 s when initiated by photobleaching. Substrate concentrations were >0.5 μM. Usually two substrates were recorded simultaneously. The number of TCs recorded in parallel was usually 30 to 40. All these numbers are starting values only. OSTR is a young technique, open for substantial additions and improvements. Some aspects are discussed in the following section.

Design and Application of Nanostructured OSTR Chips

Currently, large efforts are made to develop methods for the fabrication of nanometer-sized devices. Driven in the first line by the microelectronics and telecommunication industry, the development has strong ramifications into chemistry and biology (11). Therefore, several nano-structuring techniques, both conventional and nonconventional ones, are already available, which could be employed to create dedicated OSTR chips, i.e., transparent arrays of nanometer-sized wells. It would be highly desirable to make such chips out of glass (14, 19, 56) to promote the attachment of cellular membranes. Glass is also desirable because of its optical properties facilitating high-resolution imaging. The fabrication of OSTR chips by laser or electron beams would also permit to optimize the geometrical parameters of the TCs. Thus, the cross-sectional area and the length of the TCs could be matched to different membrane preparations, providing optimum conditions for single-transporter recording and kinetic measurements. Very small, perhaps conical, TCs might even permit scanning the surface of single large transporters. Dense arrays of TCs could be generated permitting to record one thousand or more TCs simultaneously (e.g., 2.500 TCs at a pitch of 2 μm and a scanned area of $100 \times 100\ \mu m^2$).

Combination of OSTR with Single Molecule Detection and Electrical Recording

An analytical complement to nanostructuring techniques is single molecule detection (5, 46, 97). Application of the wide-field methods for the detection and tracking of single fluorescent molecules as developed by us (36, 37, 41–43) should increase the sensitivity of OSTR measurements in terms of substrate concentration by approximately three orders of magnitude. Also, it should become possible to visualize single fluorescently labeled transporters in the membrane patches covering TCs. Fluorescence signals from these transporters may contain information (10) about conformational changes during translocation. Electrical recording under voltage clamp conditions has yielded important information on ion channels (80) and translocases (26, 85). Electrical measurements of the NPC are controversial (50). Therefore, a direct combination of optical and electrical recording will potentiate the analytical power of transport measurements. The major prerequisites are TC arrays in which individual TCs can be electrically addressed.

Generalization and Automation of OSTR

Up to now OSTR has been used almost exclusively to study nuclear transport. For the generalization of OSTR two major pathways come to mind. The first pathway begins with intact cells, when necessary freed from cell walls. In such preparations only transporters of the plasma membrane can be studied with poor control of the cellular interior. However, upon homogenization of cells and separation of subcellular fractions an abundance of membrane vesicles becomes accessible. Many methods have been worked out to separate, purify, and identify vesicles of different origin. A common feature of all these vesicles is a diameter of approximately 0.1 to 1.0 μm. The second pathway begins with transporter DNA. The transporter is expressed most frequently in *Escherichia coli* (44), yeast (7), or insect cells. After purification transporters are reconstituted in liposomes (58, 78, 82). This protocol yields unilamellar vesicles that frequently have diameters of 0.1 to 1.0 μm. But the fusion of such small proteoliposomes with giant unilamellar vesicles of ~20-μm diameter has been described (30). A central objective of our current OSTR studies is to attach membrane vesicles or reconstituted proteoliposomes to TC arrays, either directly or via fusion with preformed bilayer membranes (26, 85). Once this goal has been achieved the progression to high-throughput methods will be a matter of optimization and automation only.

CONCLUDING REMARKS

Membrane transport proteins are key elements of cellular function in health and disease. Progress in genomic analysis has prompted the discovery of a large number of new transporters, among them many translocases and ABC pumps of intracellular origin. The goals of transporter analyses are molecular models integrating structure and function. However, a *conditio sine qua non* for achieving that goal is the characterization of transporters with regard to substrate specificity, transport rate, energy requirement, and regulatory and inhibitory parameters. The most powerful approach to membrane transport kinetics is single transporter recording, previously only possible in the case of ion channels. OSTR has the potential, as shown by measurements of nuclear transport, to extend single transporter recording to a wide variety of transporters, including translocases and ABC pumps. This article has been written to stimulate further exploration of OSTR's potential.

ACKNOWLEDGMENTS

I would like to thank Jan-Peter Siebrasse, Katja Zerf, Ulrich Kubitscheck, Mathias Tschödrich-Rotter, Peter Wedekind, Thorsten Kues, and Elias Coutavas for their enthusiasm and great work. I also acknowledge productive and enjoyable discussions with Richard Wagner (Osnabrück) on translocases and their further analysis, Financial support by the Deutsche Forschungsgemeinschaft and the Volkswagenstiftung is gratefully acknowledged.

The *Annual Review of Biophysics and Biomolecular Structure* is online at http://biophys.annualreviews.org

LITERATURE CITED

1. Adam SA, Stern-Marr R, Gerace L. 1990. Nuclear protein import in permeablized mammalian cells requires soluble cytoplasmic factors. *J. Cell Biol.* 111:807–16
1a. Agarrabes FA, Dice JF. 2001. Protein translocation across membranes. *Biochim. Biophys. Acta* 1513:1–24
2. André B. 1995. An overview of membrane transport proteins in *Saccharomyces cerevisiae*. *Yeast* 11:1575–611
3. Axelrod D, Koppel DE, Schlessinger J, Elson E, Webb WW. 1976. Mobility measurement by analysis of fluorescence photobleaching recovery. *Biophys. J.* 16: 1055–69
4. Baker PF, Willis JS. 1962. Binding of the cardiac glycoside ouabain to intact cell. *J. Physiol.* 224:441–62
5. Basché T, Nie S, Fernandez JM. 2001. Single molecules. *Proc. Natl. Acad. Sci. USA* 98:10527–28
6. Bayliss B, Corbett AH, Stewart M. 2000. The molecular mechanism of transport of macromolecules through nuclear pore complexes. *Traffic* 1:448–56
7. Bill RM. 2001. Yeats—a panacea for the structure-function analysis of membrane proteins? *Curr. Genet.* 40:157–71
8. Bischoff FR, Ponstingl H. 1995. Catalysis of guanine nucleotide exchange of Ran by RCC1 and stimulation of hydrolysis of Ran-bound GTP by Ran-GAP1. *Methods Enzymol.* 257:135–44
9. Carslaw FS, Jaeger JC. 1959. *Conduction of Heat in Solids*, pp. 99–102. Oxford: Clarendon. 2nd ed.
10. Cha A, Zerangue N, Kavanaugh M, Bezanilla F. 1998. Fluorescence techniques for studying cloned channels and transporters expressed in *Xenopus* oocytes. *Methods Enzymol.* 296:566–78
11. Chen Y, Pépin A. 2001. Nanofabrication: conventional and nonconventional methods. *Electrophoresis* 22:187–207
12. Chook YM, Blobel G. 2001. Karyopherins and nuclear transport. *Curr. Opin. Struct. Biol.* 11:703–15
13. Dean M, Hamin Y, Chimini G. 2001. The human ATP-binding cassette (ABC) transporter superfamily. *J. Lipid Res.* 42:1007–17
14. Dietrich TR, Ehrfeld W, Lacher M, Krämer M, Speot B. 1996. Fabrication technologies for microsystems utilizing photoetchable glass. *Microelectron. Eng.* 30:497–504
15. Edidin M, Zagyansky Y, Lardner TJ. 1976. Measurement of membrane protein lateral diffusion in single cells. *Science* 191:466–68
16. Deleted in proof
17. Englmeier L, Fornerod M, Bishoff FR, Petosa C, Mattaj IW, Kutay U. 2001. RanBP3 influences interactions between CRM1 and its nuclear protein export substrates. *EMBO Rep.* 2:926–32
18. Englmeier L, Olivo JC, Mattaj IW. 1999. Receptor-mediated substrate translocation through the nuclear pore complex without nucleotide triphosphate hydrolysis. *Curr. Biol.* 9:1–41
19. Fertig N, Meyer C, Blick RH, Trautmann C, Behrends JC. 2001. Microstructured glass

chips for ion-channel electrophysiology. *Phys. Rev. E* 64, 040901-1-4
20. Focht DC, Spector DL, Goldman RD, Leinwand LA. 1998. Observation of live cells on the light microscope. In *Cells: A Laboratory Manual*, ed. DL Spector, RD Goldman, LA Leinwand, pp. 75:1–13. New York: Cold Spring Harbor Press
21. Görlich D, Pante N, Kutay U, Aebi U, Bischoff FR. 1996. Identification of different roles for RanGDP and RanGTP in nuclear protein import. *EMBO J.* 15:5584–94
22. Grynkiewiez G, Poenie M, Tsien RY. 1992. A new generation of Ca^{++} indicators with greatly improved fluorescence properties. *J. Biol. Chem.* 260:3440–50
23. Hamill OP, Marty A, Neher E, Sakmann B, Sigworth FJ. 1981. Improved patch-clamp techniques for high-resolution current recording from cells and cell-free membrane patches. *Pfluegers Arch.* 391: 85–100
24. Hassel I, Cezanne V, Trevino C, Schlatter H, Romero-Matuschek I, et al. 1966. Export of ribosomal subunits from resealed rat liver nuclear envelopes. *Eur. J. Biochem.* 241:32–37
25. Haupts U, Tittor J, Oesterhelt D. 1999. Closing in on bacteriorhodopsin: progress in understanding the molecule. *Annu. Rev. Biophys. Biomol. Struct.* 28:367–99
26. Hill K, Model K, Ryan M, Dietmeier K, Martin F, et al. 1998. Tom40 forms the hydrophilic channel of the mitochondrial import pore for preproteins. *Nature* 395:516–21
27. Holland B, Blight MA. 1999. ABC-ATPases, adaptable energy generators fuelling transmembrane movement of a variety of molecules in organisms from bacteria to humans. *J. Mol. Biol.* 293:381–99
28. Jacobson K, Wu E, Poste G. 1976. Measurement of the translational mobility of concanavalin A in glycerol-saline solutions and on the cell surface by fluorescence recovery after photobleaching. *Biochim. Biophys. Acta* 433:215–22
29. Jarvis SM. 2000. Assay of membrane transport in cells and membrane vesicles. In *Membrane Transport*, ed. SA Baldwin, pp. 1–20. Oxford: Oxford Univ. Press
30. Kahya N, Pécheur EI, de Boeji WP, Wierma DA, Hoekstra D. 2001. Reconstitution of membrane proteins into giant unilamellar vesicles via peptide-induced fusion. *Biophys. J.* 81:1464–74
31. Keminer O, Peters R. 1999. Permeability of single nuclear pores. *Biophys. J.* 77:217–28
32. Keminer O, Siebrasse JP, Zerf K, Peters R. 1999. Optical recording of signal-mediated protein transport through single nuclear pore complexes. *Proc. Natl. Acad. Sci. USA* 96:11842–47
33. Klein I, Sarkadi B, Varadi A. 1999. An inventory of the human ABC proteins. *Biochim. Biophys. Acta* 1461:237–62
34. Kose S, Imamato N, Tachibana T, Shimamoto T, Yoneda Y. 1997. Ran-unassisted nuclear migration of a 97-kD component of nuclear pore-targeting complex. *J. Cell Biol.* 139:841–49
35. Kubitscheck U, Heinrich O, Peters R. 1996. Continuous scanning microphotolysis: a simple laser scanning microscopic method for lateral transport measurements employing single- or two-photon excitation. *Bioimaging* 4:158–67
36. Kubitscheck U, Kückmann O, Kues T, Peters R. 2000. Imaging and tracking of single GFP molecules in solution. *Biophys. J.* 78:2170–79
37. Kubitscheck U, Kues T, Peters R. 1999. Visualization of nuclear pore complex and its distribution by confocal laser scanning microscopy. *Methods Enzymol.* 307:207–30
38. Deleted in proof
39. Kubitscheck U, Tschödrich-Rotter M, Wedekind P, Peters R. 1996. Two-photon scanning microphotolysis for three-dimensional data storage and biological transport measurements. *J. Microsc.* 182:225–33
40. Kubitscheck U, Wedekind P, Peters R. 1994. Lateral diffusion measurement at high spatial resolution by scanning

microphotolysis in a confocal microscope. *Biophys. J.* 67:948–56
41. Kubitscheck U, Wedekind P, Zeidler O, Grote M, Peters R. 1996. Single nuclear pores visualized by confocal microscopy and image processing. *Biophys. J.* 70: 2067–77
42. Kues T, Dickmanns A, Lührmann R, Peters R, Kubitscheck U. 2001. High intranuclear mobility and dynamic clustering of the splicing factor U1 snRNP observed by single particle tracking. *Proc. Natl. Acad. Sci. USA* 98:12021–26
43. Kues T, Peters R, Kubitscheck U. 2001. Visualization and tracking of single protein molecules in the cell nucleus. *Biophys. J.* 80:2954–67
44. Laage R, Langosch D. 2001. Stragtegies for prokaryotic expression of eukaryotic membrane proteins. *Traffic* 2:99–104
45. Lang I, Scholz M, Peters R. 1986. Molecular mobility and nucleocytoplasmic flux in hepatoma cells. *J. Cell Biol.* 102:1183–90
46. Lange F de, Cambi A, Huijbens R, Bakker B de, Rensen W, et al. Cell biology beyond the diffraction limit: near-field scanning optical microscopy. *J. Cell Sci.* 114:4153–60
47. Deleted in proof
48. Macara IG. 2001. Transport into and out of the nucleus. *Microbiol. Mol. Rev.* 65:570–94
49. Maul GG. 1977. The nuclear and the cytoplasmic pore complex: structure, dynamics, distribution, and evolution. *Int. Rev. Cytol.* (Suppl. 6):75–186
50. Mazzanti M, Bustamante JO, Oberleithner H. 2001. Electrical dimension of the nuclear envelope. *Physiol. Rev.* 81:1–19
51. Nachury MV, Weis K. 1999. The direction of transport through the nuclear pore can be inverted. *Proc. Natl. Acad. Sci. USA* 96:9622–27
52. Nakielny S, Dreyfuss G. 1998. Import and export of the nuclear protein import receptor transportin by a mechanism independent of GTP hydrolysis. *Curr. Biol.* 8:89–95
53. Neher E, Sakmann B. 1976. Single-channel currents recorded from membrane of denervated frog fibres. *Nature* 260:779–802
54. Nelissen B, De Warchter R. 1997. Classification of all putative permeases and other membrane plurispanners of the major facilitator superfamily encoded by the complete genome of *Saccharomyces cervisiae. FEMS Microbiol. Rev.* 21:113–34
55. Nemergut ME, Macara IG. 2000. Nuclear import of the Ran exchange factor, RCC1, is mediated by at least two distinct mechanisms. *J. Cell Biol.* 140:835–50
56. Nolte S, Momma C, Chivkov BN, Welling H. 1999. Mikrostrukturierung mit ultrakurzen Laserpulsen. *Physikal. Blätter* 55: 41–44
57. Paine PL, Moore LC, Horowitz SB. 1975. Nuclear envelope permeability. *Nature* 254:109–14
58. Palmieri L, Larosa FM, Vozza A, Agrimi G, Fiermonte G, et al. 2000. Identification and functions of new transporters in yeast mitochondria. *Biochim. Biophys. Acta* 1459:363–69
59. Paulsen IT, Sliwinski MK, Nelissen B, Goffeau A, Saier MH. 1998. Unified inventory of established and putative transporters encoded within the complete genome of *Saccharomyces cerevisiae. FEBS Lett.* 430:116–25
60. Deleted in proof
61. Peters R. 1983. Nuclear envelope permeability measured by fluorescence microphotolysis of single liver cell nuclei. *J. Biol. Chem.* 258:11427–29
62. Peters R. 1984. Nucleo-cytoplasmic flux and intracellular mobility in single hepatocytes measured by fluorescence microphotolysis. *EMBO J.* 3:1831–36
63. Peters R. 1986. Fluorescence microphotolysis to measure nucleocytoplasmic transport and intracellular mobility. *Biochim. Biophys. Acta* 864:305–59
64. Peters R, Brünger A, Schulten K. 1981. Continuous fluorescence microphotolysis: a sensitive method for study of diffusion

processes in single cells. *Proc. Natl. Acad. Sci. USA* 78:962–66
65. Peters R, Coutavas E, Siebrasse JP. 2002. Nuclear transport kinetics in microarrays of nuclear envelope patches. *J. Struct. Biol.* 140:268–78
66. Peters R, Kubitscheck U. 1999. Scanning microphotolysis: three-dimensional diffusion measurement and optical single-transporter recording. *Methods* 18:508–17
67. Peters R, Passow H. 1984. Anion transport in single erythrocyte ghosts measured by fluorescence microphotolysis. *Biochim. Biophys. Acta* 777:334–38
68. Peters R, Peters J, Tews KH, Bähr W. 1974. A microfluorimetric study of translational diffusion in erythrocyte membranes. *Biochim. Biophys. Acta* 367:282–94
69. Peters R, Sauer H, Tschopp J, Fritzsch G. 1990. Transients of perforin pore formation observed by fluorescence microscopic single channel recording. *EMBO J.* 9:2447–51
70. Deleted in proof
71. Pun YK, Lecar H. 1998. In *Cell Physiology Source Book*, ed. N Sperelakis, pp. 391–405. San Diego: Academic
72. Deleted in proof
73. Radu A, Moore MS, Blobel G. 1995. The peptide repeat domain of nucleoporin Nup98 functions as a docking site in transport across the nuclear pore complex. *Cell* 81:215–22
74. Rexach M, Blobel G. 1995. Protein import into nuclei: association and dissociation reactions involving transport substrate, transport factors, and nucleoporins. *Cell* 83:683–92
75. Ribbeck K, Görlich D. 2001. Kinetic analysis of translocation through nuclear pore complexes. *EMBO J.* 20:1320–30
76. Ribbeck K, Kutay U, Parskava E, Görlich D. 1999. The translocation of transportin-cargo complexes through the nuclear pores is independent of both Ran and energy. *Curr. Biol.* 9:47–50
77. Richards SA, Carey KL, Macara IG. 1997. Requirement of guanosine-triphosphate-bound Ran for signal-mediated nuclear protein export. *Science* 276:1842–44
78. Rigaud JL, Pitard B, Levy D. 1995. Reconstitution of membrane proteins into liposomes: application to energy-transducing membrane proteins. *Biochim. Biophys. Acta* 1231:223–46
79. Rout MP, Aitchison JD, Suprapto A, Hjertaas K, Zhao Y, Chait B. 2000. The yeast nuclear pore complex: composition, architecture and transport mechanism. *J. Cell Biol.* 148:635–51
80. Sakmann B, Neher E. 1995. *Single-Channel Recording*. New York: Plenum. 2nd ed.
81. Sauer H, Pratsch L, Fritzsch G, Bhakdi S, Peters R. 1991. Complement pore genesis observed in erythrocyte membranes by fluorescence microscopic single channel recording. *Biochem. J.* 276:395–99
82. Sharom FJ, Liu RL, Qin Q, Romsicki Y. 2001. Exploring the structure and function of the P-glycoprotein multidrug transporter using fluorescence spectroscopic tools. *Cell Dev. Biol.* 12:257–66
83. Siebrasse JP, Coutavas E, Peters R. 2002. Reconstitution of nuclear protein export in isolated nuclear envelopes. *J. Cell Biol.* 158:849–54
84. Siebrasse JP, Peters R. 2002. Rapid translocation of NTF2 through the nuclear pore of isolated nuclei and nuclear envelopes. *EMBO Rep.* 3:887–92
85. Simon SM, Blobel G, Zimmerberg J. 1989. Large aqueous channels in membrane vesicles derived from the rough endoplasmatic reticulum of canine pancreas or the plasma membrane of *Escherichia coli*. *Proc. Natl. Acad. Sci. USA* 86:6176–80
86. Talcott B, Moore MS. 1999. Getting across the nuclear pore complex. *Trends Cell Biol.* 9:312–18
87. Tschödrich-Rotter M, Kubitscheck U, Ugochukwu G, Buckley JT, Peters R. 1996. Optical single-channel analysis of the aerolysin pore in erythrocyte membranes. *Biophys. J.* 70:723–32

88. Tschödrich-Rotter M, Peters R. 1998. An optical method for recording the activity of single transporters in membrane patches. *J. Microsc.* 192:114–25
89. Tsien RY, Miyawaki A. 1998. Seeing the machinery of live cells. *Science* 280:1954–55
90. Van Belle D, André B. 2001. A genomic view of yeast membrane transporters. *Curr. Opin. Cell Biol.* 13:389–98
91. Vancurova I, Lou W, Paine TM, Paine PL. 1993. Nucleoplasmin uptake by facilitated transport and intranuclear binding. *Eur. J. Cell Biol.* 62:22–33
92. Vasu SK, Forbes DJ. 2001. Nuclear pores and nuclear assembly. *Curr. Opin. Cell Biol.* 13:363–75
93. Venosa RA, Horowicz P. 1981. Density and apparent localization of the sodium pump in frog satorius muscle. *J. Physiol.* 138:119–39
94. Verkman AS. 1995. Optical methods to measure membrane transport processes. *J. Membr. Biol.* 148:99–110
95. Verkman AS. 2000. Water permeability measurement in living cells and complex tissues. *J. Membr. Biol.* 173:73–87
96. Wedekind P, Kubitscheck U, Peters R. 1994. Scanning microphotolysis: a new photobleaching technique based on fast intensity modulation of a scanned laser beam and confocal imaging. *J. Microsc.* 176:23–33
97. Weiss S. 1999. Fluorescence spectroscopy of single biomolecules. *Science* 283:1676–83
98. Wente S. 2000. Gatekeepers of the nucleus. *Science* 288:1374–77

Annu. Rev. Biophys. Biomol. Struct. 2003. 32:69–92
doi: 10.1146/annurev.biophys.32.110601.142445

First published online as a Review in Advance on December 2, 2002

THE ROLE OF DYNAMICS IN ENZYME ACTIVITY

R.M. Daniel,[1] R.V. Dunn,[1] J.L. Finney,[2] and J.C. Smith[3]
[1]*Department of Biological Sciences, University of Waikato, Hamilton 2001, New Zealand; email: r.daniel@waikato.ac.nz; rvd@waikato.ac.nz*
[2]*Department of Physics and Astronomy, University College, London WCIE 6BT, United Kingdom; email: j.finney@ucl.ac.uk*
[3]*Lehrstuhl für Biocomputing, IWR, Universität Heidelberg, D-69120 Heidelberg, Germany; email: biocomputing@iwr.uni-heidelberg.de*

Key Words protein flexibility, enzyme stability, dynamical transition, solvation, hydration

■ **Abstract** Although protein function is thought to depend on flexibility, precisely how the dynamics of the molecule and its environment contribute to catalytic mechanisms is unclear. We review experimental and computational work relating to enzyme dynamics and function, including the role of solvent. The evidence suggests that fast motions on the 100 ps timescale, and any motions coupled to these, are not required for enzyme function. Proteins where the function is electron transfer, proton tunneling, or ligand binding may have different dynamical dependencies from those for enzymes, and enzymes with large turnover numbers may have different dynamical dependencies from those that turn over more slowly. The timescale differences between the fastest anharmonic fluctuations and the barrier-crossing rate point to the need to develop methods to resolve the range of motions present in enzymes on different time- and lengthscales.

CONTENTS

INTRODUCTION

Enzyme activity is widely regarded as being dependent on protein flexibility. But our understanding of how the dynamics of the molecule and its environment contribute to catalytic mechanisms lags far behind our knowledge of three-dimensional structures and chemical mechanisms. This is due not only to the complexity of the enzyme system but also to the wide range of internal motions exhibited by globular proteins. These cover lengthscales from 0.01 to 10 Å, timescales from 10^{-15} to >1 s (20, 87), and motions ranging from rapid local motions of individual groups to slow, collective fluctuations of large regions within the molecule. A major challenge is to understand the natures of these motions and how they contribute to enzyme function. In this article we review work relating protein dynamics to enzyme function. We focus on the global dynamics of proteins because these will inevitably have broader implications than local motions at particular active sites. Because we may expect most globular proteins to have many aspects of their global dynamics in common, we have included work relating to nonenzymic proteins where appropriate.

DYNAMICS OF PROTEINS

Experimental Techniques

A wide range of experimental techniques has been used to study the dynamics of enzymes, as well as computer simulations. Some of these are listed in Table 1, together with references for further details. Different techniques may measure different quantities, although this is not always made clear in the interpretations of the results. For example, some techniques detect the motion of a single probe nucleus (e.g., Fe in Mössbauer spectroscopy), while others report on the dynamics of more globally distributed probes (e.g., neutron scattering). Moreover, the timescales of the motions measured may differ between techniques. Most X-ray "Bragg" crystallography averages over hours, Mössbauer looks at motions on a timescale of about 10^{-7} s, while neutron scattering probes motions on a range of timescales from 10^{-12} to 10^{-8} s, depending on the instrumental resolution. Presently, computer simulations are largely limited to simulation times of no longer than $\sim 10^{-8}$ s.

TABLE 1 Dynamical transition temperatures obtained in a range of experimental studies

Technique	Protein	Preparation	Transition temperature (K)	Activity transition (K)	Reference
Mössbauer	Cytochrome *c*	Solution	~200		(58)
Mössbauer	Hemoglobin	Solution	~200		(57)
Mössbauer	Myoglobin	Solution	~240		(83)
Mössbauer	Oxymyoglobin	Frozen aqueous solution	~200		(72)
Mössbauer	Deoxymyoglobin	Crystal	~205		(96)
Mössbauer	Deoxymyoglobin	Crystal	~220		(13)
Mössbauer	Metmyoglobin	Crystal	~220		(13)
Mössbauer	Bacteriorhodopsin	In membrane	~200		(94)
Mössbauer	Chromatophore membranes	In membrane	~170	~190	(95)
Neutron scattering	Xylanase	Solution	~220	Not seen	(40)
Neutron scattering	Glutamate dehydrogenase	Solution	~230	Not seen	(32)
Neutron scattering	Amylase	Hydrated powder	200		(51)
Neutron scattering	Myoglobin	Hydrated powder	~180		(28, 37)
Neutron scattering	Superoxide dismutase	Hydrated powder	~180		(3)
Neutron scattering	Superoxide dismutase	Hydrated powder	~190		(45)
Neutron scattering	Bacteriorhodopsin	In membrane	~230	~230	(43)
Neutron scattering	Bacteriorhodopsin	In membrane	~150 ~220		(107)
Neutron scattering	Bacteriorhodopsin	In membrane	~150 ~250	Some correlations	(82)
Fluorescence	Cytochrome *c* peroxidase–cytochrome *c*	Solution	~220		(91)
Infrared	Carbonmonoxymyoglobin	Solution	~200		(68)
Infrared vibrational echo	Myoglobin	Solution	~200		(108)
Optical absorption (Soret band lineshape)	Carbonmonoxy heme proteins	Solution	~200		(35)
Time-resolved transient hole burning	Cytochrome *c*	Solution	~200		(115)
X-ray diffraction	Metmyoglobin	Crystal	~200		(62)
X-ray diffraction	Ribonuclease A	Crystal	170–220	~210	(105, 124)

The experimental technique may also constrain the nature of the sample. For example, signal/noise considerations favor neutron scattering measurements working with low hydration samples rather than solutions, crystallography requires crystals, while Mössbauer is generally restricted to solid samples. Environments can be tailored in computer simulation, although computational restrictions raise problems for working in dilute solutions or with large macromolecules. Furthermore, making correlations between measured dynamics and activity requires the enzyme to be active under the conditions in which the dynamics measurements are made.

Range of Motions

A polypeptide chain is intrinsically flexible, as many of the covalent bonds that occur in its backbone and side chains are rotationally permissive. Groups linked by rotationally permissive bonds are internally relatively rigid, possessing only high-frequency, low-displacement vibrational dynamics. These groups may therefore constitute fundamental dynamical elements in a protein molecule. Relative motions of these or analagously defined collections of rigid bodies may define principal components of protein dynamics (65).

Because proteins are densely packed, their atomic motion displays similarities to that seen in condensed phase materials. Over short times ($\leq 10^{-10}$ s) diffusive motions of rigid groups show similarities to the motions of a molecule in a liquid (74). These groups display rattling motions in a cage consisting of the neighboring protein atoms or surrounding solvent, if the group is at the protein surface. Both experiment and computer simulations have shown that substantial displacements of groups of atoms occur also over longer time intervals ($\sim 10^{-9}$ s). These may involve collective motions with a rigid-body character or local motions with significant energy barriers (i.e., fast but rare dynamics). The longer timescale internal protein dynamics deviates from liquid-like behavior owing to the additional constraint that proteins have a native molecular structure. Thus, translational and rotational diffusion of the atoms is limited and the internal atomic mean-square displacements converge with time.

Coupling Between Motions

Concerning the relationship between the various fluctuations occurring in proteins, of particular interest is the coupling between motions on different length- and timescales. If fast and slow motions were independent, then their effects would be separable and this would allow for a convenient description and classification of protein dynamics. However, although the fastest fluctuations (bond-length vibrations) probably are decoupled from the soft modes, there are indications that other displacements are not. Thus, neglect of, for example, bond-angle fluctuations in computer simulations significantly affects transition probabilities for slower motions. Theoretical activity is looking to find ways of including the faster fluctuations in computer simulations in an economic way (e.g., statistically) so as to bridge the timescale gap between what is achievable today and the holy grails of protein folding and function. The experimental counterpart to this addresses the question of whether the fast degrees of freedom are required for the slower motions that determine the rate-limiting steps in enzyme function.

ACTIVITY AND DYNAMICS

Induced Fit

A significant proportion of the catalytic power of enzymes comes from their ability to bring together substrates and the active site in favorable orientations in enzyme-substrate complexes. The specificity of substrate binding depends on the precisely defined arrangement of atoms in the active site. Fischer first proposed a lock-and-key model in 1890, whereby the substrate fitted directly into the active site (120). However, this model was found to be inadequate for a number of enzymes. In 1958, Koshland (78) proposed the idea of an induced fit mechanism of substrate binding. In this model the binding of a substrate to an enzyme is accompanied by a conformational change that aligns the catalytic groups in their correct orientations (44). The induced fit model has been shown to be a reasonable representation of substrate binding for many enzymes [see, for example, (84, 114)].

An essential aspect of the induced fit model is enzyme flexibility, and it is now widely accepted that flexibility is required for catalytic activity (7, 20, 55, 60, 66, 71). In many cases crystallography has provided direct evidence for the existence of different enzyme conformations in the presence and absence of substrate (60). Control, catalytic conversion, and product release may also require flexibility in the protein. However, the relationship between enzyme flexibility and activity is not as well understood as that between structure and activity, and the timescales and forms of the functionally important motions remain poorly characterized.

Conformational Substates

It is widely accepted that the native conformation of a protein comprises a large number of slightly different structures that correspond to local minima in the potential energy surface of the system. The presence of these substates was suggested as a means of explaining the temperature dependence of ligand binding to myoglobin (8, 55). Frauenfelder et al. (56), Ansari et al. (5), and Elber & Karplus (41) have given detailed discussions of protein conformational substates, with particular reference to myoglobin. Transitions between some of these substates may occur if the kinetic energy is sufficient and may be global (collective) or local. In principle, either kind of transition may be of biological importance (92).

In many enzymes, there are fluctuations between conformations. The equilibrium between these conformations may be perturbed on binding and release of substrate. For example, binding of an appropriate regulatory ligand may increase the population of substates with greater activity. For some enzymes, e.g., tyrosine phosphatases (121), the conformational change involves the movement of a hinged loop that folds over the active site–substrate complex upon substrate binding to promote catalysis.

Thermodynamic Aspects

Some of the important factors influencing enzyme-substrate association, such as hydrophobic, hydrogen-bonding, electrostatic, and van der Waals interactions, are well appreciated. However, a further important but poorly characterized contribution to binding thermodynamics arises from dynamical effects. Three translational and three rotational degrees of freedom are lost when two molecules associate. This is entropically unfavorable for binding, typically costing 20 to 30 kcal/mol depending on the associating species. However, this is offset by the presence of six new degrees of freedom in the complex. Moreover, the presence of the new intermolecular interactions in the complex alters the vibrational frequency distribution (density of states) relative to that of the unbound species. It is straightforward to calculate from the density of states the free energy change corresponding to the vibrational modifications. However, due to the complexity of the spectrum of protein vibrations, few estimations of this free energy change have been made. The availability of fast computers to perform normal mode analysis in the full configurational space of small proteins has now rendered this problem tractable. Using this method the vibrational effect on insulin dimerization was calculated to be 7.3 kcal/mol, favoring binding (123). In a second example, harmonic analysis of the binding of a buried water molecule to bovine pancreatic trypsin inhibitor found binding was also favored by the vibrational effect, by 3.5 kcal/mol (49, 50). These calculations confirm that both protein:protein association and small molecule:protein binding are accompanied by a highly significant free energy change due to the vibrational changes. To our knowledge, no calculation of the vibrational free energy change on enzyme-substrate binding has been reported. Moreover, no reliable experimental estimate has been made of the change in the density of states on ligand binding to a protein. Inelastic neutron scattering is the best experimental technique available for doing this. However, the few previous attempts on protein-ligand systems have been inconclusive owing to statistical inaccuracy and the lack of associated theoretical analysis (10, 69). Further progress in this important area is expected in the near future.

Specific Motions and Catalysis

The idea that enzymes might have evolved not just structurally but also dynamically to optimize function is intriguing but unproven. In this context it may be useful to distinguish conceptually between "thermal noise" background fluctuations that are present in any condensed phase system and may nonetheless be required for function and evolutionarily selected specific amplified degrees of freedom or correlated fluctuations in an enzyme active site that would enhance the rate of catalytic steps. The existence of the latter class of motions in enzymes is debated (90, 134).

In one recently studied example, conformational flexibility, including subdomain rotation and alternate loop conformations, was shown to be important for both the transformation of the substrate to product and the regeneration of the enzyme dihydrofolate reductase (DHFR) (88). Correlated motions have been found to occur over a distance of approximately 13 Å in the DHFR-folate complex, with motions in the FG loop being linked to the active site. One interpretation of the data available for DHFR is that the correlated motions are acting to increase the frequency of barrier crossing through the transition state of the enzyme-catalyzed reaction (22). Molecular dynamics simulation by Radkiewicz & Brooks (104) found evidence that the Michaelis complex exhibits correlated motions coupling distant regions of the structure. It was suggested that the coupled motions are probably necessary for the completion of the catalytic cycle.

Hydrogen Tunneling

Protein flexibility has been implicated in the efficiency of hydrogen tunneling in enzymes (75–77). Bahnson et al. (9) showed that hydrogen tunneling efficiency was reduced in a horse liver alcohol dehydrogenase mutant owing to an increase in the distance between the donor and acceptor carbons. This led to the suggestion that the flexibility in interdomain movement could influence the catalytic rate by reducing the hydrogen tunneling distance of the native enzyme. Later studies of the thermophilic alcohol dehydrogenase from *Bacillus stearothermophilus* supported this conclusion (75). Hydrogen tunneling made a significant contribution to catalysis at 65°C, and the tunneling efficiency decreased with decreasing temperature. These results led to the conclusion that thermally excited enzyme fluctuations are involved in modulating the enzyme-catalyzed hydride transfer reaction. Calculations on triose phosphate isomerase proton-transfer reactions also suggest that tunneling is important (27), although the associated factor of 10 increase in rate found for the enzyme is argued to be a relatively small contribution to the overall rate enhancement of 10^9.

Entropy Effects

An important dynamical question concerns the role of entropy in enzymic catalysis. It is conceivable that the large configurational space accessible to the reacting species in solution would be considerably reduced in an enzyme active site (93). From this it has been deduced that this configurational space reduction should lead to large entropic contributions to the difference between the activation barrier in the enzyme and in solution. Understanding the entropic contribution clearly requires an understanding of active site flexibility and how this compares with solution conditions. Recent calculations on subtilisin indicate that the residual flexibility in the active site is indeed entropically significant and should lead to revision downward of the size of the overall entropic effect (131).

Activity, Stability, and Dynamics

Dynamical arguments have been used to explain activity-stability relationships in enzymes from mesophilic and thermophilic sources (29, 128, 130, 133, 136, 144). However, recent work suggests that dynamics/activity/stability interrelationships in these systems may not be straightforward (6, 64, 70, 126). The global stability of a protein is not necessarily an accurate indication of the flexibility of its active site. For example, studies with bacteriorhodopsin showed that the dynamics of a specific region near the active site were different from the global dynamics (107). Moreover, it is unclear if there exists a simple link between active site flexibility and catalytic rate.

THE DYNAMICAL TRANSITION

Protein Dynamics Depends on Temperature

A phenomenon that has been the focus of much attention is the change in slope observed in plots of the average atomic mean-square displacement of the atoms in a protein versus temperature. An example is given in the lower line (squares) of Figure 1. A variety of proteins studied by different techniques have shown slope changes of varying degree at transition temperatures generally between 180 and 220 K. Some of the variation seems likely to be due to the variation in the nature of the protein preparation and the timescale of the technique used (the timescales of

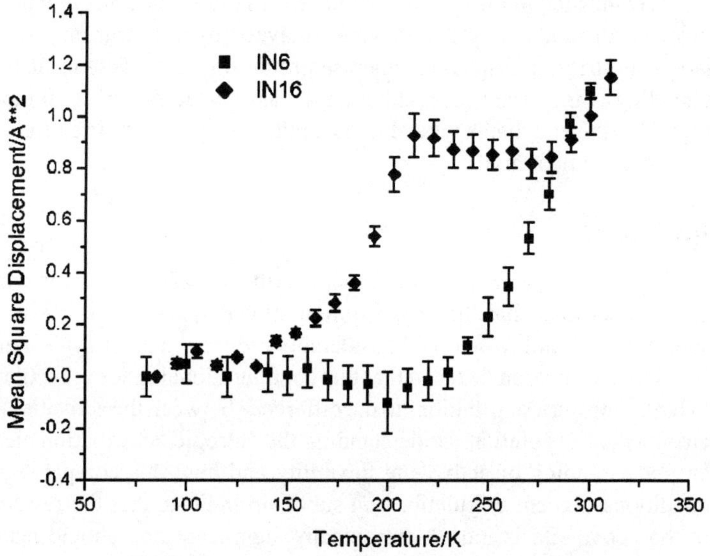

Figure 1 Mean-square displacement of GDH in 70% methanol in water, as a function of temperature, measured on two different neutron instruments at the Institut Laue-Langevin, Grenoble, France, which resolve motions on different timescales. *Squares*: measured on IN6, reporting on motions faster than about 100 ps; *diamonds*: measured on IN16, reporting on motions faster than about 5 ns. The standard dynamical transition is seen in the IN6 data at about 220 K.

the motions detected by each technique vary from approximately 10^{-7} s to 10^{-12} s). Below this "dynamical transition temperature," protein motions are argued to be purely harmonic vibrations of the atoms, while above it anharmonic motions become evident. A summary of the dynamical transition temperatures observed using various techniques for various protein preparations is given in Table 1.

Neutron scattering has been a major technique in dynamical transition research. On increasing temperature, a decrease in the elastic intensity at low-scattering factor is observed above the dynamical transition temperature. The decrease in elastic intensity is accompanied by an increase in the quasielastic scattering, which is seen as a broad peak centered on the elastic peak. Quasielastic scattering is due to stochastic processes, for example, diffusive motions that involve the crossing of energy barriers (85).

The first neutron scattering experiment reporting dynamical transition behavior in a protein was that of Doster et al. (37) on hydrated myoglobin powders. The anharmonic contribution was attributed to the onset of torsional jumps of protons among distinct sites with slightly different energy, i.e., conformational substates. A molecular dynamics simulation analysis of the neutron scattering data from myoglobin by Kneller & Smith (74) led to an alternative explanation for the nonvibrational component of the dynamics, which was attributed to liquid-like diffusive motions of the side chains rather than conformational transitions in the side chains themselves.

The change in temperature dependence of the mean-square displacements obtained from crystallographic measurements on protein crystals has also been interpreted as showing a transition in protein dynamics (55). However, as crystallographic temperature factors may include nondynamical contributions such as static disorder and refinement errors, this change cannot be rigorously assumed to have a purely dynamical origin. In metmyoglobin crystals (62) a transition in the temperature dependence of the mean-square displacement of the iron atom, and the atoms in the heme plane, was observed around 200 K. The X-ray data were compared to those obtained from Mössbauer experiments, which exhibited a dynamical transition at a similar temperature. When the overall mean-square displacement was extrapolated to 0 K, nonzero values were obtained, suggesting the existence of conformational substates of the metmyoglobin molecule that are frozen in at 0 K. Some residual nonzero value would of course also be expected from zero-point motions.

The Dynamical Transition and Protein Function

If protein function is dependent on protein flexibility and there is a significant transition from harmonic to anharmonic dynamics at the dynamical transition, a loss of function might be expected below the transition temperature. A number of studies have investigated this hypothesis. The dynamics of the membrane protein chromatophores from *Rhodospirillum rubrum* were investigated by the Mössbauer effect after incorporation of ^{57}Fe and were shown to undergo a transition at approximately 170 K (95). The temperature dependence of the efficiency of the photoinduced electron transfer was also observed to undergo a change at approximately 170 K. The function of light-driven proton pump protein bacteriorhodopsin also exhibits strong correlation with dynamical transition phenomena (43, 53, 82).

The Dynamical Transition and Enzyme Activity

One of the first studies relating enzyme binding to dynamical transition investigated the temperature dependence of inhibitor binding by crystalline ribonuclease A (105). Ribonuclease A did not bind the substrate or the inhibitor, cytidine 2′-monophosphate, at 212 K, but bound rapidly to both at 228 K. Once bound at the higher temperature, the inhibitor could not be washed off after the enzyme was cooled to 212 K. No significant structural change was seen at the two temperatures. It was suggested that thermally driven collective atomic fluctuations in proteins, present only above the transition temperature, are essential for the rapid productive binding of large ligands. However, although it is interesting to see the abovementioned qualitative difference in observed binding, the transition temperature is not clear (105, 124) (see bottom line of Table 1).

The temperature dependence of the activity of both mesophilic and thermophilic enzymes has been investigated over a wide temperature range, from approximately 190 to 360 K (32, 39, 40, 89). It was suggested that if thermophilic enzymes are less flexible than corresponding mesophilic enzymes at a given temperature, then the dynamical transition might occur at a higher temperature than that observed for mesophilic enzymes. Therefore, if anharmonic dynamics were required for activity, a significant change in activity accompanying the transition in dynamics would be seen at a higher temperature than in the mesophilic enzymes. Activity measurements were conducted down to temperatures lower than the dynamical transition temperature of ~220 K. However, no deviation from linearity in the Arrhenius plots was detected for either the thermophilic or mesophilic enzymes. This suggested that the rate-limiting step for these enzymes is independent of the fast (picosecond timescale) anharmonic dynamics determined by neutron scattering techniques. More recently, the activities of catalase and alkaline phosphatase have also been shown to exhibit no deviations in Arrhenius behavior down to near 170 K (18). This is significantly below the temperature at which the dynamical transition has been observed in a number of proteins. Therefore, although no dynamical studies have been performed specifically on these latter two enzymes, the activity results do support the observation of an independence of enzyme activity on fast anharmonic motions.

To further address the dependence of enzyme activity on anharmonic picosecond-timescale motions, the temperature dependence of the activity of glutamate dehydrogenase was compared to that of the dynamics of the enzyme cryosolution on two different timescales, as determined by neutron scattering (30). The neutron scattering experiments were performed on the instruments IN16 and IN6 at the Institut Laue-Langevin reactor in Grenoble, France, allowing the detection of motions with timescales faster than approximately 5 ns and 100 ps, respectively. The glutamate dehydrogenase-solvent solution results showed a shift of the lowest dynamical transition temperature downward as the timescale probed increased (see Figure 1). These results suggest that the temperature dependence of the transition in a protein solution is timescale dependent. The lowest dynamical transition temperature is heavily dependent upon the instrument timescale, rather than occurring at a fixed temperature. Similar results have also been observed for a xylanase enzyme (31). The timescale dependence of the dynamical transition temperature may possibly be explained simply by the anharmonic motions slowing, with motions over a given timescale progressively being replaced by slower motions, as the temperature is reduced (31).

PROTEIN SOLVATION EFFECTS

Structural Effects of Solvation

A variety of observations made as enzymes are rehydrated from the anhydrous state to the solution (23, 25, 46, 102, 109, 110, 112, 140) have associated hydration changes with changes in protein-water interactions and protein dynamics. Based mainly on detailed work on lysozyme, the process seems to involve four stages. In the first, from 0 to 0.07 h (grams of water per gram of protein), proton redistribution occurs (i.e., the ionizable groups ionize) as water molecules interact with the ionizable groups. This corresponds to a mole ratio of about 0 to 55 water molecules per enzyme molecule (typically lysozyme). If we accept that the lysozyme surface is completely covered by ~400 water molecules (47, 48), then this corresponds to 0% to 15% surface coverage of the enzyme. In stage two (0.07 to 0.30 h, corresponding to a mole ratio of 55 to 250, or 15%–65% coverage), water binds to the charged and polar sites. In stage three, between 0.30 and 0.50 h, weakly interacting (presumably largely nonpolar) surface regions are covered until water monolayer formation is complete at 0.50 h. In stage four, above 0.50 h, the hydration shell builds further, and the enzyme displays dilute solution thermodynamic properties. The hydration water has mobility close to, but slightly lower than, that of the bulk solvent. Although Rupley et al. (111) argue that no conformational changes occur during rehydration, other evidence (101, 141) suggests that fully dehydrated lysozyme has a structure slightly distorted locally from that in solution (80, 101). On rehydration, these small conformational distortions are "annealed out" by 0.2 h.

Dynamical Effects of Solvation

At physiological temperatures dry proteins are, in comparison to hydrated ones, relatively rigid and are also accepted to be inactive. A number of studies have probed the relationship between protein hydration and protein dynamics (24, 36, 43, 52, 79, 97, 99, 117, 142, 143). From related NMR measurements of exchangeability of the main chain amide hydrogens (100) and from measurements of the rate of exchange of labile hydrogens (112), it is suggested that a hydration-related increase in conformational flexibility is necessary for activity. ESR spin probe measurements on lysozyme (111) have been interpreted in terms of the protein regaining its solution dynamical state at 0.20 to 0.25 h. A break in the TEMPONE binding curve between 0.07 and 0.11 h, in light of NMR hydrogen exchange measurements, has been interpreted as a hydration-induced loosening up of the protein, an interpretation consistent with other hydrogen exchange (110) and dielectric (17) measurements.

Measurements of the Rayleigh scattering of Mössbauer radiation (79) found that at low hydration levels (~0.1 h, or 37% relative humidity), the atomic mean-square displacement in myoglobin showed no measurable motions at temperatures up to 300 K. However, at a relative humidity of 94%, an increase of the dynamical amplitudes was seen near 220 to 240 K. An increase in dynamical amplitudes was also found in subsequent work on lysozyme powders (117), and qualitatively similar results have been observed by more recent neutron scattering studies on hydrated myoglobin, lysozyme, parvalbumin, and amylase (36, 51, 99, 142). At room temperature, hydrated myoglobin and lysozyme exhibit a more pronounced quasielastic spectrum owing to diffusive motions, which was suggested to be due to modification of the amplitude, but not the timescale, of the fast protein motions (36). A later study suggested that upon hydration from dry powder to monolayer coverage, the surface side chains could diffuse locally. Further hydration increases the rate of these motions. In solution, the motions increase in amplitude compared with the fully hydrated powder and have a shorter average relaxation time than for proteins with monolayer coverage of water (99).

Solvation and the Dynamical Transition

A number of studies have suggested that the solvent strongly influences the dynamical transition. The biphasic behavior of the average mean-square displacement of ribonuclease-A was attributed to the coupling between the structure and the dynamics of the solvent shell and the protein (124). Mössbauer studies on both deoxygenated myoglobin crystals and $^{57}K_4Fe(CN)_6$ dissolved in the water of metmyoglobin crystals showed the ^{57}Fe had a similar temperature dependence of its mean-square displacement both in the protein and in the water solution.

Some studies argue that the dynamical transition is driven by a transition in the protein hydration water shell, a transition that has been seen in molecular dynamics simulation (15) and in neutron scattering experiments (51, 99). Measurements on the infrared stretching frequencies of CO bound to myoglobin indicated that the dynamical behavior of the protein is correlated with a glass transition in the surrounding solvent (68). The involvement of hydrogen-bonding interactions between the hydration water and the protein in the observed dynamical transition has been suggested in a number of studies (34, 36, 51). Molecular dynamics simulation work on ribonuclease concluded that complete structural relaxation of the protein at 300 K requires relaxation of the water hydrogen-bond network and that the short timescale water H-bond lifetime is not affected by the dynamical transition (122).

This decoupling of "rattling" motions from more global translational diffusion is a characteristic of a glass transition (4), suggesting that the water transition is glass-like.

Several studies have used environmental changes to probe the relation between solvent and the dynamical transition. By placing proteins in a room-temperature glass such as trehalose the transition to anharmonic dynamics was not observed, even up to room temperature (26). The effect of varying solution composition on the picosecond-timescale dynamics of solutions of the enzyme xylanase was investigated by dynamic neutron scattering (106). The results indicate a significant effect of the solvent, as the picosecond fluctuations of the protein solution largely follow those of the corresponding pure solvent. These results also indicate that for various protein-solvent samples the dynamical transition is strongly coupled to the melting of pure water, but is relatively invariant in cryosolvents of different compositions and melting points. One possible explanation for this observed insensitivity of the dynamical transition to cryosolvent composition may be that the structure of the solvent shell surrounding the protein is qualitatively similar in all the cryosolvents but different from that in pure water.

Although the above studies have emphasized the role of the solvent on the dynamical transition, others argue that it is dependent on the protein structure itself. For example, fluorescence quenching of the conformational dynamics in the Zn-substituted cytochrome *c* peroxidase and cytochrome *c* concluded that the transition in dynamics is an intrinsic phenomenon of the complex (91). Comparisons between the Mössbauer spectra of oxymyoglobin in aqueous solution and metmyoglobin crystals suggested that the conformational fluctuations are an intrinsic property of the protein (72).

Solvent Dynamical Mechanisms

A number of studies have suggested various mechanisms by which the solvent environment affects protein dynamics. Frictional damping of protein modes can result from intraprotein anharmonicities, protein-protein interactions (when present), and from the surrounding solvent (19, 116, 118, 127). However, frictional damping per se cannot trigger dynamical transition behavior because it affects only the time course of the exploration of normal modes and not their amplitude. Solvent molecules can also affect protein dynamics by modifying the effective potential surface of the protein. One suggestion is that the increased flexibility observed upon hydration may in part be due to the reduced interaction of charged and/or polar residues due to dielectric screening caused by water-forming clusters on the protein surface (2, 16, 113). Correspondingly, it was found that the internal mobility of chymotrypsin decreased with decreasing dielectric constant of the organic solvent, consistent with changes in the electrostatic interaction between charged residues (1).

Simulation work by Vitkup et al. (132) showed that in the hydrated state solvent fluctuations strongly influence the internal dynamics of myoglobin. Recent work (A. Tournier, J. Xu & J.C. Smith, manuscript submitted) suggests that the hydration-induced dynamical transition affects mainly external side chain atoms. This finding dovetails with the recently proposed "radially-softening" model of protein dynamics in which the average dynamical properties of a protein at 300 K vary smoothly with increasing distance from the protein core. There is a gradual increase of the diffusive amplitudes and a narrowing and shift to shorter (picosecond) times of the distribution of diffusive relaxation processes (33).

Effect of Protein Hydration on Enzyme Activity

It is generally accepted that protein hydration is essential for enzyme catalysis to occur and that dry enzymes are inactive. Although there are varying estimates of the degree of hydration required for activity, a threshold value of about 0.2 g H_2O g^{-1} protein, i.e., 0.2 h, is generally accepted (23, 46, 54, 109, 110). This fits with the view that one of the fundamental requirements for enzyme catalysis is conformational flexibility. Conformational flexibility, of the order of (sub)-nanoseconds, is expected to maximize favorable enzyme interactions with the substrate in the initial and transition state, allowing the enzyme to explore multiple conformations and therefore increase the probability that a conformational state capable of binding and converting a substrate will be achieved (112).

As mentioned previously, studies on lysozyme, subtilisin Carlsberg, α-chymotrypsin, and α-amylase show the onset of protein motions at ~0.2 h. These have been correlated with the onset of enzyme catalysis in hydrated enzyme powders (23, 38, 86,102, 109, 110, 119, 140). However, one difficulty with measuring activity in enzyme powders of low hydration is the absence of a medium for the diffusion of substrate and product, particularly in enzymes such as lysozyme that have low specific activity. The approach involving the rapid mixing and freeze-drying of enzyme and substrate solutions, for example, is likely to lead to measurements of product release rather than catalysis, and if enzyme cofactors such as NAD are involved, the rates may be limited by the local availability of these rather than by the hydration state of the enzyme.

There are two types of systems that circumvent this difficulty: enzyme powders in dry organic solvents and enzyme powders exposed to gaseous substrates. In both cases there are reports that enzyme activity may be possible at lower hydration levels.

In the work on enzyme catalysis in dry organic solvents, there is good evidence that activity may be possible at very low hydrations (73, 125). However, interpretation is complicated by the likelihood that the organic solvent is replacing water at sites on the enzyme, as well as performing a hydration function by loosening the structure (2, 21, 138).

The system that best enables the separation of those effects that are due to the structural/dynamical roles of water in catalysis from those arising from diffusional constraints is the use of gas phase substrates. Yagi et al. (139) have observed some elements of hydrogenase activity in the dry state (probably <0.03 h), and a number of groups investigating biotechnological applications of gas phase catalysis have demonstrated enzyme activities below 0.2 h (11, 12, 67, 81, 98, 103). Lind, Dunn, and Daniel (manuscript submitted) have recently demonstrated pig liver esterase activity below 0.03 h. At this hydration level, the water-to-enzyme molar ratio is about 100, allowing ~10% surface coverage and only ~15% coverage of polar groups; there are certainly too few water molecules to promote conformational flexibility. Although the effect of hydration on the dynamics of this enzyme has not been determined, there is no reason to suppose that its behavior in this respect will differ from other proteins. Every protein so far investigated shows a clear cessation of intramolecular motions below ~0.1 to 0.2 h. Because most of the proteins investigated have been fairly small, this corresponds to a water-to-enzyme molar ratio in the range of 160 to 310, and a surface coverage of 40% to 50%. These results are in keeping with those from other nondiffusion-limited systems (11, 12, 67, 73, 81, 98, 103, 125, 139). The inference is that enzyme activity is possible, independent of internal dynamics, and more generally that enzyme activity is not dependent on the properties conferred upon proteins by water above hydration levels corresponding to a surface coverage of about 10%, and only about 15% coverage of polar groups. It is also in keeping with results showing that enzymes are active at temperatures where subnanosecond anharmonic dynamics have ceased (18, 32, 40).

CONCLUSIONS

In the time since the proposal of the induced fit mechanism for enzyme-substrate binding, a considerable amount of research has been performed with the aim of determining the forms and timescales of the internal motions in enzymes that determine their catalytic function. It has become clear that the motions required will vary according to the step of the functional process considered. External, whole-molecule diffusive Brownian dynamics determines the initial enzyme-substrate encounter (59). In some cases conformational flexibility is required for substrate access (135) and/or product release (137). For the steps in between, a distinction can be drawn between reaction dynamics along the reaction coordinate and other fluctuations required for progress along the reaction coordinate to be made. A priori one would suspect that the larger the amplitude of a fluctuation in a protein, the more likely the motion is to participate in functional activity. This would certainly be expected for the motions mediating the large-scale conformational change that often accompanies enzyme activity, such as, for example, the molecular switch in GTPases (129). In contrast, however, the barrier-crossing events defining rate-limiting steps in enzyme catalysis may, in principle, be quite localized and of relatively high frequency.

Dissociating the functionally important motions from the unneccesary thermal noise is an important challenge for the future of enzymology. The dynamical transition in enzymes may have provided us with an experimental opportunity to examine which fluctuations are critical to protein function. Above the transition temperature, anharmonic fluctuations are thought to be excited. On the other hand, the transition has been observed at temperatures ranging from 170 to >230 K, generally with little reference to defining the timescale of the motions probed. When a timescale-defined window is used, the temperature of the dynamical transition appears to be timescale dependent (30). Furthermore the pattern of global dynamics may be simply explained in terms of a steady, uninterrupted, decrease in motions with temperature, rather than requiring the introduction of more complex hypotheses to explain the observed behavior (31). Whether the fluctuations that cease over 100 ps timescales at 220 K for some enzymes are required for protein function has not yet been definitively answered, but evidence from Arrhenius plots of enzyme activity down to temperatures below the dynamical transition suggests that 100 ps motions, and any motions coupled to these, are not required for enzyme function (18, 30, 32, 40). However, if the dynamical transition is to be used as an investigative tool, careful attention will need to be paid to the effects of timescale and preparation. The dependence on timescale has already been emphasized. Interpretation of data for membrane proteins such as bacteriorhodopsin is further complicated by the embedding/solution of the protein in the lipid matrix/solvent.

One possible explanation for enzyme activity in the apparent absence of global dynamics, for example at low hydrations and below the dynamical transition, could relate to the fact that most of the techniques, such as neutron beam spectroscopy, which are used to investigate protein dynamics, give a global average of all protein dynamical motions. As has been pointed out (30), the accuracy of these techniques is such that dynamics confined to the active site might not be detected. Enzyme activity is likely to be dependent on dynamics at the active site. The high reactivity of an enzyme active site is likely to be conferred by its having greater flexibility than the rest of the enzyme. The active site may thus be less dependent upon hydration, or temperature, for its flexibility than the protein as a whole.

Moreover, it may be counterproductive to discuss all enzymes/proteins as though they were the same. Proteins where the function is electron transfer, proton tunneling, or ligand binding may have different dynamical dependencies from those for enzymes; soluble enzymes may not behave in the same way as membrane-bound enzymes; and enzymes with turnover numbers of one million per second (e.g., catalase) may be expected to have different dynamical dependencies from those that have turnovers of less than 1 sec^{-1} (e.g., lysozyme).

However, the timescale difference between the fastest anharmonic fluctuations and the barrier-crossing rate, which is at least a factor of 10^6 for enzymes, points to the need to develop methods capable of resolving the range of motions present in enzymes on different time- and lengthscales. This will require new experimental developments. Examples of these are spin-echo neutron spectroscopy, which pushes the time limit for neutron scattering to 10^{-8} s (14), and thermal diffuse X-ray scattering, which enables correlated motions to be detected in X-ray crystallography (42, 65). Also potentially useful are NMR and the so-far little-developed technique of X-ray speckle interferometry using third-generation synchrotrons. Of crucial importance in understanding enzyme motions is computer simulation. Advances will certainly include improved accuracy molecular dynamics simulations over longer times than is currently feasible and the ability to deal with larger enzymes, but advances will also encompass improved quantum mechanical/molecular mechanical methods for probing enzyme reaction mechanisms together with methods for determining pathways between conformational states. Moreover, computer simulation will be increasingly useful as a tool in understanding the experimental data obtained from the enzyme systems themselves (63). However, all these methods need careful correlation with enzyme rates obtained under the same conditions, and difficulties in interpreting experimental data under such conditions need to be overcome (e.g., by careful use of computer simulation) rather than ignored.

Finally, much more work specifically targeting quasielastic and inelastic spectroscopic measurements will be needed if we are to attain the ultimate aim of seeing precisely what motions are needed for enzyme activity. Furthermore, selective isotope substitution can be used in association with neutron spectroscopy to probe the relevance of local dynamics. Such studies have begun, but they have not even scratched the surface of what is likely to be possible.

LITERATURE CITED

1. Affleck R, Haynes CA, Clark DS. 1992. Solvent dielectric effects on protein dynamics. *Proc. Natl. Acad. Sci. USA* 89:5167–70
2. Affleck R, Xu Z-F, Suzawa V, Focht K, Clark DS, Dordick JS. 1992. Enzymatic catalysis and dynamics in low-water environments. *Proc. Natl. Acad. Sci. USA* 89:1100–4
3. Andreani C, Filabozzi A, Menzinger F, Desideri A, Deriu A, Dicola D. 1995. Dynamics of hydrogen atoms in superoxide dismutase by quasielastic neutron scattering. *Biophys. J.* 68:2519–23
4. Angell CA. 1995. Formation of glasses from liquids and biopolymers. *Science* 267:1924–35
5. Ansari A, Berendzen J, Bowne SF, Frauenfelder H, Iben IET, et al. 1985. Protein states and protein quakes. *Proc. Natl. Acad. Sci. USA* 82:5000–4
6. Arnold FH. 1998. Enzyme engineering reaches the boiling point. *Proc. Natl. Acad. Sci. USA* 95:2035–36
7. Artymiuk PJ, Blake CCF, Grace DEP, Oatley SJ, Phillips DC, Sternberg MJE. 1979. Crystallographic studies of the dynamic properties of lysozyme. *Nature* 280:563–68
8. Austin RH, Beeson KW, Eisenstein L, Frauenfelder H, Gunsalus IC. 1975. Dynamics of ligand binding to myoglobin. *Biochemistry* 14:5355–73
9. Bahnson BJ, Colby TD, Chin JK, Goldstein BM, Klinman JP. 1997. A link between protein structure and enzyme catalyzed hydrogen tunneling. *Proc. Natl. Acad. Sci. USA* 94:12797–802
10. Bartunik HD, Jolles P, Berthou J, Dianoux AJ. 1982. Intramolecular low-frequency vibrations in lysozyme by neutron time-of-flight spectroscopy. *Biopolymers* 21:43–50
11. Barzana K, Karel M. 1989. Enzymatic oxidation of ethanol in the gas phase. *Biotechnol. Bioeng.* 34:1178–85
12. Barzana K, Klibanov AM, Karel M. 1987. Enzyme-catalysed gas phase reactions.

Appl. Biochem. Biotechnol. 15:25–33
13. Bauminger ER, Cohen SG, Nowik I, Ofer S, Yariv J. 1983. Dynamics of heme iron in crystals of metmyoglobin and deoxymyoglobin. *Proc. Natl. Acad. Sci. USA* 80:736–40
14. Bellissent-Funel M-C, Daniel RM, Durand D, Ferrand M, Finney JL, et al. 1998. Nanosecond protein dynamics—first detection of a neutron incoherent spin-echo signal. *J. Am. Chem. Soc.* 120:7347–48
15. Bizzarri AR, Paciaroni A, Cannistraro S. 2000. Glasslike dynamical behaviour of the plastocyanin hydration water. *Phys. Rev. E* 62:3991–99
16. Bone S. 1987. Time-domain reflectometry studies of water binding and structural flexibility in chymotrypsin. *Biochim. Biophys. Acta* 916:128–34
17. Bone S, Pethig R. 1982. Dielectric studies of the binding of water to lysozyme. *J. Mol. Biol.* 157:571–75
18. Bragger JM, Dunn RV, Daniel RM. 2000. Enzyme activity down to −100°C. *Biochim. Biophys. Acta* 1480:278–82
19. Brooks CL III, Karplus M. 1989. Solvent effects on protein motion and protein effects on solvent motion. Dynamics of the active site region of lysozyme. *J. Mol. Biol.* 208:159–81
20. Brooks CL III, Karplus M, Pettitt BM. 1988. *Proteins: A Theoretical Perspective of Dynamics, Structure and Thermodynamics*. New York: Wiley
21. Broos J, Visser AJWG, Engbersen JFJ, Verboom W, van Hoek A, Reinhoudt DN. 1995. Flexibility of enzymes suspended in organic solvents probed by time-resolved fluorescence anisotropy. Evidence that enzyme activity and enantioselectivity are directly related to enzyme flexibility. *J. Am. Chem. Soc.* 117:12657–63
22. Bruice TC, Benkovic SJ. 2000. Chemical basis for enzyme catalysis. *Biochemistry* 39:6267–74
23. Careri G, Giansanti A, Gratton E. 1979. Lysozyme film hydration events: an IR and gravimetric study. *Biopolymers* 18:1187–203
24. Careri G, Giansanti A, Rupley JA. 1986. Proton percolation on hydrated lysozyme powders. *Proc. Natl. Acad. Sci. USA* 83:6810–14
25. Careri G, Gratton E, Yang P-H, Rupley JA. 1980. Correlation of IR spectroscopic, heat capacity, diamagnetic susceptibility and enzymatic measurements on lysozyme powder. *Nature* 284:572–73
26. Cordone L, Ferrand M, Vitrano E, Zaccai G. 1999. Harmonic behaviour of trehalose-coated carbon-monoxy-myoglobin at high temperature. *Biophys. J.* 76:1043–47
27. Cui Q, Karplus M. 2002. Quantum mechanics/molecular mechanics studies of triosephosphate isomerase-catalyzed reactions: effect of geometry and tunneling on proton-transfer rate constants. *J. Am. Chem. Soc.* 124:3093–124
28. Cusack S, Doster W. 1990. Temperature dependence of the low frequency dynamics of myoglobin. *Biophys. J.* 58:243–51
29. Daniel RM, Dines M, Petach HH. 1996. The denaturation and degradation of stable enzymes at high temperatures. *Biochem. J.* 317:1–11
30. Daniel RM, Finney JL, Réat V, Dunn RV, Ferrand M, Smith JC. 1999. Enzyme dynamics and activity: time-scale dependence of dynamical transitions in glutamate dehydrogenase solution. *Biophys. J.* 77:2184–90
31. Daniel RM, Finney JL, Smith JC. 2002. The dynamic transition in proteins may have a simple explanation. *Faraday Discuss.* 122:163–69
32. Daniel RM, Smith JC, Ferrand M, Héry S, Dunn RV, Finney JL. 1998. Enzyme activity below the dynamical transition at 220 K. *Biophys. J.* 75:2504–7
33. Dellerue S, Petrescu AJ, Smith JC, Bellissent-Funel M-C. 2001. Radially softening diffusive motions in a globular protein. *Biophys. J.* 81:1666–76

34. Demmel F, Doster W, Petry W, Schulte A. 1997. Vibrational frequency shifts as a probe of hydrogen bonds: thermal expansion and glass transition of myoglobin in mixed solvents. *Eur. Biophys. J.* 26:327–35
35. Di Pace A, Cupane A, Leone M, Vitrano E, Cordone L. 1992. Protein dynamics: vibrational coupling, spectral broadening mechanisms, and anharmonicity effects in carbonmonoxy heme proteins studied by the temperature dependence of the Soret band lineshape. *Biophys. J.* 63:475–84
36. Diehl M, Doster W, Petry W, Schober H. 1997. Water-coupled low-frequency modes of myoglobin and lysozyme observed by inelastic neutron scattering. *Biophys. J.* 73:2726–32
37. Doster W, Cusack S, Petry W. 1989. Dynamical transition of myoglobin revealed by inelastic neutron scattering. *Nature* 337:754–56
38. Drapron R. 1985. Enzyme activity as a function of water activity. *NATO ASI Ser. E* 90:171–90
39. Dunn RV. 1998. *Effect of low temperature on thermophilic enzymes.* MSc thesis. Univ. Waikato, Hamilton. 124 pp.
40. Dunn RV, Réat V, Finney JL, Ferrand M, Smith JC, Daniel RM. 2000. Enzyme activity and dynamics: xylanase activity in the absence of fast anharmonic dynamics. *Biochem. J.* 346:355–58
41. Elber R, Karplus M. 1987. Multiple conformational states of proteins: a molecular dynamic analysis of myoglobin. *Science* 235:318–21
42. Faure P, Micur A, Perahia D, Doucet J, Smith JC, Benoit M. 1994. Correlated intramolecular motions and diffuse X-ray scattering in lysozyme. *Nat. Struct. Biol.* 1:124–28
43. Ferrand M, Dianoux AJ, Petry W, Zaccai G. 1993. Thermal motions and function of bacteriorhodopsin in purple membranes: effects of temperature and hydration studied by neutron scattering. *Proc. Natl. Acad. Sci. USA* 90:9668–72
44. Fersht A. 1985. *Enzyme Structure and Mechanism.* New York: Freeman
45. Filabozzi A, Deriu A, Andreani C. 1996. Temperature dependence of the dynamics of superoxide dismutase by quasi-elastic neutron scattering. *Physica B* 226:56–60
46. Finney JL. 1996. Hydration processes in biological and macromolecular systems. *Faraday Discuss.* 103:1–18
47. Finney JL, Goodfellow JM, Poole PL. 1982. The structure and dynamics of water in globular proteins. In *Structural Molecular Biology*, ed. DB Davies, S Danyluk, W Saenger, pp. 387–426. New York: Plenum
48. Finney JL, Poole PL. 1984. Protein hydration and enzyme activity: the role of hydration-induced conformation and dynamic changes in the activity of lysozyme. *Comments Mol. Cell. Biophys.* 2:129–51
49. Fischer S, Smith JC, Verma C. 2001. Dissecting the vibrational entropy change on protein/ligand binding: burial of a water molecule in bovine pancreatic trypsin inhibitor. *J. Phys. Chem. B* 105:8050–55
50. Fischer S, Verma C. 1999. Binding of buried structural water increases the flexibility of proteins. *Proc. Natl. Acad. Sci. USA* 96:9613–15
51. Fitter J. 1999. The temperature dependence of internal molecular motions in hydrated and dry α-amylase: the role of hydration water in the dynamical transition of proteins. *Biophys. J.* 76:1034–42
52. Fitter J, Lechner RE, Büldt G, Dencher NA. 1996. Internal molecular motions of bacteriorhodopsin: hydration-induced flexibility studied by quasielastic incoherent neutron scattering using oriented purple membranes. *Proc. Natl. Acad. Sci. USA* 93:7600–5
53. Fitter J, Lechner RE, Dencher NA. 1997. Picosecond molecular motions in bacteriorhodopsin from neutron scattering. *Biophys. J.* 73:2126–37
54. Franks F. 1993. Protein hydration. In

Protein Biotechnology, ed. F Franks, pp. 437–65. New Jersey: Humana
55. Frauenfelder H, Petsko GA, Tsernoglou D. 1979. Temperature-dependent X-ray diffraction as a probe of protein structural dynamics. *Nature* 280:558–63
56. Frauenfelder H, Sligar SG, Wolynes PG. 1991. The energy landscapes and motions of proteins. *Science* 254:1598–602
57. Frolov EN, Fischer M, Graffweg E, Mirishly MA, Goldanskii VI, Parak FG. 1991. Hemoglobin dynamics in rat erythrocytes investigated by Mössbauer spectroscopy. *Eur. Biophys. J.* 19:253–56
58. Frolov EN, Gvosdev R, Goldanskii VI, Parak FG. 1997. Differences in the dynamics of oxidised and reduced cytochrome *c*. *J. Biol. Inorg. Chem.* 2:710–13
59. Gabdoulline RR, Wade RC. 2002. Biomolecular diffusional association. *Curr. Opin. Struct. Biol.* 12:204–13
60. Gerstein M, Lesk AM, Chothia C. 1994. Structural mechanisms for domain movements in proteins. *Biochemistry* 33:6739–49
61. Deleted in proof
62. Hartmann H, Parak F, Steigemann W, Petsko GA, Ringe PD, Frauenfelder H. 1982. Conformational substates in a protein: structure and dynamics of metmyoglobin at 80 K. *Proc. Natl. Acad. Sci. USA* 79:4967–71
63. Hayward JA, Smith JC. 2002. Temperature dependence of protein dynamics: computer simulation analysis of neutron scattering properties. *Biophys. J.* 82:1216–25
64. Hernandez G, Jenney FEJ, Adams MWW, LeMaster DM. 2000. Millisecond time scale conformational flexibility in a hyperthermophile protein at ambient temperature. *Proc. Natl. Acad. Sci. USA* 97:3166–70
65. Héry S, Genest D, Smith JC. 1998. X-ray diffuse scattering and rigid-body motion in crystalline lysozyme probed by molecular dynamics simulation. *J. Mol. Biol.* 279:303–18
66. Huber R, Bennett WS. 1983. Functional significance of flexibility in proteins. *Biopolymers* 22:261–79
67. Hwang S, Trantolo D, Wise D. 1993. Gas phase acetaldehyde production in a continuous bioreactor. *Biotechnol. Bioeng.* 42:667–73
68. Iben IET, Braunstein D, Doster W, Frauenfelder H, Hong MK, et al. 1989. Glassy behaviour of a protein. *Phys. Rev. Lett.* 62:1916–19
69. Jacrot B, Cusack S, Dianoux AJ, Engelman D. 1982. Inelastic neutron scattering analysis of hexokinase dynamics and its modification on binding glucose. *Nature* 300:84–86
70. Jaenicke R. 2000. Do ultrastable proteins from hyperthermophiles have high or low conformational rigidity? *Proc. Natl. Acad. Sci. USA* 97:2962–64
71. Karplus M, Petsko GA. 1990. Molecular dynamics simulations in biology. *Nature* 347:631–39
72. Keller H, Debrunner PG. 1980. Evidence for conformational and diffusional mean square displacements in frozen aqueous solutions of oxymyoglobin. *Phys. Rev. Lett.* 45:68–71
73. Klibanov AM. 1989. Enzymatic catalysis in anhydrous organic solvents. *Trends Biochem. Sci.* 14:141–44
74. Kneller GR, Smith JC. 1994. Liquid-like side-chain dynamics in myoglobin. *J. Mol. Biol.* 242:181–85
75. Kohen A, Cannio R, Bartolucci S, Klinman JP. 1999. Enzyme dynamics and hydrogen tunnelling in a thermophilic alcohol dehydrogenase. *Nature* 399:496–99
76. Kohen A, Klinman JP. 1999. Hydrogen tunneling in biology. *Chem. Biol.* 6:R191–98
77. Kohen A, Klinman JP. 2000. Protein flexibility correlates with degree of hydrogen tunneling in thermophilic and mesophilic alcohol dehydrogenases. *J. Am. Chem. Soc.* 122:10738–39
78. Koshland DE. 1958. Application of a theory of enzyme specificity to protein

synthesis. *Proc. Natl. Acad. Sci. USA* 44:98–105
79. Krupyanskii YUF, Parak F, Goldanskii VI, Mössbauer RL, Gaubman EE, et al. 1982. Investigation of large intramolecular movements within metmyoglobin by Rayleigh scattering of Mössbauer radiation. *Z. Naturforsch.* 37:57–62
80. Kuntz ID, Kauzmann W. 1973. Hydration of proteins and polypeptides. *Adv. Protein Chem.* 28:239–345
81. Lamare S, Legoy MD. 1995. Working at controlled water activity in a continuous process: the gas/solid system as a solution. *Biotechnol. Bioeng.* 45:387–97
82. Lehnert U, Réat V, Weik M, Zaccai G, Pfister C. 1998. Thermal motions in bacteriorhodopsin at different hydration levels studied by neutron scattering: correlation with kinetics and light-induced changes. *Biophys. J.* 75:1945–52
83. Lichtenegger H, Doster W, Kleinert T, Birk A, Sepiol B, Vogl G. 1999. Heme-solvent coupling: a Mössbauer study of myoglobin in sucrose. *Biophys. J.* 76: 414–22
84. Lipscomb WN. 1983. Structure and catalysis of enzymes. *Annu. Rev. Biochem.* 52:17–34
85. Loncharich RL, Brooks BR. 1990. Temperature dependence of dynamics of hydrated myoglobin. *J. Mol. Biol.* 215:439–55
86. Luscher-Mattli M, Ruegg M. 1982. Thermodynamic functions of biopolymer hydration. I. Their determination by vapor pressure studies, discussed in an analysis of the primary hydration process. *Biopolymers* 21:403–18
87. McCammon JA, Harvey SC. 1987. *Dynamics of Proteins and Nucleic Acids.* Cambridge, UK: Cambridge Univ. Press
88. Miller GP, Benkovic SJ. 1998. Stretching exercises—flexibility in dihydrofolate reductase catalysis. *Chem. Biol.* 5:R105–13
89. More N, Daniel RM, Petach HH. 1995. The effect of low temperatures on enzyme activity. *Biochem. J.* 305:17–20
90. Neria E, Karplus M. 1997. Molecular dynamics of an enzyme reaction—proton transfer in TIM. *Chem. Phys. Lett.* 267: 23–30
91. Nocek JM, Stemp EDA, Finnegan MG, Koshy TI, Johnson MK, et al. 1991. Low-temperature, cooperative conformational transition within [Zn-cytochrome *c* peroxidase, cytochrome *c*] complexes: variation with cytochrome. *J. Am. Chem. Soc.* 113:6822–31
92. Ostermann A, Waschipky R, Parak F, Nienhaus U. 2000. Ligand binding and conformational motions in myoglobin. *Nature* 404:205–8
93. Page MI, Jencks WP. 1971. Entropic contributions to rate accelerations in enzymic and intramolecular reactions and the chelate effect. *Proc. Natl. Acad. Sci. USA* 68:1678–83
94. Parak F, Fischer M, Heidemeiner J, Engelhard M, Kohl K-D, et al. 1990. Investigation of the dynamics of bacteriorhodopsin. *Hyperfine Interact.* 58:2381–86
95. Parak F, Frolov EN, Kononenko AA, Mössbauer RL, Goldanskii VI, Rubin AB. 1980. Evidence for a correlation between the photoinduced electron transfer and dynamic properties of the chromophore membranes from *Rhodospirillum rubrum*. *FEBS Lett.* 117:368–72
96. Parak F, Knapp EW, Kucheida D. 1982. Protein dynamics: Mössbauer spectroscopy on deoxymyoglobin crystals. *J. Mol. Biol.* 161:177–94
97. Partridge J, Dennison PR, Moore BD, Halling PJ. 1998. Activity and mobility of subtilisin in low water organic media: Hydration is more important than solvent dielectric. *Biochim. Biophys. Acta* 1386: 79–89
98. Parvaresh F, Robert H, Thomas D, Legoy MD. 1992. Gas phase transesterification catalysed by lipolytic enzymes. *Biotechnol. Bioeng.* 39:467–73
99. Perez J, Zanotti JM, Durand D. 1999.

Evolution of the internal dynamics of two globular proteins from dry powder to solution. *Biophys. J.* 77:454–69
100. Poole PL, Finney JL. 1983. Hydration-induced conformational and flexibility changes in lysozyme at low water content. *Int. J. Biol. Macromol.* 5:308–10
101. Poole PL, Finney JL. 1983. Sequential hydration of a dry globular protein. *Biopolymers* 22:255–60
102. Poole PL, Finney JL. 1986. Solid-phase protein hydration studies. *Methods Enzymol.* 127:284–93
103. Pulvin S, Legoy MD, Lortie R, Pensa M, Thomas D. 1986. Enzyme technology and gas phase catalysis: alcohol dehydrogenase example. *Biotechnol. Lett.* 8:783–84
104. Radkiewicz J, Brooks CL. 2000. Protein dynamics in enzymatic catalysis: exploration of dihydrofolate reductase. *J. Am. Chem. Soc.* 122:225–31
105. Rasmussen BF, Stock AM, Ringe D, Petsko GA. 1992. Crystalline ribonuclease A loses function below the dynamical transition at 220 K. *Nature* 357:423–24
106. Réat V, Dunn RV, Ferrand M, Finney JL, Daniel RM, Smith JC. 2000. Solvent dependence of dynamic transitions in protein solutions. *Proc. Natl. Acad. Sci. USA* 97:9961–66
107. Réat V, Patzelt H, Ferrand M, Pfister C, Oesterhelt D, Zaccai G. 1998. Dynamics of different functional parts of bacteriorhodopsin: H-^{2}H labeling and neutron scattering. *Proc. Natl. Acad. Sci. USA* 95:4970–75
108. Rector K, Fayer MD. 1998. Myoglobin's ultrafast dynamics measured with vibrational echo experiments. *Nucl. Instr. Methods Phys. Res. B* 144:218–24
109. Rupley JA, Careri G. 1991. Protein hydration and function. *Adv. Protein Chem.* 41:37–172
110. Rupley JA, Gratton E, Careri G. 1983. Water and globular proteins. *Trends Biochem. Sci.* 8:18–22
111. Rupley JA, Yang P-H, Tollin G. 1980. Thermodynamic and related studies of water interacting with proteins. In *Water in Polymers*, ed. S Rowland, pp. 111–32. Washington, DC: Am. Chem. Soc.
112. Schinkel JE, Downer NW, Rupley JA. 1985. Hydrogen exchange of lysozyme powders. Hydration dependence of internal motions. *Biochemistry* 24:352–66
113. Schmitke JL, Wescott CR, Klibanov AM. 1996. The mechanistic dissection of the plunge in enzymatic activity upon transition from water to anhydrous solvents. *J. Am. Chem. Soc.* 118:3360–65
114. Schulz GE. 1992. Induced-fit movements in adenylate kinases. *Faraday Discuss.* 93:85–93
115. Shibata Y, Takahashi H, Kaneko R, Kurita A, Kushida T. 1999. Conformational fluctuation of native-like and molten-globule-like cytochrome *c* observed by time-resolved hole burning. *Biochemistry* 38:1802–10
116. Smith JC. 1991. Protein dynamics: comparison of simulations with inelastic neutron scattering experiments. *Q. Rev. Biophys.* 24:227–91
117. Smith JC, Cusack S, Poole PL, Finney JL. 1987. Direct measurement of hydration-related dynamic changes in lysozyme using inelastic neutron scattering spectroscopy. *J. Biomol. Struct. Dyn.* 4:583–88
118. Smith JC, Cusack S, Tidor B, Karplus M. 1990. Inelastic neutron scattering analysis of low-frequency motions in proteins: harmonic and damped harmonic models of bovine pancreatic trypsin inhibitor. *J. Chem. Phys.* 93:2974–91
119. Stevens E, Stevens L. 1979. The effect of restricted hydration on the rate of glucose-6-phosphate dehydrogenase, phosphoglucose isomerase, hexokinase and fumarase. *Biochem. J.* 179:161–67
120. Stryer L. 1988. *Biochemistry*. New York: Freeman
121. Stuckey JA, Schubert HL, Fauman EB, Zhang Z-Y, Dixon JE, Saper MA. 1994. Crystal structure of *Yersinia* protein

tyrosine phosphatase at 2.5 Å and the complex with tungstate. *Nature* 370:571–75
122. Tarek M, Tobias DJ. 2002. Role of protein-water hydrogen bond dynamics in the protein dynamical transition. *Phys. Rev. Lett.* 88:138101
123. Tidor B, Karplus M. 1994. The contribution of vibrational entropy to molecular association. The dimerization of insulin. *J. Mol. Biol.* 238:405–14
124. Tilton RF Jr, Dewan JC, Petsko GA. 1992. Effects of temperature on protein structure and dynamics: X-ray crystallographic studies of the protein ribonuclease-A at nine different temperatures from 98 to 320 K. *Biochemistry* 31:2469–81
125. Valivety RH, Halling PJ, Macrae AR. 1992. *Rhizomucor miehei* lipase remains highly active at water activity below 0.0001. *FEBS Lett.* 301:258–60
126. Van den Burg B, Vriend G, Veltman OR, Venema G, Eijsink VG. 1998. Engineering an enzyme to resist boiling. *Proc. Natl. Acad. Sci. USA* 95:2056–60
127. van Gunsteren WF, Karplus M. 1982. Protein dynamics in solution and in a crystalline environment: a molecular dynamics study. *Biochemistry* 21:2259–74
128. Varley PG, Pain RH. 1991. Relation between stability, dynamics and enzyme activity in 3-phosphoglycerate kinase from yeast and *Thermus thermophilus*. *J. Mol. Biol.* 220:531–38
129. Vetter IR, Wittinghofer A. 2001. Signal transduction—the guanine nucleotide-binding switch in three dimensions. *Science* 294:1299–304
130. Vihinen M. 1987. Relationship of protein flexibility to thermostability. *Protein Eng.* 1:477–80
131. Villa J, Strajbl M, Glennon TM, Sham YY, Chu ZT, Warshel A. 2000. How important are entropic contributions to enzyme catalysis? *Proc. Natl. Acad. Sci. USA* 97:11899–904
132. Vitkup D, Ringe D, Petsko GA, Karplus M. 2000. Solvent mobility and the protein 'glass' transition. *Nat. Struct. Biol.* 7:34–38
133. Wagner G, Wuthrich K. 1979. Correlation between the amide proton exchange rates and the denaturation temperatures in globular proteins related to the basic pancreatic trypsin inhibitor. *J. Mol. Biol.* 130:31–37
134. Warshel A. 2002. Molecular dynamics simulations of biological reactions. *Acc. Chem. Res.* 35:385–95
135. Winn PJ, Ludemann SK, Gauges R, Lounnas V, Wade RC. 2002. Comparison of the dynamics of substrate access channels in three cytochrome P450s reveals different opening mechanisms and a novel functional role for a buried arginine. *Proc. Natl. Acad. Sci. USA* 99:5361–66
136. Wrba A, Schweiger A, Schultes V, Jaenicke R, Zavodsky P. 1990. Extremely thermostable D-glyceraldehyde-3-phosphate dehydrogenase from the eubacterium *Thermotoga maritimea*. *Biochemistry* 29:7584–92
137. Wriggers W, Schulten K. 1999. Investigating a back door mechanism of actin phosphate release by steered molecular dynamics. *Proteins* 35:262–73
138. Wu J, Gorenstein DG. 1993. Structure and dynamics of cytochrome *c* in nonaqueous solvents by 2D NH-exchange NMR spectroscopy. *J. Am. Chem. Soc.* 115:6843–50
139. Yagi T, Tsuda M, Mori Y, Inokuchi H. 1969. Hydrogenase activity in the dry state. *J. Am. Chem. Soc.* 91:2801
140. Yang P-H, Rupley JA. 1979. Protein-water interactions: heat capacity of the lysozyme-water system. *Biochemistry* 18:2654–61
141. Yu N-T, Jo BH. 1973. Comparison of a protein structure in crystals and in solution by laser raman scattering. I. Lysozyme. *Arch. Biochem. Biophys.* 156:469–74
142. Zanotti J-M, Bellissent-Funel M-C, Parello J. 1997. Dynamics of a globular protein as studied by neutron scattering and

solid-state NMR. *Physica B* 234–36:228–30
143. Zanotti J-M, Bellissent-Funel M-C, Parello J. 1999. Hydration-coupled dynamics in proteins studied by neutron scattering and NMR: the case of the typical EF-hand calcium-binding parvalbumin. *Biophys. J.* 76:2390–411
144. Zavodsky P, Kardos J, Svingor A, Petsko GA. 1998. Adjustment of conformational flexibility is a key event in the thermal adaption of proteins. *Proc. Natl. Acad. Sci. USA* 95:7406–11

Annu. Rev. Biophys. Biomol. Struct. 2003. 32:93–114
doi: 10.1146/annurev.biophys.32.110601.142347
First published online as a Review in Advance on December 2, 2002

STRUCTURE AND FUNCTION OF NATURAL KILLER CELL SURFACE RECEPTORS*

Sergei Radaev and Peter D. Sun
Structural Immunology Section, Laboratory of Immunogenetics, National Institute of Allergy and Infectious Diseases, National Institutes of Health, Rockville, Maryland 20852; email: psun@nih.gov; sradaev@niaid.nih.gov

Key Words KIR, KIR/HLA complex, allotype specificity, CD94, NKG2D/ULBP complex

■ **Abstract** Since mid-1990, with cloning and identification of several families of natural killer (NK) receptors, research on NK cells began to receive appreciable attention. Determination of structures of NK cell surface receptors and their ligand complexes led to a fast growth in our understanding of the activation and ligand recognition by these receptors as well as their function in innate immunity. Functionally, NK cell surface receptors are divided into two groups, the inhibitory and the activating receptors. Structurally, they belong to either the immunoglobulin (Ig)-like receptor superfamily or the C-type lectin-like receptor (CTLR) superfamily. Their ligands are either members of class I major histocompatibility complexes (MHC) or homologs of class I MHC molecules. The inhibitory form of NK receptors provides the protective immunity through recognizing class I MHC molecules with self-peptides on healthy host cells. The activating, or the noninhibitory, NK receptors mediate the killing of tumor or virally infected cells through their specific ligand recognition. The structures of activating and inhibitory NK cell surface receptors and their complexes with the ligands determined to date, including killer immunoglobulin-like receptors (KIRs) and their complexes with HLA molecules, CD94, Ly49A, and its complex with H-$2D^d$, and NKG2D receptors and their complexes with class I MHC homologs, are reviewed here.

CONTENTS

*The U.S. Government has the right to retain a nonexclusive, royalty-free license in and to any copyright covering this paper.

INTRODUCTION

As part of innate immunity, natural killer (NK) cells are capable of killing certain tumor and virally infected cells. A balance between the activating and inhibitory receptors on the cells surface controls the cytolytic activity of NK cells. The inhibitory receptors suppress NK cell lysis of target cells that express class I MHC molecules, but allow the lysis of class I negative cells. Structurally, inhibitory receptors belong to either the immunoglobulin-like superfamily (IgSF) or the C-type lectin-like receptor (CTLR) superfamily. IgSF inhibitory receptors include the human killer cell Ig-like receptors (KIR), which recognize the $\alpha 1$ and $\alpha 2$ domains of human leukocyte antigens (HLA-A, -B, and -C) with the bound self-peptide, and the Ig-like transcripts (ILTs, also named as leukocyte Ig-like receptors or LIRs) expressed on myeloid cells that recognize the nonpolymorphic $\alpha 3$ domain of classical and nonclassical HLA molecules (38). Members of the CTLR superfamily include the CD94/NKG2A that recognize the nonclassical class I molecules HLA-E and Qa-1 in human and mouse (71), respectively, and the murine Ly49 molecules that recognize the classical class I MHC molecules (44, 71). All inhibitory receptors carry immunotyrosine inhibitory motifs (ITIM) in their cytoplasmic domain. The Ly49 receptors are found only in mice and appear to be functional orthologs to KIR, which are found in primates but not in rodents (33). The crystal structures of three KIRs (22, 46, 67), one ILT receptor, and a CD94 homodimer have been published (6, 9). Examples of MHC recognition by KIR and Ly49 inhibitory receptors were shown in the cocrystal structures of KIR2DL2/HLA-Cw3, KIR2DL1/HLA-Cw4, and Ly49A/H2-D^d (5, 21, 70).

Like inhibitory receptors, activating receptors are either IgSF or CTLR superfamily members. Activating IgSF receptors include 2B4, the natural cytotoxicity receptors NKp46, NKp30, and NKp44 (54), the noninhibitory isoforms of KIRs, and ILTs/LIRs. Activating CTLRs include CD94/NKG2C and CD94/NKG2E heterodimers, NKG2D homodimer, and activating isoforms of Ly49. All activating receptors lack activating motifs in their cytoplasmic domains. However, all of them, except 2B4, display a charged lysine or arginine residues in their transmembrane region to pair with a negatively charged residue on adaptor molecules, such as DAP12, DAP10, CD3ζ, or FcRγ, that carry signal transduction components.

In contrast to the inhibitory receptors much less is known about the structures of activating NK receptors. The only structure of the activating NK receptor determined to date is that of NKG2D (77). It is a member of CTLR superfamily and is distantly related to NKG2A, B, C, and E. NKG2D mediates NK cells' cytotoxicity against certain tumor cells and provides costimulatory signals on CD8$^+$ $\alpha\beta$ and $\gamma\delta$ T cells against virally infected cells (2, 15, 28, 72, 78). Stress-induced molecules MIC-A, MIC-B, and ULBPs in human, Rae-1, and H-60 in mice have been identified as the ligands for NKG2D (2, 8, 17). Recently, the crystal structures of a murine NKG2D in complex with Rae-1β and a human NKG2D in complex with both MIC-A and ULBP3 have been determined (40, 41, 60).

THE KILLER CELL IMMUNOGLOBULIN-LIKE RECEPTOR SUPERFAMILY

Approximately 12 genes sharing greater than 90% sequence identity encode human KIRs. They are located in the leukocyte receptor complex (LRC) region on chromosome 19q (68, 76). A similar number of KIR genes have been identified in other primate species (34). In addition, several homologous receptor families, including immunoglobulin-like transcripts (ILT) or leukocyte immunoglobulin-like receptors (LIR) (4, 11, 63), leukocyte-associated Ig-like receptors (LAIR) (50), paired Ig-like receptors (PIR), and gp49 (1), have also been identified (30, 36). They display 35%–50% sequence identity and clearly share a common fold with KIR. Together, they define a so-called KIR superfamily. A more distantly related set of proteins are the Ig-like Fc receptors (FcαR, FcγR-I, -IIa, -IIb, -III, and FcεRI), which display less than 20% sequence identity with KIR but nonetheless share a structural fold similar to it.

KIRs are type I transmembrane glycoproteins with two or three extracellular C2-type Ig-like domains (12, 14, 73). ILTs contain two or four Ig-like domains; LAIR-1 and -2 contain a single Ig-like domain; PIR-A and -B each contain six Ig-like domains. KIRs exist as either inhibitory or noninhibitory forms (20, 43, 55). The inhibitory forms of these receptors possess ITIM in their cytoplasmic tail. The noninhibitory forms of KIR have a shorter cytoplasmic tail and display a positively charged residue in their transmembrane regions through which they pair with an activating motif-containing adaptor molecule (38, 44).

The Structure of KIR

To date, the crystal structures of the extracellular domains of three members of KIR family, KIR2DL1, KIR2DL2 and KIR2DL3, have been published (22, 46, 67). Overall, the KIR fold is similar to the C2-type Ig-like fold observed in the

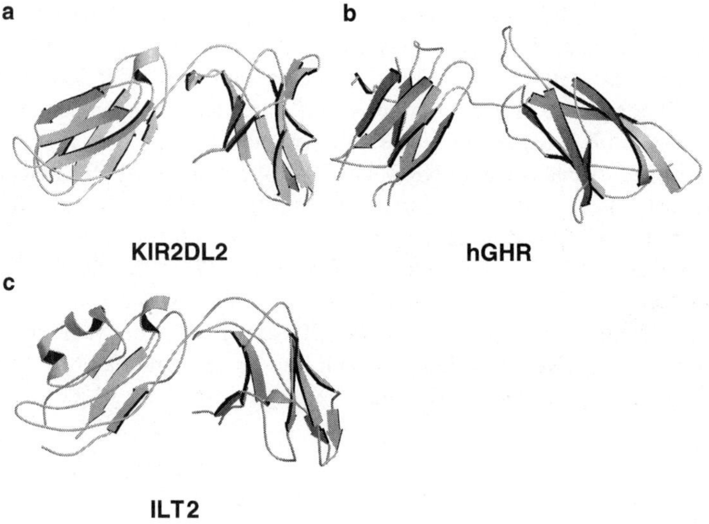

Figure 1 Comparison of KIRs with other representative two-domain Ig-like receptors. (*a*) KIR2DL2, (*b*) hGHR, (*c*) ILT-2. PDB codes: KIR2DL2, 1EFX; hGHR, 3HHR; ILT2, 1G0X.

hematopoietic receptors with the difference existing primarily in the pairing of two β strands (Figure 1). In hematopoietic receptors, β strand A pairs with strand B, whereas in KIR structures the first strand splits into two strands, A and A′, which hydrogen-bond with the B and G strands, respectively. This results in a "strand switching" in KIR that is likely attributable to the presence of a conserved *cis*-proline residue in the first strand. In addition, KIR structures possess unique tertiary packing. In particular, the hinge angle between the N-terminal D1 and C-terminal D2 domains is smaller than those observed in other two-Ig-like domain structures, such as human growth hormone receptor, the V and C domains of T cell antigen receptors, and the V and C_{H1} domains of antibodies. The hinge angle of KIR varies from 66° in KIR2DL1 to 81° in KIR2DL2 and KIR2DL3. This domain hinge angle is stabilized by a highly conserved interdomain hydrophobic core (hinge core) that consists of Leu17, Met69, Val100, Ile101, Thr102, His138, Phe178, Ser180, Pro185, Tyr186, and Trp188. An interdomain salt bridge between Asp98 and Arg149, conserved in all KIR family sequences, also helps restrict the hinge angle. Aside from the differences in the hinge angle, the structures of KIR2DL1, KIR2DL2, and KIR2DL3 are nearly identical with rms deviations less than 1 Å between the Cα atoms of their respective domains.

The Structure of LIR-1 (ILT-2)

In addition to KIR, the structure of the first two domains of LIR-1 (ILT-2), an inhibitory receptor expressed on monocytes, B cells, dendritic cells, and subsets of NK and T cells, has also been determined (9). Overall, the sequence homologies between KIR and ILT genes are about 40%, and domains 1 and 2 of LIR-1 possess the KIR type Ig fold including strand switching in the first β strands. However, distinct structural differences are observed between LIR-1 and KIRs (9). In particular, LIR-1 has two unique short 3_{10} helices in each domain. One replaces the C′ strand in the D1 domain and the C-terminal end of the C′ strand in the D2 domain found in KIR, and the other is situated between the E and F strands in the D1 domain and between the F and G strands of the D2 domain (Figure 1). A short left-handed type II polyproline-like helix is also found in the F-G loop of the D1 and D2 domains. Like KIR, primarily hydrophobic residues forming a hinge core that stabilizes the D1-D2 interdomain conformation occupy the interdomain region of LIR-1. Interestingly, a conserved interdomain salt bridge in KIR between Asp98 and Arg149 is absent in the LIR-1 structure, possibly contributing to the slightly larger hinge angle of LIR-1 (88°).

RECOGNITION OF CLASS I MHC LIGANDS BY KIR

The Structure of KIR/HLA Complexes

Class I MHC molecules were first implicated as potential ligands of NK cell receptors when an inverse relationship was identified between the susceptibility of target cells to NK cell-mediated lysis and the expression of class I on target cells (29, 69). It was further demonstrated that transfection of class I genes into a class I–deficient target cell was sufficient to protect these cells from NK cell-mediated lysis (64). Evidence for the involvement of multiple receptors that recognized distinct HLA class I molecules came from studies examining the specificity of cloned NK cells against different allogenic target cells. This led to the identification of two KIR molecules, KIR2DL1 (CD158a) and KIR2DL2 (CD158b1), that recognize HLA-Cw2, 4, 6, 15, and HLA-Cw1, 3, 7, 8 allotypes of class I MHC (12, 53, 73), respectively. Sequence comparison of HLA ligands suggested residues 77 and 80 in class I MHC heavy chains are important for their receptor specificities (10).

The first crystal structure of the KIR/HLA complex was solved between KIR2DL2 and HLA-Cw3 with a nonamer self-peptide GAVDPLLAL (GAV) derived from the importin-α1 subunit (5). More recently, a related structure of KIR2DL1/HLA-Cw4 with a bound peptide QYDDAVYKL (QYD) was determined (21). The orientation of KIR with respect to HLA is similar to that of TCRs, with the D1 and D2 domains of KIR assuming the respective positions of the Vα and Vβ domains of TCR (19) (Figure 2). The footprint of KIR on the class I HLA overlaps partially with those of TCRs on class I molecules. However, KIR contacts primarily the P7 and P8 positions of the bound peptide and associated HLA residues, whereas the TCR interacts predominantly with the P4, P5, and P6 positions of the peptide (19, 24, 74).

The KIR/HLA complex buries approximately 1600 Å^2 surface area, similar to TCR/MHC complex. Unlike TCR/MHC interfaces that consist largely of hydrogen

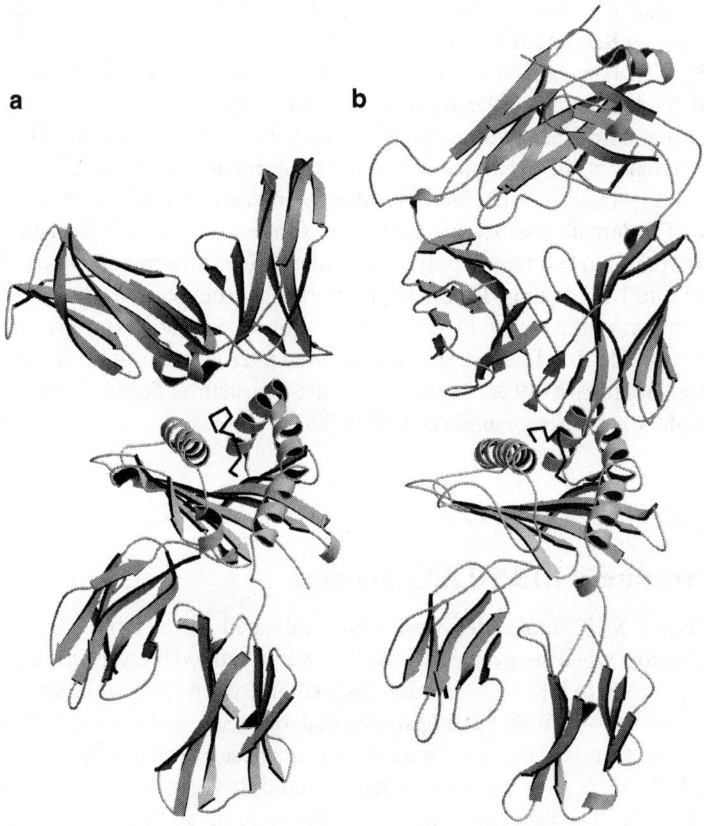

Figure 2 Comparison of KIR/MHC (PDB code 1EFX) binding to TCR/MHC (PDB code 1BD2) binding.

bonds and hydrophobic and van der Waals interactions, the KIR/HLA interface is characterized by strong charge complementarity. In all, there are six acidic residues in KIR that interact with six basic residues in HLA-Cw3, resulting in the formation of four salt bridges between E21, E106, D135, D183 of KIR and R69, R151, R145, K146 of HLA-Cw3, respectively. The dominance of charge-charge interactions in the interface resembles the interface between adhesion receptors such as CD2/CD58 (75). To evaluate the contribution of these salt bridges to the KIR/HLA recognition, three single mutations of the receptor, E106A, D135H, and D183A, were created to remove three of the four salt bridges individually, and their effects on HLA binding were measured by surface plasmon resonance (SPR). While the wild-type receptor binds HLA-Cw3 (GAV) with 30 μM affinity, E106A resulted in a sixfold reduction in the ligand binding affinity, and both D135H and

D183A mutants showed no detectable binding to the ligand. In addition to charge complementarity, the KIR/HLA interface also displays a network of six hydrogen bonds and hydrophobic interactions. The largest hydrophobic cluster includes the aliphatic portions of K44, F45, M70, and Q71 of KIR2DL2 together with the aliphatic portions of V76, R69, R75, and R79 of HLA-Cw3.

Class I Allotypic Recognition of KIR

KIRs are known to recognize multiple alleles of MHC molecules. For example, KIR2DL1 recognizes HLA-Cw2, 4, 5, 6, and 15, whereas KIR2DL2 and KIR2DL3 recognize HLA-Cw1, 3, 7, and 8 allotypes. The basis of this receptor recognition of allotypic HLA molecules was not understood until the structure of the KIR/HLA complex was solved. Among the 12 HLA-Cw3 interface residues, 11 are invariant across all HLA-C alleles despite their location within the polymorphic region of the class I heavy chain. Amino acid 80 is the only variable residue contributing to the receptor interface. In contrast, 8 of 16 HLA-A2 residues in contact with the A6 TCR are variable among the HLA-A alleles. The use of highly conserved residues within an otherwise polymorphic region of HLA enables individual KIR to recognize multiple class I HLA molecules while discriminating among various allotypes based on the identity of amino acid position 80. The recognition of conserved HLA residues by KIR has important ramifications and may reflect the functional differences between the innate and adaptive immune systems. TCRs rely on gene rearrangement and on positive selection of T cell clones to achieve exquisite specificity for specific peptides complexed to polymorphic MHC molecules. This combination of TCR maturation and the polymorphism in MHC molecules enables T cell–mediated cellular immunity to counteract virtually unlimited forms of pathogens. KIRs, as part of the innate immunity, are obligated to produce a rapid response. These receptors are germline encoded, thus they have limited ability to adapt to the evolving peptide-MHC diversities. In addition, there are significantly more MHC alleles than there are KIRs. As a result, to provide effective surveillance of MHCs by KIR requires individual KIR to recognize more than one MHC allele. By focusing recognition on conserved residues within the polymorphic regions of MHC, distinct KIR can recognize multiple related MHC molecules. Through the use of the conserved residues for KIR recognition and polymorphic residues for TCR recognition, a given class I MHC effectively accommodates the requirements for recognition by both the innate and adaptive immune systems.

As mentioned earlier, a single HLA residue at position 80 determines KIR allotype specificity. From the receptor side a single KIR residue at position 44 controls exquisite allotype specificity toward HLA. For example, KIR2DL1 with a Met at position 44 recognizes HLA-Cw2, 4, 5, 6, and 15 allotypes with Lys at position 80. KIR2DL2 has a Lys at position 44 and does not recognize HLA-C molecules with a Lys at position 80. Instead, it recognizes HLA-Cw1, 3, 7, and 8 allotypes with Asn at position 80 (Figure 3). In the crystal structure of the KIR/HLA complex, Lys44 of KIR2DL2 makes a hydrogen bond with Asn80 of HLA-Cw3. Replacing Asn80 with Lys, as in the sequence of HLA-Cw2, 4, 5, and

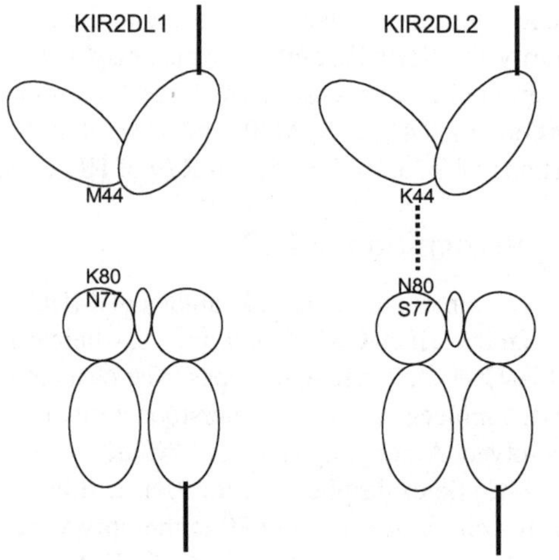

Figure 3 Schematic diagram illustrating the allotype specificity of KIR2D receptors. The *dotted line* represents the hydrogen bond between Lys44 of KIR2DL2 and Asn80 of HLA-Cw3.

6, would generate an unfavorable electrostatic interaction with Lys44 of KIR2DL2. Similarly, replacing Lys44 of KIR2DL2 with Met would result in the loss of the (KIR)Lys44-Asn80 (HLA) hydrogen bond and destabilize the KIR/HLA interface.

The Peptide Preference for KIR/HLA Binding

Peptides are involved not only in the generation of stable, properly folded MHC molecules at the cell surface but also in the direct recognition of class I MHC by NK cells. TAP-deficient cells cultured at 26°C in the absence of class I binding peptides, express "empty" class I MHC molecules on their cell surface. These class I molecules are unable to protect the target cells from NK-mediated lysis. Peptide preferences were observed in KIR2D recognition of HLA-C molecules (61, 80). The crystal structure of the KIR2DL2/HLA-Cw3 complex contains a nonamer peptide GAVDPLLAL (GAV) derived from the human importin-α1 subunit (80). KIR2DL2 makes direct contacts to the GAV peptide at both the P7 and P8 positions. The side chain of P7 leucine makes loose contacts with Leu104 and Tyr105 of the KIR. At the P8 position, Gln71 of KIR forms a hydrogen bond with the backbone amide nitrogen of the peptide. Compared to T cell receptor-MHC recognition, KIR makes less contact to the side chain of the peptide and thus is less sensitive to the peptide content. The formation of this backbone hydrogen bond may restrain the size of the P8 side chain owing to steric hindrance with the receptor. Indeed, when substitutions were introduced at the P8 position of GAV peptide the results showed that amino acids larger than Val at the P8 position completely abolished the receptor binding.

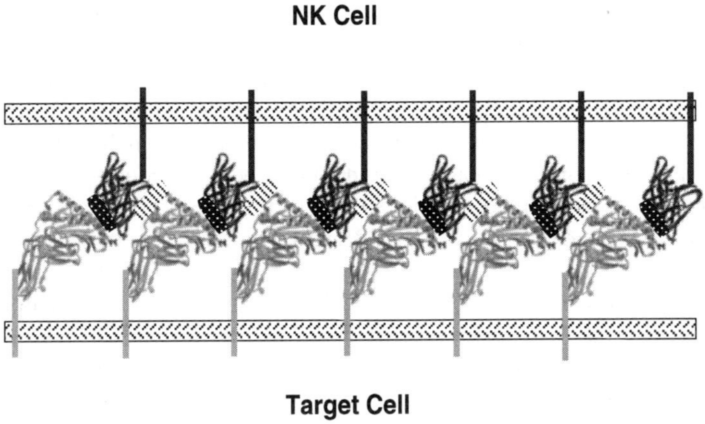

Figure 4 A lattice model for HLA-induced KIR clustering.

A Model for KIR/HLA Aggregation

Ligand-induced receptor oligomerization is presumed to be a common mechanism for initiating receptor-mediated signaling. The so-called immunological synapses at the interface of T cells and APC consist of a central cluster of TCR/MHC complexes and a peripheral ring of adhesion molecules (27, 52). In contrast, the observed NK cell immune synapse is formed with a central cluster of adhesion molecules LFA-1/ICAM-1 and a peripheral KIR/HLA cluster in the shape of a donut (16). The question of how the KIR/MHC complex forms this ordered donut-shaped oligomeric aggregate remains unclear. One possible form of oligomerization was observed in crystals of the KIR2DL2/HLA-Cw3 complex. Within these crystals, each KIR molecule makes an additional contact, apart from the functional binding interface, with a symmetry-related HLA-Cw3 molecule in a peptide-independent manner. The contact buries 530 $Å^2$ of surface area and is characterized by mostly van der Waals interactions. Interestingly, this KIR/HLA contact bridges between the adjacent pair of complexes into an oligomeric KIR/HLA aggregate (Figure 4). In this form of oligomer, the KIR/HLA complexes are all in the same orientation, and the molar ratio between receptor and ligand is maintained at 1:1. Furthermore, the putative glycosylation sites on both KIR2DL2 and HLA-Cw3 are away from the oligomerization interface. It is possible that this form of receptor-ligand oligomerization resembles the receptor clustering on the surface of NK cells.

C-TYPE LECTIN-LIKE NK RECEPTORS

The Structure of CD94 and Ly49 Receptors

C-type lectins are a family of calcium-dependent carbohydrate binding proteins, such as the mannose binding protein (MBP), E-selectins, tetranectins, and lithostathines. The carbohydrate binding domains (CRD) are approximately 120 amino acids and contain three disulfide bonds and a characteristic Ca^{2+} binding loop that is essential for carbohydrate binding. Recently, many cell surface receptor proteins have been identified that contain the homologous CRD sequences, including the conserved cysteine residues. Some of them, however, do not appear to have the conserved calcium binding residues. Collectively, these receptors are referred to as members of the C-type lectin-like receptor (CTLR) superfamily. CD94 and members of the NKG2 and Ly49 family of receptors are part of the CTLR superfamily.

The structure of CD94 showed that the putative CRD domain folds into a C-type lectin fold (6) (Figure 5). There are, however, distinct features separating CD94 from classic C-type lectins. First, the putative Ca^{2+} binding loop in CD94 lacks the conserved calcium-chelating aspartate and glutamate residues and is in a different conformation than other Ca^{2+} binding loops observed in classic C-type lectins. It is therefore predicted that the function of CD94 does not require the binding of carbohydrate. Second, there are two canonical α helices in the structure of C-lectins. Only one is observed in the structure of CD94. The residues at the position of the second helix have adopted a loop conformation at the CD94 dimer interface. The dimer of CD94 is stabilized primarily by hydrophobic interactions

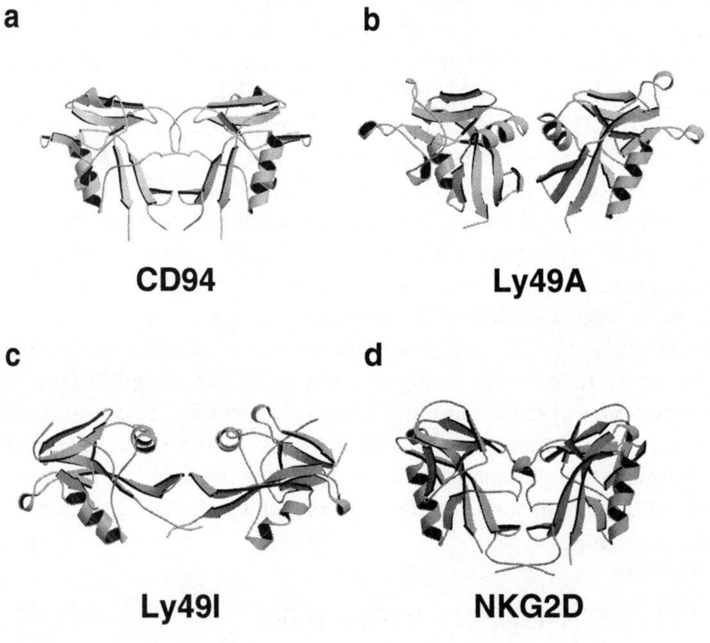

Figure 5 Structures of members of CTLR superfamily. Shown here are dimmers of (*a*) CD94 (PDB code 1B6E), (*b*) Ly49A (PDB code 1QO3), (*c*) Ly49I (PDB code 1JA3), and (*d*) NKG2D (PDB code 1KCG).

and buries about 1200 Å^2 of interface area. More recently, the crystal structure of NKG2D provided another example of the receptors from the CTLR superfamily (77). The fold of NKG2D is similar to that of CD94, including the deformed second helix. NKG2D exists as a functional homodimer whose structure can be readily superimposed onto the dimer structure of CD94.

In contrast, the structure of the Ly49A monomer showed that the murine receptor assumes a fold more similar to that of the mannose binding protein (MBP) than to that of CD94 or NKG2D (41, 70) (Figure 5). In particular, the structure of Ly49 preserves the second α helix of MBP, whereas the same region adopts a loop conformation in the structure of CD94. However, the structure of the Ly49A dimer is more similar to the structure of the CD94 homodimer than to that of the MBP. More recently, the crystal structure of Ly49I has been determined (18) (Figure 5). Overall structure of the Ly49I monomer resembles that of Ly49A. Due to the truncation in the expression construct, the crystallized Ly49I fragment existed as a monomer rather than as a dimer in solution. A noncrystallographic dimer was observed in Ly49I crystal that buries 1600 Å^2 surface area. The mode of dimerization as observed in Ly49I crystal, however, is different from that observed in Ly49A, CD94, and NKG2D.

The Structure of NKG2D

The NKG2 gene family located on human chromosome 12p13.1 encodes several similar type II lectin-like proteins known as NKG2A, B, C, D, E, and F (25, 59). NKG2A, B, C, and E show 94%–95% amino acid homology in their extracellular domain and 56% homology throughout the internal and transmembrane regions (32). NKG2D is distantly related to other members of the NKG2 family and shows only 21% overall homology to them. All NKG2 proteins except NKG2D form heterodimers with C-type lectin-like receptor CD94 (39). In contrast NKG2D forms a homodimeric structure. Recently, NKG2D generated tremendous interest owing to the finding that it activates NK cells upon the recognition of stress-inducible MIC-A, which is frequently expressed in epithelial tumors (2). NKG2D can also be found on most $\gamma\delta$ T cells, CD8+ $\alpha\beta$ T cells, and macrophages (2, 17). NKG2D associates with a signaling adapter molecule DAP10 through a charged transmembrane residue (78). NKG2D orthologs were also found in other mammals such as chimpanzee, rhesus monkey, cattle, pig, rat, and mice (3, 26, 31, 37, 65, 79).

The crystal structure of the extracellular domain of intact murine NKG2D (77) reveals a C-type lectin fold similar to CD94, Ly49A, rat MBP-A, and CD69 (6, 35, 56, 70) (Figure 5). Superposition of NKG2D with CD94 and Ly49A resulted in rms deviations of 1.4 Å and 1.7 Å, respectively, for 97 pairs of Cα atoms. Unlike Ly49A, which retains the intact $\alpha 2$ helix of classical C-type lectins, the $\alpha 2$ helix in NKG2D is deformed into only a one-turn helix. Like CD94, NKG2D also lacks the appropriate Ca^{2+} ligands in its corresponding Ca^{2+} binding loops and thus is presumed to be a non-Ca^{2+} binding C-type lectin-like receptor.

The NKG2D dimer is similar to the CD94 homodimer. Owing to an extension of the dimer interface by the N termini strands of NKG2D, the interface buries about 1900 Å^2, which is substantially more than the 1200 Å^2 dimer interface of CD94.

RECOGNITION OF CLASS I MHC BY LY49A

The structure of the C-type lectin domain of Ly49A in complex with H-2D^d is the first and the only example of class I MHC recognition by a murine inhibitory receptor Ly49A (70). In the crystal structure, Ly49A binds H-2D^d at two distinct locations, site 1 and site 2. Site 1 buries ~1000 Å^2 surface area and is located at the N-terminal end of the α1 helix of the MHC molecule (Figure 6). This interface is dominated by electrostatic interactions. Site 2 of Ly49A is located under the β sheet floor of the α1 and α2 domains interacting primarily with the α2, α3, and β2m subunits of H-2D^d. It buries more than 3000 Å^2 interface area. The interactions between Ly49A and the MHC at site 2 are made primarily by hydrogen bonds. Despite the smaller buried receptor-MHC interface, the orientation of site 1 is consistent with *trans* receptor-ligand recognition, whereas the site 2 appears to be in a *cis* orientation in which both Ly49A and H-2D^d are coming from the target cell. In addition, site 2 involves no polymorphic residues of MHC, making it difficult to explain the allelic specificity of Ly49A. Therefore, site 1 was proposed to be physiologically relevant. The crystal structure of the complex revealed no peptide contact between Ly49A and H-2D^d. This recognition model, although consistent with the nonpeptide dependence of Ly49A recognition, appears contradictory when applied to the MHC recognition by Ly49C. Subsequent mutational analysis of the binding site suggested site 2 is a more likely true binding site (49, 51). Compared to its human counterpart KIR, the recognition mode between Ly49A and H-2D^d is different from that between KIR and HLA, which resembles the TCR/MHC recognition mode.

LIGANDS OF NKG2D

Several NKG2D ligands were recently identified in both humans and mice. MIC-A and MIC-B were among the first identified ligands in humans. Structurally, MIC proteins are described as MHC class I homologs but without the association of β2-microglobulin and peptide (42) (Figure 6). Recently, a group of proteins that bind to the human cytomegalovirus glycoprotein, UL16, has been identified as ligands to human NKG2D and named ULBP. ULBP proteins contain only α1 and α2 domains similar to those of MHC class I, do not bind peptides, and are glycosyl-phosphatidylinositol (GPI) anchored to the membrane (13). In mice the retinoid acid early inducible 1 (Rae-1) family of proteins and H60 protein encoded on murine chromosome 10 were identified as ligands for murine NKG2D

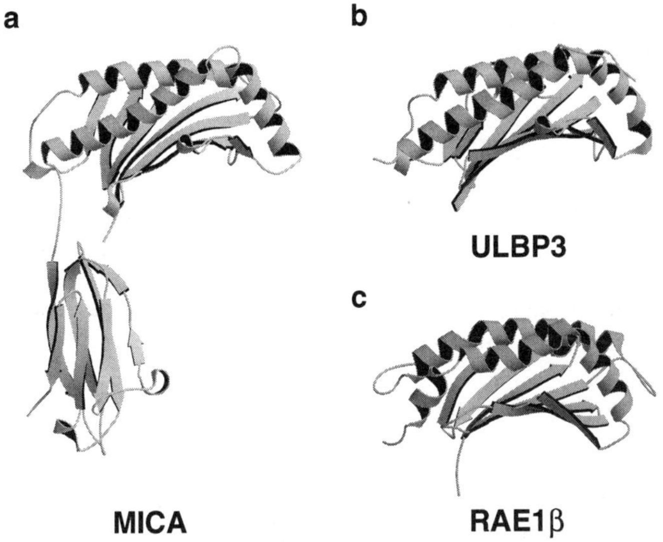

Figure 6 Ligands of NKG2D. The structures of (*a*) MIC-A (PDB code 1HYR), (*b*) ULBP3 (PDB code 1KCG), and (*c*) Rae-1β (PDB code 1JSK).

(8, 17, 47, 57). Like ULBP proteins, Rae-1 and H60 contain only $\alpha 1$ and $\alpha 2$ domains similar to those of class I MHC, do not bind peptides, and are GPI anchored to the membrane (41).

Murine and human NKG2D display significant conservation both at the amino acid sequence and at the three-dimensional structural level. The overall sequence identity of the receptor is about 60% between human and mouse. At the structural level, the superposition between the human and murine NKG2D results in an rms deviation of 0.95 Å among 233 Cα atoms. On the other hand, sequence comparison among NKG2D ligands shows limited identity between them, 24%–33%. That makes NKG2D a unique activating receptor with the ability to bind diverse MHC class I–like ligands.

A structural comparison between MIC-A, ULBP3, and Rae-1β resulted in rms differences of 2.3–2.4 Å between ULBP3 and Rae-1β or MIC-A and 3.9 Å between Rae-1β and MIC-A for 145 Cα atoms. Structural comparison between NKG2D ligands and classical MHC class I molecule HLA-Cw3 results in 2.7–3.4 Å rms differences for a similar number of Cα atoms. The conformation of the eight-stranded β sheet agrees quite well among ULBP3, MIC-A, Rae-1β, and HLA-Cw3. The differences exist primarily in the loop and helical regions. All class I/II MHC molecules present peptides, whereas no peptide has been found associating with MIC, ULBP, or Rae-1 (40, 41, 60). In addition, the spacing between the α

helices of NKG2D ligands is significantly narrower (down to 8–14 Å for ULBP3) than that of HLA-Cw3 (15–20 Å).

LIGAND RECOGNITION BY NKG2D

The crystal structures of NKG2D/MIC-A, NKG2D/ULBP3, and NKG2D/Rae-1β complexes that allow detailed insight into their binding mode were recently determined (40, 41, 60) (Figure 7). In all three complexes the mode of complex formation is similar: NKG2D uses its β strands and loops located at the end opposite to both the N and C termini to bind to the $\alpha 1/\alpha 2$ helical surface of its ligands.

The relative orientation between NKG2D and its ligands is similar to that between KIR and HLA, and between TCR and their MHC ligands. The long axis of the receptor fits diagonally across the helical axes of a ligand. The receptor footprint covers the C-terminal half of the $\alpha 1$ helix and the N-terminal half of the $\alpha 3$ helix of a ligand. Both subunits of the NKG2D bind ligand with identical receptor loops. However, the interaction between a homodimeric NKG2D and the asymmetrical ligand results in an asymmetrical receptor subunit orientation that contrasts with the perfect twofold symmetry observed in ligand-free murine receptor.

The two NKG2D subunits form a concave surface in which the convex-shaped ligand interacts with both subunits of the receptor. The interface shape complementarity value Sc is 0.65, 0.72, and 0.63 between the NKG2D and ULBP3, NKG2D and MIC-A, and NKG2D and Rae-1β, respectively, which indicates a good surface complementarity at the receptor-ligand interface. The total buried interface between NKG2D receptors and their ligands is 1900–2200 Å^2, which is larger than the 1560 Å^2 interface between KIR and HLA (5), or the 1700–1800 Å^2 between TCR and MHC (23, 24, 62). Except at the ligand binding interface, no significant conformational changes occurred in the receptor structure upon complex formation. An rms deviation of 0.48 Å among 246 Cα atoms was observed between the structure of a ligand-free and Rae-1β-bound murine NKG2D. Similarly, no large conformational adjustment occurred in the ligand structure upon receptor binding. For the structures of receptor-free and receptor-bound MIC-A and Rae-1β, the rms differences were 0.92 and 0.57 Å, respectively, for 147 Cα pairs.

Each receptor subunit contributes approximately half of the total interface area. Primarily a network of hydrogen bonds and hydrophobic interactions stabilize the receptor-ligand interface. Although the mode of the binding is similar between three different complexes, detailed examination of the binding interface reveals remarkable differences between complexes. When the three NKG2D complexes are superimposed onto each other by using only their receptors, the different ligand orientations become visible. In this case, there is about a 6° orientation difference between ULBP3 and MIC-A, 10° between ULBP3 and Rae-1β, and 10° between MIC-A and Rae-1β. These orientation differences probably reflect local adjustments of the receptor/ligand structures in an attempt to maximize their

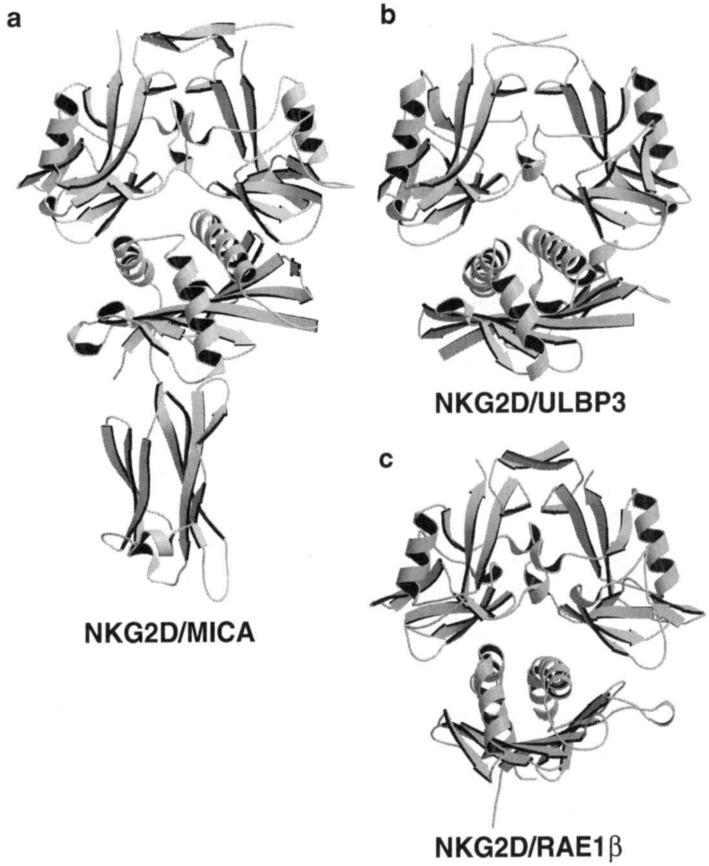

Figure 7 Structures of NKG2D in complex with (*a*) MIC-A (PDB code 1HYR), (*b*) ULBP3 (PDB code 1KCG), and (*c*) Rae-1β (PDB code 1JSK).

interaction. Detailed analysis and comparison of the hydrogen bonding networks in the complexes reveal striking differences in the topology of interacting side chains. NKG2D/MIC-A has 13 hydrogen bonds between the receptor and ligand, the largest among the three complexes; NKG2D/ULBP3 and NKG2D/Rae-1β have 9 and 7 hydrogen bonds, respectively. Three salt bridges were observed in NKG2D/ULBP3 and NKG2D/MIC-A complexes, whereas only one was observed in NKG2D/Rae-1β. Only two hydrogen bonds and one salt bridge are conserved in all three structures. Although different ligands form different hydrogen bonds with NKG2D, the signature pattern of strongest hydrogen bonds remains the same in all three complexes. Similarly, a subset of distinct but overlapping NKG2D residues participates in the hydrophobic interactions with each of the ligands and forms two large characteristic hydrophobic patches on the receptor surface whereby residues Tyr152, Ile182, Met184, and Tyr199 from each subunit make the largest contribution.

NK CELL RECEPTORS MEDIATE INNATE IMMUNE SURVEILLANCE

Cytotoxic T cells recognize viral and tumor antigens directly through peptide presentation by MHC molecules. To keep up with the evolving pathogenic world, T cell receptor repertoire is tuned in a receptor maturation process that modifies the TCR library to reflect current pathogens and to eliminate self-reacting T cells through gene recombination and somatic mutations. When MHC molecules present pathogenic peptides, T cells undergo clonal expansion to derive clones of monospecific T cells to stage an effective attack on viral infected or tumor cells. Although receptor maturation and clonal expansion renders cytotoxic T cells the most powerful arm of the immune system in combating infectious diseases, it is also a time-consuming process. It often takes days to activate T cell–mediated immunity upon infection. In contrast, components of the innate immune system, such as NK cells and macrophages, can often be activated in less than one hour. This fast activation kinetics puts the innate immune system as the first line of immune defense. The advantage in activation kinetics, however, does not alleviate the need for NK cells to distinguish foreign versus self. The dilemma is how can NK cells retain a rapid deployment status against pathogen and yet remain capable of identifying foreign and self? T cell receptors acquired this ability through an elaborate gene recombination process that is absent in NK cells. NK cell development lacks the positive/negative selection, a hallmark of T cell development.

Upon maturation, NK cells express an array of germline-encoded activating and inhibitory receptors on their surface, with the distribution of each receptor largely stochastic (66). The cytolytic profile of individual NK cells depends on the balance between the activating and inhibitory receptors expressed on that particular cell. There are about 10 KIR genes in human and over 50 class I HLA molecules that are capable of presenting over 10^3 peptides. To effectively survey HLA molecules, KIR has evolved to bind multiple HLA ligands. Structurally, KIR recognition of multiple HLA allotypes is accomplished by the receptor through binding the conserved HLA residues in the polymorphic region of HLA $\alpha 1$ and $\alpha 2$ helices. While multiple ligand recognition enables KIR to monitor more class I MHC alleles, it also has the tendency to decrease the sensitivity of immune surveillance, the ability to respond to the change in the expression of a few alleles of HLA rather than to the global change in HLA expression level. The dilemma is how to maintain the ability to recognize multiple class I MHC alleles yet still display high specificity for individual MHC alleles.

A comparison between KIR2DL2/HLA-Cw3 and KIR2DL1/HLA-Cw4 complexes showed a conserved interface. Thirteen of the 17 KIR interface residues and 11 of the 12 HLA interface residues are conserved. When the structure of the complexes was superimposed, no large conformational changes at the interface were observed. Furthermore, no significant conformational changes were observed when the structure of ligand-free KIR2DL2 was compared with that of HLA-Cw3-bound receptor. This suggests that KIR binds HLA in a lock-and-key form of recognition. The lack of conformational variation in KIR/HLA recognition is also consistent with the fast binding kinetics as measured by BIACORE (45, 48). In contrast, the activating receptor NKG2D appears to adopt an induced-fit mechanism to recognize its ligands. The induced-fit recognition of NKG2D is also consistent with a slower binding kinetics between the receptor and its ligands than that between KIR and HLA (7, 58). Furthermore, the binding affinity between NKG2D and its ligands appears to be much higher than that between the inhibitory KIR and HLA. It is important to emphasize that both KIR and NKG2D recognize multiple ligands. However, in contrast to KIR, NKG2D ligands show a low degree of sequence similarity. This suggests that KIR and NKG2D use different structural mechanisms for their ligand recognition. KIR binds HLA in a lock-and-key mode but with low affinity, whereas NKG2D binds its ligands in an induced-fit mode with higher affinity. The combination of lock-and-key recognition and low-affinity binding predicts that the KIR/HLA recognition is less tolerant of mutational changes than the induced-fit, high-affinity binding of NKG2D.

Receptor-ligand interface mutations have been created to address the issue of recognition tolerance. Five charge mutations of KIR, including K44M, Y105A, E106A, D135H, and D183A, were generated, and binding affinity between the mutant KIR and HLA-Cw3 was measured by BIACORE. Four of the five single mutants resulted in complete loss of the HLA-Cw3 binding with the dissociation constant $K_D > 400\ \mu M$. The binding of E106A to the ligand is six times lower than that of the wild-type receptor. To test the mutational tolerance of NKG2D recognition, four interface charge mutations of ULBP3 were generated, H21A, E76A, R82M, and D169A. All four retained measurable binding to NKG2D, although the individual affinity varies between 30 and 230 μM. The mutational results further support the idea that the inhibitory KIR displays greater ligand specificity than NKG2D. Indeed, despite a better than 90% sequence identity between HLA-Cw1, 3, 7, 8 and HLA-Cw2, 4, 6, 15, no cross recognition was observed between their receptors, KIR2DL2 and KIR2DL1. Furthermore, the HLA allotype specificity between KIR2DL1 and KIR2DL2 appears to be controlled by a single hydrogen bond.

As to the function that these receptors play in innate immunity, the exquisite ligand specificity of KIR indicates that the inhibitory receptors sense small changes in HLA expression on cell surface and detect the loss of a few MHC alleles rather than the global loss of MHC, which only occurs in a catastrophic event. This sensitivity is particularly important because viruses often selectively downregulate the expression of a few class I MHC alleles in order to evade T cell recognition through peptide presentation. In contrast, the function of NKG2D is to recognize and kill tumor or virally infected cells through its ligand recognition. The diversity in NKG2D ligands presumably reflects the diversity of the pathogenic world. Encoded by a single gene, NKG2D must have the ability to recognize diverse ligands and to prevent possible pathogen-induced ligand escape. High-affinity ligand binding combined with induced-fit recognition offers a structural solution to the function of NKG2D.

ACKNOWLEDGMENT

We thank Dr. J. Boyington for helpful discussions and critical reading of the manuscript.

The *Annual Review of Biophysics and Biomolecular Structure* is online at http://biophys.annualreviews.org

LITERATURE CITED

1. Arm JP, Gurish MF, Reynolds DS, Scott HC, Gartner CS, et al. 1991. Molecular cloning of gp49, a cell-surface antigen that is preferentially expressed by mouse mast cell progenitors and is a new member of the immunoglobulin superfamily. *J. Biol. Chem.* 266:15966–73
2. Bauer S, Groh V, Wu J, Steinle A, Phillips JH, et al. 1999. Activation of NK cells and T cells by NKG2D, a receptor for stress-inducible MICA. *Science* 285:727–29
3. Berg SF, Dissen E, Westgaard IH, Fossum S. 1998. Molecular characterization of rat NKR-P2, a lectin-like receptor expressed by NK cells and resting T cells. *Int. Immunol.* 10:379–85
4. Borges L, Hsu ML, Fanger N, Kubin M, Cosman D. 1997. A family of human lymphoid and myeloid Ig-like receptors, some of which bind to MHC class I molecules. *J. Immunol.* 159:5192–96
5. Boyington JC, Motyka SA, Schuck P, Brooks AG, Sun PD. 2000. Crystal structure of an NK cell immunoglobulin-like receptor in complex with its class I MHC ligand. *Nature* 405:537–43
6. Boyington JC, Riaz AN, Patamawenu A, Coligan JE, Brooks AG, Sun PD. 1999. Structure of CD94 reveals a novel C-type lectin fold: implications for the NK cell-associated CD94/NKG2 receptors. *Immunity* 10:75–82
7. Carayannopoulos LN, Naidenko OV, Kinder J, Ho EL, Fremont DH, Yokoyama WM. 2002. Ligands for murine NKG2D display heterogeneous binding behavior. *Eur. J. Immunol.* 32:597–605
8. Cerwenka A, Bakker AB, McClanahan T, Wagner J, Wu J, et al. 2000. Retinoic acid early inducible genes define a ligand family for the activating NKG2D receptor in mice. *Immunity* 12:721–27
9. Chapman TL, Heikema AP, West AP, Bjorkman PJ. 2000. Crystal structure and ligand binding properties of the D1D2 region of the inhibitory receptor LIR-1 (ILT2). *Immunity* 13:727–36
10. Colonna M, Brooks EG, Falco M, Ferrara GB, Strominger JL. 1993. Generation of allospecific natural killer cells by stimulation across a polymorphism of HLA-C. *Science* 260:1121–24
11. Colonna M, Navarro F, Bellon T, Llano M, Garcia P, et al. 1997. A common inhibitory receptor for major histocompatibility complex class I molecules on human lymphoid and myelomonocytic cells. *J. Exp. Med.* 186:1809–18
12. Colonna M, Samaridis J. 1995. Cloning of immunoglobulin-superfamily members

associated with HLA-C and HLA-B recognition by human natural killer cells. *Science* 268:405–8

13. Cosman D, Mullberg J, Sutherland CL, Chin W, Armitage R, et al. 2001. ULBPs, novel MHC class I-related molecules, bind to CMV glycoprotein UL16 and stimulate NK cytotoxicity through the NKG2D receptor. *Immunity* 14:123–33
14. D'Andrea A, Chang C, Franz-Bacon K, McClanahan T, Phillips JH, Lanier LL. 1995. Molecular cloning of NKB1. A natural killer cell receptor for HLA-B allotypes. *J. Immunol.* 155:2306–10
15. Das H, Groh V, Kuijl C, Sugita M, Morita CT, et al. 2001. MICA engagement by human Vgamma2Vdelta2 T cells enhances their antigen-dependent effector function. *Immunity* 15:83–93
16. Davis DM, Chiu I, Fassett M, Cohen GB, Mandelboim O, Strominger JL. 1999. The human natural killer cell immune synapse. *Proc. Natl. Acad. Sci. USA* 96:15062–67
17. Diefenbach A, Jamieson AM, Liu SD, Shastri N, Raulet DH. 2000. Ligands for the murine NKG2D receptor: expression by tumor cells and activation of NK cells and macrophages. *Nat. Immunol.* 1:119–26
18. Dimasi N, Sawicki MW, Reineck LA, Li Y, Natarajan K, et al. 2002. Crystal structure of the Ly49I natural killer cell receptor reveals variability in dimerization mode within the Ly49 family. *J. Mol. Biol.* 320:573–85
19. Ding YH, Smith KJ, Garboczi DN, Utz U, Biddison WE, Wiley DC. 1998. Two human T cell receptors bind in a similar diagonal mode to the HLA-A2/Tax peptide complex using different TCR amino acids. *Immunity* 8:403–11
20. Dohring C, Samaridis J, Colonna M. 1996. Alternatively spliced forms of human killer inhibitory receptors. *Immunogenetics* 44:227–30
21. Fan QR, Long EO, Wiley DC. 2001. Crystal structure of the human natural killer cell inhibitory receptor KIR2DL1-HLA-Cw4 complex. *Nat. Immunol.* 2:452–60
22. Fan QR, Mosyak L, Winter CC, Wagtmann N, Long EO, Wiley DC. 1997. Structure of the inhibitory receptor for human natural killer cells resembles haematopoietic receptors. *Nature* 389:96–100. Erratum. *Nature* 390(6657):315
23. Garboczi DN, Ghosh P, Utz U, Fan QR, Biddison WE, Wiley DC. 1996. Structure of the complex between human T-cell receptor, viral peptide and HLA-A2. *Nature* 384:134–41
24. Garcia KC, Degano M, Stanfield RL, Brunmark A, Jackson MR, et al. 1996. An alphabeta T cell receptor structure at 2.5 Å and its orientation in the TCR-MHC complex. *Science* 274:209–19
25. Glienke J, Sobanov Y, Brostjan C, Steffens C, Nguyen C, et al. 1998. The genomic organization of NKG2C, E, F, and D receptor genes in the human natural killer gene complex. *Immunogenetics* 48:163–73
26. Govaerts MM, Goddeeris BM. 2001. Homologues of natural killer cell receptors NKG2-D and NKR-P1 expressed in cattle. *Vet. Immunol. Immunopathol.* 80:339–44
27. Grakoui A, Bromley SK, Sumen C, Davis MM, Shaw AS, et al. 1999. The immunological synapse: a molecular machine controlling T cell activation. *Science* 285:221–27
28. Groh V, Rhinehart R, Randolph-Habecker J, Topp MS, Riddell SR, Spies T. 2001. Costimulation of CD8alphabeta T cells by NKG2D via engagement by MIC induced on virus-infected cells. *Nat. Immunol.* 2:255–60
29. Harel-Bellan A, Quillet A, Marchiol C, DeMars R, Tursz T, Fradelizi D. 1986. Natural killer susceptibility of human cells may be regulated by genes in the HLA region on chromosome 6. *Proc. Natl. Acad. Sci. USA* 83:5688–92
30. Hayami K, Fukuta D, Nishikawa Y, Yamashita Y, Inui M, et al. 1997. Molecular cloning of a novel murine cell-surface glycoprotein homologous to killer cell inhibitory receptors. *J. Biol. Chem.* 272:7320–27

31. Ho EL, Heusel JW, Brown MG, Matsumoto K, Scalzo AA, Yokoyama WM. 1998. Murine Nkg2d and Cd94 are clustered within the natural killer complex and are expressed independently in natural killer cells. *Proc. Natl. Acad. Sci. USA* 95:6320–25
32. Houchins JP, Yabe T, McSherry C, Bach FH. 1991. DNA sequence analysis of NKG2, a family of related cDNA clones encoding type II integral membrane proteins on human natural killer cells. *J. Exp. Med.* 173:1017–20
33. Karlhofer FM, Ribaudo RK, Yokoyama WM. 1992. MHC class I alloantigen specificity of Ly-49+ IL-2-activated natural killer cells. *Nature* 358:66–70
34. Khakoo SI, Rajalingam R, Shum BP, Weidenbach K, Flodin L, et al. 2000. Rapid evolution of NK cell receptor systems demonstrated by comparison of chimpanzees and humans. *Immunity* 12:687–98
35. Kolatkar AR, Leung AK, Isecke R, Brossmer R, Drickamer K, Weis WI. 1998. Mechanism of N-acetylgalactosamine binding to a C-type animal lectin carbohydrate-recognition domain. *J. Biol. Chem.* 273: 19502–8
36. Kubagawa H, Burrows PD, Cooper MD. 1997. A novel pair of immunoglobulin-like receptors expressed by B cells and myeloid cells. *Proc. Natl. Acad. Sci. USA* 94:5261–66
37. LaBonte ML, Levy DB, Letvin NL. 2000. Characterization of rhesus monkey CD94/NKG2 family members and identification of novel transmembrane-deleted forms of NKG2-A, B, C, and D. *Immunogenetics* 51:496–99
38. Lanier LL. 1998. NK cell receptors. *Annu. Rev. Immunol.* 16:359–93
39. Lazetic S, Chang C, Houchins JP, Lanier LL, Phillips JH. 1996. Human natural killer cell receptors involved in MHC class I recognition are disulfide-linked heterodimers of CD94 and NKG2 subunits. *J. Immunol.* 157:4741–45
40. Li P, McDermott G, Strong RK. 2002. Crystal structures of RAE-1beta and its complex with the activating immunoreceptor NKG2D. *Immunity* 16:77–86
41. Li P, Morris DL, Willcox BE, Steinle A, Spies T, Strong RK. 2001. Complex structure of the activating immunoreceptor NKG2D and its MHC class I-like ligand MICA. *Nat. Immunol.* 2:443–51
42. Li P, Willie ST, Bauer S, Morris DL, Spies T, Strong RK. 1999. Crystal structure of the MHC class I homolog MIC-A, a gammadelta T cell ligand. *Immunity* 10:577–84
43. Litwin V, Gumperz J, Parham P, Phillips JH, Lanier LL. 1994. NKB1: a natural killer cell receptor involved in the recognition of polymorphic HLA-B molecules. *J. Exp. Med.* 180:537–43
44. Long EO. 1999. Regulation of immune responses through inhibitory receptors. *Annu. Rev. Immunol.* 17:875–904
45. Maenaka K, Juji T, Nakayama T, Wyer JR, Gao GF, et al. 1999. Killer cell immunoglobulin receptors and T cell receptors bind peptide-major histocompatibility complex class I with distinct thermodynamic and kinetic properties. *J. Biol. Chem.* 274:28329–34
46. Maenaka K, Juji T, Stuart DI, Jones EY. 1999. Crystal structure of the human p58 killer cell inhibitory receptor (KIR2DL3) specific for HLA-Cw3-related MHC class I. *Struct. Fold. Des.* 7:391–98
47. Malarkannan S, Shih PP, Eden PA, Horng T, Zuberi AR, et al. 1998. The molecular and functional characterization of a dominant minor H antigen, H60. *J. Immunol.* 161:3501–9
48. Mandelboim O, Reyburn HT, Sheu EG, Vales-Gomez M, Davis DM, et al. 1997. The binding site of NK receptors on HLA-C molecules. *Immunity* 6:341–50
49. Matsumoto N, Mitsuki M, Tajima K, Yokoyama WM, Yamamoto K. 2001. The functional binding site for the C-type lectin-like natural killer cell receptor Ly49A spans three domains of its major histocompatibility complex class I ligand. *J. Exp. Med.* 193:147–58

50. Meyaard L, Adema GJ, Chang C, Woollatt E, Sutherland GR, et al. 1997. LAIR-1, a novel inhibitory receptor expressed on human mononuclear leukocytes. *Immunity* 7:283–90
51. Michaelsson J, Achour A, Rolle A, Karre K. 2001. MHC class I recognition by NK receptors in the Ly49 family is strongly influenced by the beta 2-microglobulin subunit. *J. Immunol.* 166:7327–34
52. Monks CR, Freiberg BA, Kupfer H, Sciaky N, Kupfer A. 1998. Three-dimensional segregation of supramolecular activation clusters in T cells. *Nature* 395:82–86
53. Moretta A, Bottino C, Pende D, Tripodi G, Tambussi G, et al. 1990. Identification of four subsets of human CD3-CD16+ natural killer (NK) cells by the expression of clonally distributed functional surface molecules: correlation between subset assignment of NK clones and ability to mediate specific alloantigen recognition. *J. Exp. Med.* 172:1589–98
54. Moretta A, Bottino C, Vitale M, Pende D, Cantoni C, et al. 2001. Activating receptors and coreceptors involved in human natural killer cell-mediated cytolysis. *Annu. Rev. Immunol.* 19:197–223
55. Moretta A, Sivori S, Vitale M, Pende D, Morelli L, et al. 1995. Existence of both inhibitory (p58) and activatory (p50) receptors for HLA-C molecules in human natural killer cells. *J. Exp. Med.* 182:875–84
56. Natarajan K, Sawicki MW, Margulies DH, Mariuzza RA. 2000. Crystal structure of human CD69: a C-type lectin-like activation marker of hematopoietic cells. *Biochemistry* 39:14779–86
57. Nomura M, Zou Z, Joh T, Takihara Y, Matsuda Y, Shimada K. 1996. Genomic structures and characterization of Rae1 family members encoding GPI-anchored cell surface proteins and expressed predominantly in embryonic mouse brain. *J. Biochem. (Tokyo)* 120:987–95
58. O'Callaghan CA, Cerwenka A, Willcox BE, Lanier LL, Bjorkman PJ. 2001. Molecular competition for NKG2D: H60 and RAE1 compete unequally for NKG2D with dominance of H60. *Immunity* 15:201–11
59. Plougastel B, Trowsdale J. 1997. Cloning of NKG2-F, a new member of the NKG2 family of human natural killer cell receptor genes. *Eur. J. Immunol.* 27:2835–39
60. Radaev S, Rostro B, Brooks AG, Colonna M, Sun PD. 2001. Conformational plasticity revealed by the cocrystal structure of NKG2D and its class I MHC-like ligand ULBP3. *Immunity* 15:1039–49
61. Rajagopalan S, Long EO. 1997. The direct binding of a p58 killer cell inhibitory receptor to human histocompatibility leukocyte antigen (HLA)-Cw4 exhibits peptide selectivity. *J. Exp. Med.* 185:1523–28
62. Reinherz EL, Tan K, Tang L, Kern P, Liu J, et al. 1999. The crystal structure of a T cell receptor in complex with peptide and MHC class II. *Science* 286:1913–21
63. Samaridis J, Colonna M. 1997. Cloning of novel immunoglobulin superfamily receptors expressed on human myeloid and lymphoid cells: structural evidence for new stimulatory and inhibitory pathways. *Eur. J. Immunol.* 27:660–65
64. Shimizu Y, DeMars R. 1989. Demonstration by class I gene transfer that reduced susceptibility of human cells to natural killer cell-mediated lysis is inversely correlated with HLA class I antigen expression. *Eur. J. Immunol.* 19:447–51
65. Shum BP, Flodin LR, Muir DG, Rajalingam R, Khakoo SI, et al. 2002. Conservation and variation in human and common chimpanzee CD94 and NKG2 genes. *J. Immunol.* 168:240–52
66. Smith HR, Chuang HH, Wang LL, Salcedo M, Heusel JW, Yokoyama WM. 2000. Nonstochastic coexpression of activation receptors on murine natural killer cells. *J. Exp. Med.* 191:1341–54
67. Snyder GA, Brooks AG, Sun PD. 1999. Crystal structure of the HLA-Cw3 allotype-specific killer cell inhibitory receptor KIR2DL2. *Proc. Natl. Acad. Sci. USA* 96:3864–69

68. Steffens U, Vyas Y, Dupont B, Selvakumar A. 1998. Nucleotide and amino acid sequence alignment for human killer cell inhibitory receptors (KIR), 1998. *Tissue Antigens* 51:398–413
69. Storkus WJ, Howell DN, Salter RD, Dawson JR, Cresswell P. 1987. NK susceptibility varies inversely with target cell class I HLA antigen expression. *J. Immunol.* 138:1657–59
70. Tormo J, Natarajan K, Margulies DH, Mariuzza RA. 1999. Crystal structure of a lectin-like natural killer cell receptor bound to its MHC class I ligand. *Nature* 402:623–31
71. Vance RE, Kraft JR, Altman JD, Jensen PE, Raulet DH. 1998. Mouse CD94/NKG2A is a natural killer cell receptor for the nonclassical major histocompatibility complex (MHC) class I molecule Qa-1(b). *J. Exp. Med.* 188:1841–48
72. Vivier E, Tomasello E, Paul P. 2002. Lymphocyte activation via NKG2D: towards a new paradigm in immune recognition? *Curr. Opin. Immunol.* 14:306–11
73. Wagtmann N, Biassoni R, Cantoni C, Verdiani S, Malnati MS, et al. 1995. Molecular clones of the p58 NK cell receptor reveal immunoglobulin-related molecules with diversity in both the extra- and intracellular domians. *Immunity* 2:439–49
74. Wang J, Lim K, Smolyar A, Teng M, Liu J, et al. 1998. Atomic structure of an alphabeta T cell receptor (TCR) heterodimer in complex with an anti-TCR fab fragment derived from a mitogenic antibody. *EMBO J.* 17:10–26
75. Wang JH, Smolyar A, Tan K, Liu JH, Kim M, et al. 1999. Structure of a heterophilic adhesion complex between the human CD2 and CD58 (LFA-3) counterreceptors. *Cell* 97:791–803
76. Wilson MJ, Torkar M, Trowsdale J. 1997. Genomic organization of a human killer cell inhibitory receptor gene. *Tissue Antigens* 49:574–79
77. Wolan DW, Teyton L, Rudolph MG, Villmow B, Bauer S, et al. 2001. Crystal structure of the murine NK cell-activating receptor NKG2D at 1.95 Å. *Nat. Immunol.* 2:248–54
78. Wu J, Song Y, Bakker AB, Bauer S, Spies T, et al. 1999. An activating immunoreceptor complex formed by NKG2D and DAP10. *Science* 285:730–32
79. Yim D, Jie HB, Sotiriadis J, Kim YS, Kim KS, et al. 2001. Molecular cloning and characterization of pig immunoreceptor DAP10 and NKG2D. *Immunogenetics* 53:243–49
80. Zappacosta F, Borrego F, Brooks AG, Parker KC, Coligan JE. 1997. Peptides isolated from HLA-Cw*0304 confer different degrees of protection from natural killer cell-mediated lysis. *Proc. Natl. Acad. Sci. USA* 94:6313–18

Annu. Rev. Biophys. Biomol. Struct. 2003. 32:115–33
doi: 10.1146/annurev.biophys.32.110601.142506

First published online as a Review in Advance on February 18, 2003

NUCLEIC ACID RECOGNITION BY OB-FOLD PROTEINS

Douglas L. Theobald, Rachel M. Mitton-Fry, and Deborah S. Wuttke
Department of Chemistry and Biochemistry, University of Colorado at Boulder, Boulder, Colorado 80309-0215; email: theobal@colorado.edu; fryrm@colorado.edu; deborah.wuttke@colorado.edu

Key Words single stranded, protein fold, structural alignment

■ **Abstract** The OB-fold domain is a compact structural motif frequently used for nucleic acid recognition. Structural comparison of all OB-fold/nucleic acid complexes solved to date confirms the low degree of sequence similarity among members of this family while highlighting several structural sequence determinants common to most of these OB-folds. Loops connecting the secondary structural elements in the OB-fold core are extremely variable in length and in functional detail. However, certain features of ligand binding are conserved among OB-fold complexes, including the location of the binding surface, the polarity of the nucleic acid with respect to the OB-fold, and particular nucleic acid–protein interactions commonly used for recognition of single-stranded and unusually structured nucleic acids. Intriguingly, the observation of shared nucleic acid polarity may shed light on the longstanding question concerning OB-fold origins, indicating that it is unlikely that members of this family arose via convergent evolution.

CONTENTS

INTRODUCTION

The OB-fold is a small structural motif originally named for its oligonucleotide/oligosaccharide binding properties, although it has since been observed at protein-protein interfaces as well. The nucleic acid–binding superfamily is the largest within the OB-folds, and proteins containing this motif are involved almost any time that single-stranded DNA or RNA (ssDNA/ssRNA) is present or requires manipulation. In this capacity, OB-fold proteins have been identified as critical for DNA replication, DNA recombination, DNA repair, transcription, translation, cold shock response, and telomere maintenance. We analyze a subset of these nucleic acid–binding OB-folds from a structural perspective, reviewing all the OB-fold/nucleic acid complexes for which high-resolution structures are available. The number of complex structures has nearly tripled in the past two years, and with the availability of this new structural data, it is possible to compare and contrast the topology, modularity, ligand recognition, and sequence elements featured in these diverse nucleic acid–binding OB-fold proteins.

GENERAL OB-FOLD FEATURES

OB-fold domains range between 70 and 150 amino acids in length. Although no strong sequence relationship between the disparate members of the OB-fold family can be detected, this fold is easily recognized on the basis of its distinct topology (Figure 1, see color insert). The variability in length among OB-fold domains is primarily due to dramatic differences in the length of variable loops found between well-conserved elements of secondary structure. OB-folds often occur as recognition domains in larger proteins; when seen as full proteins on their own, they frequently oligomerize or are found in large multicomponent assemblies, some examples of which are shown in Figure 2 (see color insert).

Often described as a Greek key motif, the OB-fold consists of two three-stranded antiparallel β-sheets, where strand 1 is shared by both sheets (57). As shown in Figure 1, the β-sheets pack orthogonally, forming a somewhat flattened, five-stranded β-barrel arranged in a 1-2-3-5-4-1 topology. Between strands 3 and 4, an α-helix is frequently found that packs against the bottom of the barrel, usually oriented lengthwise along the long axis of the β-barrel cross-section [also shown in Figure 3 (see color insert) for the various OB-fold domains]. Strands 3 and 5 can close the β-barrel by hydrogen bonding in a parallel arrangement. However, these strands have also been observed a full strand-width apart, which results in only a partially closed β-barrel. Several structural determinants have been identified that OB-folds share in common (17). A glycine (or other small residue) in the first half of $\beta 1$ and a β-bulge in the second half of $\beta 1$ allow this strand to contribute to both β-sheets by curving completely around the β-barrel. A second glycine residue often occurs at the beginning of strand 4 in an α_L conformation, perhaps breaking the α-helix between strands 3 and 4. Intriguingly, in the cases where the site of ligand binding is known, OB-folds tend to use a common ligand-binding interface centered on β-strands 2 and 3 (57). As shown in Figure 1, this canonical interface is augmented by the loops between $\beta 1$ and $\beta 2$ (referred to as L_{12}), $\beta 3$ and α ($L_{3\alpha}$), α and $\beta 4$ ($L\alpha 4$), and $\beta 4$ and $\beta 5$ (L_{45}). These loops define a cleft that runs across the surface of the OB-fold perpendicular to the axis of the β-barrel. The majority of nucleic acid–binding partners bind within this cleft, typically perpendicular to the antiparallel β-strands, with a polarity running $5'$ to $3'$ from strands $\beta 4$ and $\beta 5$ to strand $\beta 2$ (this orientation will hereafter be referred to as the standard polarity). As evidenced by numerous high-resolution structures, loops presented by a β-sheet appear to provide an ideal recognition surface for single-stranded nucleic acids, allowing binding through aromatic stacking, hydrogen bonding, hydrophobic packing, and polar interactions.

OVERVIEW OF STRUCTURES

OB-folds are found in eight distinct superfamilies within the SCOP (Structural Classification of Proteins) database (59). These superfamilies include staphylococcal nucleases, bacterial enterotoxins, inorganic pyrophosphatases, and nucleic acid–binding proteins, the latter of which is by far the most well-represented OB-fold superfamily in the structural database. We focus on nucleic acid–binding OB-folds that have been structurally characterized at high resolution bound to cognate nucleic acid. At this time, 11 structures of complexes have been solved, arising from four of the nine families in the SCOP nucleic acid–binding OB-fold superfamily. These OB-fold structures are divided into three categories on the basis of our current understanding of their functional recognition: (*a*) proteins that bind nucleic acids without apparent or strong sequence specificity, including human replication protein A (*hs*RPA) and *Escherichia coli* single-stranded DNA-binding protein (*Ec*SSB); (*b*) proteins that recognize specific single-stranded regions of nucleic acids, including *E. coli* Rho transcriptional terminator (*Ec*Rho), *Saccharomyces cerevisiae* Cdc13, *Oxytricha nova* telomere end-binding protein (*On*TEBP), and the *S. cerevisiae* and *E. coli* aspartyl-tRNA synthetases (AspRS); and (*c*) proteins that interact with mainly nonhelical structured nucleic acids, including *Thermatoga maritima* RecG, the ribosomal proteins *Thermus thermophilus* initiation factor 1 (IF1), *Haloarcula marismortui* L2, and *T. thermophilus* S12 and S17. The salient features of these complexes are summarized in Table 1, and a brief overview of the function of each of these systems is presented below.

TABLE 1 Summary of OB-fold protein complexes

Structure	Ligand	PDB code (bound)	Nucleic acid polarity	Buried surface area (Å)[a]	nt/OB-fold[b]	Free structure
RPA70	ssDNA	1JMC, 2.4 Å	DBD-A: standard	560	3	1FGU
			DBD-B: standard	572	5	
*Ec*SSB	ssDNA	1EYG, 2.8 Å	Reverse	1754	18	1QVC
*Ec*Rho	ssRNA	2A8V, 2.4 Å	Standard	351	2	1A8V
				379	2	
*On*TEBP	Spec. ssDNA	1JB7, 1.86 Å	α1: standard	718[c]	10	1K8G
		1KIX, 2.7 Å	α2: standard, + 3′-turn	723[c]	8	α N35
		1K8G, 2.6 Å	β: reverse	272[c]	2	
Cdc13	Spec. ssDNA	1KXL[d], NMR	Standard	1321	11	—
*Ec*AspRS	tRNA^{asp} anticodon	1ASZ (*S. cerevisiae*) 3.0 Å, inc. ATP	Standard	870[e]	8	1EQR
		1ASY (*S. cerevisiae*) 3.0 Å				
		1C0A (*E. coli*) 2.4 Å, inc. AMP				
*Sc*AspRS	Junction DNA	1GM5, 3.24 Å	Standard	569	4	—
RecG	23S rRNA	1JJ2, chain A, 2.4 Å	Standard	302	4	1RL2
L2	16S rRNA	1J5E, chain L, 3.05 Å	NA—only binds helices	1577	22	—
S12	16S rRNA	1J5E, chain Q, 3.05 Å	Standard, helices excluded	1966	31	1RIP
S17	16S rRNA	1HR0, chain W, 3.2 Å	Standard	839	10	1AH9

[a]BSA = [ASA(OB-fold + nucleic acid)−ASA(OB-fold)−ASA (nucleic acid)]/2, calculated using NACCESS v.2.1.1 with a probe size of 1.5 Å. ASA = solvent accessible surface area.

[b]Only nucleotides that are within 3.5 Å of the canonical OB-fold.

[c]For 1JB7 structure of ternary complex.

[d]Plus unpublished models w/ DNA.

[e]For *E. coli* AspRS.

Human RPA

Replication protein A (RPA) is the major eukaryotic ssDNA-binding protein and is required for many aspects of DNA metabolism, including replication, recombination, and repair [(39, 80) and references therein]. Found in eukaryotes from yeast to humans, RPA is a heterotrimeric protein composed of subunits that are roughly 70, 32, and 14 kDa (RPA70, RPA32, RPA14, respectively). The human complex includes six OB-folds, four of which are involved in DNA binding (DBD-A, -B, and -C in RPA70 and DBD-D in RPA32) and all of which have been structurally characterized in the absence of ligand, including a complex of the trimerization core with RPA14 and domains of RPA32 and RPA70 (9–11, 40). The full complex binds a ~30-nucleotide ssDNA with subnanomolar-binding affinity and prefers ssDNA over RNA or double-stranded (ds)DNA by a factor of 10^3 (44). Though usually considered a nonspecific ssDNA-binding protein, RPA displays a slight preference for binding polypyrimidine tracts. Binding is believed to occur in a sequential fashion, with the high-affinity OB-folds DBD-A and DBD-B binding weakly to a ~9-nucleotide segment, followed by conformational changes which allow DBD-C and DBD-D to interact with longer substrates (6). A fragment of RPA70 comprising DBD-A and DBD-B, shown in Figure 2*c*, has been structurally characterized in the presence of C_8 DNA (12). Each of these domains includes a helix between strands 3 and 4 that caps the bottom of the OB-fold barrel, and DBD-B has an additional helix after $\beta 5$ that may be involved in subunit trimerization.

Escherichia coli SSB

SSB is the major prokaryotic ssDNA-binding protein and, like RPA, plays essential roles in DNA replication, recombination, and repair. It forms a homotetramer of identical 19-kDa subunits capable of interacting with ssDNA in several modes with different cooperativity and a different number of nucleotides occluded by binding (51). The N-terminal 135-amino-acid chymotryptic fragment of *Ec*SSB has been structurally characterized in the presence and absence of C_{35} DNA (67, 68, 77). The monomeric OB-fold assembles into a tetramer via two distinct protein-protein interfaces. The first of these interfaces is a six-stranded β-sheet produced by interactions between two monomers along strand 1 of the sheet consisting of $\beta 1$, $\beta 4$, and $\beta 5$. The second interface is formed between two dimers interacting across this β-sheet. As seen in Figure 2*g*, each SSB monomer makes extensive contacts with DNA in the assembled tetramer (67), utilizing a large noncanonical interaction surface that includes both sides of a protracted two-stranded β-sheet extending from L_{23}. *Ec*SSB is distinguished as one of two OB-folds known to bind nucleic acid in the reverse polarity. Interestingly, archeal SSBs have been identified that contain four DNA-binding domains with sequence similarity to RPA in a single polypeptide chain, suggesting a possible evolutionary pathway for the SSB/RPA family of proteins (20, 42).

Escherichia coli Rho

The eubacterial transcriptional terminator Rho is a hexameric RNA-DNA helicase (14). Rho is believed to first bind the nascent RNA at specific sites and then actively translocate down the RNA until reaching the transcription machinery where it unwinds the DNA/RNA duplex. The Rho hexamer, which assembles into a ring, has three high-affinity ssRNA/ssDNA sites and three low-affinity ssRNA sites that exhibit a preference for poly-C substrates (76). The 47-kDa *E. coli* Rho monomer contains two major domains: a 130-amino-acid N-terminal RNA-binding domain and a C-terminal ATPase domain reminiscent of F_1 ATPase (24, 25). Structures of the N-terminal domain in the presence and absence of C_9 RNA reveal an OB-fold with a 47-amino-acid N-terminal helical extension that caps the top of the OB-fold barrel, as shown in Figure 2*b* (1, 13, 15). Rho binds two to three nucleotides of RNA with the standard polarity across the OB-fold, and it utilizes both L_{23} and the canonical OB-fold ligand-binding site for interactions with the nucleic acid (13). Interestingly, no contacts are seen to the RNA 2′-hydroxyl groups, consistent with the ability of Rho to interact with either ssDNA or ssRNA.

Saccharomyces cerevisiae Cdc13

Cdc13 is an essential yeast protein required for telomere end protection and length regulation (30, 50, 61). Cdc13 binds specifically to cognate single-stranded telomeric DNA ($TG_{1\text{-}3}$) with subnanomolar affinity (61). The protein is thought to localize to the 3′ telomeric overhang by virtue of this specific ssDNA-binding activity, recruiting relevant end-protection and telomere maintenance subcomplexes via protein-protein interactions (28, 64). Full DNA-binding activity can be found in a DNA-binding domain (residues 497–694) located centrally in the 924-amino-acid protein (2, 38). Shown in Figure 2*f*, the solution structure of this DBD in complex with an 11-nucleotide telomeric sequence reveals an OB-fold that binds DNA in an extended conformation with the standard polarity across the OB-fold (56). This OB-fold contains an unusually large (30-residue) L_{23} that folds down over the β-barrel, making critical contacts with the DNA ligand. A C-terminal helical extension may be involved in ensuring the correct orientation of this loop. The ssDNA wraps 180 degrees around the OB-fold surface, and mutagenesis of interface residues suggests that the entire contact surface is important thermodynamically for ligand binding (3).

Oxytricha nova TEBP (α/β)

The *O. nova* telomere end-binding protein (*On*TEBP) is a multimeric protein that, like Cdc13, binds with high affinity and specificity to the single-stranded 3′-overhang of macronuclear telomeric DNA (33, 65). Two different crystal structures of the α-subunit complexed with cognate ssDNA have been determined: a 35-kDa N-terminal fragment, which binds as a monomer, and the full-length 56-kDa protein, which dimerizes via a large C-terminal domain (21, 63). Intriguingly, these structures reveal three α-subunit OB-fold domains, two of which bind in concert to the ssDNA (colored green and magenta in Figure 2*a*), and one that acts as the homodimerization domain. A third crystal structure of the ternary complex (the 56-kDa α-subunit, a 28-kDa N-terminal core of the β-subunit, and a 12-nucleotide ssDNA) discloses an additional OB-fold in the β-subunit that is also involved in ssDNA recognition, colored cyan in Figure 2*a* (35, 36). These three OB-folds work together to recognize a 12-mer of ssDNA, with complete burial of the 3′ end deep within the complex. Interestingly, in the ternary complex, the C-terminal OB-fold of the α-subunit facilitates heterodimerization by recognizing a long structured loop from the β-subunit in the canonical interface. Each OB-fold presents a slightly different face to the ssDNA, and the majority of nucleotides are contacted by multiple OB-folds. Interactions are primarily mediated by L_{12} and L_{45}. Both OB-folds in the α-subunit bind the ssDNA with the standard polarity, whereas the β-subunit binds with the reverse polarity. Distinctive features of ligand recognition include two examples of arginine residues stacking face-to-face on guanine bases and an extensive aromatic stack composed of four bases and three amino acid side chains.

Saccharomyces cerevisiae and *Escherichia coli* Aspartyl-tRNA Synthetase

Aspartyl-tRNA synthetase (AspRS) is a class IIb aminoacyl-tRNA synthetase responsible for charging $tRNA^{asp}$ with aspartate. A large C-terminal catalytic domain contains the conserved class II synthetase motifs, including the enzymatic active site. tRNA recognition is achieved through five identity determinants, three of which are the anticodon bases (66). The N-terminal anticodon-binding domain of AspRS, like the other class IIb synthetases (8, 22), adopts an OB-fold, which is shown in green in Figure 2*e*. Crystal structures of cognate yeast and *E. coli* complexes (19, 27, 69) show that the tRNA anticodon bases bind across the face of this OB-fold with the standard polarity, interacting primarily via L_{45}. Complex formation induces the three anticodon bases, which stack upon each other in the free state, to bulge out and no longer stack. In *Ec*AspRS, the queuosine base (a hypermodified guanosine) in the QUC anticodon stacks on a phenylalanine (F48) projecting from strand $\beta 3$, while the pyrimidines stack on another phenylalanine (F35) from strand $\beta 2$ (see also Figure 5*a*). Interestingly, several other tRNA synthetases, including PheRS, contain an OB-fold not used for tRNA recognition (32).

Thermatoga maritima RecG

RecG, a monomeric 76-kDa multidomain bacterial protein with no known homologs in higher organisms, is a superfamily 2 helicase capable of rescuing stalled replication forks (54). Rescue occurs by RecG unwinding the nascent duplex to form chicken-foot intermediates, which are further processed to enable bypass of DNA lesions (52, 53). The crystal structure of *T. maritima* RecG in complex with a model three-way DNA junction reveals three structural domains, a large

N-terminal domain and two helicase domains (73). The center of the N-terminal domain adopts an OB-fold (shown in cyan in Figure 2*d*), referred to as the wedge domain. This OB-fold makes extensive contacts to both strands of the ssDNA at the template junction. Aromatic stacking contacts stabilize unpaired bases, forcing the parental duplex open. Because the OB-fold sits in the junction, both strands of DNA can interact with the canonical OB-fold ligand-binding face in the standard polarity.

Thermus thermophilus Ribosomal Protein S12

The recent high-resolution ribosome structures have revealed exciting examples of OB-fold recognition of structured RNAs, with the OB-fold topology observed in S12, S17, and L2 (5, 16, 79). S12, a 135-amino-acid component of the 30S ribosomal subunit, is one of the few proteins found at the interface of the small and large ribosomal subunits (represented as the small green OB-fold in Figure 2*i*). Located adjacent to the ribosomal A site, S12 is thought to be involved in tRNA decoding. Its OB-fold has a 24-amino-acid N-terminal extension that winds through the ribosomal core, terminating in a two-turn helix ~50 Å from the OB-fold center, and a short disordered C-terminal extension (16). A full one third of the protein's surface area packs against the rRNA, producing an unusually large RNA interface of greater than 3200 $Å^2$. Structures of the 30S subunit that include mRNA and a cognate tRNA anticodon stem loop in the ribosomal A site support the involvement of S12 in translational fidelity (62). Here, interaction between highly conserved residues in the S12 L_{12} loop and the conserved nucleotides A1492 and G530 allows direct interrogation of the Watson-Crick pairing status of the codon and anticodon.

Thermus thermophilus Ribosomal Protein S17

S17 is a primary assembly protein of the 30S ribosomal subunit. Depicted in blue in Figure 2*i*, S17 binds on the backside of this subunit, organizing disparate regions of the 5′ and central ribosomal domains. In *T. thermophilus*, this 105-amino-acid protein consists of an OB-fold domain that exhibits high sequence conservation among all forms of life (31) followed by a C-terminal helical extension (after $\beta5$) (16, 31, 41). Thirty-five percent of its total surface area packs against ribosomal RNA, burying over 2800 $Å^2$ of solvent-accessible surface area (16). The standard OB-fold-binding site interacts extensively with nucleotides of the 5′ domain (particularly the ribosomal RNA helices H7 and H11) via L_{45}, a long L_{23}, and L_{12}. The strikingly extended $\beta2$-$\beta3$ hairpin and the C-terminal helix protrude into the central domain, making critical contacts to helices H20 and H21.

Haloarcula marismortui Ribosomal Protein L2

L2 is one of the largest protein components of the large ribosomal subunit (237 amino acids in *H. marismortui*) and a primary binding protein for 23S rRNA (26)

(Figure 2*h*). It was long thought to be a critical part of the 50S peptidyltransferase center [(23, 43) and references therein], although recent structures show that its nearest approach to the catalytic site is over 20 Å away (5, 34). Like many ribosomal proteins, it is composed of a globular region that packs against the exterior of the ribosome and an extended region that is deeply buried within the RNA. The L2 globular region is composed of an OB-fold closely tied to an SH3-like fold by a bridging β-strand; the extended region continues both N- and C-terminally from this globular region (5, 60). The L2 OB-fold lacks both the canonical strand $\beta 5$ and the α-helix connecting strands $\beta 3$ and $\beta 4$. Strand $\beta 4$ hydrogen bonds to a fifth β-strand in a parallel orientation, giving the β-barrel an unusual "open" $\beta 1$-$\beta 2$-$\beta 3$-$\beta 5$-$\beta 4$ topology. The OB-fold makes a number of contacts to nucleotides in 23S rRNA domains IV and V primarily through L_{12}. These nucleotides do not traverse the OB-fold β-barrel, nor are they contacted solely by the OB-fold. Rather, they are sandwiched in a cleft between the two globular folds and the amino acids in both extended regions of L2.

Thermus thermophilus IF1

The prokaryotic initiation factor 1 (IF1) is a 71-amino-acid protein that functions with IF2 and IF3 at the 30S ribosomal subunit to allow translational initiation and correct start codon selection. Sequence and structural homologies have been observed between IF1 and archeal and eukaryotic translation initiation factors aIF-1A and eIF-1A (7, 49). IF1 adopts an OB-fold with a short 3_{10} helix following $\beta 3$ and no N- or C-terminal extensions (71). The crystal structure of the *T. thermophilus* 30S ribosome in complex with cognate IF1 (shown in magenta in Figure 2*i*) reveals that IF1 occupies a cleft on the surface of the 30S subunit formed by helix 44, loop 530, and protein S12, occluding the ribosomal A-site (18). The IF1 L_{12} inserts into ribosomal RNA helix H44, making hydrogen bonding interactions with the RNA backbone and triggering a striking conformational change by flipping out A1492 and A1493, which stack against conserved arginine residues between $\beta 3$ and $\beta 4$. Other conserved IF1 residues in this region, as well as in the L_{45} loop, make numerous hydrogen bonding and electrostatic interactions with nucleotides from the 530 loop. Many of these contacts are made in conjunction with S12.

COMPARISONS OF OB-FOLD COMPLEXES

Protein Side Chain Contacts with Nucleic Acid

In these OB-fold complexes, the bases are often in close contact with the protein, while the phosphodiester groups are mostly exposed to solvent, as is observed in other proteins that bind single-stranded nucleic acids and nucleic acid loop structures (4, 36). Nucleotides interact with protein primarily via stacking interactions with aromatic amino acid side chains and packing interactions with hydrophobic side chains or the aliphatic portions of more polar groups such as lysine and arginine. Such nonpolar interactions can involve both the ribose rings and the bases of the nucleic acid. Intriguingly, several examples of an arginine side chain stacking face-to-face on a base (e.g., in the *On*TEBP and IF1 complexes) suggest that cation-π interactions may be a common theme in the specific recognition of single-stranded nucleic acids by OB-folds. Additionally, hydrogen-bond donor and acceptor groups from polar side chains can satisfy hydrogen bonds, providing recognition of the edges of specific bases.

Variation in OB-Fold Loops and Nucleic Acid Recognition

In the work that initially defined the OB-fold, Murzin (57) highlighted three variable loops from the OB-fold that form the canonical ligand recognition surface: L_{12}, $L_{3\alpha}$ (or L_{34} when no capping α-helix is present), and L_{45} (Figure 1). However, subsequent studies have demonstrated that the OB-fold's loop repertoire also includes L_{23}, as seen in *Ec*Rho where the loop's length is rather modest (Figure 3*j* in green), or as seen in the greatly extended L_{23} appendages of Cdc13, *Ec*SSB, and S17, all of which are functionally used to significantly expand the ssDNA-binding surface (in green in the upper right of Figures 3*b*,*g*,*n*, respectively). As illustrated in Figure 3, the relative sizes of these loops are highly variable, and the OB-folds use them to great advantage in forming diverse nucleic acid interaction surfaces. This variability in loop size contributes to the large range of surface area buried between the OB-folds and their nucleic acid–binding partners upon binding, as listed in Table 1. The variation is roughly correlated to the number of nucleotides recognized by each fold, which varies from 2 to 11 for the single-stranded nucleic acid–binding proteins to an insuperable cluster of 31 for the ribosomal protein S17 (Figure 3*n*).

Ligand-Binding Surface

The canonical OB-fold-binding surface was initially defined based upon a careful analysis of five OB-fold proteins: staphylococcal nuclease, *Sc*AspRS, and the B-subunits of three bacterial cytotoxins (57). These five OB-folds bind their ligands centered upon β-strands 2 and 3, with additional contributions made from the C-terminal portions of $\beta 1$ and $\beta 5$. In the intervening years, 14 new complexes of nucleic acid–binding OB-folds have been solved. Although the binding surface is larger and more variable than initially characterized, its general position is remarkably constant among the various representatives of this family. In Figure 3, the 14 OB-fold complexes are shown superimposed on the N-terminal DNA-binding OB-fold domain of human RPA. As can be clearly seen, the great majority of nucleic acid partners are found on the left half of the OB-fold, situated near loops L_{23}, L_{12}, $L_{3\alpha}$, and L_{45}, and nestled against the protein face constructed primarily from $\beta 2$ and $\beta 3$. As originally noted, the N-terminal first half of $\beta 1$, before the kink that enables it to wrap around the β-barrel, rarely interacts with the ligand. Similarly, the N-terminal portion of $\beta 4$ appears to chiefly provide a scaffolding function, as it is only observed to contact ssDNA in the unusual case of the second

OB-fold in the α-subunit of *On*TEBP (β4 is shown in yellow behind the OB-fold in Figure 3*c*).

Conformational Changes upon Binding

High-resolution structures of 8 of the 11 OB-fold proteins described here have been solved in the absence of nucleic acid, allowing for an analysis of conformational changes that occur upon complex formation. A wide variety of conformational changes upon binding is observed in OB-fold complexes, ranging from virtually no change at all to dramatic movements of linked OB-fold domains, protein-loop structuring and closure, ordering of single-stranded nucleic acids, and base flipping. In nearly all these systems, either the OB-fold protein or the nucleic acid adopts a different structure depending upon the binding state, implying a cofolding event in at least one of the binding partners concurrent with complex formation. Williamson (78) has noted the ubiquity of cofolding in RNA-binding proteins and has proposed that it is a biological mechanism used to clearly distinguish free macromolecular partners from those bound in a specific complex. For example, overall the bound and unbound forms of the *hs*RPA70 OB-fold domains are similar, and the protein in the free and bound structures superimpose with an RMSD of ~1 Å. However, DNA binding causes reorientation of two RPA70 OB-folds with respect to one another, as well as conformational change in L_{12} and L_{45}. These changes tighten the DNA-binding cleft and allow the DNA to interact with the tandem OB-folds in a relatively extended conformation with the standard polarity (Figure 4*a*, see color insert). The large reorientation of the two RPA domains with respect to each other may provide a clear signal indicating that ssDNA is bound to the protein.

In the case of the *Ec*Rho transcriptional terminator, L_{12} is unstructured in the unbound form and closes down on the ssRNA upon binding (refer to Figure 4*b*). In contrast, L_{45} undergoes little conformational change, and the relative rigidity of L_{45} has been proposed to contribute to Rho's preference for pyrimidines because a larger purine could not be accommodated without significant restructuring of the loop (13).

In several cases, the nucleic acid undergoes a more pronounced conformational change. While in the AspRS and IF1 complexes the protein undergoes little conformational change, the RNA nucleotides are bulged out of helices or removed from RNA stacking interactions when recognized by the OB-folds. For example, binding by *Ec*AspRS to $tRNA^{Asp}$ removes the three anticodon bases from a stacking arrangement in the tRNA anti-codon loop and stacks them against phenylalanine side chains [see F48 marked by the gray ampersand in Figure 5*a* (see color insert)]. In addition to the obvious function of tRNA identification, this alternative bound conformation may be a signal to other translational proteins that the tRNA is unavailable for other functions, since, while bound by AspRS, it is in the process of being charged.

In the *On*TEBP complex, the ssDNA follows an irregular, contorted path, bound in a cleft formed by the junction of three OB-folds. It is highly unlikely that the ssDNA adopts a similar conformation when in the free state. Examination of the ternary complex structure reveals no clear pathway by which the ssDNA could enter its binding site unless there were also conformational changes in the protein. However, the crystal structure of the N-terminal domain of the *On*TEBP α-subunit in the absence of ssDNA shows only modest rearrangements of protein side chains compared with the bound form (RMSD = 0.43 Å), indicating that at least for α, the OB-fold-binding sites are largely preformed. In the transition between the α_2:ssDNA complex and the α:β:ssDNA ternary complex, the ssDNA undergoes a four-nucleotide register shift, indicating that β binding induces a large conformational change in the ssDNA. Unlike what is observed in the α_2:ssDNA complex, in the ternary complex, the 3′-end of the ssDNA is sequestered deep in the complex. This conformational change in the ssDNA has been proposed to be a signal that regulates telomerase activity, indicating that the chromosomal end is of the proper length (29).

Binding Modularity

The utilization of OB-folds for nucleic acid recognition is surprisingly modular (Figure 1). Often the OB-fold is just one domain of a much larger protein, as observed in RecG, Cdc13, and AspRS. In other cases, single proteins composed of several OB-fold domains are used in concert to distinguish regions of single-stranded nucleic acid, such as with the human RPA protein and the telomeric protein *On*TEBP. In the case of hexameric *Ec*Rho and tetrameric *Ec*SSB, homo-oligomers of OB-fold protein monomers coordinately bind sizeable regions of single-stranded nucleic acid. Finally, in the ribosome, compact, lone OB-folds comprise the S17, S12, and IF1 proteins, which do not multimerize but rather work as integral components within an expansive assembly fabricated from variously structured RNA domains and other relatively diminutive proteins (Figures 1*h,i*).

Structural and Sequence Conservation

The nucleic acid–binding OB-fold family is renown for the lack of any discernable sequence similarity among its members. To delineate sequence determinants that may confer common structural elements, 15 nucleic acid–binding OB-folds from the 11 complexes analyzed here were globally aligned with a multiple structural alignment algorithm (70). The OB-fold protein gp32 was also included in this analysis because the site of ssDNA binding is known from observed electron density in the crystal structure (the ssDNA is disordered and is not included in the PDB coordinate deposition) (72). The cores of the 14 OB-folds align with an RMSD of 2.1 Å over about 30 residues contained in the conserved secondary structure. As illustrated in Figure 5, even after structure-based alignment, the average sequence identity is only 12 ± 5% over the canonical OB-fold secondary structural elements (as determined by pairwise distance analysis for 52 amino acids, excluding unalignable gaps).

Nevertheless, several clear patterns emerge from the alignment. Hydrophobic residues in alternating amino acid positions are conserved for short stretches of sequence (in Figure 5*b*, completely conserved hydrophobicity is indicated by the gray columns, whereas 75% conserved hydrophobicity is indicated by yellow columns). This pattern is consistent with the structural features of OB-fold β-barrels because ideally every other residue of the component β-strands points inward and packs in the interior. This pattern is especially evident in strands $\beta1$, $\beta2$, and $\beta4$, with strand $\beta5$ being the most variable.

The interior residues of the OB-fold domain are arranged in three layers (here designated top, middle, and bottom with reference to the Murzin view of the OB-fold), with each β-strand generally donating one side chain to each layer (57). The middle layer, shown in violet in Figure 5*a*, is the most consistently conserved in terms of hydrophobicity, followed next by the bottom layer (shown in dark blue), and last by the top layer (shown in orange). Accordingly, the top layer of the OB-fold is often uncapped by secondary structures and is more exposed to solvent, in comparison to the bottom layer that is usually capped by the canonical α-helix or the β-strands that bridge strands $\beta3$ and $\beta4$ (shown in green in Figure 1). Glycines and prolines (colored green in both the structure and sequence alignment of Figure 5) are rare within β-strands yet are frequently found on either side of the strands where they may break the regular secondary structure and facilitate specific, yet irregular, conformations of the interstrand loops. As discussed previously, $\beta1$ is generally long, wrapping around the β-barrel, and contributes to both orthogonal β-sheets (shown in red in Figure 1). Consistent with previous observations (17, 57), a conspicuous kink, break, or β-bulge is usually seen in the middle of strand $\beta1$, and the characteristic glycine, which allows for this structural idiosyncrasy, is clear (marked by an asterisk). A prominent, conserved turn is found at the abrupt beginning of strand $\beta4$, indicated by the blue wedges in Figure 5*b*. Within this turn, another fairly well-conserved glycine is found (marked with a double asterisk), which again likely serves to initiate the β-strand by breaking the canonical helix found between strands $\beta3$ and $\beta4$.

Two frequently occurring hydrophobic residues, one N-terminal to the bottom residue of strand $\beta2$ and one C-terminal to strand $\beta5$, further buttress the bottom interior packing layer of the OB-fold and often pack against either the α-helix or the loop/β-strand that caps the base of the OB-barrel (these two residues are colored light blue in Figure 5*a* and indicated by light blue "xB" text in Figure 5*b*). Interestingly, the side chain found between the middle and bottom internal layer residues in strand $\beta3$ is usually hydrophobic (colored gray in the structure of Figure 5*a* and marked by an ampersand). Due to the inside/outside alternating nature of the residues in the β-strands of the OB-barrel, this nonpolar side chain must be pointing toward solvent. Being situated in the middle of the standard OB-binding surface, this residue is thus a likely candidate for hydrophobic packing interactions with the nucleic acid ligands. Indeed, in two thirds of these complexes this residue is observed interacting with nucleic acid, sometimes in a critical position such as that seen in *Ec*AspRS,where it is a phenylalanine that stacks on Q34 in the tRNA anticodon loop (shown in the highlighted structure of Figure 5*a*). Finally, the structural alignment indicates that in nearly half of the OB-folds, there is an α-helix just N-terminal to strand $\beta 1$ (shown in white in Figure 1), a common feature that has only come to light with the availability of recent structural data.

Ligand Polarity and Divergent Evolution

An intriguing aspect of OB-fold nucleic acid recognition is the extraordinary conservation of ligand-binding polarity. In 11 of 13 complexes for which the orientation is clearly discernable, the nucleic acid binds in the standard polarity, with the 5′-end directed toward strands $\beta 4$ and $\beta 5$ and the 3′-end directed toward $\beta 2$ (ignoring nonspecific contacts to only phosphodiester groups, grooves of RNA helices, and solitary nucleotides). The two exceptions to this rule are the *On*TEBP β-subunit and *Ec*SSB. A longstanding question regarding OB-folds is whether the current representatives have arisen independently during evolution or whether they are related by divergence from a common origin (57, 58, 74). Nucleic acid polarity may serve as an arbiter for this matter. No apparent biophysical reason exists for why OB-folds would prefer one polarity to the other, and the observation that *On*TEBP β and *Ec*SSB bind ssDNA with the nonstandard polarity is prima facie evidence that OB-folds are indeed physically capable of binding with either polarity. Furthermore, although possible, it is unlikely that gradual divergence could easily reverse the orientation of the nucleic acid in the OB-fold-binding cleft once a given polarity preference was set. If both polarity preferences are judged equally probable a priori, the random chance of 11 or more of 13 OB-folds arising independently with the same polarity is 0.023, which by statistical convention is a significant result against the independent origin hypothesis.

METHODS

Intermolecular distances and residue contacts were determined with MOLEMAN2 (46). Pairwise structural superimpositions were performed using LSQMAN (45), and multiple protein structural alignments were performed with STAMP v. 4.2 (70). Buried surface areas were calculated with NACCESS v. 2.1.1 (37). Figures were prepared using MolScript v. 2.1.2 (48) with Raster3D (55), and MOLMOL (47). Sequence distance analyses were performed with PAUP 4.0b10 (75).

ACKNOWLEDGMENTS

The authors are grateful to Olke Uhlenbeck for helpful discussions and Leslie Glustrom and Emily Anderson for critical reading of the manuscript. We thank the NIH (GM59414) and the Arnold and Mabel Beckman Foundation for support. R.M. Mitton-Fry is a Howard Hughes Institute Predoctoral Fellow.

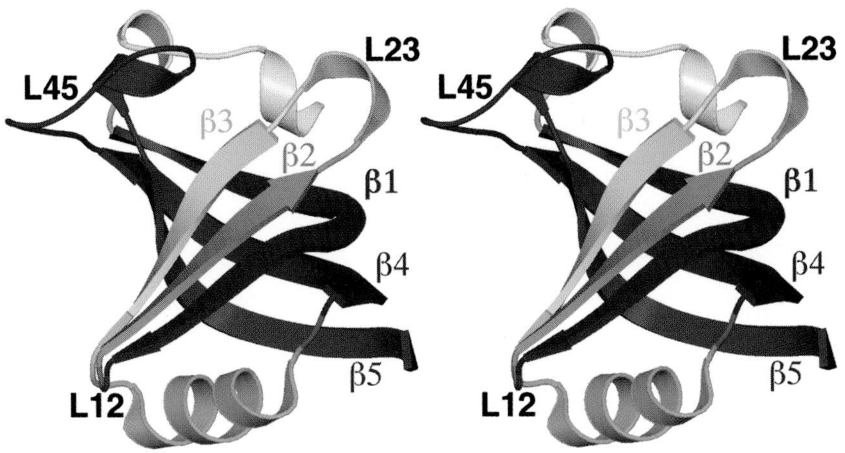

Figure 1 The canonical OB-fold domain. The OB-fold from AspRS is shown in stereo as representative of the ideal OB-fold domain. From the N terminus to the C terminus, strand β1 is shown in red, β2 in orange, β3 in yellow, the helix between β3 and β4 in green, β4 in blue, and β5 in violet. An α-helix, which is found in half of the OB-folds in these complexes, is shown in white at the top of the figure, just N-terminal to strand β1. Variable loops between strands are indicated in black text.

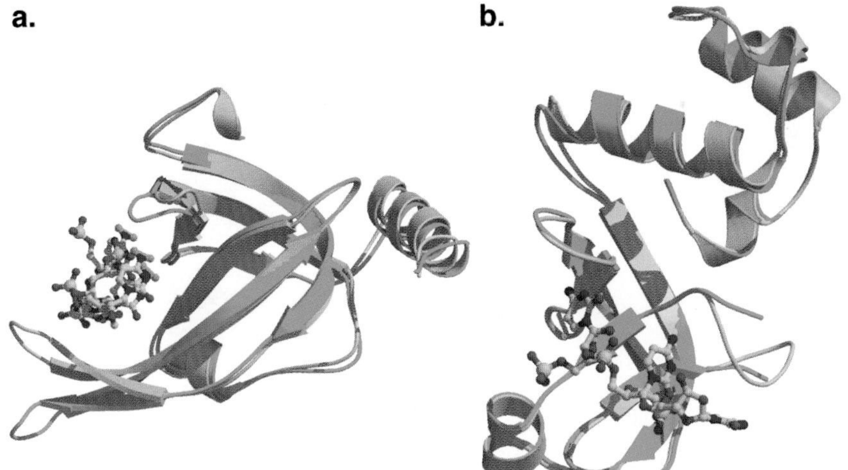

Figure 4 Conformational change upon ligand binding. The bound OB-fold complex is shown in slate blue, while the unbound protein is shown in green. (*a*) The N-terminal OB-fold of human RPA. (*b*) The OB-fold from *Ec*Rho transcriptional terminator.

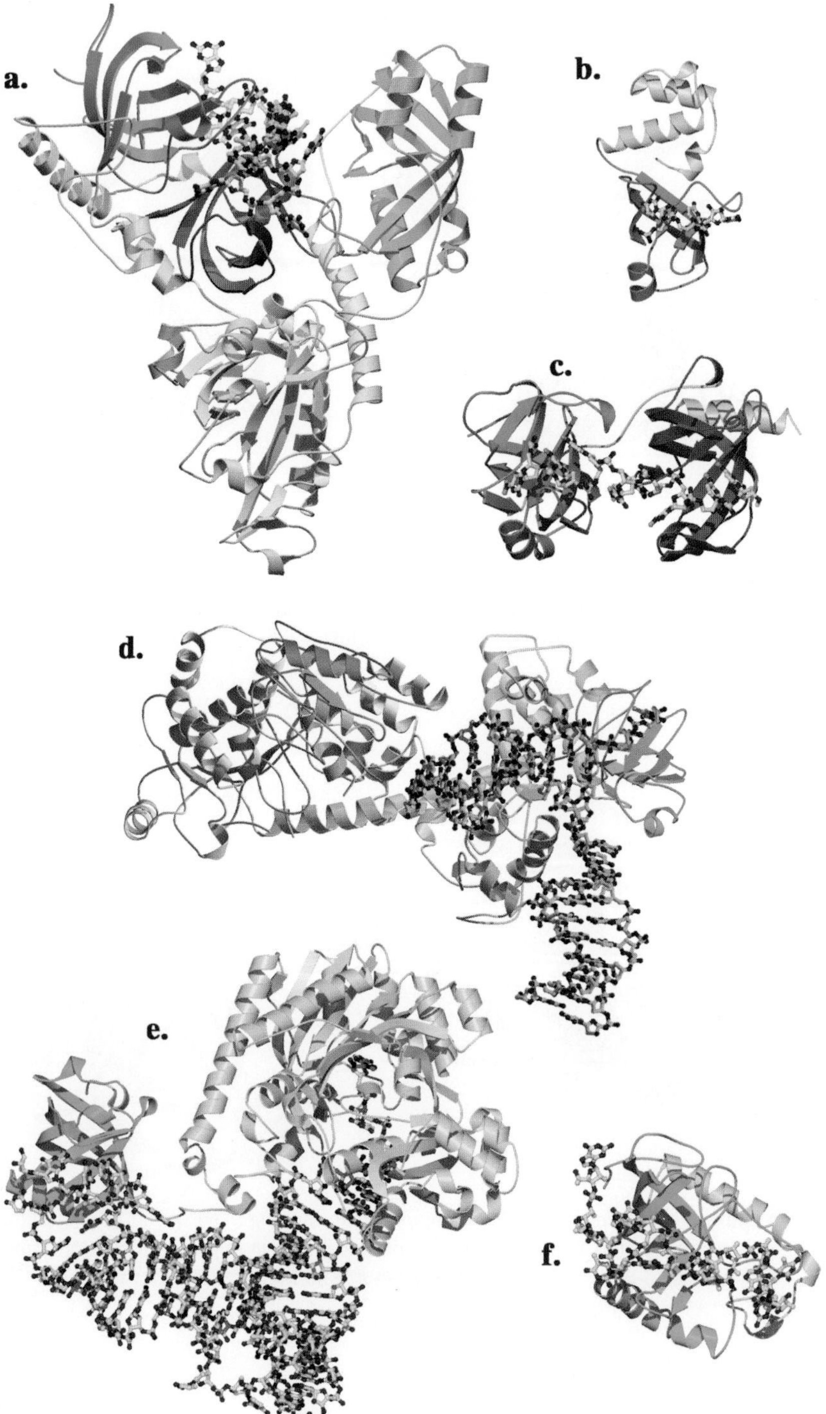
a.
b.
c.
d.
e.
f.

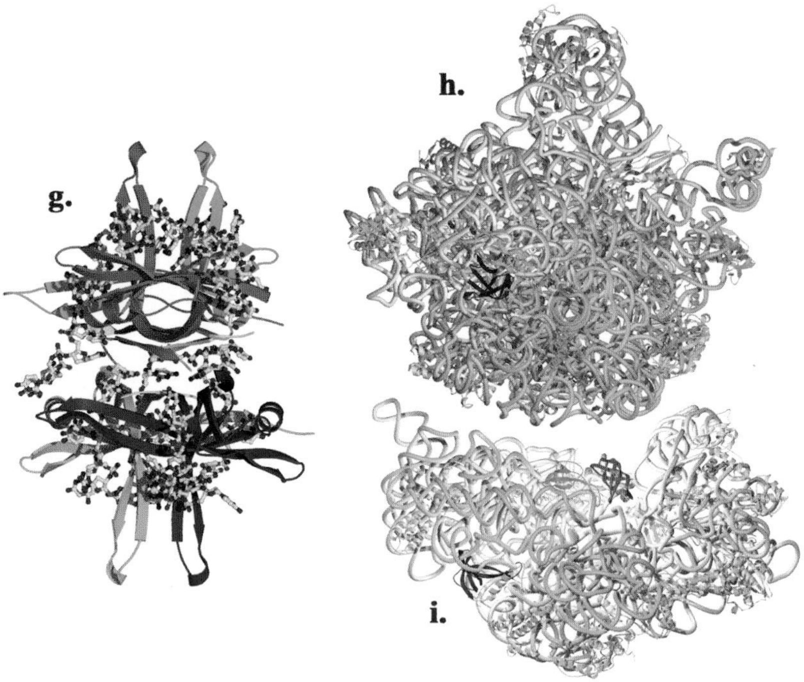

Figure 2 Structures of OB-fold/nucleic acid complexes. The high-resolution structures of several OB-fold proteins bound to nucleic acids. The individual OB-fold domains are highlighted in rainbow colors to illustrate the modularity of the domain. (*a*) *On*TEBP ternary complex, (*b*) *Ec*Rho, (*c*) human RPA, (*d*) RecG, (*e*) *Ec*AspRS, (*f*) Cdc13, (*g*) *Ec*SSB, (*h*) L2 in the large subunit of the ribosome, (*i*) S12 (green), S17 (blue), and IF1 (magenta) in the ribosomal small subunit.

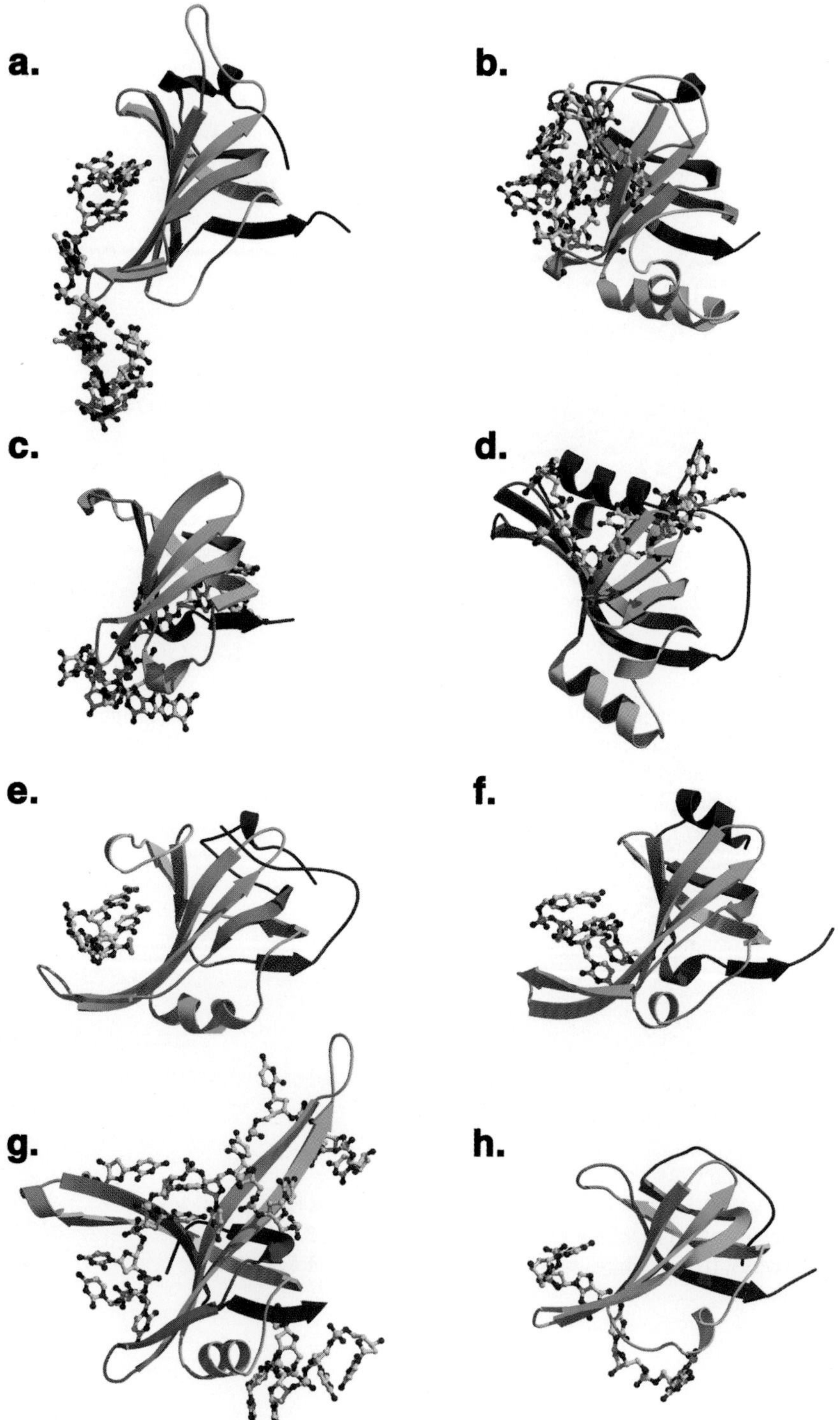
a.
b.
c.
d.
e.
f.
g.
h.

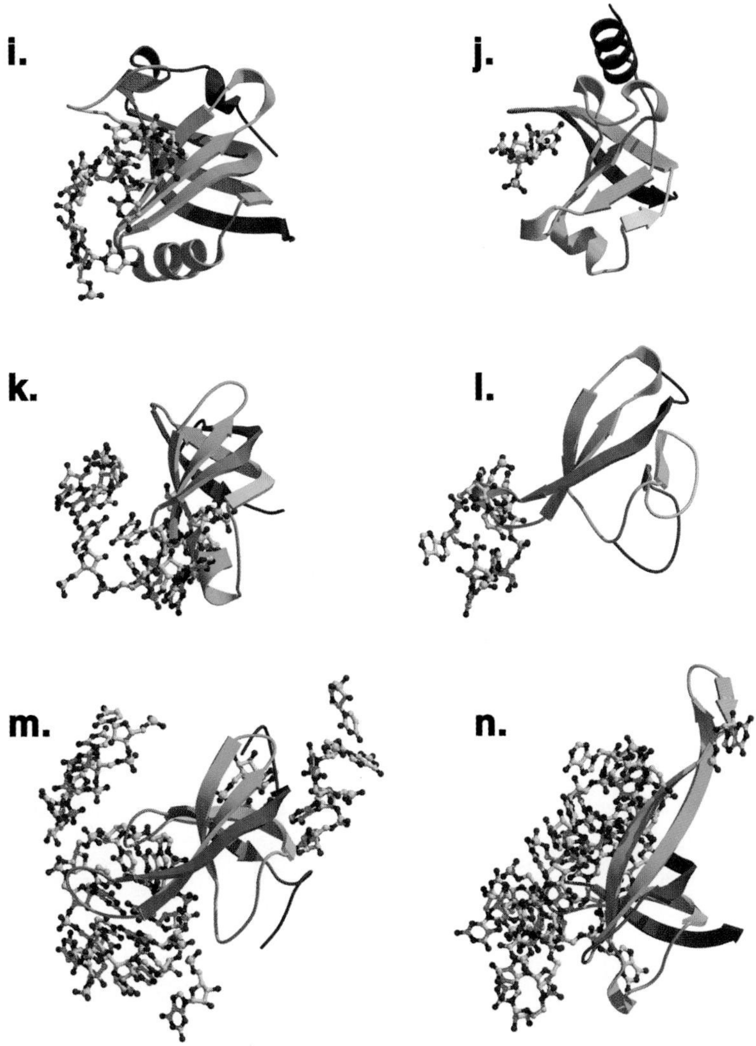

Figure 3 (*opposite, above*) Comparison of ligand binding in the OB-fold domains. The 14 independent OB-fold domains are depicted in a common orientation based on superimposition with the N-terminal OB-fold of RPA. Secondary structure is rainbow colored beginning with violet at the N terminus and ending with red at the C terminus. Nucleic acids that are within 3.5 Å of the relevant fold are rendered as ball-and-stick figures. OB-folds were aligned with LSQMAN. (*a*) *On*TEBP α1, (b) Cdc13, (*c*) *On*TEBP α2, (*d*) *On*TEBP β, (*e*) RPA-A, (*f*) RPA-B, (*g*) *Ec*SSB, (*h*) RecG, (*i*) *Ec*AspRS, (*j*) *Ec*Rho, (*k*) IF1, (*l*) L2, (*m*) S12, (*n*) S17.

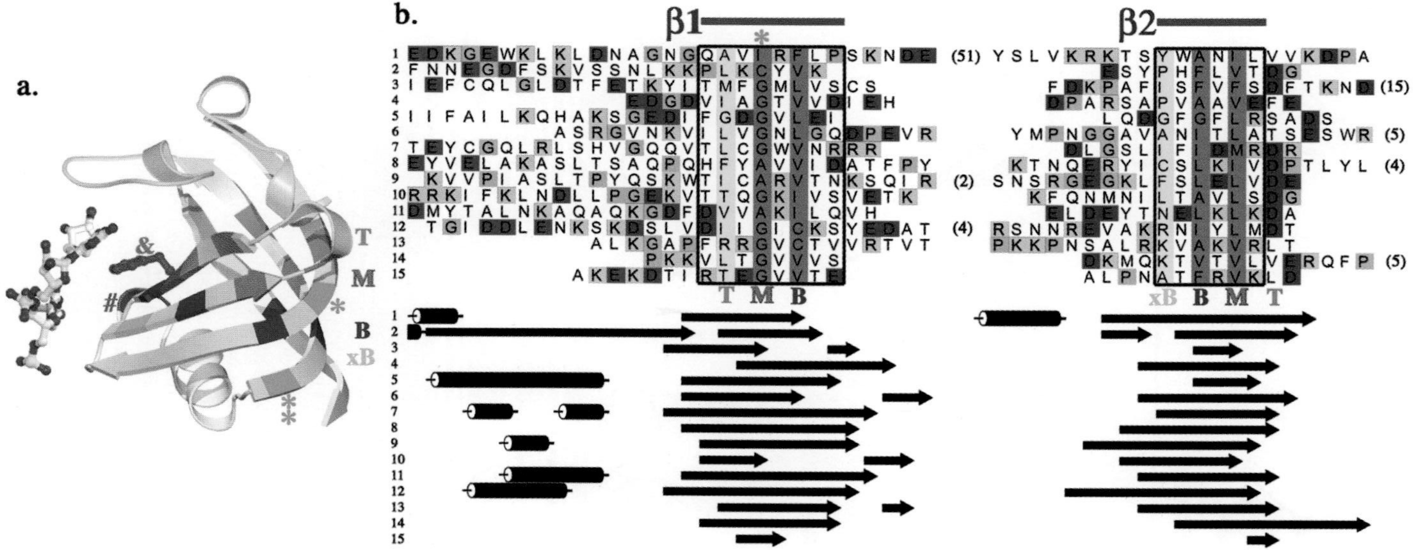

Figure 5 Structure-based sequence alignment of OB-fold domains. (*a*) The AspRS OB-fold is depicted with key residues highlighted in color. Hydrophobic residues that pack in the top layer of the barrel's interior are shown in orange, the middle layer in violet, the bottom layer in dark blue, and the "extra-bottom" layer in light blue. Two glycines are shown in green, while the conspicuous solvent-exposed hydrophobic side chain in strand β3 is shown in dark gray and rendered as a ball-and-stick figure. The DNA that interacts with this residue is rendered as ball-and-stick in CPK colors. A conserved polar residue found after strand β3 is highlighted in magenta.

b. (continued)

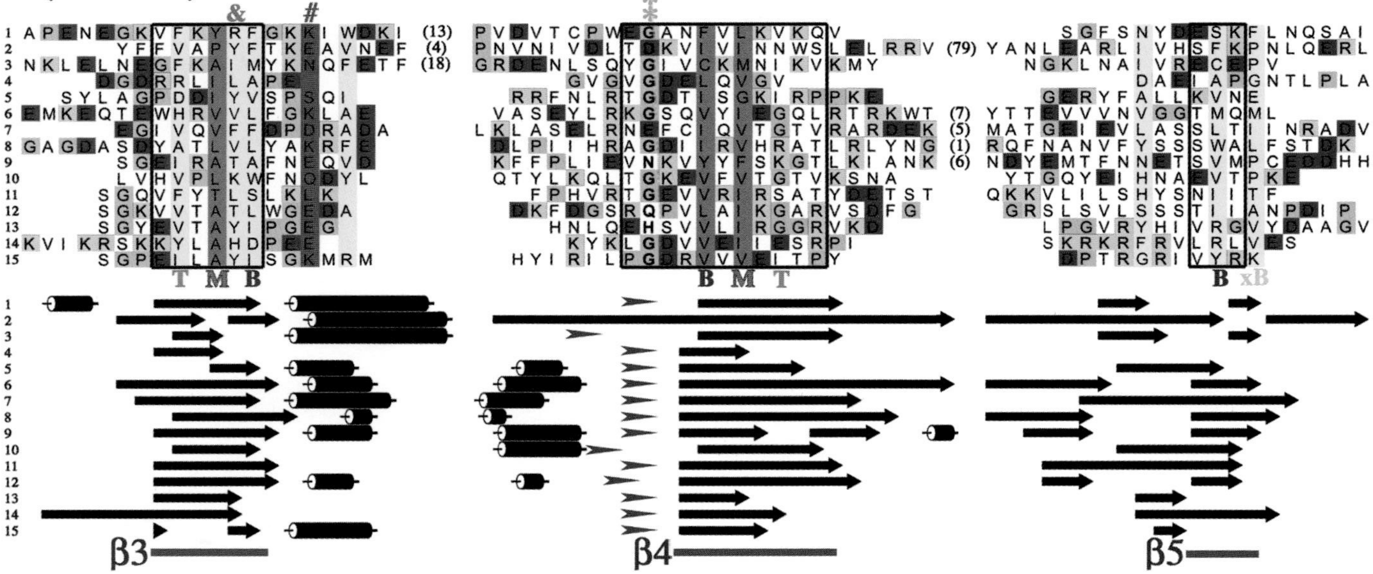

Figure 5 (continued) (*b*) In color, a structure-based sequence alignment of the OB-folds is augmented below with the corresponding secondary structures in black and white. Variable loop regions have, for the most part, been omitted and their length is indicated by number of amino acids in parentheses. Helices are shown as black cylinders, β-strands are shown as bars with arrows, and turns are shown as blue wedges. Regions of secondary structure significantly conserved among the OB-folds are boxed in thick black lines. Completely conserved hydrophobic residues are highlighted with gray columns, while positions with 75% conserved hydrophobicity are highlighted with yellow columns. A conserved polar position after strand β3 is highlighted with a magenta column. Strands are indicated above and below the alignment in blue. An orange T, a violet M, and a dark blue B indicate the top, middle, and bottom interior residues, respectively. The additional bottom residues are indicated by light blue xB text. Proline residues are colored green, acidic residues (aspartate and glutamate) are colored red, and basic residues (lysine and arginine) are colored blue. Asterisks (*) indicate two well-conserved glycines, and an ampersand (&) indicates a conspicuous solvent-exposed hydrophobic residue. [1] gp32, [2] *On*TEBP β, [3] Cdc13, [4] L2, [5] Rho, [6] *Ec*SSB, [7] *Ec*AspRS, [8] *On*TEBP α1, [9] RPA-A, [10] RecG, [11] *On*TEBP α2, [12] RPA-B, [13] S12, [14] S17, [15] IF1.

NOTE ADDED IN PROOF

Following the preparation of this review, the high-resolution structure of the conserved C-terminal domain of BRCA-2 complexed with DSS1 in the presence and absence of ssDNA was reported (Yang H, Jeffrey PD, Miller J, Kinnucan E, Sun Y, et al. 2002. BRCA2 function in DNA binding and recombination from a BRCA2-DSS1-ssDNA structure. *Science* 297:1837–48) (1MJE, 1MIU, 1IYJ). This protein contains a tandem array of three OB-folds, two of which are seen to interact with ssDNA. The DNA-binding interface is similar to that of RPA, and the ssDNA binds in the standard polarity defined here.

The *Annual Review of Biophysics and Biomolecular Structure* is online at http://biophys.annualreviews.org

LITERATURE CITED

1. Allison TJ, Wood TC, Briercheck DM, Rastinejad F, Richardson JP, Rule GS. 1998. Crystal structure of the RNA-binding domain from transcription termination factor rho. *Nat. Struct. Biol.* 5:352–56
2. Anderson EM, Halsey WH, Wuttke DS. 2002. Delineation of the high-affinity single-stranded telomeric DNA-binding domain of *S. cerevisiae* Cdc13. *Nucleic Acids Res.* 30:4305–13
3. Anderson EM, Halsey WH, Wuttke DS. 2002. Site-directed mutagenesis reveals the thermodynamic requirements for single-stranded DNA recognition by the telomere-binding protein Cdc13. *Biochemistry* 42: In press
4. Antson AA. 2000. Single stranded RNA binding proteins. *Curr. Opin. Struct. Biol.* 10:87–94
5. Ban N, Nissen P, Hansen J, Moore PB, Steitz TA. 2000. The complete atomic structure of the large ribosomal subunit at 2.4 Å resolution. *Science* 289:905–20
6. Bastin-Shanower SA, Brill SJ. 2001. Functional analysis of the four DNA binding domains of replication protein A. The role of RPA2 in ssDNA binding. *J. Biol. Chem.* 276:36446–53
7. Battiste JL, Pestova TV, Hellen CUT, Wagner G. 2000. The eIF1A solution structure reveals a large RNA-binding surface important for scanning function. *Mol. Cell* 5:109–19
8. Berthet-Colominas C, Seignovert L, Härtlein M, Grotli M, Cusack S, Leberman R. 1998. The crystal structure of asparaginyl-tRNA synthetase from *Thermus thermophilus* and its complexes with ATP and asparaginyl-adenylate: the mechanism of discrimination between asparagine and aspartic acid. *EMBO J.* 17:2947–60
9. Bochkareva E, Belegu V, Korolev S, Bochkarev A. 2001. Structure of the major single-stranded DNA-binding domain of replication protein A suggests a dynamic mechanism for DNA binding. *EMBO J.* 20: 612–18
10. Bochkarev A, Bochkareva E, Frappier L, Edwards AM. 1999. The crystal structure of the complex of replication protein A subunits RPA32 and RPA14 reveals a mechanism for single-stranded DNA binding. *EMBO J.* 18:4498–504
11. Bochkareva E, Korolev S, Lees-Miller SP, Bochkarev A. 2002. Structure of the RPA trimerization core and its role in the multistep DNA-binding mechanism of RPA. *EMBO J.* 21:1855–63
12. Bochkarev A, Pfuetzner RA, Edwards AM, Frappier L. 1997. Structure of the single-stranded-DNA-binding domain of

replication protein A bound to DNA. *Nature* 385:176–81

13. Bogden CE, Fass D, Bergman N, Nichols MD, Berger JM. 1999. The structural basis for terminator recognition by the rho transcription termination factor. *Mol. Cell* 3:487–93
14. Brennan CA, Dombroski AJ, Platt T. 1987. Transcription termination factor rho is an RNA-DNA helicase. *Cell* 48:945–52
15. Briercheck DM, Wood TC, Allison TJ, Richardson JP, Rule GS. 1998. The NMR structure of the RNA binding domain of *E. coli* rho factor suggests possible RNA-protein interactions. *Nat. Struct. Biol.* 5:393–99
16. Brodersen DE, Clemons WM Jr, Carter AP, Wimberly BT, Ramakrishnan V. 2002. Crystal structure of the 30S ribosomal subunit from *Thermus thermophilus*: structure of the proteins and their interactions with 16S RNA. *J. Mol. Biol.* 316:725–68
17. Bycroft M, Hubbard TJP, Proctor M, Freund SMV, Murzin AG. 1997. The solution structure of the S1 RNA binding domain: a member of an ancient nucleic acid-binding fold. *Cell* 88:235–42
18. Carter AP, Clemons WM Jr, Brodersen DE, Morgan-Warren RJ, Hartsch T, et al. 2001. Crystal structure of an initiation factor bound to the 30S ribosomal subunit. *Science* 291:498–501
19. Cavarelli J, Rees B, Ruff M, Thierry J-C, Moras D. 1993. Yeast $tRNA^{Asp}$ recognition by its cognate class II aminoacyl-tRNA synthetase. *Nature* 362:181–84
20. Chédin F, Seitz EM, Kowalczykowski SC. 1998. Novel homologs of replication protein A in archaea: implications for the evolution of ssDNA-binding proteins. *Trends Biochem. Sci.* 23:273–77
21. Classen S, Ruggles JA, Schultz SC. 2001. Crystal structure of the N-terminal domain of *Oxytricha nova* telomere end-binding protein α subunit both uncomplexed and complexed with telomeric ssDNA. *J. Mol. Biol.* 314:1113–25
22. Commans S, Plateau P, Blanquet S, Dardel F. 1995. Solution structure of the anticodon-binding domain of *Escherichia coli* lysyl-tRNA synthetase and studies of its interaction with $tRNA^{Lys}$. *J. Mol. Biol.* 253:100–13
23. Diedrich G, Spahn CMT, Stelzl U, Schäfer MA, Wooten T, et al. 2000. Ribosomal protein L2 is involved in the association of the ribosomal subunits, tRNA binding to A and P sites and peptidyl transfer. *EMBO J.* 19:5241–50
24. Dolan JW, Marshall NF, Richardson JP. 1990. Transcription termination factor rho has three distinct structural domains. *J. Biol. Chem.* 265:5747–54
25. Dombroski AJ, Platt T. 1988. Structure of rho factor: an RNA-binding domain and a separate region with strong similarity to proven ATP-binding domains. *Proc. Natl. Acad. Sci. USA* 85:2538–42
26. Egebjerg J, Christiansen J, Garrett RA. 1991. Attachment sites of primary binding proteins L1, L2 and L23 on 23S ribosomal RNA of *Escherichia coli*. *J. Mol. Biol.* 222:251–64
27. Eiler S, Dock-Bregeon A-C, Moulinier L, Thierry J-C, Moras D. 1999. Synthesis of aspartyl-$tRNA^{Asp}$ in *Escherichia coli*—a snapshot of the second step. *EMBO J.* 18: 6532–41
28. Evans SK, Lundblad V. 1999. Est1 and Cdc13 as comediators of telomerase access. *Science* 286:117–20
29. Froelich-Ammon SJ, Dickinson BA, Bevilacqua JM, Schultz SC, Cech TR. 1998. Modulation of telomerase activity by telomere DNA-binding proteins in *Oxytricha*. *Genes Dev.* 12:1504–14
30. Garvik B, Carson M, Hartwell L. 1995. Single-stranded DNA arising at telomeres in *cdc13* mutants may constitute a specific signal for the *RAD9* checkpoint. *Mol. Cell Biol.* 15:6128–38
31. Golden BL, Hoffman DW, Ramakrishnan V, White SW. 1993. Ribosomal protein S17: characterization of the three-dimensional structure by ^{1}H NMR and ^{15}N NMR. *Biochemistry* 32:12812–20

32. Goldgur Y, Mosyak L, Reshetnikova L, Ankilova V, Lavrik O, et al. 1997. The crystal structure of phenylalanyl-tRNA synthetase from *Thermus thermophilus* complexed with cognate $tRNA^{Phe}$. *Structure* 5: 59–68
33. Gottschling DE, Zakian VA. 1986. Telomere proteins: specific recognition and protection of the natural termini of *Oxytricha* macronuclear DNA. *Cell* 47:195–205
34. Harms J, Schluenzen F, Zarivach R, Bashan A, Gat S, et al. 2001. High resolution structure of the large ribosomal subunit from a mesophilic eubacterium. *Cell* 107:679–88
35. Horvath MP, Schultz SC. 2001. DNA G-quartets in a 1.86 Å resolution structure of an *Oxytricha nova* telomeric protein-DNA complex. *J. Mol. Biol.* 310:367–77
36. Horvath MP, Schweiker VL, Bevilacqua JM, Ruggles JA, Schultz SC. 1998. Crystal structure of the *Oxytricha nova* telomere end binding protein complexed with single strand DNA. *Cell* 95:963–74
37. Hubbard SJ, Thornton JM. 1993. NACCESS, computer program. London: Dep. Biochem. Mol. Biol., University College
38. Hughes TR, Weilbaecher RG, Walterscheid M, Lundblad V. 2000. Identification of the single-strand telomeric DNA binding domain of the *Saccharomyces cerevisiae* Cdc13 protein. *Proc. Natl. Acad. Sci. USA* 97:6457–62
39. Iftode C, Daniely Y, Borowiec JA. 1999. Replication protein A (RPA): the eukaryotic SSB. *Crit. Rev. Biochem. Mol. Biol.* 34:141–80
40. Jacobs DM, Lipton AS, Isern NG, Daughdrill GW, Lowry DF, et al. 1999. Human replication protein A: Global fold of the N-terminal RPA-70 domain reveals a basic cleft and flexible C-terminal linker. *J. Biol. NMR* 14:321–31
41. Jaishree TN, Ramakrishnan V, White SW. 1996. Solution structure of prokaryotic ribosomal protein S17 by high-resolution NMR spectroscopy. *Biochemistry* 35:2845–53
42. Kelly TJ, Simancek P, Brush GS. 1998. Identification and characterization of a single-stranded DNA-binding protein from the archaeon *Methanococcus jannaschii*. *Proc. Natl. Acad. Sci. USA* 95:14634–39
43. Khaitovich P, Mankin AS, Green R, Lancaster L, Noller HF. 1999. Characterization of functionally active subribosomal particles from *Thermus aquaticus*. *Proc. Natl. Acad. Sci. USA* 96:85–90
44. Kim C, Snyder RO, Wold MS. 1992. Binding properties of replication protein A from human and yeast cells. *Mol. Cell Biol.* 12:3050–59
45. Kleywegt GJ. 1996. Use of non-crystallographic symmetry in protein structure refinement. *Acta Crystallogr. D* 52:842–57
46. Kleywegt GJ. 1997. Validation of protein models from C^{α} coordinates alone. *J. Mol. Biol.* 273:371–75
47. Koradi R, Billeter M, Wüthrich K. 1996. MOLMOL: a program for display and analysis of macromolecular structures. *J. Mol. Graph.* 14:51–55
48. Kraulis PJ. 1991. MOLSCRIPT: a program to produce both detailed and schematic plots of protein structures. *J. Appl. Crystallogr.* 24:946–50
49. Li W, Hoffman DW. 2001. Structure and dynamics of translation initiation factor aIF-1A from the archaeon *Methanococcus jannaschii* determined by NMR spectroscopy. *Protein Sci.* 10:2426–38
50. Lin J-J, Zakian VA. 1996. The *Saccharomyces CDC13* protein is a single-strand $TG_{1\text{-}3}$ telomeric DNA-binding protein in vitro that affects telomere behavior in vivo. *Proc. Natl. Acad. Sci. USA* 93:13760–65
51. Lohman TM, Ferrari ME. 1994. *Escherichia coli* single-stranded DNA-binding protein: multiple DNA-binding modes and cooperativities. *Annu. Rev. Biochem.* 63: 527–70
52. McGlynn P, Lloyd RG. 2000. Modulation of RNA polymerase by (p)ppGpp reveals a RecG-dependent mechanism for replication fork progression. *Cell* 101:35–45
53. McGlynn P, Lloyd RG. 2001. Rescue of stalled replication forks by RecG:

Simultaneous translocation on the leading and lagging strand templates supports an active DNA unwinding model of fork reversal and Holliday junction formation. *Proc. Natl. Acad. Sci. USA* 98:8227–34

54. McGlynn P, Mahdi AA, Lloyd RG. 2000. Characterisation of the catalytically active form of RecG helicase. *Nucleic Acids Res.* 28:2324–32
55. Merritt EA, Bacon DJ. 1997. Raster3D: photorealistic molecular graphics. *Methods Enzymol.* 277:505–24
56. Mitton-Fry RM, Anderson EM, Hughes TR, Lundblad V, Wuttke DS. 2002. Conserved structure for single-stranded telomeric DNA recognition. *Science* 296:145–47
57. Murzin AG. 1993. OB (*o*ligonucleotide/*o*ligosaccharide *b*inding)-fold: common structural and functional solution for non-homologous sequences. *EMBO J.* 12:861–67
58. Murzin AG. 1998. How far divergent evolution goes in proteins. *Curr. Opin. Struct. Biol.* 8:380–87
59. Murzin AG, Brenner SE, Hubbard T, Chothia C. 1995. SCOP: a structural classification of proteins database for the investigation of sequences and structures. *J. Mol. Biol.* 247:536–40
60. Nakagawa A, Nakashima T, Taniguchi M, Hosaka H, Kimura M, Tanaka I. 1999. The three-dimensional structure of the RNA-binding domain of ribosomal protein L2; a protein at the peptidyl transferase center of the ribosome. *EMBO J.* 18:1459–67
61. Nugent CI, Hughes TR, Lue NF, Lundblad V. 1996. Cdc13p: a single-strand telomeric DNA-binding protein with a dual role in yeast telomere maintenance. *Science* 274: 249–52
62. Ogle JM, Brodersen DE, Clemons WM Jr, Tarry MJ, Carter AP, Ramakrishnan V. 2001. Recognition of cognate transfer RNA by the 30S ribosomal subunit. *Science* 292:897–902
63. Peersen OB, Ruggles JA, Schultz SC. 2002. Dimeric structure of the *Oxytricha nova* telomere end-binding protein α-subunit bound to ssDNA. *Nat. Struct. Biol.* 9:182–87
64. Pennock E, Buckley K, Lundblad V. 2001. Cdc13 delivers separate complexes to the telomere for end protection and replication. *Cell* 104:387–96
65. Price CM, Cech TR. 1987. Telomeric DNA-protein interactions of *Oxytricha* macronuclear DNA. *Genes Dev.* 1:783–93
66. Pütz J, Puglisi JD, Florentz C, Giegé R. 1991. Identity elements for specific aminoacylation of yeast $tRNA^{Asp}$ by cognate aspartyl-tRNA synthetase. *Science* 252: 1696–99
67. Raghunathan S, Kozlov AG, Lohman TM, Waksman G. 2000. Structure of the DNA binding domain of *E. coli* SSB bound to ssDNA. *Nat. Struct. Biol.* 7:648–52
68. Raghunathan S, Ricard CS, Lohman TM, Waksman G. 1997. Crystal structure of the homo-tetrameric DNA binding domain of *Escherichia coli* single-stranded DNA-binding protein determined by multiwavelength x-ray diffraction on the selenomethionyl protein at 2.9-Å resolution. *Proc. Natl. Acad. Sci. USA* 94:6652–57
69. Ruff M, Krishnaswamy S, Boeglin M, Poterszman A, Mitschler A, et al. 1991. Class II aminoacyl transfer RNA synthetases: crystal structure of yeast aspartyl-tRNA synthetase complexed with $tRNA^{Asp}$. *Science* 252:1682–89
70. Russell RB, Barton GJ. 1992. Multiple protein sequence alignment from tertiary structure comparison: assignment of global and residue confidence levels. *Proteins* 14:309–23
71. Sette M, van Tilborg P, Spurio R, Kaptein R, Paci M, et al. 1997. The structure of the translational initiation factor IF1 from *E. coli* contains an oligomer-binding motif. *EMBO J.* 16:1436–43
72. Shamoo Y, Friedman AM, Parsons MR, Konigsberg WH, Steitz TA. 1995. Crystal structure of a replication fork single-stranded DNA binding protein (T4 gp32) complexed to DNA. *Nature* 376:362–66

73. Singleton MR, Scaife S, Wigley DB. 2001. Structural analysis of DNA replication fork reversal by RecG. *Cell* 107:79–89
74. Suck D. 1997. Common fold, common function, common origin? *Nat. Struct. Biol.* 4:161–65
75. Swofford DL. 2002. *PAUP* 4.0—Phylogenetic Analysis Using Parsimony (*and Other Methods)*. Sunderland, MA: Sinauer Assoc.
76. Wang Y, von Hippel PH. 1993. *Escherichia coli* transcription termination factor rho. II. Binding of oligonucleotide cofactors. *J. Biol. Chem.* 268:13947–55
77. Webster G, Genschel J, Curth U, Urbanke C, Kang C, Hilgenfeld R. 1997. A common core for binding single-stranded DNA: structural comparison of the single-stranded DNA-binding proteins (SSB) from *E. coli* and human mitochondria. *FEBS Lett.* 411:313–16
78. Williamson JR. 2000. Induced fit in RNA-protein recognition. *Nat. Struct. Biol.* 7: 834–37
79. Wimberly BT, Brodersen DE, Clemons WM Jr, Morgan-Warren RJ, Carter AP, et al. 2000. Structure of the 30S ribosomal subunit. *Nature* 407:327–39
80. Wold MS. 1997. Replication protein A: a heterotrimeric, single-stranded DNA-binding protein required for eukaryotic DNA metabolism. *Annu. Rev. Biochem.* 66: 61–92

Annu. Rev. Biophys. Biomol. Struct. 2003. 32:135–59
doi: 10.1146/annurev.biophys.32.110601.141732

First published online as a Review in Advance on February 11, 2003

NEW INSIGHT INTO SITE-SPECIFIC RECOMBINATION FROM FLP RECOMBINASE-DNA STRUCTURES

Yu Chen and Phoebe A. Rice
Department of Biochemistry and Molecular Biology, The University of Chicago, Chicago, Illinois 60637; email: yuchen@midway.uchicago.edu; price@midway.uchicago.edu

Key Words Holliday junction, integrase, strand exchange, domain swapping, half-of-the-sites activity

■ **Abstract** The λ integrase, or tyrosine-based family of site-specific recombinases, plays an important role in a variety of biological processes by inserting, excising, and inverting DNA segments. Flp, encoded by the yeast 2-μm plasmid, is the best-characterized eukaryotic member of this family and is responsible for maintaining the copy number of this plasmid. Over the past several years, structural and biochemical studies have shed light on the details of a common catalytic scheme utilized by these enzymes with interesting variations under different biological contexts. The emergence of new Flp structures and solution data provides insights not only into its unique mechanism of active site assembly and activity regulation but also into the specific contributions of certain protein residues to catalysis.

CONTENTS

INTRODUCTION: THE TYROSINE RECOMBINASES AND FLP

The λ integrase, or tyrosine-based family of site-specific recombinases, has grown to well over 100 members with examples found in all three biological kingdoms. By catalyzing the insertion, excision, or inversion of specific DNA segments, these proteins promote a wide variety of processes such as the integration and excision of bacteriophage DNA into and out of the host genome, resolution of bacterial replicon dimers, transposition, mobilization of integron gene cassettes, and creation of antigenic diversity in some pathogens (22, 32, 53). Some of these proteins, most notably Flp and Cre, have also become increasingly useful as genetic tools. Flp, the primary subject of this review, helps maintain the copy number of the yeast 2-μm plasmid that encodes it (69).

Over the past several years, structural and biochemical studies have provided new understanding not only of the overall mechanism by which these recombinases promote the exchange of DNA strands but also of their catalytic details and activity control mechanisms. An emerging theme is a common catalytic apparatus used in a variety of contexts under different schemes of catalytic control.

The minimal requirements for tyrosine recombinase-catalyzed recombination are two DNA segments containing specific sequences recognized by the enzyme and four protein molecules. Some reactions require supplementary protein factors and DNA sequences, and some involve concerted action by two different but related proteins. The required DNA targets are usually made up of two enzyme binding sites inverted around a spacer region that is exchanged during the reaction. Our understanding of the mechanism of strand exchange reflects a variety of studies from many laboratories [reviewed in (84)]. The two strands of the DNA molecules are each cleaved and exchanged in a stepwise manner (Figure 1). After formation of an appropriate protein-DNA complex, a nucleophilic tyrosine cleaves the DNA backbone at the junction between the binding site and the spacer, forming a covalent 3′-phosphotyrosyl linkage with the DNA. The expelled 5′-hydroxyl group can then attack the protein-DNA bond on the other DNA molecule, exchanging one strand with its recombination partner and resealing the DNA backbone. This ligation step yields a characteristic DNA Holliday junction, whose isomerization is key to deactivating the two protein monomers that have already performed the cleavage reaction as well as to activating the other two. A second round of DNA cleavage and ligation reactions exchanges the other strand in the spacer region, completing the recombination process. In the absence of outside factors the reaction is isoenergetic and thus fully reversible. As shown in Figure 1, the whole reaction proceeds in a coordinated manner such that only two of four protomers are active in each step. This half-of-the-sites activity prevents double-strand cleavage and poses an interesting question of activity regulation.

In addition to the tyrosine, five other highly conserved residues have been found in the catalytic cavity of these recombinases: two arginines, a lysine, and two histidines, one of which is replaced by tryptophan in Flp and a few other family

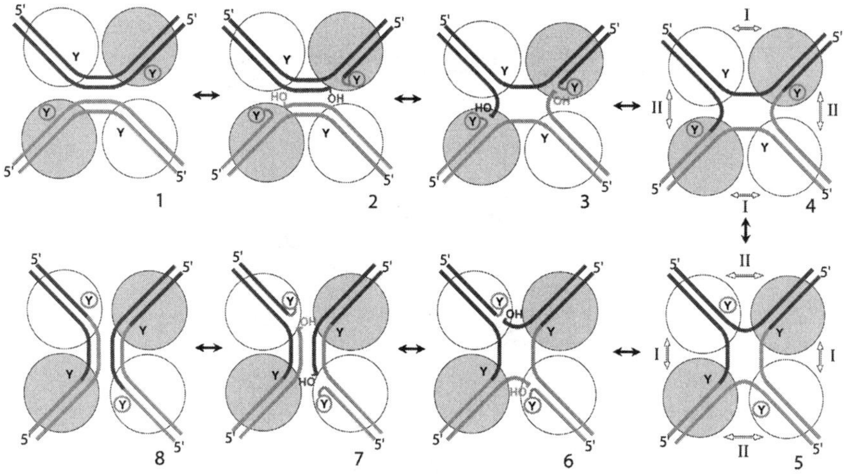

Figure 1 Site-specific recombination pathway. The two recombining DNA segments are bound by four protein monomers. Nucleophilic tyrosine residues in the active catalytic centers are shown as *circled* Ys, and the inactive ones are *uncircled*. The two different types of protein-protein interface seen in the complex are labeled explicitly in the two far right drawings. In the Flp structures, each protomer donates this tyrosine to its neighbor in a clockwise fashion, and donation across a type I interface yields an active catalytic center. Isomerization of the complex converts each type of interface to the other, activating the inactive catalytic centers and vice versa. See the text for details.

members [reviewed in (27)]. This constellation of residues may aid catalysis in several ways: They may localize the scissile phosphate group, stabilize the buildup of negative charge in the transition state, and provide a general base and general acid to deprotonate the nucleophile and protonate the leaving group (see Catalysis: The Nucleophilic Tyrosine and the Conserved Pentad, below). Despite the strong conservation of these residues among family members, two different mechanisms exist for assembling the catalytic site. For most tyrosine recombinases, all six residues in one active site originate from the same protein monomer (55). For Flp and some other fungal enzymes, the nucleophilic tyrosine and the conserved pentad come from two neighboring recombinase molecules (15, 27, 91).

In the past few years, new biochemical and structural evidence has shed light on some striking similarities between the λ integrase family of site-specific recombinases and another protein family, the type IB topoisomerases (75). Type IB topoisomerases release superhelical tension in DNA by cleaving and religating one strand. Although they function as monomers, these enzymes catalyze this reaction through a similar 3′-phosphotyrosyl intermediate and share nearly the same active site as the tyrosine recombinases with one minor exception that the highly conserved first histidine in the recombinases is replaced by a lysine in the

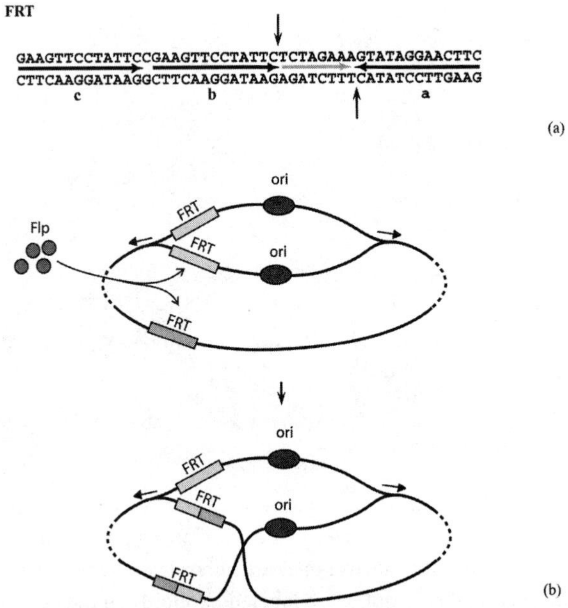

Figure 2 DNA target and Flp's function in vivo. (*a*) DNA sequence of the FRT (Flp recombination target) site. It contains three binding sites for Flp (*black horizontal arrows*) and a spacer (*gray arrow*). Subsite c is unnecessary for recombination. *Short vertical arrows* indicate where the nucleophilic tyrosine cleaves DNA. (*b*) Futcher model of 2-μm plasmid amplification through Flp-mediated recombination. The plasmid has two FRT sites (*rectangles*), one of which is closer to the origin of DNA replication ("ori") than the other (not drawn to scale). Recombination between one of the newly replicated FRT sites and the distal one changes the relative orientation of the two replication forks, whose propagating directions are shown by *short black arrows.*

topoisomerases (76). Studies on type IB topoisomerases have greatly helped the understanding of the catalytic mechanisms of the tyrosine recombinases.

The natural DNA target for Flp-mediated recombination, FRT (Flp recognition target), comprises three 13-bp Flp binding sites and one 8-bp spacer (Figure 2*a*). One of the binding sites, subsite c, has been shown to be nonessential. The length of the spacer can also vary from 7 to 9 bp without a significant effect on the recombination efficiency, an instance of flexibility rarely observed in other family members (72). In vivo, Flp can amplify the copy number of the 2-μm plasmid by recombining the two FRT sites on the plasmid (24) (Figure 2*b*). One of the FRT sites resides closer to the replication origin than the other and is thus replicated first. Recombination between one of the newly replicated FRT sites and the other one distal from the origin can alter the relative orientation of the two replication forks, resulting in rolling-circle-like replication of the plasmid. This replication is subsequently terminated and the plasmid multimers are segregated into monomers through Flp-mediated recombination. The expression of Flp, and hence the occurrence of Flp-mediated recombination, is regulated by other proteins that are also encoded by the 2-μm plasmid and might indirectly reflect the plasmid copy number [reviewed in (69)]. Through this intricate network of regulation, the maintenance of the plasmid copy number is achieved.

Like Cre, Flp does not require any accessory components for its reaction, and both are widely used in research for genetic manipulation because of the simplicity of their catalytic requirements. The application of wild type (WT) Flp, however, is limited by the enzyme's low efficiency at 37°C (9). Stewart and colleagues (8) successfully improved Flp's activity at temperatures higher than its optimal 30°C by introducing point mutations through error-prone PCR, DNA shuffling, and in vivo selection. The resulting mutant, Flpe, with four mutations (P2S, L33S, Y108N, and S294P) removed from the active site, has proven to be a more effective genetic tool (1, 68, 83).

As the best-characterized eukaryotic member of the tyrosine recombinases, Flp has been studied extensively both biochemically and structurally. Its uniqueness in assembling the active site provides useful information for elucidating the details of the catalysis by these site-specific recombinases as well as regulation of the half-of-the-sites activity. The emergence of new Flpe structures as well as solution data, combined with recent results from other related enzymes, yields some novel insights into site-specific recombination.

FLP MONOMER: CONSERVATION OF A CATALYTIC CORE

The past several years have seen an exciting number of structures determined in both the λ integrase family of site-specific recombinases and the family of type IB topoisomerases, especially a series of complexes of Cre, Flp, and human topoisomerase I with DNA (Table 1). Cre, a phage P1-derived recombinase, is involved in circularizing the phage genome and resolving plasmid multimers into monomers. Other three-dimensional models in the recombinase family include λ integrase and HP1 integrase that are responsible for integrating and excising the phage DNA into and out of the host genome, and XerD that ensures proper segregation of replicons by catalyzing recombination with another closely related enzyme, XerC.

Besides the original Flp-Holliday junction structure, two new structures of Flpe-DNA complexes have recently been solved: one with Flpe protein and the other with Flpe containing an additional mutation in a proposed catalytic residue, W330F. To distinguish between the two Flpe structures, we call the first one wild-type Flpe and the latter Flpe W330F. The Flpe protein used also carries a C-terminal $(His)_6$ tag. While the structures of the individual monomers in the Flp- and Flpe-DNA complexes are essentially identical, the Flpe-DNA complexes pack differently in the crystal because of the interaction of this tag with the end of a neighboring DNA segment, resulting in differences in some of the protein-protein interfaces within the complex. These new structures thus provide new information regarding the roles of the protein-protein interfaces and the active site residues.

TABLE 1 Crystal structures of tyrosine recombinases and type IB topoisomerases

Protein	Origin	Structure	PDB code	Reference
λ integrase	Phage λ	Catalytic domain, Y342F (aa. 170–356)	1AE9	(45)
HP1 integrase	Phage HP1	Catalytic domain (aa. 165–337)	1AIH	(35)
XerD	*E. coli*	Intact monomer	1A0P	(81)
Cre	Phage P1	WT Cre-*loxA* covalent complex Suicide DNA substrate	1CRX	(30)
		WT Cre-HJ complex Nick at scissile phosphate	2CRX	(26)
		Cre R173K-HJ complex DNA strands intact	3CRX	(26)
		Cre R173K-*loxS* synaptic complex Intact duplex DNA	4CRX	(31)
		Cre Y324F-*loxS* synaptic complex Intact duplex DNA	5CRX	(31)
		WT Cre-*loxP* HJ DNA strands intact	1KBU	(51)
		Cre Y324F-DNA trimeric complex	1F44	(88)
		Cre Y324F-DNA trimeric complex	1DRG	(88)
Flp	*S. cerevisiae*	WT Flp-HJ complex DNA shown in Figure 5	1FLO	(16)
		WT $(His)_6$-tagged Flpe-HJ covalent complex	1M6X	[a]
		$(His)_6$-tagged Flpe W330F-HJ covalent complex	[b]	
Human topoisomerase I	*H. sapiens*	Reconstituted topoisomerase[c]-DNA covalent complex	1A31	(67)
		Reconstituted topoisomerase Y723F-DNA complex	1A35	(67)
		70-kD topoisomerase (aa. 175–765) Y723F-DNA complex	1A36	(80)
		C-terminal domain (aa. 203–765) Y723F-DNA complex	1EJ9	(66)
Vaccinia topoisomerase	Vaccinia virus	Catalytic domain (aa. 81–314)	1A41	(17)

[a]A.B. Conway, Y. Chen & P.A. Rice, unpublished data.

[b]Y. Chen & P.A. Rice, unpublished data.

[c]Reconstituted topoisomerase I consisted of 58-kD core domain (aa. 175–659) complexed with 6.3-kD carboxyl-terminal domain (aa. 713–765).

WT, wild-type.

In spite of the low sequence homology among the proteins listed in Table 1, even for those within one family, the known three-dimensional models of these enzymes all resemble one another in their catalytic cores, residing in the C-terminal domain of the proteins' usually bipartite structures. As shown in Figure 3 (see color insert), a prominent feature of this catalytic core is the three-stranded β sheet and a large number of α helices. The nucleophilic tyrosine is on helix M, as labeled in Cre and Flp structures. In Flp, this helix is linked to the rest of the protein through flexible loops. A functional catalytic unit of Flp actually contains helix M donated from a neighboring molecule but positioned in the same place relative to the rest of the secondary structure elements as the equivalent helix in other enzyme protomers such as Cre. Compared with most tyrosine recombinases, Flp (and closely related fungal recombinases) has a longer linker between helices L and M, allowing it to cross between protomers in the complex. Most recombinase sequences end shortly after helix N, but Flp continues for an extra 100 or so residues after this point. These amino acids form a small extension that is tucked back to the rest of the C-terminal domain and might function as an anchor for the returning polypeptide chain after the *trans*-donated helix M. As described later, other recombinases also use helix-swapping in the assembly of a cyclic tetramer, but swap a different helix: N rather than M is packed onto a neighbor's core.

Interestingly, in most of the structures determined in the absence of DNA, the catalytic tyrosine is significantly displaced from the remainder of the active site pocket. The sole exception is the phage HP1 integrase core structure, in which a sulfate ion binds in the pocket usually occupied by the scissile phosphate. In the vaccinia virus topoisomerase, lambda integrase, and XerD structures, the majority of the catalytic domain fold matches that of the DNA-bound structures, while the polypeptide segment corresponding to helix M of Flp and Cre is drastically rearranged. This implies a conserved mobility in this region that may be tied to a common underlying theme in catalytic regulation: control of the catalytic tyrosine's orientation. In the lambda case, the contacts seen in the crystal between the rest of the protein and the displaced tyrosine-bearing segment play a regulatory role (82).

Most tyrosine recombinases and type IB topoisomerases have a domain N-terminal to the catalytic one that aids in DNA binding and, in the recombinases, protein-protein interactions. The structures of the N-terminal domains are more variable, corresponding to the less important roles this domain plays in catalysis. For Flp, the N-terminal domain contributes to protein-protein interaction and to a limited amount of largely nonspecific DNA binding capacity (58, 59). Despite differences in the overall fold of their N-terminal domains, the monomers of Flp, Cre, and human topoisomerase I all form C-shape clamps around the bound DNA. Such a two-domain structure is not universal; however, some family members, such as λ integrase, have an additional domain at the N terminus that binds to supplementary DNA sequences required for reaction, whereas others, such as FimB and FimE, function with only a catalytic domain (53).

Not surprisingly, the conserved catalytic domains interact with the DNA in similar fashions, with helix J inserted into the major groove and making direct

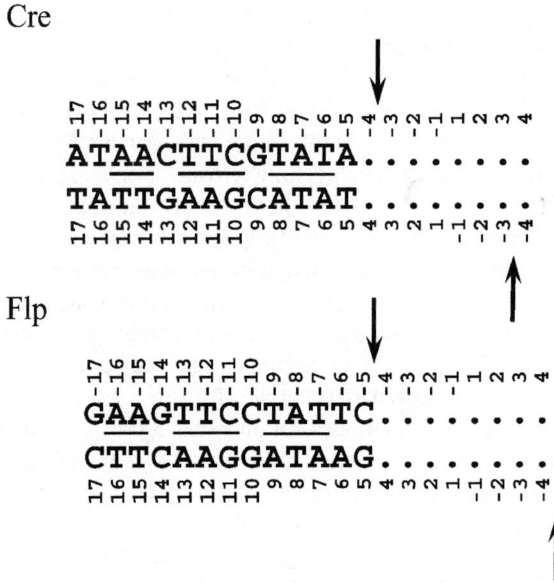

Figure 4 Comparison of Flp (*bottom*) and Cre (*top*) binding sites. The sequence of the binding site is shown and the spacer is represented as *dots*. The numbering starts in the middle of the spacer. Both binding sites contain 13 bp while both spacers are 8 bp. Cre cleaves DNA between nucleotide-3 and -4 inside the spacer, with only 6 bp between the two cleavage sites (shown by *arrow*). *Straight lines* mark the sequences shared by the two binding sites. Subsite a of FRT differs by one nucleotide from subsite b (shown here).

contact to the bases in the Flp and Cre structures, and water-mediated contacts in the nonsequence-specific topoisomerase case. The DNA binding site of Flp is close to that of Cre, sharing more than half of the sequence (Figure 4). Among the 13 base pairs of Cre binding site, mutations at A6, T7, G10, and A11 caused the most severe defects in DNA binding (34) (Figure 4, *top*). All these base pairs remain unchanged in the DNA sequence bound by Flp, of which the corresponding A7 and G11 are also among the three sites whose mutations decreased reaction rates most dramatically (73) (Figure 4, *bottom*). Both G10 for Cre and G11 for Flp are contacted directly by an arginine extended from helix J (R259 in Cre and R281 in Flp). In the Flp binding sequence, a third essential base, G5, interacts with K82, whose importance was also verified by recent solution data (86). The corresponding site on the edge of Cre's spacer region is a C4:G-4 base pair in the structure of Cre-*loxA* complex, and G-4 is also contacted by another lysine (K86) in the Cre structure. Mutations at this position of the natural Cre DNA target, *loxP*, did result in recombination deficiencies (46). The similarities between the two proteins' DNA binding sites, however, could be some evolutionary coincidence when they are compared with the more substantial variations among FRT and the DNA targets of the other fungal recombinases whose protein sequences are more closely related to Flp than any other family members. R281 of Flp is also not conserved in these fungal recombinases. Mutants with altered binding specificities have been isolated for both Flp and Cre (10, 34, 86). Such studies may broaden the usage of these two proteins as genetic tools.

HOLLIDAY JUNCTION COMPLEX: *CIS* AND *TRANS* ORGANIZATION

The DNA in the Flp and Flpe structures corresponds to the Holliday junction intermediate depicted in step 6 of Figure 1 (16). This junction, containing a 7-base spacer with a central unpaired A, resulted from a rearrangement of the initial crystallization substrate, as shown in Figure 5. Although the naturally occurring FRT

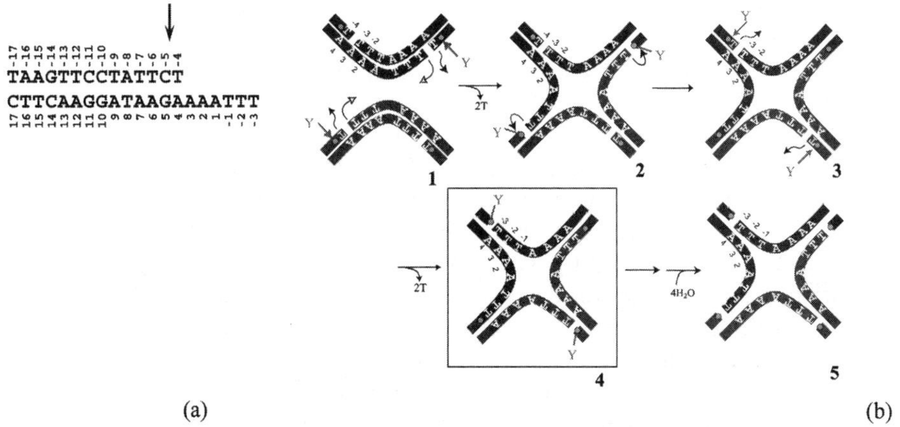

Figure 5 DNA substrate in the Flp and Flpe crystal structures. (*a*) Half-site suicide substrate, with the Flp cleavage site indicated by *arrow*. Two such substrates anneal to form a full site. Nucleotide numbering starts in the middle of the spacer. (*b*) Formation of the Holliday junction present in the crystals. Only the middle of the DNA junction is shown here. Nucleotide T-4 diffuses into solution after the cleavage reaction (stage 1). Slippage of the base pairing allows T-3 to fill the position vacated by T-4 of the opposite duplex, attack the phosphotyrosyl linkage there, and reseal the DNA strand (stage 2). After isomerization, the other two monomers attack DNA, excising two more T-4 nucleotides (stage 3). T-3 from the same strand as T-4 fits into the pocket originally occupied by the nucleotide now lost in solution, producing a covalent Flp-Holliday junction complex with 7-bp spacer containing a central unpaired A, as observed in the Flpe structures (stage 4, bounded by a *square*). Over time, the equilibrium between the noncovalent and covalent complexes is shifted by hydrolysis reactions that irreversibly sever the protein-DNA bond, eventually yielding a noncovalent nicked Holliday junction, as seen in the Flp structure (stage 5).

site contains an 8-bp spacer, Flp is a remarkably flexible enzyme and catalyzes efficient recombination of artificial sites with 7-, 8-, or 9-bp spacers (72). The covalent bond between the protein and DNA was broken in the original Flp structure due to slow hydrolysis while the crystals were stored at 4°C (39). This phosphotyrosyl linkage, however, was preserved fairly well in the Flpe structures because the crystals were flash-frozen and stored in liquid N_2 as soon as they reached a useful size. The WT Flpe structure thus represents a model closer to the reaction intermediate. It provides the first molecular model of a WT tyrosine recombinase in which the 5′-hydroxyl group is poised to attack the covalent bond between the protein and DNA. It is worth noting that the only significant hydrolysis-derived differences between the Flp and Flpe structures are observed in Y343, the scissile phosphate, and some immediately neighboring residues.

The DNA Holliday junction is held by a cyclic tetramer of Flp in a nearly square planar conformation (Figure 6, see color insert). Most of the protein-protein contacts are made by two domain-swapped helices: helix D from the N-terminal domain and helix M from the C-terminal domain. These helices pack into the hydrophobic core of a neighboring protomer but are connected to their own protomers by long loops. This creates a flexible complex and may be key to the enzyme's efficient catalysis on DNA substrates with different length spacers. The complex displays fairly accurate twofold symmetry but only inexact fourfold symmetry. Each protomer is thus approximately the same as the one opposite to it but different from the two adjacent ones. This breakdown of fourfold symmetry is in accordance with the half-of-the-sites activity of the enzyme and results in alternating types of protein-protein interfaces, referred to here as type I and type II (Figure 7, see color insert). These interfaces differ in the relative orientation of the two protomers: In type I interfaces, the C-terminal domains are closer together than in the type II interfaces. Helix M, bearing the nucleophilic tyrosine, is donated in a cyclic manner to an adjacent protomer within the complex and cleaves the DNA bound by that neighboring molecule. When this helix is donated across a type I interface an active catalytic center is formed in the recipient.

The active and inactive catalytic sites differ mainly in the location of the nucleophilic tyrosine (Figure 8, see color insert). In all four catalytic sites, the conserved pentad is stationed around the scissile phosphate. In the two active catalytic centers, the tyrosine is also present. However, in one inactive site the tyrosine and all of helix M are disordered, and in the other Y343 itself is visible in the electron density but its hydroxyl group is ~10 Å from the scissile phosphate. The observation that the other catalytic residues still surround the scissile phosphate in both types of active site is in agreement with studies of Flp-catalyzed DNA cleavage by peroxide. In these experiments, this phosphate group is unusually reactive to nucleophilic attack by peroxide even at subsite c of FRT, which is never involved in actual strand cleavage and reunion (37).

The disorder of helix M in the inactive catalytic centers stems mainly from the increased distance between the two proteins' C-terminal domains in the type II interfaces. The loops flanking helix M are simply not long enough to allow it to dock into its binding pocket on the neighboring protomer. In fact, this helix may be unraveled as well as disordered when it attempts to traverse the type II interfaces.

The type II interfaces were previously referred to as "loose" interfaces because they bury less surface area than the type I interfaces do. The disorder of the Y343-bearing helix M accounts for most of this loss in protein-protein interactions. These interactions may be essential for the correct docking of helix M in order for Y343 to attack DNA. Mutational studies have underscored the importance of this pocket in orienting helix M (28, 57).

Comparison of the Flp- and Cre-DNA complexes highlights interesting differences and similarities. The overall geometry is conserved, with good twofold symmetry and approximate fourfold symmetry. However, helix M of Cre remains in its own protomer while helix N is now donated to the neighboring molecule. In fact, the cleavage in *trans* mechanism resulting from swapping helix M is so far unique to Flp and some closely related enzymes from other yeasts. In all other recombinases where it has been tested, cleavage is in *cis*. Apart from this important difference, the active site containing the 3′-phosphotyrosyl linkage adopts almost the same configuration in the Flpe and Cre structures. Compared with Flp, the Cre-DNA complex is also more tightly packed, burying more surface area at both protein interfaces. Although there are also distinct type I and type II interfaces in the Cre complex, the differences between the two types are more subtle compared with Flp (see Protein-Protein Interfaces, below).

The *trans* cleavage mechanism used by Flp, first discovered by Jayaram's group and later visualized in the three-dimensional models, prevents it from functioning as a monomer. It is not known, however, whether this unique assembly of active sites is itself biologically useful. Although no experimental data are available, an analogy is provided by tyrosine-DNA phosphodiesterase, an enzyme that cleaves the 3′-phosphotyrosyl bond in defunct topoisomerase covalently linked to the DNA (92). The Futcher model for Flp's function in vivo (Figure 2) suggests that Flp must be active during DNA replication. Like topoisomerase, Flp is thus in danger of being rammed by a replication fork while it is covalently attached to the DNA. If the original Flp complex were damaged such that it could not detach itself, another Flp monomer would be able to release it while sealing the DNA by catalyzing a single ligation reaction. Such a rescue would be more difficult in other recombinase systems such as Cre that catalyze phosphotransfer reactions in *cis*.

CATALYSIS: THE NUCLEOPHILIC TYROSINE AND THE CONSERVED PENTAD

The conserved pentad in the active sites of tyrosine recombinases as well as type IB topoisomerases has been studied using various enzymes as model systems, recently most notably with vaccinia topoisomerase. These five residues are highly conserved throughout both families with some limited variations: The first histidine in the tyrosine recombinases (H305 for Flp) is replaced by a lysine in type IB

TABLE 2 Catalytic residues of the tyrosine recombinases and type IB topoisomerases

	Flp	Cre	λ int	HP1	XerC	XerD	V.topo[a]	H.topo[b]
R	R191	173	212	207	148	148	130	488
H/K	H305	H289	H308	H280	H240	H244	K220	K587
R	R308	292	311	283	243	247	223	590
K	K223	201	235	230	172	172	167	532
H/W	W330	W315	H333	H306	H266	H270	H265	H632
Y	Y343	324	342	315	275	279	274	723

[a]V.topo: Vaccinia topoisomerase
[b]H.topo: Human topoisomerase I

topoisomerases, and the tryptophan in Flp (W330) and Cre is a histidine in most other enzymes (Table 2).

Mutagenesis studies and sequence alignment [reviewed in (22)] initially identified the amino acids required for catalysis in addition to the tyrosine as an RHR triad, consisting of the two invariant arginines and the first histidine. The important contributions of the two arginines to catalysis have been well documented by biochemical experiments (14, 23, 57, 60, 74). Combined with structural data, these results suggest that the main functions of the two arginines might be to help position the scissile phosphate and to stabilize buildup of negative charge in the reaction transition state. In both the Flp and Cre structures, the position of H305 implies that it might serve as the general base that accepts the proton from the tyrosine during the strand cleavage reaction. However, several lines of evidence downplay the importance of its role in catalysis. This residue is not as highly conserved as the others, and mutating Flp's H305 to a variety of different residues does not fully abrogate cleavage (61). Mutagenesis of this residue in other proteins also had a relatively modest effect [reviewed in (5)]. Because the pKa of a tyrosine is lower than that of a 5′ hydroxyl, the catalytic power derived from providing a general base may be less than that derived from providing a general acid to protonate the leaving hydroxyl group.

The importance of the lysine and the second histidine (or tryptophan) has been brought to attention recently. We find that mutation of these residues in Flp to alanine has a much more drastic effect on enzyme activity than mutation of H305 to alanine (Y. Chen & P.A. Rice, unpublished data). Studies of the vaccinia topoisomerase have shown that the lysine acts as a general acid during the strand cleavage reaction, donating a proton to the leaving 5′ hydroxyl (42) (Figure 9). These experiments utilize a special DNA substrate with the 5′-bridging oxygen on the scissile phosphate replaced by a sulfur atom. With the pKa of the 5′ SH substantially lower than that of a 5′ OH, this replacement on the DNA should rescue mutations of the amino acids that are involved in general acid catalysis, without significant effects on other catalytic residues. Such experiments have not only elucidated the part K167 plays in the reaction but also implicated R130 (the first arginine) in protonation

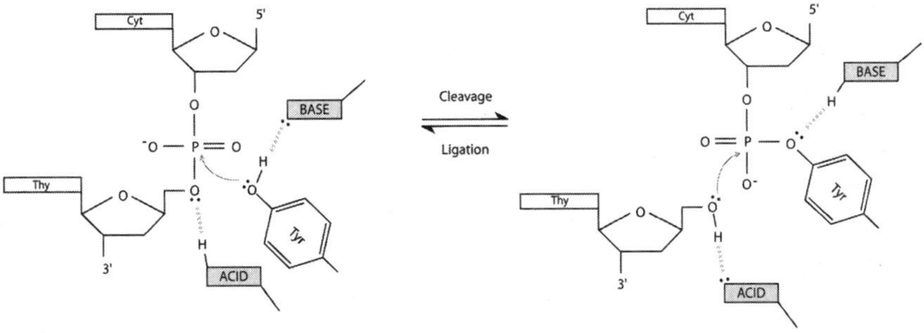

Figure 9 Schematic diagram of general acid and general base catalysis in DNA cleavage and ligation reactions. In the DNA cleavage reaction catalyzed by tyrosine recombinases and type IB topoisomerases, a general base presumably accepts a proton from the attacking tyrosine's hydroxyl group, while a general acid donates a proton to the leaving 5′-bridging oxygen on the scissile phosphate, resulting in the formation of a 3′-phosphotyrosyl bond and a free 5′-OH group. This protonation and deprotonation process is reversed in the ligation reaction. Other active site residues may help stabilize the transition state (not shown).

of the leaving hydroxyl as well (44). The structure of human topoisomerase I has previously indicated that an ordered water molecule in the active site might act as the general base to deprotonate the hydroxyl group on the tyrosine side chain (66). These observations have led to the suggestion of a proton relay network in which, during the cleavage reaction, the proton from the tyrosine is passed to the arginine, and possibly to a water, then to the lysine, and finally to the 5′-bridging oxygen (44). Although no such experiments have been performed on any tyrosine recombinase, mutations at this lysine in Flp, Cre, λ integrase, and XerD supported the significance of its contribution to catalysis [(5, 11, 84); Y. Chen & P.A. Rice, unpublished data].

The second histidine has also been well studied in the topoisomerase family, where it was mutated to alanine, glutamine, and asparagine (62, 93). H265A in vaccinia topoisomerase resulted in a 100-fold decrease in the reaction rate while H632A in human topoisomerase I caused a more severe 6000-fold drop. The glutamine and asparagine mutations in both proteins were able to restore the activity by around 30- to 60-fold, compared with the respective alanine mutant. In the structures of protein-DNA complexes, this histidine, substituted by a tryptophan in Flp and Cre, interacts with the scissile phosphate or 5′-hydroxyl group through a hydrogen bond, suggesting that it might stabilize some reaction intermediate. However, while the W330A mutant of Flpe was inactive and the W330H and W330Q mutants showed only minimal activity, the W330F protein's reaction rate was only decreased fivefold (Y. Chen & P.A. Rice, unpublished data).

The Flpe W330F structure reveals that the mutant active sites assume the same conformation as those of the wild type with only minor readjustments in the immediate vicinity of residue 330. Although the original hydrogen bond between

W330 and DNA may still play some role in catalysis, this residue's most crucial role, at least in Flp, appears to be maintaining a stable local architecture in the catalytic cavity through van der Waals interactions with nearby residues. In the structure of Flpe-Holliday junction complex, W330 contacts a number of side chains within or in immediate vicinity to the active site, including some residues from the *trans*-donated helix M. Most of these interactions are maintained in the Flpe W330F model. W330 might therefore be most important for the correct docking of helix M in Flp-mediated recombination. As also shown by the structures, W315 in Cre and H632 in human topoisomerase I are involved in many van der Waals interactions as well. These two residues could play a similar role in contributing to the stability of a local structure while possibly playing a more active role in chemical catalysis at the same time. The relative importance of this residue's hydrogen bonding and structural roles may vary among different enzymes.

PROTEIN-PROTEIN INTERFACES: FLEXIBILITY AND REGULATION OF ACTIVITY

The type I and type II interfaces, harboring the productive and unproductive active sites, respectively, hold the key to regulation of the half-of-the-sites activity. Because Flp assembles its active site in *trans*, this regulation in Flp may differ from those for other family members such as Cre. The major difference between these two interfaces in the Flp-DNA complex is that in the type I interfaces, helix M from one protomer can reach into the catalytic pocket of the other, whereas in the type II interfaces, the catalytic domains are farther apart and helix M is disordered. The Flp-DNA and Flpe-DNA complexes are packed differently in the crystal, and the ability to compare structures of the complex in different crystal packing environments sheds light on the relative importance of these interfaces. The type I interfaces remain relatively constant in both crystal forms, but the protomers in the type II interfaces rotate by $\sim 15°$ relative to one another (A.B. Conway, Y. Chen & P.A. Rice, unpublished data) (Figure 10). This is consistent with the fact that fewer protein-protein interactions occur in this interface. In solution, the type II interfaces probably exist within a range of conformations. Thus, although forming correct type I interfaces is essential for the activation of two of the catalytic centers, the inactivation of the other two does not depend on any particular conformation of the type II interfaces, but merely on the type II interfaces not being able to flex far enough to adopt the type I conformation.

What prevents the type II interfaces from also adopting the type I conformation and thus activating all four protomers within the complex? Model-building exercises show that more than two type I interfaces cannot be accommodated in a tetrameric complex without creating steric clashes (Figure 11). This is partly explained by the fact that in the type I interface, the rotation angle from one protomer to the other is $\sim 100°$ rather than $90°$. A plausible model can be constructed

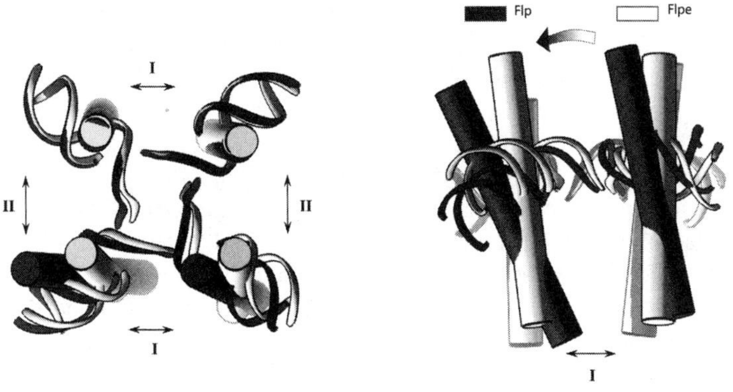

(a) (b)

Figure 10 Comparison of interfaces in Flp and Flpe structures. (*a*) Top view (as in Figure 6*a*) of the two superimposed protein-Holliday junction complex structures, with DNA represented as *ribbons* and the proteins as *rods* with one endpoint in each domain (*dark gray*, Flp; *white*, Flpe). The upper two protomers of each structure were used as guides for the superimposition. Type I and type II interfaces are as indicated. The type I interface remains approximately the same in the two structures while the type II interface flexes. (*b*) Side view of the same drawing, with the superimposed protomers in the back.

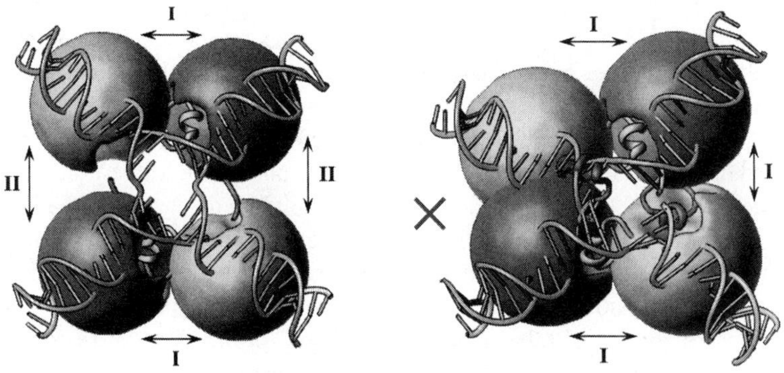

(a) (b)

Figure 11 Steric regulation of half-of-the-sites activity. (*a*) Cartoon of the normal Flp-Holliday junction complex. Proteins are represented as *spheres* with sockets for the protruding helix M. The two interface types are shown. (*b*) Model of a hypothetical synaptic complex, with three type I interfaces. Such an arrangement causes steric clash in the remaining interface.

of a trimeric complex with two adjacent type I interfaces, but this model leaves no room for the addition of a fourth protomer. Interestingly, Flp readily forms complexes with artificial substrates containing three DNA arms rather than four, and these complexes do appear to contain two adjacent active catalytic centers (48, 49, 63, 64). Trimeric complexes of Cre and λ integrase have also been studied, both in vitro and, for Cre, crystallographically (56, 88).

In the context of the naturally occurring recombination pathway, the type I interface must form first between monomers bound to the same DNA duplex. This would allow the initial strand cleavage reaction to proceed before synapsis, in agreement with solution experiments (85). When two such dimers come together to form a tetramer, they can only associate via type II interfaces because the formation of a third type I association would cause steric clash in the remaining protein interface, as shown by modeling exercises in Figure 11.

The isomerization step at the center of the recombination pathway would convert each interface type into the other and hence switch the states of the active sites within them (Figure 1). This process can be simulated by comparing the original protein-bound Holliday junction with a hypothetical complex obtained by rotating the real one by roughly 90° around the twofold symmetry axis. The isomerization would involve breaking some of the protein interactions in the old type I interfaces while establishing in the new ones those interactions required for docking helix M onto the neighboring molecule. Also the individual arms of the DNA Holliday junction rotate slightly, the angles between them change by a few degrees, and the DNA backbone at the middle of the spacer twists differently. However, the overall square planar architecture of the DNA Holliday junction is retained throughout the recombination reaction.

Cre regulates its half-of-the-sites activity through a different mechanism (84). Although the same type of steric constraints may force the complex to adopt alternating types of protein-protein interface, the structural consequences differ. The catalytic lysine (K201 in Cre) appears to play a major role in this process. In the inactive protomers, the loop containing K201 is flipped out from the active site, whereas K201 usually lies near the scissile phosphate in the productive catalytic center. The position of this loop does not significantly change in the Flp structures. The positioning of the nucleophilic tyrosine also contributes to the regulation. Although it is disordered in the inactive Flp protomers, helix M appears to be slightly less mobile in the inactive Cre protomers compared with the active ones. In vitro studies of other recombinases have shown that the region surrounding helices M and N appears to be important in the regulation of catalytic activity in those systems as well (6, 33, 78).

The Segall group (13) has recently described a series of short peptides that trap the recombination reaction at different stages. These peptides will prove useful in studying the details of the intermediates along the pathway. Preliminary experiments in our laboratory suggest that these peptides promote the growth of Holliday junction-containing crystals (A.B. Conway & P.A. Rice, unpublished data).

DIRECTIONALITY OF RECOMBINATION: RANDOM VERSUS ORDERED

As shown in Figure 12, the consequences of recombination between two target sites can vary. First, a given system produces either inversions or insertions and deletions. Second, some systems have evolved tricks for favoring one reaction direction over the other, even though the basic strand transfer reaction is isoenergetic (the product and substrate contain the same number of phosphodiester bonds).

The choice between inversion and insertion/deletion reactions reflects the relative orientations of the spacer sequences at the centers of the target sites. This stems from a requirement for Watson-Crick base pairing in the product duplexes, whose spacers contain one strand from each parent DNA (4, 47, 72) (Figure 1). The same base-pairing requirement also underlies the fact that, while the spacer sequences can be varied, two sites can only be recombined efficiently if they carry matching spacer sequences. Solution studies demonstrated that this base pairing is checked at the ligation steps of the reaction: If the incoming bases, particularly those closest to the free 5′ hydroxyl, form mismatches, ligation is inhibited (72). The structural basis for this appears to be contacts made between the N-terminal domain of Flp and the backbone of the uncleaved spacer DNA strand. An incoming DNA strand must make Watson-Crick base pairs with the uncleaved strand in order to correctly orient its 5′-hydroxyl group in the active site.

Simple recombinase systems such as Flp (and Cre) do not drive their reactions past the equilibrium dictated by entropy: an equal distribution of isomers in the inversion case and a preponderance of deletion products in the insertion/deletion case. Systems that do drive their reactions in one direction or the other require

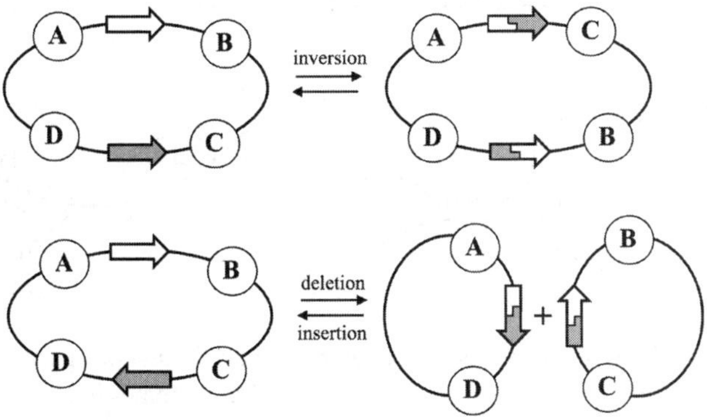

Figure 12 Directionality of recombination. *Top*: If the asymmetric spacer sequences of the target sites (*arrows*) form inverted repeats, recombination inverts the intervening segment. *Bottom*: If the spacer sequences are in direct repeat, recombination causes deletions, and in the reverse reaction, insertions.

additional factors. In some cases, such as bacteriophage integrases, the FimB and FimE invertases, and the XerC/D system, these factors are DNA-bending proteins [reviewed in (5, 6)]. The large complex formed by these accessory proteins and the recombinase on the DNA may influence the direction of recombination by acting as a topological filter, rendering assembly of an active complex dependent on the connectivity of the recombining sites (7), or it may favor the product by favoring the postisomerization conformation of the recombinase-DNA complex. XerC/D-mediated resolution of chromosomal dimers requires a different accessory factor, this time the ATPase FtsK (3, 65). Because the reversibility of the recombination reaction is a serious problem in using Flp and Cre for genetic manipulations, studies of these more complex systems will provide useful information for better controlling the reaction direction and designing more powerful genetic tools.

A related question is the order of strand exchange, i.e., whether the right or left end of the spacer is cleaved first. This too varies among recombinases: λ integrase and XerC/D show a strong preference for cleaving one strand before the other (2, 18, 38, 54), Cre shows some preference (36, 50, 51), and Flp shows little (21). Which strand is cleaved first reflects which of two possible conformations the initial DNA-bound Flp dimer assumes (Figure 13). In one, the right monomer donates its helix M to the left one thus cleaving at the left end of the spacer first, whereas in the other the left monomer donates its helix M to the right one. In the Flp case, despite the asymmetry of the spacer sequence, these two conformations must be roughly equal in free energy. This may be an advantage for a system such as the Flp/FRT, whose goal is to randomize the orientation of a DNA segment. If one conformation were significantly preferred over the other, formation of the initial

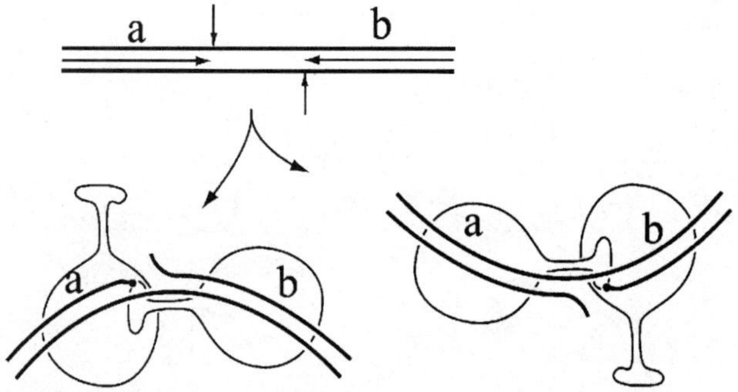

Figure 13 Order of strand cleavage. Monomers bound to a particular recombination site can interact in two alternative ways: If the monomer bound to subsite b donates its helix M (protrusion in cartoon) to the monomer bound to subsite a, the first DNA cleavage occurs at the scissile phosphate adjacent to subsite a. If the monomer at subsite a donates its helix M to the monomer at subsite b, the first cleavage occurs at the other end of the spacer. Depending on the particular recombination system, accessory factors and the sequence of the spacer itself can influence which conformer is preferred.

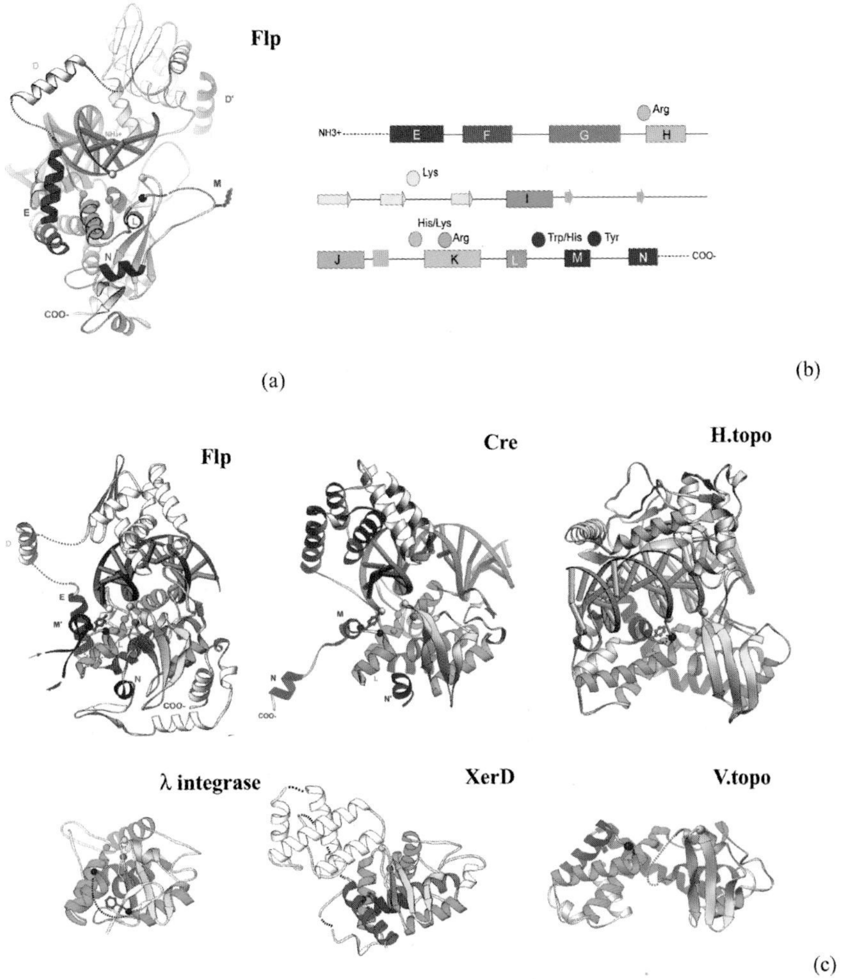

Figure 3 Flp monomer structure and comparison of related catalytic domains. (*a*) Flp monomer tertiary structure, viewed from near the center of the Holliday junction looking outward. The scissile phosphate is shown as a gray sphere and the catalytic residues as colored spheres, except for the tyrosine, which is shown as a blue stick-and-ball model. Helix M is disordered in this monomer. (*b*) Secondary structure diagram of Flp's C-terminal domain. The positions of the conserved active site amino acids are shown, and the conserved structural elements are color coded. (*c*) Comparison of the catalytic domains of Flp, Cre (PDB ID 1CRX), human topoisomerase I (H.topo, PDB ID 1EJ9), λ integrase, XerD, and vaccinia topoisomerase (V.topo), color coded as in parts (*a*) and (*b*). Helices originating from adjacent monomers are marked by the prime symbol (′). In the human topoisomerase structure, a ~75-residue insertion between the helices equivalent to L and M (dotted line) was removed. Note the variations in the position of helix M and the tyrosine in the three structures determined in the absence of DNA. This and Figures 6, 7, 8, 10, and 11 were made with Ribbons (12).

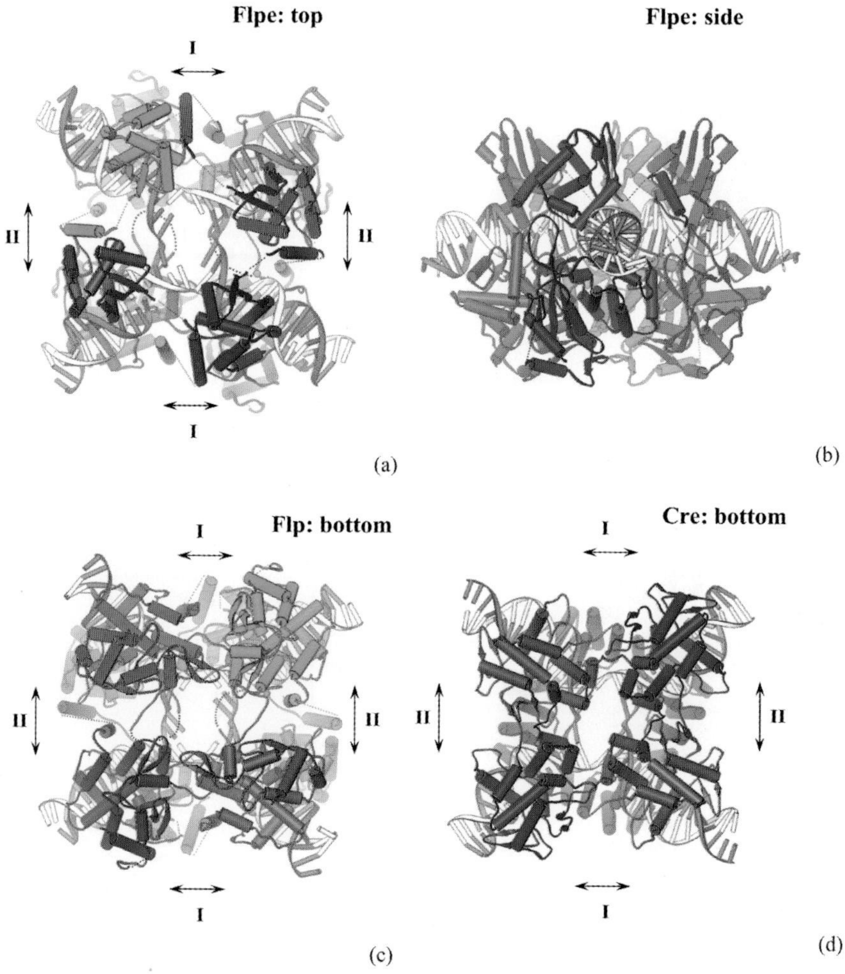

Figure 6 Structures of Flp- and Cre-DNA complexes. (*a*) Top view of the Flpe-Holliday junction complex, with the N-terminal domains in the foreground. The red and purple protomers harbor active catalytic centers; green and blue, inactive. The catalytic tyrosine (Y343) is drawn in ball-and-stick model with the Cα atom as an enlarged sphere. The continuous DNA strand is in white and the crossing strand in gray. Dotted lines represent regions that are disordered and thus not visible in the electron density map. (*b*) Side view of the Flpe-Holliday junction complex, with the N-terminal domains above the DNA and C-terminal domains below. (*c, d*) Comparison of the Flp and Cre complex structures, viewed with the C-terminal domains in the foreground. Flp (*c*) is colored as above, while in Cre (*d*) (PDB ID 3CRX), which displays exact twofold symmetry, the active protomers are red and the inactive ones blue. In Cre, the nucleophilic tyrosine on helix M attacks DNA in *cis* while the C-terminal helix N packs onto the neighboring protomer.

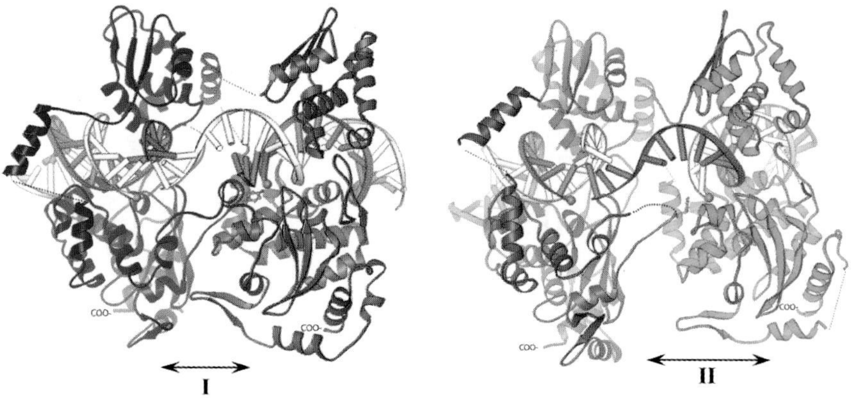

Figure 7 Two types of protein-protein interface. Flp protomers are colored as in Figure 6, and the view is from the center of the Holliday junction. (*Left*) The type I interface, also referred to as the productive interface, harbors an active catalytic center. (*Right*) The type II interface, with fewer protein-protein interactions and the nucleophilic tyrosine pulled away from the scissile phosphate as a result of the increased distance between the two C-terminal domains.

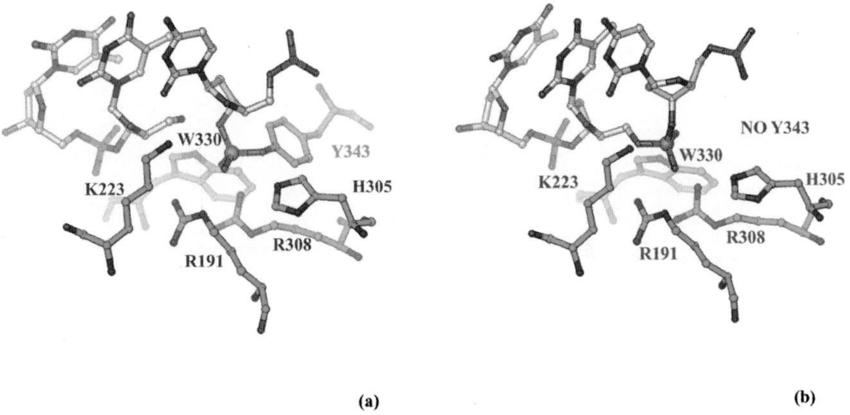

Figure 8 Two states of the active site in the Flpe-DNA complex. (*a*) The active catalytic center with the covalent 3′-phosphotyrosyl linkage. The protein is drawn with green carbons, DNA with gray carbons, and the scissile phosphate is highlighted with a gold sphere. Side chain labels are colored according to the protomer they belong to, following the color scheme in Figure 6. (*b*) The inactive catalytic center. Y343 is dislocated in this active site.

synaptic complex might be enhanced, but product formation would be inhibited because after isomerization the product DNA-dimer complex would be in the disfavored conformation.

VARIATION ON A COMMON THEME: RELATED PROTEINS

The structural and catalytic resemblances shared by the tyrosine recombinases and type IB topoisomerases extend beyond a common 3′-phosphotyrosyl reaction. While Flp can perform topoisomerase functions under some circumstances, vaccinia virus topoisomerase is capable of resolving DNA Holliday junctions and facilitating DNA synapsis, despite the fact that type IB topoisomerases normally act as monomers (70, 77, 90). Both Flp and vaccinia topoisomerase can function as DNA endonucleases under special conditions (52, 87). They also have RNase activities whose biological relevance is unknown (28, 71, 89, 90). In addition, these two enzymes can catalyze the attack of nucleophiles such as hydrogen peroxide and glycerol on a phosphodiester bond (37, 39, 40, 43). Although some details may differ between Flp and vaccinia topoisomerase, these reactions underline once again a possible evolutionary linkage between the two families that might be traced back to some DNA binding proteins. The fold of the catalytic domain can be described as two helix-turn-helix modules with multiple insertions, and in fact, the transcriptional activator MarA from *Escherichia coli*, a member of the AraC protein family, contains a similar arrangement of these modules (25). In the tyrosine recombinases, helices J and M correspond to the "recognition helices" of canonical helix-turn-helix domains. Helix J of Flp and Cre lies in the DNA major groove and is also important in sequence recognition, while helix M, from the second helix-turn-helix motif, bears the nucleophilic tyrosine. These similarities have led to a hypothesis that evolution of the helix-turn-helix modules in some ancestral DNA binding proteins conferred catalytic capabilities to the present-day enzymes (29).

This common ancestor must have evolved into the plethora of recombinases known today. The shared pseudo-fourfold geometry of the Flp and Cre complexes may reflect a common ancestor that was already a recombinase, or it may have arisen through convergent evolution from monomeric ancestors. The square planar geometry of the Holliday junction is inherently favorable because it allows the strand exchange process to proceed with a minimum of conformational changes (79). Domain swapping, reduced to the swapping of a single helix in this case, is a common method of assembling a cyclic oligomer. If a monomeric catalytic domain mutated (perhaps by an insertion in a turn) to allow the swapping of a helix among monomers, and it happened that a chain of four such domain-swapped monomers formed a closed ring, the resulting tetramer would naturally distort bound DNA duplexes so as to stabilize a Holliday junction after cleavage of two strands. The fact that Flp and Cre swap different helices to assemble cyclic tetramers implies that they may have arisen separately from a monomeric ancestor.

A new group of enzymes has recently been described that exploits the same catalytic motif in yet another context. These enzymes maintain the hairpinned ends of the linear chromosomes found in some organisms. The most thoroughly studied members of this family are TelN from phage N15 and ResT from the Lyme disease–causing spirochete *Borrelia burgdorferi* (19, 20, 41). Sequence comparisons and site-directed mutagenesis have shown that they are members of the tyrosine recombinase family, yet the product of their reaction is not a pair of rearranged duplex DNA segments but a pair of hairpinned DNA ends. Biochemical studies of these enzymes are progressing rapidly, and the details of their reaction pathway and mechanisms of catalytic control will provide a fascinating contrast to those of the more canonical tyrosine recombinases.

ACKNOWLEDGMENTS

We thank Adam Conway who carried out some of the crystallographic work discussed here, and Simone Nunes-Duby, Art Landy, and Anca Segall for sharing results before publication. We thank Kerren Swinger and Adam Conway for valuable comments on the manuscript. We are grateful to many members of the site-specific recombination community for years of interesting discussions, and in particular to Mike Cox for his generous support of our group's Flp studies. We also thank the National Institutes of Health for financial support.

The *Annual Review of Biophysics and Biomolecular Structure* is online at http://biophys.annualreviews.org

LITERATURE CITED

1. Andreas S, Schwenk F, Kuter-Luks B, Faust N, Kuhn R. 2002. Enhanced efficiency through nuclear localization signal fusion on phage PhiC31-integrase: activity comparison with Cre and FLPe recombinase in mammalian cells. *Nucleic Acids Res.* 30:2299–306
2. Arciszewska LK, Sherratt DJ. 1995. Xer site-specific recombination in vitro. *EMBO J.* 14:2112–20
3. Aussel L, Barre FX, Aroyo M, Stasiak A, Stasiak AZ, Sherratt D. 2002. FtsK is a DNA motor protein that activates chromosome dimer resolution by switching the catalytic state of the XerC and XerD recombinases. *Cell* 108:195–205
4. Azam N, Dixon JE, Sadowski PD. 1997. Topological analysis of the role of homology in Flp-mediated recombination. *J. Biol. Chem.* 272:8731–38
5. Azaro MA, Landy A. 2002. Lambda integrase and the lambda Int family. See Ref. 18a, pp. 118–48
6. Barre F-X, Sherratt DJ. 2002. Xer site-specific recombination: promoting chromosome segregation. See Ref. 18a, pp. 149–61
7. Bath J, Sherratt DJ, Colloms SD. 1999. Topology of Xer recombination on catenanes produced by lambda integrase. *J. Mol. Biol.* 289:873–83
8. Buchholz F, Angrand PO, Stewart AF. 1998. Improved properties of FLP recombinase evolved by cycling mutagenesis. *Nat. Biotechnol.* 16:657–62

9. Buchholz F, Ringrose L, Angrand PO, Rossi F, Stewart AF. 1996. Different thermostabilities of FLP and Cre recombinases: implications for applied site-specific recombination. *Nucleic Acids Res.* 24:4256–62
10. Buchholz F, Stewart AF. 2001. Alteration of Cre recombinase site specificity by substrate-linked protein evolution. *Nat. Biotechnol.* 19:1047–52
11. Cao Y, Hayes F. 1999. A newly identified, essential catalytic residue in a critical secondary structure element in the integrase family of site-specific recombinases is conserved in a similar element in eucaryotic type IB topoisomerases. *J. Mol. Biol.* 289:517–27
12. Carson M. 1997. RIBBONS. *Macromol. Crystallogr. B* 277:493–505
13. Cassell G, Klemm M, Pinilla C, Segall A. 2000. Dissection of bacteriophage lambda site-specific recombination using synthetic peptide combinatorial libraries. *J. Mol. Biol.* 299:1193–202
14. Chen JW, Evans BR, Yang SH, Araki H, Oshima Y, Jayaram M. 1992. Functional analysis of box I mutations in yeast site-specific recombinases Flp and R: pairwise complementation with recombinase variants lacking the active-site tyrosine. *Mol. Cell. Biol.* 12:3757–65
15. Chen JW, Lee J, Jayaram M. 1992. DNA cleavage in trans by the active site tyrosine during Flp recombination: switching protein partners before exchanging strands. *Cell* 69:647–58
16. Chen Y, Narendra U, Iype LE, Cox MM, Rice PA. 2000. Crystal structure of a Flp recombinase-Holliday junction complex: assembly of an active oligomer by helix swapping. *Mol. Cell* 6:885–97
17. Cheng C, Kussie P, Pavletich N, Shuman S. 1998. Conservation of structure and mechanism between eukaryotic topoisomerase I and site-specific recombinases. *Cell* 92:841–50
18. Colloms SD, McCulloch R, Grant K, Neilson L, Sherratt DJ. 1996. Xer-mediated site-specific recombination in vitro. *EMBO J.* 15:1172–81
18a. Craig NL, Craigie R, Gellert M, Lambowitz AM, eds. 2002. *Mobile DNA II.* Washington, DC: ASM. 1204 pp.
19. Deneke J, Ziegelin G, Lurz R, Lanka E. 2000. The protelomerase of temperate *Escherichia coli* phage N15 has cleaving-joining activity. *Proc. Natl. Acad. Sci. USA* 97:7721–26
20. Deneke J, Ziegelin G, Lurz R, Lanka E. 2002. Phage N15 telomere resolution. Target requirements for recognition and processing by the protelomerase. *J. Biol. Chem.* 277:10410–19
21. Dixon JE, Sadowski PD. 1993. Resolution of synthetic chi structures by the FLP site-specific recombinase. *J. Mol. Biol.* 234:522–33
22. Esposito D, Scocca JJ. 1997. The integrase family of tyrosine recombinases: evolution of a conserved active site domain. *Nucleic Acids Res.* 25:3605–14
23. Friesen H, Sadowski PD. 1992. Mutagenesis of a conserved region of thc gene encoding the FLP recombinase of *Saccharomyces cerevisiae.* A role for arginine 191 in binding and ligation. *J. Mol. Biol.* 225:313–26
24. Futcher AB. 1986. Copy number amplification of the 2 micron circle plasmid of *Saccharomyces cerevisiae. J. Theor. Biol.* 119:197–204
25. Gillette WK, Rhee S, Rosner JL, Martin RG. 2000. Structural homology between MarA of the AraC family of transcriptional activators and the integrase family of site-specific recombinases. *Mol. Microbiol.* 35:1582–83
26. Gopaul DN, Guo F, Van Duyne GD. 1998. Structure of the Holliday junction intermediate in Cre-loxP site-specific recombination. *EMBO J.* 17:4175–87
27. Grainge I, Jayaram M. 1999. The integrase family of recombinase: organization and function of the active site. *Mol. Microbiol.* 33:449–56
28. Grainge I, Lee J, Xu CJ, Jayaram M. 2001.

DNA recombination and RNA cleavage activities of the Flp protein: roles of two histidine residues in the orientation and activation of the nucleophile for strand cleavage. *J. Mol. Biol.* 314:717–33
29. Grishin NV. 2000. Two tricks in one bundle: helix-turn-helix gains enzymatic activity. *Nucleic Acids Res.* 28:2229–33
30. Guo F, Gopaul DN, Van Duyne GD. 1997. Structure of Cre recombinase complexed with DNA in a site-specific recombination synapse. *Nature* 389:40–46
31. Guo F, Gopaul DN, Van Duyne GD. 1999. Asymmetric DNA bending in the Cre-loxP site-specific recombination synapse. *Proc. Natl. Acad. Sci. USA* 96:7143–48
32. Hallet B, Sherratt DJ. 1997. Transposition and site-specific recombination: adapting DNA cut-and-paste mechanisms to a variety of genetic rearrangements. *FEMS Microbiol. Rev.* 21:157–78
33. Han YW, Gumport RI, Gardner JF. 1994. Mapping the functional domains of bacteriophage lambda integrase protein. *J. Mol. Biol.* 235:908–25
34. Hartung M, Kisters-Woike B. 1998. Cre mutants with altered DNA binding properties. *J. Biol. Chem.* 273:22884–91
35. Hickman AB, Waninger S, Scocca JJ, Dyda F. 1997. Molecular organization in site-specific recombination: the catalytic domain of bacteriophage HP1 integrase at 2.7 Å resolution. *Cell* 89:227–37
36. Hoess R, Wierzbicki A, Abremski K. 1987. Isolation and characterization of intermediates in site-specific recombination. *Proc. Natl. Acad. Sci. USA* 84:6840–44
37. Kimball AS, Lee J, Jayaram M, Tullius TD. 1993. Sequence-specific cleavage of DNA via nucleophilic attack of hydrogen peroxide, assisted by Flp recombinase. *Biochemistry* 32:4698–701
38. Kitts PA, Nash HA. 1988. Bacteriophage lambda site-specific recombination proceeds with a defined order of strand exchanges. *J. Mol. Biol.* 204:95–107
39. Knudsen BR, Dahlstrom K, Westergaard O, Jayaram M. 1997. The yeast site-specific recombinase Flp mediates alcoholysis and hydrolysis of the strand cleavage product: mimicking the strand-joining reaction with non-DNA nucleophiles. *J. Mol. Biol.* 266:93–107
40. Knudsen BR, Lee J, Lisby M, Westergaard O, Jayaram M. 1998. Alcoholysis and strand joining by the Flp site-specific recombinase. Mechanistically equivalent reactions mediated by distinct catalytic configurations. *J. Biol. Chem.* 273:22028–36
41. Kobryn K, Chaconas G. 2002. ResT, a telomere resolvase encoded by the Lyme disease spirochete. *Mol. Cell* 9:195–201
42. Krogh BO, Shuman S. 2000. Catalytic mechanism of DNA topoisomerase IB. *Mol. Cell* 5:1035–41
43. Krogh BO, Shuman S. 2000. DNA strand transfer catalyzed by vaccinia topoisomerase: peroxidolysis and hydroxylaminolysis of the covalent protein-DNA intermediate. *Biochemistry* 39:6422–32
44. Krogh BO, Shuman S. 2002. Proton relay mechanism of general acid catalysis by DNA topoisomerase IB. *J. Biol. Chem.* 277:5711–14
45. Kwon HJ, Tirumalai R, Landy A, Ellenberger T. 1997. Flexibility in DNA recombination: structure of the lambda integrase catalytic core. *Science* 276:126–31
46. Lee G, Saito I. 1998. Role of nucleotide sequences of loxP spacer region in Cre-mediated recombination. *Gene* 216:55–65
47. Lee J, Jayaram M. 1995. Role of partner homology in DNA recombination. Complementary base pairing orients the 5′-hydroxyl for strand joining during Flp site-specific recombination. *J. Biol. Chem.* 270:4042–52
48. Lee J, Jayaram M. 1997. A tetramer of the Flp recombinase silences the trimers within it during resolution of a Holliday junction substrate. *Genes Dev.* 11:2438–47
49. Lee J, Whang I, Jayaram M. 1996.

Assembly and orientation of Flp recombinase active sites on two-, three- and four-armed DNA substrates: implications for a recombination mechanism. *J. Mol. Biol.* 257:532–49

50. Lee L, Sadowski PD. 2001. Directional resolution of synthetic Holliday structures by the Cre recombinase. *J. Biol. Chem.* 276:31092–98
51. Martin SS, Pulido E, Chu VC, Lechner TS, Baldwin EP. 2002. The order of strand exchanges in Cre-LoxP recombination and its basis suggested by the crystal structure of a Cre-LoxP Holliday junction complex. *J. Mol. Biol.* 319:107–27
52. Meyer-Leon L, Inman RB, Cox MM. 1990. Characterization of Holliday structures in FLP protein-promoted site-specific recombination. *Mol. Cell. Biol.* 10:235–42
53. Nunes-Duby SE, Kwon HJ, Tirumalai RS, Ellenberger T, Landy A. 1998. Similarities and differences among 105 members of the Int family of site-specific recombinases. *Nucleic Acids Res.* 26:391–406
54. Nunes-Duby SE, Matsumoto L, Landy A. 1987. Site-specific recombination intermediates trapped with suicide substrates. *Cell* 50:779–88
55. Nunes-Duby SE, Radman-Livaja M, Kuimelis RG, Pearline RV, McLaughlin LW, Landy A. 2002. Gamma integrase complementation at the level of DNA binding and complex formation. *J. Bacteriol.* 184:1385–94
56. Nunes-Duby SE, Yu D, Landy A. 1997. Sensing homology at the strand-swapping step in lambda excisive recombination. *J. Mol. Biol.* 272:493–508
57. Pan G, Luetke K, Sadowski PD. 1993. Mechanism of cleavage and ligation by FLP recombinase: classification of mutations in FLP protein by in vitro complementation analysis. *Mol. Cell. Biol.* 13:3167–75
58. Pan G, Sadowski PD. 1993. Identification of the functional domains of the FLP recombinase. Separation of the nonspecific and specific DNA-binding, cleavage, and ligation domains. *J. Biol. Chem.* 268:22546–51
59. Pan H, Clary D, Sadowski PD. 1991. Identification of the DNA-binding domain of the FLP recombinase. *J. Biol. Chem.* 266:11347–54
60. Parsons RL, Evans BR, Zheng L, Jayaram M. 1990. Functional analysis of Arg-308 mutants of Flp recombinase. Possible role of Arg-308 in coupling substrate binding to catalysis. *J. Biol. Chem.* 265:4527–33
61. Parsons RL, Prasad PV, Harshey RM, Jayaram M. 1988. Step-arrest mutants of FLP recombinase: implications for the catalytic mechanism of DNA recombination. *Mol. Cell. Biol.* 8:3303–10
62. Petersen BO, Shuman S. 1997. Histidine 265 is important for covalent catalysis by vaccinia topoisomerase and is conserved in all eukaryotic type I enzymes. *J. Biol. Chem.* 272:3891–96
63. Qian XH, Cox MM. 1995. Asymmetry in active complexes of FLP recombinase. *Genes Dev.* 9:2053–64
64. Qian XH, Inman RB, Cox MM. 1990. Protein-based asymmetry and protein-protein interactions in FLP recombinase-mediated site-specific recombination. *J. Biol. Chem.* 265:21779–88
65. Recchia GD, Aroyo M, Wolf D, Blakely G, Sherratt DJ. 1999. FtsK-dependent and -independent pathways of Xer site-specific recombination. *EMBO J.* 18:5724–34
66. Redinbo MR, Champoux JJ, Hol WG. 2000. Novel insights into catalytic mechanism from a crystal structure of human topoisomerase I in complex with DNA. *Biochemistry* 39:6832–40
67. Redinbo MR, Stewart L, Kuhn P, Champoux JJ, Hol WG. 1998. Crystal structures of human topoisomerase I in covalent and noncovalent complexes with DNA. *Science* 279:1504–13
68. Rodriguez CI, Buchholz F, Galloway J, Sequerra R, Kasper J, et al. 2000. High-efficiency deleter mice show that FLPe

is an alternative to Cre-loxP. *Nat. Genet.* 25:139–40

69. Sadowski PD. 1995. The Flp recombinase of the 2-microns plasmid of *Saccharomyces cerevisiae. Prog. Nucleic Acid Res. Mol. Biol.* 51:53–91
70. Sekiguchi J, Seeman NC, Shuman S. 1996. Resolution of Holliday junctions by eukaryotic DNA topoisomerase I. *Proc. Natl. Acad. Sci. USA* 93:785–89
71. Sekiguchi J, Shuman S. 1997. Site-specific ribonuclease activity of eukaryotic DNA topoisomerase I. *Mol. Cell* 1:89–97
72. Senecoff JF, Cox MM. 1986. Directionality in FLP protein-promoted site-specific recombination is mediated by DNA-DNA pairing. *J. Biol. Chem.* 261:7380–86
73. Senecoff JF, Rossmeissl PJ, Cox MM. 1988. DNA recognition by the FLP recombinase of the yeast 2 mu plasmid. A mutational analysis of the FLP binding site. *J. Mol. Biol.* 201:405–21
74. Serre MC, Jayaram M. 1992. Half-site strand transfer by step-arrest mutants of yeast site-specific recombinase Flp. *J. Mol. Biol.* 225:643–49
75. Sherratt DJ, Wigley DB. 1998. Conserved themes but novel activities in recombinases and topoisomerases. *Cell* 93:149–52
76. Shuman S. 1998. Vaccinia virus DNA topoisomerase: a model eukaryotic type IB enzyme. *Biochim. Biophys. Acta* 1400:321–37
77. Shuman S, Bear DG, Sekiguchi J. 1997. Intramolecular synapsis of duplex DNA by vaccinia topoisomerase. *EMBO J.* 16:6584–89
78. Smith SG, Dorman CJ. 1999. Functional analysis of the FimE integrase of *Escherichia coli* K-12: isolation of mutant derivatives with altered DNA inversion preferences. *Mol. Microbiol.* 34:965–79
79. Stark WM, Sherratt DJ, Boocock MR. 1989. Site-specific recombination by Tn3 resolvase: topological changes in the forward and reverse reactions. *Cell* 58:779–90
80. Stewart L, Redinbo MR, Qiu X, Hol WG, Champoux JJ. 1998. A model for the mechanism of human topoisomerase I. *Science* 279:1534–41
81. Subramanya HS, Arciszewska LK, Baker RA, Bird LE, Sherratt DJ, Wigley DB. 1997. Crystal structure of the site-specific recombinase, XerD. *EMBO J.* 16:5178–87
82. Tekle M, Warren DJ, Tapan B, Ellenberger T, Lanza A, Nunes-Duby SE. 2002. Attenuating functions of the C-terminus of lambda integrase. *J. Mol. Biol.* 324:649–65
83. Umana P, Gerdes CA, Stone D, Davis JR, Ward D, et al. 2001. Efficient FLPe recombinase enables scalable production of helper-dependent adenoviral vectors with negligible helper-virus contamination. *Nat. Biotechnol.* 19:582–85
84. Van Duyne G. 2002. A structural view of tyrosine recombinase site-specific recombination. See Ref. 18a, pp. 93–117
85. Voziyanov Y, Lee J, Whang I, Lee J, Jayaram M. 1996. Analyses of the first chemical step in Flp site-specific recombination: Synapsis may not be a prerequisite for strand cleavage. *J. Mol. Biol.* 256:720–35
86. Voziyanov Y, Stewart AF, Jayaram M. 2002. A dual reporter screening system identifies the amino acid at position 82 in Flp site-specific recombinase as a determinant for target specificity. *Nucleic Acids Res.* 30:1656–63
87. Wittschieben J, Petersen BO, Shuman S. 1998. Replacement of the active site tyrosine of vaccinia DNA topoisomerase by glutamate, cysteine or histidine converts the enzyme into a site-specific endonuclease. *Nucleic Acids Res.* 26:490–96
88. Woods KC, Martin SS, Chu VC, Baldwin EP. 2001. Quasi-equivalence in site-specific recombinase structure and function: crystal structure and activity of trimeric Cre recombinase bound to a

three-way Lox DNA junction. *J. Mol. Biol.* 313:49–69
89. Xu CJ, Ahn YT, Pathania S, Jayaram M. 1998. Flp ribonuclease activities. Mechanistic similarities and contrasts to site-specific DNA recombination. *J. Biol. Chem.* 273:30591–98
90. Xu CJ, Grainge I, Lee J, Harshey RM, Jayaram M. 1998. Unveiling two distinct ribonuclease activities and a topoisomerase activity in a site-specific DNA recombinase. *Mol. Cell* 1:729–39
91. Yang SH, Jayaram M. 1994. Generality of the shared active site among yeast family site-specific recombinases. The R site-specific recombinase follows the Flp paradigm. *J. Biol. Chem.* 269:12789–96
92. Yang SW, Burgin AB Jr, Huizenga BN, Robertson CA, Yao KC, Nash HA. 1996. A eukaryotic enzyme that can disjoin dead-end covalent complexes between DNA and type I topoisomerases. *Proc. Natl. Acad. Sci. USA* 93:11534–39
93. Yang Z, Champoux JJ. 2001. The role of histidine 632 in catalysis by human topoisomerase I. *J. Biol. Chem.* 276:677–85

Annu. Rev. Biophys. Biomol. Struct. 2003. 32:161–82
doi: 10.1146/annurev.biophys.32.110601.142525

First published online as a Review in Advance on February 11, 2003

THE POWER AND PROSPECTS OF FLUORESCENCE MICROSCOPIES AND SPECTROSCOPIES

Xavier Michalet,[1] Achillefs N. Kapanidis,[1] Ted Laurence,[1] Fabien Pinaud,[1] Soeren Doose,[1] Malte Pflughoefft,[1] and Shimon Weiss[1,2]

[1]Department of Chemistry and Biochemistry, [2]Department of Physiology, UCLA, Young Hall, 607 Charles E. Young Drive East, Los Angeles, California 90095; email: michalet@chem.ucla.edu; kapanidi@chem.ucla.edu; laurence@chem.ucla.edu; fpinaud@chem.ucla.edu; sdoose@chem.ucla.edu; pflughoe@chem.ucla.edu; sweiss@chem.ucla.edu

Key Words two-photon, single-molecule, lifetime, polarization, FRET

■ **Abstract** Recent years have witnessed a renaissance of fluorescence microscopy techniques and applications, from live-animal multiphoton confocal microscopy to single-molecule fluorescence spectroscopy and imaging in living cells. These achievements have been made possible not so much because of improvements in microscope design, but rather because of development of new detectors, accessible continuous wave and pulsed laser sources, sophisticated multiparameter analysis on one hand, and the development of new probes and labeling chemistries on the other. This review tracks the lineage of ideas and the evolution of thinking that have led to the actual developments, and presents a comprehensive overview of the field, with emphasis put on our laboratory's interest in single-molecule microscopy and spectroscopy.

CONTENTS

INTRODUCTION

Optical microscopy has made continuous progress since it was born at the end of the sixteenth century, with the creation of the first two-lens microscope by the Jansens in Middleburg, Holland. Today's research microscopes and objective lenses have been perfected to achieve the diffraction limit of resolution defined by Ernst Abbe at the end of the nineteenth century (1). Since this landmark, advances in (nonfluorescence) optical microscopy have come from new imaging modes such as phase contrast, Nomarski's differential interference contrast (DIC), or Hoffman's contrast, to name a few. Despite their advantages in enhancing contrast or details due to variations in the index of refraction, these transmitted-light techniques do not provide any means to distinguish individual objects or identify small organelles other than by their shape or optical density. To circumvent this limitation, staining agents and techniques have been developed by histologists since the beginning of microscopy. However colorful, these techniques developed before the advent of molecular biology have staining specificities that rely on poorly understood interactions. They also do not allow the detection of objects that are smaller than the diffraction limit or that do not present enough contrast. Fluorescence microscopy overcomes these limitations by rejecting the excitation light, leaving visible only the sources of emission. The development of immunocytochemistry during the twentieth century gave this technique its full potential. Initially limited to fixed samples, fluorescence immunocytochemistry was extended to live cell imaging during the 1980s (133).

Confocal laser scanning microscopy (CLSM) has also steadily grown as an indispensable tool for three-dimensional imaging. CLSM capabilities have been extended to include multiphoton excitation processes and lifetime imaging, or to improve its three-dimensional resolution. Progress in detector technology, interest in the photophysics of single-quantum emitters, and questions arising from the burgeoning field of single-molecule biophysics have recently pushed fluorescence microscopy to its ultimate level of sensitivity. Single fluorescent molecules can now be detected in a living cell and localized with nanometer precision in real time. In parallel with these improvements in image versatility, sensitivity, and resolution, fluorescence has also been used as a tool to probe the dynamics, conformational changes, and interactions of single molecules. This modern development has completed the transition of fluorescence microscopy from a purely imaging technique to nanospectroscopy (spectroscopy of small volumes), extending the application of the microscope to structural biology, biochemistry, and biophysics, and providing new tools for the postgenomic (i.e., proteomic) era. Single-molecule spectroscopy (SMS) has recently shed light on inter- and intramolecular interactions, protein folding, and protein structure, as well as on the functioning of the cellular machinery.

This article gives an overview of achievements in the field of fluorescence microscopy and spectroscopy, as well as reviews recent developments focused on single-molecule sensitivity. As they take full advantage of the properties of the fluorescence emission process by individual molecules, we first present in some detail experimental results illustrating each of these properties. We then discuss how to practically take advantage of these capabilities by presenting an overview of fluorescent labeling techniques and experimental setups. The two final parts review imaging applications of fluorescence and single-molecule experiments.

FUNDAMENTALS OF FLUORESCENCE, FLUOROPHORES AND LABELING, AND FLUORESCENCE DETECTION

Fundamentals of Fluorescence

Fluorescence is the phenomenon of photon emission following absorption of one (or more) photon(s) by a molecule or material (fluorophore) that returns to its ground state. First observed for quinine by Herschel in the early nineteenth century, it was further studied by Stokes who correctly identified its main characteristics (for the case of single-photon excitation), namely that the emitted photons have a longer wavelength than the absorbed ones (70). A series of experimental and theoretical studies by several investigators in the early twentieth century uncovered most of today's known properties of fluorescence (99).

Excitation and emission processes in a typical molecule are represented by a Jablonski diagram (Figure 1) depicting the initial, final, and intermediate electronic and vibrational states of the molecule. In general, fast intramolecular vibrational relaxations result in emitted photons having a lower energy than do the incident ones (or equivalently, a larger wavelength, the difference being the so-called Stokes shift). This property is the basis of the simple separation of emitted fluorescence from excitation light, which renders fluorescence such a powerful tool. The emitted photon is detected within a typical delay (lifetime) after absorption of the excitation photon, which depends on the species studied and its local environment. Organic dyes have typical lifetimes from several tens of picoseconds to several nanoseconds. Longer lifetimes are obtained for fluorescent semiconductor nanocrystals (NCs) (tens of ns), organometallic compounds (hundreds of ns), and lanthanide complexes (up to ms). The probability distribution of emission times is usually monoexponential, characterized by a lifetime $\tau = \Gamma^{-1} = (k_r + k_{nr})^{-1}$, where k_r and k_{nr} are the radiative and nonradiative decay rates, respectively (Figure 1). The latter is dependent on the local environment via perturbations of the intramolecular transition matrix elements. For instance, the proximity to a dielectric or metallic surface markedly modifies the fluorescence lifetime (6, 141).

While in its excited state, a molecule has a probability to end up in a nonemitting triplet-state during microseconds to milliseconds, resulting in dark states (52, 95).

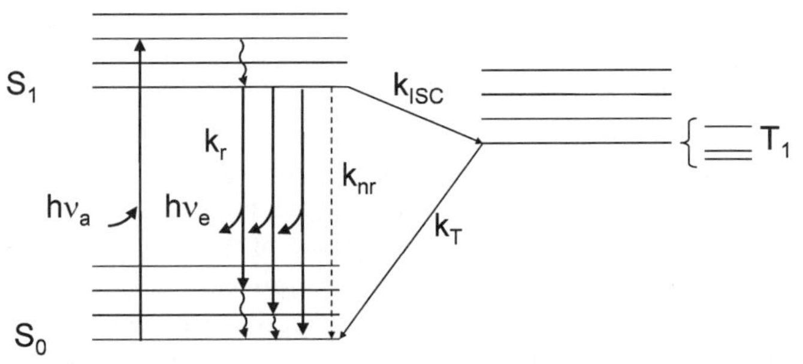

Figure 1 Jablonski diagram for fluorescence. Upon absorption of a photon of energy $h\nu_a$ close to the resonance energy $E_{S_1} - E_{S_0}$, a molecule in a vibronic sublevel of the ground singlet state S_0 is promoted to a vibronic sublevel of the lowest excited singlet state S_1. Nonradiative, fast relaxation brings the molecule down to the lowest S_1 sublevel in picoseconds. Emission of a photon of energy $h\nu_e < h\nu_a$ (radiative rate k_r) can take place within nanoseconds and bring back the molecule to one of the vibronic sublevels of the ground state. Alternatively, collisional quenching may bring the molecule back to its ground state without photon emission (nonradiative rate k_{nr}). A third type of process present in organic dye molecules is intersystem crossing to the first excited triplet state T_1 (rate k_{ISC}). Relaxation from this excited state back to the ground state is spin-forbidden, and thus the lifetime of this state is in the order of microseconds to milliseconds. Relaxation to the ground state takes place either by photon emission (phosphorescence) or nonradiative relaxation. The fluorescence lifetime is defined by $\tau = 1/\Gamma = (k_r + k_{nr})^{-1}$.

Other fluorescent molecules such as green fluorescent proteins (GFPs) (31) or semiconductor NCs (37) exhibit similar dark state intervals, although for different reasons [different long-lived dark states for GFP (31), Auger ionization or surface trapping of carriers for NC (35)].

Spectral jumps can also be observed at the single-molecule level (53). This phenomenon results from a shift of absorption and emission maxima due to the changing environment of the molecule or a sudden conformational change of the molecule itself (5, 125, 141).

The efficiency of photon absorption is proportional to $(\vec{E}.\vec{\mu})^2$, where $\vec{E}$ represents the local electric field, and $\vec{\mu}$ is the absorption dipole moment of the fluorophore (55, 70). For an immobilized fluorophore, the orientation of the molecule's absorption dipole can thus be determined by recording the emitted fluorescence as a function of the orientation of the linear polarization of the excitation light. This information in turn allows the determination of the spatial orientation of the fluorophore (51, 141). For a mobile molecule, more information is needed because the emission dipole may have time to tumble significantly (54): Fluctuations faster than the fluorescence lifetime lead to a depolarized emission; fluctuations taking place over timescales longer than the lifetime but shorter than the integration time lead to anticorrelation of the two orthogonal emission polarizations. The emission polarization is needed to fully recover the relevant information. In particular, it is important to recover the projection of the polarization on more than two orthogonal axes as illustrated in Figure 2. This can be achieved in different ways in wide-field imaging approaches as well as confocal ones (39).

Fluorescence resonance energy transfer (FRET), first described theoretically by Perrin (100) and later fully elucidated by Förster (40), is a special case of influence of the local environment on fluorescence. If a nearby molecule (acceptor) has an absorption spectrum overlapping with the emission spectrum of the studied molecule (donor), the energy absorbed by the donor can be transferred

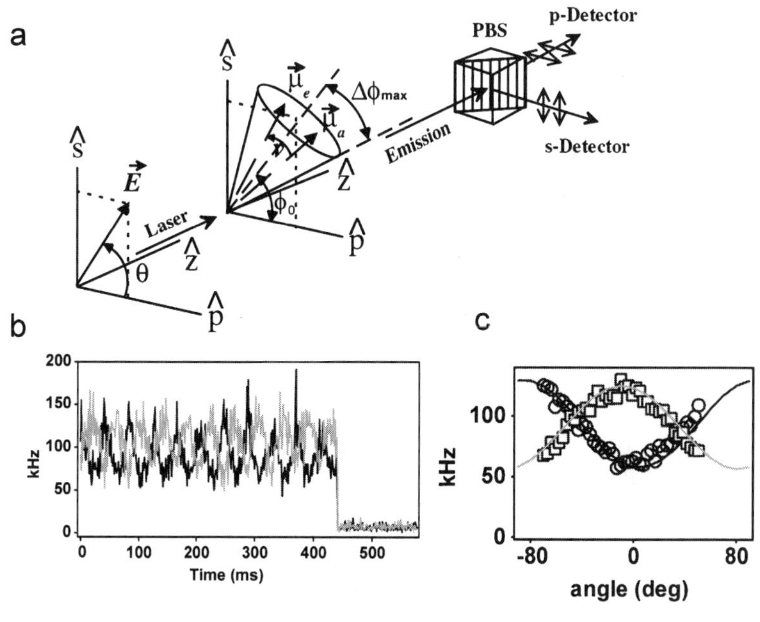

Figure 2 (*a*) Experiment schematic. $\vec{E}$: electric field, making an angle θ with the *p* polarization axis. The excitation propagates along axis *z*, which is also the collection axis. $\vec{\mu_a}$ and $\vec{\mu_e}$ are the absorption and emission dipole moments, initially aligned. ν represents the rotational diffusion of the emission dipole during the excited lifetime. The dipole is supposed to be confined in a cone positioned at an angle ϕ_0 projected on the (s, p) plane that has a half-angle $\Delta\phi_{max}$. A polarizing beam-splitter (PBS) splits the collected emission in two signals I_s and I_p, which are simultaneously recorded by APDs. (*b*) Simultaneously recorded I_s (*black*) and I_p (*gray*) of a molecule rapidly rotating in liquid. (*c*) Same data as in (*b*), but average over the 11 "on" periods. The fit corresponds to a $\Delta\phi_{max}$ close to 90° (freely rotating molecule) and permits determination of the constrained rotational diffusion parameter. Adapted with permission from Reference 54. Copyright 1998, the American Physical Society.

nonradiatively to the acceptor with an efficiency E given by:

$$E = \left(1 + (r/R_0)^6\right)^{-1}, \quad 1.$$

where r is the distance between the two emitting centers, and R_0 is the Förster radius (in Å). R_0 is of the order of a few nanometers. Stryer & Haugland (118) thus suggested using it as a molecular ruler. An example of this utilization is given in Figure 3 (see color insert). The corresponding donor-acceptor distances are measured on diffusing molecules with subnanometer precision, using the relative intensities in each color channel. Alternatively, the transfer efficiency can be measured via the fluorescence lifetimes as:$E = 1 - \tau_{D(A)}/\tau_{D(0)}$, where $\tau_{D(A)}$ and $\tau_{D(0)}$ are the donor lifetimes in the presence or absence of acceptor, respectively (49). Electron transfer can also significantly perturb the lifetime of a dye by opening a new nonradiative path in the Jablonski diagram of the molecule (Figure 1).

Although most single-molecule experiments use one-photon fluorescence excitation because of its relatively large cross-section ($\sim 10^{-16}$ cm^2), fluorescence can also be excited via a two-photon absorption process (28) using laser excitation with half the photon energy needed to attain the excited state (106). However, because of the low cross-section and the quadratic dependence on the incident power, an excitation power several orders of magnitude larger than for one-photon excitation is needed. This increases photobleaching because of the high probability of photochemical degradation in the long-lived triplet state and the interplay of multiphoton ionization processes (32). In compensation, excitation takes place in a substantially reduced volume of the sample, reducing the out-of-focus background contribution and out-of-focus bleaching.

Fluorophores and Labeling

A great variety of fluorophores are available (59, 70). Fluorescent organic molecules (13), GFPs (31) and other fluorescent proteins (73), conjugated polymers (J-aggregates) (130), light-harvesting complexes (140), dendrimers (62), or semiconductor NCs (37) are a few examples of systems that have been extensively studied at cryogenic as well as at room temperature with SMS techniques. Each of these systems exhibits fluorescence based on specific processes, which can be quite different from that illustrated on Figure 1. In NCs (4) for instance, ultraviolet-visible photon absorption by a semiconductor compound leads to the creation of an electron-hole pair (exciton). The pair recombines within few tens of nanoseconds, emitting a visible photon whose wavelength depends on the NC diameter owing to quantum confinement effects (4).

Fluorophores can be added in vitro to most proteins or other biomolecules after biosynthesis and purification (132), either statistically or specifically. Statistical labeling is mainly used for imaging purposes or restricted to preliminary stages of assay development, as for the FRET-based analysis of staphylococcal nuclease dynamics (56). However, site-specific labeling is a necessity when precise distance or orientation information is sought (27). It requires a careful choice of labeling

TABLE 1 Labeling strategies

Method	Application	References
Cysteine-specific labeling with thiol-reactive reagents	Widespread use for FRET and fluorescence polarization analysis of small proteins (<500 residues)	(2, 42, 59, 68, 107, 112, 127, 132, 135)
Bis-functional Cys-reactive fluorophore	Monitoring of orientation and dynamics of protein domains	(20, 101, 116)
Peptide ligation	Intramolecular FRET	(21, 24, 27, 88, 113, 122)
Fluorescent derivatives of the antibiotic puromycin	C terminus labeling	(90, 143)
In vitro reconstitution from purified components	Multicomponent complexes	(80, 89, 107, 135)
Genetically encoded fluorescent protein	In vivo labeling	(48, 63, 126)

chemistry, optimization of labeling reaction, and rigorous characterization of the labeled biomolecules for efficiency, site-specificity, and retention of functionality. Some of the many available methods, which are discussed in detail in (66), are presented in Table 1. Few molecules do not require labeling because of the presence of fluorescent moieties either in their own structure or in cofactors. This is the case of proteins with tryptophan residues, enzymes using NADH, or flavins as cofactors, as illustrated in Reference 73. This autofluorescence of native proteins is in fact a source of background in cell fluorescent imaging applications.

Fluorescence Detection

Fluorescence acquisition geometries can be classified according to their excitation and emission schemes. Wide-field detection schemes use either epifluorescence illumination with lamps (47), defocused laser excitation (109), or total internal reflection (TIR) excitation (42). Detectors include back-thinned charge-coupled devices (CCDs) with quantum efficiencies (QE) up to 90%, but a usable readout rate limited by readout noise to a few full frames per second (47, 109). Intensified CCDs overcome the readout noise limitation by signal amplification, allowing higher frame rates but at the price of a lower QE (<40%). The new electron-multiplying CCD technology should permit increased frame rates, with transfer rates as high as 10 MPixels/s with single-molecule sensitivity (74).

Point-detection schemes encompass confocal and near-field scanning optical microscopies (NSOM). The excitation volume has a radius of the order of the excitation wavelength for CLSM (97) and of the tapered fiber core ($\sim$100 nm) for NSOM (Figure 4, see color insert) (14, 38). Images of the sample are acquired pixel by pixel in a raster fashion using a point-detector and reconstructed by software.

Commercial CLSM uses a pair of galvanometer-mounted mirrors or acousto-optical deflectors to move the excitation laser beam across the sample and photomultiplier tubes (PMT) for photon detection. This choice has the advantage of great speed (video-rate imaging is possible) but does not provide enough sensitivity for SMS. SMS requires a slower, stage-scanning method and sensitive avalanche photodiodes (APD). Indeed, until recently, PMT had a QE $<20\%$ in the visible spectrum, against $\sim 70\%$ for silicon APD. Progress in photocathode technology (using GaAsP) should increase the QE of PMT, making them attractive detectors for beam-scanning CLSM because of their larger sensitive area.

In one version of NSOM that closely resembles the original suggestion by Synge (119) at the beginning of the twentieth century, a narrow optical fiber is brought in close proximity (tens of nm) to a sample and a raster is scanned over it (25). The underlying molecules are sensitive only to the near-field contribution of the transmitted laser electric field, which extends over distances smaller than the wavelength, resulting in a higher-resolution image collected by a microscope objective lens (38).

The use of lasers as excitation sources limits the range of accessible excitation wavelengths but provides higher intensities and allows pulsed excitation with ultrafast lasers. They can be used in the near-infrared (NIR) range to perform two-photon fluorescence excitation, whose square dependence on the intensity results in a quick decrease of the excitation away from the focus (28). Even though spatial resolution is not necessarily enhanced and fluorophore photobleaching is increased (96), two advantages result from this rather expensive and still sophisticated technique: Out-of-focus bleaching is reduced, and sample penetration is increased because of the reduced absorption of NIR radiation, allowing thick, live tissues to be imaged with little damage to the environment. The technique has thus found impressive applications in neuroimaging (77) and deep-tissue imaging (114).

In SMS, wide-field imaging is necessary for particle-tracking studies but can also be preferred over single-point detection for simultaneous observation of several spatially separated molecules (109). This reduces the amount of time needed to accumulate a statistically significant number of observations. It is especially relevant for experiments in which irreversible processes are triggered by modification of an external parameter. Point-detection geometries allow the acquisition of fluorescence time traces of immobilized molecules with high temporal resolution, as well as fluorescence lifetime information (75), but they are rather slow for imaging single molecules. CLSM is also extensively used for the study of freely diffusing molecules in solution or embedded in fluid membrane (lipid bilayer of cell membranes) by fluorescence correlation spectroscopy (FCS) (111, 139).

FLUORESCENCE IMAGING: A CLASSIC REVISITED

Imaging Modes: Intensity, Spectrum

Fluorescence allows rejection of the excitation signal and only detection of the fainter emission light using filters and dichroic mirrors. This does not lead to any

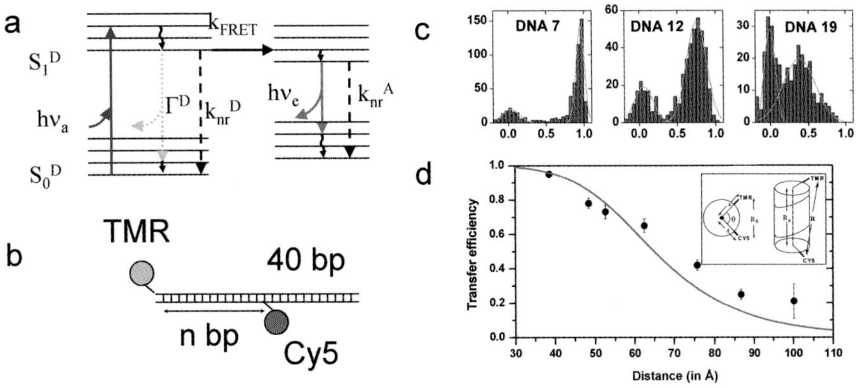

Figure 3 (*a*) Jablonski diagram for FRET. Fluorescence energy transfer involves two molecules: a donor D and an acceptor A, whose absorption spectrum overlaps the emission spectrum of the donor. Excitation of the acceptor to the lowest singlet excited state is a process identical to that described for single-molecule fluorescence (Figure 1). Energy transfer to the acceptor by dipole-dipole interaction, in the presence (within a few nm) of a nearby acceptor molecule, quenches donor fluorescence emission. The donor exhibits fluorescent emission following the rules outlined in Figure 1*a*. (*b*) DNA *n* constructs used for the FRET distance study. Tetramethylrhodamine (TMR) is attached to the 5′ end of the DNA, and Cy5 is attached to the nth base from the 5′ end ($n = 7, 12, 14, 16, 19, 24, 27$). (*c*) FRET histograms extracted from time traces for DNA 7, 12, and 19. Double Gaussian fits extract numbers for the mean (width) of the higher efficiency peak of 0.95 ± 0.05, 0.75 ± 0.13, and 0.38 ± 0.21, respectively. The peak around zero efficiency corresponds to nonfluorescent acceptor molecules. (*d*) Mean FRET efficiencies extracted from FRET histograms plotted as a function of distance for DNA 7, 12, 14, 16, 19, 24, and 27. Distances are calculated using the known B-DNA double-helix structure. The solid line corresponds to the expected Förster transfer curve for $R_0 = 65$ Å. (*b, c*) Adapted with permission from Reference 26. Copyright 1999, the National Academy of Sciences USA.

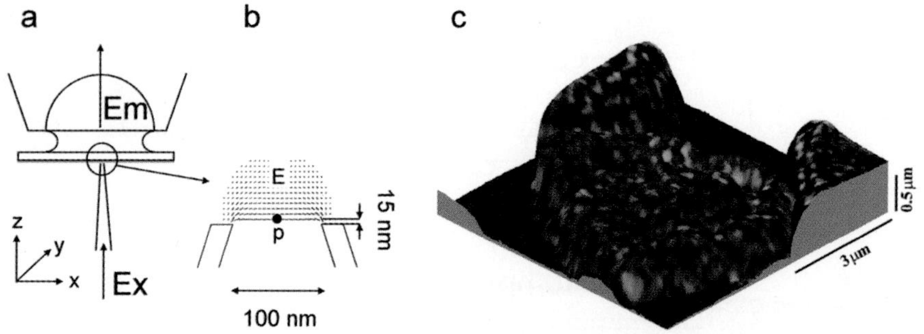

Figure 4 (*a*) Schematic of a near-field scanning optical microscope setup used for single-molecule imaging. An aluminum-coated tapered optical fiber (raster-scanned at nanometer distance from the sample) with a subwavelength aperture (50–100 nm) serves as a waveguide for laser excitation. Shear-force feedback keeps the tip at a constant distance from the sample, resulting in a signal used for nanometer-resolution topographic reconstruction of the scanned area. (*b*) The excitation volume and the corresponding local evanescent electric field are detailed in the expanded view. Fluorescence light emitted by individual molecules is collected by an oil immersion, high NA objective and recorded by an APD. (*c*) Composite image obtained by superimposing the fluorescence intensity detected in two separate channels over a topographic map of red blood cells obtained by shear force feedback of the near-field microscope. Green spots correspond to fluorescence of host proteins labeled with FITC. Red spots correspond to malaria proteins labeled with Texas Red. (*a, b*) Adapted with permission from Reference 13. Copyright 1993, the American Association for the Advancement of Science. (*c*) Adapted with permission from Reference 53. Copyright 1996, IEEE.

Figure 6 (*right*) Time-gated imaging of semiconductor NCs. (*a*) Mouse 3T3 fibroplasts were incubated with NCs and fixed in 2% formaldehyde, 0.5% glutaraldehyde. Observations were performed using a customized confocal microscope [for details see (22)]. Integration time per pixel is 10 ms, the lifetime window being 0–150 ns after the laser pulse (repetition rate: 5 MHz). This image is obtained using all the detected photons. The ellipse indicates the location of the nucleus. (*b*) The same recording, but retaining only photons arrived between 35 and 65 ns after a laser pulse. A marked decrease of the background is observed with a few bright spots clearly dominating. (*c, d*) Intensity profiles along the white-dashed line in *a* and *b*. The total signal decreases notably, but the signal-to-average-background ratio jumps from 3 to 45. Note that part of the cytoplasm fluorescence might also be due to NCs. Adapted with permission from Reference 22. Copyright 2001, Optical Society of America.

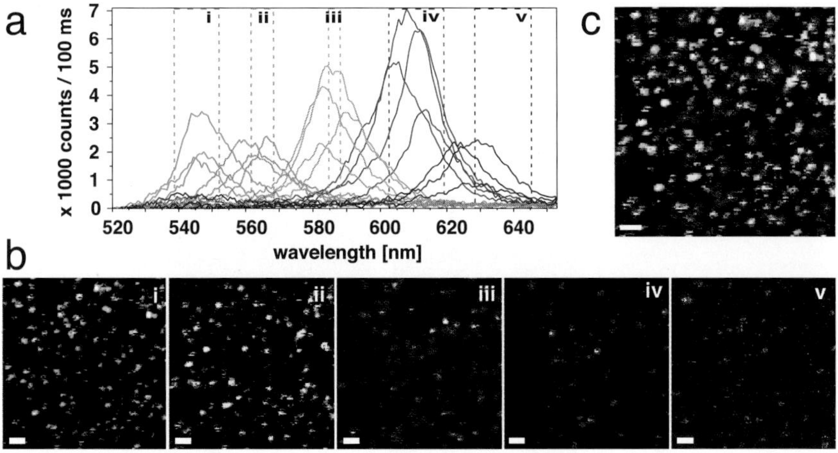

Figure 5 Spectral imaging semiconductor nanocrystals. Scans of a mixture of four NC samples (ensemble peak emissions: 540, 575, 588, and 620 nm). (*a*) A representative collection of individual NC spectra (about 20 nm FWHM) obtained from the integrated data of 3 x 3 pixels. Despite their overlap, five orthogonal spectral bands (*i* to *v*) could be defined. (*b*) Five false-color images corresponding to the spectral bands defined in (*a*). (*c*) Overlay of the five perfectly registered images of (*b*). 10 x 10 μm^2 scan (pixel size: 78 nm, scale bar: 1 μm). Adapted with permission from Reference 81. Copyright 2001, Academic Press.

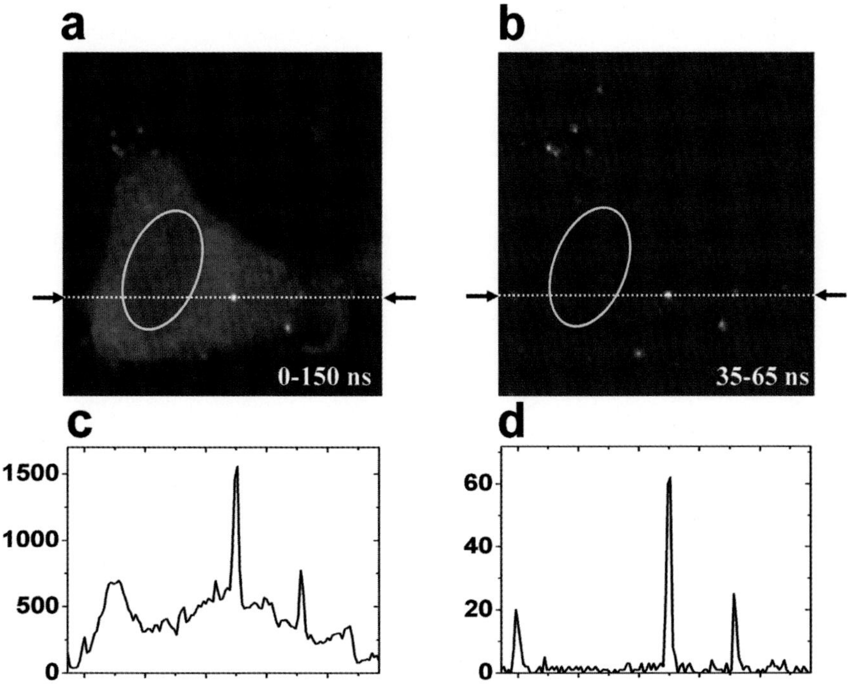

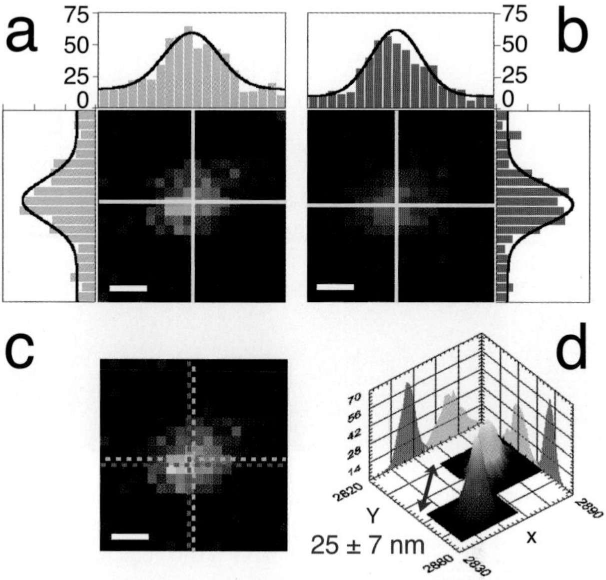

Figure 7 Ultrahigh-resolution colocalization of individual nanocrystals. Mixture of green (Em: 540 nm) and red (Em: 620 nm) NCs excited at 488 nm (excitation power: 200 nW incident or 320 W/cm^2 peak irradiance; integration time: 50 ms). (*a*, *b*) Green and red channel images of a 1 x 1 μm^2 scan obtained by raster scanning the sample through the fixed excitation PSF and recording the respective signals on two different APDs (pixel size: 50 nm; scale bar: 200 nm). As visible from the intensity profiles along two orthogonal lines passing through the PSF's centers, the count rates are similar in both channels. Black curves indicate the corresponding cross-sections of the fitted PSFs. (*c*) Overlay of the two channels with indication of the determined PSF's centers. (*d*) Bootstrap replicas of the datasets were fitted in order to estimate the uncertainty of the position determination (34). The figure shows the histograms of the fitted centers distribution obtained from 1000 simulations. The measured distance is 25 nm with a corresponding uncertainty of 7 nm (68% confidence limit). Adapted with permission from Reference 69. Copyright 2000, the National Academy of Sciences USA.

resolution improvement, but subresolution objects can now be detected as contrasted spots of fluorescent light with a diffraction-limited size (the point spread function, or PSF). With the advent of digital, high-sensitivity cameras, successive excitation and detection of different fluorophores make it possible to obtain multicolor images with good signal-to-noise ratio (SNR) (61).

Standard fluorophores have broad-emission spectra, a fact that renders their spectral separation by bandpass interference filters imperfect, and impractical in the case of several colors. It is then advantageous to use spectral imaging methods that collect all emitted light for further processing. Several of these methods have been proposed and implemented in recent years, both in wide-field (78) and point-detection geometries (30, 69). Their goal is to recover the contribution of each fluorophore to the recorded emission at different wavelengths for each individual pixel. Fluorescence being an incoherent process and emission spectra being well defined, intensities from different fluorophores simply add at each wavelength of the emission spectrum. Knowing the emission spectra, we can recover the contribution of each individual fluorophore by a simple matrix inversion. Figure 5 (see color insert) illustrates this point with a mixture of five semiconductor NC samples spin-coated on a cover-glass and imaged with a prism and ICCD-based confocal spectrometer (69, 82). These and future developments may improve the sensitivity of ion- or pH-sensitive fluorophores (59), which are characterized by spectral changes upon variation of the local concentration of an ionic species and have been used at the single-molecule level (17).

Other Imaging Modes: Lifetime, Time-Gated, and FRET Imaging

After intensity and spectrum, fluorescence lifetime is the next most useful observable fluorescence emission property for imaging. Most fluorophores excited by a pulsed laser emit fluorescence photons after a few nanoseconds. Lifetime is extremely sensitive to the molecule's environment. The purpose of fluorescence lifetime imaging (FLIM) is thus to measure the fluorescence lifetime at each point of a sample to map either the presence or absence of a species, or the environment of a known single fluorophore (45, 134). This is especially relevant for fluorescent proteins such as GFP, which cannot easily be separated spectrally (98). Two implementations of this technique are available: One uses a time-domain measurement (124), the other a frequency-domain measurement (45, 71, 134). The time-domain approach can be implemented using a time-correlated single-photon counting (TCSPC) confocal setup (124), or using a time-gated ICCD in either a multifoci excitation configuration (117) or a wide-field illumination (11). In the first case, the scanning process is time-consuming, while in the second the camera detects only photons emitted during a fixed time-window after the laser pulse, losing any information on the remaining photons. In particular, time-gated detection can only distinguish between fluorophores of well-separated lifetimes and necessitates acquiring two sets of images for that purpose. This technique has been exploited for lanthanide chelates (131), metal ligand complexes (123), and NC (22) imaging. These fluorophores have much longer lifetimes than do the autofluorescence of cell proteins, which make them attractive probes for high-sensitivity, background-free intracellular imaging (Figure 6, see color insert).

The frequency-domain approach is no less complicated and is based on radiofrequency (RF) modulation of the laser intensity and of the image intensifier gain, either in phase (homodyne) or out-of-phase (heterodyne). The acquisition of several images at different phase differences allows the calculation of an apparent lifetime and concentration for each pixel of the image, but this process can be extremely time-consuming (several minutes) (72), resulting in photobleaching. Applications to live intracellular Ca^{2+} imaging (72) or of receptor phosphorylation events (92) have illustrated the power of this imaging technique where spectral information is of little help. The advantages of two-photon microscopy can be combined with FLIM in the frequency domain (102, 115), time-gated FLIM (120), or TCSPC FLIM (10).

FRET imaging was first explored using fluorescence donor/acceptor photobleaching (23, 44) in order to extract background components of the signal. However, the irreversible modification of the sample that follows makes it far from ideal. Other methods have been proposed. FLIM can measure energy transfer using the relation $E = 1 - \tau_{D(A)}/\tau_{D(0)}$ introduced previously. It provides a concentration-independent FRET measurement and minimizes illumination (43). A combination of intensity and lifetime observations would benefit FRET study (57), giving access to dynamic interactions between FRET pairs on a cell-wide scale.

High-Resolution Imaging and Localization

Compared to an optical-sectioning technique such as DIC, standard fluorescence microscopy suffers from image blurring due to out-of-focus signal. Deconvolution algorithms have been developed during the 1980s in order to reassign out-of-focus light back to its original source location, using series of images taken at different foci (3). The reliability and ease of use of these algorithms have improved steadily with computer power, allowing in certain conditions imaging with a resolution better than the diffraction limit (super resolution) (18). Still, limited image transfer rate and processing time confine these techniques to fixed samples or slow processes. CLSM has come as a remedy to blurring by removing most ray-lights emitted out of focus with a pinhole (97).

Nevertheless, the resolution is still limited by the extension of the excitation PSF. Working toward reducing the excitation volume, several PSF engineering strategies have been proposed and demonstrated in order to reduce it at least in one direction. Interference methods have been implemented that create a narrower central excitation volume along the optical axis thus improving the vertical resolution of fluorescence microscopy (8, 41, 50, 60). A more sophisticated approach, stimulated emission depletion (STED), takes full advantage of the physics of fluorescence emission (67). Combining both approaches, an image with a three-dimensional resolution of ~30 nm has recently been obtained (33). This imaging resolution is close to the best performances of NSOM and is not limited to surface studies.

Although improvements in imaging resolution are still possible, a much better performance has already been reported in a related problem: high-resolution (co)localization. In this case, one is interested not so much in the exact geometric structure of an object than in the relative position of two objects. This problem has been investigated thoroughly in the domain of bright-field microscopy, where micrometer-sized beads were observed and localized with nanometer resolution using centroid-finding algorithms (46). In fluorescence microscopy, the image is that of the PSF. If it is detailed enough and if its SNR is large enough, precise subpixel localization can be obtained (12, 15). This idea has been successfully put in practice at the single-molecule level in near-field (53), wide-field (109, 110), and confocal microscopy (69). Reaching nanometer resolution requires a careful consideration of all sources of optical aberrations, especially chromatic if one considers distances between objects with different colors. For these reasons, NCs, which can all be excited by a single visible wavelength and detected simultaneously, represent in principle the ultimate probe for this type of application (Figure 7, see color insert) (69, 81).

SINGLE-MOLECULE SENSITIVITY: WATCHING MOLECULES ROCK 'N ROLL

The field of SMS has rapidly grown since its inception in the 1990s. The ability to watch one molecule at a time gives access to the distribution functions of observables instead of the first statistical moments obtained in ensemble measurements. It helps resolve subpopulations in heterogeneous samples or record asynchronous time trajectories of observables that would otherwise be hidden during biochemical reactions or similar processes.

SMS and microscopy experiments have already been evoked in the first parts of this article to illustrate different properties of fluorescence (Figure 2). We now point to the specific requirements needed to reach single-molecule sensitivity and present a brief overview of applications and prospects of this fast-developing methodology.

Signal-to-Noise Requirement

Single-molecule observations require a careful optimization of background, signal, and noise (83). SNR and signal-to-background ratio (SBR) can be increased by improving the collection efficiency, and the SBR can be further improved by decreasing the excitation volume V. A larger excitation power or a longer integration time improves the SNR without affecting the SBR. The value of the residual background rate can be reduced by a careful choice of buffer, embedding matrix or immersion medium, and rejection filter or use of a confocal design.

For instance, typical values of the relevant parameters in the case of a CLSM study of freely diffusing fluorescein isothiocyanate (FITC) in water may lead to count rates of a few tens to hundreds of counts/ms and SNR $\sim$10 (83). However, single molecules have a finite life span. In an oxygen-rich environment, they typically emit on the order of 10^6 photons before irreversible photobleaching. This happens after a few hundred milliseconds, larger than the typical transit time (a few hundred μs). For an immobilized molecule, however, it sets a stringent limit on the total duration of a single-molecule observation.

The Signature of a Single Fluorophore

In addition to SBR and SNR issues, care has to be taken to ensure that the collected signal originates from a single molecule. Several fluorescent molecules can indeed occupy the diffraction-limited excitation volume. If intensity fluctuations (or variations of any other spectroscopic characteristics) are observed, they could be due to single-molecule dynamics or environment changes, but they could as well reflect the stochastic mixture of emissions from nearby molecules.

Two different strategies can be envisioned to reduce this uncertainty: (*a*) Work at low concentration, such that at most one molecule is present in the excitation volume, or equivalently, minimizes the excitation volume; and (*b*) use a selective excitation or emission-detection protocol, such that only one molecule is excited or detected within the sampled volume.

The first can be used in both solution and immobilized conditions. The second strategy requires fluorophores having either separable excitation (86, 95) or emission properties (69, 124, 128) and has been illustrated with fluorescent semiconductor NCs in Figure 5.

In addition to fulfilling the above experimental criteria for single-molecule detection, a number of tests can be performed to ascertain that the observed signal actually comes from a single emitter (13). These criteria are direct consequences of the photophysical properties of fluorophores:

- The observed density of emitters varies according to the known concentration of molecules.
- The observed fluorescence intensity level is consistent with that of a single emitting molecule.
- Each immobilized emitter has a well-defined absorption or emission dipole.
- Fluorescence emission exhibits only two levels (on/off behavior due to blinking or photobleaching) over timescales where no changes in the environment are expected (93).
- If there are two or more emission levels, photophysical property changes are correlated (91).
- The emitted light exhibits antibunching, i.e., no simultaneous emission of two photons (9).

Single-Molecule Fluorescence Observation

Single-molecule fluorescence detection has undergone rapid developments in the past few years, spreading in multiple fields of science (64, 83, 85, 136, 137, 142). Here we present only a few examples of its power, with FCS methods, and studies in material science and biology.

FLUORESCENCE CORRELATION SPECTROSCOPY FCS was introduced in the early 1970s (36, 76) as a method to study thermodynamic fluctuations of freely diffusing molecules at the small ensemble level (few molecules diffusing simultaneously within the excitation volume). The method is usually implemented in a confocal detection geometry and involves the recording of the arrival time of all photons emitted in the small detection volume (~1 fl). Analysis of the correlation function of the recorded intensity permits extraction of the concentration, diffusion coefficient, photophysical characteristics, and for some of its variants, brightness of one or multiple fluorophores in solution or in fluid membranes (111). The autocorrelation function of a single-channel intensity or the cross-correlation function of multiple channels give access to fast timescales (as well as slower timescales), which are not accessible from the study of individual immobilized molecules due to shot-noise. Simultaneous consideration of other dimensions of fluorescence (polarization, lifetime) or data analysis schemes taking full advantage of the recorded arrival times have given rise to a number of applications in photophysics (138), conformational dynamics (16), macromolecular interactions (104), and study of the kinetics of enzyme-catalyzed reactions (103). Many of the recent developments in detection of molecular interactions have been focused on increasing the sensitivity to brightness to monitor ligand-protein binding equilibria (19), probe the stoichiometry of protein complexes (79), study oligonucleotide-polymer interactions (129), or to probe receptor-ligand interactions in a format compatible with ultrahigh-throughput screening (104, 108). These methods show improved sensitivity over FCS methods that rely on diffusion only.

For interactions between macromolecules of different types (for example heterodimerization of two proteins), extending the analysis to two channels improves the sensitivity over one-channel analysis. The molecules of one type are labeled with one color (e.g., yellow), and the molecules of the other type are labeled with another color (e.g., red). A complex of the two types of molecules thus has both labels. Signal from the two fluorophores is separated spectrally onto two detector channels, yellow and red. The binding of two molecules labeled with the yellow and red fluorophores is indicated by the coincident detection of simultaneous photon bursts on both channels [these applications are reviewed in (103)]. The cross-correlation technique has recently been applied in conjunction with CLSM to cellular environments (7). By imaging and subsequent placement of the confocal spot at a specific point in the cell, the endocytic pathway of bacterial cholera toxin was followed using cross-correlation analysis. Two subunits of the toxin were labeled with different fluorophores, and the subunits were colocalized until reaching the Golgi apparatus, where they separated. This study demonstrates the potential of cross-correlation analysis in living cells.

MATERIAL SCIENCE STUDIES Single-molecule photophysical properties are extremely sensitive to their local (nanometer-sized) environment (84, 87). They can for instance report on parameters such as the local pH (17) or on the local structure, as illustrated by experiments designed to study the local dynamics of a polymer matrix at the onset of the glass transition (29).

In this later experiment, molecules of the organic fluorophore Rhodamine 6G dispersed within a thin poly(methacrylate) film were observed at temperatures slightly above the melting temperature of the polymer, using fluorescence polarization CLSM (29). Each molecule exhibited a slow rotational diffusion over several hours. This study demonstrated that an individual molecule probes an increasing number of different environments over time (dynamic disorder). At long timescales, the observable characteristics (such as the autocorrelation function) are similar to those measured on an ensemble of molecules, as expected from the ergodicity hypothesis. At short timescales, each molecule reveals the peculiar local and stable characteristic of its nanoenvironment, which may be different from that of another molecule situated elsewhere (static disorder).

BIOPHYSICAL AND BIOLOGICAL STUDIES SMS has allowed a reassessment of long-standing questions in biophysics, biochemistry, and biology (64, 136) by giving scientists the possibility to study conformational dynamics and interactions of individual molecules in biological processes. From simple model systems based on DNA (54, 55) or small peptides (65), to ribozyme (145), motor-proteins (94, 116), and other biomolecules (2), or biomolecular complexes formed by association of a few molecules (121), SMS continues to contribute valuable information to biology.

An example of this versatility is provided by the labeling of the central part of a rotary motor protein, F_1-ATPase (2). In this experiment, the molecular rotor was labeled with a single fluorophore whose orientation, detected by emission polarization measurements, directly reported the angle of the rotor with respect to the shaft. The small size of the fluorophore guaranteed that the protein motion would not be hindered (contrary to previous experiments, which used many larger reporters such as micron-sized latex beads or fluorescent actin filaments). This experiment (2) reproduced the previous results, showing that the rotor performed 120° steps [more recent work revealed substeps of 90° and 30° (144)].

SMS is also the tool of choice to study protein folding (137). The transition from a denatured state to the fully folded native protein usually involves an unknown number of intermediate states, which are not accessible by ensemble measurements. Conformations of doubly labeled proteins can be monitored by SMS as they undergo folding, taking advantage of the distance dependence of FRET (Figure 3). Our laboratory has studied the enzyme chymotrypsin inhibitor 2 (CI2), believed to have two clearly distinct folding states. Their equilibrium is controlled by the concentration of denaturant guanidinium chloride (27). Using single-pair FRET (spFRET) techniques, it was possible to identify folded and unfolded molecules present at different denaturant concentrations. When averaged, these measurements yield the same denaturation curve obtained by ensemble measurements and give access to the energy landscape of the folding reaction. The additional information provided by spFRET is the number of molecules in the two different states, providing direct evidence of the two-states model derived indirectly from ensemble measurements.

Other biological systems have been successfully studied using spFRET on immobilized molecules: spFRET has revealed transient intermediate states in the *Tetrahymena* ribozyme (145), which had remained unnoticed in ensemble studies. In vivo, intermolecular spFRET permits the detection of association and dissociation events, as in the case of an epidermal growth factor receptor pair studied on cell membrane using TIR (105). For cluster formation involving larger numbers of monomers, observed with E-cadherin (63) or L-type Ca^{2+} channels (58), stoichiometric approaches relying on the quantized emission of single molecules can estimate the number of components in an aggregate.

CONCLUSION

Fluorescence microscopy is a mature field that keeps evolving toward higher sensitivity, versatility, and temporal, spectral, and spatial resolution. Its latest developments now allow reaching the level of the single molecule. In the near future, the principal limitations of SMS (low signal, limited lifespan of fluorophores, trade-off between time resolution, and the level of detail of information) will probably remain, but technical improvements toward simultaneous acquisition of all fluorescence parameters (intensity, spectrum, lifetime, polarization) are promising. New detectors will permit the combination of the high time resolution of single-photon counting devices with the large field of view and spectral resolution allowed by two-dimensional detectors. Progress is to be expected in the development of new fluorophores or in their use to probe local environmental properties.

As powerful as it is, SMS cannot replace every existing single-molecule detection or manipulation technique. Researchers start associating it with other approaches to correlate applied forces or fields and molecular conformations. New ways of controlling local fields (electric, magnetic, or others) or biochemical environments (microfluidic devices) would take advantage of the noninvasiveness, high-temporal, and spatial resolution of SMS to get a direct feedback of events at the nanometer scale in various domains of research. We thus expect that it will be possible to follow biological processes at the molecular level in individual cells.

ACKNOWLEDGMENTS

Contributions of actual and former members of the laboratory to the development of SMS and microscopy are gratefully acknowledged. Part of the work described here has been supported by NIH National Center for Research Resources grant

7 R01 RR14891-02, NIH National Institute of General Medical Sciences grant 1 R01 GM65382-01, and DOE grant DE-FG03-02ER63339.

The *Annual Review of Biophysics and Biomolecular Structure* is online at http://biophys.annualreviews.org

LITERATURE CITED

1. Abbe E. 1873. Beiträge zur Theorie der Mikroskops und der mikroskopischen Wahrnehmung. *Schultze Arch. Mikrosc.* 9: 413–68
2. Adachi K, Yasuda R, Noji H, Itoh H, Harada Y, et al. 2000. Stepping rotation of F1-ATPase visualized through angle-resolved single-fluorophore imaging. *Proc. Natl. Acad. Sci. USA* 97:7243–47
3. Agard DA, Sedat JW. 1983. Three-dimensional architecture of a polytene nucleus. *Nature* 302:676–81
4. Alivisatos AP. 1996. Semiconductor clusters, nanocrystals, and quantum dots. *Science* 271:933–37
5. Ambrose WP, Moerner WE. 1991. Fluorescence spectroscopy and spectral diffusion of single impurity molecules in a crystal. *Nature* 349:225–27
6. Ambrose WP, Goodwin PM, Martin JC, Keller RA. 1994. Alterations of single molecule fluorescence lifetimes in near-field optical microscopy. *Science* 265:364–67
7. Bacia K, Majoul IV, Schwille P. 2002. Probing the endocytic pathway in live cells using dual-color fluorescence cross-correlation analysis. *Biophys. J.* 83:1184–93
8. Bailey B, Farkas DL, Taylor DL, Lanni F. 1993. Enhancement of axial resolution in florescence microscopy by standing-wave excitation. *Nature* 366:44–48
9. Basché T, Moerner WE, Orrit M, Talon H. 1992. Photon antibunching in the fluorescence of a single dye molecule trapped in a solid. *Phys. Rev. Lett.* 69:1516–19
10. Becker W, Bergmann A, Konig K, Tirlapur U. 2001. Picosecond fluorescence lifetime microscopy by TCSPC imaging. *Proc. SPIE* 4262:414–19
11. Beeby A, Botchway SW, Clarkson IM, Faulkner S, Parker AW, et al. 2000. Luminescence imaging microscopy and lifetime mapping using kinetically stable lanthanide(III) complexes. *J. Photochem. Photobiol. B* 57:83–89
12. Betzig E. 1995. Proposed method for molecular optical imaging. *Opt. Lett.* 20: 237–39
13. Betzig E, Chichester RJ. 1993. Single molecules observed by near-field scanning optical microscopy. *Science* 262: 1422–25
14. Betzig E, Trautman JK. 1992. Near-field optics: microscopy, spectroscopy, and surface modification beyond the diffraction limit. *Science* 257:189–95
15. Bobroff N. 1986. Position measurement with a noise-limited instrument. *Rev. Sci. Instrum.* 57:1152–57
16. Bonnet G, Krichevsky O, Libchaber A. 1998. Kinetics of conformal fluctuations in DNA hairpin loops. *Proc. Natl. Acad. Sci. USA* 95:8602–6
17. Brasselet S, Moerner WE. 2000. Fluorescence behavior of single-molecule pH-sensors. *Single Mol.* 1:17–23
18. Carrington WA, Lynch RM, Moore EDW, Isenberg G, Fogarty KE, Fay FS. 1995. Superresolution three-dimensional images of fluorescence in cells with minimal light exposure. *Science* 268:1483–87
19. Chen Y, Muller JD, Tetin SY, Tyner JD, Gratton E. 2000. Probing ligand protein binding equilibria with fluorescence fluctuation spectroscopy. *Biophys. J.* 79: 1074–84

20. Corrie JE, Brandmeier BD, Ferguson RE, Trentham DR, Kendrick-Jones J, et al. 1999. Dynamic measurement of myosin light-chain-domain tilt and twist in muscle contraction. *Nature* 400:425–30
21. Cotton GJ, Muir TW. 1999. Peptide ligation and its application to protein engineering. *Chem. Biol.* 6:R247–56
22. Dahan M, Laurence T, Pinaud F, Chemla DS, Alivisatos AP, et al. 2001. Time-gated biological imaging using colloidal quantum dots. *Opt. Lett.* 26:825–28
23. Damjanovich S, Vereb G, Schaper A, Jenei A, Matko J, et al. 1995. Structural hierarchy in the clustering of HLA class I molecules in the plasma membrane of human lymphoblastoid cells. *Proc. Natl. Acad. Sci. USA* 92:1122–26
24. Dawson PE, Kent SB. 2000. Synthesis of native proteins by chemical ligation. *Annu. Rev. Biochem.* 69:923–60
25. de Lange F, Cambi A, Huijbens R, de Bakker B, Rensen W, et al. 2001. Cell biology beyond the diffraction limit: near-field scanning optical microscopy. *J. Cell Sci.* 114:4153–60
26. Deniz AA, Dahan M, Grunwell JR, Ha T, Faulhaber AE, et al. 1999. Single-pair fluorescence resonance energy transfer on freely diffusing molecules: observation of Förster distance dependence and subpopulations. *Proc. Natl. Acad. Sci. USA* 96:3670–75
27. Deniz AA, Laurence TA, Beligere GS, Dahan M, Martin AB, et al. 2000. Single-molecule protein folding: diffusion fluorescence resonance energy transfer studies of the denaturation of chymotrypsin inhibitor 2. *Proc. Natl. Acad. Sci. USA* 97:5179–84
28. Denk W, Strickler JH, Webb WW. 1990. Two-photon laser scanning fluorescence microscopy. *Science* 248:73–76
29. Deschenes LA, Vanden Bout DA. 2001. Single-molecule studies of heterogeneous dynamics in polymer melts near the glass transition. *Science* 292:255–58
30. Dickinson ME, Bearman G, Tille S, Lansford R, Fraser SE. 2001. Multi-spectral imaging and linear unmixing add a whole new dimension to laser scanning fluorescence microscopy. *Biotechniques* 31:1272–78
31. Dickson RM, Cubitt AB, Tsien RY, Moerner WE. 1997. On/off blinking and switching behaviour of single molecules of green fluorescent protein. *Nature* 388:355–58
32. Dittrich PS, Schwille P. 2001. Photobleaching and stabilization of fluorophores used for single-molecule analysis with one- and two-photon excitation. *Appl. Phys. B* 73:829–37
33. Dyba M, Hell SW. 2002. Focal spot of size $\lambda/23$ open up far-field fluorescence microscopy at 33 nm axial resolution. *Phys. Rev. Lett.* 88:163901–4
34. Efron B, Tibshirani R. 1993. *An Introduction to the Bootstrap*. New York: Chapman & Hall. 436 pp.
35. Efros AL, Rosen M. 1997. Random telegraph signal in the photoluminescence intensity of a single quantum dot. *Phys. Rev. Lett.* 78:1110–13
36. Elson EL, Magde D. 1974. Fluorescence correlation spectroscopy. I. Conceptual basis and theory. *Biopolymers* 13:1–27
37. Empedocles SA, Norris DJ, Bawendi MG. 1996. Photoluminescence spectroscopy of single CdSe nanocrystallite quantum dots. *Phys. Rev. Lett.* 77:3873–76
38. Enderle T, Ha T, Ogletree DF, Chemla DS, Magowan C, Weiss S. 1997. Membrane specific mapping and colocalization of malarial and host skeletal proteins in the *Plasmodium falciparum* infected erythrocyte by dual-color near-field scanning optical microscopy. *Proc. Natl. Acad. Sci. USA* 94:520–25
39. Forkey JN, Quinlan ME, Goldman YE. 2000. Protein structural dynamics by single-molecule fluorescence polarization. *Prog. Biophys. Mol. Biol.* 74:1–35
40. Förster. 1948. Zwischenmolekulare Energiewanderung und Fluoreszenz. *Ann. Physik* 6:55–75

41. Frohn JT, Knapp HF, Stemmer A. 2000. True optical resolution beyond the Rayleigh limit achieved by standing wave illumination. *Proc. Natl. Acad. Sci. USA* 97: 7232–36
42. Funatsu T, Harada Y, Tokunaga M, Saito K, Yanagida T. 1995. Imaging of single fluorescent molecules and individual ATP turnovers by single myosin molecules in aqueous solution. *Nature* 374:555–59
43. Gadella TW Jr, van der Krogt GN, Bisseling T. 1999. GFP-based FRET microscopy in living plant cells. *Trends Plant Sci.* 4:287–91
44. Gadella TWJ, Jovin TM. 1995. Oligomerization of epidermal growth factor receptors on A431 cells studied by time-resolved fluorescence imaging microscopy—a stereochemical model for tyrosine kinase receptor activation. *J. Cell. Biol.* 129:1543–58
45. Gadella TWJ, Jovin TM, Clegg RM. 1993. Fluorescence lifetime imaging microscopy (FLIM)—spatial resolution of microstructures on the nanosecond time scale. *Biophys. Chem.* 48:221–39
46. Gelles J, Schnapp BJ, Sheetz MP. 1988. Tracking kinesin-driven movements with nanometer-scale precision. *Nature* 331: 450–53
47. Goulian M, Simon SM. 2000. Tracking single proteins within cells. *Biophys. J.* 79:2188–98
48. Griffin BA, Adams SR, Tsien RY. 1998. Specific covalent labeling of recombinant protein molecules inside live cells. *Science* 281:269–72
49. Grinvald A, Haas E, Steinberg IZ. 1972. Evaluation of the distribution of distances between energy donors and acceptors by fluorescence decay. *Proc. Natl. Acad. Sci. USA* 69:2273–77
50. Gustafsson MGL, Agard DA, Sedat JW. 1999. I5M: 3D widefield light microscopy with better than 100 nm axial resolution. *J. Microsc.* 195:10–16
51. Ha T, Enderle T, Chemla DS, Selvin PR, Weiss S. 1996. Single molecule dynamics studied by polarization modulation. *Phys. Rev. Lett.* 77:3979–82
52. Ha T, Enderle T, Chemla DS, Selvin PR, Weiss S. 1997. Quantum jumps of single molecules at room temperature. *Chem. Phys. Lett.* 271:1–5
53. Ha T, Enderle T, Chemla DS, Weiss S. 1996. Dual-molecule spectroscopy: molecular rulers for the study of biological macromolecules. *IEEE J. Sel. Top. Quant. Elec.* 2:1115–28
54. Ha T, Glass J, Enderle T, Chemla DS, Weiss S. 1998. Hindered rotational diffusion and rotational jumps of single molecules. *Phys. Rev. Lett.* 80:2093–96
55. Ha T, Laurence TA, Chemla DS, Weiss S. 1999. Polarization spectroscopy of single fluorescent molecules. *J. Phys. Chem. B* 103:6839–50
56. Ha T, Ting AY, Liang J, Caldwell WB, Deniz AA, et al. 1999. Single-molecule fluorescence spectroscopy of enzyme conformational dynamics and cleavage mechanism. *Proc. Natl. Acad. Sci. USA* 96:893–98
57. Hanley QS, Arndt-Jovin DJ, Jovin TM. 2002. Spectrally resolved fluorescence lifetime imaging microscopy. *Appl. Spectrosc.* 56:155–66
58. Harms GS, Cognet L, Lommerse PHM, Blab GA, Kahr H, et al. 2001. Single-molecule imaging of L-type Ca^{2+} channels in live cells. *Biophys. J.* 81:2639–46
59. Haugland RP. 2002. *Handbook of Fluorescent Probes and Research Products.* Eugene, OR: Molecular Probes. 966 pp.
60. Hell S, Stelzer EHK. 1992. Properties of a 4Pi confocal fluorescence microscope. *J. Opt. Soc. Am. A* 9:2159–66
61. Hiraoka Y, Sedat JW, Agard DA. 1987. The use of charge-coupled device for quantitative optical microscopy of biological structures. *Science* 238:36–41
62. Hofkens J, Schroeyers W, Loos D, Cotlet M, Köhn F, et al. 2001. Triplet states as

non-radiative traps in multichromophoric entities: single molecule spectroscopy of an artificial and natural antenna system. *Spectrosc. Acta A* 57:2093–107
63. Iino R, Koyama I, Kusumi A. 2001. Single molecule imaging of green fluorescent proteins in living cells: E-cadherin forms oligomers on the free cell surface. *Biophys. J.* 80:2667–77
64. Ishijima A, Yanagida T. 2001. Single molecule nanobioscience. *Trends Biochem. Sci.* 26:438–44
65. Jia Y, Talaga D, Lau W, Lu H, DeGrado W, Hochstrasser R. 1999. Folding dynamics of single GCN4 peptides by fluorescence resonant energy transfer confocal microscopy. *Chem. Phys.* 247:69–83
66. Kapanidis AN, Weiss S. 2002. Fluorescent probes and bioconjugation chemistries for single-molecule fluorescence analysis of biomolecules. *J. Chem. Phys.* 117(24):10953–64
67. Klar TA, Jakobs S, Dyba M, Egner A, Hell SW. 2000. Fluorescence microscopy with diffraction resolution barrier broken by stimulated emission. *Proc. Natl. Acad. Sci. USA* 97:8206–10
68. Kunkel TA, Bebenek K, McClary J. 1991. Efficient site-directed mutagenesis using uracil-containing DNA. *Methods Enzymol.* 204:125–39
69. Lacoste TD, Michalet X, Pinaud F, Chemla DS, Alivisatos AP, Weiss S. 2000. Ultrahigh-resolution multicolor colocalization of single fluorescent probes. *Proc. Natl. Acad. Sci. USA* 97:9461–66
70. Lakowicz JR. 1999. *Principles of Fluorescence Spectroscopy*. New York: Plenum. 698 pp.
71. Lakowicz JR, Berndt KW. 1991. Lifetime-selective fluorescence imaging using an RF phase-sensitive camera. *Rev. Sci. Instrum.* 62:1727–34
72. Lakowicz JR, Szmacinski H, Nowaczyk K, Lederer WJ, Kirby MS, Johnson ML. 1994. Fluorescence lifetime imaging of intracellular calcium in COS cells using quin-2. *Cell Calcium* 15:7–27
73. Lu HP, Xun L, Xie XS. 1998. Single-molecule enzymatic dynamics. *Science* 282:1877–82
74. Mackay CD, Tubbs RN, Bell R, Burt D, Jerram P, Moody I. 2001. Sub-electron read noise at MHz pixel rates. *SPIE Proc.* 4306:289–98
75. Macklin JJ, Trautman JK, Harris TD, Brus LE. 1996. Imaging and time-resolved spectroscopy of single molecules at an interface. *Science* 272:255–58
76. Magde D, Elson EL, Webb WW. 1974. Fluorescence correlation spectroscopy. II. An experimental realization. *Biopolymers* 13:29–61
77. Mainen ZF, Maletic-Savatic M, Shi SH, Hayashi Y, Malinow R, Svoboda K. 1999. Two-photon imaging in living brain slices. *Methods* 18:231–39
78. Malik Z, Cabib D, Buckwald RA, Talmi A, Garini Y, Lipson SG. 1996. Fourier transform multipixel spectroscopy for quantitative cytology. *J. Microsc.* 182: 133–40
79. Margeat E, Poujol N, Boulahtouf A, Chen Y, Muller JD, et al. 2001. The human estrogen receptor alpha dimer binds a single SRC-1 coactivator molecule with an affinity dictated by agonist structure. *J. Mol. Biol.* 306:433–42
80. Mekler V, Kortkhonjia E, Mukhopadhyay J, Knight J, Revyakin A, et al. 2002. Structural organization of RNA polymerase holoenzyme and the RNA polymerase-promoter open complex: systematic fluorescence resonance energy transfer and distance-constrained docking. *Cell* 108: 1–20
81. Michalet X, Lacoste TD, Weiss S. 2001. Ultrahigh-resolution colocalization of spectrally resolvable point-like fluorescent probes. *Methods* 25:87–102
82. Michalet X, Pinaud F, Lacoste TD, Dahan M, Bruchez MP, et al. 2001. Properties of fluorescent semiconductor nanocrystals and their application to biological labeling. *Single Mol.* 2:261–76

83. Michalet X, Weiss S. 2002. Single-molecule spectroscopy and microscopy. *C. R. Phys.* 3:619–44
84. Moerner WE. 1994. Examining nanoenvironments in solids on the scale of a single, isolated impurity molecule. *Science* 265:46–53
85. Moerner WE. 2002. A dozen years of single-molecule spectroscopy in physics, chemistry, and biophysics. *J. Phys. Chem. B* 106:910–27
86. Moerner WE, Kador L. 1989. Optical detection and spectroscopy of single molecules in a solid. *Phys. Rev. Lett.* 62:2535–38
87. Moerner WE, Orrit M. 1999. Illuminating single molecules in condensed matter. *Science* 283:1670–76
88. Muir TW, Sondhi D, Cole PA. 1998. Expressed protein ligation: a general method for protein engineering. *Proc. Natl. Acad. Sci. USA* 95:6705–10
89. Mukhopadhyay J, Kapanidis AN, Mekler V, Kortkhonjia E, Ebright YW, Ebright RH. 2001. Translocation of S70 with RNA polymerase during transcription: fluorescence resonance energy transfer assay for movement relative to DNA. *Cell* 106:453–63
90. Nemoto N, Miyamoto-Sato E, Yanagawa H. 1999. Fluorescence labeling of the C-terminus of proteins with a puromycin analogue in cell-free translation systems. *FEBS Lett.* 462:43–46
91. Neuhauser RG, Shimizu KT, Woo WK, Empedocles SA, Bawendi MG. 2000. Correlation between fluorescence intermittency and spectral diffusion in single semiconductor quantum dots. *Phys. Rev. Lett.* 85:3301–4
92. Ng T, Squire A, Hansra G, Bornancin F, Prevostel C, et al. 1999. Imaging protein kinase Cα activation in cells. *Science* 283:2085–89
93. Nirmal M, Dabbousi BO, Bawendi MG, Macklin JJ, Trautman JK, et al. 1996. Fluorescence intermittency in single cadmium selenide nanocrystals. *Nature* 383:802–4
94. Nishiyama M, Muto E, Inoue Y, Yanagida T, Higuchi H. 2001. Substeps within the 8-nm step of the ATPase cycle of single kinesin molecules. *Nat. Cell Biol.* 3:425–28
95. Orrit M, Bernard J. 1990. Single pentacene molecules detected by fluorescence excitation in a *p*-terphenyl crystal. *Phys. Rev. Lett.* 65:2716–19
96. Patterson GH, Piston DW. 2000. Photobleaching in two-photon excitation microscopy. *Biophys. J.* 78:2159–62
97. Pawley JB, ed. 1995. *Handbook of Biological Confocal Microscopy*. New York: Plenum
98. Pepperkok R, Squire A, Geley S, Bastiaens PIH. 1999. Simultaneous detection of multiple green fluorescent proteins in live cell by fluorescent lifetime imaging microscopy. *Curr. Biol.* 9:269–72
99. Perrin F. 1929. La fluorescence des solutions. *Ann. Phys.* 12:169–275
100. Perrin F. 1932. Théorie quantique des transferts d'activation entre molécules de même espèce. Cas des solutions fluorescentes. *Ann. Phys.* 17:283–313
101. Peterman EJ, Sosa H, Goldstein LS, Moerner WE. 2001. Polarized fluorescence microscopy of individual and many kinesin motors bound to axonemal microtubules. *Biophys. J.* 81:2851–63
102. Piston DW, Sandison DR, Webb WW. 1992. Time-resolved fluorescence imaging and background rejection by two-photon excitation in laser scanning microscopy. *Proc. SPIE* 1640:379–89
103. Rarbach M, Kettling U, Koltermann A, Eigen M. 2001. Dual-color fluorescence cross-correlation spectroscopy for monitoring the kinetics of enzyme-catalyzed reactions. *Methods* 24:104–16
104. Rudiger M, Haupts U, Moore KJ, Pope AJ. 2001. Single-molecule detection technologies in miniaturized high throughput screening: binding assays for G protein-coupled receptors using fluorescence

intensity distribution analysis and fluorescence anisotropy. *J. Biomol. Screen.* 6:29–37
105. Sako Y, Minoguchi S, Yanagida T. 2000. Single-molecule imaging of EGFR signalling on the surface of living cells. *Nat. Cell Biol.* 2:168–72
106. Sánchez EJ, Novotny L, Holtom GR, Xie SX. 1997. Room-temperature fluorescence imaging and spectroscopy of single molecules by two-photon excitation. *J. Phys. Chem. A* 101:7019–23
107. Sase I, Miyata H, Ishiwata S, Kinosita K Jr. 1997. Axial rotation of sliding actin filaments revealed by single-fluorophore imaging. *Proc. Natl. Acad. Sci. USA* 94:5646–50
108. Scheel AA, Funsch B, Busch M, Gradl G, Pschorr J, Lohse MJ. 2001. Receptor-ligand interactions studied with homogeneous fluorescence-based assays suitable for miniaturized screening. *J. Biomol. Screen.* 6:11–18
109. Schmidt T, Schütz GJ, Baumgartner W, Gruber HJ, Schindler H. 1996. Imaging of single molecule diffusion. *Proc. Natl. Acad. Sci. USA* 93:2926–29
110. Schütz GJ, Pastushenko VP, Gruber H, Knaus H-G, Pragl B, Schindler H. 2000. 3D imaging of individual ion channels in live cells at 40 nm resolution. *Single Mol.* 1:25–31
111. Schwille P. 2001. Fluorescence correlation spectroscopy and its potential for intracellular applications. *Cell Biochem. Biophys.* 34:383–408
112. Selvin PR. 1995. Fluorescence resonance energy transfer. *Methods Enzymol.* 246:300–34
113. Severinov K, Muir TW. 1998. Expressed protein ligation, a novel method for studying protein-protein interactions in transcription. *J. Biol. Chem.* 273:16205–9
114. So PT, Dong CY, Masters BR, Berland KM. 2000. Two-photon excitation fluorescence microscopy. *Annu. Rev. Biomed. Eng.* 2:399–429
115. So PTC, French T, Yu WM, Berland KM, Dong CY, Gratton E. 1995. Time-resolved fluorescence microscopy using two-photon excitation. *Bioimaging* 3:49–63
116. Sosa H, Peterman EJG, Moerner WE, Goldstein SB. 2001. ADP-induced rocking of the kinesin motor domain revealed by single-molecule fluorescence polarization microscopy. *Nat. Struct. Biol.* 8:540–44
117. Straub M, Hell SW. 1998. Fluorescence lifetime three-dimensional microscopy with picosecond precision using a multifocal multiphoton microscope. *Appl. Phys. Lett.* 73:1769–71
118. Stryer L, Haugland RP. 1967. Energy transfer: a spectroscopic ruler. *Proc. Natl. Acad. Sci. USA* 58:719–26
119. Synge EH. 1928. A suggested method for extending microscopic resolution into the ultra-microscopic region. *Philos. Mag.* 6:356–62
120. Sytsma J, Vroom JM, DeGrauw CJ, Gerritsen HC. 1998. Time-gated fluorescence lifetime imaging and microvolume spectroscopy using two-photon excitation. *J. Microsc.* 191:39–51
121. Taguchi H, Ueno T, Tadakuma H, Yoshida M, Funatsu T. 2001. Single-molecule observation of protein-protein interactions in the chaperonin system. *Nat. Biotechnol.* 19:861–65
122. Tam JP, Xu J, Eom KD. 2001. Methods and strategies of peptide ligation. *Biopolymers* 60:194–205
123. Terpetschnig E, Szmacinski H, Malak H, Lakowicz JR. 1995. Metal-ligand complexes as a new class of long-lived fluorophores for protein hydrodynamics. *Biophys. J.* 68:342–50
124. Tinnefeld P, Buschmann V, Herten D-P, Han K-T, Sauer M. 2000. Confocal fluorescence lifetime imaging microscopy (FLIM) at the single molecule level. *Single Mol.* 1:215–23
125. Trautman JK, Macklin JJ, Brus LE, Betzig E. 1994. Near-field spectroscopy of single

molecules at room temperature. *Science* 369:40–42
126. Tsien RY. 1998. The green fluorescent protein. *Annu. Rev. Biochem.* 67:509–44
127. Vale RD, Funatsu T, Pierce DW, Romberg L, Harada Y, Yanagida T. 1996. Direct observation of single kinesin molecules moving along microtubules. *Nature* 380: 451–53
128. van Oijen AM, Köhler J, Schmidt J, Müller M, Brakenhoff GJ. 1998. 3-Dimensional super-resolution by spectrally selective imaging. *Chem. Phys. Lett.* 292:183–87
129. Van Rompaey E, Chen Y, Muller JD, Gratton E, Van Craenenbroeck E, et al. 2001. Fluorescence fluctuation analysis for the study of interactions between oligonucleotides and polycationic polymers. *Biol. Chem.* 382:379–86
130. Vanden Bout DA, Yip W-T, Hu D, Fu D-K, Swager TM, Barbara PF. 1997. Discrete intensity jumps and intramolecular electronic energy transfer in the spectroscopy of single conjugated polymer molecules. *Science* 277:1074–77
131. Vereb G, Jares-Erijman E, Selvin PR, Jovin TM. 1998. Temporally and spectrally resolved imaging microscopy of lanthanide chelates. *Biophys. J.* 74:2210–22
132. Waggoner A. 1995. Covalent labeling of proteins and nucleic acids with fluorophores. *Methods Enzymol.* 246:362–73
133. Wang Y-L, Taylor DL, eds. 1989. *Fluorescence Microscopy of Living Cells in Culture. Part A & B*. San Diego: Academic. Vols. 29, 30
134. Wang XF, Periasamy A, Herman B. 1992. Fluorescence lifetime imaging microscopy (FLIM): instrumentation and applications. *Crit. Rev. Anal. Chem.* 23: 369–95
135. Warshaw DM, Hayes E, Gaffney D, Lauzon AM, Wu J, et al. 1998. Myosin conformational states determined by single fluorophore polarization. *Proc. Natl. Acad. Sci. USA* 95:8034–39
136. Weiss S. 1999. Fluorescence spectroscopy of single biomolecules. *Science* 283:1676–83
137. Weiss S. 2000. Measuring conformational dynamics of biomolecules by single molecule fluorescence spectroscopy. *Nat. Struct. Biol.* 7:724–29
138. Widengren J, Mets U, Rigler R. 1995. Fluorescence correlation spectroscopy of triplet states in solution—a theoretical and experimental study. *J. Phys. Chem.* 99:13368–79
139. Widengren J, Rigler R. 1998. Fluorescence correlation spectroscopy as a tool to investigate chemical reactions in solutions and on cell surfaces. *Cell. Mol. Biol.* 44:857–79
140. Wu M, Goodwin PM, Ambrose WP, Keller RA. 1996. Photochemistry and fluorescence emission dynamics of single molecules in solution-B-phycoerythrin. *J. Phys. Chem.* 100:17406–9
141. Xie XS, Dunn RC. 1994. Probing single molecule dynamics. *Science* 265:361–64
142. Xie XS, Trautman JK. 1998. Optical studies of single molecules at room temperature. *Annu. Rev. Phys. Chem.* 49:441–80
143. Yamaguchi J, Nemoto N, Sasaki T, Tokumasu A, Mimori-Kiyosue Y, et al. 2001. Rapid functional analysis of protein-protein interactions by fluorescent C-terminal labeling and single-molecule imaging. *FEBS Lett.* 502:79–83
144. Yasuda R, Noji H, Yoshida M, Kinosita K, Itoh H. 2001. Resolution of distinct rotational substeps by submillisecond kinetic analysis of F_1-ATPase. *Nature* 410:898–904
145. Zhuang X, Bartley LE, Babcock HP, Russell R, Ha T, et al. 2000. A single-molecule study of RNA catalysis and folding. *Science* 288:2048–51

Annu. Rev. Biophys. Biomol. Struct. 2003. 32:183–206
doi: 10.1146/annurev.biophys.32.110601.141906

First published online as a Review in Advance on February 5, 2003

THE STRUCTURE OF MAMMALIAN CYCLOOXYGENASES

R. Michael Garavito and Anne M. Mulichak
Department of Biochemistry and Molecular Biology, Michigan State University, East Lansing, Michigan 48824-1319; email: garavito@msu.edu; mulichak@msu.edu

Key Words prostaglandin H_2 synthase, heme-dependent peroxidase, arachidonic acid, nonsteroidal antiinflammatory drugs, COX-2-selective inhibitors

■ **Abstract** Cyclooxygenases-1 and -2 (COX-1 and COX-2, also known as prostaglandin H_2 synthases-1 and -2) catalyze the committed step in prostaglandin synthesis. COX-1 and -2 are of particular interest because they are the major targets of nonsteroidal antiinflammatory drugs (NSAIDs) including aspirin, ibuprofen, and the new COX-2-selective inhibitors. Inhibition of the COXs with NSAIDs acutely reduces inflammation, pain, and fever, and long-term use of these drugs reduces the incidence of fatal thrombotic events, as well as the development of colon cancer and Alzheimer's disease. In this review, we examine how the structures of COXs relate mechanistically to cyclooxygenase and peroxidase catalysis and how alternative fatty acid substrates bind within the COX active site. We further examine how NSAIDs interact with COXs and how differences in the structure of COX-2 result in enhanced selectivity toward COX-2 inhibitors.

CONTENTS

CYCLOOXYGENASE

The cyclooxygenases (COXs), which are also known as prostaglandin H_2 synthases (EC 1.14.99.1), are bifunctional, membrane-bound enzymes that catalyze the committed step in prostanoid biosynthesis (45, 74, 76). Prostanoids are members of a large group of bioactive, oxygenated C_{18}–C_{22} compounds that are derived from $\omega 3$ (n-3) and $\omega 6$ (n-6) polyunsaturated fatty acids. In mammals, arachidonic acid (AA; 20:4 n-6) is the major prostanoid precursor. The heme-dependent bis-oxygenase or COX reaction converts AA to PGG_2, a 9,11-endoperoxide-15-hydroperoxide product (Figure 1). The subsequent peroxidase (POX) reaction reduces the 15-hydroperoxide of PGG_2 to form PGH_2.

The COX reaction also is the target of aspirin, ibuprofen, and other nonsteroidal antiinflammatory drugs (NSAIDs), which account for billion of dollars in sales for the pharmaceutical industry. By the end of the 1980s, the interaction of aspirin and other NSAIDs with COX had been well studied, and many felt that this area of pharmacological research had played out. However, the discovery that two isoforms of COX exist in mammals radically changed our understanding of prostanoid physiology and NSAID pharmacology (20, 76) and triggered a phenomenal increase in interest in the structure, function, and physiology of COXs during the late 1990s. Moreover, an increasing amount of evidence suggested that the COX isozymes have direct roles in many human pathologies. These include thrombosis (55, 56), inflammation, pain and fever (5), various cancers (21, 48, 82),

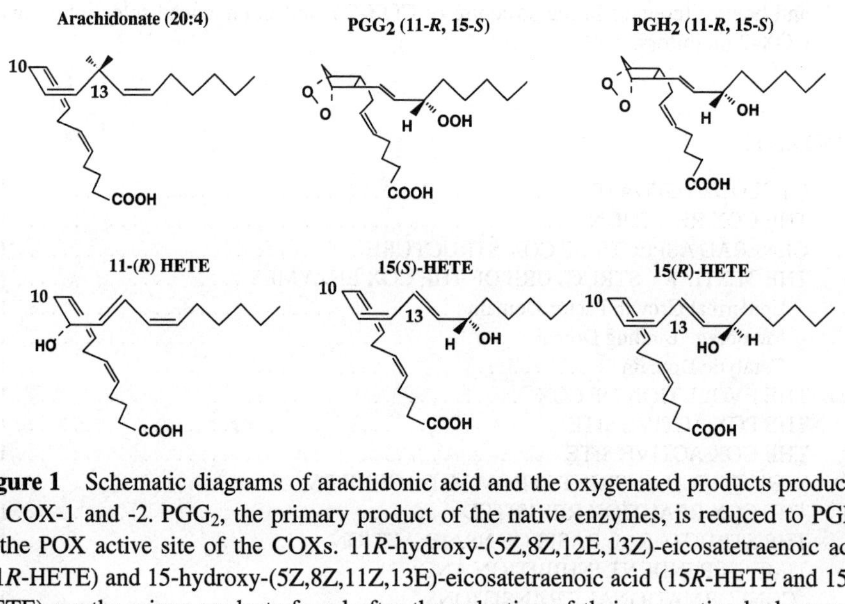

Figure 1 Schematic diagrams of arachidonic acid and the oxygenated products produced by COX-1 and -2. PGG_2, the primary product of the native enzymes, is reduced to PGH_2 at the POX active site of the COXs. 11*R*-hydroxy-(5Z,8Z,12E,13Z)-eicosatetraenoic acid (11*R*-HETE) and 15-hydroxy-(5Z,8Z,11Z,13E)-eicosatetraenoic acid (15*R*-HETE and 15*S*-HETE) are the minor products found after the reduction of their respective hydroperoxy precursors (see text).

and neurological disorders such as Alzheimer's (46) and Parkinson's (79) diseases. Finally, pharmacological research over the past decade led to the development and approval of the new COX-2-selective NSAIDs (e.g., Celebrex® and Vioxx®), which target COX-2 instead of COX-1. Armed with a panoply of new and classical COX inhibitors, researchers have begun to discover NSAID-dependent physiological effects that are seemingly not directly related to COX inhibition and the subsequent cessation of prostanoid biosynthesis. These COX-independent phenomena include the antitumor activities of some NSAIDs (63, 82), and the effects of aspirin on beta-amyloid formation in Alzheimer's disease (84) have provoked a new series of controversies about the physiological roles of the COX isozymes and the pharmacological actions of NSAIDs.

Although research on COXs has resulted in a vast amount of information about the molecular biology, pharmacology, and structural biology of these enzymes, new mysteries about the biochemistry and enzymology of the COX isoforms have arisen. The reader is encouraged to read a number of recent reviews on the physiology, enzymology, and pharmacology of these enzymes (12, 29, 43, 45, 76). In this review, we discuss the structures of the COX isoforms and their relevance to their function.

THE COX REACTION

Ruf and coworkers (9) first proposed a branched-chain, mechanistic model (Figure 2) that incorporates the requirement of heme oxidation by the COX reaction. In this scheme, a hydroperoxide reacts with the heme iron to initiate a two-electron oxidation that yields Compound I, an enzyme state with an oxyferryl-heme radical cation. Compound I quickly undergoes a single electron reduction via an intramolecular migration of the radical from the heme group to Tyr385 to create Intermediate II (77, 88). When the COX active site is occupied by an appropriate fatty acid substrate such as AA, the tyrosyl radical initiates the COX reaction by abstracting the 13*proS* hydrogen to yield an arachidonyl radical (88). The fatty acid radical then reacts with molecular O_2 to produce an 11-hydroperoxyl radical, which in turn forms the endoperoxide cyclopentane moiety of PGH_2; the addition of a second O_2 molecule at carbon 15 ultimately produces PGG_2.

The POX reaction then comes into play again to reduce the 15-hydroperoxide of PGG_2 to form PGH_2. Although the POX reaction is considered the second step in the formation of PGH_2, the COX reaction is absolutely dependent on POX activity for its activation (77, 76). Initially, the ambient hydroperoxides activate the COX reaction in a small number of COX molecules via POX turnovers. As more PGG_2 is generated, the remaining COX molecules are then activated autocatalytically. In vitro this phenomenon is exhibited as the lag period seen in COX activity assays (27).

While COX-1 synthesizes primarily PGG_2 from AA, it also produces small but significant amounts of other products (67, 85, 86) (Figure 1): 11*R*-hydroperoxy-(5Z,8Z,12E,13Z)-eicosatetraenoic acid (11*R*-HPETE) and

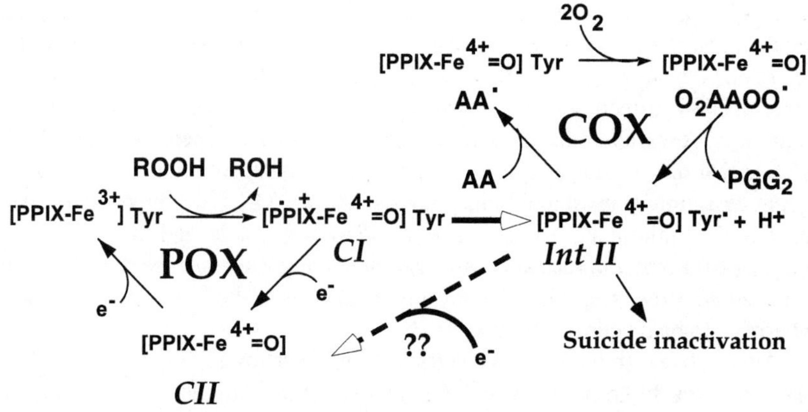

Figure 2 A schematic diagram of the branched-chain reaction mechanism for the COX enzymes that was originally proposed by Ruf and colleagues (9). In essence, one turnover of the POX reaction is needed to provide the tyrosyl radical for activating the COX reactions. Activated COX would continue to turnover, in the presence of substrate, until radical induced inactivation occurs (77, 76). The diagram was adapted from Reference 76.

15-hydroperoxy-(5Z,8Z,11Z,13E)-eicosatetraenoic acid (15*R*-HPETE and 15*S*-HPETE). The kinetics of the product formation suggests that AA may adopt up to four slightly different but catalytically competent conformers in the COX active site that then give rise to the different oxygenated products (86). Human COX-2 was also thought to form only PGG_2, 11*R*-HPETE, and 15*S*-HPETE but not 15*R*-HPETE (94), which suggested that there are only three catalytically competent conformers of AA in this isoform. However, more recent work by Schneider et al. (67) shows that wild-type COX-2 does form 15*R*-HPETE.

COX-1 and -2 also catalyze the oxygenation of other polyunsaturated fatty acids into bioactive compounds (23, 24, 40, 62, 70, 87). Both COX-1 and -2 produce the series-1 prostaglandin precursor PGH_1 from dihomo-γ-linolenic acid (DHLA; 20:3n-6) and the series-3 prostaglandin precursor PGH_3 from eicosapentaenoic acid (EPA, 20:5n-3), a dietary ω-3 fatty acid that has been linked to reduced cardiovascular disease (70). Linolenic acid (LA; 18:2n-6) is converted to 9- and 13-hydroxyoctadecadienoic acids (40). The ability of the COX enzymes to use alternative substrates demonstrates the impact of diet on physiology.

GENERAL ASPECTS OF COX STRUCTURE

After the discovery of the COX isoforms, it quickly became apparent that the isoforms were noticeably different in their expression profiles and roles in several physiological processes (20, 76). COX-1 is the constitutively expressed isoform and is apparently involved in many aspects of physiological homeostasis. On the other hand, COX-2 is the inducible isoform whose expression in a select number of cells is triggered by specific cellular events. However, both enzymes are membrane bound and are present on the lumenal surfaces of the endoplasmic reticulum and of the inner and outer membranes of the nuclear envelope (49, 83).

The primary structures of nascent COX-1 and -2 are 600–602 (depending on the species) and 604 amino acids, respectively (75), and both isoforms are then processed into mature forms by removal of signal peptides. The high degree of sequence identity between the processed isoforms and between species allows an almost one-to-one comparison (74, 76). By convention, the residues of COX are often numbered to correspond to the ovine or murine COX-1 sequence (75) to aid convenience of structural and functional comparisons across species. Sequence comparisons between COX isoforms from the same species show 60%–65% sequence identity, while sequence identity among orthologs from different species varies from 85% to 90% (75). When compared with sequences from other proteins, particularly other heme-dependent peroxidases, significant levels of similarity were detected. Clearly, the COX enzymes are members of the mammalian heme-dependent peroxidase family (61, 95), which includes myeloperoxidase (MPO) and thyroid peroxidase. Moreover, both isoforms contain an epidermal growth factor (EGF)-like domain just C-terminal to the signal peptide. The role of the EGF domain remains unclear, but it was noted that the EGF domains also occur in many cell surface membrane proteins (3).

Despite a high level of sequence homology between COX isoforms, major differences in primary structure occur in three distinct areas of the sequence. First, both isoforms have distinctly different signal peptides in terms of length and amino acid composition. Second, substantial sequence differences are found in the membrane binding domains (MBDs) between the two isoforms (54, 83), although no explanation for this phenomenon is known. Third, sequences of COX-2, but not COX-1, contain an insert of 18 amino acids that is six residues in from the C terminus. The function of this insert in COX-2 is not known, but mutations and sequence alterations at the C terminus can markedly affect the expression of active protein (17, 81). In contrast, N-terminal His-tagged versions of human COX-1 and -2 have been prepared and expressed in high yield in insect cell culture (73). In all cases, the addition of the N-terminal hexaHis-tag, between the signal peptide and the EGF domain, does not apparently impact the folding of the heterologously expressed enzyme or its activity.

Purified native COX-1 appears to be uniformly glycosylated at three sites (Asn68, Asn144, and Asn410) and appears as a single band on SDS-PAGE with a M_{app} ~67 KDa. Native COX-2, on the other hand, is more heterogeneously glycosylated at an additional site (Asn588), and multiple molecular species of COX-2 can be readily observed with SDS-PAGE with a M_{app} 68–72 KDa (53). N-glycosylation may play a role in the maturation of COXs (53), but the deglycosylation of the mature enzyme does not affect activity. Moreover, COX-1 and -2 appear as homodimers in solution after detergent solubilization (74, 76), whether glycosylated or not.

The COXs bind 1 mole of ferric-protoporphyrin IX per mole monomer for full activity, as expected for a heme-dependent peroxidase. However, native ovine COX-1 often loses much of the bound heme during detergent solubilization and purification (39, 91). For recombinant COX-2, detergent solubilization and subsequent purification tends to yield apo-enzyme (1, 14, 73), suggesting that its affinity for heme is lower than that exhibited by COX-1. Although the heme can be readily removed, active COX-1 and -2 can be easily reconstituted by the addition of hematin to the sample. This behavior makes the COXs rather unusual among the known heme-dependent peroxidases. Moreover, COXs can readily bind mangano-protoporphyrin IX or cobalt-protoporphyrin IX to create novel holo-species that are structurally native but have quite altered activity (39, 89, 90).

THE TERTIARY STRUCTURE OF THE COX ENZYMES

In 1994, Picot et al. (58) published the first three-dimensional structure of a COX enzyme, the ovine COX-1 complexed with the NSAID flurbiprofen (Figure 3*a*, see color insert). Soon afterward, the crystal structures of human (37) and murine (28) COX-2 followed. Drug interactions with COX were one of the first issues to be addressed, and complexes containing a number of different NSAIDs have been studied crystallographically with COX-1 (34–36, 69) and COX-2 (28, 37). The structural analysis of COX complexed with substrates or products was more difficult to pursue for a number of technical reasons, particularly the sensitivity of polyunsaturated fatty acids to oxidation. However, within the past three years, structures of ovine COX-1 complexed with several different fatty acid substrates have been published: with AA (20:4n-6) (38), dihomo-γ-linolenic acid (DHLA; 20:3n-6) (87), linolenic acid (LA; 18:2n-6) (40), and eicosapentaenoic acid (EPA; 20:5n-3) (40). Likewise, crystal structures of murine COX-2 complexed with AA (22) and EPA (29) have also been determined.

As was expected from the observed levels of sequence identity, the crystal structures verified that the COX isoforms are structurally homologous and quite superimposable (28, 37). The COX monomer (Figure 3*b*) consists of three structural domains: the N-terminal EGF domain, a membrane binding domain (MBD) of about 48 amino acids in length, and a large C-terminal globular catalytic domain containing the heme binding site. The C-terminal segments beyond Pro583 (17 amino acids in COX-1 and 35 amino acids in COX-2) have not been resolved crystallographically (28, 37, 58).

Epidermal Growth Factor Domain

The EGF and catalytic domains create the subunit interface in the dimer and place the two MBDs in a homodimer about 25 Å apart (Figure 3*a*). The EGF domains create a substantial portion of the dimer interface. EGF domains are common in several families of membrane proteins and secreted proteins (3). Typically, the EGF domain occurs at a position in the primary sequence N-terminal to a membrane anchor, such that these domains always occur on the extracytoplasmic face of the membrane. Garavito and colleagues (13, 57) have suggested that the EGF domains may play a role in the insertion of COX into the lipid bilayer.

Membrane Binding Domain

The MBD of COX is built up of four short consecutive amphipathic α-helices (Figure 3*b*); three of the four helices lie roughly in the same plane while the last helix angles "upward" into the catalytic domain. Hydrophobic and aromatic residues protrude from these helices to create a hydrophobic surface that would interact with only one face of the lipid bilayer (58). The MBD of COX-1 and -2 thus represents the first example of "monotopic" insertion into biological membranes. The physical and biological consequences of monotopic anchoring have been studied biochemically (33) and by computer modeling (51). The COX isozymes can only be isolated from the membrane using detergents (1, 14, 27), demonstrating that monotopic anchoring can create truly integral membrane proteins. Moreover, the COX isozymes were all crystallized in the presence of nonionic detergents, and tightly bound detergent molecules have been clearly resolved in the COX crystal structures (34, 35, 37, 38).

Catalytic Domain

The catalytic domain comprises the bulk of the COX monomer (13) and is almost entirely composed of α-helical secondary structure. The COX catalytic domain shares a great deal of structural homology with mammalian MPO (13, 58), consistent with the sequence comparisons. Structural homology between the COX catalytic domain and nonmammalian heme-dependent peroxidases is also detectable (13, 58, 61). The POX active site is in a large groove on the side opposite of the MBD (Figure 3*b*), while the entrance to the COX active site is between the helices of the MBD. The hydrophobic COX channel extends from the MBD into the interior of the catalytic domain (58), a distance of about 25 Å (Figure 3*c*). The COX channel contains several side pockets (28, 37, 58) as well as water channels (68) that extend from the COX active site near Gly533 to the dimer interface. At the interface between the MBD and catalytic domain, the COX channel narrows considerably to form an aperture that divides the COX channel into a mouth (or "foyer") and a catalytic center (Figure 3*c*). The narrowness of the aperture clearly suggests that the MBD may undergo significant conformational changes during entry of fatty acid substrates and NSAIDs (35).

THE EVOLUTION OF COX

How COX evolved from soluble heme-dependent peroxidases is an intriguing question and shows the ingenuity of Nature's biological engineering. Several crystal structures are currently available for heme-dependent peroxidases including

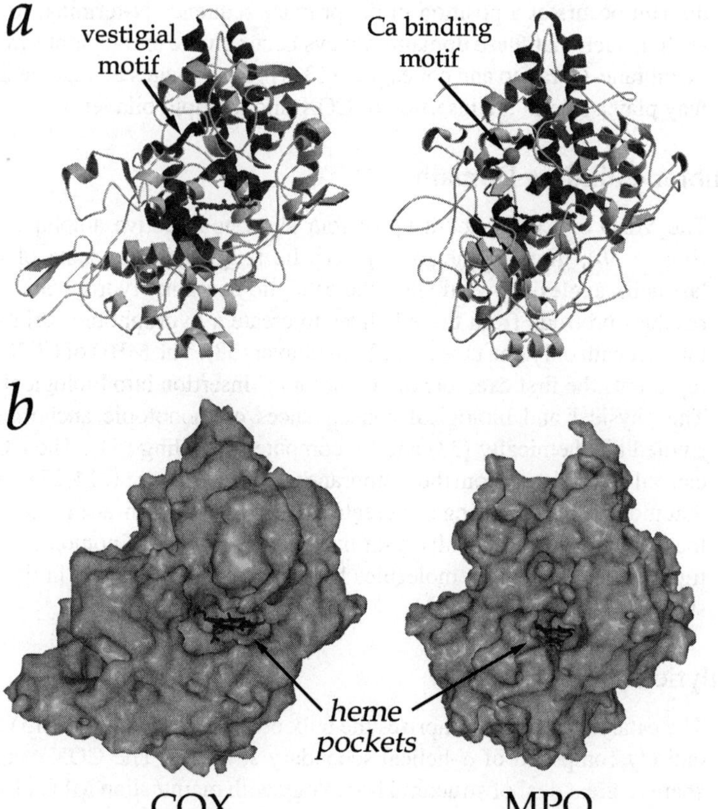

Figure 4 Views of POX active site of the COX-1 monomer with respect to myeloperoxidase (MPO) after superposition. (*a*) The structurally homologous helices that have catalytic relevance are in a *darker gray* (see text). The functioning and vestigial calcium binding motifs are *dark gray* in MPO and COX, respectively. (*b*) An accessible surface view of the POX active site shows how exposed the heme is to solvent compared to that in MPO. Molscript (25) and Raster3D (47) were used to draw (*a*).

mammalian myeloperoxidase (MPO) (6, 95), yeast cytochrome *c* peroxidase (60), and lignin peroxidase (59). Among these, MPO displays obvious close structural homology to the COX catalytic domain, with an rms deviation of 1.5 Å on C_α atoms after relative insertions/deletions are accounted for (Figure 4*a*). The most notable differences between the two enzymes are two long loops that are inserted in the MPO structure and serve to cover the heme binding pocket.

When COX is superimposed with the more distantly related nonmammalian peroxidases by aligning the proximal and distal His residues of the heme pocket, it is clear that the topologies are homologous, including a conserved spatial and topological arrangement of six helices (Figure 4*a*). Using the MPO/COX helix nomenclature (95), these helices are *helix 2*, which bears the distal His residue; *helices 5* and *6*, arranged in a distinctive helix-turn-helix motif; *helix 8*, which carries the proximal histidine residue; *helix 12*, which packs against *helix 8* in an antiparallel manner; and the long *helix 17*, which forms part of the COX channel. Optimally superimposing the conserved helices of COX with the nonmammalian peroxidases yields rms deviations on C_α atoms around 3 to 4 Å.

In addition to the series of conserved helices, COX, MPO, and several nonmammalian peroxidases retain another structural feature: a calcium-binding motif that occurs adjacent to the heme pocket (Figure 4*a*) and after *helix 2* in the sequence. This motif consists of a tight turn of the main chain and is typically composed of a (Val-X-)Gly-X-Asp-X-Ser sequence. A carboxylate oxygen of the Asp in this motif forms hydrogen bonds that bridge the Gly carbonyl and Ser amide main chain groups. The cation is coordinated by the Val and Gly carbonyl groups and the Asp and Ser side chains, as well as both carbonyl and side chain oxygen atoms from an Asp residue immediately following the distal His. This motif is highly conserved in the plant and mammalian enzymes (61); however, in both fungal peroxidases, the interaction with the initial carbonyl group is absent (59). Despite the lack of cation binding, COX retains a vestigial G-X-D-X-G calcium-binding motif (Figure 4*a*), where a Gly is substituted for the usual Ser residue, with a conserved main chain conformation and the Asp side chain making analogous hydrogen bonds with the adjacent main chain. A vestigial G-X-D-X-G calcium-binding motif also exists in the structure of yeast cytochrome *c* peroxidase, which also does not bind calcium. These features clearly demonstrate the evolutionary relationship between fungal, plant, and mammalian peroxidases.

THE POX ACTIVE SITE

The POX active sites in the COXs are quite open to the solvent in contrast to virtually all other peroxidases (Figure 4*b*). As noted above, this gives rise to the rather facile manner by which the heme can be removed and then reconstituted. Within the heme pocket, the mammalian peroxidases COX and MPO also bind the heme ring in an orientation of bound heme that is rotated 180° relative to that seen in the nonmammalian peroxidases. This results in the propionate moieties extending in the opposite direction.

In the refined crystal structure of ovine COX-1 (68), the POX active site (Figure 5, see color insert) reveals that His388 is the proximal heme ligand: The Nε nitrogen bonds to the ferric iron while the Nδ participates in a hydrogen bond network involving a water molecule and Tyr504. In COX-2, the identical arrangement is seen but the existence of a proximal water molecule has not been commented on (28, 37). In contrast, the proximal His forms a hydrogen bond with a conserved Asp side chain on an adjacent helix in the plant, yeast, and fungal peroxidases (11, 61); in MPO, the proximal His residue interacts instead with an Asn side chain (6). Thus, the proximal histidine in COX does not form an ionic bond or strong hydrogen bond, which may make the proximal His more neutral (66, 68).

On the distal side of the heme (Figure 5), His207 is predicted to be important in the deprotonation of the peroxide substrate and subsequent reprotonation of the incipient alkyloxide ion to form the alcohol during generation of Compound I (30). In both COX and MPO, a conserved Gln side chain (Gln203 in COX and Gln91 in canine MPO) is also found adjacent to the distal His in place of the Arg residue usually found in nonmammalian peroxidases. The side chain of the distal His appears to be oriented by hydrogen bonding to the Gln main chain carbonyl, as well as to the side chain immediately preceding the His (Thr206 for COX, Asp94 for MPO). Moreover, orientation of the distal His is usually stabilized by hydrogen bonding to a highly conserved Asn side chain located across the heme binding pocket. Hence, the imidazole ring of the distal His is rotated 180° in COX and MPO relative to that found in the nonmammalian peroxidases. Mutations of Gln203, His207, or His388 in ovine COX-1 and human COX-2 lead to a reduction or elimination of POX activity (30, 71).

Typical heme iron ligands (e.g., CO or CN^-) bind to the distal side of the iron with a linear or "unbent" geometry in ovine COX-1 (68). This seems to be a reasonable result, as COX has a very open active site (Figure 4*b*) and must bind large ligands such as 15-HPETE or PGG_2. However, COX-1 paradoxically exhibits reduced affinity toward small ligands such as azide, thiocyanate, and H_2O_2. In fact, H_2O_2 is a poor substrate compared to many alkyl peroxides (44). The low affinity of azide and thiocyanate for COXs probably arises from unfavorable interactions with distal "roof" residues in the POX active site (Figure 5), but the crystal structures provide little insight into how this occurs.

The proximal His is bonded directly to an Asn in MPO (95) and to an Asp in cytochrome *c* peroxidase (61), creating a more basic proximal ligand. Such strong interactions on the proximal side of the heme are considered to control the reactivity of the heme iron. In ovine COX-1, resonance Raman spectroscopy indicates that His388, the proximal ligand, is clearly more neutral in character than the corresponding proximal ligands found in other peroxidases (68). Although His388 in COX-1 hydrogen bonds to Tyr504 via a mediating buried water, mutating Tyr504 to an alanine results in only a marginal loss in POX activity (68). Thus, the proximal His in COX requires no backside hydrogen bond to catalyze peroxidation. However, this difference in acidity of the proximal histidine in native COX is clearly not reflected in the rate of POX turnover, which is comparable to those of other heme peroxidases (78).

The relatively facile removal of heme for COX-1 and -2 has allowed the creation of apo-enzyme and its reconstitution into pseudo-holoforms of COX with different metallo-protoporphyrin IX compounds. Although Fe^{3+}-protoporphyrin IX is the natural heme ligand, Mn^{3+}-protoporphyrin IX can slowly undergo the changes in redox state needed for POX catalysis and can initiate the COX reaction (89, 90). Other metals (e.g., Zn or Co) in protoporphyrin IX do not support either the POX or COX reactions (52). However, ovine COX-1 reconstituted with Co^{3+}-protoporphyrin IX does form a native-like, albeit inactive, enzyme form with

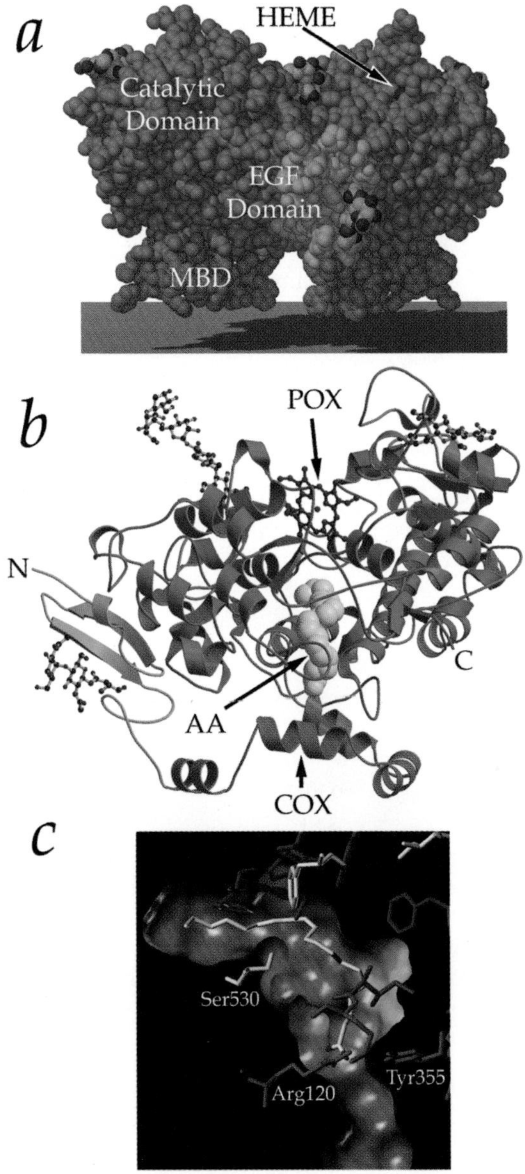

Figure 3 Structural representations of the ovine COX-1 dimer and monomer. (*a*) A space-filling view of the COX-1 dimer is shown. The EGF, MBD, and catalytic domains are colored green, gold, and blue, respectively. The locations of the heme in the POX active site are also shown, and the sites of N-linked glycosylation at asparagines N68, N144, and N410 appear as gray atoms. (*b*) A ribbon drawing shows the COX-1 monomer with bound arachidonic acid (AA in yellow) in the COX active site. The color scheme for the domains is the same as in (*a*). (*c*) A view of the COX channel with bound AA shows the aperture at the level of Tyr355 and Arg120. Views (*a*) and (*b*) were made with Molscript (25) and Raster3D (47).

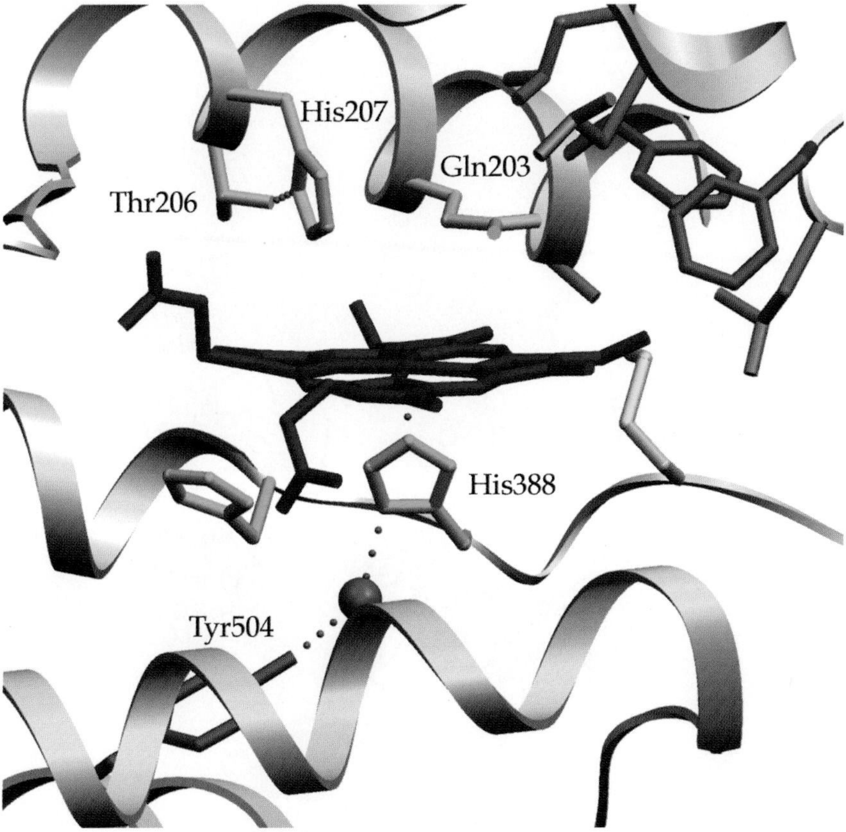

Figure 5 A view of the POX active site. The heme (red) is shown liganded to the proximal ligand His388 (green), which is hydrogen-bonded to a water molecule (blue) and Tyr504 (magenta). The distal residues His207 (green) and Gln203 (blue) are also shown. The nonpolar portion of some peroxide substrates may interact with the distal hydrophobic "roof" made up of valines and leucines (brown) and phenylalanines (magenta). Setor (10) was used to draw this figure and Figures 6 and 8.

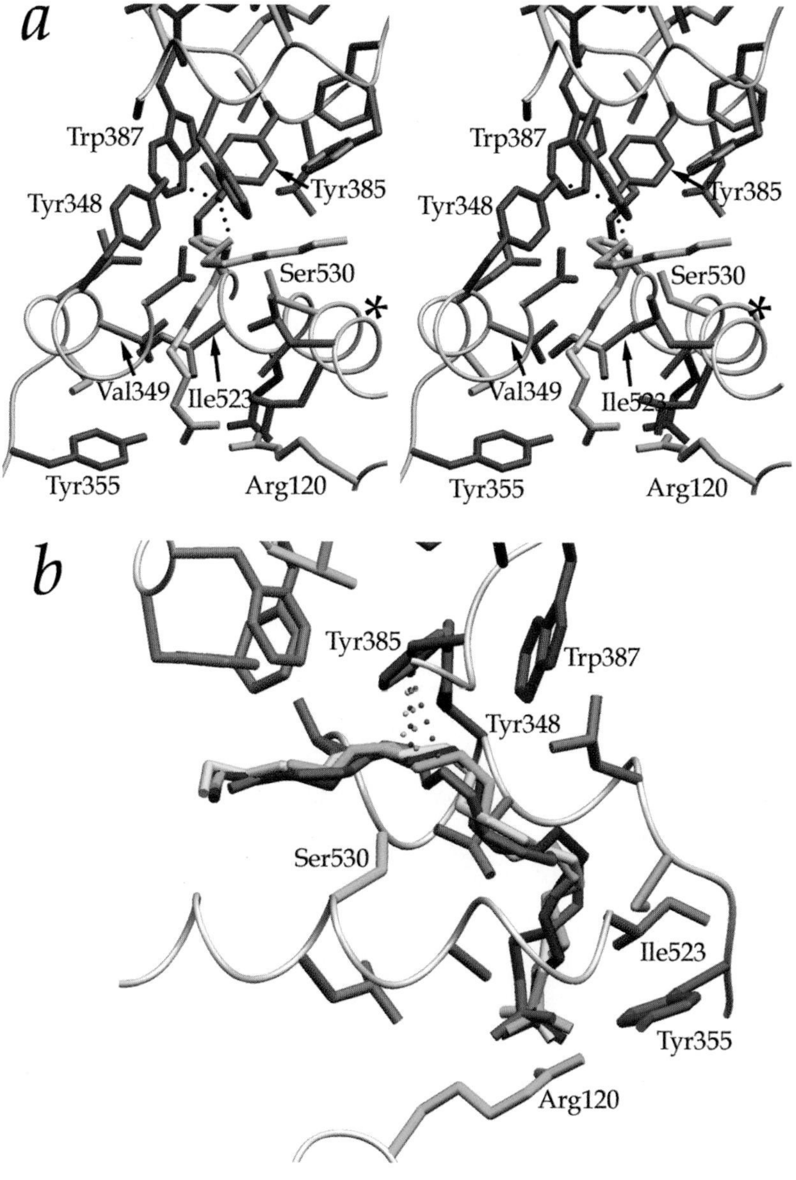

Figure 6 Fatty acid binding in ovine COX-1. (*a*) A stereo diagram of the COX active site shows the kinked L-conformation of AA (yellow). Note how Tyr385 (blue) is aligned with carbon 13 in AA, and how the carboxylate of arachidonate interacts with Arg120 (see text). Several other residues discussed in the text are also shown; the asterisk (*) refers to the position of Gly533. (*b*) A superposition of the fatty acids AA, DHLA (20:3n-6), LA (18:2n-6), and EPA (20:5n-3) in yellow, red, green, and blue, respectively, allows the comparison of their binding conformations in ovine COX-1. Note how the carboxylate ends of the fatty acids exhibit markedly different conformations, while the ω-ends adopt almost the equivalent conformations.

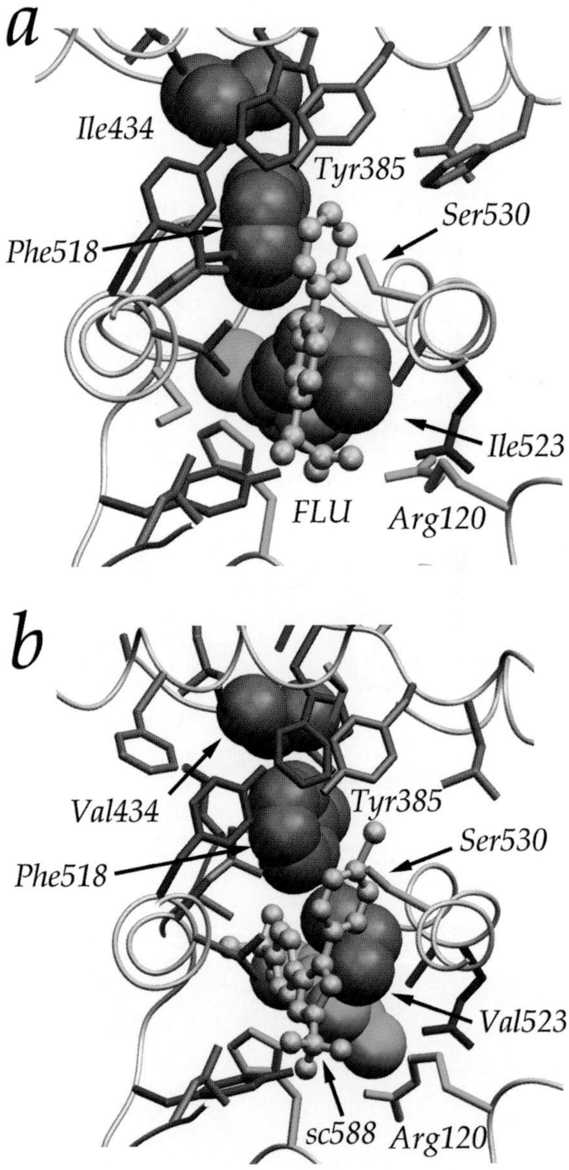

Figure 8 A view of NSAID binding in the COX active site. (*a*) Flurbiprofen (yellow) is bound in the COX active site channel in ovine COX-1 (58). Residues Ile 434 (copper), His513 (green), Phe518 (copper), and Ile523 (copper) are displayed as space-filling. (*b*) The COX-2 inhibitor SC-588 (yellow) is bound in the COX active site channel of mouse COX-2 (28). Residues Val434 (copper), Arg513 (green), Phe518 (copper), and Val523 (copper) are displayed as space-filling. The phenylsulfonamide group of SC-588 extends back into the side pocket made accessible by Val523 and interacts with Arg513. Access to the side pocket is made easier by the I434V change in COX-2, which then allows Phe518 to move out of the way when COX-2 inhibitors bind in this pocket.

the metal bound in a six-coordinate state (39). This form of the enzyme was useful for creating stable complexes of ovine COX-1 with fatty acid substrates for crystallographic analyses (38–40, 87).

THE COX ACTIVE SITE

The backbone folding in the interior of COX is homologous to that found in MPO, particularly in the region around *helix 6* through the turn and extended strand that follow it. In COX, however, this strand is shifted by ~7 Å from that seen in MPO (13). This structural alteration opens up the enzyme's interior to create the ~25 Å long COX channel (Figure 3*c*); such interior cavities are unknown in MPO and other related heme-dependent peroxidases. The COX catalytic center encompasses the upper half of a channel, extending from Arg120 to Tyr385. From an evolutionary perspective, an ancestral peroxidase must have undergone two distinct changes to create COX: (*a*) the formation of an interior channel for the COX reaction and (*b*) the acquisition of the membrane binding.

The mechanism of substrate interactions with COX-1 and -2 is becoming better understood with the growing number of COX structures containing bound substrates (22, 29, 38, 40, 87). In ovine COX-1, AA binds in an extended L-shaped but kinked conformation (Figure 3*c*). The guanidinium group of Arg120 ligands the carboxylate of the fatty acid; this interaction is a known determinant of substrate binding in COX-1 (2, 38, 64). Carbons 7 through 14 of AA form an S-shape that weaves the substrate chain around the side chain of Ser530, the residue acetylated by aspirin (8, 34). AA is positioned such that carbon 13 is oriented near the phenolic oxygen of Tyr385 (Figure 6, see color insert), where the *proS* hydrogen can be abstracted to initiate the COX reaction. The ω-end of AA (carbons 14 through 20) binds in a hydrophobic groove above Ser530, where Phe205, Phe209, Val344, Phe381, and Leu534 stabilize its conformation.

The alternative fatty acid substrates DHLA (20:3n-6), LA (18:2n-6), and EPA (20:5n-3) also bind to the COX active site in extended L-shaped conformations (Figure 6*b*) that are generally similar to those observed for AA (40, 87). These alternative substrates have their carboxylate group positioned such that it makes the critical salt bridge with Arg120, and the ω-end is placed in the hydrophobic groove above Ser530. The alternative substrates make contact with virtually the same set of residues within the active site as AA. However, the chemical differences in polyunsaturation and carbon length presented by DHLA, EPA, and LA lead to local differences in the bound conformation compared to AA. The level of conformational flexibility in each fatty acid impacts the alignment (or misalignment) of the carbon targeted for hydrogen abstraction by Tyr385 (40, 87). For example, LA (18:2n-6) is two carbons shorter and contains two fewer double bonds than AA (20:4n-6). The Arg120/carboxylate interaction is maintained, and carbons 1 to 6 of LA are positioned similarly to the equivalent carbons in AA. However, carbons 7 through 9 display a more extended stereochemistry; this positions carbon 11, instead of carbon 13, below Tyr385 at the top of the active site (40). Thus, 9- or 13-hydroperoxy octadecadienoic acids can be produced. Other fatty acids such as

18:3n-6 and 20:2n-6 should bind in a similar manner that would permit removal of the n-8 hydrogen; 18:3n-3, however, would need to be aligned for abstraction of the n-5 hydrogen (31, 65).

The attempts to study the structure of AA in COX-2 (22, 29) have yielded interesting but somewhat more equivocal results that underscore the difficulty in obtaining crystals of COX complexes with such unstable substrates. Kiefer et al. (22) reported the structure of mouse COX-2 crystallized in the presence of AA. Although they used the apo-form of the enzyme to prevent the turnover of AA during crystallization, they observed a pattern of electron density that suggested a mixture of bound substrate (AA) and product (PGG_2). After building both molecules into the COX active site, Kiefer and colleagues (29) found an almost identical set of interactions between the protein and ligand as seen in COX-1. In another attempt to prevent turnover of the substrate during crystallization, Kiefer et al. (22) made an H270A mutation to create a POX inactive version of the enzyme. The pattern of electron density again suggested that ligand had bound, but in a substantially different conformation. After careful model building at 2.4 Å resolution, Kiefer and colleagues concluded (22, 29) that AA had bound in a nonproductive, "backward" conformation, i.e., with the carboxylate hydrogen-bonded to Tyr385 and Ser530. Kiefer and colleagues (22, 29) consider this second binding mode for AA as an inhibitory binding state in COX-2 and have recently shown that EPA binds to COX-2 in an equivalent fashion. The COX active site in COX-2 is larger than in COX-1 (28, 37) and may thus accommodate a wider array of alternative conformations for many ligands than is allowed in COX-1.

While the physiological relevance of this "inhibitory" binding mode is not clear, it does, however, highlight the markedly different behavior of COX-2 toward ligands compared with that of COX-1 despite a high degree of sequence and structural homology in this region. Several interesting facts have recently come to light that suggest that the conversion of alternate substrates by the COX enzymes to oxygenated products other than PGH_2 may have significant physiological relevance. For example, COX-2 seems to be able to convert the endocannabinoids 2-arachidonylglycerol and arachidonylethanolamide into the precursors for prostaglandin glycerol esters and prostaglandin ethanolamides (23, 24, 62). Understanding the physiological roles of these alternative prostaglandins may clarify the role of COX-2 in the central nervous system. Moreover, many of the minor products from the COX reactions with arachidonate and other fatty acids have significant biological activities (70). Thus, how alternative substrates are utilized differently by COX-1 and -2 may impact the therapeutic effects of diet and NSAID use on cancer, arthritis, and cardiovascular disease.

STRUCTURAL INSIGHTS FROM MUTAGENESIS

The structures of COX complexed with fatty acids (22, 38, 40, 87) have allowed more detailed interpretations of how mutagenesis of active site residues impacts the catalytic steps leading to PGG_2 formation. Interestingly, mutagenesis of COX enzymes clearly revealed that the isoforms behave in distinctly different ways. One striking example is Arg120, which was identified as a potentially critical contributor to substrate binding via the fatty acid carboxylate (28, 37, 58). Indeed, the substitution of Arg120 with other residues can markedly decrease AA affinity by up to 1000-fold (2, 41). In sharp contrast, similar mutations in human COX-2 have little effect on K_m or V_{max} (64). These results suggest that the hydrophobic residues in the COX-2 COX channel must play a more significant role in substrate binding than in COX-1. How they compensate for the surprisingly diminished role of Arg120 in COX-2 remains unclear.

The several other residues within the COX active site, besides Arg120, have been studied by mutagenesis in both isozymes (40, 67, 85, 87). Three regions of the active site have been studied extensively: Val349 near the apical portion of AA; Ser530, Tyr348, and Trp387 near the central portion of AA; and Phe518, Leu531, and Gly533 near the ω-end of the substrate. Mutations at three of these sites (Val349, Ser530, and Gly533) (Figure 6) seem to reveal a great deal about the effects of substrate/enzyme interactions on catalysis.

When Val349 is mutated in ovine COX-1 and human COX-2, 11*R*-HPETE, not PGG_2, is the major COX product (86). The mutation of Val349 to other amino acids also has an impact on the formation of PGG_2 and 15*R/S*-HPETE by-products (67, 85). The replacement of Val349 with leucine in ovine COX-1 eliminates all 11-HPETE formations, but increases the formation of the 15*R/S*-HPETE by-products as an equimolar mixture (85). When Schneider et al. (67) replaced valine with isoleucine at residue 349 in COX-1 or -2, they observed a marked shift in product stereochemistry at carbon 15 toward the *R* stereoisomer for both PGG_2 and 15-HPETE. Thus, Val349 may play a subtle role in maintaining the proper stereochemistry of the substrate during oxygen addition at carbon 15.

Ser530 is the site of acetylation by aspirin (32, 34, 72) and makes an intimate interaction with AA at carbon 13 (38, 85). While not essential for catalysis, Ser530 may help optimally align the substrate with respect to the Tyr385 for hydrogen abstraction (38, 94) at carbon 13, as well as for subsequent oxygen addition at carbon 11 (Figure 6). When Ser530 in COX-1 is mutated to a threonine, the enzyme is active, but 15*R*-HPETE is formed almost exclusively instead of PGG_2 (85). However, the acetylation of Ser530 by aspirin in COX-1, which adds three nonhydrogen atoms completely to the serine, now blocks the binding of AA and completely inactivates the enzyme. When aspirin acetylates human COX-2, substrate turnover still occurs, but 15*R*-HPETE is produced (32, 42). The additional atoms at position 530 apparently perturb the AA conformation in COX-2 so that oxygen insertion at carbon 15, but not at carbon 11, is possible after hydrogen abstraction. As the COX active site in COX-2 is larger than in COX-1 (28, 37), COX-2 can obviously accommodate larger moieties at position 530 without inactivating. Interestingly, Schneider et al. (67) found that replacing Ser530 in COX-2 with threonine resulted in an almost complete shift in product stereochemistry at carbon 15 toward the *R* stereoisomer in PGG_2, as well. Hence, the amino acid side chain at position 530 may play a critical role in determining the stereochemistry of oxygen addition at carbon 15.

At the very end of the fatty acid binding pocket, Gly533 (Figure 6) abuts on the very last two carbons of AA in COX-1 (38). Intriguingly, the mutation of this residue in ovine COX-1 and human COX-2 markedly decreases substrate turnover (65, 72, 85). The superimposed structures of AA, DHLA, EPA, and LA (40, 87) clearly show that the ω-ends of the substrates tend to adopt a particular stereochemistry in this region of the active site (Figure 6*b*). Mutations at or near Gly533 must then perturb the conformation of the substrate away from the catalytically competent conformer (38, 65, 85). Thus, it seems that the positioning of the ω-end of a fatty acid substrate is a critical factor in its turnover (40, 87). If the tail of the substrate is not bound properly, then the appropriate hydrogen cannot be readily abstracted. On the other hand, the COX active site surrounding the carboxylate end of the fatty acids seems to better accommodate mutations without a major loss of COX activity (40, 87). This region of the active site may help compensate for any steric strain arising from the positioning of the substrate's ω-end for turnover.

THE COX REACTION REVISITED

The basic steps in the COX reaction, as originally proposed by Hamberg & Samuelsson (19), have remained virtually unchanged for over 30 years. The bulk of the biochemical and structural evidence supports all the central features of the Hamberg & Samuelsson mechanism (19, 29, 43, 45, 76), but recent structural studies have now highlighted some of the important structure-function relationships in PGG_2 formation. The COX reaction scheme can now be broken down into four basic steps (Figure 7). Initially, the arachidonyl carboxylate interacts with Arg120 (28, 34, 35, 37, 38, 58) and enters the COX channel. During this process, the enzyme appropriately positions the 13*proS* hydrogen of AA for abstraction. Carbons 8 through 12 are also positioned in a space suitable for the formation of the endoperoxide bridge and the cyclopentyl ring (38, 86). When AA is appropriately positioned, the rate-determining step begins: The radical on Tyr385 abstracts the 13-*proS* hydrogen. Subsequently, an 11*R*-peroxyl radical is quickly formed in the presence of oxygen. In the third step, the 11*R*-peroxyl radical attacks carbon 9 to form the endoperoxide, resulting in the isomerization of the radical to carbon 8. At this stage, ring closure between carbon 8 and carbon 12 cannot occur given the extended conformation exhibited by AA in the crystal structures (Figure 7). Therefore, a major reconfiguration of the substrate must occur concomitant with or immediately following formation of the endoperoxide bridge. This hypothetical conformational transition would involve a significant movement of the ω-end of the substrate toward the carboxyl half. The 11*R*-hydroperoxyl radical is hypothesized to swing "over" carbon 8 for an *R*-side attack on carbon 9 through the rotation about the carbon 10/carbon 11 bond, which brings carbon 12 closer to carbon 8 for the ring closure (Figure 7). This conformational transition would also reposition carbon 15 for the addition of the second molecule of oxygen. In the final step, the 15*S*-peroxyl radical is aligned below Tyr385 for donation of the radical to complete the catalytic cycle.

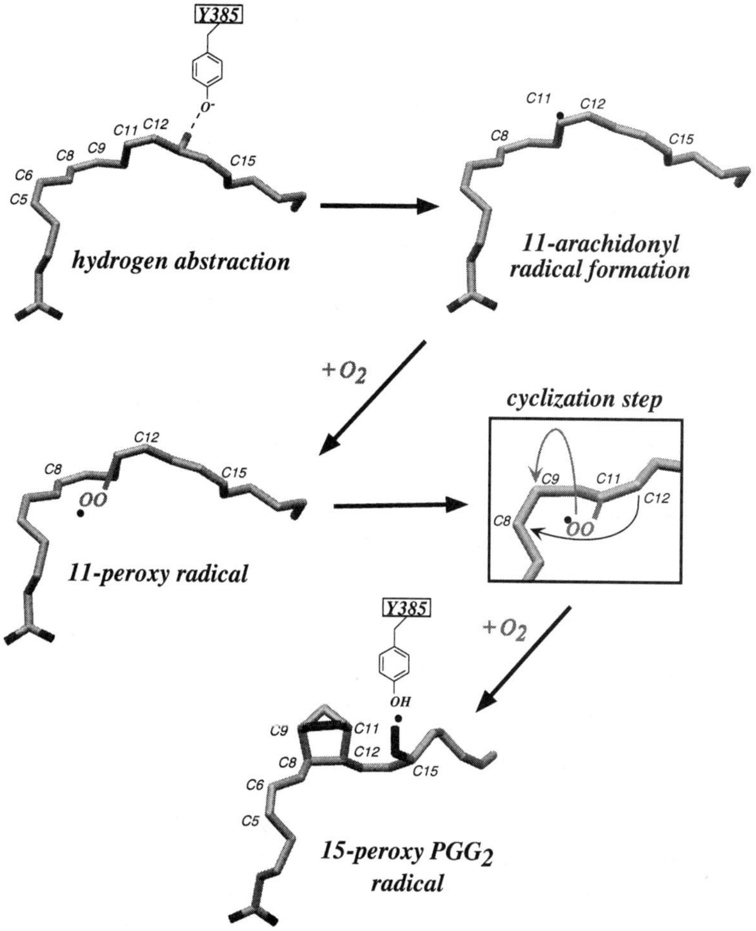

Figure 7 A schematic of the COX reaction as proposed by Malkowski et al. (38). While the formation of the 11-peroxy-arachidonate intermediate is explained by the crystal structure for the ovine COX-1/AA complex, its subsequent conversion to PGG_2 requires a major conformational transition that has not been observed or characterized.

While the formation of PGG_2 is now fairly well explained, why 11*R*-HPETE, 15*R*-HPETE, and 15*S*-HPETE are formed is unclear. Thuresson et al. (86) presented biochemical evidence that each COX product may arise from a different but catalytically competent conformer of AA. In essence, hydrogen abstraction can occur when AA is not in a conformation that allows facile ring closure, which then leads to the monooxygenation of the substrate. This may mean that the observed structure of AA bound in the COX-1 active site (38) may be a time- and space-average of more than one AA conformer, of which only the predominant conformer leads to PGG_2 formation. As little conformational variation is seen in the COX-1 and -2 crystal structures, subtle but distinct dynamic transitions in enzyme structure must be occurring to account for the different substrate conformers in the native and mutant forms of COX. As mentioned earlier, the COX channel narrows considerably to form an aperture (Figure 3*c*). The aperture, composed of parts of the catalytic center and the MBD, must open and close during entry of fatty acid substrates and NSAIDs and egress of product (35). The subtle conformational dynamics of the COX active site and the MBD may be a primary factor in determining the final COX products. Interestingly, studies on enzyme inhibition by NSAIDs have provided strong evidence that conformational variation not observed in the crystal structures occurs in COX-1 and -2.

THE STRUCTURAL BASIS OF NSAID ACTION

Several detailed reviews have been recently published on this topic (7, 12, 50), and only a summary of the major highlights of NSAID inhibition in the COX isoforms is given here. NSAIDs are usually subdivided into two classes: (*a*) classical or "nonselective" NSAIDs and (*b*) COX-2-selective or "isoform-specific" NSAID inhibitors. The classical NSAIDs inhibit both COX-1 and -2, but many tend to bind more tightly to COX-1 (7, 50). In contrast, COX-2-selective inhibitors have been designed to exhibit significantly higher selectivity toward COX-2 than toward COX-1 (7, 50). While all NSAIDs compete with arachidonate for the COX active site, each NSAID can be classified by one of three general modes of action (7, 50). An NSAID can display (*a*) rapid, reversible competitive inhibition (e.g., ibuprofen), (*b*) rapid, reversible binding followed by covalent modification (e.g., the action of aspirin on Ser530), or (*c*) rapid, lower-affinity competitive inhibition followed by a time-dependent transition to a high-affinity slowly reversible inhibitory mode (e.g., flurbiprofen). The structural basis for time-dependent inhibition is not yet well defined and may be different for different drugs. Moreover, the kinetic differences in NSAID inhibition have made simple comparisons of drug interactions between COX-1 and COX-2 difficult.

Figure 8 (see color insert) illustrates the basic features of NSAID binding to COX-1 and -2. The drugs generally bind within the upper part of the COX channel between Arg120 and Tyr385. The acidic class of NSAIDs (e.g., profens and fenamates) interact with Arg120 in both COX isozymes (7, 12) via hydrogen bonding or electrostatic interactions that provide a major portion of the binding energy and selectivity. The remaining drug/protein interactions tend to be hydrophobic (7, 12) except for potential hydrogen bonding to Ser530.

The differences between classical NSAIDs and COX-2 inhibitors arise in part from slight differences in the amino acids surrounding the active sites of COX-1 and -2. Within the catalytic center, only one structural difference is seen: Ile523 in COX-1 is substituted with Val523 in COX-2. This minor and conservative change results in a small side pocket becoming more accessible from the active site channel (28, 37). In the second shell of amino acids surrounding the COX active site, Ile434 in COX-1 is substituted by a valine in COX-2. Again, this minor substitution, coupled with the Val523, increases the accessible volume of the active site channel by enhancing the mobility of local side chains (28, 37). Hence, the combination of Val523 and Val434 in COX-2 allows a movement of Phe518 and permits access to the polar side pocket (Figure 8*b*). The larger main channel combined with this side pocket increases the volume of the COX-2 NSAID binding site by about 20% over that in COX-1 (28, 37). The larger effective size of the channel in the COX-2 may also preferentially reduce steric and ionic crowding by Arg120 in COX-2 and thus may enhance the binding of nonacidic NSAIDs to COX-2.

This extra volume is a structural feature exploited by most COX-2 inhibitors. Mutating Val523 to an isoleucine restricts access to this side pocket, and COX-2 is no longer differentially sensitive to these inhibitors (15, 18). Conversely, an I523V mutation in COX-1 increases its affinity for COX-2 inhibitors (93). The substitution of His513 in COX-1 with arginine in COX-2 also alters the chemical environment of the side pocket by placing a stable positive charge at its center (28). This arginine seemingly interacts with polar moieties entering the pocket. In combination with the I523V mutation, an H513R mutant of COX-1 becomes much more sensitive to COX-2 inhibitors (93).

TIME-DEPENDENT INHIBITION AND CONFORMATIONAL TRANSITIONS

COX is known to undergo significant conformational changes following binding of heme, fatty acid substrates, and NSAIDs (4, 26). The phenomenon of time-dependent inhibition also provides quite credible evidence of conformation changes in COXs (7, 76), where the enzymes shift from a freely reversible enzyme/ligand complex EI to a tight-binding, slowly reversible EI* complex. The fact that the new COX-2 inhibitors display time-dependent inhibition toward COX-2, but freely reversible inhibition toward COX-1, has provided a new and intriguing view of NSAID action (7, 15, 18). Mutagenesis experiments have suggested that time-dependent inhibition of COX-2 by isoform-selective inhibitors containing sulfonamide or methylsulfoxide moieties may arise from their interaction with Arg513 (28, 93). Curiously, the time-dependent inhibition displayed by the methylsulfoxide inhibitor NS-398, an early lead compound, appears to depend on interaction with Arg120 but not with Arg513. The R120E and R120Q mutations in COX-2 result in simple competitive inhibition by NS-398 (16, 64), suggesting that NS-398 binds in the COX-2 active site similarly to acidic inhibitors such as flurbiprofen.

Unfortunately, determining the precise physical basis for time-dependent inhibition has been elusive. The observed crystal structures of the COX isoforms have provided little insight into the phenomenon. In fact, the crystal structures of COX-1 with time-dependent inhibitors or competitive inhibitors are essentially indistinguishable (69). In the one case where ligand binding in human COX-2 did perturb the MBD (37), the changes were relatively minor. Why do the COX isoforms exhibit a single major conformation in crystals, regardless of whether a COX ligand is bound or not? Adding to this dilemma is the fact that the narrow aperture within the COX channel (Figure 3*c*) effectively buries all bound ligands within the catalytic domain (35, 38). As all ligands must enter the COX active site through the MBD, the COX channel region must undergo significant conformational changes during substrate entry and product exit. Recent studies (80, 92) have suggested that the reorganization of hydrogen bonding networks within the MBD (80) may play an active role in substrate binding, catalytic efficiency, and inhibition by NSAIDs. One hypothesis is that at least two conformations of the enzyme exist in equilibrium (80): an unstable ligand-free form and a more stable ligand-bound form. Conformational models have been proposed to explain time-dependent inhibition (80, 92), but how these conformational transitions are controlled and impact substrate and NSAID binding are, as yet, unanswered.

Identifying the nature of the EI* state associated with time-dependent inhibition might be easier. In the active COX, AA is rapidly converted to PGG_2 and then released. During this process, a conformational rearrangement of the AA chain must occur within the active site as the cyclopentane ring and endoperoxide bridge form (38). This structural transition may disrupt a hydrogen bond network within the COX channel and facilitate release of the newly formed product PGG_2. Thus, COX may shuttle between two conformational states as part of the catalytic mechanism for bis-oxygenation. Time-dependent inhibition may occur when NSAIDs trap the most stable conformation, i.e., that seen in the crystal structures, but then may not be able to trigger the conformational change needed for facile release. Thus, time-dependent inhibition may be a serendipitous outcome of a catalytic mechanism involving two distinct conformational steps (80, 92). Nonetheless, identification and characterization of the important conformational transitions in COX remain elusive, but these structural events will have a major impact on our understanding of how NSAIDs interact with COX. As the crystallographic research on COX continues to mature and as higher-resolution structures become available, we may be able to resolve these and the other remaining questions about the structure-function relationships in COX.

ACKNOWLEDGMENTS

NIH Grants R01 HL56773 and P01 GM57323 supported work from the author's laboratory that was mentioned in this review. The author is grateful to a number of colleagues for their personal communications regarding unpublished work on COX. The author would also like to thank Dr. Guenter Trummlitz and colleagues at Boehringer-Ingelheim for providing the raw image for Figure 3*c*.

The *Annual Review of Biophysics and Biomolecular Structure* is online at http://biophys.annualreviews.org

LITERATURE CITED

1. Barnett J, Chow J, Ives D, Chiou M, Mackenzie R, et al. 1994. Purification, characterization and selective inhibition of human prostaglandin G/H synthase 1 and 2 expressed in the baculovirus system. *Biochim. Biophys. Acta* 1209:130–39
2. Bhattacharyya DK, Lecomte M, Rieke C, Garavito RM, Smith WL. 1996. Involvement of arginine 120, glutamate 524, and tyrosine 355 in the binding of arachidonate and 2-phenylproprionic acid inhibitors to the cyclooxygenase active site of ovine prostaglandin endoperoxide H synthase-1. *J. Biol. Chem.* 271:2179–84
3. Campbell ID, Bork P. 1993. Epidermal growth factor-like modules. *Curr. Opin. Struct. Biol.* 3:385–92
4. Chen Y, Bienkowski M, Marnett L. 1987. Controlled tryptic digestion of prostaglandin H synthase. Characterization of protein fragments and enhanced rate of proteolysis of oxidatively inactivated enzyme. *J. Biol. Chem.* 262:16892–99
5. Crofford LJ, Lipsky PE, Brooks P, Abramson SB, Simon LS, van de Putte LB. 2000. Basic biology and clinical application of specific cyclooxygenase-2 inhibitors. *Arthritis Rheum.* 43:4–13
6. Davy C, Fenna R. 1996. 2.3 Å resolution X-ray crystal structure of the bisubstrate analogue inhibitor salicylhydroxamic acid bound to human myeloperoxidase: a model for a prereaction complex with hydroperoxide. *Biochemistry* 35:10967–73
7. DeWitt DL. 1999. COX-2 selective inhibitors: the new super aspirins. *Mol. Pharmacol.* 4:625–31
8. DeWitt DL, El-Harith A, Kraemer SA, Andrews MJ, Yao EF, et al. 1990. The aspirin and heme-binding sites of ovine and murine prostaglandin endoperoxide synthases. *J. Biol. Chem.* 265:5192–98
9. Dietz R, Nastainczyk W, Ruf H. 1988. Higher oxidation states of prostaglandin H synthase. Rapid electronic spectroscopy detected two spectral intermediates during the peroxidase reaction with prostaglandin G2. *Eur. J. Biochem.* 171:321–28
10. Evans SV. 1993. SETOR: hardware-lighted three-dimensional solid model representations of macromolecules. *J. Mol. Graphics* 11:134–38
11. Gajhede J, Schuller A, Henriksen A, Smith AT, Poulos TL. 1997. Crystal structure of horseradish peroxidase C at 2.15 Å resolution. *Nat. Struct. Biol.* 4:1032–38
12. Garavito RM, DeWitt DL. 1999. The cyclooxygenase isoforms: structural insights into the conversion of arachidonic acid to prostaglandins. *Biochim. Biophys. Acta* 1441:278–87
13. Garavito RM, Picot D, Loll PJ. 1994. Prostaglandin H synthase. *Curr. Opin. Struct. Biol.* 4:529–35
14. Gierse J, Hauser S, Creely D, Koboldt C, Rangwala SH, et al. 1995. Expression and selective inhibition of the constitutive and inducible forms of human cyclooxygenase. *Biochem. J.* 305:479–84
15. Gierse J, McDonald J, Hauser S, Rangwala S, Koboldt CM, Seibert K. 1996. A single amino acid difference between cyclooxygenase-1 (COX-1) and -2 (COX-2) reverses the selectivity of COX-2 inhibitors. *J. Biol. Chem.* 271:15810–14
16. Greig GM, Francis DA, Falgueyret JP, Ouellet M, Percival MD, et al. 1997. The interaction of arginine 106 of human prostaglandin G/H synthase-2 with

inhibitors is not a universal component of inhibition mediated by nonsteroidal anti-inflammatory drugs. *Mol. Pharmacol.* 52:829–38

17. Guo Q, Kulmacz RJ. 2000. Distinct influences of carboxyl terminal segment structure on function in the two isoforms of prostaglandin H synthase. *Arch. Biochem. Biophys.* 384:269–79
18. Guo Q, Wang LH, Ruan KH, Kulmacz RJ. 1996. Role of Val509 in time-dependent inhibition of human prostaglandin H synthase-2 cyclooxygenase activity by isoform-selective agents. *J. Biol. Chem.* 271:19134–39
19. Hamberg M, Samuelsson B. 1967. On the mechanism of biosynthesis of prostaglandins E-1 and F-1 alpha. *J. Biol. Chem.* 242:5336–43
20. Herschman H. 1996. Prostaglandin synthase 2. *Biochim. Biophys. Acta* 1299:125–40
21. Kalgutkar AS, Zhao Z. 2001. Discovery and design of selective cyclooxygenase-2 inhibitors as non-ulcerogenic, anti-inflammatory drugs with potential utility as anti-cancer agents. *Curr. Drug Targets* 2:79–106
22. Kiefer JR, Pawlitz JL, Moreland KT, Stegeman RA, Hood WF, et al. 2000. Structural insights into the stereochemistry of the cyclooxygenase reaction. *Nature* 405:97–101
23. Kozak KR, Crews BC, Ray JL, Tai HH, Morrow JD, et al. 2001. Metabolism of prostaglandin glycerol esters and prostaglandin ethanolamides in vitro and in vivo. *J. Biol. Chem.* 276:36993–98
24. Kozak KR, Rowlinson SW, Marnett LJ. 2000. Oxygenation of the endocannabinoid, 2-arachidonylglycerol, to glyceryl prostaglandins by cyclooxygenase-2. *J. Biol. Chem.* 275:33744–49
25. Kraulis PJ. 1991. MOLSCRIPT: a program to produce both detailed and schematic plots of protein structures. *J. Appl. Crystallogr.* 24:946–50
26. Kulmacz RJ, Lands WE. 1982. Protection of prostaglandin H synthase from trypsin upon binding of heme. *Biochem. Biophys. Res. Commun.* 104:758–64
27. Kulmacz RJ, Lands WEM. 1987. Cyclooxygenase: measurement, purification and properties. In *Prostaglandin and Related Substances: A Practical Approach*, ed. C Benedetto, RG McDonald-Gibson, S Nigan, TF Slater, pp. 209–27. Washington, DC: IRL Press
28. Kurumbail R, Stevens A, Gierse J, McDonald J, Stegeman RA, et al. 1996. Structural basis for selective inhibition of cyclooxygenase-2 by anti-inflammatory agents. *Nature* 384:644–48
29. Kurumbail RG, Kiefer JR, Marnett LJ. 2001. Cyclooxygenase enzyme: catalysis and inhibition. *Curr. Opin. Struct. Biol.* 11:752–60
30. Landino LM, Crews BC, Gierse JK, Hauser SC, Marnett LJ. 1997. Mutational analysis of the role of the distal histidine and glutamine residues of prostaglandin-endoperoxide synthase-2 in peroxidase catalysis, hydroperoxide reduction, and cyclooxygenase activation. *J. Biol. Chem.* 272:21565–74
31. Laneuville O, Breuer D, Xu N, Huang Z, Gage DA, et al. 1995. Fatty acid substrate specificities of human prostaglandin-endoperoxide H synthase-1 and -2. Formation of 12-hydroxy-(9Z, 13E/Z, 15Z)-octadecatrienoic acids from alpha-linolenic acid. *J. Biol. Chem.* 270:19330–36
32. Lecomte M, Laneuville O, Ji C, DeWitt DL, Smith WL. 1994. Acetylation of human prostaglandin endoperoxide synthase-2 (cyclooxygenase-2) by aspirin. *J. Biol. Chem.* 269:13207–15
33. Li Y, Smith T, Grabski S, DeWitt DL. 1998. The membrane association sequences of the prostaglandin endoperoxide synthases-1 and -2 isozymes. *J. Biol. Chem.* 273:29830–37
34. Loll P, Picot D, Garavito R. 1995. The structural basis of aspirin activity inferred from the crystal structure of inactivated

prostaglandin H_2 synthase. *Nat. Struct. Biol.* 2:637–43
35. Loll P, Picot D, Garavito R. 1996. The synthesis and use of iodinated non-steroidal antiinflammatory drug analogs as crystallographic probes of the prostaglandin H_2 synthase cyclooxygenase active site. *Biochemistry* 35:7330–40
36. Loll PJ, Sharkey CT, O'Connor SJ, Dooley CM, O'Brien E, et al. 2001. O-acetylsalicylhydroxamic acid, a novel acetylating inhibitor of prostaglandin H2 synthase: structural and functional characterization of enzyme-inhibitor interactions. *Mol. Pharmacol.* 60:1407–13
37. Luong C, Miller A, Barnett J, Chow J, Ramesha C, et al. 1996. Flexibility of the NSAID binding site in the structure of human cyclooxygenase-2. *Nat. Struct. Biol.* 3:927–33
38. Malkowski MG, Ginell S, Smith WL, Garavito RM. 2000. The x-ray structure of prostaglandin endoperoxide H synthase-1 complexed with arachidonic acid. *Science* 289:1933–37
39. Malkowski MG, Theisen MJ, Scharmen A, Garavito RM. 2000. The formation of stable fatty acid substrate complexes in prostaglandin H2 synthase-1. *Arch. Biochem. Biophys.* 380:39–45
40. Malkowski MG, Thuresson ED, Lakkides KM, Rieke CJ, Micielli R, et al. 2001. Structure of eicosapentaenoic and linoleic acids in the cyclooxygenase site of prostaglandin endoperoxide H synthase-1. *J. Biol. Chem.* 276:37547–55
41. Mancini J, Riendeau D, Falgueyret J, Vickers P, O'Neill G. 1995. Arginine 120 of prostaglandin G/H synthase-1 is required for the inhibition by nonsteroidal anti-inflammatory drugs containing a carboxylic acid moiety. *J. Biol. Chem.* 270:29372–77
42. Mancini JA, O'Neill GP, Bayly C, Vickers PJ. 1994. Mutation of serine-516 in human prostaglandin G/H synthase-2 to methionine or aspirin acetylation of this residue stimulates 15-R-HETE synthesis. *FEBS Lett.* 342:33–37
43. Marnett LJ. 2000. Cyclooxygenase mechanisms. *Curr. Opin. Chem. Biol.* 4:545–52
44. Marnett LJ, Maddipati KR. 1991. Prostaglandin H synthase. In *Peroxidases in Chemistry and Biology*, ed. J Everse, KE Everse, MB Grisham, pp. 293–334. Boca Raton, FL: CRC
45. Marnett LJ, Rowlinson SW, Goodwin DC, Kalgutkar AS, Lanzo CA. 1999. Arachidonic acid oxygenation by COX-1 and COX-2. *J. Biol. Chem.* 274:22903–6
46. McGeer PL, McGeer EG. 1999. Inflammation of the brain in Alzheimer's disease: implications for therapy. *J. Leukoc. Biol.* 65:409–15
47. Merrit E, Murphy M. 1994. Raster3D version 2.0—a program for photorealistic molecular graphics. *Acta Crystallogr. D* 50:869–73
48. Moore BC, Simmons DL. 2000. COX-2 inhibition, apoptosis, and chemoprevention by nonsteroidal anti-inflammatory drugs. *Curr. Med. Chem.* 7:1131–44
49. Morita I, Schindler M, Regier MK, Otto JC, Itori T, et al. 1995. Different intracellular locations for prostaglandin endoperoxide H synthase-1 and -2. *J. Biol. Chem.* 270:10902–8
50. Munroe D, Lau C. 1995. Turning down the heat: new routes to inhibition of inflammatory signaling by prostaglandin H2 synthases. *Chem. Biol.* 2:343–50
51. Nina M, Berneche S, Roux B. 2000. Anchoring of a monotopic membrane protein: the binding of prostaglandin H2 synthase-1 to the surface of a phospholipid bilayer. *Eur. Biophys. J.* 29:439–54
52. Ogino N, Ohki S, Yamamoto S, Hayaishi O. 1978. Prostaglandin endoperoxide synthetases from bovine vesicular gland microsomes. *J. Biol. Chem.* 253:5061–68
53. Otto J, Dewitt D, Smith W. 1993. N-glycosylation of prostaglandin endoperoxide synthases-1 and -2 and their orientations in the endoplasmic reticulum. *J. Biol. Chem.* 268:18234–42

54. Otto J, Smith W. 1996. Photolabeling of prostaglandin endoperoxide H synthase-1 with 3-trifluoro-3-(m-[125I]iodophenyl) diazirine as a probe of membrane association and the cyclooxygenase active site. *J. Biol. Chem.* 271:9906–10
55. Patrignani P, Panara MR, Greco A, Fusco O, Natoli C, et al. 1994. Biochemical and pharmacological characterization of the cyclooxygenase activity of human blood prostaglandin endoperoxide synthases. *J. Pharmacol. Exp. Ther.* 271:1705–12
56. Patrono C. 1994. Aspirin as an antiplatelet drug. *N. Engl. J. Med.* 330:1287–94
57. Picot D, Garavito R. 1994. Prostaglandin H synthase: implications for membrane structure. *FEBS Lett.* 346:21–25
58. Picot D, Loll PJ, Garavito RM. 1994. The X-ray crystal structure of the membrane protein prostaglandin H_2 synthase-1. *Nature* 367:243–49
59. Poulos T, Edwards S, Wariishi H, Gold M. 1993. Crystallographic refinement of lignin peroxidase at 2 Å. *J. Biol. Chem.* 268:4429–40
60. Poulos T, Freer S, Alden R, Edwards S, Skogland U, et al. 1979. The crystal structure of cytochrome *c* peroxidase. *J. Biol. Chem.* 255:575–80
61. Poulos TL, Fenna RE. 1994. Peroxidases: structure, function and engineering. See Ref. 72a, pp. 163–99
62. Prusakiewicz J, Kingsley P, Kozak K, Marnett L. 2002. Selective oxygenation of N-arachidonylglycine by cyclooxygenase-2. *Biochem. Biophys. Res. Commun.* 296:612–17
63. Raz A. 2002. Is inhibition of cyclooxygenase required for the anti-tumorigenic effects of nonsteroidal, anti-inflammatory drugs (NSAIDs)? In vitro versus in vivo results and the relevance for the prevention and treatment of cancer. *Biochem. Pharmacol.* 63:343–47
64. Rieke CJ, Mulichak AM, Garavito RM, Smith WL. 1999. The role of arginine 120 of human prostaglandin endoperoxide H synthase-2 in the interaction with fatty acid substrates and inhibitors. *J. Biol. Chem.* 274:17109–14
65. Rowlinson SW, Crews BC, Lanzo CA, Marnett LJ. 1999. The binding of arachidonic acid in the cyclooxygenase active site of mouse prostaglandin endoperoxide synthase-2 (COX-2). *J. Biol. Chem.* 274:23305–10
66. Schelvis JPM, Seibold SA, Cerda JF, Garavito RM, Arakawa T, et al. 2000. Interaction of nitric oxide with prostaglandin endoperoxide H synthase: implications for Fe-His bond cleavage in heme proteins. *J. Phys. Chem. B* 104:10844–50
67. Schneider C, Boeglin WE, Prusakiewicz JJ, Scott W, Rowlinson SW, et al. 2002. Control of prostaglandin stereochemistry at the 15-carbon by cyclooxygenases-1 and -2: a critical role for Serine 530 and Valine 349. *J. Biol. Chem.* 277:478–85
68. Seibold SA, Cerda JF, Mulichak AM, Song I, Garavito RM, et al. 2000. Peroxidase activity in prostaglandin endoperoxide H synthase-1 occurs with a neutral histidine proximal heme ligand. *Biochemistry* 39:6616–24
69. Selinsky BS, Gupta K, Sharkey CT, Loll PJ. 2001. Structural analysis of NSAID binding by prostaglandin H2 synthase: time-dependent and time-independent inhibitors elicit identical enzyme conformations. *Biochemistry* 40:5172–80
70. Serhan CN, Clish CB, Brannon J, Colgan SP, Gronert K, et al. 2000. Anti-microinflammatory lipid signals generated from dietary N-3 fatty acids via cyclooxygenase-2 and transcellular processing: a novel mechanism for NSAID and N-3 PUFA therapeutic actions. *J. Physiol. Pharmacol.* 51:643–54
71. Shimokawa T, Smith WL. 1991. Essential histidines of prostaglandin endoperoxide synthase. His-309 is involved in heme binding. *J. Biol. Chem.* 266:6168–73
72. Shimokawa T, Smith WL. 1992. Prostaglandin endoperoxide synthase: the aspirin

acetylation region. *J. Biol. Chem.* 267: 12387–92

72a. Sigal H, Sigel A, eds. 1994. *Metal Ions in Biological Systems*. New York: Marcel Dekker

73. Smith T, Leipprandt J, DeWitt DL. 2000. Purification and characterization of the human recombinant histidine-tagged prostaglandin H endoperoxide synthases-1 and -2. *Arch. Biochem. Biophys.* 374:195–200

74. Smith W, Garavito R, DeWitt D. 1996. Prostaglandin endoperoxide H synthases (cyclooxygenses)-1 and -2. *J. Biol. Chem.* 271:33157–60

75. Smith WL, DeWitt DL. 1996. Prostaglandin endoperoxide synthases-1 and -2. In *In Advances in Immunology*, ed. FJ Dixon, pp. 167–215. San Diego: Academic

76. Smith WL, DeWitt DL, Garavito RM. 2000. Cyclooxygenases: structural, cellular, and molecular biology. *Annu. Rev. Biochem.* 69:145–82

77. Smith WL, Eling TE, Kulmacz RJ, Marnett LJ, Tsai A. 1992. Tyrosyl radicals and their role in hydroperoxide-dependent activation and inactivation of prostaglandin endoperoxide synthase. *Biochemistry* 31: 3–7

78. Smith WL, Marnett LJ. 1994. Prostaglandin endoperoxide synthases. See Ref. 72a, pp. 163–99

79. Smythies J. 1996. On the function of neuromelanin. *Proc. R. Soc. London Ser. B* 263:487–89

80. So OY, Scarafia LE, Mak AY, Callan OH, Swinney DC. 1998. The dynamics of prostaglandin H synthases. Studies with prostaglandin H synthase-2 Y355F unmask mechanisms of time-dependent inhibition and allosteric activation. *J. Biol. Chem.* 273:5801–7

81. Song I, Smith WL. 1996. C-terminal Ser/Pro-Thr-Glu-Leu tetrapeptides of prostaglandin endoperoxide H synthases-1 and -2 target the enzymes to the endoplasmic reticulum. *Arch. Biochem. Biophys.* 334: 67–72

82. Song X, Lin HP, Johnson AJ, Tseng PH, Yang YT, et al. 2002. Cyclooxygenase-2, player or spectator in cyclooxygenase-2 inhibitor-induced apoptosis in prostate cancer cells. *J. Natl. Cancer Inst.* 94:585–91

83. Spencer AG, Thuresson E, Otto JC, Song I, Smith T, et al. 1999. The membrane binding domains of prostaglandin endoperoxide H synthases 1 and 2. Peptide mapping and mutational analysis. *J. Biol. Chem.* 274:32936–42

84. Thomas T, Nadackal TG, Thomas K. 2001. Aspirin and non-steroidal anti-inflammatory drugs inhibit amyloid-beta aggregation. *NeuroReport* 12:3263–67

85. Thuresson ED, Lakkides KM, Rieke CJ, Sun Y, Wingerd BA, et al. 2001. Prostaglandin endoperoxide H synthase-1: the functions of cyclooxygenase active site residues in the binding, positioning, and oxygenation of arachidonic acid. *J. Biol. Chem.* 276:10347–57

86. Thuresson ED, Lakkides KM, Smith WL. 2000. Different catalytically competent arrangements of arachidonic acid within the cyclooxygenase active site of prostaglandin endoperoxide H synthase-1 lead to the formation of different oxygenated products. *J. Biol. Chem.* 275:8501–7

87. Thuresson ED, Malkowski MG, Lakkides KM, Rieke CJ, Mulichak AM, et al. 2001. Mutational and X-ray crystallographic analysis of the interaction of dihomo-γ-linolenic acid with prostaglandin endoperoxide H synthases. *J. Biol. Chem.* 276:10358–65

88. Tsai A, Kulmacz RJ. 2000. Tyrosyl radicals in prostaglandin H synthase-1 and -2. *Prostaglandins Other Lipid Mediat.* 62: 231–54

89. Tsai A, Palmer G, Xiao G, Swinney DC, Kulmacz RJ. 1998. Structural characterization of arachidonyl radicals formed by prostaglandin H synthase-2 and

prostaglandin H synthase-1 reconstituted with mangano protoporphyrin IX. *J. Biol. Chem.* 273:3888–94
90. Tsai A, Wei C, Baek HK, Kulmacz RJ, Van Wart HE. 1997. Comparison of peroxidase reaction mechanisms of prostaglandin H synthase-1 containing heme and mangano protoporphyrin IX. *J. Biol. Chem.* 272:8885–94
91. van der Ouderaa FJ, Buytenhek M, Nugteren DH, Van Dorp DA. 1977. Purification and characterization of prostaglandin endoperoxide synthase from sheep vesicular glands. *Biochim. Biophys. Acta* 487:315–31
92. Walker MC, Kurumbail RG, Kiefer JR, Moreland KT, Koboldt CM, et al. 2001. A three-step kinetic mechanism for selective inhibition of cyclo-oxygenase-2 by diarylheterocyclic inhibitors. *Biochem. J.* 357:709–18
93. Wong E, Bayly C, Waterman HL, Riendeau D, Mancini JA. 1997. Conversion of prostaglandin G/H synthase-1 into an enzyme sensitive to PGHS-2-selective inhibitors by a double His513 → Arg and Ile523 → Val mutation. *J. Biol. Chem.* 272:9280–86
94. Xiao G, Tsai AL, Palmer G, Boyar WC, Marshall PJ, et al. 1997. Analysis of hydroperoxide-induced tyrosyl radicals and lipoxygenase activity in aspirin-treated human prostaglandin H synthase-2. *Biochemistry* 36:1836–45
95. Zeng J, Fenna RE. 1992. X-ray crystal structure of canine myeloperoxidase at 3 Å resolution. *J. Mol. Biol.* 226:185–207

Annu. Rev. Biophys. Biomol. Struct. 2003. 32:207–35
doi: 10.1146/annurev.biophys.32.110601.141709

First published online as a Review in Advance on January 16, 2003

VOLUMETRIC PROPERTIES OF PROTEINS

Tigran V. Chalikian
Department of Pharmaceutical Sciences, Leslie Dan Faculty of Pharmacy, University of Toronto, Toronto, Ontario M5S 2S2, Canada; email: chalikan@phm.utoronto.ca

Key Words thermodynamics, conformational transitions, protein binding, protein hydration, intrinsic packing

■ **Abstract** Structural and thermodynamic characterizations of a variety of intra- and intermolecular interactions stabilizing/destabilizing protein systems represent a major part of multidisciplinary efforts aimed at solving the problems of protein folding and binding. To this end, volumetric techniques have been successfully used to gain insights into protein hydration and intraglobular packing. Despite the fact that the use of volumetric measurements in protein-related studies dates back to the 1950s, such measurements still represent a relatively untapped yet potentially informative means for tackling the problems of protein folding and binding. This notion has been further emphasized by recent advances in the development of highly sensitive volumetric instrumentation that has led to intensifying volumetric investigations of protein systems. This paper reviews the volumetric properties of proteins and their low-molecular-weight analogs, in particular, discussing the recent progress in the use of volumetric data for studying conformational transitions of proteins as well as protein-ligand, protein-protein, and protein–nucleic acid interactions.

CONTENTS

INTRODUCTION

More than two millennia ago, when Archimedes immersed the golden crown of Hieron II, the king of Syracuse, into a water bath and measured the volume of displaced water, he became the first person in recorded history to determine the partial molar volume of a chemical object (golden crown) in an aqueous solution. The results of these measurements had fundamental scientific consequences while being quite lamentable for the royal jeweller: The calculated density of the crown revealed that gold had been clandestinely alloyed with some metal less dense than gold, presumably silver. Now, 22 centuries later, entering the post-genomic age, we still continue to measure partial molar volumes of chemical and biological substances, including proteins and protein complexes, in aqueous solutions. It is noteworthy that insights provided by modern volumetric measurements are often of no less fundamental importance than those provided by Archimedes' protomeasurement. With respect to proteins, partial molar volume and its pressure (compressibility) and temperature (expansibility) derivatives have proven useful in quantifying protein hydration, intrinsic packing, and conformational dynamics.

Protein Dynamics and Intrinsic Packing

To perform their biological functions, globular proteins should be in a water milieu and should also possess a unique tertiary structure with a sizeable core of water-inaccessible amino acid residues (25, 93, 135, 141). It must be noted, however, that there is a growing body of recent evidence suggesting that there are exceptions to these rules. Some proteins may be partially or fully functional in their so-called "natively unfolded" conformation (163–165, 170), while other proteins may retain significant enzymatic activity in a number of hydrogen-bonded non-aqueous liquids (9, 89–91, 130). Notwithstanding, these observations are exceptions rather than a general rule.

In native globular proteins, amino acid residues buried inside the solvent-inaccessible interior are tightly packed, although the packing is not perfect with ample intraglobular voids (25, 131, 133–136). Nearly all these voids are made up of small cavities that cause native proteins to undergo dynamic fluctuations in structure (101, 135, 166). In the course of these fluctuations, the size of intraglobular cavities varies owing to thermally activated vibrations of surrounding atoms. Such fluctuations (or mobile defects) facilitate the conformational dynamics of proteins, in particular, allowing the rotation of side chains (25).

Dynamic fluctuations of the protein between the nearly isoenergetic native-like subconformations cause fluctuations of the intrinsic volume. The mean-square fluctuations of the intrinsic volume, $\langle \delta V_M^2 \rangle$, in this case represent an effective, quantitative means of estimating protein dynamics. Significantly, δV_M, the mean amplitude of volume fluctuations, is uniquely related to the intrinsic coefficients of isothermal compressibility, β_{TM}, and thermal expansibility, α_M, of the protein molecule (28, 29):

$$\langle \delta V_M^2 \rangle = k_B T V_M \beta_{TM}, \quad 1.$$

$$\langle \delta U_M \delta V_M \rangle = k_B T V_M (T\alpha_M - P\beta_{TM}), \quad 2.$$

where k_B is Boltzmann's constant, T is the absolute temperature, P is the pressure, and δU_M is the mean amplitude of fluctuation of the intrinsic energy accompanying thermally activated subconformational transitions of the protein. Inspection of Equations 1 and 2 reveals that the value of V_M can be used in conjunction with β_{TM} and α_M for quantitative characterization of the conformational dynamics of the protein as reflected in the values of δV_M and δU_M.

Protein Hydration

Much theoretical and experimental effort has gone into thermodynamic, structural, and dynamic characterization of protein hydration, which represents a major determinant of the energetics of protein stability and recognition (93, 141). In this respect, volumetric measurements may provide useful and, in many respects, unique information about protein hydration. Interactions between solvent-exposed atomic groups of proteins and adjacent water molecules can cause the latter to exhibit thermodynamic properties distinct from those of bulk water. In particular, water of protein hydration is distinct from bulk water with respect to its partial molar volume, compressibility, and expansibility. Hence, these observables can be and have been used to discriminate between the populations of water of protein hydration and bulk water.

Protein hydration, as well as its packing and dynamics, generally changes in the course of protein folding and binding events. Consequently, volumetric measurements can be used in protein-recognition studies for identifying and quantifying changes in hydration, packing, and dynamics of a protein that are associated with its conformational transitions or complexation with other molecules. The importance of volumetric data for the thermodynamics of protein folding was emphasized more than 15 years ago by Kauzmann (79), who explicitly stated that no thermodynamic theory of protein folding and other conformational transitions can be considered valid unless it provides rationalizations for both temperature-dependent (calorimetric) and pressure-dependent (volumetric) data.

Scope of the Review

Volumetric properties of proteins and other biological compounds have been reviewed (14, 15, 20, 34, 60, 61, 65, 66, 105, 109, 143, 144, 161, 173). Therefore, this paper is not meant to be comprehensive, in particular, with respect to presenting the historical aspects of employment of volumetric data in biophysical/biochemical studies. In addition, space constraints in this review series do not allow thorough discussion of all recent protein-related investigations involving the use of volumetric measurements. Instead, this review focuses on selected aspects of some recent developments in the use of volumetric techniques in protein-related research and critical evaluation of some of the available data on protein volume, compressibility, and expansibility along with proposed microscopic interpretations of these data.

MEASUREMENTS AND INTERPRETATIONS OF VOLUMETRIC DATA

Interpretation of Volumetric Data

Recently, Blandamer et al. (5) presented an excellent review detailing the fundamentals of the use of the partial molar volume (V°), expansibility (E°), adiabatic (K_S°), and isothermal (K_T°) compressibilities of a solute in studying molecular interactions in solutions. At infinite dilution, the partial molar volumetric properties of a solute (V°, E°, K_S°, and K_T°) can be interpreted microscopically in terms of the intrinsic and hydration contributions using a general relationship (14, 15, 20, 65, 66, 143, 144):

$$X^\circ = X_M + \Delta X_h = X_M + n_h(X_h - X_0) = X_M + \sum_i n_{hi}(X_{hi} - X_0), \qquad 3.$$

where X_M is the intrinsic volume, expansibility, or compressibility of the solute, ΔX_h is the hydration-induced change in one of these properties of the solvent, n_h is the hydration number, which refers to the total number of water molecules within the solute hydration shell, X_0 is the partial molar volume, expansibility, or compressibility of the solvent, X_h is the average value of the same partial molar volumetric property of water in the solute hydration shell, n_{hi} is the hydration number of the *i*-th solvent-exposed atomic group of the solute, and X_{hi} is the value of X for water solvating the *i*-th solvent-exposed group.

For the microscopic interpretation of partial molar volume data, derivations based on scaled particle theory have been successfully employed (83, 125, 126, 152). In this presentation, the partial molar volume, V°, of a solute is considered to consist of four contributions:

$$V^\circ = V_M + V_T + V_I + \beta_{T0}RT, \qquad 4.$$

where V_M is the intrinsic volume of the solute molecule; V_T is the "thermal" volume, which results from thermally activated mutual molecular vibrations of the solute and the solvent; V_I is the "interaction volume," which represents a decrease in the solvent volume due to hydration of polar or charged atomic groups of a solute; β_{T0} is the coefficient of isothermal compressibility of the solvent; and R is the universal gas constant. The $\beta_{T0}RT$ term is small and, in macromolecular studies, is generally neglected.

Volume Measurements

The partial molar volume, V°, of proteins and their analogs or volume changes, ΔV, associated with protein transitions and ligand binding have been measured directly using the techniques of picnometry, dilatometry, falling drop method, magnetic float method, and vibrating tube densimetry (94, 173). Indirectly, a change in volume associated with a pressure-induced protein event (protein denaturation or dissociation of a protein complex) can be determined from pressure-dependent spectroscopic measurements that are based on the two-state approximation (60, 61, 137, 139, 149, 169).

The partial molar expansibility, E°, of solutes has been traditionally determined from temperature-dependent partial molar volume data [since $E^\circ = (\partial V^\circ/\partial T)_P$] using one of the abovementioned techniques of volume measurements. Recently, the so-called method of pressure-perturbation calorimetry has been developed by MicroCal, Inc. (Northampton, Massachusetts) as an alternative approach to the evaluation of the partial molar expansibility, E°, of proteins as well as expansibility changes, ΔE, associated with temperature-induced protein conformational transitions.

Compressibility Measurements

The most precise way of determining adiabatic compressibility of liquids is based on the Newton-Laplace equation, $\beta_S = \rho^{-1} U^{-2}$, where $\beta_S = -V^{-1}(\partial V/\partial P)_S$ is the coefficient of adiabatic compressibility of the medium, and ρ and U are the density and sound velocity of the medium, respectively. The partial molar adiabatic compressibility, K_S°, of a solute can be obtained from differential solution versus solvent measurements of density and sound velocity (4, 20, 119, 143, 144). Critical analyses of various techniques of ultrasonic velocity measurements have been previously presented in a number of excellent reviews (36–38, 142, 144). Perhaps, the most versatile acoustic technique that currently provides the highest precision (~10^{-4}%) of sound velocity measurements and the smallest sample volumes (~200 to ~800 μL) is the resonator method (36–38, 142, 144).

The partial molar isothermal compressibility, K_T°, of a solute can be calculated from K°_S using the relationship (5)

$$K_T^\circ = K_S^\circ + \left(T\alpha_0^2/\rho_0 c_{P0}\right)(2E^\circ/\alpha_0 - C_P^\circ/\rho_0 c_{P0}), \quad 5.$$

where α_0 is the coefficient of thermal expansion of the solvent, c_{P0} is the specific heat capacity of the solvent, and C_P° is the partial molar heat capacity of a solute.

An alternative, although less accurate, way of determining K_T° is to use direct pressure-dependent density measurements to derive K_T° (39, 146, 171). Finally, a change in isothermal compressibility, ΔK_T, associated with a pressure-induced conformational transition (such as protein denaturation) can be determined based on the two-state treatment of the transition profile monitored by some physical (most often, spectroscopic) observable (61, 127).

HYDRATION OF LOW-MOLECULAR-WEIGHT ANALOGS OF PROTEINS

In an attempt to understand the hydration of protein groups, the volumetric properties of short peptides, α-amino acids, α,ω-aminocarboxylic acids, alcohols, sugars, aminoalkanes, carboxylic acids, and other small molecules containing atomic groups identical to those in proteins have been under intensive scrutiny during the past several decades [for reviews, see (65, 66, 173)]. The main goal of such low-molecular-weight model compound-based studies is to determine group contributions of individual atomic groups and interpret these properties in terms of hydration using Equation 3. The volumetric group contributions for a number of independently hydrated aliphatic, aromatic, polar, and charged groups have been determined (10, 51, 81, 82, 97, 113). In the absence of significant intramolecular interactions, one might expect that the partial molar volume, expansibility, and adiabatic compressibility of a solute can be calculated using the additive approach based on the knowledge of the solute functional groups and their group contributions. Consistent with this expectation, analysis of literature reveals that the partial molar volumes, V°, of solutes have been satisfactorily calculated based on their chemical structures. In contrast, additive partial molar expansibility (E°) and adiabatic compressibility (K_S°) calculations often produce large errors, presumably because these parameters are exquisitely sensitive to subtle intramolecular interactions of solutes that do not significantly influence V°. Such intramolecular interactions may involve partial overlap of the hydration shells of interacting groups.

The partial molar volume contributions of amino acid residues have been reported by a number of researchers (53, 54, 84, 108, 173). There is reasonably good agreement between different sets of volume contributions of amino acid side chains. In this respect, a recent work by Häckel et al. (53) merits special mention. The authors reported extremely precise and systematic data on the partial molar volumes (V°) of GlyXGly tripeptides between 10° and 90°C. Based on these data, the volume contributions of all amino acid side chains were determined and presented as second-order polynomial functions of temperature. Analytical differentiation of these functions with respect to temperature allows one to calculate the expansibility contributions of each amino acid side chain. The volume and expansibility contributions of the glycyl unit have been evaluated based on studying gly_n and $(ala)gly_n$ oligopeptides as well as acetylated glycine ($AcglyNH_2$) and diglycine ($Acgly_2NH_2$) amides (21, 52). Taken together, the group contributions of the amino acid side chains and peptide group enable one to predict the values of V° and E° for an arbitrary oligo- or polypeptide (fully unfolded) of a given amino acid composition.

The adiabatic compressibility, K_S°, contributions of amino acid side chains and the glycyl unit have been reported by a number of researchers (18, 21, 55, 58, 59 82, 84, 113). In contrast to volume, the compressibility contributions of the amino acid side chains and glycyl unit reported by different groups (or even reported by the same group but using different model compounds) may be significantly disparate. For example, the compressibility contributions of the glycyl unit determined by Hakin et al. (55) based on the data on the gly_n and $(ala)gly_n$ oligopeptides at 25°C are equal to $-(1.08 \pm 0.08) \times 10^{-4}$ and $-(2.75 \pm 0.03) \times 10^{-4}$ $cm^3\ mol^{-1}\ bar^{-1}$, respectively. Another example is the compressibility contribution of the Phe side chain, which, if calculated by subtracting the partial molar adiabatic compressibilities of the amino acid glycine from that of phenylalanine, is equal to -7.54×10^{-4} $cm^3\ mol^{-1}\ bar^{-1}$ (113). The same contribution determined from di- and tripeptide studies equals -1.79×10^{-4} $cm^3\ mol^{-1}\ bar^{-1}$ (GlyPhe/GlyGly) (58), 5.8×10^{-4} $cm^3\ mol^{-1}\ bar^{-1}$ (GlyPheGly/GlyGlyGly) (59), 1.2×10^{-4} $cm^3\ mol^{-1}\ bar^{-1}$ (PheGlyGly/GlyGlyGly), and 0.6×10^{-4} $cm^3\ mol^{-1}\ bar^{-1}$ (GlyGlyPhe/GlyGlyGly) (18). Such disparities, which most probably originate from unaccounted intramolecular interactions within the solute molecule, currently preclude any reliable additive calculations of the partial molar adiabatic compressibility of individual solutes.

One important point related to such studies is the need to estimate changes in volume and compressibility associated with ionization/neutralization of titrable amino acid residues. To this end, several pH-dependent investigations of the volumetric properties of simple molecules containing amino, carboxyl, and imidazole groups have been conducted (18, 19, 63, 64, 67, 68, 80, 88, 174). In their pioneering study, Rasper & Kauzmann (132) discovered that, in globular proteins, neutralization of carboxyl groups results in a volume increase of 11 $cm^3\ mol^{-1}$. This value is consistent with the data on neutralization of independently hydrated carboxyl groups in small molecules. Interestingly, neutralization of amino and imidazole groups in globular proteins results in a volume increase of 16 to 18 $cm^3\ mol^{-1}$, which is only two thirds of the value observed for the same groups in simple low-molecular-weight model compounds (132). To rationalize this disparity, the authors proposed that the environment of basic groups in globular proteins is somewhat different from that of small molecules in solution. Alternatively, the observed disparity may be related to the fact that charged groups in proteins are not fully exposed to the solvent as is conventionally thought (74).

Solvent isotope effects in protein hydration (arising from substitution of H_2O with D_2O as a solvent) are believed to play an important role in the energetics of protein stability and binding (24, 107). In two recent publications, the partial molar volumes, expansibilities, and adiabatic compressibilities of a homologous series of α-amino acids, oligoglycines, and α,ω-aminocarboxylic acids were measured in D_2O and compared with similar data obtained in H_2O (99, 100). It was found that small but measurable disparities in the volumetric contributions in H_2O versus D_2O exist for various (charged, polar, and nonpolar) functional groups. For example, at 25°C, the compressibility contribution of an independently hydrated pair of $-NH_3^+$ and $-COO^-$ groups in D_2O is $(4.7 \pm 1.5) \times 10^{-4}$ cm^3 mol^{-1} bar^{-1} smaller (more negative) than that in H_2O. This observation is consistent with the data on electrolytes reported by Desrosiers & Lucas (31) and Mathieson & Conway (110), who found that the partial molar adiabatic compressibilities, K_S°, of inorganic salts (such as NaF, KBr, NaCl, and $BaCl_2$) and tetraalkylammonium halides (such as Me_4NBr and Et_4NBr) in D_2O are smaller than those in H_2O.

HYDRATION OF POLYPEPTIDES AND UNFOLDED PROTEINS

Studies of denatured protein states are as important as studies of the native state for understanding the thermodynamics of protein stability (3, 32, 42, 129, 135, 147, 167). The ideally unfolded polypeptide chain in a good solvent is a random coil with all its amino acid residues solvent accessible. It is commonly agreed that real proteins are never random coils, even under the harshest denaturing conditions (32, 96, 135, 147). Nevertheless, the hypothetical fully extended protein conformation has been widely used for modeling the extreme, limiting case of protein denaturation.

To this end, a number of investigators have described additive procedures in which the partial molar volume, expansibility, and adiabatic compressibility of a fully extended polypeptide chain are calculated based on knowledge of its amino acid composition and group contributions of amino acid residues (53, 71, 84, 108, 173). Analysis of the reported sets of group contributions of amino acid residues to partial volume reveals slight deviations among different research groups. For example, for the fully extended conformations of cytochrome *c*, lysozyme, ribonuclease A, and apomyoglobin at 25°C, Kharakoz (84) calculated partial specific volumes, v°, of 0.726, 0.713, 0.705, and 0.738 cm^3 g^{-1}, respectively. For the same proteins under the same experimental conditions, Makhatadze et al. (108) calculated the values of v° equal to 0.715, 0.716, 0.700, and 0.742 cm^3 g^{-1}, respectively, while the values obtained by Häckel et al. (53) are 0.724, 0.710, 0.705, and 0.736 cm^3 g^{-1}, respectively.

The partial specific expansibility, e°, of a fully extended polypeptide chain can be calculated from temperature-dependent data on the volume group contributions of amino acids as reported by Makhatadze et al. (108) and Häckel et al. (53). The two data sets yield somewhat different results. For example, at 25°C, the partial specific expansibilities of fully unfolded ribonuclease A and lysozyme calculated based on the data presented by Makhatadze et al. (108) are both equal to $\sim 7.5 \times 10^{-4}$ cm^3 g^{-1} K^{-1}. For the same proteins, using the data presented by Häckel et al. (53), one obtains the value of e° equal to $\sim 5.5 \times 10^{-4}$ cm^3 g^{-1} K^{-1}. It is noteworthy that these values are higher than $\sim 3 \times 10^{-4}$ cm^3 g^{-1} K^{-1}, the average partial specific expansibility of native globular proteins (22).

Iqbal & Verrall (71) and Kharakoz (84) presented the compressibility group contributions of the glycyl unit and amino acid side chains. These group contributions were used to calculate the partial specific adiabatic compressibility, k_S°, of individual fully unfolded proteins. Results of such additive calculations performed by the two research groups do not agree well. These disparities may be related, at least partially, to the fact that compressibility is significantly less "additive" than volume. The compressibility contribution of the same group derived from analyzing data on different homologous series may be significantly disparate in both magnitude and sign. These disparities reflect unaccounted effects of interactions between the side chains under investigation and the rest of the solute molecule and thereby complicate any unequivocal choice of model compounds mimicking the compressibility contributions of the amino acid residues within a polypeptide chain. Consequently, calculations/predictions of the partial specific adiabatic compressibility of an unfolded polypeptide chain of a given amino acid composition are not accurate as of yet.

In one alternative approach, no attempt was made to discriminate between the partial specific adiabatic compressibilities of individual fully unfolded proteins (161). In this approach, the average partial specific adiabatic compressibility, k_S°, of a fully unfolded polypeptide chain is taken to be equal to that of short unstructured oligopeptides. At 25°C, this value roughly equals $-(18 \pm 3) \times 10^{-6}$ cm^3 g^{-1} bar^{-1} (161). Such a highly negative value of k_S° reflects extensive hydration of unfolded, random coil-like polypeptide chains in which the majority of amino acid residues are solvent exposed and no significant solvent-inaccessible intrinsic core is preserved.

A number of research groups have measured the partial specific volumes, expansibilities, and adiabatic compressibilities of homopolymeric polypeptides including polyalanine, polyproline, polythreonine, polyglutamic acid (sodium salt), and polylysine hydrochloride [(115); G.D. Noudeh & T.V. Chalikian, unpublished data]. Excluding polyionic polyglutamic acid and polylysine, homopolymeric polyamino acids exhibit values of k_S° that are significantly larger (less negative) than -18×10^{-6} cm^3 g^{-1} bar^{-1}, the value expected for the fully extended conformation. For example, at 25°C, the partial specific adiabatic compressibilities, k_S°, of polyalanine, polyproline, and polythreonine are equal to $-(7.0 \pm 0.7) \times 10^{-6}$, $-(6.2 \pm 0.7) \times 10^{-6}$, and $-(1.3 \pm 0.7) \times 10^{-6}$ cm^3 g^{-1} bar^{-1}, respectively (G.D. Noudeh & T.V. Chalikian, unpublished data). This disparity may suggest the presence of residual structure in polyamino acids, a notion that has been confirmed by circular dichroism spectral data on polyalanine, polyproline, and polythreonine (G.D. Noudeh & T.V. Chalikian, unpublished data).

HYDRATION AND INTRINSIC PACKING OF NATIVE GLOBULAR PROTEINS

Despite the structural diversity of native proteins with respect to chain topology, folding motifs, and secondary structure content, there are certain similarities in microscopic properties of globular proteins that are characteristic of the native conformation. All globular proteins are compact and tightly packed. The packing density of globular proteins falls into a narrow range of 0.72 to 0.78 (22, 133, 135, 136). The intrinsic volume, V_M, of a globular protein (molecular volume) is uniquely related to its molecular weight, M: V_M ($Å^3$) = (1200 ± 500) + (1.04 ± 0.02) M (22). The solvent-accessible surface area, S_A, of a globular protein correlates with its molecular weight: S_A ($Å^2$) = (4.7 ± 0.2) $M^{0.76 \pm 0.03}$ (112, 161). The average contributions of charged, polar, and nonpolar atomic groups to S_A are 14 ± 4 %, 33 ± 7 %, and 53 ± 5 %, respectively (22).

These shared features result in many similarities in the average packing and hydration (if expressed per gram rather than per mole of a protein) of structurally diverse globular proteins that, in turn, bring about similarities in their partial specific volumetric properties. Specifically, near room temperature, the partial specific volumes, $v°$, of all globular proteins studied to date are between 0.69 and 0.76 $cm^3\ g^{-1}$ (22, 45–48, 93); the partial specific expansibilities, $e°$, are between 3×10^{-4} and 5×10^{-4} $cm^3\ g^{-1}\ K^{-1}$ (7, 8, 22, 47, 62, 70); and the partial specific adiabatic compressibilities, $k_S°$, are within the range of -1×10^{-6} to 10×10^{-6} $cm^3\ g^{-1}\ bar^{-1}$ (2, 22, 45–48, 70, 72, 145). The partial specific volumetric characteristics of a protein, including $v°$, $e°$, and $k_S°$, consist of the intrinsic and hydration contributions (see Equation 3). As a first step toward estimating the hydration contributions to $v°$, $e°$, and $k_S°$ from experimental volumetric results, one needs to determine the intrinsic contributions based on structural data of proteins and/or some nonthermodynamic assumptions. The intrinsic volume, V_M, of a globular protein represents the geometric volume of its solvent-inaccessible interior. The intrinsic expansibility, $E_M = \alpha_M V_M$, and intrinsic adiabatic compressibility, $K_{SM} = \beta_{SM} V_M$, of a protein represent the expansibility and adiabatic compressibility of its solvent-inaccessible interior (α_M and β_{SM} are the coefficients of expansibility and adiabatic compressibility of the protein interior, respectively). The intrinsic volume, V_M, of a protein can be determined in a relatively straightforward manner from its three-dimensional structural data. In contrast, determining the E_M and K_{SM} is not straightforward and can only be done based on some assumptions and postulated models. Below, I describe some approaches that are currently used for resolving the partial specific volumetric properties of globular proteins into the intrinsic and hydration contributions.

Volume

There are two conventional definitions of intrinsic volume that have been used in protein-related studies (98, 133). The first definition is that of Voronoi volume (133, 134), while the second definition is that of molecular volume (26, 27, 133). The Voronoi volume of a protein is calculated using an algorithm developed by Richards (133, 134) that is based on a theorem from geometry developed by Voronoi. In this algorithm, all the space within a protein is assigned to its constituent atoms by constructing around each atom a so-called Voronoi polyhedron, an irregularly shaped polyhedron of limiting size. The faces of a Voronoi polyhedron are formed by planes drawn as perpendicular bisectors of the vectors connecting each pair of atoms, while the edges of the polyhedron represent the intersections of these planes. Hence, there is a uniquely defined set of polyhedra surrounding each protein atom. These polyhedra do not overlap, nor do they produce any unaccounted void space inside the protein. The Voronoi volume of a protein is calculated as the sum of the volumes of all the constituent polyhedra.

Gerstein & Chothia (50) have compared the Voronoi volumes of amino acid residues buried inside the protein molecule with those of solvent-exposed amino acid residues on the protein surface. This comparison revealed that, on average, the Voronoi volume of an atom on the protein surface is 7% larger than that of an atom in the protein core (50). However, it should be noted that the method of Voronoi polyhedra is well defined only for buried atoms. Its application to calculating the volume of surface atoms requires the use of assumptions or molecular simulations to place water molecules around the solute. Such water molecules are needed to locate bisecting planes between surface atoms and the solvent.

Molecular volume is defined as the volume enclosed by the molecular surface of a protein. As previously defined by Richards (133), the molecular surface of a protein has two components: (*a*) the part of the protein surface that contacts a rolling probe solvent molecule, and (*b*) a reentrant surface, which corresponds to a series of patches formed by the interior-facing domain of the probe when it simultaneously contacts more than one atom on the protein surface.

The molecular volume of a globular protein is generally ~15% smaller than its Voronoi volume. The Voronoi volume ($Å^3$) of an average globular protein is related to its molecular mass (Da) via $V_M = 1.27$ M (133), while the molecular mass dependence of the molecular volume ($Å^3$) is $V_M = (1200 \pm 500) + (1.04 \pm 0.02)$ M (22). Consequently, the specific values of Voronoi volume and molecular volume of a globular protein are given by $(N_A/M)(1.27\ M) = 0.764\ (cm^3\ g^{-1})$ and $(N_A/M)(1200 + 1.04\ M) = 723/M + 0.626\ (cm^3\ g^{-1})$, respectively (where N_A is Avogadro's number).

Both definitions of intrinsic volume are physically valid and their use in protein studies is justified as long as their differences are clearly understood and taken into account. In this respect, Paci & Marchi (120) and Paci & Velikson (121) have recommended the use of Voronoi volume in computer simulations. This recommendation is based on the fact that molecular dynamics computations of protein compressibility produce results closer to experimental estimates when the computations are performed with Voronoi rather than molecular volume. However, this recommendation may be related to the specifics of the molecular dynamics simulations performed in References 120 and 121 rather than reflect any objective advantage for the use of one definition over another.

According to Equation 4, the partial specific volume of a protein consists of the intrinsic volume, v_M, thermal volume, v_T, and interaction volume, v_I. If the intrinsic volume of a protein is defined as the molecular volume, the thermal volume can be conceptually considered as consisting of an empty space of a thickness, δ, of ~1.0 Å surrounding the protein molecule (22). Consequently, the value of v_T

can be estimated by multiplying the solvent-accessible surface area of a protein by δ. Thus, for an average protein the thermal volume is given by $v_T = (N_A/M)\ \delta\ (4.7\ M^{0.76}) = 2.83\ M^{-0.24}$ ($cm^3\ g^{-1}$). The interaction volume, v_I, can be found by subtracting the intrinsic, v_M, and thermal, v_T, volumes from the partial specific volume, v°, of a protein. The average value of the partial specific volume of a globular protein is 0.72 $cm^3\ g^{-1}$. Consequently, for an average globular protein, $v_I = v^\circ - v_M - v_T = 0.094 - 0.723/M - 2.83\ M^{-0.24}$ ($cm^3\ g^{-1}$). These approximate equations can be used for rough estimates of the intrinsic, thermal, and interaction contributions to the partial specific volume of globular proteins. Such estimates yield highly negative values of v_I, which are suggestive of a smaller partial molar volume (greater density) of water solvating a protein compared to that of bulk water. This expectation is consistent with experimental observations that indicate a ~10%–15 % increase in the density of water of protein hydration relative to bulk water (11, 111, 154).

Expansibility

According to Equation 3, the partial specific expansibility, e°, of a globular protein is the sum of the intrinsic, e_M, and hydration, Δe_h, contributions. Currently, the most reliable estimates of the intrinsic expansibility of globular proteins come from temperature-dependent X-ray crystallographic measurements of Frauenfelder et al. (41), Tilton et al. (162), and Young et al. (172). These authors have determined the thermal expansion of metmyoglobin, ribonuclease A, and hen egg-white lysozyme by analysis of the refined X-ray crystallographic structures of these proteins within a wide temperature range of ~100 to ~300 K. These analyses revealed the intrinsic coefficients of thermal expansion, α_M, of 1.4×10^{-4}, 0.4×10^{-4}, and $0.9 \times 10^{-4}\ K^{-1}$ for metmyoglobin (41), ribonuclease A (162), and lysozyme (172), respectively. Significantly, the thermal expansion of a globular protein is not uniform: α-helix-rich protein domains expand more with temperature than do β-sheet-rich domains. By rearranging Equation 2, it can be shown that the intrinsic coefficient of thermal expansion of a protein system, α_M, is related to the magnitude of its thermally activated enthalpy (δH) and volume (δV) fluctuations as follows:

$$\langle \delta H \delta V \rangle = k_B T^2 V_M \alpha_M. \qquad 6.$$

Inspection of Equation 6 reveals that a larger value of local α_M for α-helices (relative to β-sheets) may indicate a larger magnitude of volume fluctuations, δV, a larger magnitude of enthalpy fluctuations, δH, or both. As discussed below, globular protein compressibility data suggest that the volume fluctuations of α-helices are more intensive than those of β-sheets. Information on the enthalpy fluctuations of α-helix- and β-sheet-rich protein domains can be potentially obtained from correlation analysis between the intrinsic heat capacity, C_{PM}, and secondary structural elements of globular proteins since $\langle \delta H^2 \rangle = k_B T^2 C_{PM}$ (28, 29).

The average value of α_M can be taken as equal to $\sim 1 \times 10^{-4}\ K^{-1}$ (41, 162, 172). Hence, the intrinsic expansibility, e_M, of a globular protein is given by

$e_M = \alpha_M v_M = \alpha_M$ (723/M + 0.626) ($cm^3\ g^{-1}\ K^{-1}$). For proteins with molecular weights between 10 and 50 kDa, the values of e_M are on the order of $0.7 \times 10^{-4}\ cm^3\ g^{-1}\ K^{-1}$. Since the partial specific expansibility, e°, of native globular proteins is between 3×10^{-4} and $5 \times 10^{-4}\ cm^3\ g^{-1}\ K^{-1}$, the hydration contribution to expansibility, $\Delta e_h = e° - e_M$, is positive and on the order of 2×10^{-4} to $4 \times 10^{-4}\ cm^3\ g^{-1}\ K^{-1}$. Based on Equation 3, this observation suggests that the partial molar expansibility, E_h, of water of protein hydration is larger than that of bulk water, E_0. To make a more quantitative estimate, it is instructive to consider molar rather than specific values of protein expansibility. In this respect, the partial molar expansibility, E°, of lysozyme is 4.29 $cm^3\ mol^{-1}\ K^{-1}$ (22). Since the intrinsic volume, V_M, of lysozyme is 9424 $cm^3\ mol^{-1}$, its molar intrinsic expansibility, E_M, is 0.85 $cm^3\ mol^{-1}\ K^{-1}$ ($0.9 \times 10^{-4} \times 9424$) (22). The hydration contribution to expansibility, ΔE_h, of lysozyme is 3.44 $cm^3\ mol^{-1}\ K^{-1}$ (4.29–0.85). Recall that $\Delta E_h = n_h (E_h - E_0)$, where n_h is the hydration number (see Equation 3). At 25°C, the partial molar expansibility of water, E_0, is $4.6 \times 10^{-3}\ cm^3\ mol^{-1}\ K^{-1}$. One estimate for the value of n_h for lysozyme is ~1800 (11). Consequently, the partial molar expansibility, E_h, of water of lysozyme hydration is $\sim 6.5 \times 10^{-3}\ cm^3\ mol^{-1}\ K^{-1}$, which is roughly 40% larger than E_0 at 25°C.

Compressibility

The intrinsic coefficients of adiabatic, β_{SM}, and isothermal, β_{TM}, compressibility of globular proteins have been determined from ultrasonic velocimetric measurements (22, 46, 48, 72, 87, 158), pressure-dependent X-ray crystallographic measurements (78, 92), spectral hole burning and fluorescence line-narrowing spectroscopy (151, 176), and molecular dynamics simulations (30, 120, 121, 124, 156). Different estimates of β_{SM} for globular proteins have yielded values between 10×10^{-6} and $30 \times 10^{-6}\ bar^{-1}$ (22, 46, 48, 72). The value of $25 \times 10^{-6}\ bar^{-1}$ appears to be the most probable estimate of the average value of β_{SM} for globular proteins (14, 85, 161). This value characterizes the interior of a globular protein as a solid-like material with elastic properties close to those of organic solids. However, the intrinsic compressibility of a globular protein is not uniform throughout the protein interior. As revealed by pressure-dependent X-ray crystallographic (92) and NMR measurements (1), as well as correlation analysis between the partial specific adiabatic compressibilities, k_S°, of globular proteins and their structural properties (46), helices and loops exhibit larger microscopic compressibilities than β-strands. Note that this observation correlates to a greater local intrinsic expansibility, α_M, of α-helix-rich protein domains relative to β-sheet-rich domains.

The intrinsic coefficient of isothermal compressibility of a protein system, β_{TM}, is related to the magnitude of its thermally activated volume, δV_M, fluctuations via Equation 1. In such analyses, the difference between β_{SM} and β_{TM} can probably be neglected. Inspection of Equation 1 reveals that a larger value of local β_{TM} or β_{SM} indicates a larger magnitude of volume fluctuations, δV_M, in α-helices relative to β-sheets. In general, it is plausible to expect that the local values of

β_{TM}, β_{SM}, and α_M might correlate to the local conformational stability of a given amino acid residue or a group of residues. Recall that, under native conditions, proteins represent statistical ensembles in which the amino acids may undergo local unfolding reactions and exhibit significantly distinct individual stabilities (69, 102).

For an average globular protein of a molecular weight of M, the intrinsic compressibility is given by $k_{SM} = \beta_{SM} v_M = \beta_{SM}\,(723/M + 0.626)$ (cm^3 g^{-1} bar^{-1}). For example, lysozyme with M of 14.3 kDa exhibits an intrinsic compressibility of 16.9×10^{-6} cm^3 g^{-1} bar^{-1}. The partial specific adiabatic compressibility, k_S°, of this protein equals $(1.3 \pm 0.5) \times 10^{-6}$ cm^3 g^{-1} bar^{-1} (22). Thus, the hydration contribution to compressibility is large and negative and equals $\Delta k_{Sh} = k_S^\circ - k_{SM} = -15.6 \times 10^{-6}$ cm^3 g^{-1} bar^{-1}. Based on Equation 3, the negative sign of Δk_{Sh} suggests that the partial molar adiabatic compressibility of water of protein hydration, K_{Sh}, is smaller than that of bulk water, K_{S0}. In agreement with this expectation, it has been estimated that the average value of K_{Sh} for native globular proteins is ~20% smaller than K_{S0} (11). At 25°C, the values of K_{S0} and K_{Sh} equal 8.1×10^{-4} and 6.5×10^{-4} cm^3 mol^{-1} bar^{-1}, respectively.

CONFORMATIONAL TRANSITIONS AND FOLDING

It is generally accepted that denatured protein species represent heterogeneous ensembles of more or less unfolded states ranging from compact intermediates to fully unfolded, random coil–like conformations (32, 135). Based on the criteria of intrinsic packing and hydration, denatured protein species can be conceptually divided into three classes: compact intermediate states, partially unfolded states, and the fully unfolded state. Compact intermediates, such as molten globules, are characterized by compact dimensions, a lack of rigid tertiary structure, a high content of secondary structure, and a sizeable core of buried amino acid residues. Partially unfolded states are not compact and do not exhibit any significant amount of secondary structure, but still retain a loosely packed core of water-inaccessible amino acid residues. The fully unfolded conformation is random coil–like with the majority of atomic groups being solvent exposed with little, if any, water-inaccessible core remaining. The compact intermediate and partially unfolded conformations exist at relatively mild denaturing conditions, such as at low pH or high temperature in the absence of denaturants. The fully unfolded conformation presumably can be obtained at high concentrations of strong denaturants, e.g., urea or guanidinium chloride (GuHCl). However, it should be noted that there is an increasing body of evidence suggesting that some proteins are not fully unfolded even under the harshest denaturing conditions (147).

Below, I review changes in volume, expansibility, and compressibility accompanying transitions of globular proteins from their native state to one of their denatured states. In general, volumetric data have provided important insights into the hydration properties and intrinsic packing of denatured protein species that may in turn have implications for protein folding.

Changes in Volume

A change in volume, Δv, associated with a conformational transition of a globular protein is generally small, being in absolute value on the order of 1 to 2% or less of the partial specific volume (13, 34, 173). Significantly, the sign of Δv measured at atmospheric pressure for native-to–molten globule (N-to-MG), native-to–partially unfolded (N-to-PU), and native-to–fully unfolded (N-to-FU) protein transitions can be either positive or negative (16, 17, 23, 40, 75, 86, 157, 160).

In contrast, changes in volume associated with pressure-induced denaturation of globular proteins at elevated pressures are always negative (6, 43, 44, 57, 60, 95, 122, 127, 140, 150, 159, 168, 175). This observation merely reflects Le Chatelier's principle and, accordingly, the fact that a protein may undergo a pressure-induced denaturation only if the partial volume of its pressure-induced denatured state is smaller than that of the native state at and above the denaturation pressure, P_M. Therefore, the volume change, Δv, associated with pressure-induced protein denaturation and determined at the P_M must be negative. If Δv at atmospheric pressure is positive, the protein still can be denatured by pressure if the change in isothermal compressibility, $\Delta k_T = -(\partial \Delta v/\partial P)_T$, accompanying protein denaturation is positive. In this scenario, the protein will denature somewhere above the pressure at which Δv changes its sign and becomes negative. Finally, Δk_T itself may be pressure dependent and its magnitude and even the sign may change as pressure increases. Clearly, for thermodynamically rigorous interpretations of "high-pressure" values of Δv, these values should be first extrapolated to atmospheric pressure. In such extrapolations, the value of Δk_T and its pressure dependence should be properly taken into account.

Conformational transitions influence the volumetric properties of proteins largely through an elimination of intramolecular voids (the intrinsic volume decreases), a loosening of the intraglobular packing (the intrinsic coefficient of adiabatic compressibility increases), and an exposure of a large number of previously buried atomic groups to the solvent (the protein hydration increases). On account of the diminution of intrinsic volume and enhancement of hydration, one might expect a large negative change in volume, Δv, to accompany protein denaturation. However, experimentally observed values of Δv are small in magnitude and, what is more important, may be either positive or negative in sign. To understand this experimental reality, Chalikian & Breslauer (13) proposed a new way of interpreting partial volume data on biopolymers. A major feature of this interpretation that heretofore has been unappreciated is the thermal volume contribution that results from the thermally induced mutual vibrational motions of solute and solvent molecules (see Equation 4). As a first approximation, the thermal volume around a protein should be proportional to its solvent-accessible surface area. When a protein unfolds, its solvent-accessible surface area increases, which leads, in turn, to an increase in the thermal volume, v_T. An increase in v_T provides the positive contribution to Δv that nearly compensates the negative contributions due to the elimination of internal voids and enhancement of hydration of protein groups.

As noted by Murphy et al. (114), however, the choice of a specific scheme to partition the solution volume into the solute and solvent components may affect the interpretation of the origin of the small volume change associated with conformational transitions of proteins. Specifically, in such analyses, Murphy et al. (114) proposed using the volume encompassed within the solvent-accessible area of the protein (the so-called excluded volume) as its intrinsic volume. An increase in the excluded volume of the protein upon its unfolding provides the positive contribution to the volume change, Δv. The authors believe that this positive contribution compensates the negative contributions owing to elimination of internal voids and enhancement in protein hydration. One ambiguous point of this partition is that the definition of excluded volume includes, in addition to the geometric volume of the protein, half the volume of water molecules incorporated within the first coordination layer of the solute. Hence, an increase in the excluded volume of the protein upon its unfolding should be offset by an equal decrease in the volume of the solvent, thereby effectively bringing to zero any change in the solution volume. Consequently, if excluded volume-based calculations were performed on the whole solution rather than on an isolated protein molecule, they would hardly produce any positive contribution to the denaturation volume, Δv.

Changes in Expansibility

There have been only a few measurements of changes in expansibility, Δe, associated with protein denaturation (8, 23, 57, 95, 146). The data accumulated to date suggest that the values of Δe are positive ranging from 0.5×10^{-4} to 1.5×10^{-4} cm^3 g^{-1} K^{-1}. Based on the foregoing discussion, the net change in expansibility, Δe, associated with a conformational transition of a globular protein should result from the changes in the intrinsic (Δe_M) and hydration (Δe_h) contributions. Recall that the values of both e_M and Δe_h for a native globular protein are positive. Upon protein unfolding, e_M is expected to decrease owing to disruption of "expandable" internal voids, whereas Δe_h is expected to increase owing to increased hydration of the unfolded state. The positive sign of the observed values of Δe suggests that an increase in Δe_h prevails over a decrease in e_M.

Changes in Compressibility

Taulier & Chalikian (161) have recently presented a comprehensive review of compressibility changes associated with conformational transitions of proteins. Compressibility measurements have been used to characterize temperature-, pH-, denaturant-, salt-, cosolvent-, and pressure-induced transitions of globular proteins. For atmospheric pressure measurements, there is a general relationship between the type of globular protein transition and the sign and magnitude of the accompanying changes in adiabatic compressibility, Δk_S (12). Specifically, all N-to-MG transitions studied to date are accompanied by small positive changes in k_S° ranging from 1×10^{-6} to 4×10^{-6} cm^3 g^{-1} bar^{-1} (16, 23, 77, 86, 118, 157). N-to-PU transitions are accompanied by small decreases in k_S° ranging from -3×10^{-6} to $-7 \times$

10^{-6} cm^3 g^{-1} bar^{-1} (16, 17, 23, 40, 160). Finally, N-to-FU transitions are accompanied by large decreases in k_S° that fall within the range of -18×10^{-6} to -20×10^{-6} cm^3 g^{-1} bar^{-1} (75, 157). These empirical observations suggest that the change in compressibility, Δk_S, associated with a protein transition is independent of the specific protein and correlates to the type of transition being monitored. To rationalize this experimental reality, Taulier & Chalikian (161) have derived analytical equations that enable one to describe and predict changes in compressibility, Δk_S, associated with N-to-MG, N-to-PU, and N-to-FU transitions. Although these equations have been derived under a number of intuition-based assumptions, they, nevertheless, account remarkably well for the experimentally observed correlations between the transition type and the sign and magnitude of Δk_S (161).

The existing body of experimental evidence suggests that pressure-induced conformational transitions are predominantly accompanied by positive changes in isothermal compressibility, k_T (6, 57, 95, 127, 146, 175). It is difficult to unequivocally rationalize the observed compressibility changes accompanying pressure-induced conformational transitions of proteins in terms of changes in hydration and intraglobular packing. Part of the problem is related to the fact that we still have only a rudimentary understanding of the pressure dependences of the partial compressibility of the native and denatured protein states. Clearly, further investigations are required in this important field. The significance of such studies has been recently emphasized by Prehoda et al. (127), who demonstrated that compressibility knowledge is fundamental for elucidating the thermodynamics of pressure stability of ribonuclease A.

PROTEIN BINDING

Interactions between proteins and other molecules play a central role in the chemistry of life. Virtually all proteins perform their biological functions by interacting with and binding to other compounds of biochemical relevance. Consequently, the thermodynamics of various forces governing protein recognition have been under intensive scrutiny [for excellent reviews see (35, 153, 155)]. In this respect, binding-induced changes in protein hydration and intrinsic packing represent major determinants of the affinity and specificity of protein binding. Note that a change in the intrinsic packing of a protein upon its complexation with another molecule reflects alterations in the configurational entropy of the reactants and the enthalpy of intramolecular interactions. Volumetric data may shed light on changes in both hydration and intrinsic packing of a protein upon its association with a low-molecular-weight ligand or another macromolecule.

Interactions of Proteins with Small Ligands

Nonspecific binding of detergent to protein was reported several years ago by Nikitina & Nikitin (117) based on ultrasonic velocimetric measurements. These authors have measured a decrease in the relative specific sound velocity increment of lysozyme upon its binding of sodium dodecyl sulfate. This observation was interpreted as suggesting binding-induced dehydration of lysozyme with concomitant melting of the protein interior. In another work, Priev et al. (128) reported that nonspecific interactions between glycerol and a number of globular proteins, including hemoglobin, BSA, β-lactoglobulin, myoglobin, and lysozyme, result in decreases in volume and adiabatic compressibility. Based on these decreases, it was proposed that glycerol induces a release of water from the protein interior with concomitant decreases in the intrinsic volume and compressibility of protein molecules. However, these results have not been confirmed in a subsequent publication by Gekko & Yamagami (49), who, in fact, observed that the partial specific volume, v°, of lysozyme changes only slightly in the presence of glycerol while the partial specific adiabatic compressibility, k_S°, significantly increases.

Volumetric properties of specific protein-ligand interactions have been determined by a number of authors. In particular, Gekko & Yamagami (49) reported decreases in volume and adiabatic compressibility associated with the binding of N-acetyl-D-glucosamine oligomers to lysozyme at 25°C. Significantly, the largest decreases in volume (equal to -0.016 ± 0.004 cm^3 g^{-1} in water and -0.010 ± 0.005 cm^3 g^{-1} in 25 mM phosphate buffer at pH 5.5) and compressibility [equal to $-(2.4 \pm 0.3) \times 10^{-6}$ cm^3 g^{-1} bar^{-1} in water and $-(1.8 \pm 0.5) \times 10^{-6}$ cm^3 g^{-1} bar^{-1} in 25 mM phosphate buffer at pH 5.5] are observed for tri-N-acetyl-D-glucosamine, which also exhibits the highest affinity for lysozyme. The authors interpreted these results in terms of binding-induced dehydration of the interacting surfaces and stiffening of the protein interior. A decrease in the intrinsic compressibility of the protein prevails over an increase in its hydration contribution to compressibility. A qualitatively similar conclusion has been drawn by Nikitin et al. (116), who employed sound velocity measurements for studying the binding of tri-N-acetyl-D-glucosamine to lysozyme.

In a study of fundamental importance, Kamiyama & Gekko (76) measured the partial specific volume, v°, and adiabatic compressibility, k_S°, of dihydrofolate reductase in the presence of various ligands, including folate, dihydrofolate, tetrahydrofolate, NADPH, NADP, methotrexate, and KCl. They found that ligand binding brings about large changes in v° and k_S°, which correlate to the binding-induced changes in the volume of internal cavities and solvent-accessible surface area of the protein. By combining volumetric results with structural information, they concluded that internal cavity effects prevail over the protein dehydration effects in determining the changes in v° and k_S°. The binding-induced change in the volume of internal cavities of dihydrofolate reductase causes a diminution in its intrinsic compressibility that, based on Equation 1, suggests a decrease in the conformational dynamics of the protein.

The first characterization of a ligand-protein binding event performed by a combination of volumetric, spectroscopic, and high-pressure measurements was reported by Dubins et al. (33). Specifically, ultrasonic velocimetry, high-precision densimetry, UV absorbance, and CD spectroscopy have been employed in conjunction with high-pressure UV melting experiments to detect and characterize the binding of ribonuclease A to cytidine 2′-monophosphate (2′-CMP) and cytidine 3′-monophosphate (3′-CMP). Small changes in volume, ΔV (between 20 and 40 cm^3 mol^{-1}, depending on temperature and ligand), and adiabatic compressibility, ΔK_S (between -20×10^{-4} and 50×10^{-4} cm^3 mol^{-1} bar^{-1}) were found to accompany the binding of the enzyme to 2′-CMP and 3′-CMP. The small magnitudes of ΔV and ΔK_S were interpreted as reflecting compensation between the large hydration and intrinsic terms. By combining the volumetric and structural data, 210 $\pm$ 40 water molecules are released to the bulk state upon the binding of 2′-CMP or 3′-CMP to ribonuclease A, while its intrinsic compressibility, k_M, decreases by about 5%. Significantly, the observed decrease in k_M correlates to the binding-induced change in the configurational entropy of the protein, which decreases by ~15%.

Protein-Protein Interactions

Many multisubunit proteins and protein-protein complexes dissociate upon an increase in hydrostatic pressure. In their seminal review, Silva & Weber (150) presented a compilation of literature data on volume changes, Δv, accompanying pressure-induced dissociation of dimeric and tetrameric proteins. Understandably, as a result of Le Chatelier's principle, all reported values of Δv for pressure-induced dissociation of multisubunit proteins are negative ranging from -0.001 to -0.008 cm^3 g^{-1}.

Taulier & Chalikian (160) reported the changes in volume and adiabatic compressibility associated with a dimer-to-monomer transition event at atmospheric pressure. Specifically, high-precision density and ultrasonic velocity measurements were carried out in an aqueous solution of β-lactoglobulin as a function of pH. The protein undergoes a pH-induced dimer-to-monomer transition between pH 2.5 and 4. This transition is accompanied by decreases in volume (-0.008 ± 0.003 cm^3 g^{-1}) and adiabatic compressibility [$-(0.7 \pm 0.4) \times 10^{-6}$ cm^3 g^{-1} bar^{-1}]. The observed changes in volume and compressibility have been interpreted, in conjunction with X-ray crystallographic data, as suggesting a 7% increase in protein hydration, with the hydration changes being localized in the area of contact between the two monomeric subunits.

Protein-DNA Interactions

Elucidation of the thermodynamics of protein-DNA complexes is central to understanding molecular mechanisms of regulation of gene expression including transcription, replication, and translation of genes (56, 73). In this respect, probing water's contribution to the energetics of specific and nonspecific protein–nucleic acid interactions attracts considerable attention among researchers (73, 148). The current state of studies of protein–nucleic acid systems under high pressure has been recently reviewed (106). Below, I focus on a few studies in which hydrostatic pressure has been used as a tool for determining the volume of protein-DNA interactions.

Macgregor (104) employed pressure-dependent gel electrophoresis assays to investigate a change in the catalytic activity of the restriction endonuclease EcoRI with pressure. The resulting data were used to estimate the volume change, ΔV, associated with the formation of the noncovalent complex between EcoRI and its target plasmid pBR 322 and hydrolysis of the phosphodiester bond. The value of ΔV was estimated to be 82 $cm^3\ mol^{-1}$. The author suggested that a large electrostatic component of the protein-DNA interaction is responsible for a large fraction of ΔV.

A similar order of ΔV was observed for the binding of the restriction endonuclease BamHI to a DNA oligonucleotide duplex that contains the cognate recognition sequence (5′-CTCGTATAATGGATCCGCAGTAAGCT-3′) (103). Specifically, gel mobility shift analysis at elevated pressure was used to study the pressure dependence of the binding of the endonuclease to its target DNA sequence (103). The determined pressure dependence of the dissociation constant allowed the determination of the volume change, ΔV, of -92 ± 8 $cm^3\ mol^{-1}$ accompanying the dissociation of the specific BamHI-DNA complex. As a possible rationalization of this result, the authors suggested that the negative value of ΔV may be brought about by increased hydration of the protein-DNA interface and structural changes of the protein.

As an alternative approach, osmotic stress measurements are sometimes employed to investigate hydration changes associated with protein–nucleic acid interactions (138, 148). In one such work, Robinson & Sligar (138) used the osmotic stress technique to investigate the binding of EcoRI to canonical (GAATTC), alternate (TAATTC), and nonspecific (TAGACG) DNA sequences. For these binding events, the authors determined the slope $-RT(\partial K_D/\partial \pi)_T$ (where K_D is the dissociation constant and π is the osmotic pressure). This slope equals -2640, -3880, and -1370 $cm^3\ mol^{-1}$ for the interaction of EcoRI with the GAATTC, TAATTC, and TAGACG sites, respectively. The value of $-RT(\partial K_D/\partial \pi)_T$ was defined as the change in equilibrium volume accompanying a protein-DNA binding event. This definition may engender significant confusion because, as shown below, $-RT(\partial K_D/\partial \pi)_T$ is not equivalent to the change in thermodynamic volume, $\Delta V = -RT(\partial K_D/\partial P)_T$ (where P is the hydrostatic pressure). The slope $-RT(\partial K_D/\partial \pi)_T$ represents the physical volume of water molecules released to the bulk from the hydration shells of the protein and DNA upon dissociation of their complex (123). In other words, $-RT(\partial K_D/\partial \pi)_T = \Delta n_h V_0^\circ$, where Δn_h is the number of water molecules released to the bulk and V_0° is the partial molar volume of water. In contrast, based on Equation 3, the change in thermodynamic volume $\Delta V = -RT(\partial K_D/\partial P)_T$ is equal to $\Delta V_M + \Delta n_h(V_h - V_0^\circ)$. In general, $\Delta n_h V_0^\circ$ is clearly distinct from $\Delta V_M + \Delta n_h(V_h - V_0^\circ)$ in both sign and magnitude. Thus, "osmotic" and thermodynamic volumes cannot be treated as equivalent, and care must be exercised when results of osmotic stress measurements are compared to those of direct volumetric measurements.

CONCLUDING REMARKS

Volumetric observables such as volume, expansibility, and compressibility can be useful in characterizing the role(s) of hydration and intrinsic packing in modulating protein stability, conformational dynamics, and recognition energetics. In particular, volumetric results can enable one to quantify changes in hydration of interacting surfaces and tightening of the protein interior. These changes, coupled with more conventional calorimetric and spectroscopic data, may be further correlated with the net energetics of protein recognition. This overview presents a review of recent progress in volumetric investigations of hydration and dynamics of globular proteins, their conformational transitions, and recognition by small ligands, nucleic acids, and other regulatory proteins.

Despite considerable progress in these investigations, volumetric measurements have not yet become routine in biophysical characterizations of protein systems. Further systematic studies are required to uncover all potential areas of biophysical applicability of volumetric measurements, in particular, defining how intraglobular packing and solute-solvent interactions influence the thermodynamics of protein folding and binding. This article provides the foundation for rationally designing future experimental strategies and for expanding biophysical/biochemical applications of volumetric techniques.

ACKNOWLEDGMENTS

Helpful discussions of specific questions with Drs. Armen P. Sarvazyan, Kenneth J. Breslauer, Robert B. Macgregor, Jr., Kunihiko Gekko, B. Montgomery Pettitt, Fumio Hirata, Roland Winter, Donald C. Rau, Kazuyuki Akasaka, and Nicolas Taulier are gratefully acknowledged. I would also like to extend special thanks to Arin Ratavosi and Arno G. Siraki for carefully reading the manuscript and their many useful comments. This work was supported by grants from the Natural Sciences and Engineering Research Council of Canada and the Canadian Institutes of Health Research. I dedicate this paper to my father Vanik Chalikian for his constant support and encouragement during my entire scientific career.

LITERATURE CITED

1. Akasaka K, Li H, Yamada H, Li R, Thoresen T, Woodward CK. 1999. Pressure response of protein backbone structure. Pressure-induced amide ^{15}N chemical shifts in BPTI. *Protein Sci.* 8:1946–53
2. Andersson GR. 1963. A study of the pressure dependence of the partial specific volume of macromolecules in solution by compression measurements in the range 1-8000 atm. *Ark. Kemi* 20:513–71
3. Arai M, Kuwajima K. 2000. Role of the molten globule state in protein folding. *Adv. Protein Chem.* 53:209–82

4. Barnatt S. 1952. The velocity of sound in electrolytic solutions. *J. Chem. Phys.* 20:278–79
5. Blandamer MJ, Davis MI, Douhéret G, Reis JCR. 2001. Apparent molar isentropic compressions and expansions of solutions. *Chem. Soc. Rev.* 30:8–15
6. Brandts JF, Oliveira RJ, Westort C. 1970. Thermodynamics of protein denaturation. Effect of pressure on the denaturation of ribonuclease A. *Biochemistry* 9:1038–47
7. Bull HB, Breese K. 1968. Temperature coefficients of protein partial volumes. *J. Phys. Chem.* 72:1817–19
8. Bull HB, Breese K. 1973. Temperature dependence of partial volumes of proteins. *Biopolymers* 12:2351–58
9. Burova TV, Grinberg NV, Grinberg VY, Rariy RV, Klibanov AM. 2000. Calorimetric evidence for a native-like conformation of hen egg-white lysozyme dissolved in glycerol. *Biochim. Biophys. Acta* 1478:309–17
10. Cabani S, Gianni P, Mollica V, Lepori L. 1981. Group contributions to the thermodynamic properties of non-ionic organic solutes in dilute aqueous solution. *J. Solut. Chem.* 10:563–95
11. Chalikian TV. 2001. Structural thermodynamics of hydration. *J. Phys. Chem. B* 105:12566–78
12. Chalikian TV, Breslauer KJ. 1996. Compressibility as a means to detect and characterize globular protein states. *Proc. Natl. Acad. Sci. USA* 93:1012–14
13. Chalikian TV, Breslauer KJ. 1996. On volume changes accompanying conformational transitions of biopolymers. *Biopolymers* 39:619–26
14. Chalikian TV, Breslauer KJ. 1998. Thermodynamic analysis of biomolecules: a volumetric approach. *Curr. Opin. Struct. Biol.* 8:657–64
15. Chalikian TV, Breslauer KJ. 1998. Volumetric properties of nucleic acids. *Biopolymers* 48:264–80
16. Chalikian TV, Gindikin VS, Breslauer KJ. 1995. Volumetric characterizations of the native, molten globule, and unfolded states of cytochrome *c* at acidic pH. *J. Mol. Biol.* 250:291–306
17. Chalikian TV, Gindikin VS, Breslauer KJ. 1996. Spectroscopic and volumetric investigation of cytochrome *c* unfolding at alkaline pH: characterization of the base induced unfolded state at 25°C. *FASEB J.* 10:164–70
18. Chalikian TV, Gindikin VS, Breslauer KJ. 1998. Hydration of diglycyl tripeptides with non-polar side chains: a volumetric study. *Biophys. Chem.* 75:57–71
19. Chalikian TV, Kharakoz DP, Sarvazyan AP, Cain CA, McGough RJ, et al. 1992. Ultrasonic study of proton-transfer reactions in aqueous solutions of amino acids. *J. Phys. Chem.* 96:876–83
20. Chalikian TV, Sarvazyan AP, Breslauer KJ. 1994. Hydration and partial compressibility of biological compounds. *Biophys. Chem.* 51:89–109
21. Chalikian TV, Sarvazyan AP, Funck T, Breslauer KJ. 1994. Partial molar volumes, expansibilities, and compressibilities of oligoglycines in aqueous solutions at 18–55°C. *Biopolymers* 34:541–53
22. Chalikian TV, Totrov M, Abagyan R, Breslauer KJ. 1996. Hydration of globular proteins as derived from the volume and compressibility measurements: cross correlating thermodynamic and structural data. *J. Mol. Biol.* 260:588–603
23. Chalikian TV, Völker J, Anafi D, Breslauer KJ. 1997. The native and the heat-induced denatured states of α-chymotrypsinogen A: a thermodynamic and spectroscopic study. *J. Mol. Biol.* 274:237–52
24. Chervenak MC, Toone EJ. 1994. A direct measure of the contribution of solvent reorganization to the enthalpy of ligand-binding. *J. Am. Chem. Soc.* 116:10533–39
25. Chothia C. 1984. Principles that determine the structure of proteins. *Annu. Rev. Biochem.* 53:537–72
26. Connolly MJ. 1983. Analytical molecular

surface calculation. *J. Appl. Crystallogr.* 16:548–58
27. Connolly MJ. 1983. Solvent-accessible surfaces of proteins and nucleic acids. *Science* 221:709–13
28. Cooper A. 1976. Thermodynamic fluctuations in protein molecules. *Proc. Natl. Acad. Sci. USA* 73:2740–41
29. Cooper A. 1984. Protein fluctuations and the thermodynamic uncertainty principle. *Prog. Biophys. Mol. Biol.* 44:181–214
30. Dadarlat VM, Post CB. 2001. Insights into protein compressibility from molecular dynamics simulations. *J. Phys. Chem. B* 105:715–24
31. Desrosiers N, Lucas M. 1974. Relation between molal volumes and molal compressibilities from viewpoint of scaled-particle theory. Prediction of apparent molal compressibilities of transfer. *J. Phys. Chem.* 78:2367–69
32. Dill KA, Shortle D. 1991. Denatured states of proteins. *Annu. Rev. Biochem.* 60:795–825
33. Dubins DN, Filfil R, Macgregor RB Jr, Chalikian TV. 2000. The role of water in protein-ligand interactions: volumetric characterizations of the binding of 2′-CMP and 3′-CMP to ribonuclease A. *J. Phys. Chem. B* 104:390–401
34. Durchschlag H. 1986. Specific volumes of biological macromolecules and some other molecules of biological interest. See Ref. 62a, pp. 45–128
35. Edgcomb SP, Murphy KP. 2000. Structural energetics of protein folding and binding. *Curr. Opin. Biotechnol.* 11:62–66
36. Eggers F, Funck Th. 1973. Ultrasonic measurements with milliliter liquid samples in the 0.5–100 MHz range. *Rev. Sci. Instrum.* 44:969–78
37. Eggers F, Kaatze U. 1996. Broad-band ultrasonic measurement techniques for liquids. *Meas. Sci. Technol.* 7:1–19
38. Eggers F, Kustin K. 1969. Ultrasonic methods. *Methods Enzymol.* 14:55–80
39. Fahey PF, Kupke DW, Beams JW. 1969. Effect of pressure on apparent specific volume of proteins. *Proc. Natl. Acad. Sci. USA* 63:548–51
40. Filfil R, Chalikian TV. 2000. Volumetric and spectroscopic characterizations of the native and acid-induced denatured states of staphylococcal nuclease. *J. Mol. Biol.* 299:829–44
41. Frauenfelder H, Hartmann H, Karplus M, Kuntz ID, Kuriyan J, et al. 1987. Thermal expansion of a protein. *Biochemistry* 26:254–61
42. Freire E. 1995. Thermodynamics of partly folded intermediates in proteins. *Annu. Rev. Biophys. Biomol. Struct.* 24:141–65
43. Frye KJ, Perman CS, Royer CA. 1996. Testing the correlation between ΔA and ΔV of protein unfolding using m value mutants of staphylococcal nuclease. *Biochemistry* 35:10234–39
44. Fuentes EJ, Wand AJ. 1998. Local stability and dynamics of apocytochrome b_{562} examined by the dependence of hydrogen exchange on hydrostatic pressure. *Biochemistry* 37:9877–83
45. Gekko K. 1991. Flexibility of globular proteins in water as revealed by compressibility. In *Water Relationships in Food*, ed. H Levine, L Slade, pp. 753–71. New York: Plenum
46. Gekko K, Hasegawa Y. 1986. Compressibility-structure relationship of globular proteins. *Biochemistry* 25:6563–71
47. Gekko K, Hasegawa Y. 1989. Effect of temperature on the compressibility of native globular proteins. *J. Phys. Chem.* 93:426–29
48. Gekko K, Noguchi H. 1979. Compressibility of globular proteins in water at 25°C. *J. Phys. Chem.* 83:2706–14
49. Gekko K, Yamagami K. 1998. Compressibility and volume changes of lysozyme due to inhibitor binding. *Chem. Lett.* pp. 839–40
50. Gerstein M, Chothia C. 1996. Packing at the protein-water interface. *Proc. Natl. Acad. Sci. USA* 93:10167–72

51. Gianni P, Lepori L. 1996. Group contributions to the partial molar volume of organic solutes in aqueous solution. *J. Solut. Chem.* 25:1–42
52. Häckel M, Hedwig GR, Hinz H-J. 1998. The partial molar heat capacity and volume of the peptide backbone of proteins in aqueous solution. *Biophys. Chem.* 73:163–77
53. Häckel M, Hinz H-J, Hedwig GR. 1999. Partial molar volumes of proteins: amino acid side-chain contributions derived from the partial molar volumes of some tripeptides over temperature range 10–90°C. *Biophys. Chem.* 82:35–50
54. Häckel M, Hinz H-J, Hedwig GR. 2000. The partial molar volumes of some tetra- and pentapeptides in aqueous solution: a test of amino acid side-chain additivity for unfolded proteins. *Phys. Chem. Chem. Phys.* 2:4843–49
55. Hakin AW, Høiland H, Hedwig GR. 2000. Volumetric properties of some oligopeptides in aqueous solution: partial molar expansibilities and isothermal compressibilities at 298.15 K for the peptides of sequence Ala(gly)$_n$, n = 1–4. *Phys. Chem. Chem. Phys.* 2:4850–57
56. Härd T, Lundbäck T. 1996. Thermodynamics of sequence-specific protein-DNA interactions. *Biophys. Chem.* 62:121–39
57. Hawley SA. 1971. Reversible pressure-temperature denaturation of chymotrypsinogen. *Biochemistry* 10:2436–42
58. Hedwig GR, Høiland H. 1991. Thermodynamic properties of peptide solutions. 7. Partial molar isentropic pressure coefficients of some dipeptides in aqueous solutions. *J. Solut. Chem.* 20:1113–27
59. Hedwig GR, Høiland H. 1994. Thermodynamic properties of peptide solutions. Part 11. Partial molar isentropic pressure coefficients in aqueous solutions of some tripeptides that model protein side-chains. *Biophys. Chem.* 49:175–81
60. Heremans K. 1982. High pressure effects on proteins and other biomolecules. *Annu. Rev. Biophys. Bioeng.* 11:1–21
61. Heremans K, Smeller L. 1998. Protein structure and dynamics at high pressure. *Biochim. Biophys. Acta* 1386:353–70
62. Hiebl M, Maksymiw R. 1991. Anomalous temperature dependence of the thermal expansion of proteins. *Biopolymers* 31:161–67
62a. Hinz H-J, ed. 1986. *Thermodynamic Data for Biochemistry and Biotechnology*. Berlin: Springer. 456 pp.
63. Høiland H. 1974. Pressure dependence of the volumes of ionization of carboxylic acids at 25, 35, 45°C and 1–200 bar. *J. Chem. Soc. Faraday Trans. 1* 70:1180–85
64. Høiland H. 1975. Volumes of ionization of dicarboxylic acids in aqueous solution from density measurements at 25°C. *J. Chem. Soc. Faraday Trans. 1* 71:797–802
65. Høiland H. 1986. Partial molar compressibilities of organic solutes in water. See Ref. 62a, pp. 129–47
66. Høiland H. 1986. Partial molar volumes of biochemical model compounds in aqueous solutions. See Ref. 62a, pp. 17–44
67. Høiland H, Vikingstad E. 1975. Partial molal volumes and volumes of ionization of hydrocarboxylic acids in aqueous solution at 25, 30, and 35°C. *J. Chem. Soc. Faraday Trans. 1* 71:2007–16
68. Høiland H, Vikingstad E. 1976. Isentropic apparent molal compressibilities and compressibilities of ionization of carboxylic acids in aqueous solution. *J. Chem. Soc. Faraday Trans. 1* 72:1441–47
69. Inoue K, Yamada H, Akasaka K, Hermann C, Kremer W, et al. 2000. Pressure-induced local unfolding of the Ras binding domain of RalGDS. *Nat. Struct. Biol.* 7:547–50
70. Iqbal M, Verrall RE. 1987. Volumetric properties of aqueous solutions of bovine serum albumin, human serum albumin, and human hemoglobin. *J. Phys. Chem.* 91:1935–41
71. Iqbal M, Verrall RE. 1988. Implications of protein folding. Additivity schemes for

volumes and compressibilities. *J. Biol. Chem.* 263:4159–65
72. Jacobson B. 1950. On the adiabatic compressibility of aqueous solutions. *Ark. Kemi* 2:177–210
73. Jen-Jacobson L. 1997. Protein-DNA recognition complexes: conservation of structure and binding energy in the transition state. *Biopolymers* 44:153–80
74. Kajander T, Kahn PC, Passila SH, Cohen DC, Lehtiö L, et al. 2000. Buried charged surface in proteins. *Structure* 8:1203–14
75. Kamiyama T, Gekko K. 1997. Compressibility and volume changes of lysozyme due to guanidine hydrochloride denaturation. *Chem. Lett.* 1063–64
76. Kamiyama T, Gekko K. 2000. Effect of ligand binding on the flexibility of dihydrofolate reductase as revealed by compressibility. *Biochim. Biophys. Acta* 1478:257–66
77. Kamiyama T, Sadahide Y, Nogusa Y, Gekko K. 1999. Polyol-induced molten globule of cytochrome *c*: an evidence for stabilization by hydrophobic interaction. *Biochim. Biophys. Acta* 1434:44–57
78. Katrusiak A, Dauter Z. 1996. Compressibility of lysozyme protein crystals by X-ray diffraction. *Acta Crystallogr. D* 52:607–8
79. Kauzmann W. 1987. Protein stabilization—thermodynamics of unfolding. *Nature* 325:763–64
80. Kauzmann W, Bodanszky A, Rasper J. 1962. Volume changes in protein reactions. II. Comparison of ionization reactions in proteins and small molecules. *J. Am. Chem. Soc.* 84:1777–88
81. Kharakoz DP. 1989. Volumetric properties of proteins and their analogues in diluted water solutions. 1. Partial volumes of amino acids at 15–55°C. *Biophys. Chem.* 34:115–25
82. Kharakoz DP. 1991. Volumetric properties of proteins and their analogues in diluted water solutions. 2. Partial adiabatic compressibilities of amino acids at 15–70°C. *J. Phys. Chem.* 95:5634–42
83. Kharakoz DP. 1992. Partial molar volumes of molecules of arbitrary shape and the effect of hydrogen bonding with water. *J. Solut. Chem.* 21:569–95
84. Kharakoz DP. 1997. Partial volumes and compressibilities of extended polypeptide chains in aqueous solution: additivity scheme and implication of protein unfolding at normal and high pressure. *Biochemistry* 36:10276–85
85. Kharakoz DP. 2000. Protein compressibility, dynamics, and pressure. *Biophys. J.* 79:511–25
86. Kharakoz DP, Bychkova VE. 1997. Molten globule of human α-lactalbumin: hydration, density, and compressibility of the interior. *Biochemistry* 36:1882–90
87. Kharakoz DP, Sarvazyan AP. 1993. Hydrational and intrinsic compressibilities of globular proteins. *Biopolymers* 33:11–26
88. King EJ. 1969. Volume changes for ionization of formic, acetic, and n-butyric acids and the glycinium ion in aqueous solution at 25°C. *J. Phys. Chem.* 73:1220–32
89. Klibanov AM. 1997. Why are enzymes less active in organic solvents than in water? *Trends Biotechnol.* 15:97–101
90. Klibanov AM. 2001. Improving enzymes by using them in organic solvents. *Nature* 409:241–46
91. Knubovets T, Osterhout JJ, Connoly PJ, Klibanov AM. 1999. Structure, thermostability, and conformational flexibility of hen egg-white lysozyme dissolved in glycerol. *Proc. Natl. Acad. Sci. USA* 96:1262–67
92. Kundrot CE, Richards FM. 1987. Crystal structure of hen egg-white lysozyme at a hydrostatic pressure of 1000 atmospheres. *J. Mol. Biol.* 193:157–70
93. Kuntz ID Jr, Kauzmann W. 1974. Hydration of proteins and polypeptides. *Adv. Protein Chem.* 28:239–345
94. Kupke DW. 1973. Density and volume change measurements. In *Physical Principles and Techniques of Protein Chemistry. Part C*, ed. SJ Leach, pp. 1–75. New York/London: Academic

95. Lassalle MW, Yamada H, Akasaka K. 2000. The pressure-temperature free energy-landscape of staphylococcal nuclease monitored by ^{1}H NMR. *J. Mol. Biol.* 298:293–302
96. Lee B. 1991. Isoenthalpic and isoentropic temperatures and the thermodynamics of protein denaturation. *Proc. Natl. Acad. Sci. USA* 88:5154–58
97. Lepori L, Giani P. 2000. Partial molar volumes of ionic and nonionic organic solutes in water: a simple additivity scheme based on the intrinsic volume approach. *J. Solut. Chem.* 29:405–47
98. Liang J, Edelsbrunner H, Fu P, Sudhakar PV, Subramaniam S. 1998. Analytical shape computation of macromolecules. 1. Molecular area and volume through alpha shape. *Proteins Struct. Funct. Genet.* 33:1–17
99. Likhodi O, Chalikian TV. 1999. Partial molar volumes and adiabatic compressibilities of a series of aliphatic amino acids and oligoglycines in D_2O. *J. Am. Chem. Soc.* 121:1156–63
100. Likhodi O, Chalikian TV. 2000. Differential hydration of a,w-aminocarboxylic acids in D_2O and H_2O. *J. Am. Chem. Soc.* 122:7860–68
101. Lumry R. 1995. The new paradigm for protein research. In *Protein-Solvent Interactions*, ed. RB Gregory, pp. 1–141. New York/Basel/Hong Kong: Marcel Dekker
102. Luque I, Leavitt SA, Freire E. 2002. The linkage between protein folding and functional cooperativity: two sides of the same coin? *Annu. Rev. Biophys. Biomol. Struct.* 31:235–56
103. Lynch TW, Kosztin D, McLean MA, Schulten K, Sligar SG. 2002. Dissecting the molecular origins of specific protein-nucleic acid recognition: hydrostatic pressure and molecular dynamics. *Biophys. J.* 82:93–98
104. Macgregor RB Jr. 1990. Reversible inhibition of EcoRI with elevated pressure. *Biochem. Biophys. Res. Commun.* 170:775–78
105. Macgregor RB Jr. 1998. Effect of hydrostatic pressure on nucleic acids. *Biopolymers* 48:253–63
106. Macgregor RB Jr. 2002. The interactions of nucleic acids at elevated hydrostatic pressure. *Biochem. Biophys. Acta* 1595:266–76
107. Makhatadze GI, Clore GM, Gronenborn AM. 1995. Solvent isotope effect and protein stability. *Nat. Struct. Biol.* 2:852–55
108. Makhatadze GI, Medvedkin VN, Privalov PL. 1990. Partial molar volumes of polypeptides and their constituent groups in aqueous solution over a broad temperature range. *Biopolymers* 30:1001–10
109. Marky LA, Kupke DW, Kankia BI. 2001. Volume changes accompanying interaction of ligands with nucleic acids. *Methods Enzymol.* 340:149–65
110. Mathieson JG, Conway BE. 1974. H_2O-D_2O solvent isotope effects in adiabatic compressibility and volume of electrolytes and non-electrolytes: relation to specificities of ionic solvation. *J. Chem. Soc. Faraday Trans. 1* 70:752–68
111. Merzel F, Smith JC. 2002. Is the first hydration shell of lysozyme of higher density than bulk water? *Proc. Natl. Acad. Sci. USA* 99:5378–83
112. Miller S, Janin J, Lesk AM, Chothia C. 1987. Interior and surface of monomeric proteins. *J. Mol. Biol.* 196:641–56
113. Millero FJ, Lo Surdo A, Shin C. 1978. The apparent molal volumes and adiabatic compressibilities of aqueous amino acids at 25°C. *J. Phys. Chem.* 82:784–92
114. Murphy LR, Matubayasi N, Payne VA, Levy RM. 1998. Protein hydration and unfolding—insights from experimental partial specific volumes and unfolded protein models. *Fold. Des.* 3:105–18
115. Nikitin SY. 1983. *Relationship between intramolecular interactions in peptides and proteins and their acoustic properties*. PhD thesis. Inst. Biol. Phys., Acad. Sci. USSR, Pushchino. 113 pp.
116. Nikitin SY, Sarvazyan AP, Kravchenko

NA. 1984. Ultrasonic velocimetry of lysozyme solutions. *Mol. Biol.* 18:685–92
117. Nikitina TK, Nikitin SY. 1990. Use of an acoustic method to study the interaction between lysozyme and sodium dodecylsulfate. *Mol. Biol.* 24:656–61
118. Nölting B, Sligar SG. 1993. Adiabatic compressibility of molten globules. *Biochemistry* 32:12319–23
119. Owen BB, Simons HL. 1957. Standard partial molal compressibilities by ultrasonics. 1. Sodium chloride and potassium chloride at 25°C. *J. Phys. Chem.* 61:479–82
120. Paci E, Marchi M. 1996. Intrinsic compressibility and volume compression in solvated proteins by molecular dynamics simulations at high pressure. *Proc. Natl. Acad. Sci. USA* 93:11609–14
121. Paci E, Velikson B. 1997. On the volume of macromolecules. *Biopolymers* 41:785–97
122. Panick G, Winter R. 2000. Pressure-induced unfolding/refolding of ribonuclease A: static and kinetic Fourier transform infrared spectroscopy study. *Biochemistry* 39:1862–69
123. Parsegian VA, Rand RP, Rau DC. 1995. Macromolecules and water: probing with osmotic stress. *Methods Enzymol.* 259:43–94
124. Phelps DK, Post CB. 1995. A novel basis for capsid stabilization by antiviral compounds. *J. Mol. Biol.* 254:544–51
125. Pierotti RA. 1965. Aqueous solutions of nonpolar gases. *J. Phys. Chem.* 69:281–88
126. Pierotti RA. 1976. A scaled particle theory of aqueous and nonaqueous solutions. *Chem. Rev.* 6:717–26
127. Prehoda KE, Mooberry ES, Markley JL. 1998. Pressure denaturation of proteins: evaluation of compressibility effects. *Biochemistry* 37:5785–90
128. Priev A, Almagor A, Yedgar S, Gavish B. 1996. Glycerol decreases the volume and compressibility of protein interior. *Biochemistry* 35:2061–66
129. Privalov PL. 1996. Intermediate states in protein folding. *J. Mol. Biol.* 258:707–25
130. Rariy RV, Klibanov AM. 1997. Correct protein folding in glycerol. *Proc. Natl. Acad. Sci. USA* 94:13520–23
131. Rashin AA, Iofin M, Honig B. 1986. Internal cavities and buried waters in globular proteins. *Biochemistry* 25:3619–25
132. Rasper J, Kauzmann W. 1962. Volume changes in protein reactions. I. Ionization reactions of proteins. *J. Am. Chem. Soc.* 84:1771–77
133. Richards FM. 1977. Areas, volumes, packing, and protein structure. *Annu. Rev. Biophys. Bioeng.* 6:151–76
134. Richards FM. 1985. Calculation of molecular volumes and areas for structures of known geometry. *Methods Enzymol.* 115:440–64
135. Richards FM. 1992. Folded and unfolded proteins—an introduction. In *Protein Folding*, ed. TE Creighton, pp. 1–58. New York: Freeman
136. Richards FM, Lim WA. 1994. An analysis of packing in the protein folding problem. *Q. Rev. Biophys.* 26:423–98
137. Robinson CR, Sligar SG. 1995. Hydrostatic and osmotic pressure as tools to study macromolecular recognition. *Methods Enzymol.* 259:395–427
138. Robinson CR, Sligar SG. 1998. Changes in solvation during DNA binding and cleavage are critical to altered specificity of the EcoRI endonuclease. *Proc. Natl. Acad. Sci. USA* 95:2186–91
139. Royer CA. 1995. Application of pressure to biochemical equilibria: the other thermodynamic variable. *Methods Enzymol.* 259:357–77
140. Royer CA. 2002. Revisiting volume changes in pressure-induced protein unfolding. *Biochim. Biophys. Acta* 1595: 201–9
141. Rupley JA, Careri G. 1991. Protein hydration and function. *Adv. Protein Chem.* 41:37–172
142. Sarvazyan AP. 1982. Development of

methods of precise ultrasonic measurements in small volumes of liquids. *Ultrasonics* 20:151–54
143. Sarvazyan AP. 1983. Ultrasonic velocimetry of biological compounds. *Mol. Biol.* 17:916–27
144. Sarvazyan AP. 1991. Ultrasonic velocimetry of biological compounds. *Ann. Rev. Biophys. Biophys. Chem.* 20:321–42
145. Sarvazyan AP, Hemmes P. 1979. Relaxational contributions to protein compressibility from ultrasonic data. *Biopolymers* 18:3015–24
146. Seemann H, Winter R, Royer CA. 2001. Volume, expansivity, and isothermal compressibility changes associated with temperature and pressure unfolding of staphylococcal nuclease. *J. Mol. Biol.* 307:1091–102
147. Shortle D, Ackerman MS. 2001. Persistence of native-like topology in a denatured protein in 8 M urea. *Science* 293:487–89
148. Sidorova NY, Rau DC. 2001. Linkage of EcoRI dissociation from its specific DNA recognition site to water activity, salt concentration, and pH: separating their roles in specific and non-specific binding. *J. Mol. Biol.* 310:801–16
149. Silva JL, Foguel D, Royer CA. 2001. Pressure provides new insights into protein folding, dynamics, and structure. *Trends Biochem. Sci.* 26:612–18
150. Silva JL, Weber G. 1993. Pressure stability of proteins. *Annu. Rev. Phys. Chem.* 44:89–113
151. Smeller L, Fidy J. 2002. The enzyme horseradish peroxidase is less compressible at high pressures. *Biophys. J.* 82:426–36
152. Stillinger FH. 1973. Structure in aqueous solutions of nonpolar solutes from the standpoint of scaled-particle theory. *J. Solut. Chem.* 2:141–58
153. Stites WE. 1997. Protein-protein interactions: interface structure, binding thermodynamics, and mutational analysis. *Chem. Rev.* 97:1233–50
154. Svergun DI, Richard S, Koch MHJ, Sayers Z, Kuprin S, Zaccai G. 1998. Protein hydration in solution: experimental observation by X-ray and neutron scattering. *Proc. Natl. Acad. Sci. USA* 95:2267–72
155. Szwajkajzer D, Carey J. 1997. Molecular and biological constraints on ligand-binding affinity and specificity. *Biopolymers* 44:181–98
156. Tama F, Myashita O, Kitao A, Go N. 2000. Molecular dynamics simulation shows large volume fluctuations of proteins. *Eur. Biophys. J.* 29:472–80
157. Tamura Y, Gekko K. 1995. Compactness of thermally and chemically denatured ribonuclease A as revealed by volume and compressibility. *Biochemistry* 34:1878–84
158. Tamura Y, Suzuki N, Mihashi K. 1993. Adiabatic compressibility of myosin subfragment-1 and heavy meromyosin with and without nucleotide. *Biophys. J.* 65:1899–905
159. Taniguchi Y, Suzuki K. 1983. Pressure inactivation of α-chymotrypsin. *J. Phys. Chem.* 87:5185–93
160. Taulier N, Chalikian TV. 2001. Characterization of pH-induced transitions of β-lactoglobulin: ultrasonic, densimetric, and spectroscopic studies. *J. Mol. Biol.* 314:873–89
161. Taulier N, Chalikian TV. 2002. Compressibility of protein transitions. *Biochim. Biophys. Acta* 1595:48–70
162. Tilton RF, Dewan JC, Petsko GA. 1992. Effects of temperature on protein structure and dynamics: X-ray crystallographic studies of the protein ribonuclease-A at nine different temperatures from 98 to 320 K. *Biochemistry* 31:2469–81
163. Uversky VN. 2002. Natively unfolded proteins: a point where biology waits for physics. *Protein Sci.* 11:739–56
164. Uversky VN. 2002. What does it mean to be natively unfolded? *Eur. J. Biochem.* 269:2–12
165. Uversky VN, Gillespie JR, Fink AL. 2000. Why are "natively unfolded" proteins

unstructured under physiologic conditions? *Proteins Struct. Funct. Genet.* 41: 415–27

166. Vanderkooi JM. 1998. The protein state of matter. *Biochim. Biophys. Acta* 1386:241–53
167. Van Gunsteren WF, Bürgi R, Peter C, Daura X. 2001. The key to solving the protein-folding problem lies in an accurate description of the denatured state. *Angew. Chem. Int. Ed.* 40:352–55
168. Vidugiris GJA, Royer CA. 1998. Determination of the volume changes for pressure-induced transitions of apomyoglobin between the native, molten globule, and unfolded states. *Biophys. J.* 75:463–70
169. Weber G. 1993. Thermodynamics of the association and the pressure dissociation of oligomeric proteins. *J. Phys. Chem.* 97:7108–15
170. Wright PE, Dyson HJ. 1999. Intrinsically unstructured proteins: re-assessing the protein structure-function paradigm. *J. Mol. Biol.* 293:321–31
171. Yayanos AA. 1972. Apparent molal volume of glycine, glycolamide, alanine, lactamide, and glycylglycine in aqueous solution at 25°C and high pressures. *J. Phys. Chem.* 76:1783–92
172. Young ACM, Tilton RF, Dewan JC. 1994. Thermal expansion of hen egg-white lysozyme. Comparison of the 1.9 Å resolution structures of the tetragonal form of the enzyme at 100 K and 298 K. *J. Mol. Biol.* 235:302–17
173. Zamyatnin AA. 1984. Amino acid, peptide, and protein volume in solution. *Annu. Rev. Biophys. Bioeng.* 13:145–65
174. Zana R. 1977. On the volume changes upon protonation of n-alkylcarboxylate ions and n-alkylamines in aqueous solutions. Extension to the interpretation of volumic effects observed in solutions of molecular ions, polyelectrolytes, micellar detergents, and proteins. *J. Phys. Chem.* 81:1817–22
175. Zipp A, Kauzmann W. 1973. Pressure denaturation of metmyoglobin. *Biochemistry* 12:4217–28
176. Zollfrank J, Friedrich J, Fidy J, Vanderkooi JM. 1991. Photochemical holes under pressure: compressibility and volume fluctuations of a protein. *J. Chem. Phys.* 94:8600–3

Annu. Rev. Biophys. Biomol. Struct. 2003. 32:237–56
doi: 10.1146/annurev.biophys.32.110601.142356

First published online as a Review in Advance on January 8, 2003

THE BINDING OF COFACTORS TO PHOTOSYSTEM I ANALYZED BY SPECTROSCOPIC AND MUTAGENIC METHODS*

John H. Golbeck
Department of Biochemistry and Molecular Biology, The Pennsylvania State University, University Park, Pennsylvania 16802; email: JHG5@psu.edu

Key Words P700, phylloquinone, iron-sulfur cluster, electron paramagnetic resonance, X-ray crystal structure

■ **Abstract** This review focuses on cofactor-ligand and protein-protein interactions within the photosystem I reaction center. The topics include a description of the electron transfer cofactors, the mode of binding of the cofactors to protein-bound ligands, and a description of intraprotein contacts that ultimately allow photosystem I to be assembled (in cyanobacteria) from 96 chlorophylls, 22 carotenoids, 2 phylloquinones, 3 [4Fe-4S] clusters, and 12 polypeptides. During the 15 years that have elapsed from the first report of crystals to the atomic-resolution X-ray crystal structure, cofactor-ligand interactions and protein-protein interactions were systematically being explored by spectroscopic and genetic methods. This article charts the interplay between these disciplines and assesses how good the early insights were in light of the current structure of photosystem I.

*Abbreviations and definitions: A_0, primary electron acceptor composed of chlorophyll monomer; A_1, secondary electron acceptor composed of phylloquinone(s); Chl, chlorophyll; ENDOR, electron-nuclear double resonance; EPR, electron paramagnetic resonance; ESEEM, electron spin echo envelope modulation; EXAFS, extended X-ray absorption fine structure; HOMO, highest occupied molecular orbital; F_A, [4Fe-4S] cluster on PsaC subunit ligated by cysteines 20, 47, 50, and 53 with *g*-values of 2.05, 1.95, and 1.85; F_B, [4Fe-4S] cluster on PsaC subunit ligated by cysteines 10, 13, 16, and 57 with *g*-values of 2.07, 1.93, and 1.88; F_X, interpolypeptide [4Fe-4S] cluster between PsaA and PsaB with *g*-values of 2.09, 1.86, and 1.75; $NADP^+$, nicotinamide adenine nucleotide phosphate; P700, primary electron donor composed of special pair of chlorophyll *a* and chlorophyll *a′* molecules; P700-F_X core, photosystem I complex stripped of PsaC, PsaD, and PsaE subunits; PhQ, phylloquinone (2-methyl-3-phytyl-1,4-naphthoquinone); PQ-9, plastoquinone; PsaA and PsaB, heterodimeric core subunits harboring the electron transfer cofactors P700, A_0, A_1, and F_X; PsaC, stromal subunit harboring F_A and F_B iron-sulfur clusters; PsaD and PsaE, stromal subunits covering PsaC responsible in part for docking site of ferredoxin; PS I, photosystem I; PS II, photosystem II; Q_A, quinone in A-site of bacterial reaction center; Q_K-A, phylloquinone on PsaA subunit; Q_K-B, phylloquinone on PsaB subunit; RC, reaction center.

1056-8700/03/0609-0237$14.00

CONTENTS

INTRODUCTION

This review examines details of cofactor-ligand and protein-protein interactions in the photosystem I (PS I) reaction center (RC) of cyanobacteria, algae, and plants. I report first on spectroscopic and mutagenesis studies that were conducted "pre-crystal" to identify the electron transfer cofactors and their ligands. I then compare these findings with the newly released 2.5 Å X-ray crystal structure (34). Where possible, I describe experimental studies conducted to identify those protein factors that modulate the functional properties of a particular cofactor. Finally, I outline strategies that have attempted to identify the amino acids involved in protein-protein contacts, and how well the outcome agrees with the current structure of PS I. The arrangement of the cofactors and the nomenclature used in this review are shown in Figure 1. Please note that unless stated otherwise the numbering of the amino acids in PsaA, PsaB, and PsaC is in reference to PS I from *Synechococcus elongatus*. This article is meant to be selective in the use of the literature to make the points enumerated above. I therefore apologize in advance to those whose work I have not cited.

THE LIGANDS TO P700

The primary electron donor, P700, was first identified as a chlorophyll (Chl) *a* "special pair" by experiments that compared the electron paramagnetic resonance (EPR) properties of $P700^+$ with those of monomeric Chl a^+ in solvent. Although both radicals have a similar electronic *g*-tensor, the narrower inhomogeneous linewidth of $P700^+$ compared to Chl a^+ is consistent with its identification as a dimer rather than as a monomer of Chl *a* (48). ^1H and ^{15}N electron-nuclear double resonance (ENDOR) spectroscopy later showed that an asymmetrical spin distribution exists between the two Chls that constitute P700 (12).

By a combination of selective isotopic labeling and ENDOR and electron spin echo envelope modulation (ESEEM) spectroscopy (41), His was identified as an axial ligand to the Chl on which the spin of $P700^+$ resides. Because the axial

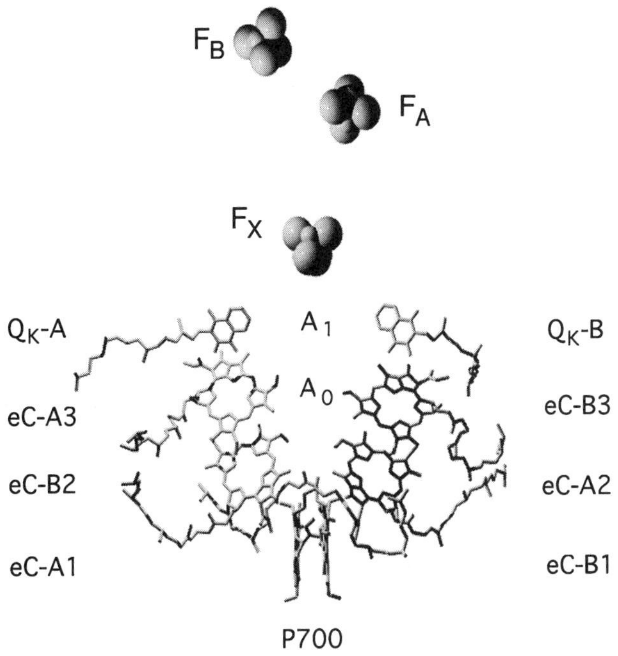

Figure 1 Placement of electron transfer cofactors in PS I derived from the crystallographic coordinates in PDB entry 1JBO. The crystallographic nomenclature for the cofactors is depicted on the left and right sides of the figure, and the spectroscopic nomenclature for the cofactors according to Reference 34 is depicted in the center of the figure.

ligands to the primary electron donor in the bacterial RC have long been known to be His, this result confirmed the long-held belief that P700 would be structurally similar. The first of the axial ligands, $His656_{PsaB}$ in *Chlamydomonas reinhardtii* ($His660_{PsaB}$ in *S. elongatus*), was identified by changes in the optical and ENDOR spectra as well as by changes in the redox potential of $P700^+$ when it was changed to Asn or Ser (79). The second axial ligand, $His676_{PsaA}$, was identified in *C. reinhardtii* ($His680_{PsaA}$ in *S. elongatus*) by a similar combination of site-directed mutagenesis and EPR/ENDOR spectroscopy (38, 53). These assignments are confirmed in the X-ray crystal structure (34), which shows that P700 is a modified dimer consisting of Chl *a*, designated as eC-B1, and Chl *a*′, designated as eC-A1 (Figure 1). $His680_{PsaA}$ provides the axial ligand to the Mg^{2+} of Chl *a*′, and $His660_{PsaB}$ provides the axial ligand to the Mg^{2+} of Chl *a*. Chl *a*′ is the $C13^2$ epimer of Chl *a*, and its presence in PS I (76), especially in highly enriched PS I complexes (43), had led to the prediction that it would be a component of P700.

There are differences in the binding sites of eC-A1 and eC-B1. The side groups of $Tyr735_{PsaA}$ and $Thr743_{PsaA}$ are close enough to form hydrogen (H)-bonds to the phytyl ester carbonyl oxygen and to the C13[1] carbonyl oxygen at ring E of the Chl a' of eC-A1, respectively. In addition, a water molecule is 3.3 Å distant from the carboxy ester oxygen of the $C13^2$ carbomethyl group of Chl a', which may be a bit long for an H-bond (this particular assignment has also been questioned in Reference 80), but interestingly, this water is also H-bonded to $Tyr603_{PsaA}$, $Ser607_{PsaA}$, and $Thr743_{PsaA}$. In contrast, there are no H-bonds to eC-B1. The H-bonds are thought to be responsible for the selective insertion of Chl a' in the PsaA protein (78). The break in C_2-symmetry in the vicinity of P700 is likely to be the reason for the asymmetric distribution of the spin density of $P700^+$, which is located on eC-B1 (38).

The redox potential of the $P700/P700^+$ couple in the wild type is +447 mV (versus H_2), which is about 420 mV more reducing than the redox potential of the monomeric Chl a/Chl a^+ couple in organic solvent (75). The factors that lead to this relatively low redox potential are only now being systematically explored (38, 79). The replacement of the axial ligand to the Mg^{2+} of either Chl a or Chl a' by uncharged polar amino acids leads to only modest increases in the redox potential of the $P700/P700^+$ couple (+18 mV when His is substituted by Gln on either PsaA or PsaB, and +18 mV and +40 mV when it is substituted by Ser on PsaA and PsaB, respectively). When Asn, Gly, or Cys substitutes His on PsaB, the redox potential increases by +40 mV, +98 mV, and +136 mV. Thus, the redox potential of P700 is more sensitive to replacement of the axial ligand to Chl a than to Chl a'. Note that with the single exception of Cys, the redox potential on the PsaB side becomes more oxidizing with polarizability and size of the amino acid side chain. This trend was rationalized by a stabilization of the $P700^+$ cation by the lone pair electron on the His, which lowers the energy of $P700^+$ and thereby increases its redox potential (38). Gly has the lowest polarizability and therefore results in the most positive potential of any substitution. Cys does not fit the pattern, and other factors such as structural changes due to the large size of the sulfur atom may come into play. His may have been selected in evolution as the axial ligand for P700 simply because it lies on one extreme of the polarizability index, thereby conferring the most reducing potential to the site.

Because an axial His ligation scheme is common to the primary electron donors in PS I, PS II, and the bacterial RC, interactions of protein with the chlorin ring must be responsible for poising the redox potentials. As pointed out in Reference 78, electrostatic effects (charges, dipoles, polarizable residues), which tend to stabilize the excited state, and H-bonds, which tend to stabilize the ground state, may function to poise the redox potential of the $P700/P700^+$ couple. Systematic studies of introducing or removing H-bonds to P700 are just now being performed. When the H-bond from $Thr739_{PsaA}$ in *C. reinhardtii* ($Thr743_{PsaA}$ in *S. elongatus*) to the C13[1] keto group of Chl a' is broken by replacement with Tyr, His, and Val, the redox potential of the $P700/P700^+$ couple decreases by −9, −9, and −32 mV, respectively (80). The nearby water molecule, which is H-bonded to the $C13^2$ carbonyl ester oxygen, is assumed to be lost in all three mutants. This would leave only one H-bond from $Tyr735_{PsaA}$ to the phytyl ester carbonyl oxygen on the Chl *a′* side. The more positive redox potential of P700/P700$^+$ that results from the addition of an H-bond may be related to a decrease in the highest occupied molecular orbital (HOMO) of Chl *a′*, thereby stabilizing the Chl *a/a′* dimer HOMO (80). The introduction of a comparable H-bond in the bacterial RC by replacing Leu160 with His on the M subunit leads to a more oxidizing redox potential when compared with the wild type (52). Thus, the presence of an H-bond to the 13^1 keto group results in a more oxidizing redox potential of the primary electron donor in both PS I and the bacterial RC. It will be interesting to probe the biophysical properties of P700 when the PsaA and PsaB sides are engineered to become even more symmetrical, either by introducing H-bonds into eC-B1 or removing H-bonds from eC-A1.

THE SECOND PAIR OF CHLOROPHYLLS: THE BRIDGING CHLOROPHYLLS

The two bridging Chl *a* molecules, eC-A2 and eC-B2 (Figure 1), were first detected in the 6 Å model of PS I (39). They had not been detected spectroscopically, and no kinetic data directly implicate them in primary charge separation. However, their very close distance to P700 and A_0 would make it highly unlikely that the bridging Chls would not be involved. The X-ray crystal structure (34) shows that the axial ligand to the Mg^{2+} of eC-A2 is a water molecule H-bonded to $Asn604_{PsaA}$. The axial ligand to the $Mg^{2+}$of eC-B2 is also a water molecule H-bonded to $Asn591_{PsaB}$. By way of comparison, a His provides the axial ligand to the Mg^{2+} in the bridging BChls of the bacterial RC (14). This is the only instance in which PsaB provides a ligand for a cofactor located on the A-branch, and in which PsaA provides a ligand for a cofactor located on the B-branch. There are no H-bonds to the chlorin ring of either of the bridging Chl *a* molecules, and unlike the asymmetry of the P700 binding pocket, there are no significant differences between the environments of eC-A2 and eC-B2. No mutants have yet been reported that attempt to displace the axially bound water by changing the identity of either $Asn604_{PsaA}$ or $Asn591_{PsaB}$.

THE THIRD PAIR OF CHLOROPHYLLS: THE PRIMARY ELECTRON ACCEPTOR(S)

The primary electron acceptor A_0 was identified as a Chl *a* molecule by EPR spectroscopy (5) and more specifically as a Chl *a* monomer by optical kinetic spectroscopy (67). In the X-ray crystal structure, one (or both) of the Chl *a* molecules that occupy the positions of eC-A3 and eC-B3 in the X-ray crystal structure (Figure 1) is assumed to be A_0. The axial ligand to eC-A3 is the sulfur atom of $Met688_{PsaA}$; similarly, the axial ligand to eC-B3 is the sulfur atom of $Met668_{PsaB}$. This mode of ligation came as a surprise because the hard acid Mg^{2+} is ligated by the soft base (methionine) sulfur, which is not predicted by the rules of inorganic chemistry. There is an H-bond between the keto oxygen of ring V of eC-A3 and $Tyr696_{PsaA}$ and an H-bond between the phytyl ester carbonyl oxygen of the phytyl group and the backbone oxygen of $Ser429_{PsaB}$. In contrast, there is only one H-bond between the keto oxygen of ring V of eC-B3 and $Tyr676_{PsaB}$. The role of the latter H-bond was probed by site-directed mutagenesis and optical spectroscopy. When $Tyr667_{PsaB}$ in *Synechocystis* sp. PCC 6803 ($Tyr676_{PsaB}$ in *S. elongatus*) was replaced with Phe (W. Xu, S. Savikhin, P. Martinsson, W. Struve & P.R. Chitnis, unpublished observations), the difference spectrum of $P700^+/P700$, the EPR spectrum of photoaccumulated A_1^-, and the A_0 to A_1 electron transfer rates were not altered, but the bleaching due to A_0 reduction was red-shifted by 3 nm. These results imply that A_0 is assigned correctly to eC-B3 in the X-ray crystal structure (the counterpart eC-A3 mutant was not constructed). However, it is difficult to reconcile the spectral bandshift with the lack of an effect on the kinetics of forward electron transfer. Perhaps this result hints that all six Chl *a* molecules are in some manner involved in the charge-separation process.

Certainly, the arrangement of three pairs of Chl *a* molecules in a bifurcating electron transfer pathway (Figure 1) raises the question of whether eC-A3, eC-B3, or both function as A_0. This, of course, is tantamount to asking whether electron transfer is unidirectional or bidirectional along the PsaA- or PsaB-ligated cofactors. One approach to answering this question involves changing the identities of the Met axial ligands to eC-A3 and eC-B3. The premise of this experiment is that a change in the axial ligand should alter the redox potential of the A_0^-/A_0 couple. This in turn should result in a change in the forward electron transfer kinetics from A_0^- to A_1 due to a change in the Frank-Condon factor in the Marcus equation, which relates the rate of electron transfer to changes in Gibbs free energy, temperature, and reorganization energy (47).

When $Met659_{PsaB}$ in *Synechocystis* sp. PCC 6803 ($Met668_{PsaB}$ in *S. elongatus*) was changed to Leu (W. Xu, R. Cohen, A. van der Est, D. Stehlik, J. Golbeck, et al., unpublished observations), the amount of $P700^+$ $[F_A/F_B]^-$ generated on a single turnover flash and the electron spin-polarized EPR spectrum of $P700^+$ A_1^- were identical to the wild type. However, when $Met684_{PsaA}$ ($Met688_{PsaA}$ in *S. elongatus*) was similarly changed to Leu, the amount of $P700^+$ $[F_A/F_B]^-$ generated was a fraction of the wild type, a spin-polarized triplet state was generated, and the electron spin-polarized EPR spectrum was strikingly different than in the wild type. The latter can be explained quantitatively by a slower rate of electron transfer from A_0^- to A_1. Given the high degree of symmetry surrounding the A_0 sites, this result implies preferential electron transfer among the PsaA-side cofactors in cyanobacterial PS I. In light of this result, it is interesting that when $Met684_{PsaA}$ in *C. reinhardtii* ($Met688_{PsaA}$ in *S. elongatus*) was changed to His, electron transfer through the PsaA branch was also inactivated (18). Moreover, when $Met664_{PsaB}$ in *C. reinhardtii* ($Met668_{PsaB}$ in *S. elongatus*) was changed to His and Ser, growth on light was severely limited, yet the rate of primary charge separation was unaltered (51). These are interesting preliminary results, and additional work on these mutants is needed to arrive at a consensus on the issue of directionality in PS I.

ENVIRONMENT OF THE TWO PHYLLOQUINONES

Dating from its initial discovery, the secondary electron acceptor A_1 in PS I has been surrounded by controversy. In the 1980s, the identity of A_1 as phylloquinone (PhQ) was called into question, but its role was slowly clarified through a combination of optical kinetic and EPR studies [reviewed in (21)]. In recent years, the controversy has shifted to whether one or both of the PhQs participate in electron transfer. A straightforward solution to this problem is complicated by the near C_2-symmetry of the PS I cofactors, particularly beyond the Chl *a/a′* special pair. As a consequence, there are no unique spectroscopic markers for the PsaA-side or PsaB-side cofactors, as there are, for example, in the bacterial RC.

The redox potential of PhQ (oxidized versus semiquinone) in the A_1 site is estimated to be <–800 mV (versus H_2), making it one of the most reducing quinones in biology (73). Although the factors that lead to this low reduction potential are only now being systematically explored, they most certainly lie in protein environment rather than in the chemical identity of the quinone. This was illustrated in mutants of *Synechocystis* sp. PCC 6803 in which the biosynthetic pathway of PhQ was interrupted (32, 33, 86). Although plastoquinone-9 (PQ-9) rather than PhQ is present in the Q_K-A and Q_K-B sites of these mutants, forward electron transfer occurs with high efficiency (62). What makes this interesting is that PQ-9 has a redox potential of approximately 0 to -150 mV when it is present in the Q_A site in PS II (24). In PS I, the replacement of PhQ by PQ-9 induces a decrease in the free energy gap between A_1 and F_A/F_B from ~-170 mV to ~-35 mV (63). Thus, PQ-9 in the A_1 site of PS I has a redox potential of ~-650 mV. The difference of $\sim$700 mV between the redox potential of PQ-9 in PS I and in PS II is due solely to the influence of protein. It should be noted that when demethyl-PhQ is incorporated into the quinone binding site, the redox potential of A_1 is driven 50 to 60 mV more oxidizing than the wild type, but there is no diminution in either the growth rate or the quantum yield of ferredoxin reduction (59).

The A_1 site was initially probed in a series of replacement experiments in which chemically extracted PS I complexes were reconstituted with various benzoquinones, naphthoquinones, and anthraquinones (31). From these studies it was proposed that the PS I quinone is stabilized by the hydrocarbon chain, by hydrophobic interactions, and by $\pi-\pi$ interactions (31). Magnetic resonance spectroscopy provided additional structural details of the quinone binding site. ENDOR and ESEEM studies placed a Trp and/or a His close to the semiquinone radical (28). Transient EPR studies showed that the PhQ anion radical is oriented with its carbonyl bonds parallel to the vector joining $P700^+$and A_1^- (66, 69). The angle between $P700^+$-A_1^- and the membrane normal was found to be 27 ± 5 degrees in ESEEM measurements of PS I crystals (4), a value that agrees with the 25°–30°

angle determined by transient EPR studies of PS I crystals (35). The distance between the P700$^+$ and A_1^- radicals was found to be 25.4 ± 0.3 Å by ESEEM measurement of the dipolar coupling between the two radicals (4). The angle and distance [see also (42)] define a circle, which when superimposed on the 4 Å model of PS I (61) results in the localization of the EPR-detectable PhQ at the intersection of a surface α-helix and a transmembrane α-helix (84).

These predictions were confirmed in the current X-ray crystal structure of PS I (34); the PhQs are located where expected and π-stacked with Trp697$_{PsaA}$ and with Trp677$_{PsaB}$. In this context, it is interesting that quantum chemical calculations show that unlike neutral quinones, which prefer a π-stacked geometry with indoles, semiquinone radical anions prefer a T-stacked geometry (36). It is possible that the hydrocarbon tail is required so that the semiquinone radical retains a π-stacked orientation with the tryptophan in the A_1 site. This should destabilize the radical, thereby contributing to the highly negative redox potential that allows the semiquinone to pass the electrons forward to F_X. The X-ray crystal structure further shows that the distance between eC-A1/Q_K-B and eC-B1/Q_K-A is 26.0 ± 0.3 Å, that is, between the A-side quinone and the B-side Chl of P700 and visa versa; and 24.2 ± 0.3 Å between eC-A1/Q_K-A and 24.4 ± 0.3 Å between eC-B1/Q_K-B, that is, between quinones and Chls on the same branches. Hence, the data support the localization of P700$^{\cdot+}$ and $A_1^{\cdot-}$ on cofactors that reside on different subunits, which is also the case in the bacterial RC. These findings represent an impressive application of X-ray crystallography and magnetic resonance spectroscopy to solve a problem that neither is capable of solving alone.

Although ENDOR studies had implicated two strong H-bonds to the EPR-visible PhQ (54), electron spin-polarized EPR studies indicated only one highly asymmetrical H-bond (69), quite unlike the two H-bonds to Q_A in the bacterial RC. This agrees qualitatively with high-field EPR studies in which the g-anisotropy of the A_1^- radical was noted to be larger than that of the Q_A^- radical in the bacterial RC and was consistent with less H-bonding or stronger π-bonding to an aromatic amino acid (or both) (42, 69). This assessment is confirmed in the X-ray crystal structure, which shows that only the carbonyl oxygen ortho to the phytyl tail is strongly H-bonded to a backbone peptide NH from Leu722$_{PsaA}$ and Leu706$_{PsaB}$. There does exist a Ser OH approximately 3.1 Å distant from the carbonyl oxygen ortho to the methyl group of both quinones; however, the angle is not consistent with a bona fide H-bond. Moreover, this Ser692$_{PsaA}$ is H-bonded to Trp697$_{PsaA}$, which is π-stacked with PhQ, as well as to the peptide carbonyl from Met688$_{PsaA}$, which provides the axial ligand to eC-A3. Similarly, this Ser672$_{PsaB}$ is H-bonded to Trp677$_{PsaB}$, which is π-stacked with PhQ, as well as to the peptide carbonyl from Met668$_{PsaB}$, which provides the axial ligand to eC-B3. Thus, there exists a near-continuous series of H-bonds from A_0 to F_X that may constitute a favorable electron transfer pathway in PS I (Figure 2, see color insert).

Given the high degree of symmetry in PS I, it is relevant to ask whether Q_K-A, Q_K-B, or both are active in electron transfer. The participation of Q_K-A in forward electron transfer was first demonstrated in a PsaE/PsaF deletion mutant of

Synechococcus sp. PCC 7002 (82). EPR studies showed that A_0^- rather than A_1^- became photoaccumulated when PS I complexes from this mutant were illuminated at 220 K in the presence of sodium dithionite and Triton X-100. This result was explained by the opening of a water channel to the quinone; the availability of a proton would lead to the facile double reduction of PhQ, thereby forcing A_0^- to become accumulated. Because PsaE and PsaF are associated with the PsaA subunit, this experiment shows that Q_K-A functions in forward electron transfer. Because the spectrum was devoid of A_1^-, this result also implies that Q_K-B is not involved in forward electron transfer.

It has long been known from optical kinetic measurements that there exist fast (~10 ns) and slow (~200 ns) kinetic phases in the forward electron transfer from A_1^- to F_X (8). Transient EPR measurements confirm the presence of the slow phase, but the risetime of the spectrometer prevents direct measurement of the fast phase (68). The two kinetic phases are due to either (*a*) bidirectional electron transfer up the PsaA-side and PsaB-side cofactors, or (*b*) unidirectional electron transfer involving either two populations of A_1 or an equilibrium between A_1^- and F_X (8). To decide between these possibilities, the π-stacked Trp residues ($W697_{PsaA}$ and $W677_{PsaB}$ in *S. elongatus*) were changed to Phe in *C. reinhardtii* (26, 50) and in *Synechocystis* sp. PCC 6803 (25). The idea was to selectively alter the environment of the quinones in order to cause a change in the rate of electron transfer rate from A_1^- to the iron-sulfur clusters. In the PsaA-side mutant, the optically measured fast kinetic phase was unaltered but the slow kinetic phase slowed in both studies. Similarly, in the PsaB-side mutant, the optically measured slow kinetic phase was unaltered but the fast kinetic phase slowed. In a second set of mutants, the H-bonded Ser residues ($S692C_{PsaA}$, $S672C_{PsaB}$ in *S. elongatus*) were changed to Cys in *Synechocystis* sp. PCC 6803. The idea was to alter the strength of the H-bond to the π-stacked Trp, thereby indirectly affecting the properties of the quinones. In general, the optically measured kinetic phases of the Ser mutants showed a pattern similar to the Trp mutants (W. Xu, T. Shen, K. Brettel, M. Guergova-Kuras, P. Chitnis & J.H. Golbeck, unpublished observations). All of the above results can be interpreted in terms of a bidirectional electron transfer scheme.

The Trp and Ser mutants in *Synechocystis* sp. PCC 6803 were also studied by transient EPR and ENDOR spectroscopy (25). The PsaA-side mutants show the expected slowing of the slow kinetic component of A_1^- oxidation, but the PsaB-side mutant remained identical to the wild type in spite of the fact that PsaB-side electron transfer was expected to have a detectable influence on the spin polarization patterns (25). The combined effect of low temperature and mutations would be expected to slow the fast kinetic phase to times that are within the window accessible by transient EPR, yet at 260 K, the spectra and kinetic curves remained unaltered. Additionally, the PsaA-side Trp mutation in *C. reinhardtii* resulted in the inability to photoaccumulate A_1^- and produced spectral changes in the transient EPR spectrum of the $P700^+ A_1^-$ radical pair, whereas there was no effect in the PsaB-side mutant (7). This result confirms that the quinone measured by photoaccumulation is the same quinone measured by transient EPR spectroscopy, i.e., Q_K-A. Hence, while transient EPR studies are consistent with PsaA-side electron transfer, there is no evidence yet for PsaB-side electron transfer. More work is obviously needed to reconcile the optical and magnetic resonance results.

THE INTERPOLYPEPTIDE F_X CLUSTER

The electron acceptor F_X was discovered by EPR spectroscopy (16) and identified as a [4Fe-4S] cluster by Mössbauer and EXAFS spectroscopy (15, 45, 49). Its proposed function as an intermediate electron carrier between A_1 and the F_A/F_B iron-sulfur clusters was proven by transient EPR studies of P700-F_X cores (46, 68). The low number of Cys residues (four on PsaA and two on PsaB in maize PS I (20), along with the presence of the highly conserved sequences ^{576}FPCDGPGRGGTCQ^{588} on PsaA and ^{563}FPCDGPGRGGTCD^{575} on PsaB (*S. elongatus* numbering), led to the proposal that F_X is an interpolypeptide [4Fe-4S] cluster positioned between PsaA and PsaB, with ligands provided by both subunits (23, 30). The predicted interpolypeptide location of F_X was confirmed in the 6 Å model of PS I (39). However, because Asp as well as Cys residues can ligate [4Fe-4S] clusters (11) and because three Asp residues are also present in the conserved sequences, the identity of the ligands to F_X remained uncertain.

The ligands to F_X were probed by site-directed mutagenesis and EPR spectroscopy. When the PsaB-side residues Cys556$_{PsaB}$ and Cys565$_{PsaB}$ in *Synechocystis* sp. PCC 6803 (Cys565$_{PsaB}$ and Cys574$_{PsaB}$ in *S. elongatus*) were changed to Ser (64, 74), the EPR spectrum of F_X changed from a highly anisotropic spectrum with *g*-values of 2.09, 1.86, and 1.75 in the wild type to a less anisotropic spectrum with *g*-values of 2.05, 1.94, and 1.81 in the mutants (74). The identical EPR spectrum of the two Ser mutants is probably due to the high degree of symmetry surrounding the F_X site. In both mutants, the mixed-ligand (3 Cys•1 Ser) cluster transferred electrons less efficiently than the wild type to F_A and F_B, resulting in backreactions from the earlier acceptors, F_X^- and A_1^- (71). Asp and His residues do not appear capable of providing ligands for F_X: A Cys565Asp$_{PsaB}$ mutant and a Cys565His$_{PsaB}$ mutant of *Synechocystis* sp. PCC 6803 (Cys574$_{PsaB}$ in *S. elongatus*) resulted in no detectable F_A/F_B and reduced levels of PS I (64), and a Cys561His$_{PsaB}$ mutant of *C. reinhardtii* (Cys565$_{PsaB}$ in *S. elongatus*) did not accumulate detectable levels of PS I (77). The 2.5 Å crystal structure (34) confirmed that Cys578$_{PsaA}$, Cys587$_{PsaA}$, Cys565$_{PsaB}$, and Cys574$_{PsaB}$ provide the ligands to F_X.

The redox potential of F_X was measured to be −705 mV by electrochemical titration and EPR detection (10), making it one of the most reducing [4Fe-4S]$^{1+/2+}$ clusters in biology. The factors that lead to this highly reducing potential are largely unknown. One idea was that if the Asp residues that flank three of the four Cys residues would harbor an uncompensated charge, the electrostatic interaction with the [4Fe-4S]$^{2+/1+}$ cluster might drive F_X more electronegative. This idea was tested using site-directed mutagenesis and EPR spectroscopy. When Asp576$_{PsaA}$ in *C. reinhardtii* (Asp579$_{PsaA}$ in *S. elongatus*) was changed to Leu, the organism grew on light, PS I assembled, and the g-tensor of F_A/F_B was altered (27). In contrast, the PS I complex did not assemble when $Asp562_{PsaB}$ in *C. reinhardtii* (Asp566 in *S. elongatus*) was changed to Asn (58). When $Asp557_{PsaB}$ and $Asp566_{PsaB}$ in *Synechocystis* sp. PCC 6803 ($Asp566_{PsaB}$ and $Asp575_{PsaB}$ in *S. elongatus*) were changed to Ala and Lys either separately or in pairs (72), the organism grew on light at impaired rates, yet the rates of $NADP^+$ photoreduction were only slightly lower than the wild type. It was concluded that the Asp residues do not significantly influence the redox properties of F_X. Instead, the impaired growth was found to be related to a reduced level of PS I. Most likely, during biogenesis, protein turnover machinery targets the altered PS I for proteolysis, and the resulting imbalance between synthesis and turnover lowers a steady-state concentration in the cell.

In hindsight, the phenotypes of these mutants are not surprising (65). The X-ray crystal structure shows that $Asp579_{PsaA}$ is salt-bridged internally to $Arg583_{PsaA}$ and externally to $Arg52_{PsaC}$ (Figure 3, see color insert). Similarly, $Asp566_{PsaB}$ is salt-bridged internally to $Arg570_{PsaB}$ and externally to $Arg65_{PsaC}$ and $Lys51_{PsaC}$. Cys-proximal $Asp575_{PsaB}$ (which has no counterpart on PsaA) ties together the stromal end of the C-terminal transmembrane α-helix with the F_X loop. Thus, the purpose of $Asp579_{PsaA}$ and $Asp566_{PsaB}$ is to bind PsaC, and the purpose of $Asp575_{PsaB}$ is to confer internal stability. The change of $Asp566_{PsaB}$ to either Ala or Lys should disrupt salt-bridges to $Lys51_{PsaC}$ and to $Arg65_{PsaC}$, which should affect PsaC stability. Yet, the PS I complex functions in electron transfer, indicating that the loss of these bonds is not sufficient to disrupt the binding of PsaC.

In nearly every known [4Fe-4S] protein that functions in electron transfer, a Pro is located adjacent to one of the Cys residues. The ubiquity of this motif might indicate an indispensable role for Pro in stabilizing a [4Fe-4S] cluster. A conserved Pro is also located at the beginning the F_X binding loop on both PsaA and PsaB; however, when $Pro560_{PsaB}$ in *C. reinhardtii* ($Pro564_{PsaB}$ in *S. elongatus*) was changed to Leu, there was no effect on the amount or function of PS I (58, 77). A second Pro residue occurs in the F_X binding loop between the Cys ligands. When $Pro564_{PsaB}$ in *C. reinhardtii* ($Pro568_{PsaB}$ in *S. elongatus*) was changed to Ala or Leu, there was likewise no loss of amount or function of PS I. The only phenotype displayed in the Pro mutants was that PsaC dissociated more readily in urea, and in vitro reconstitution using mutant PS I cores and PsaC was incomplete (58).

The X-ray crystal structure shows that $Pro564_{PsaB}$ is located at the base of the F_X binding loop and is H-bonded through the backbone O to $Arg712_{PsaB}$, thereby tying together two transmembrane α-helices. $Pro568_{PsaB}$ is located near the apex of the F_X loop and has no bonds to any other residue. Pro residues are frequently found on the outside of proteins, and the four Pro residues present on the F_X binding loop may provide a large surface that, among other things, protects F_X from dioxygen. These loops may be motionally constrained owing to the presence of the [4Fe-4S] cluster, the Pro residues, the internal salt-bridge between $Arg583_{PsaA}$ ($Arg570_{PsaB}$) and $Asp579_{PsaA}$ ($Asp566_{PsaB}$), and the H-bonds between the loop residues. Hence, the F_X binding region is different from a traditional [4Fe-4S] binding motif in that each loop is best considered as a facet of a surface on which PsaC is bound. Indeed, PsaC does not bind to P700-F_X cores in the absence of F_X (J. Golbeck, unpublished data).

On the basis of the primary amino acid sequence, Arg570$_{PsaB}$ and Arg583$_{PsaA}$ were initially proposed to be involved in promoting interaction between the PsaA/PsaB heterodimer and PsaC by virtue of being located in the middle of the F_X binding loop (56). Altering the charge on these residues tested this idea. Although no PS I was found in the Arg566Glu$_{PsaB}$ mutant of *C. reinhardtii* (Arg570Glu$_{PsaB}$ in *S. elongatus*) (58), PS I did assemble in the analogous Arg561Glu$_{PsaB}$ mutant of *Synechocystis* sp. PCC 6803 (57). The stromal subunits, including PsaC, were more easily removed by urea, and reconstitution experiments with PsaC were more efficient in the presence of divalent cations. Thus, the data supported the proposal that the Glu replacement offered an unfavorable electrostatic interaction between surface-exposed residues on PsaC and on PS I cores (57).

The idea that PsaA/PsaB and PsaC are bound by electrostatic interactions is nicely supported by the X-ray crystal structure. However, opposite to prediction, Arg570$_{PsaB}$ forms an internal salt-bridge to Asp566$_{PsaB}$, which in turn is salt-bridged to Arg65$_{PsaC}$ and Lys51$_{PsaC}$ (Figure 3), thereby providing a part of the interface between PsaC and the PsaA/PsaB heterodimer. The PsaA subunit has a comparable set of contacts; Arg583$_{PsaA}$ forms an internal salt-bridge to Asp579$_{PsaA}$, which in turn is salt-bridged to Arg52$_{PsaC}$. The function of the basic Arg residues on PsaA/PsaB may be to position the acidic Asp residues on PsaA/PsaB so that they are in the proper place to form salt-bridges with the basic Arg residues on PsaC. Although the proposal that the Arg residues are directly involved in binding PsaC through carboxylate groups has not proven to be correct, the X-ray data show that these residues do have an indirect role in the binding of PsaC. The substitution of an acidic Glu residue at position 570 on PsaB could influence the orientation of Asp566$_{PsaB}$ and thereby diminish the strength of binding to the appropriate residues on PsaC. A weakening of this bond might explain why urea is able to remove the PsaC subunit easily in the Arg570Glu$_{PsaB}$ mutant, and why divalent cations are able to stabilize the binding of PsaC on mutant PS I cores.

STUDIES OF STROMAL SUBUNIT PsaC

F_A and F_B were discovered in the early 1970s (3, 17) and were shown to be [4Fe-4S] clusters (9) bound to a single 9-kDa peripheral membrane polypeptide (29) termed PsaC (6). The [4Fe-4S] clusters are spectroscopically distinct when PsaC is bound to PS I; the *g*-tensor of F_A is characterized by resonances at $g = 2.05$, 1.95, and 1.85, and that of F_B by resonances at $g = 2.07$, 1.93, and 1.88. When both clusters are reduced, a so-called interaction spectrum results with resonances at $g = 2.05$, 1.94, 1.92, and 1.86. This new spectrum is the result of an additional term in the spin Hamiltonian that is necessary due to the magnetic coupling of two closely spaced clusters.

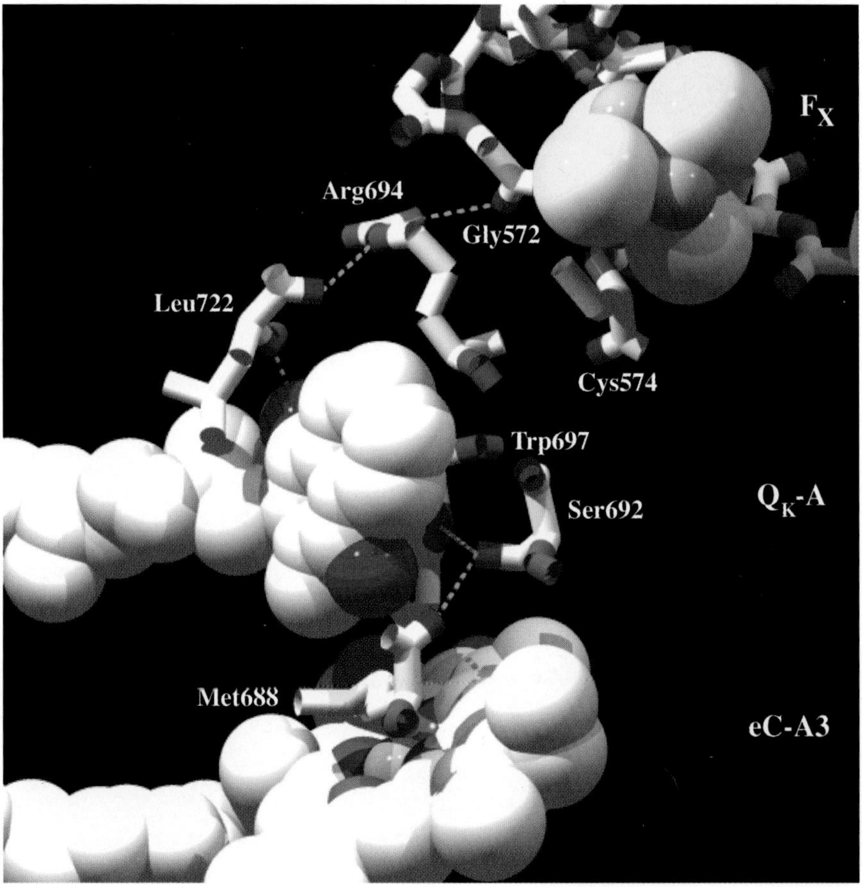

Figure 2 The electron transfer cofactors eC-A3, Q_K-A, and F_X (in CPK form) and surrounding amino acids (in stick form) on the PsaA side of PS I. H-bonds are depicted in green and are proposed to be involved in a favorable electron transfer pathway from A_0 to F_X.

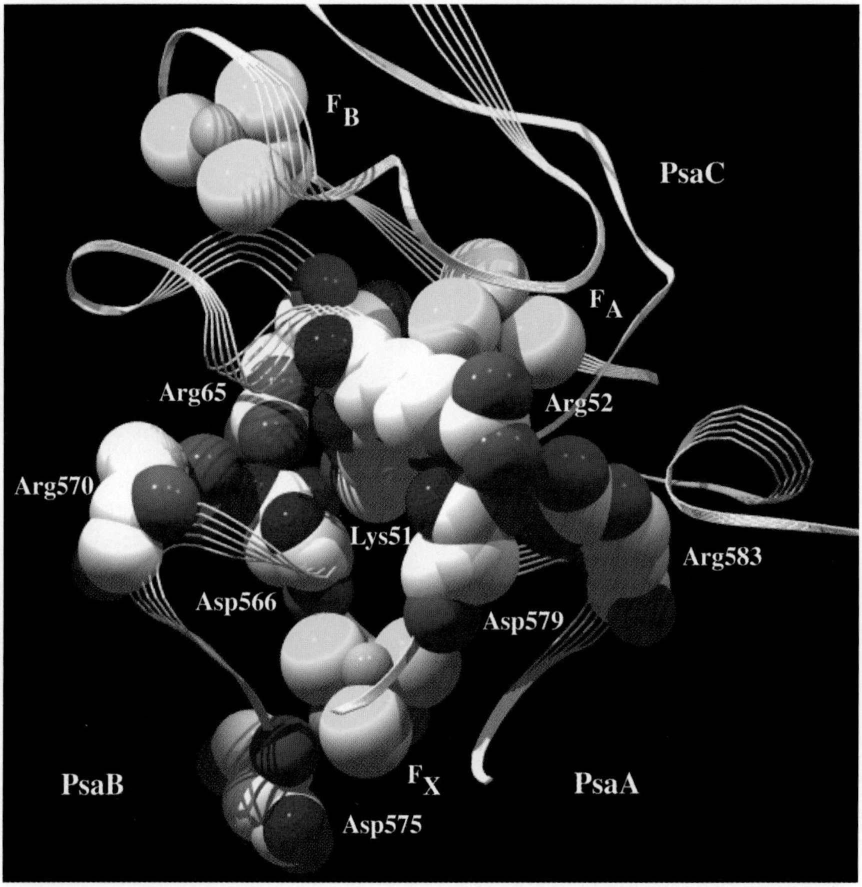

Figure 3 Amino acids involved in the contact area between PsaC and PsaA/PsaB. Note that $Asp555_{PsaB}$, which is located on a neighboring α-helix, is additionally salt-bridged to $Arg65_{PsaC}$, and $Asp568_{PsaA}$, which is located on a neighboring α-helix, is additionally salt-bridged to $Arg52_{PsaC}$. These residues are not shown for reasons of clarity.

PsaC belongs to the family of low-molecular-weight bacterial dicluster ferredoxins that contain two classic [4Fe-4S] binding motifs CxxCxxCxxxCP separated by the one-and-one-half-turn α-helices that rigidly separate the [4Fe-4S] clusters by 12 Å. The F_A and F_B clusters in PsaC have the task of vectoring the electron from the interpolypeptide [4Fe-4S] center F_X to the soluble electron carrier ferredoxin. It is tempting to speculate that F_X may have been the original terminal acceptor and that a bacterial ferredoxin was later recruited to serve as an additional helper protein. This protein ultimately became PsaC. Its function may have been to increase the lifetime of the charge separation when proteins such as the [2Fe-2S] ferredoxin evolved to serve as the soluble electron acceptor. Interruption mutants of PsaC in *Synechocystis* sp. PCC 6803 do not live on light, and although PS I complexes contain cofactors up to and including F_X, there is no detectable reduction of $NADP^+$ (83). The two additional stromal proteins PsaD and PsaE are also missing from PsaC deletion mutants (44, 83). PsaC and PsaD can be rebound to P700-F_X cores without addition of energy and without the participation of chaperones (83). These results can be understood in the context of the X-ray crystal structure (2, 65). PsaC is bound to the PsaA/PsaB heterodimer primarily through ionic contacts. PsaD binds to PsaB, but a long stretch of amino acids, termed a clamp, runs over the entire surface of PsaC and fastens with an ionic contact on the opposite side on PsaA. Thus, the three-dimensional structure of PsaD is dependent on PsaC. PsaD may fall into the classification of natively unfolded protein (81), which attains a proper three-dimensional structure only when it is bound to PS I.

The three-dimensional structure of PsaC is known both in the unbound state (1, 2) and when bound to the PS I complex (34, 65). Prior to the determination of these structures, certain central structural features common to all dicluster ferredoxins were assumed to apply to PsaC (13); these include a local pseudo-C_2-symmetry around both the iron-sulfur cluster binding motifs and, to a lesser extent, other regions of the protein. In light of the known structures for PsaC, bacterial ferredoxins have proven to be good models. However, the pseudo-C_2-symmetry of PsaC leads to an interesting problem, i.e., which cluster represents F_A and which cluster represents F_B. This problem was solved using site-directed mutagenesis and EPR spectroscopy. PsaC was engineered so that the second Cys in each [4Fe-4S] binding motif was changed to Asp, which has the effect of altering the ground spin state from $S = 1/2$ to $S \geq 3/2$. The mutant PsaC proteins were reconstituted onto P700-F_X cores in the presence of PsaD, and the EPR spectrum of F_A and F_B was measured. Because the unaltered clusters retain a net ground spin state of $S = 1/2$, it could be deduced that Cys10, 13, 16, and 57 provide the ligands to F_B, and Cys47, 50, 53, and 20 provide the ligands to F_A (85). This is yet another instance in which structural information combined with magnetic resonance data answered a question that neither could have addressed separately.

The 6 Å model of PS I (39) revealed yet another symmetry-related issue: the twofold pseudorotation axis, which relates PsaA and PsaB, intersects the local twofold C_2 axis of PsaC at an angle of 62 degrees, and places one of the two iron-sulfur clusters, F_A or F_B, closer to F_X and the other cluster closer to the stromal surface (Figure 3). In principle, the orientation of PsaC could be solved by X-ray crystallography, but the resolution of the 6 Å and 4 Å maps (37, 40) was insufficient to resolve the nonsymmetrical elements in PsaC such as the extended loop structure between the iron-sulfur clusters or the C-terminal extension. In the end, this problem was solved using biophysical techniques. One approach used the known ability of $HgCl_2$ to selectively denature the F_B cluster (60). Optical kinetics, photovoltage, and EPR saturation studies of the F_B-deficient PS I complex all indicated that F_A is proximal to F_X [reviewed in (70)]. A second approach compared the EPR spin states of site-directed mutants of PsaC prepared in vivo (in *Synechocystis* sp. PCC 6803) with those prepared in vitro (PsaC expressed in *Escherichia coli* and rebound in vitro to P700-F_X cores). This analysis [summarized in (22)] also concluded that F_A is proximal to F_X. The X-ray crystal structure confirms the orientation PsaC deduced from the biochemical and biophysical studies: PsaC binds to PsaA/PsaB such that the iron-sulfur cluster ligated by cysteines 20, 47, 50, and 53 (known to be F_A from the mutagenesis studies) is proximal to F_X, and the iron-sulfur cluster ligated by cysteines 10, 13, 16, and 57 (known to be F_B from the mutagenesis studies) is proximal to the ferredoxin binding site.

A few mutants were constructed prior to the X-ray crystal structure that examined possible binding contacts between PsaC and PsaA/PsaB (Figure 3). With the premise that charged amino acids are involved, a series of eight Asp-to-Arg and Glu-to-Arg changes were made in PsaC, and the efficiency of PsaC rebinding was measured. The mutants $Asp8Arg_{PsaC}$, $Asp8Ala_{PsaC}$, $Glu26Arg_{PsaC}$, and $Asp31Arg_{PsaC}$ resulted in a lower level of PsaC binding. These amino acids were therefore implicated in electrostatic interactions with the PsaA/PsaB core (55). However, the X-ray crystal structure shows that none of these basic amino acids are salt-bridged to acidic residues on PsaA and PsaB (Figure 3). The reason for the lower efficiency of reconstitution of some mutants of PsaC is not entirely clear; for example, $Asp8_{PsaC}$ forms bonds only with $Tyr34_{PsaE}$ and $Arg117_{PsaD}$ and has no contacts with PsaA/PsaB. In retrospect, the idea that the binding contacts between PsaC and PsaA/PsaB are electrostatic in nature (56) has proven to be entirely correct; only the details of the residues involved were mistaken.

An attempt was made to alter the redox potential of the iron-sulfur clusters by targeting the charged pair, $Lys51_{PsaC}$ and $Arg52_{PsaC}$, located near F_A in PsaC. The background to these experiments is that Ser and Ala are present in comparable positions in the PscB protein of *Chlorobium limicola*, and the higher redox potential of F_B is higher than that of the comparable cluster in PsaC. To test whether basic residues in PsaC are involved in poising the redox potential of the iron-sulfur clusters, $Lys51Ser/Arg52Asp_{PsaC}$ and $Lys51Ser/Arg52Ala_{PsaC}$ mutants were constructed in *C. reinhardtii* (19). Both mutants grew under anaerobic conditions, and in contrast to the wild type, F_B was preferentially photoreduced in the latter mutant. The latter was proposed to be due to changes in the redox potential of F_A and/or structural modification of the PsaC subunit. However, the X-ray crystal structure (Figure 3) shows that $Lys51_{PsaC}$ is involved in a salt-bridge with $Asp566_{PsaB}$ and $Arg52_{PsaC}$ is involved in a salt-bridge with $Asp579_{PsaA}$ and $Asp568_{PsaA}$ (65). These contacts would obviously be lost. The salt-bridges remaining would be the contacts between $Arg65_{PsaC}$ and $Asp566_{PsaB}$, and those between $Arg65_{PsaC}$ and $Asp555_{PsaB}$ [three H-bonds between C-terminal amino acids on PsaC and PsaB would also be retained (65)]. It is thus remarkable that a large number of ionic contacts between PsaC and the PsaA/PsaB heterodimer can be broken, and there still remains a population of PS I with bound and functional PsaC. This speaks to a high degree of redundancy in the binding of this PS I subunit.

CLOSING OBSERVATIONS

One of the lessons learned in PS I is that mutagenesis experiments yield the greatest insights when an analog exists with known structures. In the absence of structural clues, these types of experiments have less success in uncovering important structural details. Now that the atomic-resolution X-ray crystal structure is available, the challenge changes to one of unraveling how the protein environment modulates the thermodynamic and kinetic properties of the electron transfer cofactors.

LITERATURE CITED

1. Antonkine ML, Bentrop D, Bertini I, Luchinat C, Shen G, et al. 2000. Paramagnetic 1H NMR spectroscopy of the reduced, unbound photosystem I subunit PsaC: sequence-specific assignment of contact-shifted resonances and identification of mixed- and equal-valence Fe-Fe pairs in [4Fe-4S] centers FA- and FB-. *J. Biol. Inorg. Chem.* 5:381–92
2. Antonkine ML, Liu G, Bentrop D, Bryant DA, Bertini I, et al. 2002. Solution structure of the unbound, oxidized photosystem I subunit PsaC, containing [4Fe-4S] clusters F_A and F_B: A conformational change occurs upon binding to photosystem I. *J. Biol. Inorg. Chem.* 7:461–72
3. Bearden AJ, Malkin R. 1972. Quantitative EPR studies of the primary reaction of photosystem I in chloroplasts. *Biochim. Biophys. Acta* 283:456–68
4. Bittl R, Zech SG, Fromme P, Witt HT, Lubitz W. 1997. Pulsed EPR structure analysis of photosystem I single crystals: localization of the phylloquinone acceptor. *Biochemistry* 36:12001–4
5. Bonnerjea J, Evans MCW. 1982. Identification of multiple components in the intermediary electron carrier complex of photosystem I. *FEBS Lett.* 148:313–16
6. Bordier Høj P, Svendsen I, Vibe Scheller H, Lindberg Møller B. 1987. Identification of a chloroplast-encoded 9-kDa polypeptide as a 2[4Fe-4S] protein carrying centers A and B of photosystem I. *J. Biol. Chem.* 262:12676–84
7. Boudreaux B, MacMillan F, Teutloff C, Agalarov R, Gu F, et al. 2001. Mutations in both sides of the photosystem I reaction center identify the phylloquinone observed by electron paramagnetic resonance spectroscopy. *J. Biol. Chem.* 276:37299–306
8. Brettel K, Leibl W. 2001. Electron transfer in photosystem I. *Biochim. Biophys. Acta* 1507:100–14
9. Cammack R, Evans MCW. 1975. EPR spectra of iron-sulfur proteins in dimethyl

sulfoxide solutions. Evidence that chloroplast photosystem I particles contain 4Fe-4S centers. *Biochem. Biophys. Res. Commun.* 67:544–49
10. Chamorovsky SK, Cammack R. 1982. Direct determination of the midpoint potential of the acceptor X in chloroplast photosystem I by electrochemical reduction and ESR spectroscopy. *Photobiochem. Photobiophys.* 4:195–200
11. Conover RC, Kowal AT, Fu WG, Park JB, Aono S, et al. 1990. Spectroscopic characterization of the novel iron-sulfur cluster in *Pyrococcus furiosus* ferredoxin. *J. Biol. Chem.* 265:8533–41
12. Davis IH, Heathcote P, Maclachlan DJ, Evans MCW. 1993. Modulation analysis of the electron spin echo signals of *in vivo* oxidised primary donor N-14 chlorophyll centres in bacterial, P870 and P960, and plant photosystem-I, P700, reaction centres. *Biochim. Biophys. Acta* 1143:183–89
13. Dunn PPJ, Gray JC. 1988. Localization and nucleotide sequence of the gene for the 8 kDa subunit of photosystem I in pea and wheat chloroplast DNA. *Plant Mol. Biol.* 11:311–19
14. Ermler U, Fritzsch G, Buchanan SK, Michel H. 1994. Structure of the photosynthetic reaction centre from *Rhodobacter sphaeroides* at 2.65 Å resolution: cofactors and protein-cofactor interactions. *Structure* 2:925–36
15. Evans EH, Dickson DPE, Johnson CE, Rush JD, Evans MCW. 1981. Mössbauer spectroscopic studies of the nature of center X of photosystem I reaction centers from the cyanobacterium *Chlorogloea fritschii*. *Eur. J. Biochem.* 118:81–84
16. Evans MCW, Sihra CK, Cammack R. 1976. The properties of the primary electron acceptor in the photosystem I reaction center of spinach chloroplasts and its interaction with P700 and the bound ferredoxin in various oxidation-reduction states. *Biochem. J.* 158:71–77
17. Evans MCW, Telfer A, Lord AV. 1972. Evidence for the role of a bound ferredoxin as the primary electron acceptor of photosystem I in spinach chloroplasts. *Biochim. Biophys. Acta* 267:530–37
18. Fairclough WV, Evans MCW, Purton S, Jones S, Rigby SEJ, Heathcote P. 2001. *Site-directed mutagenesis of PsaA:M684 in* Chlamydomonas reinhardtii. Presented at PS2001 12th Int. Congr. Photosynthesis, Melbourne, Aust.
19. Fischer N, Sétif P, Rochaix JD. 1997. Targeted mutations in the psaC gene of *Chlamydomonas reinhardtii*: Preferential reduction of F-B at low temperature is not accompanied by altered electron flow from photosystem I ferredoxin. *Biochemistry* 36:93–102
20. Fish LE, Kuck U, Bogorad L. 1985. Two partially homologous adjacent light-inducible maize chloroplast genes encoding polypeptides of the P700 chlorophyll a-protein complex of photosystem I. *J. Biol. Chem.* 260:1413–21
21. Golbeck JH. 1987. Structure, function and organization of the photosystem I reaction center complex. *Biochim. Biophys. Acta* 895:167–204
22. Golbeck JH. 1999. A comparative analysis of the spin state distribution of *in vivo* and *in vitro* mutants of PsaC. A biochemical argument for the sequence of electron transfer in photosystem I as $F_X \rightarrow F_A \rightarrow F_B \rightarrow$ ferredoxin/flavodoxin. *Photosyn. Res.* 61: 107–49
23. Golbeck JH, Cornelius JM. 1986. Photosystem I charge separation in the absence of centers A and B. I. Optical characterization of center 'A_2' and evidence for its association with a 64-kDa polypeptide. *Biochim. Biophys. Acta* 849:16–24
24. Golbeck JH, Kok B. 1979. Redox titration of electron acceptor Q and the plastoquinone pool in photosystem II. *Biochim. Biophys. Acta* 547:347–60
25. Golbeck JH, Xu W, Zybailov B, van der Est A, Pushkar J, et al. 2001. *Electron transfer through the quinone acceptor in cyanobacterial photosystem I.* Presented at

PS2001 12th Int. Congr. Photosynthesis, Melbourne, Aust.

26. Guergova-Kuras M, Boudreaux B, Joliot A, Joliot P, Redding K. 2001. Evidence for two active branches for electron transfer in photosystem I. *Proc. Natl. Acad. Sci. USA* 98:4437–42
27. Hallahan BJ, Purton S, Ivison A, Wright D, Evans MCW. 1995. Analysis of the proposed Fe-Sx binding region of photosystem I by site-directed mutation of PsaA in *Chlamydomonas reinhardtii*. *Photosynth. Res.* 46:257–64
28. Hanley J, Deligiannakis Y, MacMillan F, Bottin H, Rutherford AW. 1997. ESEEM study of the phyllosemiquinone radical $A_1^{\cdot-}$ in N-14- and N-15-labeled photosystem I. *Biochemistry* 36:11543–49
29. Hayashida N, Matsubayashi T, Shinozaki K, Sugiura M, Inoue K, Hiyama T. 1987. The gene for the 9 kd polypeptide, a possible apoprotein for the iron-sulfur centers A and B of the photosystem I complex, in tobacco chloroplast DNA. *Curr. Genet.* 12:247–50
30. Høj PB, Møller BL. 1986. The 110-kDa reaction center protein of photosystem I, P700-chlorophyll a-protein 1, is an iron-sulfur protein. *J. Biol. Chem.* 261:14292–300
31. Itoh S, Iwaki M, Ikegami I. 2001. Modification of photosystem I reaction center by the extraction and exchange of chlorophylls and quinones. *Biochim. Biophys. Acta* 1507:115–38
32. Johnson TW, Shen G, Zybailov B, Kolling D, Reategui R, et al. 2000. Recruitment of a foreign quinone into the A_1 site of photosystem I. I. Genetic and physiological characterization of phylloquinone biosynthetic pathway mutants in *Synechocystis* sp. PCC 6803. *J. Biol. Chem.* 275:8523–30
33. Johnson TW, Zybailov B, Jones AD, Bittl R, Zech S, et al. 2001. Recruitment of a foreign quinone into the A_1 site of photosystem I. *In vivo* replacement of plastoquinone-9 by media-supplemented naphthoquinones in phylloquinone biosynthetic pathway mutants of *Synechocystis* sp. PCC 6803. *J. Biol. Chem.* 276:39512–21
34. Jordan P, Fromme P, Witt HT, Klukas O, Saenger W, Krauß N. 2001. Three dimensional structure of photosystem I at 2.5 Å resolution. *Nature* 411:909–17
35. Kamlowski A, Altenberg-Greulich B, van der Est A, Zech SG, Bittl R, et al. 1998. The quinone acceptor A_1 in photosystem I: binding site, and comparison to Q_A in purple bacterial reaction centers. *J. Phys. Chem. B* 102:8278–87
36. Kaupp M. 2002. The function of photosystem I. Quantum chemical insight into the role of tryptophan-quinone interactions. *Biochemistry* 41:2895–900
37. Klukas O, Schubert WD, Jordan P, Krauß N, Fromme P, et al. 1999. Photosystem I, an improved model of the stromal subunits PsaC, PsaD, and PsaE. *J. Biol. Chem.* 274:7351–60
38. Krabben L, Schlodder E, Jordan R, Carbonera D, Giacometti G, et al. 2000. Influence of the axial ligands on the spectral properties of P700 of photosystem I: a study of site-directed mutants. *Biochemistry* 39:13012–25
39. Krauß N, Hinrichs W, Witt I, Fromme P, Pritzkow W, et al. 1993. 3-Dimensional structure of system-I of photosynthesis at 6 Å resolution. *Nature* 361:326–31
40. Krauß N, Schubert WD, Klukas O, Fromme P, Witt HT, Saenger W. 1996. Photosystem I at 4 Å resolution represents the first structural model of a joint photosynthetic reaction centre and core antenna system. *Nat. Struct. Biol.* 3:965–73
41. Mac M, Tang XS, Diner BA, McCracken J, Babcock GT. 1996. Identification of histidine as an axial ligand to $P700^+$. *Biochemistry* 35:13288–93
42. MacMillan F, Hanley J, van der Weerd L, Knupling M, Un S, Rutherford AW. 1997. Orientation of the phylloquinone electron acceptor anion radical in photosystem I. *Biochemistry* 36:9297–303
43. Maeda H, Watanabe T, Kobayashi M,

Ikegami I. 1992. Presence of 2 chlorophyll *a′* molecules at the core of photosystem I. *Biochim. Biophys. Acta* 1099:74–80

44. Mannan RM, Pakrasi HB, Sonoike K. 1994. The PsaC protein is necessary for the stable association of the PsaD, PsaE, and PsaL proteins in the photosystem I complex: analysis of a cyanobacterial mutant strain. *Arch. Biochem. Biophys.* 315:68–73
45. McDermott AE, Yachandra VK, Guiles RD, Sauer K, Klein MP, et al. 1989. EXAFS structural study of F_X, the low-potential iron-sulfur center in photosystem I. *Biochemistry* 28:8056–59
46. Moenne-Loccoz P, Heathcote P, Maclachlan DJ, Berry MC, Davis IH, Evans MCW. 1994. Path of electron transfer in photosystem 1: direct evidence of forward electron transfer from A_1 to Fe-S_X. *Biochemistry* 33:10037–42
47. Moser CC, Keske JM, Warncke K, Farid RS, Dutton PL. 1992. Nature of biological electron transfer. *Nature* 355:796–802
48. Norris RJ, Uphaus RA, Crespi HL, Katz JJ. 1971. Electron spin resonance of chlorophyll and the origin of signal I in photosynthesis. *Proc. Natl. Acad. Sci. USA* 68:625–28
49. Petrouleas V, Brand JJ, Parrett KG, Golbeck JH. 1989. A Mössbauer analysis of the low-potential iron-sulfur center in photosystem I: spectroscopic evidence that F_X is a iron-sulfur [4Fe-4S] cluster. *Biochemistry* 28:8980–83
50. Purton S, Stevens DR, Muhiuddin IP, Evans MC, Carter S, et al. 2001. Site-directed mutagenesis of PsaA residue W693 affects phylloquinone binding and function in the photosystem I reaction center of *Chlamydomonas reinhardtii*. *Biochemistry* 40:2167–75
51. Ramesh VM, Gibasiewicz K, Lin S, Webber AN. 2001. *Specific mutation of the methionine axial ligand to chlorophyll A_0 of PSI in* Chlamydomonas reinhardtii. Presented at PS2001 12th Int. Congr. Photosynthesis, Melbourne, Aust.
52. Rautter J, Lendzian F, Schulz C, Fetsch A, Kuhn M, et al. 1995. ENDOR studies of the primary donor cation radical in mutant reaction centers of *Rhodobacter sphaeroides* with altered hydrogen-bond interactions. *Biochemistry* 34:8130–43
53. Redding K, MacMillan F, Leibl W, Brettel K, Hanley J, et al. 1998. A systematic survey of conserved histidines in the core subunits of photosystem I by site-directed mutagenesis reveals the likely axial ligands of P700. *EMBO J.* 17:50–60
54. Rigby SE, Evans MC, Heathcote P. 1996. ENDOR and special triple resonance spectroscopy of A_1^- of photosystem 1. *Biochemistry* 35:6651–56
55. Rodday SM, Do LT, Chynwat V, Frank HA, Biggins J. 1996. Site-directed mutagenesis of the subunit PsaC establishes a surface-exposed domain interacting with the photosystem I core binding site. *Biochemistry* 35:11832–38
56. Rodday SM, Jun SS, Biggins J. 1993. Interaction of the F_A/F_B-containing subunit with the photosystem-1 core heterodimer. *Photosynth. Res.* 36:1–9
57. Rodday SM, Schulz R, McIntosh L, Biggins J. 1994. Structure-function studies on the interaction of PsaC with the photosystem 1 heterodimer. The site directed change R561E in PsaB destabilizes the subunit interaction in *Synechocystis* sp. PCC 6803. *Photosynth. Res.* 42:185–90
58. Rodday SM, Webber AN, Bingham SE, Biggins J. 1995. Evidence that the F_X domain in photosystem I interacts with the subunit PsaC: Site-directed changes in PsaB destabilize the subunit interaction in *Chlamydomonas reinhardtii*. *Biochemistry* 34:6328–34
59. Sakuragi Y, Zybailov B, Shen G, Jones AD, Chitnis PR, et al. 2002. Insertional inactivation of the *menG* gene, encoding 2-phytyl-1,4-naphthoquinone methyltransferase of *Synechocystis* sp. PCC 6803, results in the incorporation of 2-phytyl-1, 4-naphthoquinone into the A_1 site and alteration of the equilibrium constant between

A_1 and F_X in photosystem I. *Biochemistry* 41:394–405
60. Sakurai H, Inoue K, Fujii T, Mathis P. 1991. Effects of selective destruction of iron-sulfur center B on electron transfer and charge recombination in photosystem I. *Photosynth. Res.* 27:65–71
61. Schubert WD, Klukas O, Krauss N, Saenger W, Fromme P, Witt HT. 1997. Photosystem I of *Synechococcus elongatus* at 4 Å resolution: comprehensive structure analysis. *J. Mol. Biol.* 272:741–69
62. Semenov AY, Vassiliev IR, van Der Est A, Mamedov MD, Zybailov B, et al. 2000. Recruitment of a foreign quinone into the A_1 site of photosystem I. Altered kinetics of electron transfer in phylloquinone biosynthetic pathway mutants studied by time-resolved optical, EPR, and electrometric techniques. *J. Biol. Chem.* 275:23429–38
63. Shinkarev VP, Zybailov B, Vassiliev IR, Golbeck J. 2002. Modeling of the $P700^+$ charge recombination kinetics with phylloquinone and plastoquinone-9 in the A_1 site of photosystem I. *Biophys. J.* 83:2885–97
64. Smart LB, Warren PV, Golbeck JH, McIntosh L. 1993. Mutational analysis of the structure and biogenesis of the photosystem-I reaction center in the cyanobacterium *Synechocystis* sp. PCC 6803. *Proc. Natl. Acad. Sci. USA* 90:1132–36
65. Stehlik D, Antonkine ML, Golbeck JH, Jordan P, Fromme P, Krauß N. 2001. *Assembly of PsaC subunit within stromal ridge of photosystem I core. Comparison of PsaC bound and solution structure*. Presented at PS2001 12th Int. Congr. Photosynthesis, Melbourne, Aust.
66. Stehlik D, Bock CH, Petersen J. 1989. Anisotropic electron spin polarization of correlated spin pairs in photosynthetic reaction centers. *J. Phys. Chem.* 93:1612–19
67. Swarthoff T, Gast P, Amesz J, Buisman HP. 1982. Photoaccumulation of reduced primary electron acceptors of photosystem I of photosynthesis. *FEBS Lett.* 146:129–32
68. van der Est A, Bock C, Golbeck J, Brettel K, Sétif P, Stehlik D. 1994. Electron transfer from the acceptor A_1 to the iron-sulfur centers in photosystem I as studied by transient EPR spectroscopy. *Biochemistry* 33:11789–97
69. van der Est A, Prisner T, Bittl R, Fromme P, Lubitz W, et al. 1997. Time-resolved X-, K- and W-band EPR of the radical pair state $P700^+ A_1^-$ of photosystem I in comparison with $P865^+ Q_A^-$ in bacterial reaction centers. *J. Phys. Chem. B* 101:1437–43
70. Vassiliev IR, Antonkine ML, Golbeck JH. 2001. Iron-sulfur clusters in type I reaction centers. *Biochim. Biophys. Acta* 1507:139–60
71. Vassiliev IR, Jung YS, Smart LB, Schulz R, Mcintosh L, Golbeck JH. 1995. A mixed-ligand iron-sulfur cluster ($C556S_{PsaB}$ or $C565S_{PsaB}$) in the F_X-binding site leads to a decreased quantum efficiency of electron transfer in photosystem I. *Biophys. J.* 69:1544–53
72. Vassiliev IR, Yu JP, Jung YS, Schulz R, Ganago AO, et al. 1999. The cysteine-proximal aspartates in the F_X-binding niche of photosystem I—effect of alanine and lysine replacements on photoautotrophic growth, electron transfer rates, single-turnover flash efficiency, and EPR spectral properties. *J. Biol. Chem.* 274:9993–10001
73. Vos MH, Van Gorkom HJ. 1990. Thermodynamic and structural information on photosynthetic systems obtained from electroluminescence kinetics. *Biophys. J.* 58: 1547–55
74. Warren PV, Smart LB, McIntosh L, Golbeck JH. 1993. Site-directed conversion of cysteine-565 to serine in PsaB of photosystem-I results in the assembly of [3Fe-4S] and [4Fe-4S] clusters in F_X—A mixed-ligand [4Fe-4S] cluster is capable of electron transfer to F_A and F_B. *Biochemistry* 32:4411–19
75. Wasielewski MR, Norris JR, Shipman LL, Lin C-P, Svec WA. 1981. Monomeric chlorophyll *a* enol: evidence for its possible role as the primary electron donor in photosystem I of plant photosynthesis. *Proc. Natl. Acad. Sci. USA* 78:2957–61

76. Watanabe T, Kobayashi M. 1988. Chlorophylls as functional molecules in photosynthesis. Molecular composition *in vivo* and physical chemistry *in vitro*. *Nippon Kagaku Kaishi* 4:383–95
77. Webber AN, Gibbs PB, Ward JB, Bingham SE. 1993. Site-directed mutagenesis of the photosystem-I reaction center in chloroplasts—the proline-cysteine motif. *J. Biol. Chem.* 268:12990–95
78. Webber AN, Lubitz W. 2001. P700: the primary electron donor of photosystem I. *Biochim. Biophys. Acta* 1507:61–79
79. Webber AN, Su H, Bingham SE, Kass H, Krabben L, et al. 1996. Site-directed mutations affecting the spectroscopic characteristics and midpoint potential of the primary donor in photosystem I. *Biochemistry* 35:12857–63
80. Witt H, Schlodder E, Teutloff C, Niklas J, Bordignon E, et al. 2002. Hydrogen bonding to P700: site-directed mutagenesis of threonine A739 of photosystem I in *Chlamydomonas reinhardtii*. *Biochemistry* 41:8557–69
81. Xia Z, Broadhurst RW, Laue ED, Bryant DA, Golbeck JH, Bendall DS. 1998. Structure and properties in solution of PsaD, an extrinsic polypeptide of photosystem I. *Eur. J. Biochem.* 255:309–16
82. Yang F, Shen G, Schluchter WM, Zybailov B, Ganago AO, et al. 1998. Deletion of the PsaF polypeptide modifies the environment of the redox-active phylloquinone. Evidence for unidirectionality of electron transfer in photosytem I. *J. Phys. Chem.* 102:8288–99
83. Yu JP, Smart LB, Jung YS, Golbeck J, McIntosh L. 1995. Absence of PsaC subunit allows assembly of photosystem I core but prevents the binding of PsaD and PsaE in *Synechocystis* sp. PCC 6803. *Plant Mol. Biol.* 29:331–42
84. Zech SG, Hofbauer W, Kamlowski A, Fromme P, Stehlik D, et al. 2000. A structural model for the charge separated state $P700^+ A_1^-$ in photosystem I from the orientation of the magnetic interaction tensors. *J. Phys. Chem. B* 104:9728–39
85. Zhao J, Li N, Warren P, Golbeck J, Bryant D. 1992. Site-directed conversion of a cysteine to aspartate leads to the assembly of a [3Fe-4S] cluster in PsaC of photosystem-I—The photoreduction of F_A is independent of F_B. *Biochemistry* 31:5093–99
86. Zybailov B, van der Est A, Zech SG, Teutloff C, Johnson TW, et al. 2000. Recruitment of a foreign quinone into the A_1 site of photosystem I. II. Structural and functional characterization of phylloquinone biosynthetic pathway mutants by electron paramagnetic resonance and electron-nuclear double resonance spectroscopy. *J. Biol. Chem.* 275:8531–39

Annu. Rev. Biophys. Biomol. Struct. 2003. 32:257–83
doi: 10.1146/annurev.biophys.32.110601.142439

First published online as a Review in Advance on January 16, 2003

THE STATE OF LIPID RAFTS: From Model Membranes to Cells

Michael Edidin
Biology Department, The Johns Hopkins University, Baltimore, Maryland 21218; email: edidin@jhu.edu

Key Words sphingolipids, cholesterol, membrane domains, fluidity, trafficking

■ **Abstract** Lipid raft microdomains were conceived as part of a mechanism for the intracellular trafficking of lipids and lipid-anchored proteins. The raft hypothesis is based on the behavior of defined lipid mixtures in liposomes and other model membranes. Experiments in these well-characterized systems led to operational definitions for lipid rafts in cell membranes. These definitions, detergent solubility to define components of rafts, and sensitivity to cholesterol deprivation to define raft functions implicated sphingolipid- and cholesterol-rich lipid rafts in many cell functions. Despite extensive work, the basis for raft formation in cell membranes and the size of rafts and their stability are all uncertain. Recent work converges on very small rafts <10 nm in diameter that may enlarge and stabilize when their constituents are cross-linked.

CONTENTS

INTRODUCTION

> *"... it is not as sure as both chymists and Aristotelians are wont to think it, that every seemingly similar or distinct substance that is separated from a body ... was pre-existent in it..."*
>
> *Robert Boyle*, The Sceptical Chymist, *1661 (13)*

Domain models of cell membranes, which view lipid bilayers as more mosaic than fluid, have long been proposed based on the properties of lipids in liposome membranes (54, 63). Indeed, even the model that summed up early work on membrane organization and emphasized bilayer fluidity, the fluid-mosaic model of Singer & Nicolson (129), considered the possibility of small membrane domains (~100 nm at most) in the fluid cell membrane bilayer. However, all models of cell membranes comprising domains were rather general and did not focus on specific biological functions that required domain formation.

Lipid rafts were conceived as functional lipid microdomains less than 15 years ago (128, 143). They were proposed as the solution to a particular biological problem, the selective delivery of lipids to the apical and basolateral surfaces of morphologically polarized epithelial cells. The lipid raft microdomain model drew on the physical chemistry of model lipid membranes to create a mechanism for segregating and sorting newly synthesized lipids for traffic to the cell surface. In turn, it suggested experiments in models to test aspects of the raft model.

This review begins with a history of the way in which physical chemistry, biochemistry, and cell biology came together to create and expand the raft model for lipid trafficking to encompass other areas of membrane biology, including signaling and invasion by intracellular pathogens. This survey of the early work in several areas maps some of the streams that feed the lipid microdomain model. These streams have disappeared in successive reviews of the field, but they are important to understanding the prospects and limits of lipid raft microdomain models. Other streams, in particular that of caveolae and caveolin, are the subject of excellent recent reviews (1, 2) and are not mapped here.

The fruitful interplay between results in model membranes and those on cell membranes continues to this day. Accordingly, after the historical introduction, this review will cover data from model membranes, monolayers, supported bilayers and liposomes, and cell membranes. We will see that the results with models demand rethinking of many of the procedures defining lipid rafts in cell membranes. My reading of this literature is that procedures create artifacts that do not reflect the state of lipids in native membranes. On the other hand, both model and cell experiments appear to be converging on the scale of lipid raft domains in native membranes and on mechanisms for organizing lipid rafts. I expect that further work will show that protein-lipid interactions rather than lipid-lipid interactions are important in organizing membrane lipids for trafficking and in signaling.

PREHISTORY AND EARLY HISTORY OF LIPID RAFTS

Epithelial Cells and Membrane Sphingolipids

> *"... in the plane of the membrane there are domains enriched in sphingomyelin and cholesterol (or both)" (47).*

Epithelial cells define and separate functional compartments in tissues and organs. They are morphologically and functionally polarized. The apical surface of a typical epithelial cell is chemically and functionally distinct from the basolateral surface. The demarcation between the two surfaces is a band of tight junctions binding adjacent cells in the epithelial sheet. This band of tight junctions allows selective fractionation of apical from basolateral plasma membranes. Apical lipids of apical membrane fractions are enriched, relative to basolateral membranes, in glycosphingolipids (GSL). This enrichment was confirmed using enveloped viruses to sample apical or basolateral host cell membranes (111, 142) and by other approaches (132). This difference in composition appeared to be due to the sorting of proteins and lipids in the Golgi complex and *trans*-Golgi membranes and directed trafficking of these molecules to the cell surface. Simons and van Meer (128, 143) based the lipid microdomain model for these cell biological observations on model membrane and other physical chemical studies that showed a high propensity for hydrogen bonding between GSL (11, 94), as well as on observations that neutral GSL, but not charged GSL, were found in small clusters, about 15 molecules maximum, in model and native bilayer membranes (108, 140). Thus in the proto-lipid raft model, trafficking microdomains were seen as clusters of hydrogen-bonded GSL, with associated proteins, either trapped in the hydrogen-bonded sphingolipid phase or specifically associated with sphingolipids. The proteins were presumed to carry the sorting signals that would result in trafficking entire microdomains to the apical surface.

Lipid-Anchored Proteins

Glycosylphosphatidylinositol (GPI)-anchored membrane proteins were found to concentrate at the apical surfaces of polarized epithelial cells both in vivo and in vitro (71). Their distribution was often more highly polarized than that of GSL, and the GPI-anchor appeared to be the sorting tag. Substituting a GPI modification sequence for the transmembrane region and cytoplasmic tail of a basolateral membrane marker redirected it to the apical surface (14, 70).

The parallel between polarity of GSL and that of GPI-anchored proteins suggested that the two classes of membrane molecules trafficked in the same unit, the lipid microdomain of van Meer and Simons. A scattering of biochemical data indicated that sphingolipids (1, 36) and GPI-anchored proteins (48) were insoluble in the neutral detergent Triton X-100, especially when cholesterol was present. These data were pulled together in a systematic study by Hooper & Turner (50) that showed that GPI-anchored ectoenzymes of renal epithelial apical membranes were insoluble in 5.9 mM Triton X-100 at 4°C, but soluble in low concentrations of other detergents. In a landmark paper, Brown & Rose (15) used this insolubility criterion to follow the membrane organization of a GPI-anchored alkaline phosphatase as the protein traveled through the endoplasmic reticulum and Golgi complex on its way to the apical surface of MDCK cells. Ninety percent of nascent alkaline phosphatase molecules were Triton X-100 insoluble by the time their N-oligosaccharide chains had matured, but had not been sialylated. This placed the point of their conversion from detergent-soluble to detergent-insoluble forms and entry into lipid microdomains for sorting somewhere prior to the medial Golgi complex. When detergent extracts were fractionated by density, mature alkaline phosphatase was found in lipid-rich, low-density fractions, whereas a control transmembrane protein fractionated to lipid-poor, high-density fractions.

The low-density fractions enriched in alkaline phosphatase appeared to be membrane vesicles of varied sizes, but typically hundreds of nanometers in diameter. Relative to whole cell lipids, the vesicles were depleted of the glycerophospholipids phosphatidylethanolamine (PE) and phosphatidylcholine (PC) and enriched in cholesterol and sphingolipids. The molar ratio of glycerophospholipid:sphingolipid:cholesterol was 34:36:32 or ~1:1:1. This ratio has become the canonical "lipid raft" mixture—the composition used for model membrane studies of raft properties.

Two other observations further defining a sorting lipid raft followed that of Brown & Rose (15). Le Bivic and colleagues (32) used a brief extraction (5 min in 1% Triton X-100, rather than the 20 min used by Brown & Rose) and reported that 50%–60% of marker GPI-anchored proteins were Triton X-100 insoluble. Although it was expected from the composition of detergent-insoluble rafts, and from some observations on clustering of GPI-anchored proteins (112), Le Bivic et al. found no effect of cholesterol-binding drugs on insolubility of GPI-anchored proteins. Another group, using a longer extraction time (30 min), did find that GPI-anchored alkaline phosphatase was almost all soluble if cells were treated with saponin, to complex cholesterol, before extraction (20).

Together, the experiments summarized above redefined lipid rafts as requiring both sphingolipids and cholesterol. They also created two operational definitions for lipid rafts. One definition, for constituents of rafts, was fractionation of proteins of interest to a lipid-rich complex insoluble in cold Triton X-100. A second definition, for raft requirements in cell function, was perturbation of some cellular process by cholesterol depletion, either acute or chronic. These two operational definitions opened the way for a vigorous pursuit of lipid rafts in trafficking newly synthesized lipids and proteins to cell surfaces. A consensus emerged that cholesterol/sphingolipid-rich membrane rafts are central to this traffic (126); however, this consensus is strained and challenged by the plasticity and range of lipid-sorting patterns in a variety of cells (149) and by the adaptability of cells to cholesterol or GSL depletion (37, 93).

Signaling Tyrosine Kinases and Lipid Rafts

Cross-linking GPI-anchored proteins can activate lymphocytes. Because activation is signaled through a cascade of intracellular kinases, signaling by GPI-anchored proteins raises the same formal problem that is raised by their vectorial transport, i.e., communication between the outer leaflet of the membrane bilayer and the cell cytoplasm. In three closely linked papers, Horejsi, Stockinger, and their colleagues showed that GPI-anchored proteins of human lymphocytes could be isolated in cold Triton X-100–insoluble complexes that also contained src family tyrosine kinases (22, 134, 135). These papers are the basis of a large literature that generally shows association of ligated receptors, signaling kinases and phosphatase, and G proteins with raft fractions to form signaling complexes (127). The three types of receptors most explored are the so-called IgE receptor (FcεR1) of mast cells and basophils (124), the T-lymphocyte receptor for antigen (150), and the B-lymphocyte receptor for antigen (95). Although the case for some involvement of rafts or raft lipids in receptor-mediated signaling and in regulation of this signaling (57) is compelling, work in this area is bedeviled by a lack of quantitative precision. The model for the function of rafts in T cell receptor signaling has also received a vigorous challenge from a detailed consideration of the biochemistry, rather than the physical chemistry, of receptor signaling (33).

More recently, operational definitions have pointed to lipid rafts as sites of entry and exit for intracellular pathogens (141). These may prove quite useful in sorting out raft function because they offer the possibility of using mutant viral and bacterial proteins as probes of native cell membranes.

THE LIPID RAFT RUBRIC

Together, the data of the last section give us the rubric—the outline and direction—for the rest of this review, a discussion of model membranes, lipid monolayers and bilayers, and cell membranes. First, we examine the ways in which sphingolipids, cholesterol, and glycerolipids behave in model membranes to gain an understanding of the way they could interact to form rafts in cell membranes. Models that begin as rather general investigations into the properties of a class of lipids, for example, studies of pure sphingolipid monolayers, can evolve into highly specific systems, such as mixtures of 1:1:1 phosphoglycerolipids:sphingolipids:cholesterol, which approximate the composition of detergent-insoluble domains isolated from cells. Most of these studies are consistent with the inferences about lipid rafts in intact cells. They also shed new light on the way in which bilayer leaflets could be coupled in lipid raft–mediated signaling and on the minimum size of cholesterol/lipid complexes. On the other hand, we will see that some studies of the solubilization of membranes by Triton X-100 may subvert the operational definition of lipid rafts in terms of detergent insolubility.

I also argue on cell biological grounds that the second operational definition of lipid rafts, cholesterol depletion, is also unsound. This leaves us with the problem of probing for lipid rafts in native cell membranes, without biochemical modification or detergent extraction. Optical methods, such as fluorescence resonance energy transfer (FRET), single molecule and single particle tracking, and laser trapping, have all proved useful here. These optical methods appear to be converging on a size for a unit or core lipid raft that is small indeed, consisting of at most half a dozen molecules, a far cry from the 100-nm rafts first defined as detergent-resistant membrane vesicles.

Although core lipid rafts shrink to a few molecules, we are left with the problems in cell biology that led to the raft model in the first place. These imply the reorganization of core rafts into larger structures. The trafficking of lipids, formation of signaling complexes, and virus interaction with cells all report segregation and concentration of raft lipids. I suggest that the common denominator in all these events is protein interaction with lipids, either by recruiting subsets of lipids to clustered transmembrane domains or by clustering of lipids interacting with protein exodomains apposed to the membrane surface. In short, proteins recruit lipids, rather than the opposite.

INTERACTION OF PHOSPHOLIPIDS, GLYCOLIPIDS, AND CHOLESTEROL IN MODEL MEMBRANES

Model 1: Sphingolipids

As we noted earlier, the lipid raft model was based on studies of purified sphingolipids. As a class, the acyl chains of sphingolipids are more saturated than those of glycerolipids. Typical plasma membrane glycerolipids bear saturated acyl chains at the glycerol *sn-1* position and unsaturated acyl chains at *sn-2*, for example, 16:0/18:1. In contrast, the sphingoid backbone of a sphingolipid is 18:0, and the acyl chain is also saturated, 18:0 to 24:0 (16, 106). A mixture of fluid glycerophospholipid and sphingolipid might be expected to segregate sphingolipid-rich domains, optimizing acyl chain packing. This tendency is enhanced by the presence of both hydrogen bond donor (hydroxyl) and acceptor (fatty acyl carbonyl) groups in sphingomyelin (SM). It is opposed by sphingolipid headgroups, the phosphocholine of SM, and the neutral or charged oligosaccharides of GSL. The bulk and charge of these headgroups limit the packing density of pure sphingolipid species, particularly of the negatively charged gangliosides (76). On the other hand, hydrogen bonding between sugars of the headgroups can stabilize glycolipid interactions. Indeed, as noted earlier (108, 140), clusters of neutral glycolipids may be present in native membranes. In contrast, clustering of gangliosides is not observed, although they can be induced to cluster by Ca^{++} or by multivalent toxins, such as cholera toxin (35).

Mixtures of sphingolipids with different size headgroups segregate from glycerophospholipids, forming complexes that can locally develop into gel phases in a liquid crystalline phospholipid matrix. Thus the gel to liquid crystalline phase transition temperature of either 16:0/16:0 PC or SM in a matrix of fluid native PC (16:1, 18:0) is raised significantly by the addition of 10% ceramide (acylated sphingoid base) to the mixture. The ceramide has a larger effect on the melting temperature of SM than on that of dipalmitoylphosphatidylcholine (DPPC), presumably because it can interact with SM through both acyl chain interactions and hydrogen bonding (78). In this system, where the acyl chain length of all lipids was similar, there was no evidence of formation of ceramide-rich domains. Only interactions with other lipids led to segregation of a gel domain. The author speculates that local changes in ceramide concentration, due to local sphingomyelinase activity, could create small gel domains in cell membranes, perhaps changing local permeability and activating membrane-associated enzymes.

SM also forms complexes with glycolipids, consistent with the co-isolation of these molecules in raft fractions (79). The preferential segregation of GSL and SM from glycerophospholipids, independently of other lipids or cholesterol, may be important in organizing the apical surfaces of renal and intestinal epithelial cells, which stabilizes them against breakdown by the hostile fluids (for example, bile salt–containing intestinal fluids) that bathe them. This is suggested by the observation that apical brush border lipids form detergent-resistant domains even in the absence of cholesterol (38, 85). This is also suggested by recent work comparing the permeability of apical and basolateral membrane vesicles from guinea pig colon (18). In some sections of the colon, the cholesterol:phospholipids ratio is twofold higher in basolateral membrane vesicles than in apical membrane vesicles, yet the permeability of the latter is much lower than that of the former.

Model 2: Cholesterol-Containing Lipid Mixtures

> *"... in biomembranes containing various phospholipids species one would expect that certain lipids (saturated species for example) would tend to accumulate in the vicinity of cholesterol..." (53).*

The lipids of mammalian cell plasma membranes are roughly one-third cholesterol. This rigid sterol molecule intercalates between acyl chains of phospho- and glycolipids whether these are in the gel or the liquid crystalline phase. In the former, because the rigid sterol cannot conform to pack with all-*trans* acyl chains, cholesterol decreases acyl chain order and this is reflected in the mixture melting at a slightly lower temperature and over a much broader range than the pure lipid. In the liquid crystalline phase—the matrix of cell membranes—cholesterol orders acyl chains, decreasing the area per molecule. It has been well put that "sterols appear to have evolved to fill the flickering spaces among acyl chains in membrane bilayers" (92). The great range of cholesterol interactions is evidenced both by model membrane studies and by the finding that cell metabolism of cholesterol and choline phospholipids, both glycerolipids and SM, is coordinately regulated. These interactions are discussed in an excellent recent review (92). Here we focus on cholesterol-containing membranes whose composition is modeled on lipid rafts, as defined by detergent extraction. As noted above, raft lipid composition is approximately 1:1:1 phosphoglycerolipids:sphingolipids (both SM and GSL):cholesterol. In rafts prepared with detergent, the phospholipid population is enriched in saturated acyl chains relative to the average for whole cell phospholipids (30, 97) [although this is not the case for sphingolipid-rich raft fractions prepared without detergent (9)]. This finding converged with studies of cholesterol interactions with saturated chain PCs, particularly DPPC, to suggest that the lipids of rafts, as defined by detergent insolubility, were in a liquid-ordered phase, l_o. This phase was defined in theoretical work based on five different experimentally determined phase diagrams for DPPC cholesterol (52). It is one in which the unfavorable free energy of cholesterol solvation in an ordered gel phase is balanced by specific short-range interactions of cholesterol with the conformationally ordered acyl chains of this phase. The result is decoupling of acyl chain conformational order from PC crystalline order. Thus, while acyl chains are largely in the all-*trans* conformation, rotational and translational diffusion of lipid molecules is almost equal to that in a liquid crystalline phase. Moderate (~10–20 mol %) concentrations of cholesterol result in a mixture of gel and liquid-ordered phases below the PC melting temperature (41°C) and a mixture of two immiscible liquid crystalline phases, liquid ordered and liquid disordered, above the melting temperature. Further theoretical work has reinforced the view that cholesterol and related sterols with a smooth, unbroken hydrophobic surface (but not sterols such as lanosterol with this surface made "bumpy" by methylene groups) are uniquely suited to interact with acyl chains to create a liquid-ordered state (84).

Not only is the lipid composition of detergent-insoluble lipid rafts consistent with the existence of a liquid-ordered state, but liposomes made of a binary mixture of DPPC and cholesterol 2:1, proportions that are in the liquid-ordered region of the phase diagram, were completely insoluble under the conditions used to define lipid rafts by detergent insolubility. Liposomes with compositions approximating that of lipid rafts, rich in cholesterol and sphingolipids, were also largely detergent insoluble (121). In a series of experiments comparing lipid mixtures capable of forming liquid-ordered domains with those that are not, excellent correlations were made between detergent insolubility, ordering of acyl chains, and formation of ordered domains detected by dequenching a fluorescence probe. The probe, diphenylhexatriene (DPH), could partition into more ordered phases, whereas the short-range quencher, a spin-labeled lipid, partitioned in disordered phases. Hence, increases in fluorescence were taken to indicate at least partial segregation of the fluorescent probe into liquid-ordered domains. These experiments are summarized in a recent review (72). They speak to lateral variation in the surface of liposomes of appropriate composition but only explore the phase diagram of the systems studied to a limited extent. They also do not speak to the size of the liquid-ordered domains, an issue of paramount interest in studying native cell membranes. Short-range quenching could be impeded if donor fluorophores were in small domains and separated from the quencher by a few phospholipid radii, as well as if the fluorophores were in large domains and sequestered many molecule diameters away from the quencher.

Fluorescence resonance energy transfer (FRET) measurements can be used to estimate the size of lipid domains, if the apparent partition coefficients of donors or acceptors determined by FRET can be compared with the partition coefficients measured by some distance-independent method. Prieto and colleagues (74) have taken this approach with the classic liquid-ordered/liquid-disordered model membrane, DPPC/cholesterol. They find that liquid-ordered domains dispersed in liquid-disorder phase are small, $\leq$20 nm, the R_o for their donor-acceptor fluorophore pair.

The interpretation of the DPPC/cholesterol phase diagram in terms of immiscible liquid phases l_o and l_d is not unique. In another view, these mixtures comprise immiscible phases, one of which is rich in specific complexes of cholesterol with phospholipids. Complex formation is treated as a reversible reaction characterized by an equilibrium constant, K_{eq}, with a stoichiometry that appears to be ~1:5 cholesterol:glycerophospholipid or sphingolipid in monolayers (104) and 1:12 cholesterol:phospholipid in vesicle bilayers [(3), reviewed in (82)]. A complex-rich phase separates from cholesterol-rich and liquid crystalline phases [of dimyristoylphosphatidylcholine (DMPC), which does not form complexes efficiently] in monolayers at low pressures (82). At higher pressures, comparable to those in a bilayer membrane, the complexes appear to be miscible with the other phases.

The cholesterol/lipid complex model can be used to interpret thermodynamic phase diagrams for mixtures of cholesterol and disaturated acyl chain phospholipids, for example, dipalmitoylphosphatidylethanolamine (DPPE), which were used to infer the coexistence of l_o and l_d phases, without specifying a liquid-ordered phase or indeed specifying anything about the high-cholesterol region of the phase diagram, except that it contains cholesterol/phospholipid complexes. It contrasts with the l_o/l_d model, but each model has a different emphasis and different uses. A membrane model with immiscible l_o domains is useful for thinking about the ways bilayer leaflets can be coupled by cholesterol and interdigitating acyl chains (116). It also sets a direction for modeling immiscible liquid-state lipid phases and for interpreting cell experiments where cross-linked membrane receptors change their association with particular lipids. This sort of model tempts one into thinking about l_o phases as stable regions of cell membranes, i.e., available to cross-linked receptors. It also neglects the chemical activity of cholesterol, i.e., its availability for specific chemical reactions with proteins and lipids.

A membrane model in which cholesterol is in a stoichiometric complex with either phospho- or glycolipids emphasizes the regulation of cholesterol activity in membranes and points to the possibilities of rapid local remodeling of cell membranes by available cholesterol. Changes in cholesterol activity can in turn lead to changes in membrane enzyme activity (in response to small changes in bilayer thickness or chain order) (73, 136), changes in activity of lipases (118) and accessibility of their substrates (69), and changes in receptor (96) and antigen display (10, 125), as well as global responses by cells.

The canonical raft lipid mixture and its variants yield multimicrometer-diameter domains in both monolayers and supported bilayers below the melting temperature of all their component lipids (24, 25, 114). Overall, studies of these domains yield convincing evidence for the possibility of liquid/liquid immiscibility in cell membranes and also highlight the mechanical and physical properties of l_o phases. One group (114) emphasized the mechanics and dynamics of domains excluding R-DOPE (dioleylphosphatidylethanolamine), showing that properties of these dark domains are those of fluids, not gels. The behavior of the domains under shear and the dynamics of their fusion may speak to the way small, functional l_o domains form in cell membranes. Their behavior when the bathing solution was stirred indicated that they were thicker than the surrounding bilayer, consistent with atomic force microscopy (AFM) (107, 117) and X-ray diffraction measurements (31). Another group (24, 25) explored variables of probe, form of model membrane, and composition. The probes for raft domains in monolayers and supported bilayers were a series of headgroup-tagged DPPE molecules: NBD-PE, flourescein (FL)-PE, and Texas Red-PE. NBD-PE, the probe with the smallest headgroup, entered l_o domains in supported monolayers at lateral pressures of ~30 dyne/cm similar to those inferred for cell membrane bilayers. However, when supported bilayers were probed, FL-PE partitioned about equally well into l_o and l_d phases. The l_o domains could not be resolved after cholesterol depletion and reappeared when cholesterol was added back to the depleted bilayer.

When monolayers were formed from renal epithelial cell brush border lipids, the domains were often ovoid with rough contours rather than circular shapes with smooth perimeters, as expected for a liquid phase. Similar shapes were seen for dark regions (excluding TR-DPPE) in monolayers prepared from cholesterol-depleted brush border lipids. This suggests the possibility that the domains are gel, not l_o. Indeed, although the dark domains in complete brush border lipids appeared to allow some penetration of TR-DPPE, the rough-contoured domains in cholesterol-depleted brush border lipids were impermeable to TR-DPPE, consistent with their being in the gel state.

Lipid monolayers were also used to visualize two other important parts of the lipid raft model: (*a*) partition of GPI-anchored proteins into l_o domains and (*b*) the effects of cross-linking on partition of a probe molecule between l_o and l_d. Results for both visualizations were not as clear-cut as might have been expected. The GPI-anchored Thy-1 molecule was present in both the l_d matrix lipids and the l_o domains of dioleylphosphatidylcholine (DOPC)/SM/cholesterol or brush border lipid monolayers, with more intensity in putative l_o domains. A similar result for bilayers was reported for GPI-anchored alkaline phosphatase (117). Surprisingly, less Thy-1 partitioned into l_o if a small amount (<1 mol %) of ganglioside GM1 was present. Because rafts defined by detergent extraction are enriched in Thy-1, gangliosides, and other glycolipids, it is not clear why Thy-1 and GM1 did not colocalize.

Another surprise was found when FL-DOPE or FL-DPPC was cross-linked. FL-DOPE partitioned mainly into the l_d matrix lipids of the monolayer regardless of whether it was cross-linked, consistent with its unsaturated acyl chains, and FL-DPPE partitioned into l_o domains only after cross-linking by antifluorescein antibody. The partition of FL-DPPE between l_d and l_o is different in supported bilayers than in monolayers. This suggests that subtle changes in molecular interactions could shift it more toward the l_o phase even in monolayers. It would be interesting to know if TR-DPPE also partitions into the l_o phase after cross-linking.

Although l_d and l_o domains can be detected by microscopy, wide-angle X-ray diffraction studies detected only a single lamellar phase in multilamellar vesicles of 1:1:1 DOPC:SM:cholesterol. Gel and fluid phases could be distinguished in 1:1 DOPC:SM vesicles in which SM forms a gel phase (31). The authors discuss possible reasons for this difference, pointing out variations in boundary conditions and composition among the model membranes used in References 24, 25, 31, and 114, though citing other explanations as well.

Model 3: Triton X-100 and the Solubility of Model Lipid Mixtures

Membrane lipid raft domains are often defined operationally in terms of detergent insolubility in cold Triton X-100, although not exclusively (for example see Reference 26 and a discussion of lipid solubilization by different classes of detergent in Reference 7). However, few of these experiments explore a wide range of Triton concentrations or follow the timecourse of membrane solubilization by detergent. Indeed, much of the published work on detergent-resistant domains in cell membranes does not even specify the ratio of detergent to membrane lipid. A few experiments on model membranes do explore a large sample of the variable space. These show that SM/cholesterol monolayers and bilayers are more resistant to solubilization by Triton X-100 than are pure SM membranes. Hydrogen bonding between cholesterol and SM may be important for this. However, Triton X-100 at low concentrations penetrates SM-containing membranes more readily than chain-matched phospholipid DPPC membranes, and SM membranes are solubilized at lower concentrations of detergent than are phosphoglycerolipid membranes [see (90) and references therein].

The modeling work also suggests that detergent-resistant lipid domains are created when low concentrations of Triton X-100 induce membrane blebbing and fusion, possibly scrambling membrane lipids (47, 131). Addition of small, sublytic amounts of detergent to vesicles made of raft lipids increases their light scattering, implying an increase in size, owing to bilayer fusion. Electron microscopy of vesicles treated this way shows that multilayer vesicles are indeed converted to larger structures containing ramified branched bilayers (47), as are cell membranes (130). Furthermore, it appears that inclusion of GSL in model membranes reduces the detergent concentration required for membrane solubilization. The authors note that this is consistent with the tendency for GSL to form micelles in water.

A recent study suggests that the detergent could induce domain formation in otherwise miscible lipid mixtures (45). NMR and three different types of calorimetric measurements—isothermal titration, differential scanning, and pressure perturbation—were used to follow micellization of lipids, an indication of membrane solubilization, as a function of temperature and Triton X-100 concentration and, in parallel, thermal transitions. When Triton X-100 was titrated into a POPC:SM:cholesterol raft model membrane system, a sharp change in heat of transfer occurred at detergent:lipid ratios that did not solubilize membrane. This was interpreted as a detergent-driven transition within the lamellar state. The author argues that the observed changes represent the balance between enthalpically favored associations of SM and cholesterol and entropically driven dispersion of SM and cholesterol in the liquid-disordered membrane phase. When detergent:lipid ratios reach some critical value, in this system about 0.3 Triton X-100/POPC, the enthalpic penalty of mixing becomes so high that SM and cholesterol segregate from the POPC and form detergent-resistant domains. These results suggest that the large SM/cholesterol-rich domains isolated from detergent do not reliably report on the organization of these lipids in the cell membrane.

MODELS MEET CELLS AND RAISE PROBLEMS

Problem 1: How Are the Two Monolayers of a Cell Lipid Bilayer Membrane Coupled to Form a *Trans*-Bilayer Raft?

One might expect that rafts would form independently in each leaflet. In the event, the two leaflets appear to be coupled in both gel/fluid glycerophospholipid mixtures (4, 64) and raft mixtures (24, 25). This might be due to interdigitation of longer acyl chains (106) and/or cholesterol (116) of the two monolayers. However, understanding the coupling of raft lipids between the inner and outer leaflets of cell plasma membranes is vexed by a number of factors, such as the high degree of unsaturation of inner leaflet phospholipid acyl chains, the asymmetry of SM distribution, with only 15% of the total in the inner leaflet, and the unresolved distribution of cholesterol between the inner and outer leaflet. Estimates for cholesterol distribution range from about 65% in the outer leaflet (23) to 75%–80% in the inner leaflet (120). Thus, the global ratio of cholesterol to sphingolipids may be as high as 10:1 in the inner leaflet and as low as 0.3:1 in the outer leaflet, with a consensus ratio of 2:1. There is also evidence of multiple pools of membrane cholesterol extractable with methyl-β-cyclodextrin (43, 91) or accessible to cholesterol oxidase (28, 99). It is not clear what fraction of membrane cholesterol can associate with sphingolipids to form rafts.

In a model mixture of PCs with unsaturated (20:4 and 18:0) acyl chains, ^{2}H NMR spectra suggest that cholesterol segregates into cholesterol-rich regions (17); however, these are not likely to be in a raft-like l_o state. No l_o domains are detected in bilayers made of cholesterol and lipids characteristic of the inner leaflet, PS and PE with unsaturated acyl chains (147). The authors speculate that any inner leaflet rafts must be organized by factors, such as proteins, extrinsic to the inner leaflet. Immiscible liquid phases do form in monolayers made from erythrocyte membrane inner leaflet lipids, about 5% in SM (60), but these could be cholesterol/lipid-complex-rich domains, not l_o domains.

Coupling of inner and outer leaflets in native cell membranes has been shown by qualitative (41, 55) and quantitative (103) colocalization (at the resolution of the light microscope) of receptors, GPI-anchored proteins, or glycolipids clustered in the outer leaflet with acylated cytoplasmic signaling and other proteins found in detergent-defined raft fractions. The range of such lipidated proteins is summarized in the introduction to a recent study of the importance of lipidation for their colocalization with membrane cholesterol and gangliosides (81). Green fluorescent protein (GFP) fused to specific sequences from Yes, Fyn, Lck, Src, and other proteins that contained acylation sites and, in some cases, a polybasic sequence colocalized well with cholesterol labeled by fillipin, and also with GM1 labeled by cholera toxin. This is a useful model, although incomplete, because it is clear that the isoforms and conformations of lipidated proteins play a role in localizing them to cholesterol-rich, raft-like regions of the inner leaflet (89, 101).

Raft localization of cyan fluorescent protein (CFP) or yellow fluorescent protein (YFP) fused to consensus sequences for acylation or prenylation has also been detected in terms of FRET between the fluorophores (151). The FRET data were analyzed using an approach that can distinguish FRET in a randomly distributed population of fluorophores from FRET among clustered fluorophores (61) by the relationship between FRET efficiency (E) and acceptor concentration (A) (measured in terms of YFP fluorescence). The data are extremely scattered and noisy, but the change in E versus A after cholesterol is extracted with methyl-β-cyclodextrin is consistent with cholesterol-dependent clustering of doubly acylated GFP. The fluorescence quenching method described earlier (72) also showed that dual acylation is sufficient to associate peptides with l_o fractions in model membranes (146).

Once directed to the bilayer by acylation, a protein may explore it, rather than localize to one lipid environment. The cytosolic signaling protein Raf, a serine/threonine kinase, has an affinity for both anionic lipids and cholesterol (46), opening the possibility of its moving from one region of membrane to another. If this sort of dynamic is true of all or most associations of lipidated proteins with the inner leaflet, they could sample a number of local lipid environments, with their dwell time in a given environment dependent upon local biochemical reactions (81, 122). As nicely put in a recent review, lipidated proteins are in "a perpetual search for short relationships" (51).

Problem 2: What Do Detergent Extraction and Insolubility Tell Us About the State of Cell Membranes?

As we have seen, solubility in cold Triton X-100 detergent is a principal biochemical definition of lipid rafts. Results with model membranes and cell membranes weaken these operational definitions of lipid rafts.

The model membrane and early cell membrane experiments cited in an earlier section make the case that cell membranes are vesiculated and fused by low, sublytic concentrations of Triton X-100. This suggests that raft components defined in detergent-resistant membranes came together by vesicle fusion rather than because they were proximally associated in cell membranes. The careful analysis of the way in which detergent interacts with raft mixtures (45) suggests that detergent can drive reorganization of lipids to form detergent-insoluble phases. This suggestion is reinforced by comparison of the lipid composition of membrane rafts isolated with or without detergent (97). The former are rich in lipids with saturated acyl chains, whereas the latter are rich in highly unsaturated acyl chains. If the rafts defined in detergent extracts existed prior to detergent addition, then we might expect that the lipid compositions of raft fractions isolated in different ways would be grossly similar.

A review of the large literature on lipid rafts as detergent-insoluble, lipid-rich membrane fractions also highlights the imprecision of most extractions. Detergent:lipid ratios are not well controlled and often not defined. Also, partition of proteins of interest between detergent-insoluble and detergent-soluble fractions is seldom quantified and rarely absolute, with all of a protein of interest in one fraction or the other. Even the starting material for extraction varies considerably. Comparing papers on the association of lipidated cytoplasmic proteins with raft fractions highlights the possible importance of controlling this factor. Thus McCabe & Berthiaume (81) extracted monolayers of intact COS-7 cells with 1% Triton X-100 at 4°C for 20 min and found that GFP with the Lck-derived N-terminal acylation sequence (myristate + palmitate) was totally soluble in detergent, while Wang et al. (146) found that 60% of the same construct, also expressed in COS-7, was detergent insoluble when isolated membranes, rather than whole cells, were extracted. The same extent of insolubility was found for GFP with the Lyn acylation sequence, when MDCK membranes were extracted (46). Another discordance is between two reports on the effects of brefeldin A treatment on the detergent insolubility of GPI-anchored proteins (in the lumenal leaflet of the Golgi complex) compared with that of G proteins (associated with the cytoplasmic leaflet of the Golgi complex). Brefeldin A treatment results in collapse of the Golgi into the endoplasmic reticulum. One group finds that GPI-anchored proteins of the Golgi complex are detergent soluble after brefeldin A treatment (20). Another group finds that a detergent-insoluble membrane fraction containing G protein subunits and caveolin (but not probed for GPI-anchored proteins) could be isolated from whole cell extracts after brefeldin A treatment (34). Still another discordance is the recent observation that although the IL-1 receptor β chain and GPI-anchored proteins are co-isolated in detergent-insoluble membrane fractions, they traffic separately from the cell surface (113).

Problem 3: How Directed Are the Effects of Cholesterol Depletion on Cell Function?

Cholesterol is required to form l_o domains to maintain clusters of GPI-anchored proteins and caveolae and for formation of detergent-insoluble fractions in cell and membrane extracts. However, cell requirements for cholesterol extend far beyond this. I have already noted the effects of membrane lipids (including cholesterol) on the function of membrane proteins and that levels of membrane cholesterol affect turnover of other lipids, including SM, and the activity of the cholesterol storage and synthetic pathways (29, 67, 92). These may come into play when growing cells in LDL-deficient media chronically depletes cell cholesterol. Because most studies of cholesterol's role in raft function involve acute depletion with either cholesterol oxidase or nonionic surfactant methyl-β-cyclodextrin, which forms inclusion complexes with cholesterol [briefly reviewed in (91)], it might be expected that they would be specific for rafts. However, acute cholesterol depletion disrupts the clusters of snares required for exocytosis (66), blocks formation of clathrin-coated endocytic vesicles (109, 137), and, perhaps most important of all, delocalizes a plasma membrane signaling phospholipid, phosphatidylinositol(4,5)-bisphosphate [PIP(4,5)P2], from the plasma membrane (98). PIP(4,5)P2 is a major regulator of the actin cytoskeleton (19, 56, 105) and is intimately involved in endocytosis (12). Hence, cholesterol depletion, dispersing PIP(4,5)P2 from sites of functional interaction with cell proteins, can disrupt many cell functions. Indeed, recently, we have found (65) that cholesterol depletion, like sequestration of PIP(4,5)P2, alters cell actin organization and inhibits lateral diffusion of membrane proteins. Overall, it would seem that a correlation between reduction in cell cholesterol, loss of detergent-resistant membrane fractions, and loss of a particular function, for example, virus budding or receptor-mediated signaling, cannot necessarily be taken to show that lipid rafts are directly involved in the cellular function that is measured.

CELLS INSIST ON LIPID RAFTS

Lipid Rafts and Function

Despite great reservations about the interpretation of "classical" operational definitions of lipid raft components and functions, we are left with cells stubbornly insisting that lipids, lipid-anchored proteins, and acylated cytoplasmic signaling proteins are selectively trafficked and associated, and left with viruses that selectively sample host lipids to enrich their envelopes with sphingolipids and cholesterol. In the signaling field, perhaps the most striking finding is that receptors found in raft fractions of mature B cells are not found in these fractions from B cells at an earlier stage of development (133) and that raft localization of receptors is different in functionally different T cells (5). There are also a number of examples of viruses selectively incorporating GPI-proteins, cholesterol, and sphingolipids in their membranes (8, 44, 58, 87, 119).

There is abundant evidence that lipids and lipid-anchored proteins are segregated along the endocytic pathway. It was early noted that GPI-anchored proteins are internalized in a pathway separate from that for transmembrane proteins (6, 59) and, more recently, that fluorescent lipid analogs, dialkylindocarbocyanines (DiIs) with unsaturated or saturated acyl chains, were sorted away from each other after endocytosis with the unsaturated chain DiI going to recycling endosomes, returning to the plasma membrane, and saturated chain DiI going to late endosomes (86). In the past few years, focus has returned to the trafficking of GPI-anchored proteins and sphingolipids. Although SM analogs enter by both clathrin-dependent and clathrin-independent pathways (102), GPI-anchored proteins and glycolipids are internalized by a nonclathrin fluid-phase endocytic process (88, 102, 113). Once internalized, GPI-anchored proteins return to the surface more slowly than other recycling membrane components (80). This appears to be because they are retained in recycling endosomes by association with components of lipid rafts; their retention is sensitive to depletion of cell cholesterol. Recycling time is shortened if cell cholesterol or SM levels are reduced (21, 80). Recently, the Pagano laboratory (123) has pushed the limits of resolution in the light microscope and imaged the segregation of GSL analogs in the early endosome membrane. Separate domains of high and low concentrations of lactosylceramide form in endosomes. The high-concentration domains, about half the total, recycle to the cell surface, while the low-concentration domains traffic to the Golgi complex. It will be interesting to connect these observations to the cholesterol-dependent domains recently visualized on the surface of cultured cells (39).

Detecting Lipid Rafts in Intact Cell Membranes

The raft model implies that component molecules are concentrated and clustered. The lipid composition defined for rafts implies a higher viscosity for raft lipids than for the membrane average. The first implication has been tested by cross-linking and FRET measurements; the second by a variety of diffusion measurements. Measurements through 2000 are summarized in a recent review (2). They converge on raft diameters <100 nm. Some FRET measurements (61) detected no clustering at all, implying that rafts were vanishingly small or that only a fraction ($<20\%$) of the molecules of interest were in rafts. The most refined recent diffusion measurements used a laser trap to confine molecules labeled with beads to small areas (<100-nm diameter) and then followed the thermal motion of the beads around their position (100). Local membrane viscosity was inferred from this motion; the distribution of viscosities appeared to report on the diffusion of proteins within a raft and, as well, on the diffusion of an entire raft. The average raft diameter was estimated as 26 ± 13 nm with a raft lifetime of >1 min. Another group using the same logic, but conventional fluorescence recovery after photobleaching (FRAP) for measuring diffusion, was unable to detect diffusion of rafts as opposed to individual components of rafts (62). However, it is hard to compare the two results because of the difference in time and spatial scales of the methods used.

Fresh approaches to FRET (S. Mayor, personal communication) and single particle tracking techniques (138) have yielded a new picture of lipid rafts in unperturbed membranes. Mayor and colleagues (144) had earlier used fluorescence polarization as a measure of energy transfer between labeled GPI-anchored proteins. In their new work, they returned to this technique but combined it with time-resolved anisotropy measurements and theoretical modeling of the way in which the relationship between fluorescence anisotropy and fluorophore concentration would vary as a function of raft size. Using this analysis they concluded that units containing GPI-anchored proteins, probably associated with cholesterol, are <10 nm in diameter and contain perhaps five molecules of GPI-anchored protein. This size converges with the cholesterol/sphingolipid complexes discussed earlier (3, 82, 104).

Kusumi and colleagues (138) used high-speed particle tracking to compare the Brownian motion of GPI-anchored proteins with a di-unsaturated phosphatidylcholine, DOPC, which should not enter rafts. With a time resolution of 25 μs they found that GPI-anchored proteins and DOPC both diffused freely within compartments $\sim$110 nm in diameter, hopping to a new compartment every 25 ms. From the identical behavior of the two classes of molecules, they concluded that at steady state these core lipid rafts consist of only a few molecules and persist for $\sim$1 ms or less. However, if the GPI-anchored protein CD59 was cross-linked with antibody or its native ligand, larger rafts developed with a diffusion coefficient about eightfold smaller than that of core rafts. Cross-linked CD59 was frequently immobilized, and these immobilized rafts were associated with actin and signaling kinases. Thus membrane rafts appear to be small and unstable, but they are capable of developing into larger functional structures when their constituents are ligated.

HOW DOES IT ALL WORK?

Model membrane data and the stubborn insistence by cells that raft lipids can be organized and segregated into membrane domains can be integrated if we consider the ways in which proteins can interact with raft lipids and lipid-anchored proteins. Proteins can either recruit raft lipids to their transmembrane domains or bind their exodomains to the membrane surface [reviewed in (77, 148)]. A recent review emphasizes the formation of lipid shells around proteins integrated into the bilayer (2) and as well lists a baker's dozen of proteins that specifically bind cholesterol or sphingolipids.

An important interaction of raft components with cytoplasmic proteins is with actin (40), which is implicated in raft-mediated signaling (42, 49, 110, 138). This interaction is coming into new focus as important in surface organization of rafts (27) and for the mechanical properties of cell membranes (139). Actin does not bind directly to membranes, but rather to proteins, such as annexins, and polybasic proteins, such as MARCKS and GAP43. It has been suggested that annexin binding to raft lipids would hinder assembling of signaling proteins in a raft [though it would order lipid acyl chains, promoting association of raft lipids (25, 147)]. However, another model was developed recently (83) in which polybasic proteins with single acyl chains bind negatively charged PIP(4,5)P2, creating an oligomer with multiple saturated acyl chains that could recruit cholesterol and SM and so consolidate raft components with a signaling lipid that has highly unsaturated acyl chains. This can be generalized to a mechanism in which patches of charges on monoacylated proteins bind counter-charged lipids (PE and PS) and create a region of ordered acyl chains that can recruit SM and cholesterol. Chains might also be ordered by insertion of hydrophobic regions of a protein; this is suggested by the finding that prion protein binds negatively charged lipids by a combination of electrostatic and hydrophobic interactions (115). Still another possibility is that local sphingomyelinase activity could create small foci of highly ordered lipids (78). Whatever the mechanism for clustering inner leaflet lipids, it could be coupled to the outer leaflet and so recruit raft components such as signaling kinases and G proteins in response to receptor cross-linking.

Binding of proteins in the outer leaflet may organize lipids there. An energy transfer study of the proximity of the exodomain of a GPI-anchored protein to the bilayer surface shows that the protein is close enough to the surface to interact with the bilayer lipids, suggesting two-way transmission of conformational changes in the interacting partners (69). More specific interactions between raft glycolipid components and exogenous toxins, for example, cholera toxin and shigatoxin, have long been recognized (35). There is also a hint of specific lipid binding in the fact that influenza membrane proteins found in rafts (hemagglutinin and neuraminidase) no longer target there if their cytoplasmic domains are truncated (152). Recently, a common sphingolipid-binding domain (which binds SM as well as a glycolipid) has been identified on Alzheimer β-amyloid protein, prion protein, and HIV-1 glycoprotein (75). This suggests another way in which membrane integral and peripheral proteins could interact with lipids to organize larger domains from core rafts. Interestingly, we have found the homologous sequence in a resident membrane integral protein, class I MHC (D. Fooksman, personal communication). Finally there is some evidence that a tetraspan protein, EMP2, modulates the surface expression of GPI-anchored proteins (145a), which hints at a function for EMP2 as an organizer of raft lipids.

CONCLUDING REMARKS

The interaction of proteins with membrane lipids to enlarge and stabilize transient rafts offers a perspective on the function of lipid rafts that combines physical chemistry and biochemistry. We can imagine cell membranes rich in raft components whose interactions with one another depend upon changes in the properties of small regions of membrane. Enzyme activation, receptor cross-linking, recruitment of lipids by proteins, and local changes in lipid concentrations could all work toward recruiting raft components into stable, relatively long-lived functional complexes, or the opposite, dispersing interacting components and quenching reactions. All this involves a mix of two-dimensional and three-dimensional diffusion as well. Model membranes can be used to investigate these interactions, but we should keep in mind the difference between the conveniences of scale required for observing the models and the actual scale of creating and dispersing functional signaling or trafficking rafts. Whatever we learn from models should be tempered by the fact that the cells are always right.

ACKNOWLEDGMENTS

I thank my colleagues Ira Probodh, Kalina Hristova, Anne Kenworthy, and Adrian Parsegian for reading and commenting on an early draft of this review. I also thank authors who sent me manuscripts ahead of publication. Our research on rafts and the writing of this review are supported by grants from the National Institutes of Health.

The *Annual Review of Biophysics and Biomolecular Structure* is online at http://biophys.annualreviews.org

LITERATURE CITED

1. Anderson RGW. 1998. The caveolar membrane system. *Annu. Rev. Biochem.* 67:199–225
2. Anderson RGW, Jacobson K. 2002. A role for lipid shells in targeting proteins to caveolae, rafts and other lipid domains. *Science* 296:1821–25
3. Anderson TG, McConnell HM. 2001. Condensed complexes and the calorimetry of cholesterol-phospholipid bilayers. *Biophys. J.* 81:2774–85
4. Bagatolli LA, Gratton E. 2000. Two photon fluorescence microscopy of coexisting lipid domains in giant unilamellar vesicles of binary phospholipids mixtures. *Biophys. J.* 78:290–305
5. Balamuth F, Leitenberg D, Unternaehrer J, Mellman I, Bottomly K. 2001. Distinct patterns of membrane microdomain partitioning in Th1 and Th2 cells. *Immunity* 15:729–38
6. Bamezai A, Goldmacher VS, Rock KL. 1992. Internalization of glycosylphosphatidylinositol (GPI)-anchored lymphocyte proteins. II. GPI-anchored and transmembrane molecules internalize through distinct pathways. *Eur. J. Immunol.* 22:15–21
7. Banerjee P, Joo JB, Buse JT, Dawson G. 1995. Differential solubilization of lipids along with membrane proteins by different classes of detergents. *Chem. Phys. Lipids* 77:65–78
8. Bavari S, Bosio CM, Wiegand E, Ruthel G, Will AB, et al. 2002. Lipid raft microdomains: a gateway for compartmentalized trafficking of Ebola and Marburg viruses. *J. Exp. Med.* 195:593–602
9. Blom TS, Koivusalo M, Kuismanen E, Kostiainen, Somerharju P, Ikonen E. 2001. Mass spectrometric analysis reveals an increase in plasma membrane polyunsaturated phospholipids species upon cellular cholesterol loading. *Biochemistry* 40:14635–44
10. Bodnar A, Jenei A, Bene L, Damjanovich S, Matko J. 1996. Modification of membrane cholesterol affects expression and clustering of class I HLA molecules at the surface of JY human lymphoblasts. *Immunol. Lett.* 54:221–26
11. Boggs JM. 1987. Lipid intermolecular hydrogen bonding: influence on structural organization and membrane function. *Biochim. Biophys. Acta* 906:353–404
12. Botelho RJ, Teruel M, Dierckman R, Anderson R, Wells A, et al. 2000. Localized biphasic changes in phosphatidylinositol-4,5-bisphosphate at sites of phagocytosis. *J. Cell Biol.* 151:1353–68
13. Boyle R. 1964. *The Sceptical Chymist*

(1661), ed. EO Moelwyn-Hughs, p. 63. London: Dent
14. Brown D, Crise B, Rose JK. 1989. Mechanism of membrane anchoring affects polarized expression of two proteins in MDCK cells. *Science* 245:1499–501
15. Brown DA, Rose JK. 1992. Sorting of GPI-anchored proteins to glycolipids-enriched membrane subdomains during transport to the apical cell surface. *Cell* 68:533–44
16. Brown RE. 1998. Sphingolipid organization in biomembranes: what physical studies of model membranes reveal. *J. Cell Sci.* 111:1–9
17. Brzustowicz MR, Cherezov V, Caffrey M, Stillwell W, Wassall SR. 2002. Molecular organization of cholesterol in polyunsaturated membranes: microdomain formation. *Biophys. J.* 82:285–98
18. Busche R, Dittman J, Meyer zu Duttingdorf H-D, Glockenthor U, von Engelhardt W, Sallmann H-P. 2002. Permeability properties of apical and basolateral membranes of the guinea pig caecal and colonic epithelia for short-chain fatty acids. *Biochim. Biophys. Acta* 1565:55–63
19. Caroni P. 2001. Actin cytoskeleton regulation through modulation of PI(4,5)P2 rafts. *EMBO J.* 20:4332–36
20. Cerneus DP, Ueffing E, Posthuma G, Strous GJ, van der Ende A. 1993. Detergent insolubility of alkaline phosphatase during biosynthetic transport and endocytosis. *J. Biol. Chem.* 268:3150–55
21. Chatterjee S, Smith ER, Hanada K, Stevens VL, Mayor S. 2001. GPI-anchoring leads to sphingolipid-dependent retention of endocytosed proteins in the recycling endosomal compartment. *EMBO J.* 20:1583–92
22. Cinek T, Horejsi V. 1992. The nature of large noncovalent complexes containing glycosyl-phosphatidylinositol-anchored membrane glycoproteins and protein tyrosine kinases. *J. Immunol.* 149:2262–70
23. Clejan S, Bittman R, Rottem S. 1981. Effects of sterol structure and exogenous lipids on the transbilayer distribution of sterols in the membrane of *Mycoplasma capricolum*. *Biochemistry* 20:2200–4
24. Dietrich C, Bagatolli LA, Volvyk ZN, Thompson NL, Levi M, et al. 2001. Lipid rafts reconstituted in model membranes. *Biophys. J.* 80:1417–28
25. Dietrich C, Volvyk ZN, Levi M, Thompson NL, Jacobson K. 2001. Partitioning of Thy-1, GM1, and cross-linked phospholipid analogs into lipid rafts reconstituted in supported model membrane monolayers. *Proc. Natl. Acad. Sci. USA* 98:10642–47
26. Drevot P, Langlet C, Guo X-J, Bernard A-M, Collard O, et al. 2002. TCR signal initiation machinery is pre-assembled and activated in a subset of membrane rafts. *EMBO J.* 21:1899–902
27. Dustin M. 2002. Shmoos, rafts and uropods—the many facets of cell polarity. *Cell* 110:13–18
28. El Yandouzi EH, Le Grimellec C. 1992. Cholesterol heterogeneity in the plasma membrane of epithelial cells. *Biochemistry* 31:547–51
29. Fielding CJ, Fielding PE. 2000. Cholesterol and caveolae: structural and functional relationships. *Biochim. Biophys. Acta* 1529:210–22
30. Fredriksson EK, Shipkova PA, Sheets ED, Holowka D, Baird B, McLafferty FW. 1999. Quantitative analysis of phospholipids in functionally important membrane domains from RBL-2H3 mast cells using tandem high-resolution mass spectrometry. *Biochemistry* 38:8056–63
31. Gandhavadi M, Allende D, Vidal A, Simon SA, McIntosh SA. 2002. Structure, composition, and peptide binding properties of detergent soluble bilayers and detergent resistant rafts. *Biophys. J.* 82:1469–82
32. Garcia M, Mirre C, Quaroni A, Reggio H, Le Bivic A. 1993. GPI-anchored proteins associate to form microdomains

during their intracellular transport in Caco-2 cells. *J. Cell Sci.* 104:1281–90

33. Germain RN. 2001. The T cell receptor for antigen: signaling and ligand discrimination. *J. Biol. Chem.* 276:35223–26
34. Gkantiragas I, Brugger B, Stuven E, Kaloyanova D, Li X-Y, et al. 2001. Sphingomyelin-enriched microdomains at the Golgi complex. *Mol. Biol. Cell* 12:1819–33
35. Goins B, Masserini M, Barisas BG, Freire E. 1986. Lateral diffusion of ganglioside G_{M1} in phospholipid bilayer membranes. *Biophys. J.* 49:849–56
36. Hagman J, Fishman PH. 1982. Detergent extraction of cholera toxin and gangliosides from cultured cells and isolated membranes. *Biochim. Biophys. Acta* 720:181–87
37. Hannan LA, Edidin M. 1996. Traffic, polarity and detergent solubility of a glycosylphosphatidylinositol-anchored protein after LDL-deprivation of MDCK cells. *J. Cell Biol.* 133:1265–76
38. Hansen GH, Immerdal L, Thorsen E, Niels-Christiansen L-L, Nystrom BT, et al. 2001. Lipid rafts exist as stable cholesterol-independent microdomains in the brush border membrane of enterocytes. *J. Biol. Chem.* 276:32338–44
39. Hao M, Mukherjee S, Maxfield FR. 2001. Cholesterol depletion induces large scale domain segregation in living cell membranes. *Proc. Natl. Acad. Sci. USA* 98:13072–77
40. Harder T, Kellner R, Parton RG, Gruenberg J. 1997. Specific release of membrane-bound annexin II and cortical cytoskeletal elements by sequestration of membrane cholesterol. *Mol. Biol. Cell* 8:533–45
41. Harder T, Scheiffele P, Verkade P, Simons K. 1998. Lipid domain structure of the plasma membrane revealed by patching of membrane components. *J. Cell Biol.* 141:929–42
42. Harder T, Simons K. 1999. Clusters of glycolipid and glycosylphosphatidylinositol-anchored proteins in lymphoid cells: accumulation of actin regulated by local tyrosine phosphorylation. *Eur. J. Immunol.* 29:556–62
43. Haynes MP, Phillips MC, Rothblat GH. 2000. Efflux of cholesterol from different cellular pools. *Biochemistry* 39:4508–17
44. Hedayati SM, Atkinson JP, Holguin MH, Parker CJ, Spear GT. 1997. Human immunodeficiency virus type 1 incorporates both glycosyl phosphatidylinositol-anchored CD55 and CD59 and integral membrane CD46 at levels that protect from complement-mediated destruction. *J. Gen. Virol.* 78:1907–11
45. Heerklotz H. 2002. Triton promotes domain formation in lipid raft mixtures. *Biophys. J.* 83:2693–701
46. Hekman M, Hamm H, Villar AV, Bader B, Kuhlmann J, et al. 2002. Associations of B- and C-Raf with cholesterol, phosphatidylserine and lipid second messengers. *J. Biol. Chem.* 277:14067–76
47. Hertz R, Barenholz Y. 1977. The relations between the composition of liposomes and their interaction with Triton X-100. *J. Colloid Interface Sci.* 60:188–200
48. Hoessli D, Rungger-Brandle E. 1985. Association of specific cell-surface glycoproteins with a Triton X-100-resistant complex of plasma membrane proteins isolated from T-lymphoma cells. *Exp. Cell Res.* 156:239–50
49. Holowka D, Sheetz ED, Baird B. 2000. Interactions between FcεRI and lipid raft components are regulated by the actin cytoskeleton. *J. Cell Sci.* 113:1009–19
50. Hooper NM, Turner AJ. 1988. Ectoenzymes of the kidney microvillar membrane. *Biochem. J.* 250:865–69
51. Hurley JH, Meyer T. 2001. Subcellular targeting by membrane lipids. *Curr. Opin. Cell Biol.* 13:146–52
52. Ipsen JH, Karlstrom G, Mouritsen OG, Wennerstrom H, Zuckermann MJ. 1987.

Phase equilibria in the phosphatidylcholine-cholesterol system. *Biochim. Biophys. Acta* 905:162–72
53. Jain MK. 1975. Role of cholesterol in biomembranes and related systems. *Curr. Top. Membr. Trans.* 7:1–57
54. Jain MK, White HB 3rd. 1977. Long-range order in biomembranes. *Adv. Lipid Res.* 15:1–60
55. Janes PW, Ley SC, Magee AI. 1999. Aggregation of lipid rafts accompanies signaling via the T cell antigen receptor. *J. Cell Biol.* 147:447–61
56. Janmey PA, Xian W, Flanagan LA. 1999. Controlling cytoskeleton structure by phosphoinositide-protein interactions: phosphoinositide binding protein domains and effects of lipid packing. *Chem. Phys. Lipids* 101:93–107
57. Kabouridis PS, Janzen J, Magee AL, Ley SC. 2000. Cholesterol depletion disrupts lipid rafts and modulates the activity of multiple signaling pathways in T lymphocytes. *Eur. J. Immunol.* 30:954–63
58. Kawasaki K, Yin J-J, Subczynski WK, Hyde JS, Kusumi A. 2001. Pulse EPR detection of lipid exchange between protein-rich raft and bulk domains in the membrane: methodology development and its application to studies of influenza viral membrane. *Biophys. J.* 80:738–48
59. Keller GA, Siegel MW, Caras IW. 1992. Endocytosis of glycosphingolipid-anchored and transmembrane forms of CD4 by different endocytic pathways. *EMBO J.* 11:863–74
60. Keller SL, Pitcher WH III, Huestis WH, McConnell HM. 1998. Red blood cell lipids form immiscible liquids. *Phys. Rev. Lett.* 81:5019–22
61. Kenworthy AK, Edidin M. 1998. Distribution of a GPI-anchored protein at the apical surface of MDCK cells examined at a resolution of <100 Å using imaging fluorescence resonance energy transfer. *J. Cell Biol.* 142:69–84
62. Kenworthy AK, Nichols BJ, Bekiranov S, Kumar M, Zimmerberg J, et al. 2001. Large-scale lateral diffusion measurements of plasma membrane proteins reveal uncorrelated diffusion of lipid raft markers. *Mol. Biol. Cell* 12S:471 (Abstr.)
63. Klausner RD, Kleinfeld AM, Hoover RL, Karnovsky MJ. 1980. Lipid domains in membranes. Evidence derived from structural perturbations induced by free fatty acids and lifetime heterogeneity analysis. *J. Biol. Chem.* 255:1286–95
64. Korlach J, Schwille P, Webb WW, Feigenson G. 1999. Characterization of lipid bilayer phases by confocal microscopy and fluorescence correlation spectroscopy. *Proc. Natl. Acad. Sci. USA* 96:8461–66
65. Kwik, Margolis, Boyle S, Sheetz MS, Edidin M. 2002. Membrane cholesterol, lateral diffusion of membrane proteins and the PI(4,5)P2-dependent organization of cell actin. *Mol. Biol. Cell* 13S: 142 (Abstr.)
66. Lang T, Bruns D, Wenzel D, Riedel D, Holroyd P, et al. 2001. SNAREs are concentrated in cholesterol-dependent clusters that define docking and fusion sites for exocytosis. *EMBO J.* 20:2202–13
67. Lange Y, Steck TL. 1996. The role of intracellular cholesterol transport in cholesterol homeostasis. *Trends Cell Biol.* 6:205–8
68. Lehto MT, Sharom FJ. 2002. PI-specific phospholipase C cleavage of a reconstituted GPI-anchored protein: modulation by the lipid bilayer. *Biochemistry* 41:1398–408
69. Lehto MT, Sharom FJ. 2002. Proximity of the protein moiety of a GPI-anchored protein to the membrane surface: a FRET study. *Biochemistry* 41:8368–76
70. Lisanti MP, Caras IW, Davitz MA, Rodriguez-Boulan E. 1989. A glycophospholipid membrane anchor acts as an apical targeting signal in polarized epithelial cells. *J. Cell Biol.* 109:2145–56
71. Lisanti MP, Sargiacomo M, Graeve L, Saltiel AR, Rodriguez-Boulan E. 1988.

Polarized apical distribution of glycosyl-phosphatidylinositol-anchored proteins in a renal epithelial cell line. *Proc. Natl. Acad. Sci. USA* 85:9557–61
72. London E. 2002. Insights into lipid raft structure and formation from experiments in model membranes. *Curr. Opin. Struct. Biol.* 12:480–86
73. Los DA, Murata N. 2000. Regulation of enzymatic activity and gene expression by membrane fluidity. *Science's STKE* http://www.stke.org/cgi/content/full/OC_sigtrans;2000/62/pe1
74. Loura LMS, Fedorov A, Prieto M. 2001. Fluid-fluid membrane microheterogeneity: a fluorescence resonance energy transfer study. *Biophys. J.* 80:776–88
75. Mahfoud R, Garmy N, Maresca M, Yahi N, Puigserver A, Fantini J. 2002. Identification of a common sphingolipid-binding domain in Alzheimer, prion, and HIV-1 proteins. *J. Biol. Chem.* 277:11292–96
76. Majewski J, Kuhl TL, Kjaer K, Smith GS. 2001. Packing of ganglioside-phospholipid monolayers: an X-ray diffraction and reflectivity study. *Biophys. J.* 81:2707–15
77. Marsh D, Horvath LI. 1998. Structure, dynamics and composition of the lipid-protein interface. Perspectives from spin labeling. *Biochim. Biophys. Acta* 1376:267–96
78. Massey J. 2001. Interaction of ceramides with phosphatidylcholine, sphingomyelin and sphingomyelin/cholesterol bilayers. *Biochim. Biophys. Acta* 1510: 167–84
79. Mattjus P, Kline A, Pike HM, Molotkovsky JG, Brown RE. 2002. Probing for preferential interactions among sphingolipids in bilayer vesicles using the glycolipids transfer protein. *Biochemistry* 41:266–73
80. Mayor S, Sabharanjak S, Maxfield FR. 1998. Cholesterol-dependent retention of GPI-anchored proteins in endosomes. *EMBO J.* 17:4626–38
81. McCabe JB, Berthiaume LG. 2001. N-terminal protein acylation confers localization to cholesterol, sphingolipid-enriched membranes but not to lipid rafts/caveolae. *Mol. Biol. Cell* 12:3601–17
82. McConnell HM, Vrljic M. 2003. Liquid-liquid immiscibility in membranes. *Annu. Rev. Biophys. Biomol. Struct.* In press
83. McLaughlin S, Wang J, Gambhir A, Murray D. 2002. PIP_2 and proteins: interactions, organization, and information flow. *Annu. Rev. Biophys. Biomol. Struct.* 31:151–75
84. Mia L, Nielsen M, Thewalt J, Ipsen JH, Bloom M, et al. 2002. From lanosterol to cholesterol: structural evolution and differential effects on lipid bilayers. *Biophys. J.* 82:1429–44
85. Milhiet PE, Giocondi M-C, Le Grimellec C. 2002. Cholesterol is not crucial for the existence of microdomains in kidney brush-border membrane models. *J. Biol. Chem.* 277:875–78
86. Mukherjee S, Soe TT, Maxfield FR. 1999. Endocytic sorting of lipid analogues differing solely in the chemistry of their hydrophobic tails. *J. Cell Biol.* 144:1271–84
87. Nguyen DH, Hildreth JEK. 2000. Evidence for budding of human immunodeficiency virus type 1 selectively from glycolipid-enriched membrane lipid rafts. *J. Virol.* 74:3264–72
88. Nichols BJ, Kenworthy AK, Polishchuk RS, Lodge R, Roberts TH, et al. 2001. Rapid cycling of lipid raft markers between the cell surface and the Golgi complex. *J. Cell Biol.* 153:529–41
89. Niv H, Gutman O, Kloog Y, Henis YI. 2002. Activated K-Ras and H-Ras display different interactions with saturable nonraft sites at the surface of live cells. *J. Cell. Biol.* 157:865–72
90. Nyholm T, Slotte JP. 2001. Comparison of Triton X-100 penetration into phosphatidylcholine and sphingomyelin

mono- and bilayers. *Langmuir* 17:4724–30
91. Ohvo H, Olsio C, Slotte JP. 1997. Effects of sphingomyelin and phosphatidylcholine degradation on cyclodextrin-mediated cholesterol efflux in cultured fibroblasts. *Biochim. Biophys Acta* 1349:131–34
92. Ohvo-Rekila H, Ramstedt B, Leppimaki P, Slotte JP. 2002. Cholesterol interactions with phospholipids in membranes. *Prog. Lipid Res.* 41:66–97
93. Ostermeyer AG, Beckrich BT, Ivarson KA, Grove KE, Brown DA. 1999. Glycosphingolipids are not essential for formation of detergent-resistant membrane rafts in melanoma cells. *J. Biol. Chem.* 274:34459–66
94. Pascher I. 1976. Molecular arrangements in sphingolipids. Conformation and hydrogen bonding of ceramide and their implication on membrane stability and permeability. *Biochim. Biophys. Acta* 455:433–51
95. Pierce SK. 2001. Lipid rafts and B-cell activation. *Nat. Rev. Immunol.* 2:96–104
96. Pike LJ, Casey L. 2002. Cholesterol levels modulate EGF receptor-mediated signaling by altering receptor function and trafficking. *Biochemistry* 41:10315–22
97. Pike LJ, Han X, Chung K-N, Gross RW. 2002. Lipid rafts are enriched in arachidonic acid and plasmenylethanolamine and their composition is independent of caveolin-1 expression: a quantitative electrospray ionization/mass spectrometric analysis. *Biochemistry* 41:2075–88
98. Pike LJ, Miller JM. 1998. Cholesterol depletion delocalizes phosphatidylinositol bisphosphate and inhibits hormone-stimulated phosphostidylinositol turnover. *J. Biol. Chem.* 273:22298–304
99. Pörn MI, Slotte JP. 1995. Localization of cholesterol in sphingomyelinase-treated fibroblasts. *Biochem. J.* 308:269–74
100. Pralle A, Keller P, Florin E-L, Simons K, Hörber JKH. 2000. Sphingolipid-cholesterol rafts diffuse as small entities in the plasma membrane of mammalian cells. *J. Cell Biol.* 148:997–1007
101. Prior IA, Harding A, Yan J, Sluimer J, Parton RG, Hancock JF. 2001. GTP-dependent segregation of H-ras from lipid rafts is required for biological activity. *Nat. Cell* 3:868–72
102. Puri V, Watanabe R, Singh RD, Dominguez M, Brown JC, et al. 2001. Clathrin-dependent and -independent internalization of plasma membrane sphingolipids initiates two Golgi targeting pathways. *J. Cell Biol.* 154:535–47
103. Pyenta PS, Holowka D, Baird B. 2001. Cross-correlation analysis of inner-leaflet-anchored green fluorescent proteins co-distributed with IgE receptors and outer leaflet raft components. *Biophys. J.* 80:2120–32
104. Radhakrishnan A, Anderson TG, McConnell HM. 2000. Condensed complexes, rafts, and the chemical activity of cholesterol in membranes. *Proc. Natl. Acad. Sci. USA* 97:12422–27
105. Raucher D, Stauffer T, Chen W, Shen K, Guo S, et al. 2000. Phosphatidylinositol 4,5-bisphosphate functions as a second messenger that regulates cytoskeleton-plasma membrane adhesion. *Cell* 100:221–28
106. Rietveld A, Simons K. 1998. The differential miscibility of lipids as the basis for the formation of functional membrane rafts. *Biochim. Biophys. Acta* 1376:467–79
107. Rinia HA, Snel MME, van der Erden JPJM, de Kruijff B. 2001. Visualizing detergent resistant domains in model membranes with atomic force microscopy. *FEBS Lett.* 501:92–96
108. Rock P, Allietta M, Young WW Jr, Thompson TE, Tillack TW. 1990. Organization of glycosphingolipids in phosphatidylcholine bilayers: use of antibody molecules and Fab fragments as morphologic markers. *Biochemistry* 29:8484–90

109. Rodal SK, Skretting G, Garred Ø, Vilhardt F, van Deurs B, Sandvig K. 1999. Extraction of cholesterol with methyl-β-cyclodextrin perturbs formation of clathrin-coated endocytic vesicles. *Mol. Biol. Cell* 10:961–74
110. Rodgers W, Zavzavadjian J. 2001. Glycolipid-enriched membrane domains are assembled into membrane patches by associating with the actin cytoskeleton. *Exp. Cell Res.* 267:173–83
111. Rodriguez-Boulan E, Sabatini DD. 1978. Asymmetric budding of viruses in epithelial monlayers: a model system for study of epithelial polarity. *Proc. Natl. Acad. Sci. USA* 75:5071–75
112. Rothberg KG, Ying Y-S, Kamen BA, Anderson RGW. 1990. Cholesterol controls the clustering of the glycophospholipid-anchored membrane receptor for 5-methyltetrahydrofolate. *J. Cell Biol.* 111:2931–38
113. Sabharanjak S, Sharma P, Parton RG, Mayor S. 2002. GPI-anchored proteins are delivered to recycling endosomes via a distinct cdc42-regulated, clathrin-independent pinocytic pathway. *Dev. Cell* 2:411–23
114. Samsonov AV, Mihalyov I, Cohen FS. 2001. Characterization of cholesterol-sphingomyelin domains and their dynamics in bilayer membranes. *Biophys. J.* 81:1486–1500
115. Sanghera N, Pinheiro TJ. 2002. Binding of prion protein to lipid membranes and implications for prion conversion. *J. Mol. Biol.* 315:1241–56
116. Sankaram MB, Thompson TE. 1990. Modulation of phospholipids acyl chain order by cholesterol: a solid-state ^{2}H nuclear magnetic resonance study. *Biochemistry* 29:10676–84
117. Saslowsky DE, Lawrence J, Ren X, Brown DA, Henderson RM, Edwardson JM. 2002. Placental alkaline phosphatase is efficiently targeted to rafts in supported lipid bilayers. *J. Biol. Chem.* 277:26966–70
118. Scarlata S. 2002. Regulation of the lateral association of phospholipase $C^{\beta}2$ and G protein subunits by lipid rafts. *Biochemistry* 41:7092–99
119. Scheiffle P, Rietveld A, Wilk T, Simons K. 1999. Influenza viruses select ordered lipid domains during budding from the plasma membrane. *J. Biol. Chem.* 274:2038–44
120. Schroeder F, Nemecz G, Wood WG, Joiner C, Morrot G, et al. 1991. Transmembrane distribution of sterol in the human erythrocyte. *Biochim. Biophys. Acta* 1066:183–92
121. Schroeder R, London E, Brown D. 1994. Interactions between saturated acyl chains confer detergent resistance on lipids and glycosylphosphatidylinositol (GPI)-anchored proteins: GPI-anchored proteins in cells and liposomes show similar behavior. *Proc. Natl. Acad. Sci. USA* 91:12130–34
122. Shahinian S, Silvius JR. 1995. Doubly-lipid-modified sequence motifs exhibit long-lived anchorage to lipid bilayer membranes. *Biochemistry* 34:3813–22
123. Sharma DK, Choudhury AK, Singh R, Wheatley CL, Marks DL, Pagano RE. 2003. Glycosphingolipids internalized *via* non-clathrin endocytosis rapidly merge with the clathrin pathway in early endosomes and form "lipid rafts" for recycling. *J. Biol. Chem.* 278(9):7564–72
124. Sheets ED, Holowka D, Baird B. 1999. Critical role for cholesterol in Lyn-mediated tyrosine phosphorylation of FcεR1 and their association with detergent-resistant complexes. *J. Cell Biol.* 145:877–87
125. Shinitzky M, Souroujon M. 1979. Passive modulation of blood-group antigens. *Proc. Natl. Acad. Sci. USA* 76:4438–40
126. Simons K, Ikonnen E. 1997. Functional rafts in cell membranes. *Nature* 389:569–72
127. Simons K, Toomre D. 2000. Lipid rafts

and signal transduction. *Nat. Rev. Mol. Cell Biol.* 1:31–41
128. Simons K, van Meer G. 1988. Lipid sorting in epithelial cells. *Biochemistry* 27:6197–202
129. Singer SJ, Nicolson GL. 1972. The fluid mosaic model of the structure of cell membranes. *Science* 175:720–31
130. Sohn R, Marinetti GV. 1974. Effect of detergents on the enzyme activities of the rat liver plasma membrane. *Chem. Phys. Lipids* 12:17–30
131. Sot J, Collado MI, Arrondo JLR, Alonso A, Goni FM. 2002. Triton X-100-resistant bilayers: effect of lipid composition and relevance to the raft phenomenon. *Langmuir* 18:2828–35
132. Spiegel S, Blumenthal R, Fishman PH, Handler JS. 1985. Gangliosides do not move from apical to basolateral plasma membrane in cultured epithelial cells. *Biochim. Biophys. Acta* 821:310–18
133. Sproul TW, Malapati S, Kim J, Pierce SK. 2000. B cell antigen receptor signaling occurs outside lipid rafts in immature B cells. *J. Immunol.* 165:6020–23
134. Stefanova I, Horejsi V. 1991. Association of the CD59 and CD55 cell surface glycoproteins with other membrane molecules. *J. Immunol.* 147:1587–92
135. Stefanova I, Horejsi V, Ansotegui IJ, Knapp W, Stockinger H. 1991. GPI-anchored cell-surface molecules complexed to protein tyrosine kinases. *Science* 254:1016–19
136. Stubbs CD. 1983. Membrane fluidity: structure and dynamics of membrane lipids. *Essays Biochem.* 19:1–39
137. Subtil A, Gaidarov I, Kobylarz K, Lampson MA, Keen JH, McGraw TE. 1999. Acute cholesterol depletion inhibits clathrin-coated pit budding. *Proc. Natl. Acad. Sci. USA* 96:6775–80
138. Suzuki K, Sanematsu F, Fujiwara T, Edidin M, Kusumi A. 2001. Rapid continual formation/dispersion of raft-like domains in the resting cell membrane. *Mol. Biol. Cell* 12S:470 (Abstr.)
139. Suzuki K, Sheetz MP. 2001. Binding of cross-linked glycosylphosphatidylinositol-anchored proteins to discrete actin-associated sites and cholesterol-dependent domains. *Biophys. J.* 81:2181–89
140. Thompson TE, Tillack TW. 1985. Organization of glycosphingolipids in bilayers and plasma membranes of mammalian cells. *Annu. Rev. Biophys. Biophys. Chem.* 14:361–86
141. van der Groot FG, Harder T. 2001. Raft membrane domains: from liquid-ordered membrane phase to a site of pathogen attack. *Semin. Immunol.* 13:89–97
142. van Meer G, Simons K. 1982. Viruses budding from either the apical or the basolateral plasma membrane domain of MDCK cells have unique phospholipid compositions. *EMBO J.* 1:847–52
143. van Meer G, Simons K. 1988. Lipid polarity and sorting in epithelial cells. *J. Cell. Biochem.* 36:51–58
144. Varma R, Mayor S. 1998. GPI-anchored proteins are organized in submicron domains at the cell surface. *Nature* 394:798–801
145. Wadhera M, Goodglick L, Braun J. 2002. The tatraspan protein EMP2 modulates the surface expression of caveolins and GPI linked proteins. *FASEB J.* 16:A1178 (Abstr.)
146. Wang TY, Leventis R, Silvius JR. 2001. Partitioning of lipidated peptide sequences into liquid-ordered lipid domains in model and biological membranes. *Biochemistry* 40:13031–40
147. Wang T-Y, Silvius JR. 2001. Cholesterol does not induce segregation of liquid-ordered domains in bilayers modeling the inner leaflet of the plasma membrane. *Biophys. J.* 81:2762–73
148. Watts A. 1998. Solid-state NMR approaches for studying the interaction of peptides and proteins with membranes. *Biochim. Biophys. Acta* 1376:297–318
149. Weimbs T, Low SH, Chapin SJ, Mostov KE. 1997. Apical targeting in polarized

epithelial cells: There's more afloat than rafts. *Trends Cell Biol.* 7:393–99

150. Werlen G, Palmer E. 2002. The T-cell receptor signalosome: a dynamic structure with expanding complexity. *Curr. Opin. Immunol.* 14:299–305
151. Zacharias DA, Violin JD, Newton AC, Tsien RY. 2002. Partitioning of lipid-modified monomeric GFPs into membrane microdomains of live cells. *Science* 296:913–16
152. Zhang J, Pekosz A, Lamb RA. 2000. Influenza virus assembly and lipid raft microdomains: a role for the cytoplasmic tails of spike glycoproteins. *J. Virol.* 74:4634–44

Annu. Rev. Biophys. Biomol. Struct. 2003. 32:285–310
doi: 10.1146/annurev.biophys.32.110601.142516

First published online as a Review in Advance on February 10, 2003

X-RAY CRYSTALLOGRAPHIC ANALYSIS OF LIPID-PROTEIN INTERACTIONS IN THE BACTERIORHODOPSIN PURPLE MEMBRANE*

Jean-Philippe Cartailler[1] and Hartmut Luecke[1,2]
[1]Department of Molecular Biology and Biochemistry, and [2]Departments of Physiology and Biophysics, and Information and Computer Science, University of California Irvine, Irvine, California 92697-3900; email: hudel@uci.edu

Key Words bilayer membrane, proteolipid, archeol, crystallization, light-driven ion pump

■ **Abstract** The past decade has witnessed increasingly detailed insights into the structural mechanism of the bacteriorhodopsin photocycle. Concurrently, there has been much progress within our knowledge pertaining to the lipids of the purple membrane, including the discovery of new lipids and the overall effort to localize and identify each lipid within the purple membrane. Therefore, there is a need to classify this information to generalize the findings. We discuss the properties and roles of haloarchaeal lipids and present the structural data as individual case studies. Lipid-protein interactions are discussed in the context of structure-function relationships. A brief discussion of the possibility that bacteriorhodopsin functions as a light-driven inward hydroxide pump rather than an outward proton pump is also presented.

CONTENTS

*Abbreviations: BR, bacteriorhodopsin; CP, cytoplasmic; EC, extracellular; HR, halorhodopsin; PM, purple membrane; PG, phosphatidylglycerol (diphytanylglycerol ether analog); PGP-Me, phosphatidylglycerophosphate methyl ester (diphytanylglycerol ether analog); PGS, phosphatidylglycerosulfate (diphytanylglycerol ether analog); S-TGA-1 (a.k.a. S-TGD-1), 3-HSO_3-Galp-β1,6-Manp-α1,2-Glcp-α1,1-sn-2,3-diphytanylglycerol; BPG (bisphosphatidylglycerol), *sn*-2,3-di-*O*-phytanyl-1-phosphoglycerol-3-phospho-*sn*-2,3-di-*O*-phytanylglycerol; GlyC, archaeal glycocardiolipin or 3-HSO_3-Galp-β1,6-Manp-α1, 2-Glcp-α1,1-[sn-2,3-di-*O*-phytanylglycerol] -6 -[phospho - sn-2,3-di-O-phytanylglycerol]; SRII, sensory rhodopsin II.

INTRODUCTION

All living cells are separated from the surrounding environment by a cell membrane that provides more than just a simple barrier. Highly specialized, cell membranes are metabolically active in processes such as respiration, photosynthesis, protein transport, signal transduction, motility, and solute transport. Despite varying physicochemical composition, the basic structural unit of virtually all biomembranes is the phospholipid bilayer. The bilayer is typically composed of two leaflets of phospholipid molecules whose apolar acyl chains form the 3-nm-thick hydrophobic interior while leaving the polar headgroups on both sides solvent exposed. Within this bilayer structure, multitudes of integral and peripheral membrane proteins are scattered, together with the bilayer comprising the biological cell membrane. Aside from providing structural and functional integrity to the membrane, the lipids allow membrane proteins to reside within the bilayer and allow for sufficient flexibility required for critical cellular events such as vesicle budding and cellular division while maintaining an electrical seal. Molecular shape, flexibility, and charge of individual lipids determine the physical properties of the membrane such as thickness and curvature. In the cell, a variety of lipids are required to establish the biophysical properties of a given membrane, in contrast to the simple bilayer structures that can be formed in vitro with the prototypic lipid phosphatidylcholine. Membrane lipids self-assemble into different aggregate structures depending on their environment, a phenomenon called lipid polymorphism. Lipid assemblies other than the common bilayer also include nonbilayer structures, such as the inverted hexagonal phase and associated intermediate structures, typically formed from (cone-shaped) nonbilayer-forming lipids. Much research has been performed on nonbiological bilayers (103), and the emerging picture is that nonbilayer-forming lipids confer special structural features on membranes and their functional attributes, as seen with lactose permease (phosphatidylethanolamine) (9) and cytochrome *c* oxidase (cardiolipins, phosphatidylethanolamine, and phosphatidylcholine) (89). A high degree of complementarity between the hydrophobic protein surface and the neighboring lipids is generally believed to be important for the stable integration of integral membrane proteins into the bilayer, as previously reviewed elsewhere (24, 37). After a brief introduction on light-driven ion pumping, this review focuses on the structural arrangement of the protein and lipid components of the *Halobacterium salinarum* purple membrane (PM).

BACTERIORHODOPSIN

The Photocycle

Microbial rhodopsins comprise a family of photoactive, seven-transmembrane-helical retinal proteins found in phylogenetically diverse microorganisms, including archaea, proteobacteria, cyanobacteria, fungi, and algae (3, 91, 92). The family can be divided into two classes: (*a*) light-driven ion pumps, such as bacteriorhodopsin (BR) and halorhodopsin (HR) in archaea, and proteorhodopsin in marine bacteria, with a rapid (>50 Hz) photocycle, and (*b*) photosensory receptors, such as sensory rhodopsins I and II (SRI and SRII) in archaea, *Chlamydomonas* rhodopsins CsoA and CsoB, and *Anabaena* rhodopsin, with a slower photocycle (<20 Hz).

Bacteriorhodopsin, the most studied member of this family, is a light-driven transmembrane ion pump found in extreme halophilic archaea such as *H. salinarum* (74). Photoisomerization of the all-*trans* retinal chromophore, covalently attached to Lys216 through a protonated Schiff base, to the 13-*cis*,15-*anti* configuration results in ion translocation across the cell membrane and establishes an electrochemical gradient for ATP synthesis and other energy-requiring membrane processes. The cyclic reaction, know as the photocycle, that follows the photoisomerization of the retinal without further input of energy produces distinct, spectroscopically identifiable intermediates (Figure 1, see color insert). Figure 2*A* (see color insert) depicts the key ion translocation events during the photocycle, and the following list summarizes the salient features of each of the photocycle states BR, K, L, M, N, and O:

- The BR (ground/resting/light-adapted) state ($\lambda_{max} = 568$ nm) is the best-defined state in terms of the structure as well as side chain ionization states. The retinal chromophore is in a nonstrained all-*trans* configuration, and Lys216, to which it is covalently attached, is part of a so-called π-bulge in the middle of helix G (64). This π-bulge is thought to impart greater flexibility to this region, which was shown to undergo large conformational changes during the photocycle (58, 59). Toward the extracellular (EC) side, an extensive hydrogen-bonded network leads from the Schiff base via key water W402, deprotonated Asp85, W406, Arg82, and additional waters to the initially protonated proton release group (Glu204/Glu194 and nearby waters) and on to the EC surface. In marked contrast, the cytoplasmic (CP) side is hydrophobic, with no polar residues or ordered waters between the Schiff base and the CP surface, with the exception of a pair of uncharged polar side chains, protonated Asp96 and Thr46, about equidistant from the Schiff base and the surface (64).

- The K state (λ_{max} = 590 nm) arises within a few picoseconds of photon absorption, which deposits about 50 kcal/mol of energy into the retinal. The K state has a reported ΔH of 11.6 kcal/mol, thus about 20% of the photon energy is converted to enthalpy and a substantial portion of this enthalpy gain is due to charge separation (6, 7). Spectroscopic methods have determined a highly strained (twisted) 13-*cis*,15-*anti* configuration of the retinal when prepared at 77 K (denoted K_{LT} state) as evidenced by large-amplitude hydrogen out-of-plane (HOOP) vibrations (10, 82, 90). In contrast, when illumination is carried out at ambient temperature to prevent low-temperature artifacts, the retinal in K adopts a less twisted, near planar configuration, only to become twisted again in the transition to the early L state (17a, 67a). Although the K_{LT} state can be cryo-trapped at temperatures below 140 K, spectral overlap with the ground state (Figure 1*B*) and other factors limit its maximal occupancy to about 50% (1), making analysis of K_{LT} by traditional structural methods difficult (59).
- In the L state (λ_{max} = 550 nm) the retinal is less twisted than in the K_{LT} state produced at 77 K, but more so than in a K state produced at ambient temperature (17a, 67a), and hydrogen bonds of the retinal, protein groups and bound waters begin to change (66). The conformation at this point in the photocycle is of particular importance as the active site is now primed for the decisive event in the photocycle, the protonation of Asp85 and the deprotonation of the Schiff base (L to M reaction). The structural mechanism underlying these protonation changes is currently under intense investigation. However, as in the case of the K_{LT} state, spectral overlap (Figure 1*B*) and photoinduced interconversion between states limit L_{LT} state occupancies to about 40% or less (1).
- The M state (λ_{max} = 412 nm) is defined by a deprotonated Schiff base with a 13-*cis* chromophore and consists of at least two substates, often referred to as M_1 and M_2 (98). The M state is the only state other than the ground state that can be populated to and cryo-trapped at 100% occupancy owing to its large spectral blueshift of over 150 nm. The deprotonated Schiff base nitrogen is now pointing into a hydrophobic pocket on the CP side, while Asp85 on the EC side has been protonated, resulting in the elimination of the light-induced charge separation present in the K and L states. This in turn is thought to allow the charged guanidinium of Arg82 to approach the proton release group, causing proton release at the EC surface. On the CP side, a chain of ordered water molecules is starting to extend from protonated Asp96 toward the deprotonated Schiff base (61), presumably to bring about reprotonation of the Schiff base from the CP side in the M to N transition.
- In the N state (λ_{max} = 560 nm) the Schiff base has been reprotonated, whereas Asp96 to the CP side has deprotonated (25). With Asp85 still protonated

and Asp96 now deprotonated (charge reversal with respect to ground state), the retinal binding site now preferentially accommodates the 13-*cis*,15-*anti* configuration of the chromophore (17). During the M to N transition, the CP channel also opens up considerably (95, 100), presumably increasing hydration of this otherwise hydrophobic region. Clearly, this hydration has to be orchestrated in such a way as to prevent deprotonation of Asp96 to the bulk solvent.

- The O state (λ_{max} = 630 nm) occurs after both reprotonation of Asp96 from the CP surface and thermal reisomerization of the retinal to all-*trans*. The driving force for its relatively slow transition back to the ground state is not clear, but in this essentially unidirectional step of the photocycle the initially low pK_a of Asp85 is reestablished, causing this residue to deprotonate in a strongly downhill reaction. Concurrently, the EC proton release group reprotonates (42, 81), or, when a proton is not released from this site in the L to M reaction, a proton is released directly to the surface. Recent studies of the D85S mutant of BR in the absence of chloride have suggested a large transient opening of the EC side in the O state (83).

Outward Proton Pump or Inward Hydroxide Pump?

Historically, BR has been described as an outward proton pump, i.e., pumping one proton from the CP side to the EC side per photocycle. The reason for assuming the pumped ionic species to be protons is based on the circumstance that most time-resolved visible (55) and infrared (65) spectroscopic studies observe protonation/deprotonation events, such as the deprotonation of the Schiff base during the L to M transition, or the reprotonation of Asp96 in the N to O transition.

In contrast, the genetically, structurally, and presumably functionally related light-driven ion pump HR is a bona fide anion pump, which is known to pump the much heavier chloride anion into cells. Furthermore, BR can be converted into an inward chloride anion pump either through a single mutation that neutralizes the charge of Asp85 (Figure 2*B*, D85S or D85T) (53) or by lowering the pH to affect protonation of Asp85. And vice versa, the inward anion pump HR can be converted into what is presumably an outward cation (proton) pump by the simple addition of azide (2).

The overall photocycle of HR (and D85S/T BR) when pumping chloride is strikingly similar to that of wild-type BR, with the exception that the Schiff base never deprotonates, which leads to ion pumping without M intermediate formation. Furthermore, because HR is a net chloride pump, during the course of each pump cycle one large and relatively heavy anion must pass through the whole length of the molecule, including passing the seal formed by the retinal in its binding pocket (Figure 2*B*).

Thus, one could argue by analogy that wild-type BR might function as an inward hydroxide (OH^-) anion pump rather than as an outward proton (H^+) cation pump. Although the possibility that BR might be an inward anion pump has been discussed privately in the research community for decades, the first such proposal based on an atomic resolution structure was put forward only recently (64), followed by a more detailed mechanism (58). Since then, other groups have also started to discuss net OH^- pumping mechanisms (5a, 18, 38).

Net OH^- transport across the bilayer into a cell can be achieved by proton transport in the outward direction, coupled to the transport of water molecules into the cell ($OH^- \uparrow = H^+ \downarrow + H_2O \uparrow$). Thus, most of the inferred proton transfer events of the BR photocycle could equally be part of a net hydroxide pumping mechanism (58), with the exception of the specific ion movements that take place during the L to M transition, when the Schiff base deprotonates and Asp85 protonates. Unfortunately, experimental evidence of net cation versus net anion pumping is difficult to obtain because of the high abundance of water and the rapid exchange of water protons combined with the high water permeability of most bilayers.

The remainder of this review focuses on the functional significance of the interaction between BR and the lipid bilayer that it is embedded in, both of which are integral parts of PM.

Functional Significance of Purple Membrane Lipids

Are lipids really that important in BR function? There is disagreement about the importance of the two-dimensional crystalline state of BR. Hartmann et al. (30) showed that the crystalline state is important for in vivo physiology of BR. However, in vitro experiments hinted that the arrangement of BR trimers into a hexagonal array (two-dimensional crystalline state) is not essential for its function (12) (Figure 3, see color insert). Nonetheless, delipidated BR reconstitutes into the PM only with native lipids (39), marking their importance in specific lipid-protein contacts. Specific lipids can influence the photocycle; in particular a combination of phosphatidylglycerophosphate (PGP-Me) and squalene is required to maintain normal photocycle behavior (40).

PHYSICAL AND CHEMICAL PROPERTIES OF ARCHAEAL LIPIDS

Unique Properties of Archaeal Lipids

The membrane lipids of archaebacteria [now designated archaea (104)] consist uniquely of diphytanylglycerol diether (43) and its dimer dibiphytanyldiglycerol tetraether (15), both derivatives of a C_{20} fatty acid. In contrast, eubacteria and plants mostly have diacylglycerol-derived membrane lipids, and eukarya contain mostly diacylglycerol lipids and some monoacylglycerol-derived lipids (44). These peculiar lipids appear to have properties that are well suited to the rather harsh environment in which extreme halophiles such as *Halobacteria*, *Haloarcula*, *Haloferax*, and *Halococci* live. The branched alkyl ester structure, in contrast to the acyl ester found in eukarya and eubacteria, imparts stability to the lipids over a wide range of pH and temperatures, and the saturated alkyl chains impart stability toward oxidative degradation for those species exposed to air and sunlight (41). An elevated proportion of acidic lipids in archaeal membranes creates a high negative charge density on the membrane surface (44), balanced by counterions and shielding by the high salt concentrations present. Such a highly negatively charged membrane surface is likely required for the survival of halophiles in such high salt concentration (3–4 M). Furthermore, the phosphoryl group is attached to the glycerol moiety of the diether lipid via an *sn*-1 stereoconfiguration, not the *sn*-3 configuration typically found in bacteria, thereby imparting resistance to foreign phospholipases (44). The chemical nature of these lipids contributes significantly to the robust nature of archaeal life.

Purple Membrane Lipids

Within the CP membrane of *H. salinarum*, the bilayer is composed of the prototypic lipid archaeol (Figure 4*A*) and contains the highly specialized PM (74), which is composed of protein and lipid components. The major protein component is the 248-residue bacterioopsin (BO) (46) that, together with a covalently attached all-*trans* retinal, forms the photoactive ion pump BR. A minor component is composed of incomplete BR precursors (71, 88, 105). The lipid component was initially determined to be composed of (in molar ratios to BR/retinal) PGP-Me as the major lipid (45) (3 to 4), S-TGD-1 (2), PG (0.3), PGS (0.3), squalene (0.6), and traces of vitamin MK_8 for a total of 7 to 8 lipids per molecule of BR (45). However, a more recent study using high-precision instrumentation revealed the additional presence of archaeal glycocardiolipin (GlyC) and archaeal cardiolipin (BPG), leading to a correction of the molar ratios of all the lipids in the PM as shown in Table 1 (14). Overall, there is a 3:1 weight-to-weight ratio of BR to lipids.

Interestingly, the lipid composition of PM differs markedly from that of the surrounding CP membrane. Carotenoids found in the CP membrane are excluded from the PM. In contrast, the sulfated triglycosylarchaeol (S-TGA-1) is exclusively associated with the PM (49), as confirmed by mass spectrometry (102). This glycolipid was also proposed to be located entirely in the outer leaflet of the PM (34).

Functional Roles of Lipids in the Purple Membrane

These results spawned proposals describing highly specific interactions between BR and lipid molecules, which are essential for lattice formation. For example, it was proposed that S-TGA-1, in association with the major phospholipid PGP-Me, is directly involved in the energy-producing ion conductance pathway (97). The polar headgroup sulfate of the S-TGA-1 and the phosphate groups of PGP-Me would serve to transport the protons pumped by light-activated BR across the outer surface of the PM toward the red membrane, where the PGP-Me headgroup phosphates would conduct the protons to the sites of H^+-ATPase molecules situated in the red membrane, to drive ATP synthesis (22). Experimental support for this mechanism

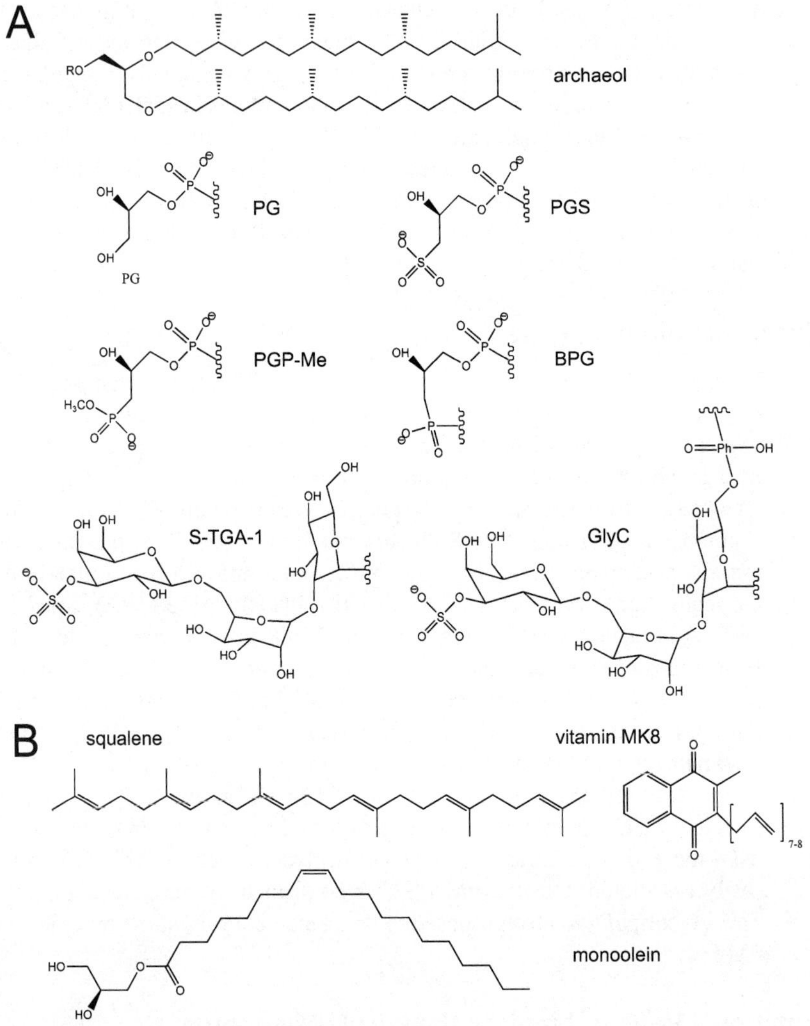

Figure 4 Lipid components of purple membrane. *Panel A* shows archaeol, the core hydrophobic moiety of purple membrane phospho- and glycolipids. The R groups, or headgroups, are shown for the phospholipids PG, PGP-Me, PGS, and BPG (cardiolipin analog), and for the glycolipids S-TGA-1 and GlyC. *Panel B* shows monoolein, the most commonly used lipid for cubic lipid phase crystallization of integral membrane proteins. It contains a characteristic *cis* double bond at the C9 position. Squalene and vitamin MK8 are also shown.

TABLE 1 Lipid:retinal molar ratios from PM total lipid extract [adapted from (14)]

Lipid	Corcelli[a]	Kates[b]
S-TGD-1[c]	3	2
GlyC	1	—
PGP-Me	2.4	3–4
PG	1.2	0.3
PGS + BPG	0.37	0.3[d]
Squalene	2	0.6
Vitamin MK-8	Traces	Traces
Total	10	7–8

[a]NMR data (14).

[b]From Reference 45.

[c]Same as S-TGA-1.

[d]Refers only to PGS, as the cardiolipin BPG was not known at that time.

comes from studies that showed that reconstituting delipidated BR in PGP-Me vesicles containing S-TGA-1 results in increased rates of proton pumping (56).

Recently, novel glycolipids and phospholipids were discovered: GlyC, a phosphosulfoglycolipid, and BPG, a glycerol diether analog of bisphosphatidylglycerol (cardiolipin) (13). Subsequently, data obtained from ^{31}P-NMR and ^{1}H-NMR spectra of the total lipid extract of PM were combined to redetermine the molar ratio of lipid to BR (14). Their data show that in total 10 lipids per BR/retinal are present in the PM (Table 1). Of these, five are phospholipids, three are glycolipid sulfate, and two are squalene. Among the phospholipid cardiolipin analogs, one GlyC per BR was found, while BPG cardiolipin was a minor component. Whereas these novel cardiolipins bind tightly to BR, and possibly play an important role in stability (57), their location in the PM lattice remains unknown. In contrast, eukaryotic cardiolipins have been localized in crystals of the photosynthetic reaction center of *Rhodobacter sphaeroides* (68, 69, 101). Furthermore, tightly bound cardiolipin plays a role in stabilizing the quaternary structure of bovine cytochrome *bc1* (27).

However, it has been long predicted that the glycolipids of the PM are specifically located on the PM EC side (34). Recent neutron diffraction experiments of PM containing deuterium-labeled S-TGA-1 revealed that there are two S-TGA-1 locations per BR monomer: in the inter- and intratrimer space (102). Unlike the abundant S-TGA-1, the minor phospholipid BPG [BPG:retinal molar ratio < 0.37 (14)] is likely to be located in intertrimer spaces, mediating specific interactions between trimers, a role in overall PM structural integrity. Likewise, PGS, a minor phospholipid, allows BR trimer reconstitution into a hexagonal lattice and may indeed also be present in the intertrimer space (93).

EMERGING STRUCTURAL DETAILS OF PURPLE MEMBRANE LIPID-PROTEIN INTERACTIONS

From Electron Microscopy to Three-Dimensional X-ray Crystallography

The various studies of PM composition and structure suggest a tight association between unique lipids and protein. The locations of lipids in PM were initially reported with limited accuracy from electron diffraction studies of the PM (2BRD) (28, 29, 30a). A more recent 3.0 Å electron crystallography structure of the two-dimensional PM (2AT9) (72) also revealed lipid locations. Of the 31 BR models currently available from the Protein Data Bank (5) (Table 2), 21 are X-ray diffraction structures with 16 containing some form of lipid moiety. In order to review

TABLE 2 Listing of all electron diffraction (top section) and X-ray diffraction (bottom section) models of bacteriorhodopsin currently deposited in the Protein Data Bank (PDB), in order of publication in the respective category. Cryo-trapped intermediates are rendered in italics.

PDB Code	Description	Resolution [Å]	Method
1BRD	First atomic bacteriorhodopsin structure	3.5	ED
2BRD	Refinement of 1BRD	3.5	ED
1AT9	Wild type	3.0	ED
2AT9	Refinement of 1AT9	3.0	ED
1FBB	Native, based on 2BRD	3.2	ED
1FBK	Triple mutant with cytoplasmically open conformation	3.2	ED
1AP9*	First CLP structure, merohedral twinning not recognized	2.35	X-RAY, CLP
1BRX	Wild type, ground state, merohedral twinning	2.3	X-RAY, CLP
1BM1	Wild type, ground state	3.5	X-RAY, via spherical vesicles
1BRR	Wild type, ground state, trimer in the asymmetric unit	2.9	X-RAY, detergent
1AP9	Further refinement of 1AP9*, taking twinning into account	2.35	X-RAY, CLP
1QHJ	Wild type, ground state	1.9	X-RAY, CLP
1C3W	Wild type, ground state, twinning	1.55	X-RAY, CLP
1C8R	D96N mutant, ground state	1.8	X-RAY, CLP
1C8S	*D96N mutant, late M intermediate generated at ambient with 100% occupancy*	*2.0*	*X-RAY, CLP*
1QKO/1QKP	*Wild type, low temperature K intermediate (K_{LT}) with 35% occupancy*	*2.1*	*X-RAY, CLP*

(Continued)

TABLE 2 *(Continued)*

PDB Code	Description	Resolution [Å]	Method
1CWQ	*Wild type, mixture of early & late M, and N intermediates with 35% occupancy, twinning*	*2.25*	*X-RAY, CLP*
1QM8	Based on 1BM1, wild type, ground state	2.5	X-RAY, via spherical vesicles
1DZE#	*Wild type, M intermediate with 100% occupancy*	*2.5*	*X-RAY, via spherical vesicles*
1EOP%	*Wild type, low temperature L intermediate (L_{LT}) with 36% occupancy, twinning*	*2.1*	*X-RAY, CLP*
1F50	E204Q mutant, ground state	1.7	X-RAY, CLP
1F4Z	*E204Q mutant, early M intermediate generated at ambient temperature with 100% occupancy*	*1.8*	*X-RAY, CLP*
1JV7	D85S mutant without halide, O-like state, head-to-tail dimers	2.25	X-RAY, CLP
1JV6	D85S/F219L double mutant without halide, O-like state, head-to-tail dimers	2.0	X-RAY, CLP
1KGB	Wild type, ground state	1.65	X-RAY, CLP
1KG8	*Wild type, low temperature early M intermediate (M_{LT}) with 100% occupancy*	*2.0*	*X-RAY, CLP*
1KME	PDB status WAIT, crystallized from bicelles, head-to-tail dimers	2.0	X-RAY, from bicelles

1AP9*: As published in Reference 78.

1DZE#: No publication since deposition in February 2000, when the title was "Sliding of G-helix in bacteriorhodopsin during proton transport."

1EOP%: For which the original publication and the PDB file specify 70% L occupancy, an estimate that was later reduced to 36% by the same group, along with 12% K and 12% M contamination.

ED: Determined by electron diffraction.

CLP: Grown in cubic lipid phase.

the interaction of the lipid with BR in its resting state, we only investigate those lipids found in wild-type resting state BR. As shown in Table 3, those resting state structures that reside in the $P6_3$ spacegroup are 1C3W (64) and 1QHJ (4). Those found in spacegroups other than $P6_3$ are 1BRR (C2) (19), 1QM8 (P622) (96), and 1KME ($P2_1$) (21). The $P6_3$-based structures all adopt a PM-like arrangement of stacked bilayers with unit cell constants nearly identical with those of the native PM in the *ab*-plane, making them an attractive crystal form. To date, all the structural information available on PM lipids is based on the work of Kates and coworkers (45, 49, 51). Future studies will have to rely on this information as well as more recent studies that have shown the inclusion of new lipids within the PM (13, 14).

TABLE 3 Overview of bacteriorhodopsin structures with refined lipids reviewed here

PDB	Spacegroup	*a* (Å)	*b* (Å)	*c* (Å)	R-value	Res. (Å)	Reference
1C3W	$P6_3$	60.631	60.631	108.156	0.158	1.55	(64)
1QHJ	$P6_3$	60.800	60.800	110.520	0.224	1.90	(4)
1KME	$P2_1$	45.000	108.900	55.900	0.263	2.00	(21)
1BM1	P622	104.70	104.70	114.10	0.260	3.50	(86)
1BRR	C2	120.50	105.960	80.190	0.256	2.90	(19)
2BRD	P3	62.45	62.45	100.90	0.280	3.50	(29)
2AT9	P3	62.45	62.45	100.00	0.237	3.00	(72)

Structural biology of membrane proteins started in 1975 with the landmark work of Henderson & Unwin (36), who determined a 7 Å projection map of BR PM, demonstrating the presence of hexagonally packed trimers (spacegroup P3), which are composed of monomers each containing seven-transmembrane helices. Electron density profiles of the PM also showed the BR molecules to be asymmetrically embedded in the bilayer, along with the bilayer leaflets containing different amounts of lipids (8). An orthorhombic two-dimensional crystal form of PM was obtained, and its 6.5 Å projection structure determined by electron microscopy and diffraction shows an identical molecular arrangement to that of the P3 form (36, 70). X-ray and electron diffraction of lipid-depleted PM later showed that trimers move closer together in response to the decrease in unit cell size (62.4→57.3 Å), showing the first structural significance of lipids in the packing of trimers within the PM (26). In 1990, the three-dimensional electron cryo-microscopy structure of BR was solved to 3.5 Å and provided the first details about the location of the side chains involved in ion transport (11, 29, 32, 33, 35). In 1997, following the groundbreaking development of cubic lipid phase (CLP) three-dimensional crystallization of BR (52), the structure of BR was determined from three-dimensional microcrystals to 2.5 Å (78). The following year, the crystals were discovered to be strongly merohedrally twinned and the structure was redetermined to 2.35 Å taking twinning into account (60). In recent years, there has been a multitude of crystal structures based on the CLP crystallization method for ground state and photointermediates (4, 20, 61, 63, 64, 83–85). Recent reviews on the structure of BR and its photocycle intermediates are available (54, 58, 59).

Structural Details of Lipids

CASE STUDY 2BRD In the refined 3.5 Å electron crystallographic structure of BR, the first lipids were modeled as phosphatidylglycerophosphate with dihydrophytol chains, five in each leaflet of the bilayer (29). However, only four of the EC leaflet lipids and two of the CP leaflet lipids had electron density. Other lipids were added to fill the space with similar molecules despite the lack of electron density.

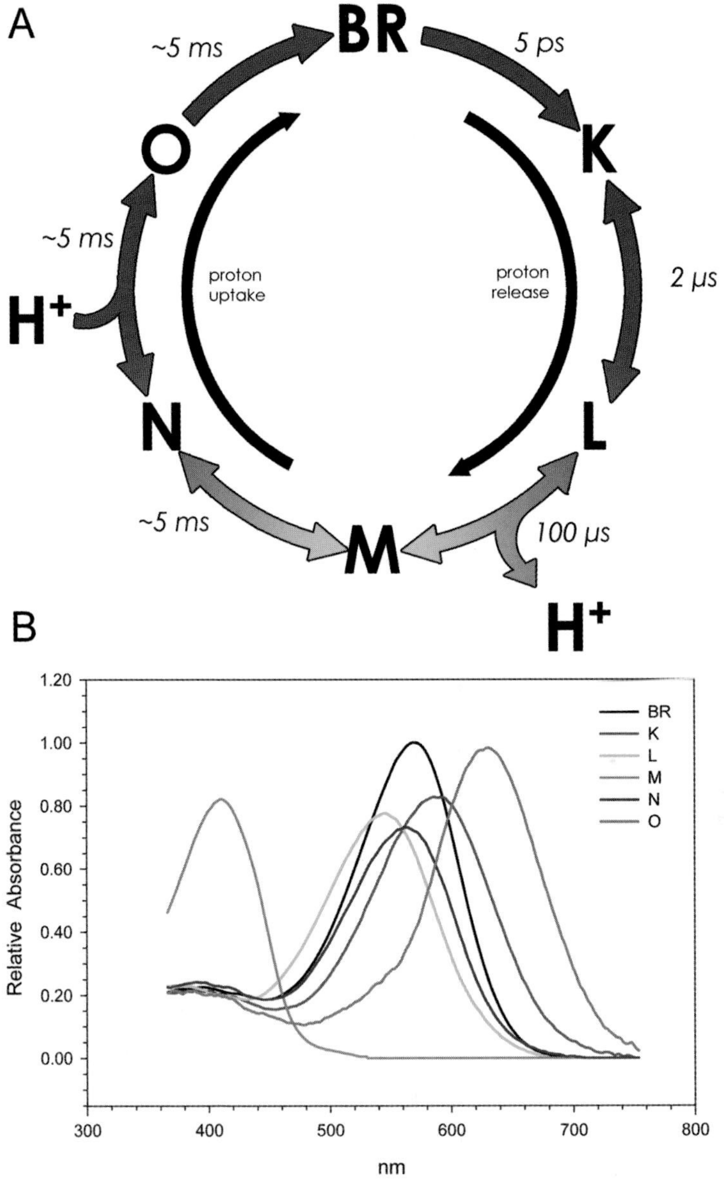

Figure 1 (*A*) The bacteriorhodopsin photocycle is composed of distinct spectroscopically identifiable states named BR, K, L, M, N, and O. Double-ended arrows denote the reversibility of these steps. Proton release and uptake occur to the extracellular and from the cytoplasmic bulk solvent, respectively. (*B*) Visible absorption spectra of the ground state and various intermediates scaled relative to ground state. There is significant spectral overlap between the ground state and the intermediates with the exception of the M intermediate (on the left, colored orange in this plot), which is strongly blue-shifted.

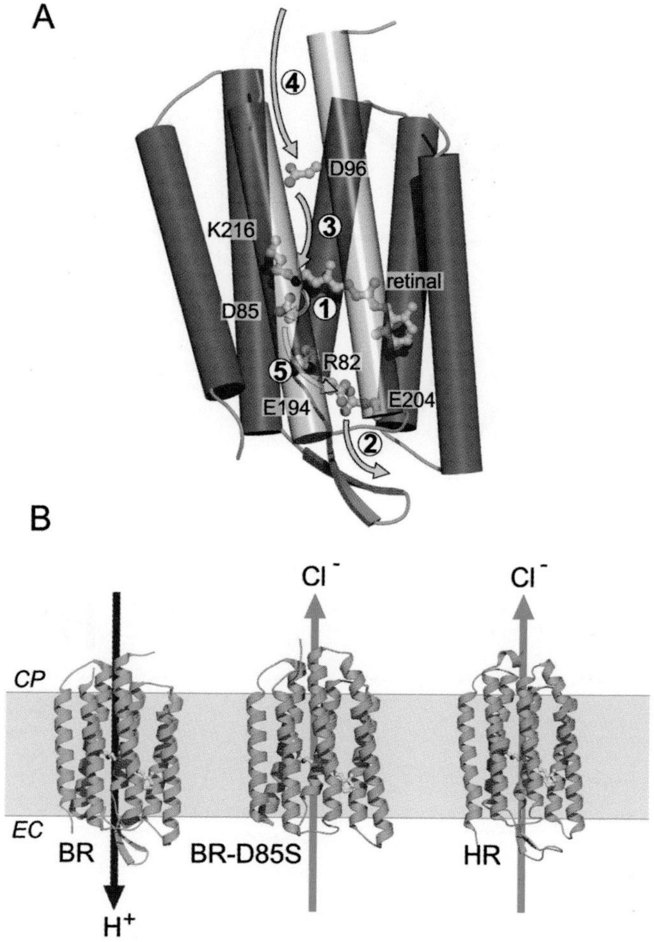

Figure 2 (*A*) Bacteriorhodopsin with its retinal chromophore and residues directly implicated in ion transport (cytoplasmic side on the *top*). The arrows (numbered 1 through 5) indicate sequential proton transfer steps of the wild-type photocycle, following the initial photoisomerization of the retinal: [1] deprotonation of the Schiff base, protonation of Asp85, yielding the early M intermediate; [2] coupled to step 1 via motion of the Arg82 guanidinium, proton release from the proton release group to the extracellular surface, yielding the late M intermediate; [3] deprotonation of Asp96, reprotonation of the Schiff base, yielding the N intermediate; [4] reprotonation of Asp96 from the cytoplasmic surface and thermal reisomerization of the retinal, yielding the O intermediate; [5] deprotonation of Asp85, reprotonation of the proton release group, regenerating the ground state. (*B*) Commonly reported net pumping activities of wild-type BR (*left*), the BR single mutant D85S (*center*), and HR (*right*). The structural similarity of all three proteins has prompted proposals that there might also be functional similarity that wild-type BR might function as an inward hydroxide (OH^-) anion pump rather than as an outward proton (H^+) cation pump.

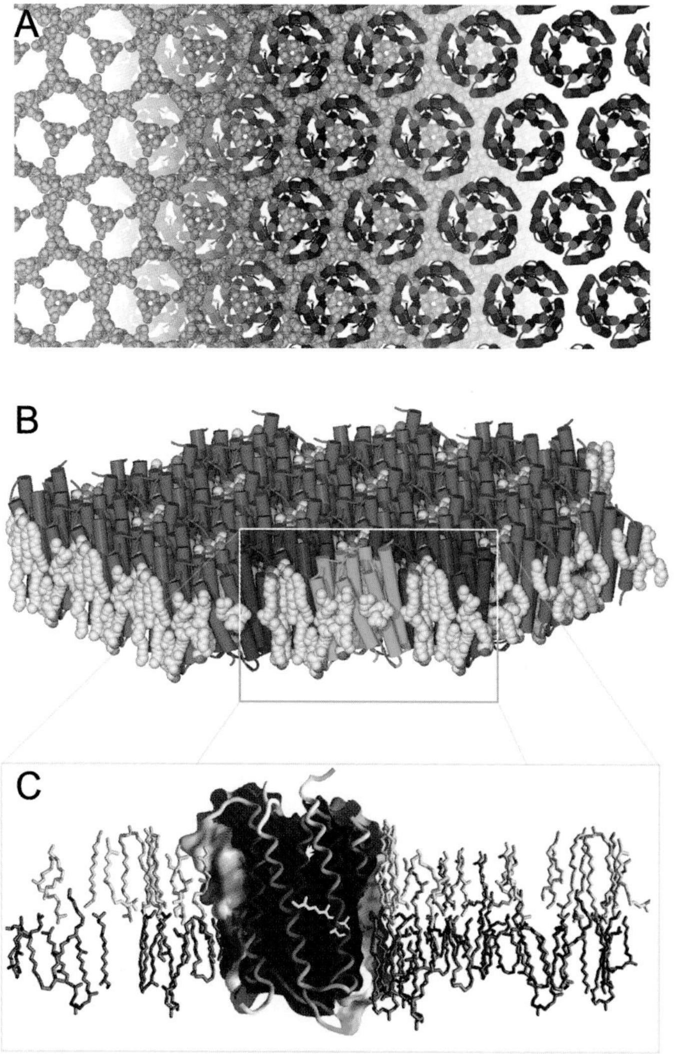

Figure 3 Arrangement of bacteriorhodopsin and lipids in three-dimensional crystals grown in the cubic lipid phase. The three-dimensional crystals are stacked layers of the naturally occurring purple membrane two-dimensional crystals. Complementarity between the protein trimers (purple) and lipids (beige) is illustrated in (*A*). Both the lipids (shown by themselves on the *far left*) and the bacteriorhodopsin trimers (shown by themselves on the *far right*) are essential for formation of the crystal lattice (*center*). (*B*) View (*A*) tilted slightly to show the edge of the bilayer, with (*C*) providing a magnified view of one bacteriorhodopsin molecule embedded in the surrounding lipid bilayer. The surface colors of this protein molecule correspond to the refined temperature or disorder factors (blue and red denote low and high temperature factor, respectively). The retinal chromophore is shown in yellow.

Even though a particular lipid (269) was located in a crevice between adjacent BR monomers within a trimer and was supposedly well resolved, its B-factor was 212 Å^2. In fact, the range of B-factors for the lipids in this study was 154–223 Å^2. Extremely high values for the loop regions (up to 375 Å^2) were also present, whereas the helical segments had more customary B-factors (10–60 Å^2). We do not believe that the accuracy of the positions of these lipids is enough to warrant further analysis. Shortly prior to this study, a projection electron microscopy study on deoxycholate-treated PM revealed that, aside from the loss of half the lipids, there was a 7% reduction in unit cell dimension and a 2° counterclockwise rotation of the trimers, which are otherwise indistinguishable from the nontreated PM (28). However, three lipids were observed after deoxycholate treatment, one in the CP crevice between monomers in the trimer, one on the EC side of the trimer's central pore, and one on the EC threefold axis II (28).

CASE STUDY 1BRR The next major step was the epitaxial growth of BR three-dimensional crystals on organic benzamidine sulfate crystals that diffract to 2.9 Å (PDB 1BRR) (19, 87). Despite the monoclinic arrangement (spacegroup C2), the monomers still packed into trimers, the latter forming the asymmetric unit. However, these trimers packed into a non-PM-like arrangement. The authors observed that they had to lower the detergent concentration of their stock BR solutions in order to obtain crystals with low mosaicity. The trimer packing contacts are likely to be stabilized by native archaeal lipids bound to the protein and an overabundance of detergent probably stripped some contact lipid away, giving rise to higher mosaicity. Mass spectrometry analysis of dissolved crystals revealed the presence of archaeal PG, PGS, PGP-ME, and S-TGA-1. Of the six lipids that were observed in the resulting electron density maps, three were modeled as single phytanol chains that were sandwiched into the CP side hydrophobic crevice between helices AB and DE of neighboring monomers within the trimer. The presence of this lipid is in accordance with that found in the deoxycholate-treated PM (28) and the refined 2BRD model (29). The other three lipids were found on the EC side of the central trimer compartment and form a continuous patch. The electron density reveals that the 3-phytanol (the lipid tail attached to the glycerol 3-OH position) has an elongated conformation and fits neatly in a channel formed by three consecutive glycine residues on helix D (Gly113, Gly116, Gly120) lined by hydrophobic residues. The results of mutating Gly113 or Gly116 to bulkier leucine residues, with the aim of filling in this channel, indicated an accumulation of BR monomers or small aggregates from a disruption in PM lattice stability (48). The 2-phytanol moiety of this lipid is bent sharply where they contact W80 (base of helix C), and these three do not assume perfect threefold noncrystallographic symmetry. Mutating Trp80 to smaller amino acids with smaller side chains disrupts the native PM-like arrangement in this crystal form (H. Besir & D. Oesterhelt, unpublished data) (19). Of these three lipids located in the central pore of the trimer, two were identified to be S-TGA-1 from the presence of their triglycoside headgroups intercalated with the BC loop residues. Earlier neutron diffraction studies of PM labeled with deuterated glycolipids indicated the presence of three S-TGA-1 lipids in the central compartment as described above (102). The interactions between the triglycoside and the protein consist of four hydrogen bonds and one salt bridge. Additional interactions likely occur with intervening water molecules, but the resolution of this study prevented their identification. Unlike the well-ordered EC central compartment lipids, there was no observable electron density on the CP side. The boundary of the EC bilayer leaflet is 5 Å shallower in the central compartment (S-TGA-1 lipids) than in the bulk lipid phase, yet leaves sufficient room for lipids to pack in the CP central compartment. The "membrane thinning" in the isolated central compartment is most likely an artifact of the packing interactions between the trimer and S-TGA-1.

During this time, a revolution in the field of three-dimensional BR crystallization was well underway. Based on the groundbreaking work of Landau & Rosenbusch (52), the first of many X-ray structures to be determined based on the CLP crystallization method was published (79) and thereafter improved (60). These X-ray structures revealed that charged residues and water molecules are involved in forming the ion translocation pathway (60, 79), with one key water (W402) situated between the protonated Schiff base and the anionic proton acceptor Asp85 (60). However, owing to their limited resolution these structures did not yet reveal details on the lipids that are a major component of the stacked PM bilayer sheets that constitute these crystals.

CASE STUDY 2AT9 With the aim of determining the charge states of ionizable residues involved in ion pumping, as identified by these early X-ray structures (60, 79), an electron crystallography study provided additional insight into lipid-protein interactions (72). The experimental maps revealed the presence of eight lipid molecules per asymmetric unit, including partial headgroups. However, there were not enough data to identify the glycolipid sulfate or squalene. All the lipids were modeled as PGP-Me, although it is likely that some of these are interchangeable with other headgroups. Five of these lipids were located in the EC region (lipids 261–265) while three were present in the CP region (lipids 266–269). Lipid 263, the only lipid to be located in the inter-trimer region, is most likely S-TGA-1. Furthermore, lipid 269 is located between helices AB and DE of neighboring monomers as previously observed (19, 28, 29). The bilayer thickness was reported here to be 31.6 Å with a standard deviation of 3.5 Å on the CP side and 1.6 Å on the EC side [the calculation did not include the central compartment lipids because they are isolated and at a different height, as previously observed (19)]. The position of BR within the bilayer is asymmetric with the CP region protruding from the membrane surface while the EC region is buried.

CASE STUDY 1BM1 Despite the advent of the CLP crystallization method, other techniques were still being developed. Three-dimensional crystals of BR were initially obtained that contain hexagonally packed spherical protein clusters with a diameter of approximately 50 nm (16, 47). Unfortunately, these diffracted weakly to approximately 40 Å. More recently, by varying the temperature of concentration and incubation of the spherical vesicles, fusion of these yielded hexagonal crystals that diffracted to 3 Å resolution (96). Data was collected on this crystal form to 3.5 Å, and the spacegroup was determined to be P622 with one monomer in the asymmetric unit. BR trimers are stacked in a planar honeycomb lattice, each layer stacking in opposite orientation. The distance between adjacent trimer centers is 60.5 Å, slightly less than for the native PM unit cell dimension of 62.3 Å (36). Only one lipid was observed and located at the crevice between adjacent monomers within the trimer between helices AB and DE as previously seen in numerous BR structures (19, 28, 29, 72). Its exact identity is unknown because the headgroup region is not fully resolved; however, it is believed to be PG, PGP, or PGS. Nonetheless, the interactions of the lipid with the protein are composed of strong electrostatic and van der Waals contacts. The phosphate group is in contact with the ε-amino group of Lys40 (helix B, CP side) to form a salt bridge, whereas the phytanyl chain is tightly packed between helices A/B and D/E of adjacent monomers. It was reported that a BR homolog (port-BR), in which Lys40 is substituted for a neutral residue, cannot self-assemble in vivo to form PM (76). Results also demonstrate that BR, devoid of all native lipids, behaves as a monomer in a pure dimyristoylphosphatidylcholine (DMPC) bilayer, and addition of PGP or PGS results in the formation of two-dimensional crystals (93). The observation that the phosphate group of the lipid described above forms a salt bridge with Lys40 suggests that PG (or a derivative) acts as the "glue" for the tight association of BR monomers into trimers.

CASE STUDY 1QHJ Following up on their earlier work plagued with undetected merohedral twinning (79), the first sub-2 Å structure was determined using the CLP crystallization method (4). Crystals free of merohedral twinning were obtained that diffracted to 1.9 Å. The electron density maps revealed the presence of lipids tails and connecting glycerol groups that were modeled as nine archaeol moieties (Figure 4*A*), four in the CP region and five in the EC region. Interestingly, the authors report that refining these lipids led to B-factors on the order of 40–80 Å^2, higher than the protein's average B-factor (30.2 Å^2) (4), which might not be surprising considering that bilayer lipids are generally more mobile than folded proteins. In turn, this study did not refine the lipids and assigned a B-factor of 57.3 Å^2 to all lipid atoms, which is unfortunate as atomic B-factors are a guide to coordinate error. Even though the crystallization matrix is composed of a monoolein (MO or 1-monooleoyl-*rac*-glycerol, a single chain lipid), none of the MO was detected in the crystals as determined by matrix-assisted laser desorption/ionization mass spectrometry, which revealed the presence of only endogenous PM lipids PGS, PGP-Me, and S-TGA-1. Their presence in the crystals, which are grown via incorporation of detergent-solubilized BR into the CLP, corroborates the evidence that endogenous lipids reside in protein-detergent-mixed micelles used in NMR (77) and in non-PM-like structures (19). Aside from the lipid chains found in the bulk phase of the PM within van der Waals contact to BR monomers, lipid 506 was found on the EC side, sandwiched between helices AB and DE of neighboring monomers as previously described (19, 28, 29, 72, 86). Only one lipid chain (500) was found in the central compartment on the EC side, likely to correspond to S-TGA-1. Aside from the majority of hydrophobic contacts between the lipid chains and the protein surface, there were some polar interactions with a glycerol moiety oxygen atom and Tyr64-OH. As predicted from neutron diffraction studies (102), two glycolipids are believed to reside nearby tryptophan side chains. In the lipids presented in this structure, Trp80 is in van der Waals contact with lipid 500, and it is reported that Trp10 is in the vicinity of the putative headgroup of lipid 504 (4).

CASE STUDY 1C3W Shortly after the elucidation of the BR structure at 1.9 Å (4), the resolution was pushed to 1.55 Å, again from CLP crystals (64). Aside from providing a wealth of new information about irregularities in the alpha helices, locations of side chains, and the retinal and key water molecules involved in the photocycle of BR, 18 individual phytanyl moieties were fitted into electron density and refined, 4 of which were modeled as the native diether archaeol. Lipid headgroup electron densities were observed in many instances but were not modeled because of their lower quality and consequent ambiguity. Again, the presence of the archaeol and phytanyl moieties, detected by mass spectrometry, indicated that native PM lipids were carried through the BR detergent extraction and purification, whereas the monoolein used in the CLP crystallization was not incorporated into the three-dimensional crystals. Attempts to further purify the protein after detergent extraction actually inhibited crystal growth. In retrospect, this finding is interpreted as the result of the loss of native lipids essential to crystal formation. The refined average B-factor for lipid atoms was 57 Å^2, compared to the average protein B-factor of 26 Å^2.

On the EC side, four single phytanyl moieties (lipids 603–606) and three archaeol moieties (lipids 601, 602, 607) were modeled. One of these, diether lipid 607, is located inside the central chamber of the trimer and is likely to be S-TGA-1. On the CP side, five single phytanyl moieties (lipids 608, 609, 611–613) and one archaeol moiety (lipid 610) were located. Lipid 612 is specifically wedged between helices AB and DE of adjacent monomers as commonly observed (4, 19, 28, 29, 72, 86). Whereas most of the lipid chains are only slightly bent or curved, one (lipid 701) adopts an S-shape, as evident from omit electron density maps. This feature near the center of the bilayer was tentatively modeled as squalene, and is tightly associated with the protein surface where a complementary groove exists near the π-bulge of helix G (64). Mass spectrometry revealed the presence of squalene in our dissolved crystals in an approximate 1:1 molar ratio with BR (H.T. Richter & J.K. Lanyi, unpublished data). Its location at the distorted and functionally active region of helix G would be consistent with the observation that squalene affects reprotonation of the retinal Schiff base by Asp96 during the photocycle (40). The presence of squalene, as well as the other native lipids, is suggested also by the fact that the photochemical cycle is essentially the same in these three-dimensional crystals as in the BR-containing two-dimensional membranes before their solubilization (31, 60).

Furthermore, as seen in other structures of BR in PM-like environments, there exists minimal contact between trimers within the same layer, computed here to be 1.2% of the buried accessible surface area. Of the total trimer accessible surface area, 39.6% is buried by the lipid bilayer, thus making up the majority of crystal-packing contacts. Specifically, 8.1% of the trimers are covered by the lipids (608 and 618) found.

More recently, the 1C3W 1.55 Å model was used as a starting model for the refinement of the BR resting state at 1.6 Å (PDB 1KGB) (20). The RMSD between the two models is 0.31 Å for the backbone atoms and 0.38 Å for all protein atoms, with the loop regions accounting for most of the deviation. The lipids that were previously reported (64) were all included in this model as well, except for the squalene. The locations and B-factors for these are nearly identical to those reported previously, and no additional information was obtained by this study on lipid-protein interactions.

CASE STUDY 1KME Most recently, another novel method of membrane protein crystallization was developed consisting of a bicelle-forming lipid/detergent mixture (21). Bacteriorhodopsin was used as the candidate protein and crystals were obtained in spacegroup $P2_1$ that consist of stacked two-dimensional crystalline layers. Each monomer is surrounded by three other monomers packed in antiparallel fashion with alternating orientation along the axes, which is different from a native P3 arrangement obtained within $P6_3$ crystals. Using the 1.55 Å BR structure (64) as a search model for molecular replacement, the refined structure had an overall RMSD of 0.72 Å. The authors state that because the structure of BR is largely unaffected by packing, it is not really influenced by the trimer formation, bound lipids, or the overall lamellar assembly of the PM (21). However, there exists an overwhelming wealth of evidence that packing and surrounding lipids do influence the structure and thereby function of BR. When BR is exhaustively purified away from its endogenous lipids and reconstituted at a wide range of protein densities (concentrations) into DMPC, bilayers do not arrange into two-dimensional arrays like those found in PM (93). Also, when PM-derived lipids are partially removed from the PM by nonsolubilizing amounts of detergent, a loss of hexagonal BR patches occurs, suggesting that lipids have an influence on array formation (70). Furthermore, the integrity of the trimer structure was assayed using circular dichroism, and short exposure to non-ionic detergent adversely affects the photocycle while not altering the overall trimer structure (73). These observations support the view that lipid-protein interactions are important for a normal photocycle. Unfortunately, it was not shown that BR is still active in the bicelle crystal form (21). Other than the extensive monomer-to-monomer packing interactions, the study reported one lipid per monomer, bound distant from the crystal-packing contacts. It is therefore apparent that most of the lipids were stripped from BR prior to crystallization. The lipid was labeled as squalene but no additional information was reported, nor does it look S-shaped, as expected from the conjugated system. Furthermore, this squalene is not in the same position as the one previously observed in the native packing arrangement (64).

Assembly of the Lipid Mosaic

As seen in the case studies presented here, which include only ground-state wild-type structures, there are many occurrences of lipids. However, there is no standard convention for naming and numbering these lipids, thereby making comparisons difficult. Perhaps the convention used by Essen and coworkers (19) is best, in that they separated the lipids into (*a*) the phytanyl tails, common to all PM archaeol-based lipids, and (*b*) the headgroups, thereby simplifying the nomenclature until all lipids have been identified and located in the PM.

Nonetheless, the structures presented here do differ from one another by the type of technique used to determine them, with ensuing differences in quality. It is therefore difficult to combine the results discussed here into a structural consensus. However, there are commonalities between many of these structures that are worth reiterating. Most of the structures reviewed here contain, in the EC central trimer region, a glycolipid most likely to be S-TGA-1. S-TGA-1 is the major glycophospholipid of the PM and is exclusively associated with it (49), as first suggested (36). It has been proposed that the polar headgroup sulfate of S-TGA-1 and the phosphate groups of PGP-Me would serve to transport the protons pumped by light-activated BR across the outer surface of the PM to the red membrane for ensuing ATP synthesis (22, 97). Reconstitution of delipidated BR in PGP-Me vesicles containing S-TGA-1 results in increased rates of proton pumping, thereby supporting the theory outlined above (56).

Another interesting lipid feature is the PG lipid found in all the structures presented here [except for the bicelle crystallization-based nontrimeric structure (21)] between the AB and DE helices of neighboring monomers within an intact trimer (4, 19, 20, 28, 29, 64, 72, 86). To be so prevalent in these structures implies that it is tightly bound and that it likely provides structural and functional integrity to the trimer. This diether lipid is oriented such that one of its lipid tails is wedged between helices A and B of one monomer and D and E of another adjacent monomer on the CP side, where helices B and D are within van der Waals contact. Without the lipid, there would be a deep cavity between monomers on the CP side. Having a lipid at this position, a molecule with such high flexibility is likely to indicate a functional role in the latter part of the photocycle when the CP side undergoes pronounced opening and closing (86). Likewise, it may also play a role in photocycle cooperativity among the monomers within a trimer. Also, common hydrogen bonds and salt bridges formed between lipid headgroups and the protein are found in most of these structures (4, 19, 20, 28, 29, 64, 72, 86). Unfortunately, each of these interactions is uniquely described from a structural standpoint, preventing an overall consensus. Aside from strong polar or electrostatic interactions, it has been shown that replacing the phytanyl tails with alkyl tails results in a large blue shift of BR, implying that the bulkiness of the phytanyl tails, and possibly specific binding to surface cavities of BR, is important in providing additional structural stability (80). As evident from the location of this lipid, sandwiched between adjacent monomers in a well-defined crevice, it is plausible that such a replacement would have a significant effect on BR.

Generally, it is also shown that BR is asymmetrically oriented within the bilayer, with its CP region more solvent exposed than its EC counterpart by ~5 Å. The higher B-factors in the CP region, as previously observed (72), along with the positional asymmetry have been prescribed to the large-scale conformational changes that occur in the CP portion during the photochemical cycle (94, 99). Similar conformational changes, described as rigid-body motions of the same transmembrane helices as in BR, have been detected by spin-spin distance measurements in visual rhodopsin (23). In the crystal structure of the D85S mutant of BR, we observed large conformational changes on the EC side that were also accompanied by elevated B-factors (83).

Overall, the protein is more compact and rigid in the regions exposed to the bilayer interior, much like the hydrophobic core of soluble proteins. On the hydrophobic protein surface, grooves are formed by specific arrangements of the side chains and provide a highly structured interface for lipid binding. The lipid tails on the whole are aligned in these grooves (64) and imply specific interactions between the lipid chains and the protein groups. Theoretical models had predicted the existence of such intimate lipid-protein contacts in the form of van der Waals–London forces (103).

CONCLUSIONS

Generally, membrane protein surfaces have two obvious features: (*a*) a neutral, apolar belt that forms the surface that interacts with the hydrophobic interior of the bilayer and (*b*) protruding water-soluble regions on either side of the bilayer surface that may interact with substrate, protein, or other factors. As seen in the case of BR, the intramembrane protein surface is irregular, providing grooves and channels, some of which are formed by the turns of the alpha helices. The shape of the protein is such that lipids will most likely undergo significant rearrangement from a perfect bilayer to fit into these grooves and channels in order to form an electrically sealed membrane. A good match between the hydrophobic protein surface and the neighboring lipids is thought to be important for the stable integration of integral membrane proteins into the bilayer. Additional examples include the tight interaction between cardiolipin and (*a*) *R. sphaeroides* photosynthetic reaction center (68, 69) and (*b*) cytochrome *c* oxidase (75). Nonbilayer-forming lipids, as well as lipids with branched tails and unsaturated tails, are likely to play a role, as they may be nonlinear, less compact, and/or more flexible than bilayer-forming lipids.

Spin-label electron paramagnetic resonance spectroscopy carried out on a variety of membrane proteins indicates that the first shell of lipids that interacts with the protein is motion restricted, presumably as a result of the specific interactions of these lipids with the surface of the protein (67). A single layer of spatially constrained "annular" lipids surrounds the protein, and these lipids are distinguishable from the bulk phase membrane lipids, despite an exchange rate on the microsecond timescale (67). Such an annulus exists in BR, and one of the tightest-bound lipids is likely to be the one located in the aforementioned crevice in the CP region.

The CLP-based three-dimensional crystals (52) are especially useful for the study of the PM because the arrangement of BR inside the crystals is in a hexagonal lattice of trimers, indistinguishable from that found in the two-dimensional PM lattice with the same unit cell dimensions. Furthermore, and perhaps most importantly, the process by which BR is overexpressed and purified is such that native *Halobacterium* lipids are carried through the purification and crystallization and are not stripped away by the detergents used. This is also interesting from the point of view that these lipids appear to be strongly bound to BR, since the use of such conditions on other membrane proteins typically strips most lipids, therefore adding to the notion of lipids being an integral part of BR structure/function. In the case of a related protein, SRII, the addition of native lipids to the CLP setup was necessary to obtain crystals (62, 84), again demonstrating their crucial role in structural stability.

Even with dozens of structures available, there are many details of lipid-protein interactions left to be determined. Higher-resolution work, coupled with other biophysical methods, is necessary to identify the lipid headgroups, thereby identifying the lipids themselves. Because crystallographic techniques rely on the repeating unit cells within the crystal lattice, it is likely that we will never determine exactly which lipids are present at the special positions of the threefold axes where we observe electron density. Likewise, in the case of rare lipids present in the PM such as vitamin MK_8, if these are solely incorporated in a small fraction of trimers, we are unlikely to observe them by structural methods that rely on lattice assemblies. Is the squalene molecule really present near the π-bulge of helix G? How and with what PGP-Me would S-TGA-1 communicate to deliver protons to the red membrane for subsequent ATP synthesis? How does the bilayer structure change throughout the photocycle? Does it play a specific role in photocooperativity? These and many other questions remain to be answered. Nonetheless, the PM is likely the best structurally understood biological membrane to date.

The *Annual Review of Biophysics and Biomolecular Structure* is online at http://biophys.annualreviews.org

LITERATURE CITED

1. Balashov SP, Ebrey TG. 2001. Trapping and spectroscopic identification of the photointermediates of bacteriorhodopsin at low temperatures. *Photochem. Photobiol.* 73:453–62
2. Bamberg E, Tittor J, Oesterhelt D. 1993. Light-driven proton or chloride pumping by halorhodopsin. *Proc. Natl. Acad. Sci. USA* 90:639–43
3. Beja O, Aravind L, Koonin EV, Suzuki

MT, Hadd A, et al. 2000. Bacterial rhodopsin: evidence for a new type of phototrophy in the sea. *Science* 289:1902–64

4. Belrhali H, Nollert P, Royant A, Menzel C, Rosenbusch JP, et al. 1999. Protein, lipid and water organization in bacteriorhodopsin crystals: a molecular view of the purple membrane at 1.9 Å resolution. *Struct. Fold Des.* 7:909–17
5. Berman HM, Westbrook J, Feng Z, Gilliland G, Bhat TN, et al. 2000. The Protein Data Bank. *Nucleic Acids Res.* 28:235–42

5a. Betancourt FM, Glaeser RM. 2000. Chemical and physical evidence for multiple functional steps comprising the M state of the bacteriorhodopsin photocyle. *Biochim. Biophys. Acta* 1460(1):106–18

6. Birge RR, Cooper TM, Lawrence AF, Masthay MB, Zhang CF, Zidovetzki R. 1991. Assignment of energy storage in the primary photochemical event in bacteriorhodopsin. *J. Am. Chem. Soc.* 113:4327–28
7. Birge RR, Cooper TM, Lawrence AF, Masthay MB, Vasilakis C, et al. 1989. A spectroscopic, photocalorimetric, and theoretical investigation of the quantum efficiency of the primary event in bacteriorhodopsin. *J. Am. Chem. Soc.* 111:4063–74
8. Blaurock AE, King GI. 1977. Asymmetric structure of the purple membrane. *Science* 196:1101–4
9. Bogdanov M, Dowhan W. 1995. Phosphatidylethanolamine is required for in vivo function of the membrane-associated lactose permease of *Escherichia coli*. *J. Biol. Chem.* 270:732–39
10. Braiman M, Mathies R. 1982. Resonance Raman spectra of bacteriorhodopsin's primary photoproduct: evidence for a distorted 13-*cis* retinal chromophore. *Proc. Natl. Acad. Sci. USA* 79:403–7
11. Ceska TA, Henderson R. 1990. Analysis of high-resolution electron diffraction patterns from purple membrane labelled with heavy-atoms. *J. Mol. Biol.* 213:539–60
12. Cherry RJ, Muller U. 1978. Temperature-dependent aggregation of bacteriorhodopsin in dipalmitoyl- and dimyristoylphosphatidylcholine vesicles. *J. Mol. Biol.* 121:283–98
13. Corcelli A, Colella M, Mascolo G, Fanizzi FP, Kates M. 2000. A novel glycolipid and phospholipid in the purple membrane. *Biochemistry* 39:3318–26
14. Corcelli A, Lattanzio VM, Mascolo G, Papadia P, Fanizzi F. 2002. Lipid-protein stoichiometries in a crystalline biological membrane: NMR quantitative analysis of the lipid extract of the purple membrane. *J. Lipid Res.* 43:132–40
15. De Rosa M, Gambacorta A, Gliozzi A. 1986. Structure, biosynthesis, and physicochemical properties of archaebacterial lipids. *Microbiol. Rev.* 50:70–80
16. Denkov ND, Yoshimura H, Kouyama T, Walz J, Nagayama K. 1998. Electron cryomicroscopy of bacteriorhodopsin vesicles: mechanism of vesicle formation. *Biophys. J.* 74:1409–20
17. Dioumaev AK, Richter HT, Brown LS, Tanio M, Tuzi S, et al. 1998. Existence of a proton transfer chain in bacteriorhodopsin: participation of Glu-194 in the release of protons to the extracellular surface. *Biochemistry* 37:2496–506

17a. Doig SJ, Reid PJ, Mathies RA. 1991. Picosecond time-resolved resonance Raman spectroscopy of bacteriorhodopsin's J, K, and KL intermediates. *J. Phys. Chem.* 5(16):6372–79

18. Essen L. 2002. Halorhodopsin: light-driven ion pumping made simple? *Curr. Opin. Struct. Biol.* 12:516–22
19. Essen L, Siegert R, Lehmann WD, Oesterhelt D. 1998. Lipid patches in membrane protein oligomers: crystal structure of the bacteriorhodopsin-lipid complex. *Proc. Natl. Acad. Sci. USA* 95:11673–78
20. Facciotti MT, Rouhani S, Burkard FT, Betancourt FM, Downing KH, et al. 2001. Structure of an early intermediate in the M-state phase of the bacteriorhodopsin photocycle. *Biophys. J.* 81:3442–55

21. Faham S, Bowie JU. 2002. Bicelle crystallization: a new method for crystallizing membrane proteins yields a monomeric bacteriorhodopsin structure. *J. Mol. Biol.* 316:1–6
22. Falk KE, Karlsson KA, Samuelsson BE. 1980. Structural analysis by mass spectrometry and NMR spectroscopy of the glycolipid sulfate from *Halobacterium salinarium* and a note on its possible function. *Chem. Phys. Lipids* 27:9–21
23. Farrens DL, Altenbach C, Yang K, Hubbell WL, Khorana HG. 1996. Requirement of rigid-body motion of transmembrane helices for light activation of rhodopsin. *Science* 274:768–70
24. Fyfe PK, McAuley KE, Roszak AW, Isaacs NW, Cogdell RJ, Jones MR. 2001. Probing the interface between membrane proteins and membrane lipids by X-ray crystallography. *Trends Biochem. Sci.* 26:106–12
25. Gerwert K, Hess B, Soppa J, Oesterhelt D. 1989. Role of aspartate-96 in proton translocation by bacteriorhodopsin. *Proc. Natl. Acad. Sci. USA* 86:4943–47
26. Glaeser RM, Jubb JS, Henderson R. 1985. Structural comparison of native and deoxycholate-treated purple membrane. *Biophys. J.* 48:775–80
27. Gomez B Jr, Robinson NC. 1999. Phospholipase digestion of bound cardiolipin reversibly inactivates bovine cytochrome bc1. *Biochemistry* 38:9031–38
28. Grigorieff N, Beckmann E, Zemlin F. 1995. Lipid location in deoxycholate-treated purple membrane at 2.6 A. *J. Mol. Biol.* 254:404–15
29. Grigorieff N, Ceska TA, Downing KH, Baldwin JM, Henderson R. 1996. Electron-crystallographic refinement of the structure of bacteriorhodopsin. *J. Mol. Biol.* 259:393–421
30. Hartmann R, Sickinger HD, Oesterhelt D. 1977. Quantitative aspects of energy conversion in halobacteria. *FEBS Lett.* 82:1–6
30a. Hayward SB, Stroud RM. 1981. Projected structure of purple membrane determined to 3.7 Å resolution by low temperature electron microscopy. *J. Mol. Biol.* 151:491–517
31. Heberle J, Buldt G, Koglin E, Rosenbusch JP, Landau EM. 1998. Assessing the functionality of a membrane protein in a three-dimensional crystal. *J. Mol. Biol.* 281:587–92
32. Henderson R, Baldwin JM, Ceska TA, Zemlin F, Beckmann E, Downing KH. 1990. An atomic model for the structure of bacteriorhodopsin. *Biochem. Soc. Trans.* 18:844–48
33. Henderson R, Baldwin JM, Ceska TA, Zemlin F, Beckmann E, Downing KH. 1990. Model for the structure of bacteriorhodopsin based on high-resolution electron cryo-microscopy. *J. Mol. Biol.* 213: 899–929
34. Henderson R, Jubb JS, Whytock S. 1978. Specific labelling of the protein and lipid on the extracellular surface of purple membrane. *J. Mol. Biol.* 123:259–74
35. Henderson R, Schertler GF. 1990. The structure of bacteriorhodopsin and its relevance to the visual opsins and other seven-helix G-protein coupled receptors. *Philos. Trans. R. Soc. London Sci. Ser. B* 326:379–89
36. Henderson R, Unwin PN. 1975. Three-dimensional model of purple membrane obtained by electron microscopy. *Nature* 257:28–32
37. Hendler RW, Dracheva S. 2001. Importance of lipids for bacteriorhodopsin structure, photocycle, and function. *Biochemistry* 66:1311–14
38. Herzfeld J, Lansing JC. 2002. Magnetic resonance studies of the bacteriorhodopsin pump cycle. *Annu. Rev. Biophys. Biomol. Struct.* 31:73–95
39. Huang KS, Bayley H, Khorana HG. 1980. Delipidation of bacteriorhodopsin and reconstitution with exogenous phospholipid. *Proc. Natl. Acad. Sci. USA* 77:323–27
40. Joshi MK, Dracheva S, Mukhopadhyay AK, Bose S, Hendler RW. 1998.

Importance of specific native lipids in controlling the photocycle of bacteriorhodopsin. *Biochemistry* 37:14463–70
41. Kamekura M, Kates M. 1988. Lipids of halophilic archaebacteria. In *Halophilic Bacteria*, ed. F Rodriquez-Velera, 2:25–54. Boca Raton, FL: CRC. 336 pp.
42. Kandori H, Yamazaki Y, Hatanaka M, Needleman R, Brown LS, et al. 1997. Time-resolved Fourier transform infrared study of structural changes in the last steps of the photocycles of Glu-204 and Leu-93 mutants of bacteriorhodopsin. *Biochemistry* 36:5134–41
43. Kates M. 1978. The phytanyl ether-linked polar lipids and isoprenoid neutral lipids of extremely halophilic bacteria. *Prog. Chem. Fats Other Lipids* 15:301–42
44. Kates M. 1993. Membrane lipids of Archaea. In *The Biochemistry of Archaea (Archaeabacteria)*, ed. M Kates, DJ Kushner, AT Matheson, 6:261–96. Amsterdam: Elsevier. 582 pp.
45. Kates M, Kushwaha SC, Sprott GD. 1982. Lipids of purple membrane from extreme halophiles and of methanogenic bacteria. *Methods Enzymol.* 88:98–111
46. Khorana HG, Gerber GE, Herlihy WC, Gray CP, Anderegg RJ, et al. 1979. Amino acid sequence of bacteriorhodopsin. *Proc. Natl. Acad. Sci. USA* 76:5046–50
47. Kouyama T, Yamamoto M, Kamiya N, Iwasaki H, Ueki T, Sakurai I. 1994. Polyhedral assembly of a membrane protein in its three-dimensional crystal. *J. Mol. Biol.* 236:990–94
48. Krebs MP, Li W, Halambeck TP. 1997. Intramembrane substitutions in helix D of bacteriorhodopsin disrupt the purple membrane. *J. Mol. Biol.* 267:172–83
49. Kushwaha SC, Kates M, Martin WG. 1975. Characterization and composition of the purple and red membrane from *Halobacterium cutirubrum*. *Can. J. Biochem.* 53:284–92
50. Deleted in proof
51. Kushwaha SC, Kates M, Stoeckenius W. 1976. Comparison of purple membrane from *Halobacterium cutirubrum* and *Halobacterium halabium*. *Biochim. Biophys. Acta* 426:703–10
52. Landau EM, Rosenbusch JP. 1996. Lipidic cubic phases: a novel concept for the crystallization of membrane proteins. *Proc. Natl. Acad. Sci. USA* 93:14532–35
53. Lanyi JK. 1995. Bacteriorhodopsin as a model for proton pumps. *Nature* 375:461–63
54. Lanyi JK, Luecke H. 2001. Bacteriorhodopsin. *Curr. Opin. Struct. Biol.* 11:415–19
55. Lanyi JK, Váró G. 1995. The photocycles of bacteriorhodopsin. *Isr. J. Chem.* 35:365–86
56. Lind C, Hojeberg B, Khorana HG. 1981. Reconstitution of delipidated bacteriorhodopsin with endogenous polar lipids. *J. Biol. Chem.* 256:8298–305
57. Lopez F, Lobasso S, Colella M, Agostiano A, Corcelli A. 1999. Light-dependent and biochemical properties of two different bands of bacteriorhodopsin isolated on phenyl-sepharose CL-4B. *Photochem. Photobiol.* 69:599–604
58. Luecke H. 2000. Atomic resolution structures of bacteriorhodopsin photocycle intermediates: the role of discrete water molecules in the function of this light-driven ion pump. *Biochim. Biophys. Acta* 1460:133–56
59. Luecke H, Lanyi J. 2003. Structural clues to the mechanism of ion pumping in bacteriorhodopsin. *Adv. Prot. Chem.* 63:111–30
60. Luecke H, Richter HT, Lanyi JK. 1998. Proton transfer pathways in bacteriorhodopsin at 2.3 angstrom resolution. *Science* 280:1934–37
61. Luecke H, Schobert B, Cartailler JP, Richter HT, Rosengarth A, et al. 2000. Coupling photoisomerization of retinal to directional transport in bacteriorhodopsin. *J. Mol. Biol.* 300:1237–55
62. Luecke H, Schobert B, Lanyi JK, Spudich EN, Spudich JL. 2001. Crystal structure of sensory rhodopsin II at 2.4 angstroms: insights into color tuning and transducer interaction. *Science* 293:1499–503

63. Luecke H, Schobert B, Richter HT, Cartailler JP, Lanyi JK. 1999. Structural changes in bacteriorhodopsin during ion transport at 2 angstrom resolution. *Science* 286:255–61
64. Luecke H, Schobert B, Richter HT, Cartailler JP, Lanyi JK. 1999. Structure of bacteriorhodopsin at 1.55 Å resolution. *J. Mol. Biol.* 291:899–911
65. Maeda A. 1995. Application of FTIR spectroscopy to the structural study on the function of bacteriorhodopsin. *Isr. J. Chem.* 35:387–400
66. Maeda A, Kandori H, Yamazaki Y, Nishimura S, Hatanaka M, et al. 1997. Intramembrane signaling mediated by hydrogen-bonding of water and carboxyl groups in bacteriorhodopsin and rhodopsin. *J. Biochem.* 121:399–406
67. Marsh D, Horvath LI. 1998. Structure, dynamics and composition of the lipid-protein interface. Perspectives from spin-labelling. *Biochim. Biophys. Acta* 1376:267–96
67a. Mathies RA, Lin SW, Ames JB, Pollard WT. 1991. From femtoseconds to biology: mechanism of bacteriorhodopsin's light-driven proton pump. *Annu. Rev. Biophys. Biophys. Chem.* 20:491–98
68. McAuley KE, Fyfe PK, Ridge JP, Cogdell RJ, Isaacs NW, Jones MR. 2000. Ubiquinone binding, ubiquinone exclusion, and detailed cofactor conformation in a mutant bacterial reaction center. *Biochemistry* 39:15032–43
69. McAuley KE, Fyfe PK, Ridge JP, Isaacs NW, Cogdell RJ, Jones MR. 1999. Structural details of an interaction between cardiolipin and an integral membrane protein. *Proc. Natl. Acad. Sci. USA* 96: 14706–11
70. Michel H, Oesterhelt D, Henderson R. 1980. Orthorhombic two-dimensional crystal form of purple membrane. *Proc. Natl. Acad. Sci. USA* 77:338–42
71. Miercke LJ, Ross PE, Stroud RM, Dratz EA. 1989. Purification of bacteriorhodopsin and characterization of mature and partially processed forms. *J. Biol. Chem.* 264:7531–35
72. Mitsuoka K, Hirai T, Murata K, Miyazawa A, Kidera A, et al. 1999. The structure of bacteriorhodopsin at 3.0 Å resolution based on electron crystallography: implication of the charge distribution. *J. Mol. Biol.* 286:861–82
73. Mukhopadhyay AK, Dracheva S, Bose S, Hendler RW. 1996. Control of the integral membrane proton pump, bacteriorhodopsin, by purple membrane lipids of *Halobacterium halobium*. *Biochemistry* 35:9245–52
74. Oesterhelt D, Stoeckenius W. 1971. Rhodopsin-like protein from the purple membrane of *Halobacterium halobium*. *Nat. New Biol.* 233:149–52
75. Ostermeier C, Harrenga A, Ermler U, Michel H. 1997. Structure at 2.7 Å resolution of the *Paracoccus denitrificans* two-subunit cytochrome *c* oxidase complexed with an antibody FV fragment. *Proc. Natl. Acad. Sci. USA* 94:10547–53
76. Otomo J, Urabe Y, Tomioka H, Sasabe H. 1992. The primary structures of helices A to G of three new bacteriorhodopsin-like retinal proteins. *J. Gen. Microbiol.* 138(Pt. 11):2389–96
77. Patzelt H, Ulrich AS, Egbringhoff H, Dux P, Ashurst J, et al. 1997. Towards structural investigations on isotope labelled native bacteriorhodopsin in detergent micelles by solution-state NMR spectroscopy. *J. Biomol. NMR* 10:95–106
78. Pebay-Peyroula E, Neutze R, Landau EM. 2000. Lipidic cubic phase crystallization of bacteriorhodopsin and cryotrapping of intermediates: towards resolving a revolving photocycle. *Biochim. Biophys. Acta* 1460:119–32
79. Pebay-Peyroula E, Rummel G, Rosenbusch JP, Landau EM. 1997. X-ray structure of bacteriorhodopsin at 2.5 angstroms from microcrystals grown in lipidic cubic phases. *Science* 277:1676–81
80. Pomerleau V, Harvey-Girard E, Boucher F. 1995. Lipid-protein interactions in the

purple membrane: structural specificity within the hydrophobic domain. *Biochim. Biophys. Acta* 1234:221–24
81. Richter HT, Needleman R, Kandori H, Maeda A, Lanyi JK. 1996. Relationship of retinal configuration and internal proton transfer at the end of the bacteriorhodopsin photocycle. *Biochemistry* 35:15461–66
82. Rothschild KJ, Roepe P, Ahl PL, Earnest TN, Bogomolni RA, et al. 1986. Evidence for a tyrosine protonation change during the primary phototransition of bacteriorhodopsin at low temperature. *Proc. Natl. Acad. Sci. USA* 83:347–51
83. Rouhani S, Cartailler JP, Facciotti MT, Walian P, Needleman R, et al. 2001. Crystal structure of the D85S mutant of bacteriorhodopsin: model of an O-like photocycle intermediate. *J. Mol. Biol.* 313:615–28
84. Royant A, Nollert P, Edman K, Neutze R, Landau EM, et al. 2001. X-ray structure of sensory rhodopsin II at 2.1-Å resolution. *Proc. Natl. Acad. Sci. USA* 98:10131–36
85. Sass HJ, Buldt G, Gessenich R, Hehn D, Neff D, et al. 2000. Structural alterations for proton translocation in the M state of wild-type bacteriorhodopsin. *Nature* 406:649–53
86. Sato H, Takeda K, Tani K, Hino T, Okada T, et al. 1999. Specific lipid-protein interactions in a novel honeycomb lattice structure of bacteriorhodopsin. *Acta Crystallogr. D* 55(Pt. 7):1251–56
87. Schertler GF, Bartunik HD, Michel H, Oesterhelt D. 1993. Orthorhombic crystal form of bacteriorhodopsin nucleated on benzamidine diffracting to 3.6 Å resolution. *J. Mol. Biol.* 234:156–64
88. Seehra JS, Khorana HG. 1984. Bacteriorhodopsin precursor. Characterization and its integration into the purple membrane. *J. Biol. Chem.* 259:4187–93
89. Seelig A, Seelig J. 1985. Phospholipid composition and organization of cytochrome *c* oxidase preparations as determined by 31P-nuclear magnetic resonance. *Biochim. Biophys. Acta* 815:153–58
90. Siebert F, Mantele W. 1983. Investigation of the primary photochemistry of bacteriorhodopsin by low-temperature Fourier-transform infrared spectroscopy. *Eur J. Biochem.* 130:565–73
91. Sineshchekov OA, Jung KH, Spudich JL. 2002. Two rhodopsins mediate phototaxis to low- and high-intensity light in *Chlamydomonas reinhardtii. Proc. Natl. Acad. Sci. USA* 99:8689–94
92. Spudich JL, Yang CS, Jung KH, Spudich EN. 2000. Retinylidene proteins: structures and functions from archaea to humans. *Annu. Rev. Cell Dev. Biol.* 16:365–92
93. Sternberg B, L'Hostis C, Whiteway CA, Watts A. 1992. The essential role of specific *Halobacterium halobium* polar lipids in 2D-array formation of bacteriorhodopsin. *Biochim. Biophys. Acta* 1108:21–30
94. Subramaniam S, Gerstein M, Oesterhelt D, Henderson R. 1993. Electron diffraction analysis of structural changes in the photocycle of bacteriorhodopsin. *EMBO J.* 12:1–8
95. Subramaniam S, Henderson R. 2000. Crystallographic analysis of protein conformational changes in the bacteriorhodopsin photocycle. *Biochim. Biophys. Acta* 1460:157–65
96. Takeda K, Sato H, Hino T, Kono M, Fukuda K, et al. 1998. A novel three-dimensional crystal of bacteriorhodopsin obtained by successive fusion of the vesicular assemblies. *J. Mol. Biol.* 283:463–74
97. Teissie J, Prats M, LeMassu A, Stewart LC, Kates M. 1990. Lateral proton conduction in monolayers of phospholipids from extreme halophiles. *Biochemistry* 29:59–65
98. Varo G, Lanyi JK. 1991. Effects of the crystalline structure of purple membrane on the kinetics and energetics of the bacteriorhodopsin photocycle. *Biochemistry* 30:7165–71

99. Vonck J. 1996. A three-dimensional difference map of the N intermediate in the bacteriorhodopsin photocycle: Part of the F helix tilts in the M to N transition. *Biochemistry* 35:5870–78
100. Vonck J. 2000. Structure of the bacteriorhodopsin mutant F219L N intermediate revealed by electron crystallography. *EMBO J.* 19:2152–60
101. Wakeham MC, Sessions RB, Jones MR, Fyfe PK. 2001. Is there a conserved interaction between cardiolipin and the type II bacterial reaction center? *Biophys. J.* 80:1395–405
102. Weik M, Patzelt H, Zaccai G, Oesterhelt D. 1998. Localization of glycolipids in membranes by in vivo labeling and neutron diffraction. *Mol. Cell* 1:411–19
103. White SH, Wimley WC. 1999. Membrane protein folding and stability: physical principles. *Annu. Rev. Biophys. Biomol. Struct.* 28:319–65
104. Woese CR, Kandler O, Wheelis ML. 1990. Towards a natural system of organisms: proposal for the domains Archaea, Bacteria, and Eucarya. *Proc. Natl. Acad. Sci. USA* 87:4576–79
105. Wolfer U, Dencher NA, Buldt G, Wrede P. 1988. Bacteriorhodopsin precursor is processed in two steps. *Eur. J. Biochem.* 174:51–57

Annu. Rev. Biophys. Biomol. Struct. 2003. 32:311–34
doi: 10.1146/annurev.biophys.32.110601.142536

First published online as a Review in Advance on February 21, 2003

ACETYLCHOLINE BINDING PROTEIN (AChBP): A Secreted Glial Protein That Provides a High-Resolution Model for the Extracellular Domain of Pentameric Ligand-Gated Ion Channels

Titia K. Sixma[1] and August B. Smit[2]
[1]Division of Molecular Carcinogenesis, Netherlands Cancer Institute, Plesmanlaan 121, 1066 CX Amsterdam, The Netherlands; email: t.sixma@nki.nl
[2]Department of Molecular and Cellular Neurobiology, Research Institute Neurosciences Vrije Universiteit, Faculty of Biology, De Boelelaan 1087, 1081 HV Amsterdam, The Netherlands; email: absmit@bio.vu.nl

Key Words nicotinic acetylcholine receptor, $GABA_A$, $5HT_3$, glycine receptor, Cys-loop

■ **Abstract** Acetylcholine binding protein (AChBP) has recently been identified from molluskan glial cells. Glial cells secrete it into cholinergic synapses, where it plays a role in modulating synaptic transmission. This novel mechanism resembles glia-dependent modulation of glutamate synapses, with several key differences. AChBP is a homolog of the ligand binding domain of the pentameric ligand-gated ion-channels. The crystal structure of AChBP provides the first high-resolution structure for this family of Cys-loop receptors. Nicotinic acetylcholine receptors and related ion-channels such as $GABA_A$, serotonin $5HT_3$, and glycine can be interpreted in the light of the 2.7 Å AChBP structure. The structural template provides critical details of the binding site and helps create models for toxin binding, mutational effects, and molecular gating.

CONTENTS

INTRODUCTION

In this review we discuss the remarkable physiology of acetylcholine binding protein (AChBP) in the great pond snail, *Lymnaea stagnalis*, its secretion by glial cells, and its possible role in molluskan cholinergic synapses. Current views of the presence of related proteins in other organisms are discussed. We use the crystal structure of AChBP to shed light on structure/function aspects of pentameric ligand-gated ion-channels (LGICs), which mediate and modulate chemical synaptic transmission. This family of transmembrane receptors includes the nicotinic acetylcholine (nAChR), serotonin $5HT_3$, γ-aminobutyric acid ($GABA_A$ and $GABA_C$), and glycine receptors (86) as well as invertebrate glutamate (27) and histamine (130) channels. These receptors are, apart from their endogenous neuronal ligands, receptive to diverse compounds such as nicotine, alcohol, and various snake and snail venoms. In addition, they are prime targets for pharmaceuticals such as barbiturates, benzodiazepines, and anti-emetics. Mutations in these receptors are involved in diseases such as congenital myasthenia gravis, epilepsy, startle syndrome, and in sensitivity to alcohol (120). Also, nAChRs mediate nicotine addiction in chronic tobacco users. Because LGICs are involved in important aspects of brain functioning and brain diseases, they are considered prime targets for novel drug discovery programs.

AChBP IN *LYMNAEA STAGNALIS*

Lymnaea stagnalis was analyzed because it has large easily identifiable neurons that have been characterized extensively. This precise knowledge allowed in vitro culture of specific synapses (128). The AChBP protein was found in a study that concentrated on the role of glial cells in cholinergic synapses in *L. stagnalis* (111).

The Role of Glial Cells in Synaptic Transmission

Glia are the most numerous cells in the central nervous system (CNS). Their main role was seen as providing metabolic and trophic support to neurons. In recent years this classical view on the role of glia has been challenged, and recent findings indicate active glial involvement in the modulation of synaptic transmission. Various studies have supported the view that the neuron-glia communication is bidirectional: Glial cells receive neuronal input and may also release transmitter onto neurons thereby affecting neuronal excitation and synaptic transmission (13, 123). Compelling examples of this come from glutamatergic signaling between neurons and glia in the hippocampus, from GABAergic synapses, where GABA released by astrocytes potentiates inhibitory synaptic transmission (54), and from the cholinergic neuromuscular junction, where acetylcholine (ACh) released by perisynaptic glia (53, 95) is postulated to feed back onto the presynaptic nerve terminal (94). As such, glial and neuronal cells can form integral modulatory components of synaptic function (115).

The insight that glial cells are part of the functional brain circuitry stimulated a study of their precise modulatory role in well-characterized individual synapses. Individually identifiable neurons isolated from the CNS of the mollusk *L. stagnalis* have the advantage to readily form synapses in vitro that in turn can be cocultured with or without glial cells (36, 44, 127, 128). These glial cells were found to produce the AChBP protein (111). In reconstructed excitatory cholinergic synapses in vitro, cocultured synaptic AChBP-producing glia abolished the presynaptically induced facilitatory excitatory postsynaptic potentials (EPSPs) and action potentials in the postsynaptic cell. Together, various types of experiments demonstrated that glial modulation of cholinergic synaptic efficacy was mediated by AChBP (111).

Production and Release of AChBP from Molluskan Glia

The AChBP gene is specifically expressed in about 50% of the glial cells located between neuronal cell bodies and axon terminals of the *Lymnaea* CNS. The AChBP protein was visualized in the endoplasmic reticulum and the Golgi apparatus and found in granule-like compartments (111), which might represent large, dense core vesicles, all indicating that AChBP is synthesized in a regulated protein secretory pathway. Transmitter-containing vesicles have not been demonstrated so far in mammalian glial cells. Secretion of transmitters from mammalian glia is similar to that from neurons, e.g., glutamate vesicles produced in astrocytes are released upon a glutamate stimulus in a Ca^{2+}-dependent manner, involving the SNARE vesicle release machinery (5, 88). Release of AChBP from glia was also stimulus dependent, i.e., dependent on exposure to ACh, and mediated by a glial nAChR. Whether AChBP secretion results from calcium influx coupled to nAChR activation is not yet known.

AChBP-Mediated Modulation of Synaptic Transmission

Astrocyte-mediated modulation of glutamatergic synaptic transmission is via glutamate and involves a bidirectional mode of neuron-glia signaling (Figure 1*a*). Glia-mediated modulation of cholinergic synaptic transmission via AChBP is mechanistically different, as it involves direct interaction of a glia-derived protein and synaptic transmitter (111) (Figure 1*b*).

Models of the kinetics of AChBP in synaptic modulation require as yet absent data on the concentrations of ACh, acetylcholinesterase (AChE), and postsynaptic nAChRs, the localization and individual affinities of these components, as well as diffusion characteristics of ACh (59, 64, 126). Using a simple equilibrium equation, however, the concentration of free ACh can be calculated for given concentrations of AChBP and of total ACh (Figure 1*b*). In two scenarios of cholinergic transmission, under low and high concentration of AChBP (e.g., 80 μM versus 600 μM

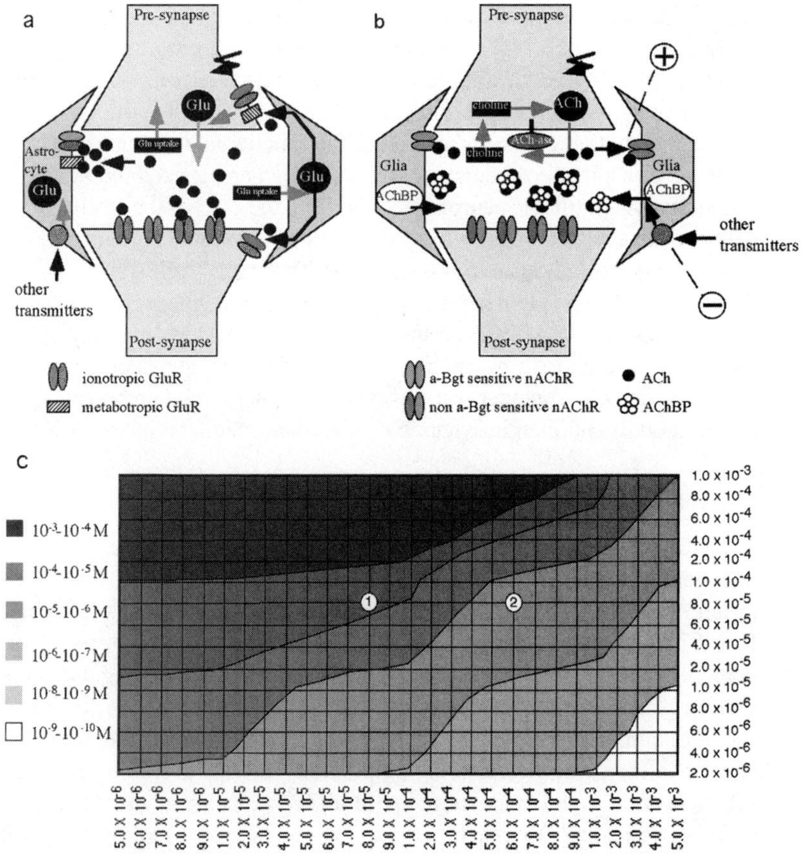

Figure 1 The glia-derived glutamate and AChBP have similar roles in synaptic transmission. (*a*) Schematic of glutamate-dependent modulation of synaptic transmission in astrocytes. Presynaptically released glutamate activates postsynaptic AMPA- or NMDA-type receptors and, in parallel, activates AMPA- of metabotropic glutamate receptors (12) on synapse-ensheathing astrocytes. The astrocytes subsequently release glutamate onto the synaptic elements, which activates NMDA and metabotropic glutamate receptors (5) and thereby modulates synaptic transmission (9). (*b*) Schematic of AChBP-dependent modulation of cholinergic synaptic transmission in molluskan glia. Under conditions of active presynaptic transmitter release (61), ACh activates both postsynaptic receptors and nAChRs on synaptic glial cells (EC_{50} is in the μM range). The latter enhances AChBP secretion thereby increasing its concentration in the synaptic cleft, which either diminishes or terminates the ongoing ACh transmission or the subsequent postsynaptic responses to ACh. (*c*) Two-dimensional plot representing the calculated concentration of free ACh in the synaptic cleft under varying concentrations of ACh and AChBP (given as $Kd = 4\ \mu$M and based on the equilibrium of AChBP + ACh ⇔ AChBP-ACh). At point 1, the concentration of free Ach = 16 μM, at point 2 it is 0.6 μM. Postsynaptic receptors in *Lymnaea* typically do not activate below 1 micromolar and thus would be inactive in condition 2.

in Figure 1*c*), only the low AChBP titer would yield a free ACh concentration of $\sim$16 μM, which is probably adequate to activate postsynaptic receptors. As such, the actual synaptic concentration of AChBP might critically determine whether transmission is either fully active or suppressed.

Various transmitter or hormone systems modulate glial cells in the mammalian brain. Many receptor types on glia, e.g., amino-methyl proprionic acid (AMPA)- (31), $GABA_B$- (54), α1-adrenoreceptors (63), and also different nAChR subunits, have been found in oligodendrocytes (96) and astrocytes (51). Thus molluskan synaptic glial cells might integrate various signals. These might adjust the synaptic concentration of AChBP at which modulation of synaptic transmission will occur (Figure 1*c*).

THE ACETYLCHOLINE BINDING PROTEIN

The AChBP sequence encodes a 210-residue mature protein, including two disulfide bonds. An N-terminal signal sequence is cleaved off in the mature protein. There is one glycosylation site at residue Asn66. The recombinant AChBP protein, expressed in the yeast *Pichia pastoris*, assembles into stable homopentamers as shown by gel filtration (111) and analytical ultracentrifugation (K. Brejc, M.H. Lamers & and T.K. Sixma, unpublished data).

AChBP in Other Species

Related AChBP genes have been found in the mollusk *Aplysia californica* (Acc. nr: AF364899) and in the leech *Haementeria ghilianii* (98). Based on the absence of typical α-subunit sequence features, the latter protein is named β-subunit-like. *Lymnaea* and *Aplysia* AChBP have 33% sequence identity, *Lymnaea* and *Haementeria* only 18.2%. Thus, sequence conservation is low, even within the molluskan phylum, in which leeches are nowadays often included. Although ligand binding characteristics for the *Aplysia* protein are not yet available, it is probably a functional homolog of *Lymnaea* AChBP, but with different ligand binding characteristics. The *Lymnaea* and *Aplysia* AChBP have lower sequence identity to the two α-subunits of nAChRs identified in *Aplysia* (25% and 26%, respectively) than to each other (33%). An important question remains whether orthologs of AChBP will be found in other animal phyla. The cloning of AChBP orthologs in more distantly related species is obviously hindered by the low degree of sequence conservation. Database searches in the genomes of *Caenorhabditis elegans*, *Drosophila*, and human have not yielded AChBP orthologs so far.

AChBP Structure

The crystal structure of AChBP was determined to 2.7 Å resolution (18). The AChBP homopentamer forms a doughnut-like structure with a radius of 80 Å and a height of 62 Å (Figure 3, see color insert). Each AChBP monomer folds into an N-terminal helix and a curled extended β-sandwich with modified immunoglobulin topology. Where a conserved "tyrosine corner" is found in different immunoglobulin families (45, 49), AChBP has a disulfide bridge. The electron density shows evidence of calcium ions and a HEPES buffer molecule, both present at ~100 mM during crystallization. The HEPES molecule was bound in the ligand binding site, whereas Asp161 and 175 and the main chain of residue 176 liganded the calcium ion.

THE LIGAND-GATED ION CHANNEL SUPERFAMILY

The superfamily of Cys-loop LGICs has a highly conserved structure (65, 86). These receptors form homo- and heteropentamers in which each monomer has a conserved extracellular N-terminal domain and four transmembrane domains (M1–M4). Ligand binding takes place in the N-terminal domain, and a signal is conveyed to a gate that is created by the M2 transmembrane helix. The channels are cation or anion selective, dependent on residues in the M2 domain. The extracellular M2/M3 loop has a possible role in channel gating (41) while the cytoplasmic M3/M4 loop of variable length has a role in regulation by posttranslational modification such as phosphorylation or ubiquitination. In general the other superfamily members have not been as extensively studied as the nAChRs, but their importance as drug targets has recently sparked a lot of interest.

AChBP Sequence Compared with LGIC N-Terminal Domains

AChBP aligns well with the ligand binding domains of all LGICs (18). It has sequence identities of 15%–20% with $GABA_A$, glycine, and serotonin $5HT_3$ receptors, and it shows greatest similarity to the nicotinic receptor subunits, between 20%–25% identity (Figure 2*c*).

In the crystal structure the conserved residues map to the monomer core, indicating that the fold of the other family members will resemble AChBP. The only conserved region in the LGIC superfamily that is different in AChBP is the Cys-loop [AChBP: 123–136, *Torpedo* nAChR: 128–142], which has a conserved length and hydrophobic character in the receptors (22) but is one residue shorter and mostly hydrophilic in AChBP. It is found on the membrane-facing side (bottom) (Figure 2) and probably interacts with the transmembrane domain in the receptors.

Many different alignments have been published of AChBP to various ligand-gated ion channels (18, 25, 66, 81, 93, 99, 110). The main variations reside in two regions that are difficult to align, the region around the so-called F-loop and around the C-loop (see Ligand Binding Site, below). Both regions affect (F-loop) or are critical (C-loop) for ligand binding, and it is likely that informed analysis of particular substructures leads to optimized alignments. However, owing to large insertions, in particular within the F-loop region, alignment will remain difficult.

Nicotinic Acetylcholine Receptors

The nAChRs are found in the CNS and in neuromuscular junctions and function in either synaptic transmission or as modulators of neurotransmitter release. The muscle nAChR with the complex subunit stoichiometry $(\alpha_1)_2\beta_1\delta\gamma$ is found at the

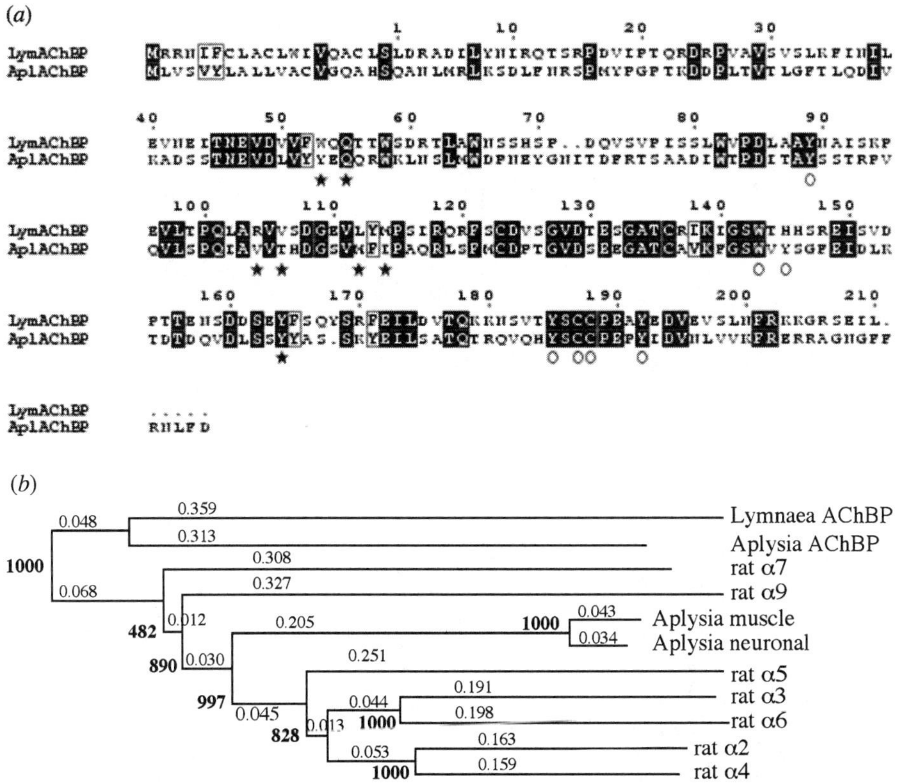

Figure 2 (*a*) Alignment of the sequences of *Lymnaea* and *Aplysia* AChBPs (acc.nrs AF322877 and AF364899). Identical amino acids are boxed. (*b*) Dendrogram of the sequence relatedness of *Lymnaea* and *Aplysia* AChBPs, the extracellular domains of the *Aplysia* ACh receptor subunits (neuronal AF467898; muscle AF467899), and the rat alpha2-9 (respectively, acc.nrs NM_133420, NM_052805, NM_024354, NM_017078, L08227, NM_012832, NM_22930). Note the distinct group that is formed by the AChBP sequences. Alignments and dendrogram were made using default settings in Clustal-X. Bootstrap values (*bold*) and branch lengths are indicated. (*c*) Alignment of *Lymnaea* AChBP with ligand-gated ion channels (18). Secondary structure and sequence conservation are indicated (Blosum62 matrix); residues contributing to ligand binding site are indicated by an asterisk (*_ (principal side) or O (complementary side). Figures (*a*) and (*c*) were prepared with ESPript (39).

neuromuscular junction in vertebrates and in the electric organs of fish (e.g., electric ray, *Torpedo californica*). Neuronal nAChRs exist in many different arrangements, with a combination of α_4 and β_2 as a major subtype. Stoichiometries vary but heteropentameric neuronal receptors contain at least two α-subunits. In addition α_7–α_9 can also form homopentamers (74).

Structural data for the nicotinic receptor are available thanks to the discovery of tubular crystals of ACh receptors (59a), X-ray structure analysis (60a), and more

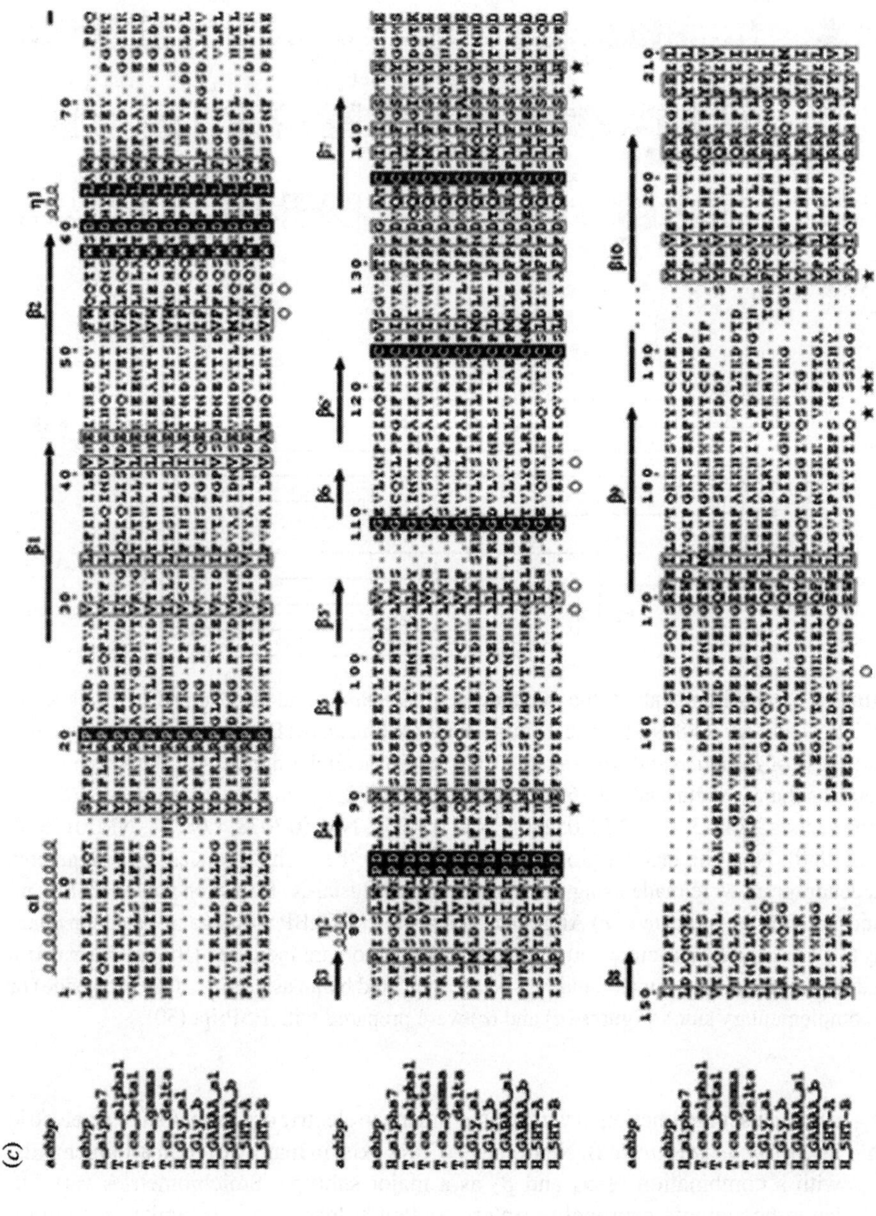

Figure 2 *(Continued)*

recently electron microscopy (EM) studies by Unwin and coworkers using *Torpedo* receptors (10, 77, 117). Comparison of the 4.6 Å EM data with the AChBP crystal structure shows that the proteins are indeed similar and that subunit conformations are especially comparable to the non-α-subunits (118). The α-subunits themselves resemble AChBP once ligand is bound.

Several nAChR models have been created based on the AChBP structure (37, 66, 81, 99, 110). Because the homology is weak (<25% identity) such models are intrinsically inaccurate. However, they will be relatively good where the similarities are largest such as the ligand binding site. Details of the models with respect to ligand and toxin binding will be discussed below. An interesting approach was followed by Sine et al. (110), who used lysine scanning as an additional constraint on the modeling of a nAChR ε-subunit.

ACHBP PHARMACOLOGY RESEMBLES A NICOTINIC RECEPTOR AChBP contains all the amino acid residues needed for ligand binding. It has a pharmacological profile that resembles a homopentameric receptor (111), such as the neuronal α_7 type (3). Thus it has a relatively weak ACh affinity and high affinity for nicotine and especially epibatidine. Many antagonists of the nicotinic receptor also bind to AChBP, such as α-bungarotoxin, d-tubocurarine, and gallamine. In addition, some muscarinic antagonists bind to AChBP, a mixed pharmacology that was also observed for the homopentameric α_9 receptor (33). The observed Hill coefficients do not indicate cooperativity between ligand binding sites in AChBP. Interestingly ligand binding to AChBP results in intrinsic fluorescence quenching (45a). Stopped-flow analysis using this effect showed that AChBP has rapid association and dissociation kinetics for agonists and slow rates for α-neurotoxins. Ligand stoichiometry was approximately five sites per pentamer, and no cooperativity between sites could be observed. A two-step binding process was observed for the neurotoxin, with fast binding, followed by a slower conformational change.

LIGAND BINDING SITE Affinity labeling, mutagenesis, and the substituted cysteine accessibility method (SCAM) (57) have identified residues involved in ligand binding in the muscle and *Torpedo* nAChR subtypes. The binding site is found at the interface between an α-subunit and a neighboring subunit (89), forming principal and complementary parts of the binding pocket, respectively (6, 23, 56). Some ligands, in particular d-tubocurarine and various neurotoxins, show differential binding dependent on the complementary subunit (14, 20, 89, 105–107).

The α-subunit contributes the principal binding site, which consists of residues clustered in loop regions A (38), B (29), and C (29, 55, 109). The neighboring subunit, γ or δ, contributes loops D (20), E (20, 21, 105), and F (28). In homopentamers the equivalent loops will be contributed by α-subunits. No negatively charged residues are found in the binding site, to compensate the ACh positive charge, and the electrostatic contribution is expected to derive from cation-π interactions (30, 131), as seen in the AChE structure (47).

In the AChBP structure all loop regions form a single pocket region at the subunit interface, with loops A through C contributed from one subunit and D through F from another (Figure 4, see color insert). Loops A, B, C, and F are indeed structural loops in the protein, but the residues contributed from loops D and E are found on β-strands. Aromatic residues of loop A through D line the pocket, and the hydrophobic components of the residues in loop E form the lid of the pocket. The F-loop region is not conserved in sequence or length, and the density is relatively weak. Different conformations of this region are likely in other family members. The functionally important vicinal disulfide bond (55) in the C-loop resembles the major *trans* conformer observed in model peptides (24), but higher-resolution data are needed to confirm this.

The binding site is located toward the outside of the AChBP pentamer, in a location that disagrees with suggestions based on interpretation of electron microscopy data (77). The site suggested by the EM data does not contain any residues shown to be important for ligand binding.

A HEPES (N-2-hydroxyethylpiperazine-N9-2-ethanesulphonic acid) buffer molecule was found in the ligand binding site (18). This weak ligand (IC_{50} = 100 mM) has a positive-charged nitrogen atom, similar to ACh. It stacks on Trp143, forming cation-π interactions (30, 131). The position of the HEPES molecule was unclear in the 2.7 Å data (18), and recent 2.0 Å data (P. Celie, W. van Dijk & T.K. Sixma, unpublished data) show that it was incorrectly oriented (Figure 4). The reorientation does not affect the cation position.

Models of ACh and nicotine binding were created using an AChBP-based α_7 model (66). Schapira et al. (99) created a more cautious model of the binding site only, with ACh and nicotine bound. These models are available on websites of the authors (65, 99), which are important because they allow comparisons between them. It will also be interesting to compare these models with crystallographic studies of ligand-bound AChBP.

TOXIN BINDING Toxins from snakes and snails behave as competitive antagonists for nAChR ligand binding and have been used to probe the active site and to determine the relative differences between subtypes of nicotinic receptors (6, 23). Chimeras between receptor subtypes were analyzed, using double-mutant cycles, in which nonadditive free-energy changes are taken as a measure of proximity of residues (4, 19, 73, 87, 92). Several different classes of toxins have been used for these purposes, the α-neurotoxins, the conotoxins, and the waglerins. The three-fingered toxins include the long-chain α-neurotoxins (e.g., α-bungarotoxin) and the short-chain α-neurotoxins [e.g., erabutoxin, α-cobratoxin, *Naja mossambica mossambica* I (NmmI)]. A major determinant for bungarotoxin binding is the C-loop region. Peptide binding studies with a synthetic C-loop peptide have an affinity of 10^{-4} M (82). Using an iterative approach of phage display techniques with combinatorial libraries (8), peptides with affinity for bungarotoxin in the nanomolar range were found (58).

The structure of α-bungarotoxin (68a) has also been determined in complex with a high-affinity peptide (46). A comparison of the 1.8 Å high-affinity peptide structure with the C-loop region of AChBP showed a remarkable similarity

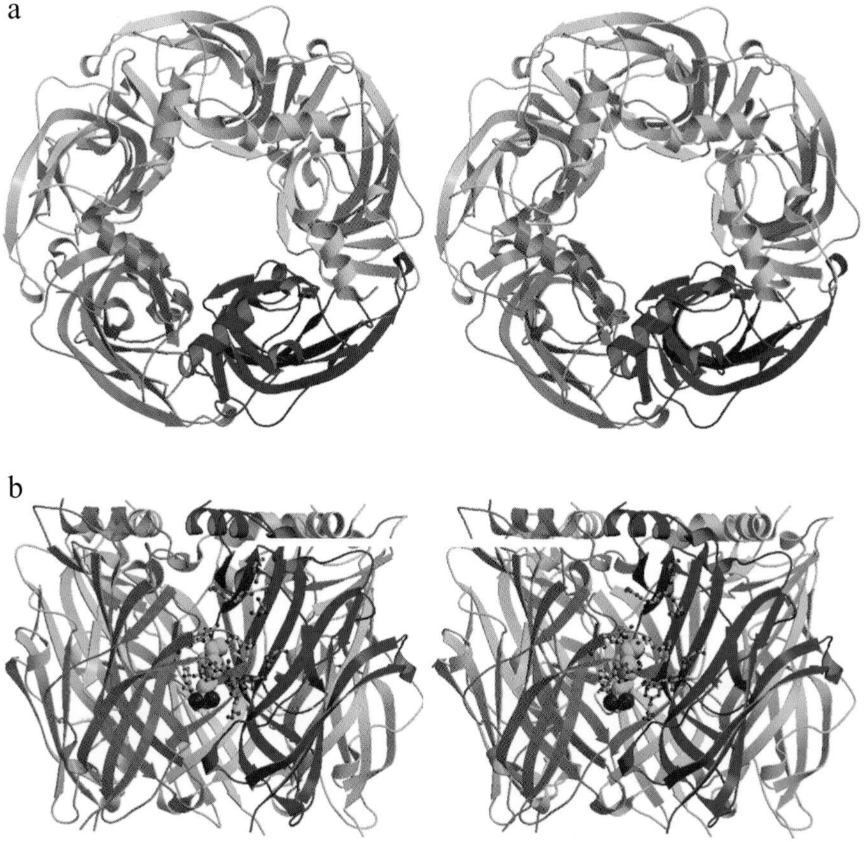

Figure 3 (*a*) Stereoview of main chain of AChBP with two monomers separately colored, viewed toward the membrane. (*b*) Side view, with ligand binding residues in ball-and-stick configuration (principal binding site colored yellow, complementary binding site colored blue) with bound HEPES (green), shown as CPK model. In this orientation the membrane will be at the bottom of the figure.

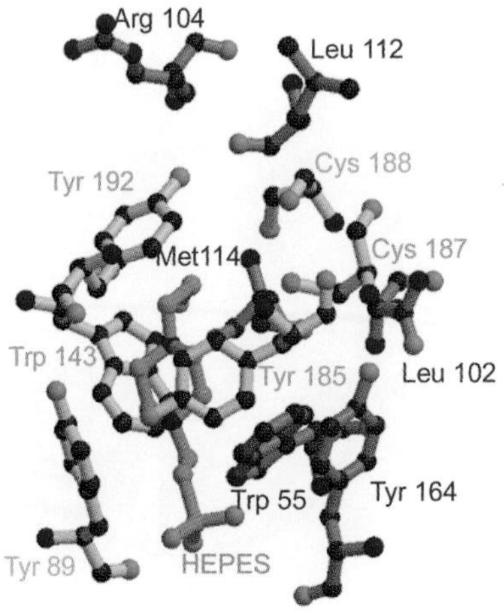

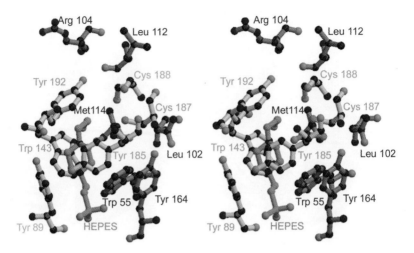

Figure 4 The aromatic cage of AChBP ligand binding site with HEPES bound as determined in high-resolution structure (P. Celie & T.K. Sixma, unpublished data). Color scheme as in Figure 3*b*.

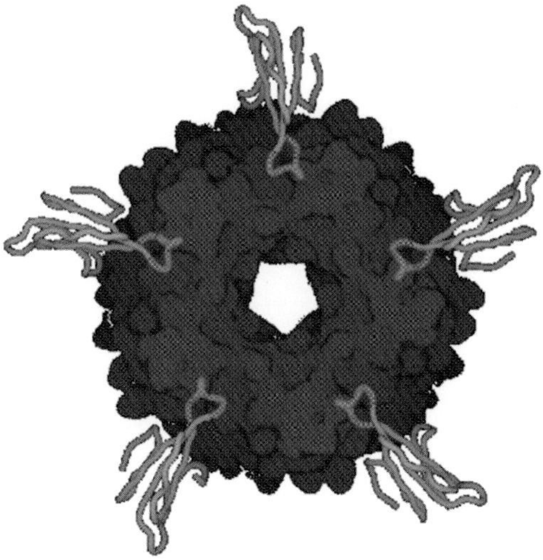

Figure 5 Proposed binding mode of α-bungarotoxin (yellow) to AChBP (red) (46) provides a model for the toxin binding to nAChRs.

between these β-hairpins (RMS deviation of 0.76 Å for 12 Cαs) (46). Based on this superposition a good model of bungarotoxin binding to AChBP and, by inference, to the nicotinic receptor was obtained. The few minor clashes can easily be resolved, by conformational changes. The toxin binds halfway between membrane and top of the ligand binding domain, angled outward from the center (Figure 5, see color insert). It makes close contacts to the C-loop and inserts its finger 2 into the ligand binding site, thus precluding ligand binding completely. The toxin contacts both the principal and the complementary subunits. In the complementary subunit, it interacts with the F-loop region (which is difficult to align) as well as residue δSer36 or γLys 34 in region I [or the D-loop region in the seven-loop model of Taylor et al. (107)] (91), confirming its importance for ligand binding (107). A similar toxin-AChBP model was produced (79) from a nuclear magnetic resonance (NMR) structure of bungarotoxin with an authentic 19-mer peptide of the α_7 receptor. The proposed binding locus is identical, but the toxin is tilted more toward the membrane.

Using double-mutant cycle studies (73), which involve change reversal of residues, proximity data between Nmm toxin residues and both the α- and γ-subunit interfaces were obtained. A model for α-cobratoxin binding to an AChBP-based α_7 nAChR was based on nine distance constraints derived from similar double-mutant cycles (37, 66). Complex formation required adjustments to the F-loop region, but then the biochemical constraints could be satisfied with only one exception. The model was validated by an a posteriori double-mutant experiment involving α7-W54 and cobratoxin F29. The final model is similar to the bungarotoxin complexes with respect to the second toxin finger inserted into the binding site, positioning an arginine side chain (R33) for cation-π interactions (37), but cobratoxin makes no contacts with the E-loop region, in agreement with mutant data. All toxin complex models agree well with prior data on α-neurotoxin binding, except with cross-linking data of neurotoxin II from *Naja naja oxiana* to δ268 (70).

Conotoxins are an amazingly variable series of short ~20-residue peptide toxins produced by the *Conus* species of snails that target neuronal receptors (75). Double-mutant cycles of the M1 α-conotoxin with nAChRs revealed selectivity for the α-δ interface over the α-γ interface (107). The residues involved are in region I (δS36 versus γK34) and the loop E (δY113 versus γS111) and loop F (δI179 versus γF172). In AChBP the equivalent residues are S32, E110, and approximately Ser160, although the latter region is difficult to define. In the α-subunit residue Tyr198 (AChBP: Y192) is important for the interaction, while double-mutant cycles showed that hydrophobic interactions are important for M1 binding (19). All these experiments agree with the observed binding site in the AChBP crystal structure.

Waglerins have specificity for the α-ε interface of nAChR muscle subtypes. Recent mutant studies established the residues contributing to interaction (80, 81). Binding is strongly dependent on the presence of Asp residue in the F-loop region (ε173, γ172). Modeling of the residues using the AChBP structure (80, 81) helped delineate a number of intrasubunit contacts, in particular with residues ε57 and

ε173 or in the δ-subunit between δ178 and the δ36, δ59, and δ61. Residues ε173 or δ178 are found in the F-loop region, which has a large insertion in the γ-, δ-, and ε-subunits. In the crystal structure this region has badly defined electron density, and the structure is somewhat unusual. The observed weak calcium binding site is close to this site. Lack of a good model may be overcome by constraints from biochemical data.

An interesting twist has been given to the toxin binding studies with the finding of the lynx1 protein (52, 76). This mammalian protein is predicted to have a structure that resembles the three-fingered α-neurotoxins. Lynx1 modulates the function of nAChRs and physically associates with the receptor. It will be interesting to see whether the details of the lynx1-nAChR binding resemble those of the snake neurotoxins.

NONCOMPETITIVE MODULATORS Several noncompetitive modulators (1) have been described for the nAChRs, but many of these are thought to act on the transmembrane portion of the receptor. However, there is increasing interest in the role of galantamine and related compounds, which may bind to the ligand binding domain (72). These inhibitors of AChE have clinical benefits for patients with Alzheimer's disease. Because the effect on AChE does not fully explain the clinical efficacy of these compounds, further research has identified nAChRs as additional targets. The binding site for these compounds is mostly unclear, although human α_1-K125 is involved in binding physostigmine (102). This residue is found in strand β6, close to the start of the Cys-loop. It is not yet clear whether AChBP will bind to these noncompetitive agonists.

FOLDING AND ASSEMBLY The assembly of the muscle type nicotinic receptor $\alpha\gamma\alpha\delta\beta$ has been controversial, but it can be resolved by the crystal structure of AChBP. The subunits must be arranged in anticlockwise order (18), in agreement with an analysis of the possible coexpression patterns (87). An earlier incorrect assignment (71) was due to a wrong assumption about the orientation of a toxin.

The subunit interface can be analyzed on the basis of the AChBP structure. Subunit interfaces are different between the different LGIC families, although they are well preserved within each family (18). Modeling studies already started to indicate potential contacts in other interfaces (110).

The structure of AChBP gives no direct information about the transmembrane portion of LGICs because that region does not exist in AChBP, but it positions the start of the M1 transmembrane domains at ~31 Å from each other. In a comparison of AChBP with the 4.6 Å EM data (118), this localization helped to map the synaptic side of the four transmembrane regions. Because the M2 regions should form the pore and because the M4 region is most accessible, total mapping is possible. This resulted in a prediction that the M2/M3 extracellular loop interacts with the Cys-loop region, which is interesting because the M2/M3 region is involved in gating (41). The AChBP crystal structure also provides insight in the surface area of the ligand binding domain that could contact the transmembrane portion.

ACTIVATION STATE An interesting question is whether the AChBP structure resembles the resting, the activated, or the desensitized state of the nAChR. Several arguments have been put forward in favor of a desensitized state: AChBP has no apparent cooperativity, a ligand is bound to the structure (43), and the current position of the F-loop precludes toxin binding (37). The high affinity of AChBP for acetylcholine presented another argument for the activated or desensitized state because the resting state has low affinity for agonist (114). Upon comparison of AChBP with the EM data of the *Torpedo* nAChR, Unwin et al. (118) concluded that in the resting state AChBP resembled the non-α-subunits but not the α-subunits, whereas in the 9 Å ACh-bound structure the latter had changed conformation to look like AChBP. A movement of the inner β-sheet with respect to the outer β-sheet was invoked to explain the movement between resting and ligand-bound states (118).

GATING Analysis of chimeras comprising the N-terminal domain of the nAChR and the transmembrane domain of the $5HT_3$ receptor has shown that ligand binding and gating are located on different domains: The chimeras have nAChR-like ligand binding and $5HT_3$-like gating (32). Thus ligand binding and receptor gating are uncoupled in this superfamily, and transmission of the signal is conserved between receptor subtypes. Therefore each ligand binding domain should undergo similar conformational changes that lead to opening of the gate.

Based on an extensive series of mutants, using single-channel experiments as readout, Auerbach and colleagues (42) derived a series of linear free-energy relationships for the relative importance for gating of various residues in the nAChRs. This led to the suggestion that the changes in conformation behave as a propagated conformational wave traveling from binding site to channel opening. The largest changes were seen close to the binding site and were smoothly becoming smaller toward the position of the gate.

Several suggestions have been made on the conformational changes in the extracellular domain that lead to gating. The allosteric model of Monod-Wyman-Changeux would suggest that changes in conformation occur symmetrically in the different subunits. A rigid body rotation of the ligand binding subunits has been suggested as a mechanism for transmitting the signal from the ligand binding site to the gate (43). The changes that were seen between ligand-free and ligand-bound state in *Torpedo* nAChR led to the suggestion (117, 118) that the inner and outer β-sheet of the α-subunit could move with respect to each other, to accomplish the observed changes in conformation. Examination of the AChBP ligand binding site led Karlin (56) to the suggestion that ligand size could be important for effective gating because antagonists are generally larger. Thus gating could be initiated by a contraction of the aromatic residues around the ligand. This suggestion fits with a difference in conformation that was observed for the C-loop region (the $\beta 9$-$\beta 10$ hairpin) in the comparison between AChBP and the EM structure of the nAChR (118). Such a contracting mechanism was concluded for the glutamate receptor subunit 2 (GluR2) channel after comparison between the ligand-bound and ligand-free state (7). In this tetrameric homolog of the human AMPA, receptor contraction around the ligand binding site leads to a conformational change that opens the receptor. Interestingly for the muscarinic ACh receptor a similar contraction of the ACh binding site was recently proposed (69). In this G protein–coupled receptor (GPCR) the result of ACh binding is to expose the G protein binding site. The muscarinic receptor also seems to rely on an aromatic cage for ligand binding, but otherwise the two types of ACh receptor appear to have little in common.

MYASTHENIA GRAVIS The disease myasthenia gravis is clinically known for muscle weakness and fatigue (68). It can be caused by an autoimmune response to proteins involved in muscle activation such as the nicotinic receptor. More than half of these antibody types are directed against the main immunogenic region (MIR) on the α1-subunit [residues 67–76] (116). The MIR was located by EM (10) at the top of the molecule, angled outward. In AChBP the loop is not conserved, but its position confirms the localization (18). Comparison of an MIR peptide NMR structure (85) to the AChBP equivalent region shows a reasonably good superposition for the important N terminus of the peptide. Superposition of AChBP 66–70 (NSSHS) on residues 2–6 of the MIR peptide (NPDDY) showed an RMS deviation of 1.1 Å. A model of the MIR peptide-antibody complex has been published (60, 90). Combination of the modeled complex with the above superposition results in a rough orientation of the myasthenia antibody relative to AChBP. This model shows no clashes and gives a general idea of the MIR-antibody binding to the receptor. It fits with data showing that the antibodies do not occlude the binding site and could act by cross-linking receptors to each other.

Congenital myasthenic syndrome A hereditary form of muscle weakness, congenital myasthenia, results from mutations in proteins involved in muscle activation, including the nicotinic receptor (34, 124). These mutations can result in either a slow-channel or a fast-channel response depending on their effect on the receptor. Missense mutations in the ligand binding domains that affect channel kinetics can now be interpreted in light of the crystal structure of AChBP. A εP121L mutation (84), combined with a null-mutation in the other allele, results in a fast-channel response, with diminished postsynaptic response to ACh. The homologous residue in AChBP, S116, is found at the interface between subunits not far from the ligand binding site. The proline mutation in the ε-subunit could change the main chain conformation affecting ligand binding and gating. Two slow-channel missense mutations that do more than affect expression levels are found in the α-subunit, G153S (108), and V156M (26). Mapping to the AChBP structure does not immediately explain the effect of these mutations. The G153-related residue S147 is involved in interaction with the C-loop through a water molecule and could possibly affect C-loop movements during gating. The V156 related residue, I150, is buried in the subunit; the effect of this mutation is difficult to understand without additional data.

$5HT_3$ Receptors

The serotonin $5HT_3$ receptor responds to 5-hydroxytryptamine (5-HT, serotonin). The six other 5HT receptors are GPCRs. The $5HT_3$ receptor can make homopentamers of α-subunits and heteropentamers in which a recently identified β-subunit participates (93). $5HT_3$ receptors are mostly cation channels. Some of the ligands for the $5HT_3$ receptor and the nAChRs cross-react (93). The ligand binding involves residues from the A-loop (Phe130) (15, 113), B-loop (Trp183) (112), C-loop (50, 78, 112) and D-loop regions (W90, R92, Y94) (129, 112). Adjustment of the published alignment (18) (Figure 2*c*) is necessary for the A-loop region: If F130 is equivalent to Y89 in AchBP, there has to be a two-residue insertion at the start of strand β4. $5HT_3$ receptors are modulated by a variety of different substances, but little is yet known of how they act. Calcium is an inhibitor of $5HT_3$ receptors (93), and location of the binding site is unclear. Comparison with the AChBP calcium binding site might be interesting.

GABA Receptors

GABA receptors form chloride channels in response to stimulation with γ-aminobutyric acid. They are the major inhibitory LGICs in the CNS. $GABA_A$ and $GABA_C$ channels are LGICs, whereas the $GABA_B$ receptors are GPCRs. $GABA_A$ receptors have many different subunits, and their stoichiometry is under debate (103). $GABA_A$ receptors can be modulated by, for example, benzodiazepines that act as noncompetitive activators. The $GABA_C$ receptors consist of homo- or heteropentamers of ρ-subunits and have distinct pharmacological properties (35). The number of binding sites for GABA on the ρ-subunits was determined to be four or five, although it is unknown how many need to be occupied to generate maximal activation (35). In contrast to the $GABA_A$ family, no allostery was detected in the $GABA_C$ receptors.

A major binding site for GABA was determined at the β1-α interface, with β as principal subunit, and α as the complementary. Ligand binding residues have been found for loop B (βY157), loop C (βF200, βY205), and loop D (αF65), identified by mutational studies, covalent labeling, and/or SCAM studies (2, 104, 125). The A-loop residue Y89 in AChBP is replaced by a leucine (βL99) in $GABA_A$, while nearby βY97 is also important for binding (17, 25). The AChBP structure has been used to produce a model of a $GABA_A$ ligand binding domain, with two GABA binding sites and one benzodiazepine binding site (25). Notable differences in the GABA binding sites compared with AChBP are the likely relevance of residue βY97 as well as βE155 in the B-loop region. Because the tryptophans in loop B and D in AChBP are replaced by tyrosine, cation-π interactions are probably less important and hydrogen bonding to the tyrosine hydroxyl more important for recognizing GABA. In addition, αR67 may play a role in binding the carboxyl group of GABA, a suggestion that fits with mutational analysis (16, 48). Residues involved in benzodiazepine binding are thought to comprise loop A (αH102), loop B (αY160), loop C (αY210, S205), loop D (γF77), and loop

E (γT142). In comparison to the GABA site, the benzodiazepine site is more open thanks to a serine in place of Y (S205). The many additional residues that have an effect on benzodiazepine binding show the complexity of this binding site (25).

Glycine Receptors

The glycine receptors form inhibitory receptors that respond to glycine. They are close homologs of the GABA receptors (11, 67). These receptors convey strychnine sensitivity. Mutations in these receptors lead to defects in movements, such as the *spa*, *spd*, and *spdloc* animals, and to hyperekplexia or startle disease in humans, a disease characterized by stiff joints.

The binding site for glycine has been partially characterized (67). The A-loop (119), B-loop (F159, Y161) (100, 121), and C-loop regions (K200–K206) (122) play a role in ligand binding. Subtle differences were found between C-loop mutations for different ligands, which have been interpreted as separate binding sites (83, 121). These residues map to the β9-β10 hairpin in AchBP, and the differential effects are more likely to reflect subtleties in the binding mode. Antagonist binding affects a region around K104, F108, and T112 (101). In AChBP the equivalent residues are in loops on both the plus and minus side of the subunit interface and probably have an indirect effect on ligand binding. Interestingly the *spd* mutation, a missense mutation at α-subunit A52 (97) giving rise to a spasmodic phenotype in mice, aligns to AChBP N42. This residue, found at the tip of strand β1, is located at the interface with the transmembrane domain and may well affect the gating properties of the receptor.

Analysis of assembly of different glycine receptor subunits has led to assignment of so-called assembly boxes (62). Individual mutations were mapped that would affect assembly of hetero-oligomerization (40). Mutants were expected to form part of the subunit interface, but sequence alignment of the glycine receptor against AChBP (18) shows that most of these regions are found in surface loops. This is surprising; possibly the mutations have a subtle effect on subunit folding that is only noticeable in the hetero-oligomers.

CONCLUSIONS

AChBP provides a good starting point for understanding ligand binding in nAChRs and in the more remote pentameric LGICs. However, to what extent will these models be valid? Although the generalized fold and the details of the binding site are present in AChBP, it must be remembered that the conservation is always less than 25% identity. This is low for homology modeling, and although AChBP has confirmed many data and answered various questions, it will not address every issue. Any conclusion drawn from AChBP modeling needs to be validated in the receptor and/or AChBP. All in all there is still urgent need for a good high-resolution structure of a real pentameric ligand-gated receptor.

ACKNOWLEDGMENTS

We thank Palmer Taylor and colleagues for sending us unpublished material. We thank Katjusa Brejc, Patrick Celie, and Remco Klaassen for their input and many discussions. We thank them and Palmer Taylor, Henry Lester, Sara Fuchs, Steen Pedersen, and Meindert Lamers for critically reading the manuscript. Funding for this work was provided by NWO-CW and STW.

The *Annual Review of Biophysics and Biomolecular Structure* is online at http://biophys.annualreviews.org

LITERATURE CITED

1. Albuquerque EX, Alkondon M, Pereira EF, Castro NG, Schrattenholz A, et al. 1997. Properties of neuronal nicotinic acetylcholine receptors: pharmacological characterization and modulation of synaptic function. *J. Pharmacol. Exp. Ther.* 280: 1117–36
2. Amin J, Weiss DS. 1993. GABAA receptor needs two homologous domains of the beta-subunit for activation by GABA but not by pentobarbital. *Nature* 366:565–69
3. Anand R, Bason L, Saedi MS, Gerzanich V, Peng X, Lindstrom J. 1993. Reporter epitopes: a novel approach to examine transmembrane topology of integral membrane proteins applied to the alpha 1 subunit of the nicotinic acetylcholine receptor. *Biochemistry* 32:9975–84
4. Antil-Delbeke S, Gaillard C, Tamiya T, Corringer PJ, Changeux JP, et al. 2000. Molecular determinants by which a long chain toxin from snake venom interacts with the neuronal alpha 7-nicotinic acetylcholine receptor. *J. Biol. Chem.* 275:29594–601
5. Araque A, Parpura V, Sanzgiri RP, Haydon PG. 1999. Tripartite synapses: glia, the unacknowledged partner. *Trends Neurosci.* 22:208–15
6. Arias HR. 2000. Localization of agonist and competitive antagonist binding sites on nicotinic acetylcholine receptors. *Neurochem. Int.* 36:595–645
7. Armstrong N, Gouaux E. 2000. Mechanisms for activation and antagonism of an AMPA-sensitive glutamate receptor: crystal structures of the GluR2 ligand binding core. *Neuron* 28:165–81
8. Balass M, Katchalski-Katzir E, Fuchs S. 1997. The alpha-bungarotoxin binding site on the nicotinic acetylcholine receptor: analysis using a phage-epitope library. *Proc. Natl. Acad. Sci. USA* 94:6054–58
9. Bergles DE, Roberts JD, Somogyi P, Jahr CE. 2000. Glutamatergic synapses on oligodendrocyte precursor cells in the hippocampus. *Nature* 405:187–91
10. Beroukhim R, Unwin N. 1995. Three-dimensional location of the main immunogenic region of the acetylcholine receptor. *Neuron* 15:323–31
11. Betz H, Kuhse J, Schmieden V, Laube B, Kirsch J, Harvey RJ. 1999. Structure and functions of inhibitory and excitatory glycine receptors. *Ann. NY Acad. Sci.* 868:667–76
12. Bezzi P, Carmignoto G, Pasti L, Vesce S, Rossi D, et al. 1998. Prostaglandins stimulate calcium-dependent glutamate release in astrocytes. *Nature* 391:281–85
13. Bezzi P, Volterra A. 2001. A neuron-glia signalling network in the active brain. *Curr. Opin. Neurobiol.* 11:387–94
14. Blount P, Merlie JP. 1989. Molecular basis of the two nonequivalent ligand binding sites of the muscle nicotinic acetylcholine receptor. *Neuron* 3:349–57
15. Boess FG, Steward LJ, Steele JA, Liu D,

Reid J, et al. 1997. Analysis of the ligand binding site of the 5-HT3 receptor using site directed mutagenesis: importance of glutamate 106. *Neuropharmacology* 36:637–47

16. Boileau AJ, Evers AR, Davis AF, Czajkowski C. 1999. Mapping the agonist binding site of the GABAA receptor: evidence for a beta-strand. *J. Neurosci.* 19:4847–54
17. Boileau AJ, Newell JG, Czajkowski C. 2002. GABA(A) receptor beta 2 Tyr97 and Leu99 line the GABA-binding site. Insights into mechanisms of agonist and antagonist actions. *J. Biol. Chem.* 277:2931–37
18. Brejc K, van Dijk WJ, Klaassen RV, Schuurmans M, van Der Oost J, et al. 2001. Crystal structure of an ACh-binding protein reveals the ligand-binding domain of nicotinic receptors. *Nature* 411:269–76
19. Bren N, Sine SM. 2000. Hydrophobic pairwise interactions stabilize alpha-conotoxin MI in the muscle acetylcholine receptor binding site. *J. Biol. Chem.* 275:12692–700
20. Chiara DC, Cohen JB. 1997. Identification of amino acids contributing to high and low affinity d-tubocurarine sites in the Torpedo nicotinic acetylcholine receptor. *J. Biol. Chem.* 272:32940–50
21. Chiara DC, Xie Y, Cohen JB. 1999. Structure of the agonist-binding sites of the Torpedo nicotinic acetylcholine receptor: affinity-labeling and mutational analyses identify gamma Tyr-111/delta Arg-113 as antagonist affinity determinants. *Biochemistry* 38:6689–98
22. Cockcroft VB, Lunt GG, Osguthorpe DJ. 1990. Modelling of binding sites of the nicotinic acetylcholine receptor and their relation to models of the whole receptor. *Biochem. Soc. Symp.* 57:65–79
23. Corringer PJ, Le Novere N, Changeux JP. 2000. Nicotinic receptors at the amino acid level. *Annu. Rev. Pharmacol. Toxicol.* 40:431–58
24. Creighton CJ, Reynolds CH, Lee DH, Leo GC, Reitz AB. 2001. Conformational analysis of the eight-membered ring of the oxidized cysteinyl-cysteine unit implicated in nicotinic acetylcholine receptor ligand recognition. *J. Am. Chem. Soc.* 123:12664–69
25. Cromer BA, Morton CJ, Parker MW. 2002. Anxiety over GABA(A) receptor structure relieved by AChBP. *Trends Biochem. Sci.* 27:280–87
26. Croxen R, Newland C, Beeson D, Oosterhuis H, Chauplannaz G, et al. 1997. Mutations in different functional domains of the human muscle acetylcholine receptor alpha subunit in patients with the slow-channel congenital myasthenic syndrome. *Hum. Mol. Genet.* 6:767–74
27. Cully DF, Vassilatis DK, Liu KK, Paress PS, Van der Ploeg LH, et al. 1994. Cloning of an avermectin-sensitive glutamate-gated chloride channel from *Caenorhabditis elegans*. *Nature* 371:707–11
28. Czajkowski C, Karlin A. 1995. Structure of the nicotinic receptor acetylcholine-binding site. Identification of acidic residues in the delta subunit within 0.9 nm of the 5 alpha subunit-binding. *J. Biol. Chem.* 270:3160–64
29. Dennis M, Giraudat J, Kotzyba-Hibert F, Goeldner M, Hirth C, et al. 1988. Amino acids of the *Torpedo marmorata* acetylcholine receptor alpha subunit labeled by a photoaffinity ligand for the acetylcholine binding site. *Biochemistry* 27:2346–57
30. Dougherty DA, Stauffer DA. 1990. Acetylcholine binding by a synthetic receptor: implications for biological recognition. *Science* 250:1558–60
31. Dzubay JA, Jahr CE. 1999. The concentration of synaptically released glutamate outside of the climbing fiber-Purkinje cell synaptic cleft. *J. Neurosci.* 19:5265–74
32. Eisele JL, Bertrand S, Galzi JL, Devillers-Thiery A, Changeux JP, Bertrand D. 1993. Chimaeric nicotinic-serotonergic receptor combines distinct ligand binding and channel specificities. *Nature* 366:479–83
33. Elgoyhen AB, Johnson DS, Boulter J,

Vetter DE, Heinemann S. 1994. Alpha 9: an acetylcholine receptor with novel pharmacological properties expressed in rat cochlear hair cells. *Cell* 79:705–15
34. Engel AG, Ohno K, Sine SM. 1999. Congenital myasthenic syndromes: recent advances. *Arch. Neurol.* 56:163–67
35. Enz R. 2001. GABA(C) receptors: a molecular view. *Biol. Chem.* 382:1111–22
36. Feng ZP, Klumperman J, Lukowiak K, Syed NI. 1997. In vitro synaptogenesis between the somata of identified *Lymnaea* neurons requires protein synthesis but not extrinsic growth factors or substrate adhesion molecules. *J. Neurosci.* 17:7839–49
37. Fruchart-Gaillard C, Gilquin B, Antil-Delbeke S, Le Novere N, Tamiya T, et al. 2002. Experimentally based model of a complex between a snake toxin and the alpha 7 nicotinic receptor. *Proc. Natl. Acad. Sci. USA* 99:3216–21
38. Galzi JL, Revah F, Black D, Goeldner M, Hirth C, Changeux JP. 1990. Identification of a novel amino acid alpha-tyrosine 93 within the cholinergic ligands-binding sites of the acetylcholine receptor by photoaffinity labeling. Additional evidence for a three-loop model of the cholinergic ligands-binding sites. *J. Biol. Chem.* 265:10430–37
39. Gouet P, Courcelle E, Stuart DI, Metoz F. 1999. ESPript: analysis of multiple sequence alignments in PostScript. *Bioinformatics* 15:305–8
40. Griffon N, Buttner C, Nicke A, Kuhse J, Schmalzing G, Betz H. 1999. Molecular determinants of glycine receptor subunit assembly. *EMBO J.* 18:4711–21
41. Grosman C, Salamone FN, Sine SM, Auerbach A. 2000. The extracellular linker of muscle acetylcholine receptor channels is a gating control element. *J. Gen. Physiol.* 116:327–40
42. Grosman C, Zhou M, Auerbach A. 2000. Mapping the conformational wave of acetylcholine receptor channel gating. *Nature* 403:773–76
43. Grutter T, Changeux JP. 2001. Nicotinic receptors in wonderland. *Trends Biochem. Sci.* 26:459–63
44. Hamakawa T, Woodin MA, Bjorgum MC, Painter SD, Takasaki M, et al. 1999. Excitatory synaptogenesis between identified *Lymnaea* neurons requires extrinsic trophic factors and is mediated by receptor tyrosine kinases. *J. Neurosci.* 19:9306–12
45. Hamill SJ, Cota E, Chothia C, Clarke J. 2000. Conservation of folding and stability within a protein family: the tyrosine corner as an evolutionary cul-de-sac. *J. Mol. Biol.* 295:641–49
45a. Hansen SB, Radic Z, Talley TT, Molles BE, Deerinck T, et al. 2002. Tryptophan fluorescence reveals conformational changes in the acetylcholine binding protein. *J. Biol. Chem.* 277:41299–302
46. Harel M, Kasher R, Nicolas A, Guss JM, Balass M, et al. 2001. The binding site of acetylcholine receptor as visualized in the X-ray structure of a complex between alpha-bungarotoxin and a mimotope peptide. *Neuron* 32:265–75
47. Harel M, Schalk I, Ehret-Sabatier L, Bouet F, Goeldner M, et al. 1993. Quaternary ligand binding to aromatic residues in the active-site gorge of acetylcholinesterase. *Proc. Natl. Acad. Sci. USA* 90:9031–35
48. Hartvig L, Lukensmejer B, Liljefors T, Dekermendjian K. 2000. Two conserved arginines in the extracellular N-terminal domain of the GABA(A) receptor alpha(5) subunit are crucial for receptor function. *J. Neurochem.* 75:1746–53
49. Hemmingsen JM, Gernert KM, Richardson JS, Richardson DC. 1994. The tyrosine corner: a feature of most Greek key beta-barrel proteins. *Protein Sci.* 3:1927–37
50. Hope AG, Belelli D, Mair ID, Lambert JJ, Peters JA. 1999. Molecular determinants of (+)-tubocurarine binding at recombinant 5-hydroxytryptamine3A receptor subunits. *Mol. Pharmacol.* 55:1037–43
51. Hosli E, Ruhl W, Hosli L. 2000. Histochemical and electrophysiological evidence for estrogen receptors on cultured

astrocytes: colocalization with cholinergic receptors. *Int. J. Dev. Neurosci.* 18:101–11

52. Ibanez-Tallon I, Miwa JM, Wang HL, Adams NC, Crabtree GW, et al. 2002. Novel modulation of neuronal nicotinic acetylcholine receptors by association with the endogenous prototoxin lynx1. *Neuron* 33:893–903
53. Jahromi BS, Robitaille R, Charlton MP. 1992. Transmitter release increases intracellular calcium in perisynaptic Schwann cells in situ. *Neuron* 8:1069–77
54. Kang J, Jiang L, Goldman SA, Nedergaard M. 1998. Astrocyte-mediated potentiation of inhibitory synaptic transmission. *Nat. Neurosci.* 1:683–92
55. Kao PN, Dwork AJ, Kaldany RR, Silver ML, Wideman J, et al. 1984. Identification of the alpha subunit half-cystine specifically labeled by an affinity reagent for the acetylcholine receptor binding site. *J. Biol. Chem.* 259:11662–65
56. Karlin A. 2002. Emerging structure of the nicotinic acetylcholine receptors. *Nat. Rev. Neurosci.* 3:102–14
57. Karlin A, Akabas MH. 1998. Substituted-cysteine accessibility method. *Methods Enzymol.* 293:123–45
58. Kasher R, Balass M, Scherf T, Fridkin M, Fuchs S, Katchalski-Katzir E. 2001. Design and synthesis of peptides that bind alpha-bungarotoxin with high affinity. *Chem. Biol.* 8:147–55
59. Katz B, Miledi R. 1973. The binding of acetylcholine to receptors and its removal from the synaptic cleft. *J. Physiol.* 231:549–74

59a. Kistler J, Stroud RM. 1981. Crystalline arrays of membrane-bound acetylcholine receptor. *Proc. Natl. Acad. Sci. USA* 78:3678–82

60. Kleinjung J, Petit MC, Orlewski P, Mamalaki A, Tzartos SJ, et al. 2000. The third-dimensional structure of the complex between an Fv antibody fragment and an analogue of the main immunogenic region of the acetylcholine receptor: a combined two-dimensional NMR, homology, and molecular modeling approach. *Biopolymers* 53:113–28

60a. Klymkowsky MW, Stroud RM. 1979. Immunospecific identification and three-dimensional structure of a membrane-bound acetylcholine receptor from *Torpedo californica*. *J. Mol. Biol.* 128:319–34

61. Kuffler SW, Yoshikami D. 1975. The number of transmitter molecules in a quantum: an estimate from iontophoretic application of acetylcholine at the neuromuscular synapse. *J. Physiol.* 251:465–82
62. Kuhse J, Laube B, Magalei D, Betz H. 1993. Assembly of the inhibitory glycine receptor: identification of amino acid sequence motifs governing subunit stoichiometry. *Neuron* 11:1049–56
63. Kulik A, Haentzsch A, Luckermann M, Reichelt W, Ballanyi K. 1999. Neuron-glia signaling via alpha(1) adrenoceptor-mediated Ca(2+) release in Bergmann glial cells in situ. *J. Neurosci.* 19:8401–8
64. Land BR, Salpeter EE, Salpeter MM. 1981. Kinetic parameters for acetylcholine interaction in intact neuromuscular junction. *Proc. Natl. Acad. Sci. USA* 78:7200–4
65. Le Novere N, Changeux JP. 2001. LGICdb: the ligand-gated ion channel database. *Nucleic Acids Res.* 29:294–95
66. Le Novere N, Grutter T, Changeux JP. 2002. Models of the extracellular domain of the nicotinic receptors and of agonist- and Ca2+-binding sites. *Proc. Natl. Acad. Sci. USA* 99:3210–15
67. Legendre P. 2001. The glycinergic inhibitory synapse. *Cell Mol. Life Sci.* 58:760–93
68. Lindstrom JM. 2000. Acetylcholine receptors and myasthenia. *Muscle Nerve* 23:453–77

68a. Lover RA, Stroud RM. 1986. The crystal structure of α-bungarotoxin and its relation to solution structure and binding to AcCh receptor. *Protein Eng.* 1:37–46

69. Lu ZL, Saldanha JW, Hulme EC. 2002. Seven-transmembrane receptors: crystals clarify. *Trends Pharmacol. Sci.* 23:140–46
70. Machold J, Utkin Y, Kirsch D, Kaufmann R, Tsetlin V, Hucho F. 1995. Photolabeling reveals the proximity of the alpha-neurotoxin binding site to the M2 helix of the ion channel in the nicotinic acetylcholine receptor. *Proc. Natl. Acad. Sci. USA* 92:7282–86
71. Machold J, Weise C, Utkin Y, Tsetlin V, Hucho F. 1995. The handedness of the subunit arrangement of the nicotinic acetylcholine receptor from *Torpedo californica. Eur. J. Biochem.* 234:427–30
72. Maelicke A, Albuquerque EX. 2000. Allosteric modulation of nicotinic acetylcholine receptors as a treatment strategy for Alzheimer's disease. *Eur. J. Pharmacol.* 393:165–70
73. Malany S, Osaka H, Sine SM, Taylor P. 2000. Orientation of alpha-neurotoxin at the subunit interfaces of the nicotinic acetylcholine receptor. *Biochemistry* 39:15388–98
74. McGehce DS. 1999. Molecular diversity of neuronal nicotinic acetylcholine receptors. *Ann. NY Acad. Sci.* 868:565–77
75. McIntosh JM, Santos AD, Olivera BM. 1999. Conus peptides targeted to specific nicotinic acetylcholine receptor subtypes. *Annu. Rev. Biochem.* 68:59–88
76. Miwa JM, Ibanez-Tallon I, Crabtree GW, Sanchez R, Sali A, et al. 1999. lynx1, an endogenous toxin-like modulator of nicotinic acetylcholine receptors in the mammalian CNS. *Neuron* 23:105–14
77. Miyazawa A, Fujiyoshi Y, Stowell M, Unwin N. 1999. Nicotinic acetylcholine receptor at 4.6 Å resolution: transverse tunnels in the channel wall. *J. Mol. Biol.* 288: 765–86
78. Mochizuki S, Miyake A, Furuichi K. 1999. Identification of a domain affecting agonist potency of meta-chlorophenylbiguanide in 5-HT3 receptors. *Eur. J. Pharmacol.* 369:125–32
79. Moise L, Piserchio A, Basus VJ, Hawrot E. 2002. NMR structural analysis of alpha-bungarotoxin and its complex with the principal alpha-neurotoxin-binding sequence on the alpha 7 subunit of a neuronal nicotinic acetylcholine receptor. *J. Biol. Chem.* 277:12406–17
80. Molles BE, Rezai P, Kline EF, McArdle JJ, Sine SM, Taylor P. 2002. Identification of residues at the alpha and epsilon subunit interfaces mediating species selectivity of Waglerin-1 for nicotinic acetylcholine receptors. *J. Biol. Chem.* 277:5433–40
81. Molles BE, Tsigelny I, Nguyen PD, Gao SX, Sine SM, Taylor P. 2002. Residues in the epsilon subunit of the nicotinic acetylcholine receptor interact to confer selectivity of waglerin-1 for the alpha-epsilon subunit interface site. *Biochemistry* 41:7895–906
82. Neumann D, Barchan D, Fridkin M, Fuchs S. 1986. Analysis of ligand binding to the synthetic dodecapeptide 185–196 of the acetylcholine receptor alpha subunit. *Proc. Natl. Acad. Sci. USA* 83:9250–53
83. O'Connor V, Phelan PP, Fry JP. 1996. Interactions of glycine and strychnine with their receptor recognition sites in mouse spinal cord. *Neurochem. Int.* 29:423–34
84. Ohno K, Wang HL, Milone M, Bren N, Brengman JM, et al. 1996. Congenital myasthenic syndrome caused by decreased agonist binding affinity due to a mutation in the acetylcholine receptor epsilon subunit. *Neuron* 17:157–70
85. Orlewski P, Marraud M, Cung MT, Tsikaris V, Sakarellos-Daitsiotis M, et al. 1996. Compared structures of the free nicotinic acetylcholine receptor main immunogenic region (MIR) decapeptide and the antibody-bound [A76]MIR analogue: a molecular dynamics simulation from two-dimensional NMR data. *Biopolymers* 40:419–32
86. Ortells MO, Lunt GG. 1995. Evolutionary history of the ligand-gated ion-channel superfamily of receptors. *Trends Neurosci.* 18:121–27
87. Osaka H, Malany S, Kanter JR, Sine SM,

Taylor P. 1999. Subunit interface selectivity of the alpha-neurotoxins for the nicotinic acetylcholine receptor. *J. Biol. Chem.* 274:9581–86

88. Parpura V, Basarsky TA, Liu F, Jeftinija K, Jeftinija S, Haydon PG. 1994. Glutamate-mediated astrocyte-neuron signalling. *Nature* 369:744–47
89. Pedersen SE, Cohen JB. 1990. d-Tubocurarine binding sites are located at alpha-gamma and alpha-delta subunit interfaces of the nicotinic acetylcholine receptor. *Proc. Natl. Acad. Sci. USA* 87:2785–89
90. Poulas K, Eliopoulos E, Vatzaki E, Navaza J, Kontou M, et al. 2001. Crystal structure of Fab198, an efficient protector of the acetylcholine receptor against myasthenogenic antibodies. *Eur. J. Biochem.* 268:3685–93
91. Prince RJ, Sine SM. 1996. Molecular dissection of subunit interfaces in the acetylcholine receptor. Identification of residues that determine agonist selectivity. *J. Biol. Chem.* 271:25770–77
92. Quiram PA, Jones JJ, Sine SM. 1999. Pairwise interactions between neuronal alpha7 acetylcholine receptors and alpha-conotoxin ImI. *J. Biol. Chem.* 274:19517–24
93. Reeves DC, Lummis SC. 2002. The molecular basis of the structure and function of the 5-HT3 receptor: a model ligand-gated ion channel (review). *Mol. Membr. Biol.* 19:11–26
94. Robitaille R. 1998. Modulation of synaptic efficacy and synaptic depression by glial cells at the frog neuromuscular junction. *Neuron* 21:847–55
95. Robitaille R, Jahromi BS, Charlton MP. 1997. Muscarinic Ca2+ responses resistant to muscarinic antagonists at perisynaptic Schwann cells of the frog neuromuscular junction. *J. Physiol.* 504:337–47
96. Rogers SW, Gregori NZ, Carlson N, Gahring LC, Noble M. 2001. Neuronal nicotinic acetylcholine receptor expression by O2A/oligodendrocyte progenitor cells. *Glia* 33:306–13
97. Ryan SG, Buckwalter MS, Lynch JW, Handford CA, Segura L, et al. 1994. A missense mutation in the gene encoding the alpha 1 subunit of the inhibitory glycine receptor in the spasmodic mouse. *Nat. Genet.* 7:131–35
98. Salgueiro S, Potts S, McIlgorm EA, Ansell KH, Hemberger J, Gussow D. 1999. A protein from the salivary glands of the giant Amazon leech with high sequence homology to a nicotinic acetylcholine receptor subunit. *Z. Naturforsch.* 54:963–71
99. Schapira M, Abagyan R, Totrov M. 2002. Structural model of nicotinic acetylcholine receptor isotypes bound to acetylcholine and nicotine. *BMC Struct. Biol.* 2:1
100. Schmieden V, Kuhse J, Betz H. 1993. Mutation of glycine receptor subunit creates beta-alanine receptor responsive to GABA. *Science* 262:256–58
101. Schmieden V, Kuhse J, Betz H. 1999. A novel domain of the inhibitory glycine receptor determining antagonist efficacies: further evidence for partial agonism resulting from self-inhibition. *Mol. Pharmacol.* 56:464–72
102. Schrattenholz A, Coban T, Schroder B, Okonjo KO, Kuhlmann J, et al. 1993. Biochemical characterization of a novel channel-activating site on nicotinic acetylcholine receptors. *J. Recept. Res.* 13:393–412
103. Sieghart W, Fuchs K, Tretter V, Ebert V, Jechlinger M, et al. 1999. Structure and subunit composition of GABA(A) receptors. *Neurochem. Int.* 34:379–85
104. Sigel E, Baur R, Kellenberger S, Malherbe P. 1992. Point mutations affecting antagonist affinity and agonist dependent gating of GABAA receptor channels. *EMBO J.* 11:2017–23
105. Sine SM. 1997. Identification of equivalent residues in the gamma, delta, and epsilon subunits of the nicotinic receptor that

contribute to alpha-bungarotoxin binding. *J. Biol. Chem.* 272:23521–27

106. Sine SM, Claudio T. 1991. Gamma- and delta-subunits regulate the affinity and the cooperativity of ligand binding to the acetylcholine receptor. *J. Biol. Chem.* 266:19369–77
107. Sine SM, Kreienkamp HJ, Bren N, Maeda R, Taylor P. 1995. Molecular dissection of subunit interfaces in the acetylcholine receptor: identification of determinants of alpha-conotoxin M1 selectivity. *Neuron* 15:205–11
108. Sine SM, Ohno K, Bouzat C, Auerbach A, Milone M, et al. 1995. Mutation of the acetylcholine receptor alpha subunit causes a slow-channel myasthenic syndrome by enhancing agonist binding affinity. *Neuron* 15:229–39
109. Sine SM, Quiram P, Papanikolaou F, Kreienkamp HJ, Taylor P. 1994. Conserved tyrosines in the alpha subunit of the nicotinic acetylcholine receptor stabilize quaternary ammonium groups of agonists and curariform antagonists. *J. Biol. Chem.* 269:8808–16
110. Sine SM, Wang HL, Bren ND. 2002. Lysine scanning mutagenesis delineates structural model of the nicotinic receptor ligand binding domain. *J. Biol. Chem.* 277:29210–23
111. Smit AB, Syed NI, Schaap D, van Minnen J, Klumperman J, et al. 2001. A glia-derived acetylcholine-binding protein that modulates synaptic transmission. *Nature* 411:261–68
112. Spier AD, Lummis SC. 2000. The role of tryptophan residues in the 5-hydroxytryptamine(3) receptor ligand binding domain. *J. Biol. Chem.* 275:5620–25
113. Steward LJ, Boess FG, Steele JA, Liu D, Wong N, Martin IL. 2000. Importance of phenylalanine 107 in agonist recognition by the 5-hydroxytryptamine(3A) receptor. *Mol. Pharmacol.* 57:1249–55
114. Sullivan D, Chiara DC, Cohen JB. 2002. Mapping the agonist binding site of the nicotinic acetylcholine receptor by cysteine scanning mutagenesis: antagonist footprint and secondary structure prediction. *Mol. Pharmacol.* 61:463–72
115. Theodosis DT, Poulain DA. 1993. Activity-dependent neuronal-glial and synaptic plasticity in the adult mammalian hypothalamus. *Neuroscience* 57:501–35
116. Tzartos SJ, Changeux JP. 1983. High affinity binding of alpha-bungarotoxin to the purified alpha-subunit and to its 27,000-dalton proteolytic peptide from *Torpedo marmorata* acetylcholine receptor. Requirement for sodium dodecyl sulfate. *EMBO J.* 2:381–87
117. Unwin N. 1995. Acetylcholine receptor channel imaged in the open state. *Nature* 373:37–43
118. Unwin N, Miyazawa A, Li J, Fujiyoshi Y. 2002. Activation of the nicotinic acetylcholine receptor involves a switch in conformation of the alpha subunits. *J. Mol. Biol.* 319:1165–76
119. Vafa B, Lewis TM, Cunningham AM, Jacques P, Lynch JW, Schofield PR. 1999. Identification of a new ligand binding domain in the alpha1 subunit of the inhibitory glycine receptor. *J. Neurochem.* 73:2158–66
120. Vafa B, Schofield PR. 1998. Heritable mutations in the glycine, GABAA, and nicotinic acetylcholine receptors provide new insights into the ligand-gated ion channel receptor superfamily. *Int. Rev. Neurobiol.* 42:285–32
121. Vandenberg RJ, French CR, Barry PH, Shine J, Schofield PR. 1992. Antagonism of ligand-gated ion channel receptors: two domains of the glycine receptor alpha subunit form the strychnine-binding site. *Proc. Natl. Acad. Sci. USA* 89:1765–69
122. Vandenberg RJ, Handford CA, Schofield PR. 1992. Distinct agonist- and antagonist-binding sites on the glycine receptor. *Neuron* 9:491–96
123. Vesce S, Bezzi P, Volterra A. 2001. Synaptic transmission with the glia. *News Physiol. Sci.* 16:178–84

124. Vincent A, Palace J, Hilton-Jones D. 2001. Myasthenia gravis. *Lancet* 357: 2122–28
125. Wagner DA, Czajkowski C. 2001. Structure and dynamics of the ABA binding pocket: a narrowing cleft that constricts during activation. *J. Neurosci.* 21:67–74
126. Wathey JC, Nass MM, Lester HA. 1979. Numerical reconstruction of the quantal event at nicotinic synapses. *Biophys. J.* 27:145–64
127. Woodin MA, Hamakawa T, Takasaki M, Lukowiak K, Syed NI. 1999. Trophic factor-induced plasticity of synaptic connections between identified *Lymnaea* neurons. *Learn Mem.* 6:307–16
128. Woodin MA, Munno DW, Syed NI. 2002. Trophic factor-induced excitatory synaptogenesis involves postsynaptic modulation of nicotinic acetylcholine receptors. *J. Neurosci.* 22:505–14
129. Yan D, Pedersen SE, White MM. 1998. Interaction of D-tubocurarine analogs with the 5HT3 receptor. *Neuropharmacology* 37:251–57
130. Zheng Y, Hirschberg B, Yuan J, Wang AP, Hunt DC, et al. 2002. Identification of two novel *Drosophila melanogaster* histamine-gated chloride channel subunits expressed in the eye. *J. Biol. Chem.* 277:2000–5
131. Zhong W, Gallivan JP, Zhang Y, Li L, Lester HA, Dougherty DA. 1998. From ab initio quantum mechanics to molecular neurobiology: a cation-pi binding site in the nicotinic receptor. *Proc. Natl. Acad. Sci. USA* 95:12088–93

Annu. Rev. Biophys. Biomol. Struct. 2003. 32:335–73
doi: 10.1146/annurev.biophys.32.110601.142532

First published online as a Review in Advance on January 24, 2003

MOLECULAR RECOGNITION AND DOCKING ALGORITHMS

Natasja Brooijmans[1] and Irwin D. Kuntz[2]
[1]Chemistry and Chemical Biology Graduate Program and [2]Department of Pharmaceutical Chemistry, University of California San Francisco, San Francisco, California 94143-2240; email: nbrooij@itsa.ucsf.edu; kuntz@cgl.ucsf.edu

Key Words scoring functions, search algorithms, thermodynamics of binding, structure-based drug design, virtual screening

■ **Abstract** Molecular docking is an invaluable tool in modern drug discovery. This review focuses on methodological developments relevant to the field of molecular docking. The forces important in molecular recognition are reviewed and followed by a discussion of how different scoring functions account for these forces. More recent applications of computational chemistry tools involve library design and database screening. Last, we summarize several critical methodological issues that must be addressed in future developments.

CONTENTS

INTRODUCTION

As biological research has become increasingly data intensive, biomedical projects require informatics tools. As an example, in drug discovery research, high-throughput screening often requires the screening of millions of compounds for a particular protein target. Important tools that can enhance such screens are molecular docking and database mining.

Molecular docking can be defined as the prediction of the structure of receptor-ligand complexes, where the receptor is usually a protein or a protein oligomer and the ligand is either a small molecule or another protein. Different simplifications are used to make molecular docking tractable in different applications. Initially, molecular docking was used to predict and reproduce protein-ligand complexes (75, 80, 119, 145). Successes of these early studies led to the exploration of molecular docking as a tool in drug discovery to find and optimize lead compounds, often by database screening. Since the development of combinatorial chemistry, molecular docking is applied to aid in the design of combinatorial libraries and to prescreen "real" or "virtual" compound databases in silico.

There are two key parts to any docking program, namely a search of the configurational and conformational degrees of freedom and the scoring or evaluation function. The search algorithm must search the potential energy landscape in enough detail to find the global energy minimum. In rigid docking this means that the search algorithm explores different positions for the ligand in the receptor active site using the translational and rotational degrees of freedom. Flexible ligand docking adds exploration of torsional degrees of freedom of the ligand to this process. The scoring function has to be realistic enough to assign the most favorable scores to the experimentally determined complex. Usually, the scoring function assesses both the steric complementarity between the ligand and the receptor as well as their chemical complementarity. We focus on developments of concepts in both these areas. Our review is divided into six parts: introduction, molecular recognition, molecular docking, molecular docking experiments in which different methods were used sequentially, database methods, and critical issues.

In the introduction we discuss different docking applications that illustrate major algorithmic changes. Second, we discuss the implementation of scoring grids and molecular surfaces because they have had a large influence on the development and utility of docking tools. Third, ligand flexibility and receptor flexibility algorithms are discussed.

The first half of the review considers the physical basis for molecular recognition and discusses different scoring functions, which all try to account for the forces involved in protein-ligand recognition. In the second half we examine experiments in which docking was one of several tools used sequentially to solve a particular docking problem, focus on library design and database screening methods, and provide our view on critical limitations of current methods.

Overview of Docking Applications

The first docking procedures directly followed the development of interactive molecular graphics programs, which allowed researchers to manipulate two molecules to find different configurations that appeared complementary, both geometrically and chemically. Molecular docking programs allowed automation of the search and more objective ways of assessing the fit between two molecules. The number of ways two molecules can interact is large, and in order to make the search efficient, the molecular docking problem is simplified and only certain degrees of freedom are explored.

The first molecular dockings of protein-protein complexes used the rigid body approximation, fixing all internal degrees of freedom, except for the three translations and three rotations. Protein atoms were represented explicitly and simple potential energy functions were used to evaluate the complementarity of proposed complex configurations (80, 119). Experimental biochemical data helped filter the highest scoring configurations.

A later application of molecular docking attempted to reproduce the known crystal structure of the BPTI-trypsin protein-protein complex (145). Different configurations of the complex were generated using simplified models for the proteins. The energy function consisted of a nonbonded term and a surface area term. In a second step, the best scoring models were further refined with a potential energy function consisting of a nonbonded term and a solvent term. Although the crystal structure could be regenerated, the potential energy function that was used to refine the models could not distinguish the crystal structure complex from certain non-native complexes (145). Greer & Bush (44) further exploited the notion that protein-protein complexes are formed by fitting complementary areas of their surfaces, using the solvent-accessible surface definition as proposed by Richards (117).

While the protein docking experiments used positive images (atomic representations) of both molecules involved, the first molecular docking study of protein–small molecule complexes used a negative image of the receptor—spheres that fill pockets and grooves on the receptor surface (75). Such spheres describe potential interaction sites on the receptor surface for ligands and can subsequently be used to match to ligand atoms. Again, the only degrees of freedom explored were the six translational and rotational degrees of freedom. The interactions were evaluated by assessing overlap between atoms and the potential to form hydrogen bonds in the complex. Several protein-ligand crystal structures were correctly reproduced by this method, but, as in the study of Wodak & Janin (145), non-native configurations sometimes scored better than the crystal structure configuration.

The rigid body approximation of protein-ligand interactions has clear limitations: It does not account for induced-fit (72), especially when a cryptic site in the uncomplexed receptor "opens" under the influence of the ligand. A simple approach to sampling some of the ligand degrees of freedom was described by DesJarlais et al. (27). Since then, multiple algorithms have been developed that explore ligand flexibility. Limited receptor flexibility has also been considered. These advances are described later in this section.

The ability to reproduce known crystal structures of pharmaceutically relevant protein-ligand complexes allowed the exploration of these methods as tools in structure-based drug design. The first application screened a library of small molecule ligands against two different protein receptors. Chemically different ligands were found that matched the receptor sites geometrically, but their chemical complementarity was not assessed (28). HIV-1 protease was studied using in silico library screening after the crystal structure was solved, and several interesting nonpeptide inhibitors were found. The most promising ones were screened experimentally and inhibit the protease selectively (26). A recent review of structure-based drug design is available (67).

Combinatorial chemistry allows for the rapid synthesis of large compound libraries, but experimental screening of these libraries can be time-consuming and expensive. Computational chemistry can be used to prescreen compound databases to decrease the number of compounds to be tested, and assist in the design of libraries by using chemical informatics methods to select a highly diverse subset of compounds from the original database for screening. Molecular docking has also been applied to design directed libraries against specific targets. Those compounds that have the most favorable scores are subsequently synthesized and tested for their experimental binding affinity (28, 66).

Perspective on Algorithmic Developments

For molecular docking programs to be useful tools in (high-throughput) drug discovery, accuracy and speed are key factors. These are two contradictory requirements, and compromises have to be made. The speed of the elementary steps has increased many fold owing to the much higher clock speeds of modern computers, the larger word size, and the representation of the surface of a protein by the molecular surface and the use of scoring grids. Recent advances in parallel computing have also improved the performance of docking algorithms significantly. This improved performance has permitted more realistic scoring functions, and the exploration of ligand flexibility in the active site so that the basic throughput of ~1 molecule per minute per processor has not changed much over the years.

Initially, docking programs represented both the receptor and the ligand with explicit atoms (80, 119, 144). These representations make numerous complex configurations possible, which all have to be evaluated. The development of molecular surface calculations (25) and their subsequent use in docking applications significantly reduces the number of possible complexes and thus increases the speed of docking algorithms (35, 59, 75, 103).

The next major development in molecular docking came from the application of grids to store physico-chemical properties of the receptor. The grid only needs to be calculated once, assuming the receptor is rigid. The interactive docking program DOCKER mapped the extended van der Waals radii onto a grid, which could be used to explore the unoccupied surface/volume with a ligand (16). In GRID, for each of the grid points the potential energy of the interaction between the receptor and a probe atom/group is calculated (41). Some of the probe groups that are used are amino, methyl, and hydroxyl groups. The grids thus display the kind of interactions (polar/nonpolar) that are preferred at each position in the binding site.

The disadvantages of the GRID and DOCKER implementations are that interaction energies of the protein and ligand cannot be easily estimated. Two different implementations (110, 129) do allow for direct energy evaluation while manipulating the system. The program from Langridge and coworkers (110) maps the van der Waals and Coulomb contributions of the receptor onto the grid. Another implementation builds on Goodford's GRID and maps the interactions of different probes with the receptor on the grid, which are stored in a look-up table. The interaction energy of the ligand with the receptor can be estimated by summing up the interaction energies of each probe atom that corresponds to a ligand atom. In addition to the van der Waals and Coulomb interactions, a grid is also made to map the hydrogen bonding properties (129). The first docking programs to implement the use of grids in an automated way were AutoDock (42) and DOCK (93). The significant time-savings obtained with the use of grids resulted in the implementation of more sophisticated scoring functions (42, 93).

Ligand Flexibility Algorithms

Conformational searching during docking is necessary because usually it is not known which conformation of a ligand interacts most favorably with a receptor. Ligand flexibility algorithms can be divided in three types of searches, namely systematic, stochastic, and deterministic searches. Some algorithms use more than one of these approaches. Systematic search algorithms are based on a grid of values for each formal degree of freedom, and each of these grid values is explored in a combinatorial fashion during the search. As the number of degrees of freedom increases, the number of evaluations needed increases rapidly. To deal with this problem, termination criteria are inserted to prevent the algorithm from sampling space that is known to lead to the wrong solution. An example of a systematic search is the anchor-and-grow or incremental construction algorithm. Stochastic search algorithms make random changes, usually changing one degree of freedom of the system at a time. One of the major concerns with stochastic searches is the uncertainty of convergence. In order to improve convergence, multiple, independent runs can be performed. Examples of stochastic searches are Monte Carlo (MC) methods and evolutionary algorithms. In deterministic searches the initial state determines the move that can be made to generate the next state, which generally has to be equal to or lower in energy than the initial state. Deterministic searches performed on exactly the same starting system (including each degree of freedom) with the same parameters will generate exactly the same final state.

A problem with deterministic algorithms is that they often get trapped in local minima because they cannot traverse barriers. Different approaches will be discussed that try to increase the ability to cross barriers or that try to decrease the height of the energy barriers. Examples of deterministic methods are energy minimization methods and molecular dynamics (MD) simulations.

INCREMENTAL CONSTRUCTION Incremental construction algorithms, or anchor-and-grow methods, generally divide a ligand into rigid and flexible regions. One or more rigid "anchors" with flexible parts are defined by perception of rotatable bonds. Typical procedures dock the anchor first with the flexible parts added sequentially, with systematic scanning of the torsion angles. Implementations differ in the way the anchors are docked in the active site and in pruning procedures when the number of rotatable bonds increases. The anchor-and-grow method assumes that the anchor position observed in the intact ligand is among the lowest *n* anchor positions when evaluated independently of the rest of the ligand. The parameter, *n*, controls the breadth of the search tree. This assumption might not be correct, especially for smaller anchors, because the anchor can potentially fit well at numerous positions on the receptor surface. Systematically taking the lowest *n* poses to the next level of growth is often called the "greedy" algorithm.

The first incremental construction algorithm for docking was described by DesJarlais et al. (27) and was part of the program DOCK. In that implementation different rigid fragments of the ligand were docked independently of each other. In a second step, each combination of the rigid fragments was reattached as in the parent ligand if the atoms to be connected were within certain distances from each other. Later implementations of this algorithm search the degrees of freedom of the flexible part more explicitly during the docking. Leach & Kuntz (78) docked the anchor rigidly first, and the flexible part of the ligand was added later. Each of the dihedral angles was sampled systematically, but the number of dihedral angles allowed per angle was kept low to prevent combinatorial explosions. The method also implemented a backtracking procedure to reoptimize angles to prevent serious steric clashes in other parts of the flexible part, and to prevent combinatorial explosions from occurring (78). The DOCK4.0 algorithm docks the anchor based on steric complementarity. The flexible parts are grown incrementally; the dihedral degrees of freedom are explored and minimized. Pruning occurs at each step of the growth to ensure diversity. When the molecule is complete, it is reminimized and the final score is calculated (32). Minimization was possible since the scoring function was analytic in form, and differentiable.

The program FlexX docks the anchor based on modeling chemical interactions, instead of just steric complementarity, using an algorithm called "pose clustering." Fewer matches should be accepted with chemical complementarity. A clustering algorithm is implemented to merge similar transformations of the ligand into the active site (114). It is critical to choose anchor atoms that have directional interactions with the receptor (e.g., hydrogen bonds, salt bridges). Flexible fragments are added incrementally, after placement of the anchor in the active site. Each preferred angle is constructed, and those angles that overlap with the receptor or clash internally are rejected. A simple overlap function is used to assess clashes. The angles are optimized for interactions with matching atoms in the receptor. Energy calculations are performed and the partially built molecules are ranked. This procedure is repeated until the ligand is completely built.

Hammerhead uses a different approach than the anchor-and-grow scheme. The program breaks the ligand into fragments and each fragment is docked into the binding site. The high-affinity fragments ("heads") are used to rebuild the complete ligand. Instead of systematically searching torsion angles, the ligands are optimized after each fragment is added using energy minimization. The algorithm was specifically developed for database screening (141).

MONTE CARLO ALGORITHMS In MC implementations, the ligand is considered as a whole and random changes are made to change the translation and rotation of the ligand, as well as the torsion angles. After each move, the structure is minimized, and the energy of the new structure is determined. Minimization before evaluating the Metropolis criterion was shown to increase convergence (131).

In this procedure the ligand is usually placed randomly in the binding site. To increase the chance of reaching the global energy minimum, the simulation may consist of several cycles. The first cycle is performed at high temperature and later cycles are done at increasingly lower temperatures (simulated annealing MC). Usually each cycle starts with the lowest energy from the previous cycle.

AutoDock was the first docking program to implement simulated annealing MC methods. Usually tens of thousands of steps are performed during each cycle. At the beginning of each new cycle, the temperature is reduced (42).

Biased probability MC was implemented in the program ICM. Translational and rotational degrees of freedom are sampled by pseudo-Brownian motion, while torsion angles are sampled using biased probability moves. After the MC moves, a local energy minimization is performed, which is followed by a surface-based solvation energy and entropy calculation, and these terms are added to the gas-phase energy from the minimization. The sum of the three energy terms is used in the Metropolis selection criterion to determine whether the new structure is rejected or accepted (1, 2, 130). The biased probabilities can be derived from known structures (e.g., Ramachandran plots of amino acid dihedral angles), and moves likely sample high probability zones, which are approximated by Gaussian distributions (1).

Caflisch and coworkers (18) applied Scheraga's MC methods procedure (81) to docking. First, a set of docked conformations is generated based upon a simple energy function that removes bad contacts between the ligand atoms and the receptor. In later work, the overlap function is replaced by a shifted Lennard-Jones potential (4). Second, the docked conformations are subsequently subjected to random MC movements of the dihedral angles only, followed by a minimization procedure. The energy function used in the second step is similar to a molecular mechanics force field with a hydrogen bonding term (18). It is critical to ensure adequate coverage of the binding site in the first step to make sure the global minimum can be found. The method was later improved by allowing translational and rotational MC moves as well (17).

The MCDOCK package was similar to the original MC methods procedure in that it first generated protein-ligand configurations based on an overlap function, followed by MC using a scoring function. The difference lies in the fact that no energy minimization is done until after the energy-based MC (85).

Prodock, developed by Scheraga & Trosset (132), uses the scaled collective variable Monte Carlo method (SCVMCM) with energy minimizations after each MC step for docking. SCVMC is biased MC because it uses information about the topology of the potential energy "hypersurface" (132).

PRO_LEADS implements a "Tabu search," similar to MC. It uses random moves to explore conformational space, and records the conformations already sampled. When a new solution is not lower in energy, it is only kept if it is not similar to anything in the "Tabu" list. This procedure stimulates sampling of space that has not been sampled before (98).

EVOLUTIONARY ALGORITHMS Evolutionary algorithms are stochastic methods similar to MC methods and are used to find the global energy minimum. Evolutionary algorithms demand that the fittest "individuals" are carried on to the next generation, and random or biased mutations can be made to increase genetic diversity and prevent premature convergence. Crossover, a process that swaps large regions of the "parents," is permitted in genetic algorithms (GAs). Complex, non-differentiable scoring functions may be used. Size of population, mutation rates, crossover rates, and number of evolutionary rounds are parameters that can influence results. Generally, minimization is not applied until convergence is reached.

The first two applications of GAs to the molecular docking problem appeared in 1995; one uses an anchor-and-grow mechanism. After the growth phase has ended, the GA runs for several more generations (63). A GA was also implemented in the program DOCK, which docks either the complete ligand into the active site or a rigid fragment of the ligand first. The sphere-based GA gives broad sampling within the receptor site of a rigid ligand, whereas the explicit-orientation-based method gives detailed conformational sampling within a subsection of the active site for a flexible ligand. Rigid body minimizations are performed after the GA runs for only the low-scoring solutions. The explicit-orientation-based method gave significantly better results (106).

The docking program GOLD uses multiple subpopulations of the ligand, rather than a single large population, and manipulates these simultaneously. Migration of members from one subpopulation to another one can occur, which increases the efficiency of the GA. The GOLD program was the first docking algorithm to be tested on a large set (>100) of protein-ligand complexes (61).

The AutoDock added a GA and a "Lamarckian" GA (LGA) to improve convergence for ligands with more than eight rotatable bonds. The LGA switches between "genotypic space" and "phenotypic space." Mutation and crossover occur in genotypic space, while phenotypic space is determined by the energy function to be optimized. Energy minimization (local sampling) is performed after genotypic changes have been made to the population (global sampling) in phenotypic space, which is conceptually similar to MC minimization. The phenotypic changes from the energy minimization are mapped back onto the genes (by changing the ligand coordinates in the chromosome). The LGA was shown to be the most efficient, reliable, and successful of the three search algorithms (96).

An evolutionary programming algorithm involves generation of an initial population of ligand conformations, and no crossover occurs during evolution. This population competes for survival against each other. Upon convergence, minimization is performed on the best member (36). It would be useful to have definitive tests of the efficiency and efficacy of ligand flexibility algorithms.

PREGENERATED CONFORMATIONAL LIBRARIES Pregenerated conformational libraries are an efficient way of taking ligand flexibility into account because the cost of generating multiple conformers per molecule only has to be incurred once and the internal energy of the conformers can be assessed prior to docking. Each of the generated conformers will be docked rigidly into the receptor, and its fit with the binding site will be determined.

FLOG is a docking program that generates and docks conformational libraries called Flexibases. The Flexibases are generated using distance geometry. The docking algorithm used by FLOG is similar to the one used in DOCK (65, 94).

Shoichet and coworkers (122) added a ligand-ensemble docking method to DOCK. An ensemble of ligand conformations is pregenerated for each molecule in the compound library. Using the rigid fragment of each molecule, the complete ensemble of ligand conformations is placed rigidly into the receptor active site at once, instead of rigidly docking each conformer individually. For each ligand conformer in the ensemble the fit with the receptor is evaluated (86).

EUDOC uses conformational searches of the ligand to generate different ligand structures. Each of the structures is subsequently placed in the receptor active site using a systematic search. For each receptor-ligand complex, an energy evaluation is performed (109). There is currently no consensus on whether pregenerated conformations or "on-the-fly" flexibility is the better technology.

MOLECULAR DYNAMICS METHODS Unlike MC methods, molecular dynamics (MD) cannot cross energy barriers larger than 1–2 kT within reasonable simulation times and thus can get easily trapped in local minima when the potential energy landscape is rugged. Different approaches have been developed that allow for better exploration of the binding free energy landscape. One approach increases the temperature of the simulation. Other methods manipulate the potential energy surface and make it smoother (100, 101, 108), allowing for full exploration of the modified energy surface. All the degrees of freedom of the receptor can, in principle, be explored. The simulation time per complex, however, limits the use to the study of single protein-ligand complexes at this time.

The MD docking algorithm couples different degrees of freedom of the system to different temperatures. During the simulation, the two temperatures that determine the flexibility of the ligand are varied, which help prevent the system from getting trapped in a local minimum (30, 88).

IMPLICIT LIGAND FLEXIBILITY The algorithms described above take ligand flexibility into account explicitly, but ligand flexibility can also be taken into account implicitly by allowing for overlap between the protein and the ligand through a "soft" scoring function. Soft scoring functions have proven to be especially useful in protein-protein docking.

Jiang & Kim (59) developed the soft docking approach, which allows for overlap by counting favorable and unfavorable matches based on the physico-chemical properties of a pair of matching surfaces.

The FTDOCK algorithm generates configurations for the protein-protein complex using Fourier transformations of the coordinates. Electrostatic interactions in the protein-protein interface are estimated by calculating the electric field for the receptor and storing this on a grid. The strength of the interaction is then calculated using the explicit atoms of the ligand and the electrostatic field of the grid points of the receptor. Using the electrostatic field for the receptor disperses the point charges, allowing implicit side chain flexibility (35).

COMPARISON STUDIES OF SEARCH ALGORITHMS Two studies implement different search algorithms with a single scoring function, which allows for comparison between the success rates and efficiencies of the different search methods. The study by Westhead and coworkers (143) compares a simulated annealing MC method, an evolutionary programming algorithm, a GA, and a Tabu search. As a control, a random search algorithm was tested as well. The Tabu search appears to search the landscape more broadly, but it is not as good at local optimizations as the GA (143).

The Brooks group (134, 135) has also investigated different search algorithms. Their study compares a GA, an MD-simulated annealing method, and the AutoDock MC algorithm. The MD algorithm was the most efficient in docking structures in a large search space, whereas the GA worked best in a small search space. Furthermore, the MD method gives the lowest mean energies for the docked structures, while the GA gives the largest fraction of structures within 3 Å of the crystal structure. Simulated annealing MC appears to work well for both small and large search spaces. All three algorithms are more efficient than AutoDock, but AutoDock is better at finding the global minimum (134).

Receptor Flexibility Methods

Allowing ligand flexibility was a huge step forward, but changes in receptor structure upon ligand binding are observed frequently as well. In cross-docking experiments, a ligand A from a receptor-ligand A complex is docked into a receptor from receptor-ligand B complex, where the receptor is the same protein. Success of cross-docking depends critically on structural similarity in the protein active site between the two receptors. Murray and coworkers (98) compared the success rate of PRO_LEADS in regenerating crystal structures using the receptor structure from either the complex or an alternative complex structure. The success rate dropped from 75% to 50%, indicating the importance of receptor flexibility (98). Similar observations were reported by Knegtel et al. (68).

Some methods that sample ligand flexibility are also suited to sample receptor flexibility, namely GAs, MD methods, and MC methods. Usually, MC methods are restricted to side chain torsional degrees of freedom. Many applications only study the receptor active site. Methods that specifically deal with receptor flexibility are discussed first, followed by a brief mention of those ligand flexibility methods that have been used to explore receptor flexibility as well.

CONFORMER LIBRARIES Two different approaches use conformational libraries to sample receptor flexibility. The first one relies on rotamer libraries for receptor side chains in the active site, and the second uses pregenerated receptor conformers. To make rotamer libraries suitable for docking, algorithms need to be implemented that can exhaustively search all conformer combinations. The most favorable receptor structure for a particular ligand configuration needs to be determined within a reasonable amount of time. As with sampling ligand flexibility, algorithms need to be implemented that search through the rotamers in such a manner that no combinatorial explosions occur.

Andrew Leach (77) implemented rotamer libraries for the receptor and used two search algorithms to determine the most favorable receptor structure for a ligand configuration. A Dead End Elimination (DEE) algorithm quickly removes those rotamers for each side chain that are incompatible with the receptor backbone or with all possible rotamers of a particular amino acid residue. The remaining rotamers are then searched using the A* algorithm to find the lowest energy for the protein structure in the complex (77).

Sternberg's (55) self-consistent mean field approach to optimizing receptor side chain conformations is similar in concept. The algorithm determines the most probable set of rotamers from an ensemble of rotamers by determining the interaction of a particular amino acid rotamer with all other rotamers of other residues weighted by their probabilities. The probability of a particular rotamer for a residue can be calculated using the Boltzmann principle. The probabilities for each residue rotamer are recalculated until the value reaches convergence (55).

The EUDOC program and McCammon and coworkers have implemented more implicit ways of sampling protein flexibility. Both EUDOC and McCammon's "relaxed complex scheme" use MD to generate different protein structures. These different structures are subsequently used for docking. In the case of EUDOC, all pregenerated ligand conformers are rigidly docked against all pregenerated receptor structures (109). The relaxed complex scheme uses the flexible ligand docking program AutoDock with the LGA algorithm (82).

ENSEMBLE GRIDS Precalculating receptor contributions to the binding free energy on a grid significantly increases the speed of docking algorithms. However, the grid also constricts docking to a rigid receptor because the grid is only calculated once. Different approaches try to overcome this problem by incorporating multiple receptor structures into one grid.

The first implementation of an ensemble grid was implemented in DOCK. Different crystal structures of the same receptor were used, and two different weighting schemes to construct the ensemble grid were tested. The energy-weighted grid was constructed by weighting the potential energy of the van der Waals and Coulomb interactions between each receptor and ligand. The geometry-weighted averages considered the variance in the atom positions. Either the averages were taken or the atoms were represented without volume when too disordered. The attractive portion of the van der Waals energy of the disordered atoms was represented on the grid but scaled by the number of copies of the side chain. There was no difference in success rates among the different weighting schemes (68). The energy-weighted ensemble grid (68) performed equally well as an energy-weighted grid that does not use arbitrary thresholds and applies a Boltzmann assumption (107).

FlexE incorporates a similar scheme. The main difference lies in the representation of dissimilar parts of the receptor active site. Instead of making a single representation as above, these parts of the structure are treated as separate alternatives. It can thus be considered as a combination of an ensemble grid and a conformer library (24).

MONTE CARLO METHODS In theory MC methods can allow protein flexibility, although this can make docking a time-consuming process. Three programs have been used in flexible protein and flexible ligand docking, namely ICM (130), Prodock (132), and MCM (4). Only selected receptor side chains are allowed to move. MCM reported timings of approximately five days for one docking experiment (4), which makes these methods too time-consuming for high-throughput drug discovery applications.

MOLECULAR DYNAMICS METHODS Two MD docking algorithms have been used for flexible protein-ligand docking, namely the multicanonical MD algorithm of Nakajima et al. (101) and the Pak & Wang algorithm (108). Again, only selected receptor side chains are allowed to move during the docking. The molecular mechanics/Poisson-Boltzmann surface area (MM-PBSA) also explores receptor flexibility using MD "snapshots" (126).

EVOLUTIONARY ALGORITHMS None of the evolutionary algorithms have been applied to flexible protein docking, although partial flexibility of the receptor can be treated with the GOLD algorithm (60, 61).

MOLECULAR RECOGNITION

Different molecular recognition models are described and followed by a discussion of the physical basis for molecular recognition and how binding free energies can be calculated with "master-equation" methods. These methods are applied in later stages of drug discovery such as lead optimization.

Molecular Recognition Models

The molecular recognition model proposed by Emil Fischer in 1894 is the lock-and-key model. In this model molecular recognition occurs when the shape of the receptor active site is exactly complementary to the ligand shape and the ligand fits as a key in a lock (33). The early docking programs relied heavily on the lock-and-key model because both the receptor and ligand were treated as rigid bodies.

The Koshland "induced fit" model proposed an optimization in which both the receptor and ligand adapt their structures to bind to each other (72). The induced fit model was implemented in docking algorithms by allowing for ligand flexibility or both receptor and ligand flexibility during docking.

In the "conformation selection" model, different conformations exist of both the receptor and the ligand, and binding preferentially occurs between those conformers that are complementary. They may not be the lowest energy conformers in solution, but they are part of the conformational ensemble of the molecule in solution. Kinetic experiments by Foote & Milstein (34) on hapten recognition by antibodies support this model. The EUDOC docking program uses the conformation selection model in docking experiments by pregenerating different receptor and ligand structures and performing all-against-all docking experiments (109).

Physical Basis for Molecular Recognition

Pauling & Delbrück (111) postulated in 1940 that intermolecular Van der Waals interactions, electrostatic interactions, and hydrogen bonds stabilize molecules and complexes. Furthermore, complexes are formed between molecules that are complementary in structure as well as in physical-chemical properties (111). Kauzmann (64) first described in 1959 the hydrophobic effect, an entropy-based attraction.

Before binding, both molecules interact with the solvent. Desolvating polar and charged parts of the molecule upon complex formation especially comes at a high desolvation penalty. The desolvation penalty is generally only partially offset by the Coulomb interactions and/or hydrogen bonds that are formed in the bound complex because these interactions are usually not as favorable as the interactions made with water molecules. Desolvating nonpolar parts of the surface releases water molecules, which increases entropy. This entropy gain is favorable to complex formation. The increase in solvent entropy upon burial of nonpolar surfaces is the hydrophobic effect (64). The current view is that the hydrophobic effect is a major stabilizing term for biomolecular complexes, while the Coulomb interactions and hydrogen bonds provide specificity to protein-ligand interactions. Besides the increase in entropy of the solvent, there are also changes in entropy of the solutes. First, there are translational and rotational entropy losses because two molecules will become one complex. Second, there are changes in vibrational entropy due to complex formation. Third, there are changes in conformational entropy due to the restriction of dihedral angles upon complex formation.

The internal energy of the molecules has to be taken into account as well because the internal energy of the receptor and/or ligand may be higher in the complex than it is in the uncomplexed states, and this "strain energy" destabilizes the complexes.

The stability of a particular complex can be measured by determining the equilibrium binding constant, K_{eq}, or by determining both the on-rate k_{on} and the off-rate k_{off} of the reaction. The binding constant K_{eq} is directly related to the Gibbs free energy, as shown in Equation 1:

$$\Delta G_{bind} = -RT \ln K_{eq} = -RT \ln\left(\frac{k_{on}}{k_{off}}\right). \qquad 1.$$

The Gibbs binding free energy is the experimental quantity that is of interest to the computational chemist because it can be obtained with theoretical methods. The binding free energy consists of both enthalpic and entropic contributions (Equation 2).

The experimentally determined equilibrium binding constant is always determined with respect to a reference state, usually 1 M concentration and 25°C for biological preparations. The measured binding constant depends critically on the chosen reference state and on variables such as temperature, ionic strength, pH, pressure, and concentrations of solutes. The influence of these variables on measured binding constants often makes it difficult to compare experimental results on the system that have been performed in different labs. Comparing theoretically predicted binding free energies with the experimental values is difficult as well because it is not possible to model all the variables correctly.

Although it is possible to compute the binding free energy directly using, for example, the predominant states approximation to the configuration integral (48), this is too time-consuming for high-throughput screening. Usually, the binding free energy is estimated through separate determination of the enthalpic and entropic contributions to complex stability. This is done by determining the free energy of the complex, receptor, and ligand separately and by taking the difference between these terms (Equation 2).

$$\Delta G_{bind} = \Delta H - T\Delta S = G_{complex} - (G_{receptor} + G_{ligand}). \qquad 2.$$

The current most rigorous way of determining the binding free energy is through free energy perturbation (FEP) or thermodynamic integration (TI) methods. In most FEP/TI applications the free energy difference between two similar states is determined by slowly mutating one state into the other ("chemical alchemy"). This is done for both the complex and the ligand free in solution, if the free energy difference between two similar ligands is to be determined [see (70) for a review]. Although it is possible to calculate the absolute binding free energy by slowly growing a ligand into the binding site from a "dummy" ligand, this is time-consuming (49). Comparisons of calculated binding free energy differences with FEP/TI have shown that the errors are usually within 1 kcal/mol. Improvements in accuracy beyond this value are expected to require quadratically longer simulations and force field improvements.

Recent methods directly use the molecular mechanics force fields to estimate the enthalpic contributions to the free energy of binding, and entropic effects are calculated separately for the solute and the solvent using different methods. These "master-equation" methods estimate the binding free energy by only considering the initial and final states, as opposed to FEP/TI methods, which consider numerous unphysical intermediate states.

Most molecular mechanics force fields use a functional form similar to the one shown in Equation 3:

$$E_{MM} = \sum_{bonds} K_r(r - r_{eq})^2 + \sum_{angles} K_\vartheta(\vartheta - \vartheta_{eq})^2 + \sum_{dihedrals} \frac{V_n}{2}[1 + \cos(n\phi - \gamma)] + \sum_{i<j}\left[\frac{A_{ij}}{R_{ij}^{12}} - \frac{B_{ij}}{R_{ij}^{6}} + \frac{q_i q_j}{\varepsilon R_{ij}}\right]. \tag{3}$$

The internal energy of the molecule is determined by the first three terms in Equation 3, which estimates the internal energy by penalizing deviations from ideal bond lengths, bond angles, and dihedral angles. Enthalpic contributions to the free energy are estimated through the Lennard-Jones (or 6-12) potential, which estimates the van der Waals interactions of the system, and the Coulomb equation, which determines the strength of charge-charge interactions. Equation 3 is used with additive force fields, which do not take into account any polarization effects. There are developments under way in several laboratories to develop polarizable, nonadditive force fields (7, 22).

Translational and rotational entropies of solutes are usually determined separately with statistical mechanics approaches (92), and the vibrational entropy can be determined with either normal mode analysis or quasi-harmonic analysis. The configurational entropy loss can be estimated with the side chain entropy scale developed by Pickett & Sternberg (112) based on statistical thermodynamics, which requires calculating the solvent-accessible surface areas of amino acid side chains before and after binding. Only the vibrational entropy calculation is based on rigorous physical principles.

Bulk solvent effects arise from screening of electrostatic interactions, which attenuate their strength. The solvent free energy term is usually divided into two terms, nonpolar and polar solvent effects, which are calculated separately. The nonpolar solvent term is generally estimated using the surface area of the solute. This term accounts for the energy cost of making a cavity in the solvent that can contain the solute (accounting for freezing of the water molecules around the solute) as well as the favorable van der Waals interaction energy of the solute with the solvent. The difference in nonpolar solvation free energy between the complex and its components is usually favorable and thus increases complex stability. The polar solvation term accounts for the difference in electrostatic energy between the solute embedded in a low-dielectric medium and a high-dielectric medium, which arises from a difference in polarization effects between low-dielectric and high-dielectric solvents. This term can be calculated with continuum solvent approximations such as the Poisson-Boltzmann equation (102) or the generalized Born equation (8, 128). Continuum solvent models model the bulk properties of aqueous solvents well, but they cannot correctly account for effects owing to interactions specific water molecules make at the solute-solvent interface.

All of the abovedescribed terms together determine the strength of the interaction between proteins and ligands. These physical forces lead to the following master equation for the binding free energy shown in Equation 4:

$$\Delta G_{bind} = E_{MM} - T\Delta S_{solute} + \Delta G_{solvent}. \qquad 4.$$

E_{MM} is given in Equation 3; the solute entropy consists of four terms, namely translational, rotational, vibrational, and conformational entropy. The solvent free energy consists of the two terms described above, a nonpolar and a polar term. The solvent term is a free energy because it takes into account both enthalpic and entropic effects of the solvent. This equation has recently been applied to the study of a number of different systems, where the calculated free energies are averages over a number of different structures obtained through explicit solvent MD simulations (MM-PBSA or ES/IS approach) (91, 126, 136). The success of these methods compared with results using a single structure obtained through energy minimization underscores the importance of taking into account flexibility of protein systems (52, 116). In the best cases, the accuracy of these methods can be similar to FEP/TI methods, around ±1 kcal/mole (52, 53).

MOLECULAR DOCKING: SCORING FUNCTIONS, VALIDATION, AND SYSTEM REPRESENTATION

Scoring Functions

Estimating binding free energies accurately is a time-consuming process. State-of-the-art efforts are represented by the FEP/TI methodology. Although the MM-PBSA and explicit solvent/implicit solvent (ES/IS) methods can achieve similar accuracy at a smaller computational cost, these methodologies cannot currently be used in screening large numbers of ligands against a protein target. Furthermore, the effects of using continuum solvent models for a wide range of ligands need further exploration (43). The need for a fast, yet accurate, scoring function for docking studies has led to a number of different functions that can be divided into four main classes, namely first-principles methods, semiempirical scoring functions, empirical methods, and knowledge-based potentials. Different scoring functions are presented for each class of function, and approximations and assumptions made by each are discussed.

Besides simplified scoring functions, docking programs make other approximations as well. While FEP/TI and MM-PBSA/ES/IS calculate the binding free energy based on structural ensembles, docking algorithms assume the binding free energy can be approximated using a single structure. Often this is a good approximation because the lowest energy structure makes the largest contribution to the partition function. In addition, most functions assume that the binding free energy can be approximated by a linear combination of pairwise terms, while many forces involved in complex formation are nonadditive (e.g., solvent screening). Furthermore, as expressed in Equation 1, the binding free energy is a function of the interactions between the receptor and the ligand in the complex, as well as of the interactions between the receptor and the solvent, and the ligand and the solvent. Docking programs usually only consider the complexed state explicitly, whereas the uncomplexed molecules are often accounted for implicitly.

FIRST-PRINCIPLES METHODS Both DOCK (93) and AutoDock (42) implemented scoring functions use the Coulomb and van der Waals terms of force field functions. To account for the screening effect of the solvent on electrostatic interactions, a distance-dependent dielectric constant is used. Internal ligand energies and entropic terms are completely ignored. The program EUDOC uses a force field function as well, without a grid (109). These force field–based scoring functions ignore most solvent effects as well as solute entropies, and the calculated scores are just energies or enthalpies rather than free energies.

Shoichet et al. (122) later added the effects of the solvent on protein-ligand interactions using implicit solvent models. The van der Waals interactions were calculated using the Lennard-Jones potential; the electrostatic interaction between the ligand and the protein was estimated using a precalculated receptor potential determined with DelPhi (102), which solves the Poisson-Boltzmann equation. Ligand desolvation penalties were calculated with HYDREN (115), which estimates the polar desolvation term with the Born equation, and the nonpolar desolvation term is obtained using surface area calculations. The solvent-corrected scores were closer to experimental binding free energies than the regular DOCK scores, but still too favorable. The overestimation of complex stability can be partially ascribed to the neglect of solute entropic terms (122). This energy function thus takes into account the interactions the unbound receptor and ligand make with the solvent, although van der Waals interactions with the solvent or solute entropy changes are not calculated.

SEMIEMPIRICAL METHODS Most central processing unit (CPU) time required to perform FEP/TI calculations arises from sampling physically irrelevant states with MD or MC simulations because only small mutations can be made at a time.

Åqvist and coworkers proposed a method in which only the initial and final states need to be sampled, namely the ligand free in solution and bound to the receptor. The linear interaction energy (LIE) method is based on the linear response of electrostatic forces due to changes in electric fields, as for example described in the Born equation for ion solvation. The hydration free energy of ions and small polar groups was equal to half of the solute-solvent electrostatic energy, and this observation forms the basis for the LIE method. Changes in nonpolar interactions are modeled with an empirically determined coefficient, α, derived by fitting to experimental binding data. The scoring function is

$$\Delta G_{bind} = \frac{1}{2}\left\langle \Delta V_{i-s}^{el} \right\rangle + \alpha \left\langle \Delta V_{i-s}^{vdw} \right\rangle, \quad 5.$$

where the first term is the difference in the electrostatic interaction energy of the inhibitor i with the surroundings s, which consist either of solvent or of solvent and receptor. The second term is the empirically derived nonpolar contribution to the binding free energy. The brackets indicate that the energy differences are ensemble averages, which are obtained through MD or MC simulations of the unbound ligand and the bound ligand, usually both in explicit solvent. Due to the empirical coefficient α, the applicability of this scoring function depends on the training set from which α was derived. The initial results were promising, with errors less than 1.5 kcal/mol, at considerable less computational cost than FEP/TI (5, 6).

Later applications of the LIE method showed that the linear response approximation is not valid for nonpolar groups, and the coefficient of $\frac{1}{2}$ for the electrostatic response has therefore become an empirically determined variable, β, making this method empirical rather than semiempirical. Furthermore, Jorgensen and coworkers (91a) added a third term, γ, to Equation 5 to account for the cost of forming a cavity in the solvent. This term is linearly related to the solvent-accessible surface area, and thus basically a constant (91a, 124). Åqvist and coworkers (47, 89) further explored the LIE method and allowed the variable β to have different values depending on the chemical composition of the ligand. The parameter α is derived empirically again. Interestingly, in this model, the parameter γ is almost zero, which means only the two terms from the original work are needed (Equation 5), although the $\frac{1}{2}$ is replaced by the variable β, which varies depending on chemical groups (47, 89). The LIE method and its derivatives thus explicitly account for the unbound state of the ligand, which most scoring functions in docking programs do not do. Because either MD or MC simulations in explicit or implicit solvents need to be performed, this method is still quite time-consuming, and it has not been used as a scoring function in docking. Replacement of explicit waters with a continuum solvent model, for example generalized Born, decreases the CPU time needed for a single binding free energy calculation (147).

The GOLD scoring function consists of three terms, a hydrogen bonding term, a van der Waals term, and an internal energy term. The original scoring function used a 6-12 potential for the van der Waals interaction energy and the internal energy of the ligand (60). Later the van der Waals interaction energy was replaced with a 4-8 potential (61). The hydrogen bonding term is based on empirical values for the strength of hydrogen bonds between different atom types. The hydrogen bond energy is weighted based on the angle and the bond length between the donor and acceptor. The total energy is a weighted sum of the three terms, making this a semiempirical scoring function (60).

The Kuntz group recently developed SDOCK, which combines the van der Waals force field score and the generalized Born-surface area (GB-SA) model to account for polar and nonpolar solvation effects. Although based on first-principles methods, each term in the scoring function is multiplied by an empirically derived scaling factor as in Equation 6:

$$\Delta G_{binding} = \sigma_1 \Delta(SA_{hp}) + \beta VDW - \sigma_2 \Delta(SA) + G_{pol}. \qquad 6.$$

The first term is linearly related to the hydrophobic surface area and is the cavity term, which measures the energetic cost of making a cavity for the solute in the solvent. The second term is the van der Waals interaction energy between the receptor and the ligand, calculated with the Lennard-Jones potential. The third term is another surface area term and accounts for the favorable van der Waals interactions between the solute and the solvent. This term thus accounts for nonpolar desolvation effects because surface area is lost upon complex formation. The last term is calculated using the generalized Born equation, which takes into account the effect of the solvent on Coulomb interactions in the complex (screening effect), and is used to estimate the polar desolvation penalties for the receptor and the ligand as well. The parameters σ_1, σ_2, and β were derived empirically by fitting calculated binding free energies to experimental values for a test set consisting of the receptors dihydrofolate reductase and trypsin with several ligands (148).

The docking program AutoDock later implemented a semiempirical scoring function as in Equation 7:

$$\Delta G = \Delta G_{vdW} \sum_{i,j} \left(\frac{A_{ij}}{R_{ij}^{12}} - \frac{B_{ij}}{R_{ij}^{6}} \right) + \Delta G_{hbond} \sum_{i,j} E(t) \left(\frac{C_{ij}}{R_{ij}^{12}} - \frac{D_{ij}}{R_{ij}^{6}} \right)$$
$$+ \Delta G_{elec} \sum \frac{q_i q_j}{\varepsilon(r_{ij}) r_{ij}} + \Delta G_{tor} N_{tor} + \Delta G_{sol} \sum_{i,j} (S_i V_j + S_j V_i) \cdot e^{(-r_{ij}^2/2\sigma^2)}. \qquad 7.$$

The first three terms are force field terms and each of these terms is scaled empirically by the ΔG term. The hydrogen bonding term has directionality owing to the E(t) factor, which is a function of the angle. The last term accounts for desolvation effects. Autodock uses a pairwise, volume-based method to estimate the buriedness of the atom, which is multiplied by the atomic solvation parameter for that atom. This function was evaluated based on precalculated grids for the receptor contributions, and was derived based on 30 protein-ligand complexes (96).

EMPIRICAL METHODS Empirical scoring functions are logical extensions of the structure-activity relationships first developed in the 1960s. The linear–free energy relationships advanced by, for example, Hammett (σ function) and Hansch (π substituent constant) predicted chemical reactivity and octanol-water partitions of compounds, respectively, based on the physical-chemical properties of the solute alone. Hansch later combined octanol-water coefficients and the Hammett substituent constant σ to predict biological activities based on ligand properties only. This early work led to the field of quantitative structure activity relationships (QSAR), and QSAR is still used in drug discovery in the absence of knowledge of the target structure. Empirical scoring functions in docking are based on receptor-ligand structure properties rather than on ligand properties alone.

The first disadvantage of empirical scoring functions is that it is difficult to know what each term exactly accounts for and to assess where errors come from. Second, binding free energy predictions can only be successful if the molecules make similar interactions to the ones in the training set complexes (transferability issues). Third, pH, salt concentration, and temperature can influence the measured binding constants significantly. These different conditions are generally ignored when calculating binding free energies from experimental binding constants, and this limits both training of empirical functions and accuracy of predicted binding free energies.

The first empirical scoring function developed to predict binding free energies was implemented in LUDI and was derived using experimental binding free energies and protein-ligand crystal structures for 45 complexes.

$$\Delta G_{bind} = \Delta G_o + \Delta G_{hb} \sum_{h-bonds} f(\Delta R, \Delta\alpha) + \Delta G_{ionic} \sum_{ionic\ int.} f(\Delta R, \Delta\alpha) + \Delta G_{lipo} \left|A_{lipo}\right| + \Delta G_{rot} NROT. \qquad 8.$$

As shown in Equation 8, the binding free energy is modeled using hydrogen bonds, salt bridges, the hydrophobic effect, and solute entropy terms. The hydrogen bond and salt bridge terms are modified by a penalty function, $f(\Delta R, \Delta\alpha)$, that accounts for large deviations from ideal hydrogen bond and salt bridge geometry. Lipophilic interactions are based on the lipophilic contact surface between the receptor and the ligand. Entropy loss of the ligand upon complex formation is based on the number of rotatable bonds NROT in the ligand. The first term in Equation 8, ΔG_o, accounts for the reduction in translational and rotational entropy (10).

The LUDI function was extended to include the difference in strength in hydrogen bonds based on physico-chemical properties of the donor and acceptor and solvent effects on interaction strength. The hydrogen bond and salt bridge terms are represented by three functions instead of one (Equation 8). The two functions added account for buriedness of the hydrogen bond/salt bridge and the amount of polar contact surface divided by the number of hydrogen bonds/salt bridges. The first new penalty function (buriedness) accounts for solvent effects on the strength of hydrogen bonds and salt bridges ("screening effect"), while the second function distinguishes between strong interactions and weak interactions. Besides these modifications in the hydrogen bond and salt bridge terms, three new terms were added to the function shown in Equation 8. The first term added penalizes long-range electrostatic repulsions. This term is calculated for polar interactions that are not in direct contact. The second new term accounts for so-called π-π interactions between aromatic rings. Its form is similar to the one used for hydrogen bonds in the original LUDI function with a penalty function to account for deviations from ideal geometry (Equation 8). The final new term accounts for explicit water molecules in the active site in contact with the nonpolar surface of the binding site, which will be replaced by the ligand upon binding. This term thus accounts for nonpolar desolvation effects (11).

A similar function is implemented in the docking program FlexX, shown in Equation 9:

$$\Delta G = \Delta G_o + \Delta G_{rot} N_{rot} + \Delta G_{hb} \sum_{neutralh-bonds} f(\Delta R, \Delta\alpha) + \Delta G_{io} \sum_{ionic\ int.} f(\Delta R, \Delta\alpha)$$
$$+ \Delta G_{aro} \sum_{aro\ int.} f(\Delta R, \Delta\alpha) + \Delta G_{lipo} \sum_{lipo.\ cont.} f^*(\Delta R). \qquad 9.$$

As in the LUDI function, $f(\Delta R, \Delta\alpha)$ is a scaling function that penalizes deviations from ideal geometries, and the scaling factor $f^*(\Delta R)$ accounts for contacts with "ideal" distance but penalizes for close contacts (113).

The last empirical scoring function we discuss was specifically developed for protein-protein docking and is based on atomic solvation parameter (ASP) models. ASP parameters for amino acids have been derived from experimental free energies of transfer from octanol to water. DeLisi and coworkers (142) define the binding free energy of protein-protein complex formation as the sum of two terms, namely a constant term (ΔG_{const}) and a variable term (ΔG_{var}). The constant term accounts for translational and rotational free energies. The variable term encompasses all other terms playing a role in complex formation, including solvent effects and side chain entropies. This term is the sum of three terms (Equation 10):

$$\Delta G_{var} = E_d^{rl} + \Delta G_s - T\Delta S_c \qquad 10.$$

The first term is the Coulomb interaction between the receptor and the ligand, and the last term accounts for the loss of side chain entropy upon complex formation, which is obtained through the Pickett and Sternberg scale described earlier. The ΔG_s term accounts for the free energy of desolvation, which is calculated based on solvent-accessible surface areas of amino acids and the experimentally determined ASPs of amino acids (142).

KNOWLEDGE-BASED POTENTIALS/POTENTIALS OF MEAN FORCE Knowledge-based potentials are derived using observed frequencies of atom-atom interactions in known structures of protein-ligand complexes. If these frequencies are converted into free energies using Boltzmann distributions, the potentials are generally called potentials of mean force (PMF). PMFs have been used in the protein-folding field for quite some time and have recently been developed for protein-ligand structure prediction. The main difference with empirical potentials is that no binding data are needed, which has the advantage of easily devising relatively large training sets. Both knowledge-based functions and PMFs are used as scoring functions in docking.

While PMFs for protein folding studies are usually derived on only one or two interaction points per amino acid, complex structure prediction requires the use of an all-atom potential. The Helmholtz free energy of interaction is then calculated by summing over all protein-ligand atom pairs:

$$\Delta A_{bind} = \sum_{kl} A_{ij}(r), \tag{11.}$$

where A_{ij} is the potential of mean force between atom type i and atom type j, and the sum is over all protein-ligand atom pairs kl and all interactions of type ij. The potential of mean force is defined as follows:

$$A_{ij}(r) = -kT \ln \frac{g_{ij}(r)}{g(r)}, \tag{12.}$$

where k is Boltzmann's constant and $g_{ij}(r)$ is the normalized probability of atom pair ij to be in contact at distance r, while $g(r)$ is the normalized reference probability. The reference state is defined as the state where protein and ligand do not interact with each other. There are two important aspects in deriving PMFs. First, the definition of the reference state is critical. Approximations to the correct reference state are made because it is not feasible to generate receptor and ligand structures in the unbound conformations for all complexes. The reference state is either set to a specific (empirical) value or an average potential is calculated over all data using an appropriate volume correction. Second, the manner in which the interaction statistics are collected is important. This can be done as coarse-grained bins, where the potentials are collected over a single interval and the interaction is either on or off based on cutoff distances. In smooth-grained PMFs, the potentials are dependent on the distance between atoms i and j in a continuous manner.

In theory, PMFs include all forces that play a role in complex formation, although some effects, such as solvent effects, may be underestimated. As with empirical functions, the uncertainty in what forces are represented in PMFs makes it difficult to assess where problems come from when experimental results cannot be reproduced correctly.

We begin a more detailed look at knowledge-based potentials with the simple piecewise linear potential (PLP), based on four atom types, developed by the group at Agouron. The PLP accounts for steric and hydrogen bond interactions, and a simple torsional potential calculates the internal energy of the ligand. By optimizing its ability to reproduce known crystal structures (36), the parameters for the PLP were derived. The hydrogen bond potential was later modified to include an angular term (120).

Shakhnovich's group at Harvard developed a coarse-grained PMF for de novo design called SMoG, which was derived from protein-ligand crystal complexes. A cutoff of 5 Å is used for any protein-ligand atom pair. The reference state is defined as a complex of randomly connected protein atoms and randomly connected ligand atoms that do not interact. The reference potential is approximated by the average potential over all atom types (29). This SMoG potential was recently improved by introducing two different potentials for each protein-ligand atom pair (two different distance intervals) and a new definition of the reference state. The new reference state potential accounts for statistical effects due to the composition of the training set and is again an average over all atom pair types. While the number of contacts are calculated from a training set of 725 protein-ligand complexes, the four free parameters are optimized by assessing the correlation between predicted and experimental binding free energies of 119 protein-ligand complexes (54, 121).

The BLEEP potential is a smooth-grained potential with a cutoff distance of 8 Å, which was derived either with or without a layer of water around the complex. The potential that included the explicit waters was shown to be more effective. The potential of the reference state was approximated by an imported function based on neon-neon interactions (95).

PMFScore is another smooth-grained potential. Here only ligand-protein atom pairs of the same type define the reference state. The PMF was derived using a cutoff of 12 Å to include all solvent effects, but for scoring a 9 Å cutoff was used for noncarbon protein-ligand atom pairs while a 6 Å cutoff was used for carbon-carbon interactions. A linear scaling factor was introduced to correlate the PMFScore with experimental binding free energies (97).

The developers of FlexX implcmented DrugScore, which is a smooth-grained PMF. The potentials were derived over short distances only to make sure specific interactions dominate the potential. As a consequence, solvent effects are not properly included, and these are calculated separately based on another PMF that uses surface areas of atoms. For the SAS calculations, the reference state is the solvated state and the SAS potentials are based on atom types rather than atom type pairs (40).

The last potential derived in this section was derived specifically for protein-protein itneractions by Lai and coworkers (58). The derivation is similar to the BLEEP potential, except for the use of only four atom types. Similar to PMFScore, a scaling factor is used to relate the PMF score to the binding free energy. The reference potential was set to an empirical value (58).

Validation of Docking Algorithms

There are three main applications of docking programs: binding mode prediction, lead optimization, which includes directed-library design, and database screening. Different measures are used to evaluate the performance of a docking algorithm in a specific application.

BINDING MODE PREDICTION This is the oldest application of docking algorithms, and the ability of a particular program to predict the correct geometry of the protein-ligand complex is usually assessed with the root-mean-square derivative (RMSD) of the predicted ligand configuration versus the crystal structure. The current standard in the field is a 2 Å cutoff, beyond which the prediction is considered a docking failure (32, 135). In most tests no independent assessment of the ligand conformation is generated, although it would be straightforward to report displacements in the dihedral angles either from the reference structure or from database values (3). With the increase in available crystal structures, large test sets of over 100 protein-ligand structures are becoming the norm to assess the ability of a docking algorithm to predict binding modes (61, 73, 109).

One attempt to assess the prediction in the absence of a reference structure is the use of Z-scores, which take into account the energy distribution of the predicted geometries. The larger the differences, the more selective the scoring function (135). One could use the selectivity measure to assess whether one or several possible binding modes should be taken into account based on the energy gap between them.

Docking failures arise due to either incomplete sampling ["soft failures" (133)] or inaccuracies in the scoring function ["hard failures" (133)]. A comparison of the score of the lowest energy docked conformation with the score of the minimized crystal score can indicate whether a soft or hard failure is occurring. If the score of the docked complex is lower in energy than the crystal structure, a hard failure has occurred. If the score of the docked complex is higher (less favorable) than the crystal structure, a search algorithm (or soft) failure has occurred because the search has not found the global minimum yet and the energy landscape was thus sampled insufficiently (133). A relatively easy remedy for soft failures is to increase sampling. Solving hard failures requires improving or changing the scoring function used.

It is worth pointing out that the hard/soft distinction depends on the assumption that the energy landscape near the experimental structure is adequately sampled and/or that the experimental structure is, in fact, a global free energy minimum. While this assumption might seem a truism, it must be remembered that even the tightest molecular complexes are, thermodynamically speaking, ensembles of structures, not one of which exactly represents the free energy minimum.

Brooks and coworkers (135) generated 144 unique scoring functions by varying the parameter values and used selectivity and efficiency to assess their ability to reproduce crystal structures. The selectivity measure determines how large the energy gap is between the lowest docked structure and the lowest energy misdocked structure. The efficiency measure assesses the kinetic accessibility of the global minimum. Scoring functions with a hard core potential perform best in terms of selectivity, while soft core potentials allow for high efficiency (135).

LEAD OPTIMIZATION For lead optimization, the docking program has to be able to rank correctly chemically similar compounds. In either lead optimization or directed-library design the compounds to be ranked can be similar in chemical structure, but their binding affinity can be different. A good scoring function limits the number of both false positives and false negatives.

To test the ability of a docking program to rank order similar compounds, experimental binding data are needed. Ideally, a crystal structure is available for at least one protein-ligand complex in the test set if the ligands are similar, or for each different ligand scaffold. Muegge and coworkers (45) had 61 IC_{50} values of compounds with two similar scaffolds against a matrix metalloproteinase protein (MMP-3). Crystal structures for six protein-ligand complexes were available as well. They compared three different docking/scoring functions, and the DOCK4/PMF combination performed best at both binding mode prediction and ranking ligands (45).

The Kuntz and Ellman groups designed and tested a library against cathepsin D in which two different libraries and their hit rates were compared. The directed library (based on the target structure) had both higher hit rates (6%–7% versus 2%–3%) and higher affinity hits (three- to fourfold more potent) than the diverse library, based on libraries of 1000 compounds (66). Target-based libraries were also designed against three different serine proteases. The correct side chain preferences, known from experimental studies, were reproduced by the library design method (76).

VIRTUAL DATABASE SCREENING In lead optimization similar compounds have to be ranked correctly. In database screening, different compounds are compared. The docking algorithm must identify the small number of active compounds in a database of mostly inactive compounds. There are two metrics in common used to evaluate the success in virtual screening.

The first metric is the hit rate, where the hit rate is defined as the recovery of known actives among the top-scoring x compounds of the library. The higher the hit rate, the fewer compounds have to be screened experimentally to find active compounds. The second measure of success in database screening is the enrichment factor (EF), which is defined as

$$EF = \frac{a/n}{A/N}, \qquad 13.$$

where a is the number of active databases in the top n compounds, and A is the total number of active compounds in the database of N compounds. Hit rates of scoring functions can differ significantly from target to target (9, 21, 127).

Only two studies have published EFs. Knegtel & Wagener (69) used a 1000-compound database with known activities for thrombin and the estrogen receptor and assessed EFs using rigid and flexible docking in DOCK4.0. The influence of different scoring functions and different amounts of sampling during flexible docking on the EF was assessed. For the progesterone receptor the largest enrichment (∼6) was retrieved with the DOCK force field score, while for thrombin the DOCK chemical score gave the largest enrichment (∼9) (69). Diller & Merz (31) studied a library containing more than 10,000 compounds, which were tested experimentally against plasmepsin. They found a maximum enrichment of 3, depending on the scoring function used. Besides increased hit rates, computational screening can also ensure drug-likeness of the hits because these can be easily eliminated before any screening is done, based, for example, on Lipinski's rule of 5 (83).

Different combinations of scoring functions ("consensus scoring") and their hit rates on different targets have been assessed (9, 21, 23, 127). If one function is much worse than all the others, it will strongly degrade the consensus results.

Differences Between Protein-Protein and Protein-Ligand Docking

Several search algorithms and scoring functions have been developed specifically for protein-protein docking (46). In protein-protein docking the rigid body approximation is still the standard because of the large number of degrees of flexibility and because it is much harder to predict where the protein interaction site is. The scoring function should be rather soft because some atom clashes are likely to occur even at near-native configurations. When both protein surfaces are sampled fully, the number of generated complexes is huge, requiring efficient sampling and scoring functions. Often, the generated complexes are scored in two stages. The first stage is a geometric filter that eliminates conformers not likely to be the complex structure. In the second stage, the geometrically reasonable configurations are scored with a more elaborate energy function. Surfaces are represented in a reduced manner such as description of knobs and holes rather than explicit atom representations. More success is obtained using the complexed conformations for both the protein and the protein ligand ("bound" docking). The "unbound" docking problem is much more difficult to solve and requires more extensive sampling of amino acid side chains to account for induced fit.

DOCKING AS PART OF A MULTISTEP PROCESS

Hierarchical experiments for binding mode prediction and lead optimization typically use docking as a first step to sample the binding free energy landscape broadly. MD or MC simulations are subsequently used to sample particular binding modes more thoroughly and to obtain a more accurate estimate of the binding free energy through the use of a more sophisticated scoring function. Such binding modes can be picked either by score (138) or by clustering all the binding modes and taking the cluster heads forward (133, 139).

Application 1: Binding Mode Prediction

Several multistep studies have been published. Two reports have appeared that combine docking and MD simulations to difficult protein-ligand complexes. Both studies show improvement over docking alone (133, 139).

Both the Jorgenson group (118) and the Kollman group (139), independently, correctly predicted the binding mode of Sustiva, an HIV-1 reverse transcriptase (HIV-1 RT) inhibitor, before the structure was published. The Jorgenson group used the known binding modes for nevirapine, HEPT, and 9-CL TIBO to test the docking procedure. To validate the predicted binding mode for Sustiva, the known influence of receptor mutations on Sustiva activity were used as a comparison to predicted binding free energies between the wild-type receptor-ligand complex and several mutant receptor-complexes (118). The Kollman group used DOCK4.0 to generate binding modes for Sustiva to HIV-1 RT, which were subsequently clustered. Each of the five cluster heads was simulated using explicit water MD simulations, and binding free energies for each binding mode was predicted. The binding mode with the lowest predicted binding free energy had had the lowest RMSD to the crystal structure (139).

Application 2: Lead Optimization

In lead optimization, similar compounds have to be compared and correctly ranked according to their affinity for the target structure. Kollman and coworkers (53) predicted the binding free energy for the cathepsin D inhibitors found in the library design study of Kuntz and coworkers (66) with the MM-PBSA method. The average error between predicted binding free energy and the experimental value was 1 kcal/mol and the correlation between prediction and experiment was 0.98 (53).

DATABASE METHODS

It is now possible to easily synthesize tens of thousands of compounds and test them for activity against a target of interest. Rather than making compounds randomly, three different library design approaches have been developed to decide which compounds to synthesize. Furthermore, many compound databases are available, either commercially or in-house, that can be screened for activity. Rather than selecting random compounds, the database is usually filtered based on physicochemical properties of the compounds.

Ligand-Based Library Design

Ligand-based library design can be done either diversity or similarity based. When designing a library from scratch, diversity is a useless concept because chemical space is so vast [10^{100} compounds (137)], but when selecting compounds from an existing database, diversity is useful. Selecting diverse compounds should increase the chance of finding compounds with biological activity because similar compounds usually have similar biological activity (90). In initial screens, when no leads are available against a particular target, the chance of finding a lead can be increased by screening as diverse a set of compounds as possible.

Similarity-based library design can be used when one or more compounds bind the target of interest. Of interest are more potent compounds, or compounds with similar biological activity, but with, for example, different absorption properties. Similarity can be based on many properties, and both two-dimensional and three-dimensional molecular descriptors can be used. Brown & Martin (13) tested different descriptors and clustering methods to determine which performed best at distinguishing active from inactive compounds based on several datasets of active and inactive compounds. Two-dimensional descriptors perform better at distinguishing actives from inactives. The MACCS structural key descriptor and the Ward clustering method were the most successful. The success of two-dimensional descriptors is especially encouraging because clustering based on two-dimensional descriptors is much less time-consuming than three-dimensional descriptors (13). In a followup paper, Brown & Martin (14) investigated the ability to predict different physical-chemical properties important for biological activity with these structural descriptors. Again, two-dimensional descriptors are much better at predicting these different physical-chemical properties. The success of distinguishing active from inactive compounds by two-dimensional descriptors can thus be attributed to the fact that these descriptors capture properties important for ligand activity (14).

Receptor-Based Library Design

Receptor-based design can be applied when the structure of the target is known, ideally in the presence of a ligand. In receptor-based library design one can either screen an existing library in order to prioritize compounds for high-throughput screening (often called virtual screening) or build a combinatorial library based on a known ligand scaffold (target-based library design).

Reagent-Based Library Design

The last approach to library design is based on reagents available for making the combinatorial library. Usually, two or more sets of R groups are selected. If we consider a reaction between two reagent groups, each compound in the R1 group undergoes the same reaction with each compound in the R2 group, but because the compounds in the R1 and R2 groups all have different substituents, different compounds arise that all have a similar scaffold. Reagent-based libraries are usually built from diverse R1 and R2 reagents to increase the diversity of the library of products. Several studies investigated reagent-based and ligand-based diverse libraries, and libraries built using diverse product selection rather than diverse reactants led to more diversity (39, 57). In this context, diversity is again a useful concept because a particular scaffold can usually only be constructed from a limited set of reactants, and there is a limited set of products based on the scaffold from which one wants to choose as diverse a set as possible.

Database Filtering

Filtering methods eliminate compounds based on their physico-chemical properties and are used to increase drug-likeness of lead compounds. Lipinski (83) performed the first large-scale study investigating the properties of commercially available drugs. The "rule of 5" is based on four properties obeyed by most drug molecules (83, 84). There are certain functional groups that give rise to toxicity problems, and compounds containing these functional groups can also be eliminated using chemoinformatics tools. One can also exclude compounds that lack functional groups known to commonly occur in marketed drugs.

In early-stage drug discovery the search is for a good lead rather than for a good drug molecule. A comparative study of the physico-chemical properties of drug molecules versus lead compounds has shown that lead compounds are generally smaller and more polar than drug molecules, and as potency and drug-likeness are optimized throughout the drug discovery process leads will become larger and more hydrophobic (105). While Lipinski's "rule of 5" might be appropriate for selecting compounds in later stages of drug discovery, in an earlier stage it might be better to restrict acceptable compounds even more.

CRITICAL ISSUES

Our first concern is whether current protocols are effective enough to be of practical use in the discovery/design of new ligands. High-throughput screening is well-suited to process large numbers of compounds. Virtual screening can parallel this process either with or without structural information on the target. Using structure-based docking, compounds of appropriate activity (usually defined as inhibitory constants below micromolar) can be carried forward in a discovery pipeline, often through a coupling with chemical strategies (20, 21, 66). Several groups have reported EFs of detecting known inhibitors from databases using computational methods (31, 69). There are also reports of quite novel inhibitors of processes as diverse as viral fusion and kinesin binding to microtubules (50, 51). Computational efforts, especially in conjunction with coordinated chemical strategies, can add efficiency and quality to ligand discovery and design (66).

We next turn to several technical issues dealing with the quality of the scoring and sampling procedures, and the representation of the solvent. We focus our attention on the physics-based methods. First, we ask if the approximate potential energy expressions are adequate for calculating the binding affinities of receptor-ligand complexes. Molecular mechanics force fields, usually parameterized against experimental or quantum mechanical data, certainly have limitations: They completely omit entropic considerations, are restricted to pairwise potentials, and assume transferability at an atomic level. There is no exact way to "add entropy" to a molecular configuration because entropy is a distributive quantity containing components from the conformational choices of ligand and receptor and from the configurations of the solvent. The two basic strategies to solve this problem are (*a*) to emulate the physical situation by using MD or MC simulations to generate an ensemble of thermally equilibrated structures that include large numbers of solvent molecules and counter ions, and (*b*) to use the continuum solvent models (Poisson-Boltzmann, generalized Born) along with an empirical correction for hydrophobic interactions based on buried surface area. The major concern with the first class of approaches is that it is difficult to assure sufficient sampling to obtain good averaging of thermodynamic properties. The only criteria used routinely to track sampling are to monitor the geometric and energetic stability of a simulation. These are inadequate to demonstrate that the system is at equilibrium or that thermodynamic properties have converged. Running longer simulations is inefficient. With parallel processor facilities, multiple trajectories are likely to be more productive (125). The problem with continuum solvent models is that most implementations are too slow to calculate the desolvation effects at every step during docking (2). The accuracy of the representation is also an issue.

What can be said about the accuracy of these calculations in reproducing experimental results? FEP or TI methods are presently the most accurate calculations available to assess binding free energies. In favorable cases, average errors of less than ± 1 kcal/mole have been reported (70). These errors contain contributions from inaccuracies in the force field, sampling, and in experimental uncertainties. They can unlikely be reduced much further without careful attention to each of these areas.

Using perturbation methods as the standard, we can assess the more approximate free energy calculations using continuum solvent models (147) or the admixture of molecular solvent and continuum solvent in MM-PBSA (126). In several papers, results comparable to the perturbation methods are achieved (53, 71, 139). The authors note anticipated limitations: failure to account for specific water interactions, sensitivity to trajectories, and sensitivity to induced-fit.

The overall impression is that the force field/simulation procedures may be as good as we can get without expanding the number of parameters and may not be the limit in accurate experiment/theory comparisons. It will be interesting to see if the newer "polarizable" force fields provide significant improvements. Another fruitful test of the quality of force fields is provided by the advent of high-resolution protein crystal structures that point to ways to improve current methods (56).

Second, we consider the accuracy of different solvent representations. Studies of water are well advanced, with a variety of models that simulate thermodynamic and kinetic properties with considerable accuracy (62, 87). There are two issues. First, no single water model captures all the thermodynamic and kinetic properties equally well; so one must accept some compromises. Second, the shift from pure water to macromolecular solutions places additional demands on the water model charges and geometric properties. The free energies, enthalpies, and entropies of solution for small solutes in water are available (19, 140), but equivalent data for protein solutions are harder to obtain. It should be possible to use the improved accuracy in placing water molecules (and other solvents) in high-resolution protein structures. Calculations of occupancy, B factors, and order parameters can now all be compared with new experimental results.

Continuum solvent models are of interest because of their efficiency and their generally acceptable quality for solvation phenomena. The Poisson-Boltzmann formulation provides a more physically compelling picture than the generalized Born implementation, but the quantitative features of the two approaches are similar (104). Because these approaches do not contain the frictional forces of the explicit water models, they should offer more sampling per unit time step (146). They will fail when specific water-ligand or water-receptor interactions are important.

Third, we assess the quality of sampling of the conformational and configurational space associated with macromolecular complexes. One concern, in the docking area, is to assess the frequency of soft docking errors. We estimate that ~15% of the cases in which docking does not reproduce known geometries are due to sampling limitations (12). Sampling errors also arise in perturbation and MM-PBSA calculations. In the former case, one can run the perturbation in the "forward" and "reverse" directions, testing, at least, for numerical convergence on the same solution. A similar test is not available for MM-PBSA, though Kuhn & Kollman (74) examined the reproducibility of results from two different trajectories, and this idea can be generalized to multiple trajectories.

In summary, the time has come to use the relatively large expansion in computer resources to provide a more quantitative assessment of the degree of sampling in molecular recognition calculations.

The last topic we consider is the comparison between experiment and calculation. In many areas of chemistry and physics this activity is straightforward. However, both the measurement and the calculation of the thermodynamic properties of macromolecular solutions are complex procedures with many underlying assumptions that make this comparison challenging. To begin at the beginning: The concept of a biological reference state for these solutions involves (hypothetical) macromolecular concentrations of 1 molar! While such a state could be simulated, it is never achieved in practice. Instead, all experiments are carried out in a differential fashion, typically using one ligand as the standard for others. Although such procedures yield legitimate results for comparisons within the protocol, it is nontrivial to compare different enzymes, different mutations, and different phases (crystal versus solution). The other difficulty is in transferring these conditions to the calculation. For example, most simulations are carried out in concentrated solutions, most structures are determined in the crystal state, and most binding experiments are conducted in dilute solutions. The pH and ionic strength for the binding experiments are generally well regulated at precise values within a (typically) narrow range near pH 7; NMR structures are often determined at acidic pH and crystals often grow at biologically extreme pHs. It is possible to simulate a particular pH (15), but it requires a much more elaborate procedure than one can use for database screening or even for most dynamic trajectories. Instead, the normal practice is to fix the ionization states of (all) protonatable groups, most especially those near the active site based on the pH and the pKas of the groups. Even here it is rare to take into account individual pKas generated by the local environment of the target or the ligand-target complex, although it is not uncommon to find perturbations of 2 pH units or even more in such systems (38). Enzyme active sites offer many such challenges (e.g., aspartyl proteases, particularly HIV protease) and require a prior determination of protonation of the active site aspartic acids; polymerases use divalent metals whose positions are not precisely known. The high charge density of nucleic acid systems raises similar concerns about the location of monovalent and divalent cations. Even a perfect potential function cannot overcome errors in initial charge distribution.

A closely related issue is the comparison of the physical conditions for obtaining the structural data with those for the binding data. Obviously, there will be shifts in the structure as the pH, salt, and solvent conditions are changed. Further, the diffraction structures are derived under concentrated conditions with physical contact among the macromolecular constituents while the binding experiments are in highly dilute solutions. Both the experimental and computational communities would be well served if there were direct measurements of the change in binding as a function of these variables, especially pH, macromolecular concentration, cosolvent, and ionic strength for some representative systems. Conversely, if it were possible to measure structural properties as a function of the same variables, perhaps by NMR, one could build up more powerful tests of theory and experiment.

In spite of these concerns, there is general consensus that structures are (relatively) insensitive to "crystal" artifacts. In fact, the agreement between NMR and diffraction structures is often remarkably good, especially around active sites of enzymes. From our own experience, the area that benefits the most in the short term takes care in the calculations to generate an appropriate protonation of critical side chains (53).

Most of our attention has been on experimental and theoretical pursuit of free energies. Enthalpies and entropies of binding are considerably more difficult to obtain (79). The extra effort is worthwhile because of the many ligand binding relationships that are dominated by entropic considerations (79), and design protocols must do a better job of accounting for these terms.

The most realistic and quantitatively most difficult tests would be regular CASP-like ligand docking "contests" where the unliganded form of the receptor is supplied along with the chemical formula of the ligand. Also useful would be structural and binding data obtained under similar pH and salt conditions for families of ligands and receptors. Such data would be of great value for parameter development.

Computational approaches to molecular recognition are moving forward rapidly because of the conjunction of improved experimental data, the advent of combinatorial chemistry, greatly enhanced computer resources, and theoretical advances. The utility of these techniques, particularly when arrayed in hierarchical protocols, will define the future.

ACKNOWLEDGMENTS

This work was supported by grants from the National Institute of General Medical Sciences. We thank Dr. Robert Stroud for his generous advice.

The *Annual Review of Biophysics and Biomolecular Structure* is online at http://biophys.annualreviews.org

LITERATURE CITED

1. Abagyan R, Totrov R. 1994. Biased probability Monte Carlo conformational searches and electrostatic calculations for peptides and proteins. *J. Mol. Biol.* 235:983–1002
2. Abagyan R, Totrov R, Kuznetsov D. 1994. ICM—a new method for protein modeling and design: applications to docking and structure prediction from the distorted native conformation. *J. Comput. Chem.* 15: 488–506
3. Allen FH. 2002. The Cambridge

Structural Database: a quarter of a million crystal structures and rising. *Acta Crystallogr. B* 58:380–88
4. Apostolakis J, Plueckthun A, Caflisch A. 1998. Docking small ligands in flexible binding sites. *J. Comput. Chem.* 19:21–37
5. Åqvist J. 1996. Calculation of absolute binding free energies for charged ligands and effects of long-range electrostatic interactions. *J. Comput. Chem.* 17:1587–97
6. Åqvist J, Medina C, Samuelsson J-E. 1994. A new method for predicting binding affinity in computer-aided drug design. *Protein Eng.* 7:385–91
7. Banks JL, Kaminski GA, Zhou RH, Mainz DT, Berne BJ, Friesner RA. 1999. Parametrizing a polarizable force field from ab initio data. I. The fluctuating point charge model. *J. Chem. Phys.* 110:741–54
8. Bashford D, Case DA. 2000. Generalized Born models of macromolecular solvation effects. *Annu. Rev. Phys. Chem.* 51:129–52
9. Bissantz C, Folkers G, Rognan D. 2000. Protein-based virtual screening of chemical databases. 1. Evaluation of different docking/scoring combinations. *J. Med. Chem.* 43:4759–67
10. Boehm H-J. 1994. The development of a simple empirical scoring function to estimate the binding constant for a protein-ligand complex of known three-dimensional structure. *J. Comput. Aid. Mol. Des.* 8:243–56
11. Boehm H-J. 1998. Prediction of binding constants of protein ligands: a fast method for the prioritization of hits obtained from de novo design or 3D database search programs. *J. Comput. Aid. Mol. Des.* 12:309–23
12. Brooijmans N. 2003. *Theoretical studies of molecular recognition.* PhD thesis. Univ. Calif., San Francisco. 250 pp.
13. Brown RD, Martin YC. 1996. Use of structure-activity data to compare structure-based clustering methods and descriptors for use in compound selection. *J. Chem. Inf. Comput. Sci.* 36:572–84
14. Brown RD, Martin YC. 1997. The information content of 2D and 3D structural descriptors relevant to ligand-receptor binding. *J. Chem. Inf. Comput. Sci.* 37:1–9
15. Buergi R, Kollman PA, van Gunsteren WF. 2002. Simulating proteins at constant pH: an approach combining molecular dynamics and Monte Carlo simlution. *Proteins* 47:469–80
16. Busetta B, Tickle IJ, Blundell TL. 1983. DOCKER, an interactive program for simulating protein receptor and substrate interactions. *J. Appl. Crystallogr.* 16:432–37
17. Caflisch A, Fischer S, Karplus M. 1997. Docking by Monte Carlo minimization with a solvation correction: application to an FKBP-substrate complex. *J. Comput. Chem.* 18:723–43
18. Caflisch A, Niederer P, Anliker M. 1992. Monte Carlo docking of oligopeptides to proteins. *Proteins* 13:223–30
19. Chambers CC, Hawkins GD, Cramer DJ, Truhlar DG. 1996. Model for aqueous solvation based on class IV atomic charges and first solvation shell effects. *J. Phys. Chem.* 100:16385–98
20. Charifson PS, ed. 1997. *Practical Application of Computer-Aided Drug Design.* New York: Marcel Dekker
21. Charifson PS, Corkery JJ, Murcko MA, Walters WP. 1999. Consensus scoring: a method for obtaining improved hit rates from docking databases of three-dimensional structures into proteins. *J. Med. Chem.* 42:5100–9
22. Cieplak P, Caldwell JW, Kollman PA. 2001. Molecular mechanical models for organic and biological systems going beyond the atom centered two body additive approximation: aqueous solution free energies of methanol and N-methyl acetamide, nucleic acid base, and amide hydrogen bonding and chloroform/water partition coefficients of the nucleic acid bases. *J. Comput. Chem.* 22:1048–57
23. Clark RD, Strizhev A, Leonard JM, Blake

JF, Matthew JB. 2002. Consensus scoring for ligand/protein interactions. *J. Mol. Graph.* 20:281–95
24. Claussen H, Buning C, Rarey M, Lengauer T. 2001. FlexE: efficient molecular docking considering protein structure variations. *J. Mol. Biol.* 308:377–95
25. Connolly ML. 1983. Analytical molecular surface calculation. *J. Appl. Crystallogr.* 16:548–58
26. DesJarlais RL, Seibel GL, Kuntz ID, Furth PS, Alvaraz JC, et al. 1990. Structure-based design of nonpeptide inhibitors specific for human immunodeficiency virus 1 protease. *Proc. Natl. Acad. Sci. USA* 87: 6644–48
27. DesJarlais RL, Sheridan RP, Dixon JS, Kuntz ID, Venkataraghavan R. 1986. Docking flexible ligands to macromolecular receptors by molecular shape. *J. Med. Chem.* 29:2149–53
28. DesJarlais RL, Sheridan RP, Seibel GL, Dixon JS, Kuntz ID, Venkataraghavan R. 1988. Using shape complementarity as an initial screen in designing ligands for a receptor binding site of known three-dimensional structure. *J. Med. Chem.* 31: 722–29
29. DeWitte RS, Shaknovich EI. 1996. SMoG: de novo design method based on simple, fast, and accurate free energy estimates. 1. Methodology and supporting evidence. *J. Am. Chem. Soc.* 118:11733–44
30. Di Nola A, Roccatano D, Berendsen HJC. 1994. Molecular dynamics simulation of the docking of substrates to proteins. *Proteins* 19:174–82
31. Diller DJ, Merz KM Jr. 2001. High throughput docking for library design and library prioritization. *Proteins* 43:113–24
32. Ewing TJA, Makino S, Skillman AG, Kuntz ID. 2001. DOCK 4.0: search strategies for automated molecular docking of flexible molecule databases. *J. Comput. Aid. Mol. Des.* 15:411–28
33. Fischer E. 1894. Einfluss der Configuration auf die wirkung der Enzyme. *Ber.* 27:2985–93
34. Foote J, Milstein C. 1994. Conformational isomerism and the diversity of antibodies. *Proc. Natl. Acad. Sci. USA* 91:10370–74
35. Gabb J, Jackson RM, Sternberg MJE. 1997. Modelling protein docking using shape complementarity, electrostatics and biochemical information. *J. Mol. Biol.* 272:106–20
36. Gehlhaar DK, Verkhivker GM, Rejto PA, Sherman CJ, Fogel DB, et al. 1995. Molecular recognition of the inhibitor AG-1343 by HIV-1 protease: conformationally flexible docking by evolutionary programming. *Chem. Biol.* 2:317–24
37. Deleted in proof
38. Georgescu RE, Alexov EG, Gunner MR. 2002. Combining conformational flexibility and continuum electrostatics for calculating pK_as in proteins. *Biophys. J.* 83:1731–48
39. Gillet VJ, Willett P, Bradshaw J. 1997. The effectiveness of reactant pools for generating structurally-diverse combinatorial libraries. *J. Chem. Inf. Comput. Sci.* 37:731–40
40. Gohlke H, Hendlich M, Klebe G. 2000. Knowledge-based scoring function to predict protein-ligand interactions. *J. Mol. Biol.* 295:337–56
41. Goodford PJ. 1985. A computational procedure for determining energetically favorable binding sites on biologically important macromolecules. *J. Med. Chem.* 28:849–57
42. Goodsell DS, Olson AJ. 1990. Automated docking of substrates to proteins by simulated annealing. *Proteins* 8:195–202
43. Gouda H, Kuntz ID, Case DA, Kollman PA. 2002. Free energy calculations for theophylline binding to an RNA aptamer: comparison of MM-PBSA and thermodynamic integration methods. *Biopolymers* 68:16–34
44. Greer J, Bush BL. 1978. Macromolecular shape and surface maps by solvent exclusion. *Proc. Natl. Acad. Sci. USA* 75:303–7

45. Ha S, Andreani R, Robbins A, Muegge I. 2000. Evaluation of docking/scoring approaches: a comparative study based on MMP3 inhibitors. *J. Comput. Aid. Mol. Des.* 14:435–48
46. Halperin I, Ma B, Nussinov R. 2002. Principles of docking: an overview of search algorithms and a guide to scoring functions. *Proteins* 47:409–43
47. Hansson T, Marelius J, Åqvist J. 1998. Ligand binding affinity prediction by linear interaction energy methods. *J. Comput. Aid. Mol. Des.* 12:27–35
48. Head MS, Given JA, Gilson MK. 1997. "Mining minima": direct computation of conformational free energy. *J. Phys. Chem. A* 101:1609–18
49. Helms V, Wade RC. 1998. Computational alchemy to calculate absolute protein-ligand binding free energy. *J. Am. Chem. Soc.* 120:2710–13
50. Hoffman LR, Kuntz ID, White JM. 1997. Structure-based identification of an inducer of the low-pH conformational change in the influenza virus hemagglutinin: irreversible inhibition of infectivity. *J. Virol.* 71:8808–20
51. Hopkins SC, Vale RD, Kuntz ID. 2000. Inhibitors of kinesin activity from structure-based computer screening. *Biochemistry* 39:2805–14
52. Huo S, Massova I, Kollman PA. 2002. Computation alanine scanning of the 1:1 human growth hormone-receptor complex. *J. Comput. Chem.* 23:15–27
53. Huo S, Wang J, Cieplak P, Kollman PA, Kuntz ID. 2002. Molecular dynamics and free energy analyses of cathepsin D-inhibitor interactions: insight into structure-based ligand design. *J. Med. Chem.* 45:1412–19
54. Ishchenko AV, Shaknovich EI. 2002. SMall Molecule Growth 2001 (SMoG-2001): an improved knowledge-based scoring function for protein-ligand interactions. *J. Med. Chem.* 45:2770–80
55. Jackson RM, Gabb HA, Sternberg MJE. 1998. Rapid refinement of protein interfaces incorporating solvation: application to the docking problem. *J. Mol. Biol.* 276:265–85
56. Jacobson MP, Friesner RA, Xiang Z, Honig B. 2002. On the role of crystal environment in determining protein side-chain conformations. *J. Mol. Biol.* 320:597–608
57. Jamois EA, Hassan M, Waldman M. 2000. Evaluation of reagent-based and product-based strategies in the design of combinatorial library subsets. *J. Chem. Inf. Comput. Sci.* 40:63–70
58. Jiang F, Gao Y, Mao F, Liu Z, Lai L. 2002. Potential of mean force for protein-protein interaction studies. *Proteins* 46:190–96
59. Jiang F, Kim S-H. 1991. "Soft docking": matching of molecular surface cubes. *J. Mol. Biol.* 219:79–102
60. Jones G, Willett P, Glen RC. 1995. Molecular recognition of receptor sites using a genetic algorithm with a description of desolvation. *J. Mol. Biol.* 245:43–53
61. Jones G, Willett P, Glen RC, Leach AR, Taylor R. 1997. Development and validation of a genetic algorithm for flexible docking. *J. Mol. Biol.* 267:727–48
62. Jorgensen WL, Jenson C. 1998. Temperature dependence of TIP3P, SPC, and TIP4P water from NPT Monte Carlo simulations: seeking termperatures of maximum density. *J. Comput. Chem.* 19:1179–86
63. Judson RS, Tan YT, Mori E, Melius C, Jaeger EP, et al. 1995. Docking flexible molecules: a case study of three proteins. *J. Comput. Chem.* 16:1405–19
64. Kauzmann W. 1959. Some factors in the interpretation of protein denaturation. *Adv. Protein Chem.* 14:1–63
65. Kearsley SK, Underwood DJ, Sheridan RP, Miller MD. 1994. Flexibases: a way to enhance the use of molecular docking methods. *J. Comput. Aid. Mol. Des.* 8:565–82
66. Kick EK, Roe DC, Skillman AG, Liu G, Ewing TJA, et al. 1997. Structure-based design and combinatorial chemistry yield

low nanomolar inhibitors of cathepsin D. *Chem. Biol.* 4:297–307
67. Klebe G. 2000. Recent developments in structure-based drug design. *J. Mol. Med.* 78:269–81
68. Knegtel RMA, Kuntz ID, Oshiro CM. 1997. Molecular docking to ensembles of protein structures. *J. Mol. Biol.* 266:424–40
69. Knegtel RMA, Wagener M. 1999. Efficacy and selectivity in flexible database docking. *Proteins* 37:334–45
70. Kollman PA. 1993. Free energy calculations: applications to chemical and biochemical phenomena. *Chem. Rev.* 93:2395–417
71. Kollman PA, Massova I, Reyes C, Kuhn B, Huo S, et al. 2000. Calculating structures and free energies of complex molecules: combining molecular mechanics and continuum models. *Acc. Chem. Res.* 33:889–97
72. Koshland DE Jr. 1958. Application of a theory of enzyme specificity to protein synthesis. *Proc. Natl. Acad. Sci. USA* 44:98–104
73. Kramer B, Rarey M, Lengauer T. 1999. Evaluation of the FlexX incremental construction algorithm for protein-ligand docking. *Proteins* 37:228–41
74. Kuhn B, Kollman PA. 2000. A ligand that is predicted to bind better to avidin than biotin: insights from computational fluorine scanning. *J. Am. Chem. Soc.* 122:3909–16
75. Kuntz ID, Blaney JM, Oatley SJ, Langridge R, Ferrin TE. 1982. A geometric approach to macromolecule-ligand interactions. *J. Mol. Biol.* 161:269–88
76. Lamb ML, Burdick KW, Toba S, Young MM, Skillman AG, et al. 2001. Design, docking, and evaluation of multiple libraries against multiple targets. *Proteins* 42:296–88
77. Leach AR. 1994. Ligand docking to proteins with discrete side-chain flexibility. *J. Mol. Biol.* 235:345–56
78. Leach AR, Kuntz ID. 1992. Conformational analysis of flexible ligands in macromolecular receptor sights. *J. Comput. Chem.* 13:730–48
79. Leavitt S, Freire E. 2001. Direct measurement of protein binding energetics by isothermal titration calorimetry. *Curr. Opin. Struct. Biol.* 11:560–66
80. Levinthal C, Wodak SJ, Kahn P, Dadivanian AK. 1975. Hemoglobin interaction in sickle cell fibers. I. Theoretical approaches to the molecular contacts. *Proc. Natl. Acad. Sci. USA* 72:1330–34
81. Li Z, Scheraga HA. 1987. Monte Carlo-minimization approach to the multiple-minima problem in protein folding. *Proc. Natl. Acad. Sci. USA* 84:6611–15
82. Lin J-H, Perryman AL, Schames JR, McCammon JA. 2002. Computational drug design accommodating receptor flexibility: the relaxed complex scheme. *J. Am. Chem. Soc.* 124:5632–33
83. Lipinski CA. 2000. Drug-like properties and the causes of poor solubility and poor permeability. *J. Pharmacol. Toxicol.* 44:235–49
84. Lipinski CA, Lombardo F, Dominy BW, Feeney PJ. 2001. Experimental and computational approaches to estimate solubility and permeability in drug discovery and development settings. *Adv. Drug Deliv. Rev.* 46:3–26
85. Liu M, Wang S. 1999. MCDOCK: a Monte Carlo simulation approach to the molecular docking problem. *J. Comput. Aid. Mol. Des.* 13:435–51
86. Lorber DM, Shoichet BK. 1998. Flexible ligand docking using conformational ensembles. *Protein Sci.* 7:938–50
87. Mahoney MW, Jorgensen WL. 2001. Diffusion constant of the TIP5P model of liquid water. *J. Chem. Phys.* 114:363–66
88. Mangoni M, Roccatano D, Di Nola A. 1999. Docking of flexible ligands to flexible receptors in solution by molecular dynamics simulation. *Proteins* 35:153–62
89. Marelius J, Hansson T, Åqvist J. 1998. Calculation of ligand binding free energies from molecular dynamics simulations. *Int. J. Quantum Chem.* 69:77–88

90. Martin YC, Kofron JL, Traphagen LM. 2002. Do structurally similar molecules have similar biological activity? *J. Med. Chem.* 45:4350–58
91. Massova I, Kollman PA. 1999. Computational alanine scanning to probe protein-protein interactions: a novel approach to evaluate binding free energies. *J. Am. Chem. Soc.* 121:8133–43
91a. McDonald NA, Carlson HA, Jorgensen WL. 1997. Free energies of solvation in chloroform and water from a linear response approach. *J. Phys. Org. Chem.* 10: 563–76
92. McQuarrie DA. 2000. *Statistical Mechanics*. New York: Univ. Sci. Books
93. Meng EC, Shoichet BK, Kuntz ID. 1992. Automated docking with grid-based energy evaluation. *J. Comput. Chem.* 13: 505–24
94. Miller MD, Kearsley SK, Underwood DJ, Sheridan RP. 1994. FLOG: a system to select 'quasi-flexible' ligands complementary to a receptor of known three-dimensional structure. *J. Comput. Aid. Mol. Des.* 8:153–74
95. Mitchell JBO, Laskowski RA, Alex A, Thornton JM. 1999. BLEEP—potential of mean force describing protein-ligand interactions. I. Generating potential. *J. Comput. Chem.* 20:1165–76
96. Morris GM, Goodsell DS, Halliday RS, Huey R, Hart WE, et al. 1998. Automated docking using a Lamarckian genetic algorithm and an empirical free energy function. *J. Comput. Chem.* 19:1639–62
97. Muegge I, Martin YC. 1999. A general and fast scoring function for protein-ligand interactions: a simplified potential approach. *J. Med. Chem.* 42:791–804
98. Murray CW, Baxter CA, Frenkel AD. 1999. The sensitivity of the results of molecular docking to induced effects: application to thrombin, thermolysin and neuraminidase. *J. Comput. Aid. Mol. Des.* 13:547–62
99. Deleted in proof
100. Nakajima N, Higo J, Kidera A, Nakamura H. 1997. Flexible docking of a ligand peptide to a receptor protein by multicanonical molecular dynamics simulation. *Chem. Phys.* 278:297–301
101. Nakajima N, Nakamura H, Kidera A. 1997. Multicanonical ensemble generated by molecular dynamics simulation for enhanced conformational sampling of peptides. *J. Phys. Chem. B* 101:817–24
102. Nicholls A, Honig B. 1991. A rapid finite difference algorithm, utilizing successive over-relaxation to solve the Poisson-Boltzmann equation. *J. Comput. Chem.* 12:435–45
103. Norel R, Lin SL, Wolfson HJ, Nussinov R. 1994. Shape complementarity at protein-protein interfaces. *Biopolymers* 34:933–40
104. Onufriev A, Case DA, Bashford D. 2002. Effective Born radii in the generalized Born approximation: the importance of being perfect. *J. Comput. Chem.* 23:1297–304
105. Oprea TI. 2002. Current trends in lead discovery: Are we looking for the appropriate properties? *J. Comput. Aid. Mol. Des.* 16:325–34
106. Oshiro CM, Kuntz ID, Dixon JS. 1995. Flexible ligand docking using a genetic algorithm. *J. Comput. Aid. Mol. Des.* 9:113–30
107. Osterberg F, Morris GM, Sanner MF, Olson AJ, Goodsell DS. 2002. Automated docking to multiple target structures: incorporation of protein mobility and structural water heterogeneity in AutoDock. *Proteins* 46:34–40
108. Pak Y, Wang S. 2000. Application of a molecular dynamics simulation method with a generalized effective potential to the flexible molecular docking problems. *J. Phys. Chem. B* 104:354–59
109. Pang Y-P, Perola E, Xu K, Prendergast FG. 2001. EUDOC: a computer program for identification of drug interaction sites in macromolecules and drug leads from chemical databases. *J. Comput. Chem.* 22: 1750–71

110. Pattabiraman N, Levitt M, Ferrin TE, Langridge R. 1985. Computer graphics in real-time docking with energy calculation and minimization. *J. Comput. Chem.* 6:432–36
111. Pauling L, Delbrück M. 1940. The nature of the intermolecular forces operative in biological processes. *Science* 92:77–79
112. Pickett SD, Sternberg MJE. 1993. Empirical scale of side-chain conformational entropy in protein folding. *J. Mol. Biol.* 231:825–39
113. Rarey M, Kramer B, Lengauer T, Klebe G. 1996. A fast flexible docking method using an incremental construction algorithm. *J. Mol. Biol.* 261:470–89
114. Rarey M, Wefing S, Lengauer T. 1996. Placement of medium-sized molecular fragments into active sites of proteins. *J. Comput. Aid. Mol. Des.* 10:41–54
115. Rashin AA. 1990. Hydration phenomena, classical electrostatics and the boundary element method. *J. Phys. Chem.* 94:1725–33
116. Reyes C, Kollman PA. 2000. Investigating the binding specificity of U1A-RNA by computational mutagenesis. *J. Mol. Biol.* 295:1–6
117. Richards FM. 1977. Areas, volumes, packing, and protein structure. *Annu. Rev. Biophys. Bioeng.* 6:151–76
118. Rizzo RC, Wang D-P, Tirado-Rives J, Jorgensen WL. 2000. Validation of a model for the complex of HIV-1 reverse transcriptase with sustiva through computation of resistance profiles. *J. Am. Chem. Soc.* 122:12898–900
119. Salemme FR. 1976. An hypothetical structure for an intermolecular electron transfer complex of cytochromes *c* and b_5. *J. Mol. Biol.* 102:563–68
120. Schaffer L, Verkhivker GM. 1998. Predicting structural effects in HIV-1 protease mutant complexes with flexible ligand docking and protein side-chain optimization. *Proteins* 33:295–310
121. Shimada J, Ishchenko AV, Shaknovich EI. 2000. Analysis of knowledge-based protein-ligand potentials using a self-consistent method. *Protein Sci.* 9:765–75
122. Shoichet BK, Leach AR, Kuntz ID. 1999. Ligand solvation in molecular docking. *Proteins* 34:4–16
123. Deleted in proof
124. Smith RH Jr, Jorgensen WL, Tirado-Rives J, Lamb ML, Janssen PAJ, et al. 1998. Prediction of binding affinities for TIBO inhibitors of HIV-1 reverse transcriptase using Monte Carlo simulations in a linear response method. *J. Med. Chem.* 41:5272–86
125. Snow CD, Nguyen H, Pande VS, Gruebele M. 2002. Absolute comparison of simulated and experimental protein-folding dynamics. *Nature* 420:102–6
126. Srinivasan J, Cheatham TE III, Cieplak P, Kollman PA, Case DA. 1998. Continuum solvent studies of the stability of DNA, RNA, and phosphoramidate-DNA helices. *J. Am. Chem. Soc.* 120:9401–9
127. Stahl M, Rarey M. 2001. Detailed analysis of scoring functions for virtual screening. *J. Med. Chem.* 44:1035–42
128. Still WC, Tempczyk A, Hawley RC, Hedrickson T. 1990. Semianalytical treatment of solvation for molecular mechanics and dynamics. *J. Am. Chem. Soc.* 112:6127–29
129. Tomioka N, Itai A, Iitaka Y. 1987. A method for fast energy estimation and visualization of protein-ligand interaction. *J. Comput. Aid. Mol. Des.* 1:197–210
130. Totrov R, Abagyan R. 1997. Flexible protein-ligand docking by global energy optimization in internal coordinates. *Proteins* (Suppl.) 1:215–20
131. Trosset J-Y, Scheraga HA. 1998. Reaching the global minimum in docking simulations: a Monte Carlo energy minimization approach using Bezier Splines. *Proc. Natl. Acad. Sci. USA* 95:8011–15
132. Trosset J-Y, Scheraga HA. 1999. Prodock: software package for protein modeling and docking. *J. Comput. Chem.* 20:412–27

133. Verkhivker GM, Bouzida D, Gehlhaar DK, Rejto PA, Arthurs S, et al. 2000. Deciphering common failures in molecular docking of ligand-protein complexes. *J. Comput. Aid. Mol. Des.* 14:731–51
134. Vieth M, Hirst JD, Dominy BN, Daigler H, Brooks CL. 1998. Assessing search strategies for flexible docking. *J. Comput. Chem.* 19:1623–31
135. Vieth M, Hirst JD, Kolinski A, Brooks CL. 1998. Assessing energy functions for flexible docking. *J. Comput. Chem.* 19:1612–22
136. Vorobjev YN, Almagro JC, Hermans J. 1998. Discrimination between native and intentionally misfolded conformations of proteins: ES/IS, a new method for calculating conformational free energy that uses both dynamics simulations with explicit solvent and a continuum solvent models. *Proteins* 32:399–413
137. Walters WP, Stahl M, Murcko MA. 1998. Virtual screening—an overview. *Drug Discov. Today* 3:160–78
138. Wang J, Kollman PA, Kuntz ID. 1999. Flexible ligand docking: a multistep strategy approach. *Proteins* 36:1–19
139. Wang J, Morin P, Wang W, Kollman PA. 2001. Use of MM-PBSA in reproducing the binding free energies of HIV-1 RT of TIBO derivatives and predicting the binding mode to HIV-1 RT of Efivirenz by docking and MM-PBSA. *J. Am. Chem. Soc.* 123:5221–30
140. Wang J, Wang W, Huo S, Lee MR, Kollman PA. 2001. Solvation model based on weighted solvent accessible surface area. *J. Phys. Chem. B* 105:5055–67
141. Welch W, Ruppert J, Jain AN. 1996. Hammerhead: fast, fully automated docking of flexible ligands to protein binding sites. *Chem. Biol.* 3:449–62
142. Weng Z, Vajda S, DeLisi C. 1996. Prediction of protein complexes using empirical free energy functions. *Protein Sci.* 5:614–26
143. Westhead DR, Clark DE, Murray CW. 1997. A comparison of heuristic search algorithms for molecular docking. *J. Comput. Aid. Mol. Des.* 11:209–28
144. Wodak SJ, De Crombrugghe M, Janin J. 1987. Computer studies of interactions between macromolecules. *Prog. Biophys. Mol. Biol.* 49:29–63
145. Wodak SJ, Janin J. 1978. Computer analysis of protein-protein interactions. *J. Mol. Biol.* 124:323–42
146. Xia B, Tsui V, Case DA, Dyson HJ, Wright PE. 2002. Comparison of protein solution structures refined by molecular dynamics simulation in vacuum, with a generalized Born model, and with explicit water. *J. Biomol. NMR* 22:317–31
147. Zhou R, Friesner RA, Ghosh A, Rizzo RC, Jorgensen WL, Levy RM. 2001. New linear interaction method for binding affinity calculations using a continuum solvent model. *J. Phys. Chem. B* 105:10388–97
148. Zou X, Sun Y, Kuntz ID. 1999. Inclusion of solvation in ligand binding free energy calculations using the generalized-Born model. *J. Am. Chem. Soc.* 121:8033–43

Annu. Rev. Biophys. Biomol. Struct. 2003. 32:375–97
doi: 10.1146/annurev.biophys.32.110601.142520

First published online as a Review in Advance on February 5, 2003

THE CRYSTALLOGRAPHIC MODEL OF RHODOPSIN AND ITS USE IN STUDIES OF OTHER G PROTEIN–COUPLED RECEPTORS

Slawomir Filipek,[1,6,7,8] David C. Teller,[2,6]
Krzysztof Palczewski,[3,4,5] and Ronald Stenkamp,[1,6]
Departments of [1]Biological Structure, [2]Biochemistry, [3]Ophthalmology, [4]Chemistry, [5]Pharmacology, and [6]Biomolecular Structure Center, University of Washington, Seattle, Washington 98195; email: teller@u.washington.edu; palczews@u.washington.edu; stenkamp@u.washington.edu
[7]International Institute of Molecular and Cell Biology and [8]Faculty of Chemistry, University of Warsaw, 02-109 Warsaw, Poland; email: sfilipek@iimcb.gov.pl

Key Words transmembrane protein, signal transduction, homology models, vision, phototransduction

■ **Abstract** G protein–coupled receptors (GPCRs) are integral membrane proteins that respond to environmental signals and initiate signal transduction pathways activating cellular processes. Rhodopsin is a GPCR found in rod cells in retina where it functions as a photopigment. Its molecular structure is known from cryo-electron microscopic and X-ray crystallographic studies, and this has reshaped many structure/function questions important in vision science. In addition, this first GPCR structure has provided a structural template for studies of other GPCRs, including many known drug targets. After presenting an overview of the major structural elements of rhodopsin, recent literature covering the use of the rhodopsin structure in analyzing other GPCRs will be summarized. Use of the rhodopsin structural model to understand the structure and function of other GPCRs provides strong evidence validating the structural model.

CONTENTS

INTRODUCTION

G protein–coupled receptors (GPCRs) are a large group of integral membrane proteins that provide molecular links between extracellular signals and intracellular processes (15, 36, 54, 61, 117, 121). Proteins in this class respond to stimuli such as molecular ligands or light and couple these to internal signal transduction systems involving G proteins. The receptors are sensors in molecular level communication systems connecting external signals with intracellular functions and pathways.

Several hundred GPCRs are found in tissues throughout the human body, where they respond to a range of signals and ligands. For instance, rhodopsin and the cone opsins respond to light, other receptors respond to small molecular signals such as dopamine, histamine, or serotonin, and yet other receptors bind to larger ligands such as angiotensins or chemokines. The roles of these GPCRs in fundamental cellular processes make them important drug targets. A large fraction of the clinically important drugs in use today are targeted toward GPCRs (6).

One of the most widely studied GPCRs is rhodopsin, the photopigment found in the visual system (32, 44, 72, 78, 81, 85, 94, 95, 102). This protein (with its retinal chromophore) is found in the outer segments of retinal rod cells. Absorption of a photon causes the retinal chromophore to change its conformation from *11-cis* to all-*trans*. This is accompanied by conformational changes in the protein that result in a binding site on its cytoplasmic surface for its cognate G protein, transducin (Gt) (30, 43, 71, 90). Once activated, the α subunit of Gt activates a phosphodiesterase that converts cyclic-GMP to GMP. Ion channels gated by cyclic-GMP then close, leading to a hyperpolarized cell that can initiate a nerve signal from the retina to the brain.

Biological, chemical, and physical studies of rhodopsin, both as a vision protein and as a prototypical GPCR, have taken advantage of its high abundance in bovine eyes. Many of the methods used to probe GPCR structure and function were first applied to rhodopsin because of its availability. This also contributed to the efforts leading to a three-dimensional crystal structure determination of rhodopsin (86).

The first part of this article provides an overview of the structural features of ground-state rhodopsin determined using X-ray crystallographic methods. More detailed summaries of the structure and function of rhodopsin are found in References 32 and 95. The second part of this review examines the application of the rhodopsin structural model in studies of other GPCRs.

STRUCTURAL STUDIES OF RHODOPSIN

The seven transmembrane helices in the structural core of GPCRs were first observed in electron microscopic studies of two-dimensional crystals of bovine rhodopsin (97). In 1997, a low-resolution view of the helices was obtained from cryo-electron microscopic studies of two-dimensional crystals of frog rhodopsin (114).

These basic structures, along with the higher-resolution bacteriorhodopsin models (26, 65, 66, 87, 110), supported efforts to computationally model many GPCRs (5, 89). In addition, other biophysical techniques, including NMR (1, 20, 27, 126), spin-label EPR (2, 3), and disulfide formation rates (18, 57), were also used to obtain structural information about these molecules.

Tetsuji Okada, in Palczewski's laboratory, succeeded in obtaining three-dimensional crystals of bovine rhodopsin (80) using protein purified with a protocol he developed (82). One important contribution to his success was his ability to stabilize the protein (and the crystal) by addition of Zn^{2+} in the purification. This is consistent with recent ideas coupling membrane protein stability with successful structural studies (93). Whether the techniques successful for crystallization of rhodopsin can be applied to other GPCRs is an open question.

The crystallographic structure was solved using multi-wavelength anomalous dispersion techniques on a mercury derivative (86). Crystals with high twinning ratios provided diffraction data sufficient to show the helical core of the structure, but not the connecting loops on either end of the molecule. Miyano and coworkers at the SPring8 synchrotron identified a crystal with a low twinning ratio and used it to obtain phases showing the entire molecule.

The structure was refined using standard crystallographic techniques for twinned crystals. Three coordinate sets are currently (June 2002) available from the Protein Data Bank (13). The set with identification code 1F88 is the model reported in the initial structure description (86). Model 1HZX was obtained by further refinement of 1F88 (112). Model 1L9H was obtained by refinement at slightly higher resolution, 2.6 Å (79). Small conformational differences between the models are well within the error estimates for these resolution limits. In the case of 1HZX and 1L9H, various nonprotein components of the structure have been added to the model: zinc, mercury, water, heptane-1,2,3-triol, β-nonyl glucoside, palmitoyl groups, and carbohydrate. These structures were determined at a higher resolution than previously seen for GPCRs, but these are not "high-resolution" structures. There is a serious lack of structural detail available at 2.8 or 2.6 Å resolution. The structural models can be expected to change as higher-resolution data become available.

The residue numbering scheme used in this article combines the generic system of Ballesteros & Weinstein (9) and the residue numbers for the particular receptor being discussed. As an example, consider K7.43[296]. The residue is number 7.43 in the generic numbering system and number 296 in the bovine rhodopsin sequence. In the former, the first digit denotes which helix contains the residue and the second number is an index locating the residue in the helix. The most conserved residue in each helix is assigned an index of 50, so residue Lys7.43 is located in helix 7 and is 7 (= 50–43) residues before the most conserved residue.

MOLECULAR STRUCTURE OF RHODOPSIN

Overview of the Structure

The crystallographic model of rhodopsin (Figure 1, see color insert) confirmed many of the previously known or assumed structural features for GPCRs. The seven transmembrane helices are arranged as seen in the cryo-electron microscopy studies, with the same topology as found in bacteriorhodopsin (46).

The N terminus of the protein is on the extracellular side of the membrane. This is also the intradiscal side in the membrane disks found in rod cells in the retina. The helices then traverse the membrane with odd-numbered helices going from the extracellular/intradiscal side to the intracellular/cytoplasmic side when passing along the polypeptide from the N terminus to the C terminus. There are three extracellular loops connecting helices on that side of the membrane. Along with the N-terminal tail, these make up the extracellular surface. Three cytoplasmic loops as well as the C-terminal region form the intracellular surface of the protein.

Cytoplasmic Loops

The temperature factors for the refined model show an interesting trend in that their values are higher toward the cytoplasmic side of the protein. The cytoplasmic surface of the protein is where its Gt will bind. Accordingly, the major conformational shifts initiating the signaling cascade should occur on this face. The structure is consistent with this in that one of the cytoplasmic loops (C-III) and part of the C-terminal tail are not seen in the electron density maps. The structural disorder accounting for this might be either static or dynamic. These parts of the protein appear to be more flexible, a characteristic that could be useful for rearranging a binding site for Gt.

Bent Helices

The transmembrane helices in rhodopsin are not regular α-helices. The helices are bent and contain segments with 3_{10}- or π-helical conformations. Detailed analyses of these distortions (92) have shown that in many of the helices, Pro residues are associated with the bends. The conformational restraints imposed on the backbone atoms of Pro by its side chain ring do not keep this residue from being located in helical regions of Φ-Ψ space. The major effect of Pro on a helical structure is removal of a hydrogen bond due to the lack of a hydrogen atom on the imino nitrogen. Although this destroys the continuity of the hydrogen-bonding pattern for a helix, it is not sufficient to cause a distortion in the helical structure. The methylene carbons of Pro account for that. In a potential α-helix, the carbonyl oxygen of a residue four positions before the Pro would be its hydrogen-bond acceptor. The δ-methylene of Pro is located where the hydrogen atom of an amino acid would normally be found. This is too close to the carbonyl oxygen, so to accommodate the packing of the atoms, the carbonyl group tilts away from the helical axis. This introduces a distortion in the backbone conformational angles that results in a bending of the entire helix. This type of distortion was pointed out in the past (67, 91, 96, 119), and it and related distortions are found in 6 of the 10 helical bends in rhodopsin.

The bend of helix II is different in that it is associated with an introduction of an extra residue in one turn of the helix. The residues involved in this turn take on backbone conformational angles characteristic of π-helices. This is a rare conformational state for residues found in protein structures, but the presence of an extra residue in this turn of helix is not a resolution artifact.

The turn of π-helix in helix II contains two adjacent Gly residues (G2.56[89] and G2.57[90]). This Gly-Gly pattern is conserved in many GPCRs (49). Gly residues, with no bulky side chains, can take on conformational angles that other residues cannot. However, the reason adjacent Gly residues are associated with an insertion of an extra residue in the helix is not apparent.

The bends in the helices, and the twists associated with them, might be hinge-points for relevant conformational changes associated with GPCR activation (119). The hydrogen-bond patterns are already broken at the bends, so further bending of the helices should have minor enthalpic effects. Several investigators have suggested that the helical bending has functional significance in these molecules, so it will be interesting to see if that idea is supported by structural studies of complexes of the receptors and their G proteins.

Chromophore Conformation and Binding Site

The importance of the conformational change in the chromophore has made it the object of extensive biophysical studies (59, 70). In the ground-state inactive form of the protein, the polyene segment of the chromophore is in the *11-cis* conformation. NMR and cross-linking studies (16, 17, 104, 106, 111) indicate that the β-ionone ring is oriented to give the *6-s-cis* conformer. This is consistent with the X-ray diffraction pattern.

The interactions of the chromophore with the surrounding protein residues are of interest because they influence the absorption spectrum of the protein/chromophore complex. In addition, they might serve as a model of ligand binding sites in other GPCRs. Hydrophobic groups largely cover the surface of the retinal binding cavity. The β-ionone ring is packed between two Phe residues (F5.47[212] and F6.44[261]), while the side chain of W6.48[265] is located in the center of the site with the chromophore bent around it in the *11-cis* conformation. Although the chromophore is completely buried inside the protein with no accessibility to the aqueous or membrane environment, there are several polar or charged protein side chains nearby. Burial of polar groups within the protein is thermodynamically costly, so their presence near the retinal is likely due to important structural or functional reasons. T3.33[118] and Y5.61[268] are located near the polyene tail, but the major charged group associated with the retinal is E3.28[113]. This residue is located near the Schiff base linkage of the chromophore with K7.43[296] and likely serves as a counterion when the Schiff base is protonated.

Major candidates for comparative structure/function studies with rhodopsin are the cone opsins, the related molecules in cone cells responsible for wavelength discrimination and color vision (102, 108). Description of those comparisons lies outside the scope of this review.

Activation and Interactions with G Proteins

After the quick *cis-trans* isomerization of the chromophore, rhodopsin passes through a set of conformational states that can be characterized spectroscopically (102). One of these photostates, Meta II, is the form of the protein capable of activating Gt. The α subunit of Gt then extends the signal transduction cascade. The structural changes leading to Meta II are believed to be similar in the activation of most GPCRs.

In the case of rhodopsin, structural information concerning activation is limited because we have a static, time-averaged view of only the initial state in the crystals. However, investigators have used mutagenesis to introduce Cys residues into rhodopsin for use in disulfide formation, accessibility, and spin-label experiments probing the structure of the activated and ground-state molecules (2, 3, 18, 57, 72, 78). These studies and others have shown which transmembrane helices must move during activation, and by how much. The interresidue distances obtained using these techniques are affected by chemical modifications and perturbations of the protein structure, but the experiments provide the best information currently available about the activation process, albeit without a lot of atomic level detail.

Several papers have appeared illustrating and summarizing the major structural principles connected with GPCR activation (3, 32, 43, 71, 78). The response of GPCRs to a signal is to expand and open up their structures on the cytoplasmic side of the membrane. This is accomplished by movement of the cytoplasmic ends of helices II, VI, and VII to create a binding crevice across the cytoplasmic surface of the protein. No reports are yet available describing detailed models of an activated receptor or a receptor/G protein complex.

MODELING OF OTHER GPCRs

The molecular model of bovine rhodopsin provides a useful structural framework for understanding the large amounts of structural and functional data compiled for the vision photopigments. The 2.8 Å resolution structural model has also been important for researchers focusing on the chemistry and biology of GPCRs. Because three-dimensional structures for membrane proteins are not readily available, computational modeling of GPCRs has been an active research area (5, 35, 47, 89, 118). Several important issues complicate modeling of GPCRs based on the model of rhodopsin. One is the low sequence identity between rhodopsin and other receptors (116, 118). Alignment of the amino acid sequences for these proteins is also complicated by variation in the lengths of the interhelical loops (84). In addition, the retinal binding pocket is completely buried in the protein and may make it an unsuitable model for the binding of ligands to most other GPCRs (69). If the dynamics of the protein permit occasional access to the binding site from the aqueous or membrane environment, this might not be an issue. Finally, the currently available structure is for the inactive form of the receptor, and this should be remembered if activated receptors are being modeled.

The next sections summarize how the three-dimensional crystal structure of bovine rhodopsin has been used to understand functional or structural characteristics of other GPCRs. Some of the studies cited record use of the rhodopsin model itself for understanding a particular GPCR while others describe homology models built to explain chemical and biochemical properties of the receptor of interest. Energy minimization and molecular dynamics calculations have been applied to some models, while in a few cases, energy functions were not used owing to concerns about whether those actions improve or degrade models (75) or whether the energy functions are appropriate for proteins in lipid environments (10).

Several of the papers describe comparisons of models with the X-ray crystallographic model. One measure of the fit of two models is the rms distance between equivalent atoms. Some of the models have large rms distances above 3 or 4 Å (31, 76, 99), which are substantially larger than rms distances calculated for structures of homologous soluble proteins obtained from X-ray diffraction studies.

The primary goal of the studies reviewed in the remainder of this article has been to understand a particular GPCR in terms of the rhodopsin structure. Another thing to learn from these studies is whether the structural model of rhodopsin is representative of other GPCRs. All molecular models, regardless of source, need to be consistent with information obtained from complementary approaches (76). Extrapolation of the structure to explain the properties of other GPCRs provides one means of validating the crystallographic model.

Characterization of Ligand Binding Sites

Knowing where ligands bind to a receptor is useful for designing new agonists or antagonists. Much is known about the binding sites in most GPCRs, but the more detailed crystal structure of rhodopsin has provided new views of the site for investigating ligand binding. As expected, the residues implicated in ligand binding are located near the extracellular or intradiscal surface of the receptor. Figure 2 (see color insert) shows the binding pocket residues identified for the angiotensin I receptor located on the rhodopsin crystal structure, as well as residues involved in the protein conformational change and G protein interactions.

One experimental approach for determining which residues contribute to a binding site is the substituted Cys-accessibility method (SCAM). Using mutagenesis techniques, Cys can be substituted at positions in the polypeptide chain, and their accessibility can then be determined by reaction with sulfhydryl reagents in the absence and presence of receptor ligands. In the case of human A_1 adenosine receptor (22), residues T7.35[270], A7.38[273], I7.39[274], T7.42[277], H7.43[278], N7.49[284], and Y7.53[288] were identified as ligand binding residues. The authors then noted the positions of these residues in the rhodopsin structure and concluded that they make up part of a contiguous binding pocket. SCAM was also used to investigate residues of helix VI of the μ, δ, and κ opioid receptors (124). Similarities in the accessibility patterns of these receptors and that of dopamine D2 receptor for helix VI (53) indicate that the secondary and tertiary structures of these receptors are quite similar.

Mapping studies of receptors with multiple binding sites have also been carried out. Allosteric effectors of M_1 muscarinic receptor are known, and two residues have been identified that affect binding of the effector gallamine (14). W3.28[101] and W7.35[400] (muscarinic receptor residue numbers) affect gallamine binding. W7.35[400] is located on the edge of a cleft in the extracellular part of a model obtained by threading the muscarinic receptor sequence onto the rhodopsin structure. W3.28[101] is buried inside the protein, on the other side of the β-strands. It likely plays some role in stabilizing the structure rather than in interacting directly with the allosteric effector.

Models based on the rhodopsin crystal structure have also been reported for GPCRs that respond to small molecule ligands. In the case of α_1-adrenergic receptor (122), an antagonist binding site has been shown to involve residues F7.35[308] and F7.39[312].

Also, three papers have described differences in the binding of agonists to human and rodent histamine receptors. In one comparing the binding characteristics of human and guinea pig histamine H_2-receptor (55), a homology model based on the rhodopsin structure helped predict that a nonconserved Asp in helix VII was responsible for the binding differences. Mutagenesis of Asp to Ala confirmed this idea. In another study comparing human and rat histamine H_3-receptors (107), an Asp on helix III could account for the difference in the binding profiles for the two receptors. A homology model with docked ligands, including histamine, helped in understanding the differences in binding. Finally, the contributions of residues in helix V to the binding of agonists to human H_3-receptor were investigated (115). In the homology model in this study, helix V had to be manually rotated and positioned to allow interactions between histamine and E5.46[206] that were known from other experiments.

Combinations of mutagenesis techniques, homology modeling, and other biochemical approaches have mapped out the binding sites and characterized several peptide binding GPCRs. Studies making use of the rhodopsin crystal structure have been reported for rat angiotensin II receptor of type 1 (in complex with angiotensin II) (77), bradykinin receptors (68, 88), opioid receptors (19), the cholecystokinin receptors (24, 29, 37), complement factor 5a receptor (33), human formyl peptide receptor (73), mouse gastrin-releasing peptide receptor (113), parathyroid hormone receptor (74), and human gonadotropin-releasing hormone receptor (50).

In all these cases, the biochemical data were reconciled with the homology models. Access to the ligand binding site and the structure of the extracellular loops are major issues for models of the peptide binding GPCRs because the peptides will likely come from the aqueous phase and the binding site might be large. The authors of these papers dealt with these in different ways. In some of the cases, only the transmembrane helices were included in the homology model. In others, special attention was given to one or more of the loops. Some were modeled by looking for equivalent structures in the Protein Data Bank for the loop peptides, some were modeled using ab initio approaches, and some were modeled using NMR-derived structures for peptide segments of the receptors.

The rhodopsin model was used in two studies of melanocortin-4 receptor. Hydropathy plots based on the sequences of human melanocortin receptors were used to map mutations onto the rhodopsin model to ensure that the sites were in the transmembrane region (125). Residues D3.25[122] and D3.29[126] in helix III and F6.51[261] and H6.54[264] in helix VI decreased the binding of the melanocyte-stimulating hormone. In a study of the mouse receptor (45), several residues involved in agonist and antagonist binding were identified using a model built before the crystallographic structure became available as well as one built using the crystal structure. E2.59[92], D3.25[114], and D3.29[118] are involved in melanocortin-based peptide binding. E2.59[92] and D3.29[118] interact with antagonists. F4.60[176], Y4.63[179], F6.52[254], and F6.57[259] interact with Phe on a melanocortin-based ligand.

Properties and Design of Ligands

Several research groups have used homology models based on the rhodopsin structure for structure/activity studies of GPCR ligands. The receptors involved in these studies include human A_3 adenosine receptor (10, 11), rat serotonin 5-HT_4 receptor (63, 64), and turkey $P2Y_1$ receptor (56). In the case of the adenosine receptor, a homology model was used in the design of a new ligand (10) that was experimentally proven to be an antagonist of the receptor. While the focus of these kinds of studies is on the properties of the ligands, they also provide insight into the effects of the ligands on receptor structure and signaling. A model of human CXCR4 chemokine receptor (missing the extracellular loop between helices IV and V) was generated for a study of cyclam and cicyclam antagonists (34). This chemokine receptor is a coreceptor for HIV, and the authors proceeded to test the binding of antiviral compounds to it. Docking of AMD3100, an antiviral bicyclam, into the model placed one cyclam close to D4.60[171] and one close to D6.58[262]. The agonist activity is felt to be due to the ligand preventing the receptor from altering its conformation into an active one. This idea of restraining or constraining the structure of the receptor will return in the next portion of this review.

Another example showing the use of a rhodopsin-based homology model in a molecular design process concerns the development of mutant receptors (neoceptors) with selective binding properties for synthetic ligands (52). The approach used identifies positions in a receptor binding site where an altered ligand can be accommodated by a mutated receptor. In the case reported, a mutant adenosine A_3 receptor (H7.43[272]E) was developed, and the binding of amine-modified nucleosides was measured. A homology model of the neoceptor was extensively used in this study, which hoped to bind a 3′-amino-3′-deoxyadenosine via a direct electrostatic interaction between the ligand and the substituted glutamic acid. The derivative bound well, but its binding was not explained by the proposed interaction. Instead, loss of repulsion from His is felt to be a more important contribution to the altered binding.

The DRY Motif

Sequence comparisons and other studies have pointed out the existence of a tripeptide sequence, DRY, in helix III of a large number of GPCRs (the residue numbers for these amino acids are D3.49, R3.50, and Y3.51). If the motif is expanded to be (E/D)R(Y/W), the pattern's conservation is greatly extended. The location of the ERY sequence in rhodopsin is shown in Figure 3 (see color insert). The overall picture common to many GPCRs is that R3.50 is hydrogen bonded to a carboxylate side chain at position 3.49 and to one or two residues in helix VI. In rhodopsin, those are E6.30 and T6.34. Removal of these interactions often results in constitutive activation of the receptor.

Several studies combining mutagenesis and homology modeling have attempted to identify the residue(s) interacting with R3.50 and what their physiological effects might be. Groups have investigated this motif in rat μ opioid receptor (51, 62), β_2-adrenergic receptor (7), α_{1b}-adrenergic receptor (41), lutropin/choriogonadotropin receptor (4), and 5-hydroxytryptamine 2A serotonin receptor (100). Mutagenesis of D3.49 in the opioid receptor results in activation, and neutralization of the charge on E6.30 in the adrenergic receptors does likewise. The current favorite interpretation of these results is that the electrostatic–hydrogen bonding interactions restrain the position of R3.50. Removal of the restraints results in straightening of helix VI and a separation of helices III and VI to form an activated receptor.

As attractive as this bit of molecular machinery is, there are two or three suggestions from other modeling studies that need to be kept in mind when considering the biological function of the DRY motif. First, the charged groups interacting with R3.50 will not be found in all GPCRs. In the case of lutropin receptor (98), mutagenesis methods have ruled out an interaction between R3.50 and Asp meant to substitute for E6.30 in rhodopsin. Second, the Kaposi's sarcoma–associated herpesvirus GPCR (48) has no negatively charged side chains for R3.50. This receptor has high levels of basal activity, and the authors interpret this to mean the receptor can serve as a model of the activated receptor. Third, in the M-3 muscarinic acetylcholine receptor (120), helices V and VI move closer together upon

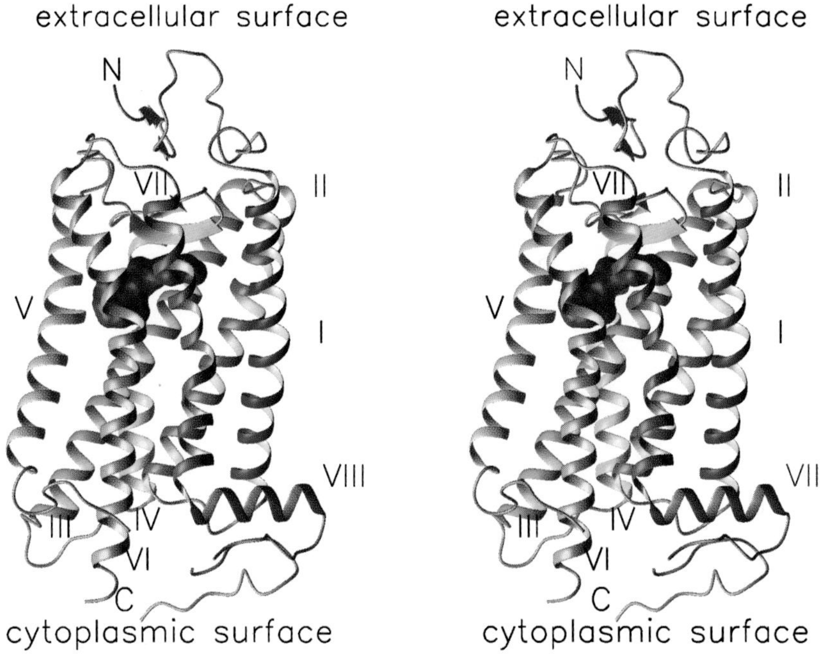

Figure 1 Stereoview showing the secondary and tertiary structure of rhodopsin. The helices are labeled I–VIII. The retinal chromophore is shown in red. The molecule is drawn in the orientation usual for GPCRs with the extracellular surface at the top. Figure drawn using MOLMOL (60).

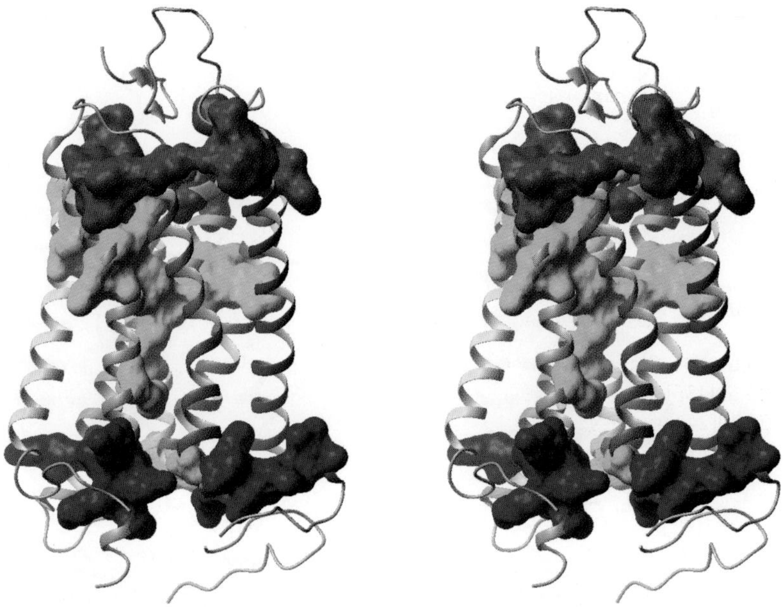

Figure 2 Stereoview showing regions of the at1 angiotensin receptor involved in ligand binding (red), signal propagation (green), and G protein binding (blue). The residues with atoms shown as spheres are from the crystallographic model of rhodopsin, but their positions were chosen on the basis of a tabulation of important residues in the at1 angiotensin receptor (77). Figure drawn using MOLMOL (60).

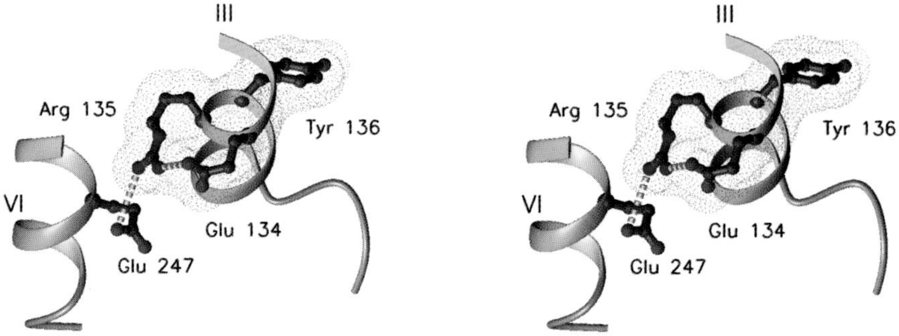

Figure 3 Stereoview of the ERY motif in rhodopsin. E3.49[134], R3.50[135], and Y3.51[136] are shown in ball-and-stick mode enclosed in a van der Waals envelope. E6.30[247] is also shown with its hydrogen bond to R3.50[135]. Figure drawn using MOLMOL (60).

activation. This, instead of a separation of those helices, may be the real activation movement, but further structural studies are needed to confirm this.

Modeling of Bent Helices

As expected from the extensive sequence comparisons and alignments of GPCRs, the receptors share several three-dimensional structural elements that correlate with similarities in their functions as signaling proteins. One such structural feature concerns the bent helices that help differentiate GPCRs from the other seven transmembrane helical proteins. As described above, six of the transmembrane helices in rhodopsin are bent, with the bending often associated with Pro or a pair of adjacent Gly residues. These residues are often conserved in the GPCR sequences (9, 12, 54, 123), and it is anticipated that as a consequence, the bent helices will also be conserved. Several modeling and computational exercises support this view.

Modeling studies of β_2-adrenergic receptor (7) included computer simulations of a bend in helix VI of the receptor. While the presence of Pro causes the helix to bend, the amount of bending in the simulations is less than that found in the rhodopsin crystal structure. The authors point out other interactions in the inactive, nonsignaling form of the protein that can constrain the helix to be more bent.

Further simulations of helices containing Pro residues were carried out employing the chemokine receptor CCR5 (39). In this receptor, helix II contains a TXP sequence, and molecular dynamics calculations indicate that the helix could bend at these residues. In fact, the Thr located two residues before Pro increases the bend of the helix by 10° in simulations, presumably owing to hydrogen bonding possibilities between the side chain hydroxyl and the main chain atoms. The corresponding sequence in bovine rhodopsin is GGF, which is another sequence motif associated with bending of helices. It is interesting that the two receptors have different amino acid sequences in this region, both of which support a helical bend. However, the equivalent sequences, although located in the same positions, bend the helices in different directions. In rhodopsin, the helix is bent toward helix I. In the adrenergic receptor, the helix bends toward helix III. It would be interesting to know if this difference is real and how it fits with the idea of "structural mimicry" proposed by Ballesteros et al. (8) as a mechanism for generating selective receptors with a common tertiary structure and signaling mechanism.

A study of human $VPAC_1$ receptor (58) pointed out another issue about the Pro-kinked helices that requires attention in modeling GPCRs. This receptor is a member of the secretin class of GPCRs. The conserved Pro residues in this class of receptors are located in helices IV, V, and VI, not helices V, VI, and VII as in rhodopsin. This requires special attention in generating homology models because the bent helices are likely to be functionally important. In the case of $VPAC_1$ receptor, mutations of Pro residues in helices IV, V, and VI affected the properties of the receptor. Alteration of Pro in helix IV reduced the activity of the receptor, while the mutations in helices V and VI increased the ability of the receptor to initiate cAMP production. In a related study of human prostacyclin receptor (109), it was discovered that mutation of Pro and Ala residues in helices IV, V, and VI reduced ligand binding while similar mutations for Pro residues in helices I, II, III, IV, VI, and VII affected activation of the receptor. Sorting out why these mutations have their different effects requires more research.

For some modeling exercises, straightening of the helices will require consideration when the amino acid sequence contains no special sequence motifs that support bends. Dopamine D2 receptor has no Pro in helix I, and simulations were undertaken to see how the sequence might affect the dynamics of that structural element (101). The Gly residue equivalent to Pro in helix I in rhodopsin gives the dopamine receptor's helix more flexibility than would other amino acid substitutions, but it is not clear whether Gly would permit as much bending as found in rhodopsin.

Transmembrane Helices III, VI, and VII

As described in the discussion of the DRY motif, interactions between helices III and VI are involved in restraining the structure of GPCRs in the inactive, nonsignaling state. The set of interacting residues can be expanded beyond R3.50 and its hydrogen-bonded partners to include additional residues, some of which are found in helix VII. The following examples present experimental evidence for interactions among the three helices and how they are associated with the activity of the receptors. What is uncertain is whether these studies on different GPCRs provide information relevant to all GPCRs or whether variation exists among the receptors in the structural elements and interactions giving rise to signaling. At this time, there is no strong evidence requiring radically different activation schemes for different receptors.

One part of the basic interaction between helices III, VI, and VII is illustrated by a study of human B_2 bradykinin receptor (68). In this receptor, W6.48[256] controls the balance between active and inactive conformations. Through its interactions with inverse agonists, it can control the relative motion of the three helices. It was hypothesized that an interaction between N3.35[113] and W6.48[256] would stabilize an inactive conformation. W6.48[256]F and Q mutants were constitutively active, consistent with the hypothesized interaction in the wild-type protein. A lack of constitutive activity for the W6.48[256]A mutant was taken as an indication that additional residues and interactions might be involved in activation of the receptor.

Additional interactions between helix III and helix VI have been described for human lutropin receptor (103). The natural mutant L3.43[457]R for this receptor is constitutively active. This leucine is highly conserved and is located in helix III. Mutagenesis of this residue shows that activation is associated with a positively charged amino acid at this position. Modeling shows that R, K, or H at 3.43[457] interact electrostatically with D6.44[578] in helix VI. This perturbs the interhelical interactions between III, VI, and VII and gives rise to the activity of the receptor.

Further examples can be found where individual parts of helices III, VI, and VII are implicated in activation. In opioid receptors, the major residues involved in antagonist binding are located in helix VI (124). In thyrotropin receptor (40), an interaction between D6.44[633] and N7.49[674] serves as an on/off switch for activation. Constitutive activity in muscarinic receptors can be generated by mutations of residues in helices III and VI (105). Finally, for C5a receptor, it has been proposed that binding of agonists via interactions with residues on helices III and VII causes them to change in their relative orientations and allow helix VI to separate from them (33).

Activation and Binding of G Proteins and Other Macromolecules

As was stated in the first part of this article, movement of the transmembrane helices is a major component in the transition from an inactive ground state to an active signaling form of the receptor. However, it is not sufficient for binding of G proteins. The cytoplasmic loops of the receptors are also involved in recognition and binding of the G proteins. Using the bovine rhodopsin structure to model the interaction with the G proteins is difficult. The cytoplasmic loops are flexible and disordered in the crystals, enough so that residues in the loop are not seen in the electron density maps. The resulting protein structure cannot provide much information about the structure of the loops relevant for the function of the receptor. Nevertheless, two studies have been reported [one for the V2 vasopressin receptor (28) and one for the α_{1b}-adrenergic receptor (42)] where modeling of the receptors with the cytoplasmic loops started with the rhodopsin crystal structure. The models found some use in the design of mutants, but the range of structures available for the loops complicated the analysis of the mutants.

While interactions between GPCRs and their G proteins are of prime importance, interactions between neighboring receptors have become the object of much speculation and experiment. One group has used modeling techniques to identify regions on the hydrophobic surfaces of GPCRs that might participate in protein-protein interactions leading to the formation of dimers and other oligomers (23, 38). Two sets of surfaces on the proteins, one containing helices II and III and the other containing V and VI, are suggested as regions that might be involved in the formation of hetero- and homodimers.

Investigators are also using molecular structure information in approaching questions concerning expression and localization of GPCRs. For example, differences in expression of murine and human vasopressin V_2 receptors in COS cells are due to a single amino acid change at position 100 at the interface between helix II and the second extracellular loop (83). Examination of a homology model shows that Lys at this position can make hydrogen bonds with residues in helices I, II, and III, whereas Asp (the change leading to reduced expression) would destabilize the protein structure via electrostatic clashes with other carboxylates.

Assessment of Alternative Models

The 2.8 Å resolution crystal structure of bovine rhodopsin has had a major impact on GPCR research, both in providing structural information for modeling and structure analysis, and in showing that structural studies of proteins in this family are possible. The large number of papers using the experimental structure for modeling studies shows the level of interest in using structural information for understanding GPCR function. In the absence of experimentally determined structures, modeling provides a means of generating that base of structural information. However, it is often difficult to assess the quality and relevance of molecular models. Concerns about the structures of the loops in rhodopsin have been expressed (21, 69), as have questions about the validity of the rhodopsin structure as a general template for GPCRs (25, 76, 77).

There are five papers available at this time (June 2002) containing comparisons between models based on the rhodopsin crystal structure and models obtained using other approaches. In a modeling study of cholecystokinin receptor, the rhodopsin crystal structure was compared with an earlier receptor model generated by the authors (25). Imposition of the receptor sequence on the rhodopsin template introduced many structural clashes between residues within the transmembrane region, and these made its use for further studies unattractive.

Another instance allowing comparison of alternative models is in a paper describing structure/function studies of melanocortin receptor (45). This particular study was already underway before the rhodopsin crystal structure was reported. The investigators were using a homology model to generate lists of possible ligand-receptor interactions. They built another homology model based on the rhodopsin crystal structure and could see no significant differences in the possible ligand-protein interactions. Each model was adequate for this purpose.

Another example where two GPCR models were directly compared was in an investigation of human B_2 bradykinin receptor (68). Again, two models were built. The "experimental model," was built using a previous model as a template. In addition, two of the helices, VI and VII, were rotated slightly in response to binding information from mutant receptors. The "rhodopsin-like model" was built using the crystal structure as a template. Molecular dynamics simulations of ligand binding resulted in different binding modes for each ligand/model pair, along with differences in the predictions about which residues were in contact with the ligand. Resolution of the differences did not favor either model. However, in another part of the study, the proximity of two residues was predicted on the basis of earlier work. Replacement of the two residues with His, followed by exposure of the protein to zinc ions, showed that the residues were close to one another. The results were more in line with the rhodopsin-like model than with earlier published rhodopsin models.

Finally, two examples are available where differences between two structural models were clarified using mutagenesis techniques. In the case of α_{1b}-adrenergic receptor (41), two models were generated, one being a homology model based on the rhodopsin crystal structure and the other being generated ab initio. The prediction from the first model is that R3.50 (in the DRY motif) would interact with E6.30. The ab initio model predicts that R3.50 interacts with D2.50 and Y7.53. Mutagenesis experiments were consistent with the prediction from the homology model based on the rhodopsin structure.

Studies of a possible sodium binding site on dopamine D_2 receptor (75) also provide strong evidence supporting the validity of the rhodopsin crystal structure as a template for GPCRs. An older model of the receptor indicated that D2.50[80], S3.39[121], N3.42[124], and S7.46[420] form the base of a pyramidal sodium binding site with N7.49[423] at the apex. A new model based on the rhodopsin structure placed N7.45[419] (backbone oxygen) at the apex, with D2.50[80], S3.39[121], N7.45[419], and S7.46[420] at the vertices. Mutagenesis of S3.39[121] and S7.46 [420] alters the receptor's dependence on sodium, but mutations of N3.42[124] do not.

CONCLUSIONS

Rhodopsin remains a molecule of importance and interest to two large research areas. The X-ray crystallographic model has been useful for understanding the biochemical and biophysical properties of rhodopsin in the vision system, but it has also proven important for extrapolations into the structure and function of GPCRs. This is consistent with the idea that the basic structure and properties of rhodopsin are representative of proteins in the entire GPCR family. The secondary and tertiary structures of rhodopsin are consistent with experimental information available for other receptors, and the tertiary structure is robust and not overly sensitive to the packing forces it encounters in two-dimensional or three-dimensional crystals. The homology models based on the crystal structure indicate that the details of the helical bends likely apply to many proteins in the family, as do other structural details.

The major structural feature of rhodopsin that likely cannot be carried over directly into studies of other GPCRs is the ligand binding site. Currently, computational techniques appear to be providing suitable alterations in the binding site, but additional three-dimensional crystal structures of GPCRs are needed to determine how varied the binding sites might be. Progress in obtaining additional crystal structures might also provide the strongest evidence that the rhodopsin structure is typical for this class of proteins. One crystallographic technique capable of differentiating between seemingly similar structures such as those of rhodopsin and bacteriorhodopsin is that of molecular replacement. The method provides an orientation and position of a structural model in a new crystal form. It is a powerful technique for determining the structure of a protein related to one with a known structure. In the case of rhodopsin, bacteriorhodopsin and prior models of rhodopsin were inadequate templates for solving the structure. If those techniques can be used with the current rhodopsin model to solve new crystal structures of GPCRs, it will be strong proof of the generality of the structure.

ACKNOWLEDGMENTS

This research was supported in part by U.S. Public Health Service grants EY01730 and EY0861 from the National Eye Institute, National Institutes of Health, grant GM63020 from the NIGMS, an unrestricted Grant from Research to Prevent Blindness, Inc., to the Department of Ophthalmology at the University of Washington, and a grant from the E.K. Bishop Foundation.

The *Annual Review of Biophysics and Biomolecular Structure* is online at http://biophys.annualreviews.org

LITERATURE CITED

1. Albert AD, Watts A, Spooner P, Groebner G, Young J, Yeagle PL. 1997. A distance measurement between specific sites on the cytoplasmic surface of bovine rhodopsin in rod outer segment disk membranes. *Biochim. Biophys. Acta* 1328:74–82
2. Altenbach C, Cai KW, Klein-Seetharaman J, Khorana HG, Hubbell WL. 2001. Structure and function in rhodopsin: mapping light-dependent changes in distance between residue 65 in helix TM1 and residues in the sequence 306–319 at the cytoplasmic end of helix TM7 and in helix H8. *Biochemistry* 40:15483–92
3. Altenbach C, Klein-Seetharaman J, Cai KW, Khorana HG, Hubbell WL. 2001. Structure and function in rhodopsin: mapping light-dependent changes in distance between residue 316 in helix 8 and residues in the sequence 60–75, covering the cytoplasmic end of helices TM1 and TM2 and their connection loop CL1. *Biochemistry* 40:15493–500
4. Ascoli M, Fanelli F, Segaloff DL. 2002. The lutropin/choriogonadotropin receptor, a 2002 perspective. *Endocr. Rev.* 23: 141–74
5. Baldwin JM, Schertler GFX, Unger VM. 1997. An alpha-carbon template for the transmembrane helices in the rhodopsin family of G-protein-coupled receptors. *J. Mol. Biol.* 272:144–64
6. Ballesteros J, Palczewski K. 2001. G protein-coupled receptor drug discovery: implications from the crystal structure of rhodopsin. *Curr. Opin. Drug Discov. Dev.* 4:561–74
7. Ballesteros JA, Jensen AD, Liapakis G, Rasmussen SGF, Shi L, et al. 2001. Activation of the beta(2)-adrenergic receptor involves disruption of an ionic lock between the cytoplasmic ends of transmembrane segments 3 and 6. *J. Biol. Chem.* 276:29171–77
8. Ballesteros JA, Shi L, Javitch JA. 2001. Structural mimicry in G protein-coupled receptors: implications of the high-resolution structure of rhodopsin for structure-function analysis of rhodopsin-like receptors. *Mol. Pharmacol.* 60:1–19
9. Ballesteros JA, Weinstein H. 1995. Integrated methods for the construction of three-dimensional models and computational probing of structure-function relations in G protein coupled receptors. *Methods Neurosci.* 25:366–428
10. Baraldi PG, Cacciari B, Moro S, Romagnoli R, Ji XD, et al. 2001. Fluorosulfonyl- and bis-(beta-chloroethyl)amino-phenylamino functionalized pyrazolo 4,3-e1, 2,4-triazolo 1,5-c pyrimidine derivatives: irreversible antagonists at the human A(3) adenosine receptor and molecular modeling studies. *J. Med. Chem.* 44:2735–42
11. Baraldi PG, Cacciari B, Moro S, Spalluto G, Pastorin G, et al. 2002. Synthesis, biological activity, and molecular modeling investigation of new pyrazolo 4,3-e-1,2,4-triazolo 1,5-c pyrimidine derivatives as human A(3) adenosine receptor antagonists. *J. Med. Chem.* 45:770–80
12. Barlow DJ, Thornton JM. 1988. Helix geometry in proteins. *J. Mol. Biol.* 201:601–19
13. Berman HM, Westbrook J, Feng Z, Gilliland G, Bhat TN, et al. 2000. The Protein Data Bank. *Nucleic Acids Res.* 28:235–42
14. Birdsall NJM, Lazareno S, Popham A, Saldanha J. 2001. Multiple allosteric sites on muscarinic receptors. *Life Sci.* 68: 2517–24
15. Bockaert J, Claeysen S, Becamel C, Pinloche S, Dumuis A. 2002. G protein-coupled receptors: dominant, players in

cell-cell communication. *Int. Rev. Cytol.* 212:63–132

16. Borhan B, Souto ML, Imai H, Shichida Y, Nakanishi K. 2000. Movement of retinal along the visual transduction path. *Science* 288:2209–12
17. Buss V, Kolster K, Terstegen F, Vahrenhorst R. 1998. Absolute sense of twist of the C12–C13 bond of the retinal chromophore in rhodopsin—semiempirical and nonempirical calculations of chiroptical data. *Angew. Chem. Int. Ed.* 37:1893–95
18. Cai KW, Klein-Seetharaman J, Altenbach C, Hubbell WL, Khorana HG. 2001. Probing the dark state tertiary structure in the cytoplasmic domain of rhodopsin: proximities between amino acids deduced from spontaneous disulfide bond formation between cysteine pairs engineered in cytoplasmic loops 1, 3, and 4. *Biochemistry* 40:12479–85
19. Chaturvedi K, Christoffers KH, Singh K, Howells RD. 2000. Structure and regulation of opioid receptors. *Biopolymers* 55:334–46
20. Chopra A, Yeagle PL, Alderfer JA, Albert AD. 2000. Solution structure of the sixth transmembrane helix of the G-protein-coupled receptor, rhodopsin. *Biochim. Biophys. Acta* 1463:1–5
21. Chung DA, Zuiderweg ERP, Fowler CB, Soyer OS, Mosberg HI, Neubig RR. 2002. NMR structure of the second intracellular loop of the alpha 2A adrenergic receptor: evidence for a novel cytoplasmic helix. *Biochemistry* 41:3596–604
22. Dawson ES, Wells JN. 2001. Determination of amino acid residues that are accessible from the ligand binding crevice in the seventh transmembrane-spanning region of the human A(1) adenosine receptor. *Mol. Pharmacol.* 59:1187–95
23. Dean MK, Higgs C, Smith RE, Bywater RP, Snell CR, et al. 2001. Dimerization of G-protein-coupled receptors. *J. Med. Chem.* 44:4595–614
24. Ding XQ, Miller LJ. 2001. Characterization of the type A cholecystokinin receptor hormone-binding domain: use of contrasting and complementary methodologies. *Peptides* 22:1223–28
25. Ding XQ, Pinon DI, Furse KE, Lybrand TP, Miller LJ. 2002. Refinement of the conformation of a critical region of charge-charge interaction between cholecystokinin and its receptor. *Mol. Pharmacol.* 61:1041–52
26. Edman K, Nollert P, Royant A, Belrhali H, Pebay-Peyroula E, et al. 1999. High-resolution X-ray structure of an early intermediate in the bacteriorhodopsin photocycle. *Nature* 401:822–26
27. Eilers M, Ying WW, Reeves PJ, Khorana HG, Smith SO. 2002. Magic angle spinning nuclear magnetic resonance of isotopically labeled rhodopsin. *Methods Enzymol.* 343:212–22
28. Erlenbach I, Kostenis E, Schmidt C, Serradeil-Le Gal C, Raufaste D, et al. 2001. Single amino acid substitutions and deletions that alter the G protein coupling properties of the V2 vasopressin receptor identified in yeast by receptor random mutagenesis. *J. Biol. Chem.* 276:29382–92
29. Escrieut C, Gigoux V, Archer E, Verrier S, Maigret B, et al. 2002. The biologically crucial C terminus of cholecystokinin and the non-peptide agonist SR-146,131 share a common binding site in the human CCK1 receptor—evidence for a crucial role of Met-121 in the activation process. *J. Biol. Chem.* 277:7546–55
30. Fain GL, Matthews HR, Cornwall MC. 1996. Dark adaptation in vertebrate photoreceptors. *Trends Neurosci.* 19:502–7
31. Fanelli F, Themmen APN, Puett D. 2001. Lutropin receptor function: insights from natural, engineered, and computer-simulated mutations. *IUBMB Life* 51: 149–55
32. Filipek S, Stenkamp RE, Teller DC, Palczewski K. 2003. G-protein-coupled receptor rhodopsin: a prospectus. *Annu. Rev. Physiol.* 65:851–79

33. Gerber BO, Meng EC, Dotsch V, Baranski TJ, Bourne HR. 2001. An activation switch in the ligand binding pocket of the C5a receptor. *J. Biol. Chem.* 276:3394–400
34. Gerlach LO, Skerlj RT, Bridger GJ, Schwartz TW. 2001. Molecular interactions of cyclam and bicyclam non-peptide antagonists with the CXCR4 chemokine receptor. *J. Biol. Chem.* 276:14153–60
35. Gershengorn MC, Osman R. 2001. Minireview: insights into G protein-coupled receptor function using molecular models. *Endocrinology* 142:2–10
36. Gether U, Kobilka BK. 1998. G protein-coupled receptors. II. Mechanism of agonist activation. *J. Biol. Chem.* 273:17979–82
37. Giragossian C, Mierke DF. 2002. Intermolecular interactions between cholecystokinin-8 and the third extracellular loop of the cholecystokinin-2 receptor. *Biochemistry* 41:4560–66
38. Gouldson PR, Dean MK, Snell CR, Bywater RP, Gkoutos G, Reynolds CA. 2001. Lipid-facing correlated mutations and dimerization in G-protein coupled receptors. *Protein Eng.* 14:759–67
39. Govaerts C, Blanpain C, Deupi X, Ballet S, Ballesteros JA, et al. 2001. The TXP motif in the second transmembrane helix of CCR5—a structural determinant of chemokine-induced activation. *J. Biol. Chem.* 276:13217–25
40. Govaerts C, Lefort A, Costagliola S, Wodak SJ, Ballesteros JA, et al. 2001. A conserved Asn in transmembrane helix 7 is an on/off switch in the activation of the thyrotropin receptor. *J. Biol. Chem.* 276:22991–99
41. Greasley PJ, Fanelli F, Rossier O, Abuin L, Cotecchia S. 2002. Mutagenesis and modelling of the alpha(1b)-adrenergic receptor highlight the role of the helix 3/helix 6 interface in receptor activation. *Mol. Pharmacol.* 61:1025–32
42. Greasley PJ, Fanelli F, Scheer A, Abuin L, Nenniger-Tosato M, et al. 2001. Mutational and computational analysis of the alpha(1b)-adrenergic receptor. Involvement of basic and hydrophobic residues in receptor activation and G protein coupling. *J. Biol. Chem.* 276:46485–94
43. Hamm HE. 2001. How activated receptors couple to G proteins. *Proc. Natl. Acad. Sci. USA* 98:4819–21
44. Hargrave PA, McDowell JH. 1992. Rhodopsin and phototransduction—a model system for G-protein-linked receptors. *FASEB J.* 6:2323–31
45. Haskell-Luevano C, Cone RD, Monck EK, Wan YP. 2001. Structure activity studies of the melanocortin-4 receptor by in vitro mutagenesis: identification of agouti-related protein (AGRP), melanocortin agonist and synthetic peptide antagonist interaction determinants. *Biochemistry* 40:6164–79
46. Henderson R, Baldwin JM, Ceska TA, Zemlin F, Beckmann E, Downing KH. 1990. Model for the structure of bacteriorhodopsin based on high-resolution electron cryomicroscopy. *J. Mol. Biol.* 213:899–929
47. Herzyk P, Hubbard RE. 1998. Combined biophysical and biochemical information confirms arrangement of transmembrane helices visible from the three-dimensional map of frog rhodopsin. *J. Mol. Biol.* 281:741–54
48. Ho HH, Ganeshalingam N, Rosenhouse-Dantsker A, Osman R, Gershengorn MC. 2001. Charged residues at the intracellular boundary of transmembrane helices 2 and 3 independently affect constitutive activity of Kaposi's sarcoma-associated herpesvirus G protein-coupled receptor. *J. Biol. Chem.* 276:1376–82
49. Horn F, Vriend G, Cohen FE. 2001. Collecting and harvesting biological data: the GPCRDB and NucleaRDB information systems. *Nucleic Acids Res.* 29:346–49
50. Hovelmann S, Hoffmann SH, Kuhne R, ter Laak T, Reilander H, Beckers T. 2002. Impact of aromatic residues within transmembrane helix 6 of the human

gonadotropin-releasing hormone receptor upon agonist and antagonist binding. *Biochemistry* 41:1129–36
51. Huang P, Li J, Chen CG, Visiers I, Weinstein H, Liu-Chen LY. 2001. Functional role of a conserved motif in TM6 of the rat mu opioid receptor: Constitutively active and inactive receptors result from substitutions of Thr6.34(279) with Lys and Asp. *Biochemistry* 40:13501–9
52. Jacobson KA, Gao ZG, Chen AS, Barak D, Kim SA, et al. 2001. Neoceptor concept based on molecular complementarity in GPCRs: a mutant adenosine A(3) receptor with selectively enhanced affinity for amine-modified nucleosides. *J. Med. Chem.* 44:4125–36
53. Javitch JA, Shi L, Liapakis G. 2002. Use of the substituted cysteine accessibility method to study the structure and function of G protein-coupled receptors. *Methods Enzymol.* 343:137–56
54. Ji T-H, Grossmann M, Ji I. 1998. G Protein-coupled receptors. I. Diversity of receptor-ligand interactions. *J. Biol. Chem.* 273:17299–302
55. Kelley MT, Burckstummer T, Wenzel-Seifert K, Dove S, Buschauer A, Seifert R. 2001. Distinct interaction of human and guinea pig histamine H-2-receptor with guanidine-type agonists. *Mol. Pharmacol.* 60:1210–25
56. Kim HS, Barak D, Harden TK, Boyer JL, Jacobson KA. 2001. Acyclic and cyclopropyl analogues of adenosine bisphosphate antagonists of the P2Y(1) receptor: structure-activity relationships and receptor docking. *J. Med. Chem.* 44:3092–108
57. Klein-Seetharaman J, Hwa J, Cai KW, Altenbach C, Hubbell WL, Khorana HG. 2001. Probing the dark state tertiary structure in the cytoplasmic domain of rhodopsin: proximities between amino acids deduced from spontaneous disulfide bond formation between Cys316 and engineered cysteines in cytoplasmic loop 1. *Biochemistry* 40:12472–78
58. Knudsen SM, Tams JW, Fahrenkrug J. 2001. Functional roles of conserved transmembrane prolines in the human VPAC-(1) receptor. *FEBS Lett.* 503:126–30
59. Kochendoerfer GG, Lin SW, Sakmar TP, Mathies RA. 1999. How color visual pigments are tuned. *Trends Biochem. Sci.* 24:300–5
60. Koradi R, Billeter M, Wüthrich K. 1996. MOLMOL: a program for display and analysis of macromolecular structures. *J. Mol. Graph.* 14:51–55
61. Lefkowitz RJ. 1998. G protein-coupled receptors. III. New roles for receptor kinases and beta-arrestins in receptor signaling and desensitization. *J. Biol. Chem.* 273:18677–80
62. Li J, Huang P, Chen CG, de Riel JK, Weinstein H, Liu-Chen LY. 2001. Constitutive activation of the mu opioid receptor by mutation of D3.49(164), but not D3.32(147): D3.49(164) is critical for stabilization of the inactive form of the receptor and for its expression. *Biochemistry* 40:12039–50
63. Lopez-Rodriguez ML, Benhamu B, Viso A, Murcia M, Pardo L. 2001. Study of the bioactive conformation of novel 5-HT4 receptor ligands: influence of an intramolecular hydrogen bond. *Tetrahedron* 57:6745–49
64. Lopez-Rodriguez ML, Murcia M, Benhamu B, Viso A, Campillo M, Pardo L. 2001. 3-D-QSAR/CoMFA and recognition models of benzimidazole derivatives at the 5-HT4 receptor. *Bioorg. Med. Chem. Lett.* 11:2807–11
65. Luecke H, Richter HT, Lanyi JK. 1998. Proton transfer pathways in bacteriorhodopsin at 2.3 Å resolution. *Science* 280:1934–37
66. Luecke H, Schobert B, Richter HT, Cartailler JP, Lanyi JK. 1999. Structure of bacteriorhodopsin at 1.55 Å resolution. *J. Mol. Biol.* 291:899–911
67. Macarthur MW, Thornton JM. 1991. Influence of proline residues on protein conformation. *J. Mol. Biol.* 218:397–412
68. Marie J, Richard E, Pruneau D, Paquet

JL, Siatka C, et al. 2001. Control of conformational equilibria in the human B-2 bradykinin receptor—modeling of nonpeptidic ligand action and comparison to the rhodopsin structure. *J. Biol. Chem.* 276:41100–11
69. Marshall GR. 2001. Peptide interactions with G-protein coupled receptors. *Biopolymers* 60:246–77
70. Mathies RA, Lugtenburg J. 2000. The primary photoreaction of rhodopsin. *Handb. Biol. Phys.* 3:55–90
71. Meng EC, Bourne HR. 2001. Receptor activation: What does the rhodopsin structure tell us? *Trends Pharmacol. Sci.* 22:587–93
72. Menon S, Han M, Sakmar TP. 2001. Rhodopsin: structural basis of molecular physiology. *Physiol. Rev.* 81:1659–88
73. Mills JS, Miettinen HM, Cummings D, Jesaitis AJ. 2000. Characterization of the binding site on the formyl peptide receptor using three receptor mutants and analogs of Met-Leu-Phe and Met-Met-Trp-Leu-Leu. *J. Biol. Chem.* 275:39012–17
74. Monticelli L, Mammi S, Mierke DF. 2002. Molecular characterization of a ligand-tethered parathyroid hormone receptor. *Biophys. Chem.* 95:165–72
75. Neve KA, Cumbay MG, Thompson KR, Yang R, Buck DC, et al. 2001. Modeling and mutational analysis of a putative sodium-binding pocket on the dopamine D-2 receptor. *Mol. Pharmacol.* 60:373–81
76. Nikiforovich GV, Galaktionov S, Balodis J, Marshall GR. 2001. Novel approach to computer modeling of seven-helical transmembrane proteins: current progress in the test case of bacteriorhodopsin. *Acta Biochim. Pol.* 48:53–64
77. Nikiforovich GV, Marshall GR. 2001. 3D model for TM region of the AT-1 receptor in complex with angiotensin II independently validated by site-directed mutagenesis data. *Biochem. Biophys. Res. Commun.* 286:1204–11
78. Okada T, Ernst OP, Palczewski K, Hofmann KP. 2001. Activation of rhodopsin: new insights from structural and biochemical studies. *Trends Biochem. Sci.* 26:318–24
79. Okada T, Fujiyoshi Y, Silow M, Navarro J, Landau EM, Shichida Y. 2002. Functional role of internal water molecules in rhodopsin revealed by X-ray crystallography. *Proc. Natl. Acad. Sci. USA* 99:5982–87
80. Okada T, Le Trong I, Fox BA, Behnke CA, Stenkamp RE, Palczewski K. 2000. X-ray diffraction analysis of three-dimensional crystals of bovine rhodopsin obtained from mixed micelles. *J. Struct. Biol.* 130:73–80
81. Okada T, Palczewski K. 2001. Crystal structure of rhodopsin: implications for vision and beyond. *Curr. Opin. Struct. Biol.* 11:420–26
82. Okada T, Takeda K, Kouyama T. 1998. Highly selective separation of rhodopsin from bovine rod outer segment membranes using combination of divalent cation and alkyl(thio)glucoside. *Photochem. Photobiol.* 67:495–99
83. Oksche A, Leder G, Valet S, Platzer M, Hasse K, et al. 2002. Variant amino acids in the extracellular loops of murine and human vasopressin V2 receptors account for differences in cell surface expression and ligand affinity. *Mol. Endocrinol.* 16:799–813
84. Otaki JM, Firestein S. 2001. Length analyses of mammalian G-protein-coupled receptors. *J. Theor. Biol.* 211:77–100
85. Palczewski K, ed. 2000. *Methods in Enzymology*. Vol. 315: *Vertebrate Phototransduction and the Visual Cycle. Part A.* SanDiegeo, CA: Academic
86. Palczewski K, Kumasaka T, Hori T, Behnke CA, Motoshima H, et al. 2000. Crystal structure of rhodopsin: a G protein-coupled receptor. *Science* 289:739–45
87. Pebay-Peyroula E, Rummel G, Rosenbusch JP, Landau EM. 1997. X-ray structure of bacteriorhodopsin at 2.5 Å from

microcrystals grown in lipidic cubic phases. *Science* 277:1676–81
88. Piserchio A, Prado GN, Zhang R, Yu J, Taylor L, et al. 2002. Structural insight into the role of the second intracellular loop of the bradykinin 2 receptor in signaling and internalization. *Biopolymers* 63:239–46
89. Pogozheva ID, Lomize AL, Mosberg HI. 1997. The transmembrane 7-alpha-bundle of rhodopsin: distance geometry calculations with hydrogen bonding constraints. *Biophys. J.* 72:1963–85
90. Polans A, Baehr W, Palczewski K. 1996. Turned on by Ca^{2+}! The physiology and pathology of Ca^{2+}-binding proteins in the retina. *Trends Neurosci.* 19:547–54
91. Ri Y, Ballesteros JA, Abrams CK, Oh S, Verselis VK, et al. 1999. The role of a conserved proline residue in mediating conformational changes associated with voltage gating of Cx32 gap junctions. *Biophys. J.* 76:2887–98
92. Riek RP, Rigoutsos I, Novotny J, Graham RM. 2001. Non-alpha-helical elements modulate polytopic membrane protein architecture. *J. Mol. Biol.* 306:349–62
93. Rosenbusch JP. 2001. Stability of membrane proteins: relevance for the selection of appropriate methods for high-resolution structure determinations. *J. Struct. Biol.* 136:144–57
94. Sakmar TP. 2002. Structure of rhodopsin and the superfamily of seven-helical receptors: the same and not the same. *Curr. Opin. Cell Biol.* 14:189–95
95. Sakmar TP, Menon ST, Marin EP, Awad ES. 2002. Rhodopsin: insights from recent structural studies. *Annu. Rev. Biophys. Biomol. Struct.* 31:443–84
96. Sankararamakrishnan R, Vishveshwara S. 1992. Geometry of proline-containing alpha-helices in proteins. *Int. J. Pept. Protein Res.* 39:356–63
97. Schertler GFX, Villa C, Henderson R. 1993. Projection structure of rhodopsin. *Nature* 362:770–72
98. Schulz A, Bruns K, Henklein P, Krause G, Schubert M, et al. 2000. Requirement of specific intrahelical interactions for stabilizing the inactive conformation of glycoprotein hormone receptors. *J. Biol. Chem.* 275:37860–69
99. Shacham S, Topf M, Avisar N, Glaser F, Marantz Y, et al. 2001. Modeling the 3D structure of GPCRs from sequence. *Med. Res. Rev.* 21:472–83
100. Shapiro DA, Kristiansen K, Weiner DM, Kroeze WK, Roth BL. 2002. Evidence for a model of agonist-induced activation of 5-hydroxytryptamine 2A serotonin receptors that involves the disruption of a strong ionic interaction between helices 3 and 6. *J. Biol. Chem.* 277:11441–49
101. Shi L, Simpson MM, Ballesteros JA, Javitch JA. 2001. The first transmembrane segment of the dopamine D2 receptor: accessibility in the binding-site crevice and position in the transmembrane bundle. *Biochemistry* 40:12339–48
102. Shichida Y, Imai H. 1998. Visual pigment: G-protein-coupled receptor for light signals. *Cell. Mol. Life Sci.* 54:1299–315
103. Shinozaki H, Fanelli F, Liu XB, Jaquette J, Nakamura K, Segaloff DL. 2001. Pleiotropic effects of substitutions of a highly conserved leucine in transmembrane helix III of the human lutropin/choriogonadotropin receptor with respect to constitutive activation and hormone responsiveness. *Mol. Endocrinol.* 15:972–84
104. Singh D, Hudson BS, Middleton C, Birge RR. 2000. Conformation and orientation of the retinyl chromophore in rhodopsin: A critical evaluation of recent NMR data on the basis of theoretical calculations results in a minimum energy structure consistent with all experimental data. *Biochemistry* 40:4201–4
105. Spalding TA, Burstein ES. 2001. Constitutively active muscarinic receptors. *Life Sci.* 68:2511–16
106. Spooner PJR, Sharples JM, Verhoeven MA, Lugtenburg J, Glaubitz C, Watts A.

2002. Relative orientation between the β-ionone ring and the polyene chain for the chromophore of rhodopsin in native membranes. *Biochemistry* 41:7549–55

107. Stark H, Sippl W, Ligneau X, Arrang JM, Ganellin CR, et al. 2001. Different antagonist binding properties of human and rat histamine H-3 receptors. *Bioorg. Med. Chem. Lett.* 11:951–54
108. Stenkamp RE, Filipek S, Driessen CAGG, Teller DC, Palczewski K. 2002. Crystal structure of rhodopsin: a template for cone visual pigments and other G protein-coupled receptors. *Biochim. Biophys. Acta Biomembr.* 1565:168–82
109. Stitham J, Martin KA, Hwa J. 2002. The critical role of transmembrane prolines in human prostacyclin receptor activation. *Mol. Pharmacol.* 61:1202–10
110. Subramaniam S, Henderson R. 2000. Crystallographic analysis of protein conformational changes in the bacteriorhodopsin photocycle. *Biochim. Biophys. Acta Bioenerg.* 1460:157–65
111. Tan Q, Lou JH, Borhan B, Karnaukhova E, Berova N, Nakanishi K. 1997. Absolute sense of twist of the C12-C13 bond of the retinal chromophore in bovine rhodopsin based on exciton-coupled CD spectra of 11,12-dihydroretinal analogues. *Angew. Chem. Int. Ed.* 36:2089–93
112. Teller DC, Okada T, Behnke CA, Palczewski K, Stenkamp RE. 2001. Advances in determination of a high-resolution three-dimensional structure of rhodopsin, a model of G-protein-coupled receptors (GPCRs). *Biochemistry* 40:7761–72
113. Tokita K, Katsuno T, Hocart SJ, Coy DH, Llinares M, et al. 2001. Molecular basis for selectivity of high affinity peptide antagonists for the gastrin-releasing peptide receptor. *J. Biol. Chem.* 276:36652–63
114. Unger VM, Hargrave PA, Baldwin JM, Schertler GFX. 1997. Arrangement of rhodopsin transmembrane alpha-helices. *Nature* 389:203–6
115. Uveges AJ, Kowal D, Zhang YX, Spangler TB, Dunlop J, et al. 2002. The role of transmembrane helix 5 in agonist binding to the human H3 receptor. *J. Pharmacol. Exp. Ther.* 301:451–58
116. Vakser IA, Jiang SL. 2002. Strategies for modeling the interactions of transmembrane helices of G protein-coupled receptors by geometric complementarity using the GRAMM computer algorithm. *Methods Enzymol.* 343:313–28
117. Vaughan M. 1998. G protein-coupled receptors minireview series. *J. Biol. Chem.* 273:17297
118. Visiers I, Ballesteros JA, Weinstein H. 2002. Three-dimensional representations of G protein-coupled receptor structures and mechanisms. *Methods Enzymol.* 343:329–71
119. Visiers I, Braunheim BB, Weinstein H. 2000. Prokink: a protocol for numerical evaluation of helix distortions by proline. *Protein Eng.* 13:603–6
120. Ward SDC, Hamdan FF, Bloodworth LM, Wess J. 2002. Conformational changes that occur during M-3 muscarinic acetylcholine receptor activation probed by the use of an in situ disulfide cross-linking strategy. *J. Biol. Chem.* 277:2247–57
121. Watson S, Arkinstall S. 1994. *The G-Protein Linked Receptor Factsbook*. New York: Academic
122. Waugh DJJ, Gaivin RJ, Zuscik MJ, Gonzalez-Cabrera P, Ross SA, et al. 2001. Phe-308 and Phe-312 in transmembrane domain 7 are major sites of alpha(1)-adrenergic receptor antagonist binding—imidazoline agonists bind like antagonists. *J. Biol. Chem.* 276:25366–71
123. Woolfson DN, Williams DH. 1990. The influence of proline residues on alpha-helical structure. *FEBS Lett.* 277:185–88
124. Xu W, Li J, Chen CG, Huang P, Weinstein H, et al. 2001. Comparison of the amino acid residues in the sixth transmembrane domains accessible in the binding-site crevices of mu, delta, and kappa opioid receptors. *Biochemistry* 40:8018–29
125. Yang Y, Fong TM, Dickinson CJ, Mao

C, Li JY, et al. 2000. Molecular determinants of ligand binding to the human melanocortin-4 receptor. *Biochemistry* 39:14900–11

126. Yeagle PL, Choi G, Albert AD. 2001. Studies on the structure of the G-protein-coupled receptor rhodopsin including the putative G-protein binding site in unactivated and activated forms. *Biochemistry* 40:11932–37

Annu. Rev. Biophys. Biomol. Struct. 2003. 32:399–424
doi: 10.1146/annurev.biophys.32.110601.141854
First published online as a Review in Advance on January 24, 2003

PROTEOME ANALYSIS BY MASS SPECTROMETRY*

P. Lee Ferguson and Richard D. Smith
Biological Sciences Division, Pacific Northwest National Laboratory, Richland, Washington 99352; email: patrick.ferguson@pnl.gov; rds@pnl.gov

Key Words proteomics, FTICR, protein identification, chromatography, two-dimensional HPLC

■ **Abstract** The coupling of high-performance mass spectrometry instrumentation with highly efficient chromatographic and electrophoretic separations has enabled rapid qualitative and quantitative analysis of thousands of proteins from minute samples of biological materials. Here, we review recent progress in the development and application of mass spectrometry–based techniques for the qualitative and quantitative analysis of global proteome samples derived from whole cells, tissues, or organisms. Techniques such as multidimensional peptide and protein separations coupled with mass spectrometry, accurate mass measurement of peptides from global proteome digests, and mass spectrometric characterization of intact proteins hold great promise for characterization of highly complex protein mixtures. Advances in chemical tagging and isotope labeling techniques have enabled quantitative analysis of proteomes, and highly specific isolation strategies have been developed aimed at selective analysis of posttranslationally modified proteins.

CONTENTS

INTRODUCTION

The increasing availability of fully or partially sequenced genomes for a variety of organisms (including humans) has ushered in a new era of biological research aimed toward the holistic understanding of life processes. This new science, sometimes termed systems biology, has as its goal an eventual understanding of the information flow from gene to biological function, with explicit treatment of an organism's response to perturbations (chemical, environmental, or otherwise) at the basic molecular level. One potential promise of such research is an eventual understanding of disease processes in humans, leading to development of novel and targeted drugs and therapies aimed at improving the quality of human life (14, 61). This ambitious undertaking relies heavily on the capability to examine in a massively parallel fashion the identity, concentration, function, and interaction of a wide variety of biological macromolecules. High-throughput DNA sequencing techniques have now matured as stable technologies and have enabled our access to fully characterized organism genomes. These genomes take the form of large databases containing the nucleotide sequence code derived from an organism's full DNA complement, providing a basic foundation for studies of biological systems.

While the genome of an organism may be considered to be static over short timescales, this is certainly not the case for the protein complement for which an organism's genes code through open reading frames (ORFs), collectively known as the proteome. Proteins are the penultimate functional macromolecules in cellular systems, and the fraction of potentially expressed proteins actually present in a cell or tissue at a given time is strongly dependent on the organism's environment and physiological state. While some information about relative protein expression levels may be obtained from analysis of an organism's transcriptome (or mRNA complement), e.g., via high-throughput cDNA microarray analysis (50), measured mRNA levels do not necessarily correlate strongly with the corresponding activity or abundance of proteins in biological tissues (23). In addition, the function of proteins may be extensively modified by posttranslational modifications, such as phosphorylation (6), or by complexation with other biomolecules or small molecules (16, 27), and this information is not accessible by transcriptome analysis.

It is therefore not surprising that proteome analysis has become a key enabling technology in the emerging science of systems biology. Proteomics may be defined as the direct qualitative and quantitative analysis of the full complement of proteins present in an organism, tissue, or cell under a given set of physiological or environmental conditions. The advantage of proteomics lies in the ability to directly examine the biomolecules and assemblies of biomolecules that are most responsible for the function of biological systems. For many years, protein chemists have focused on the sequential isolation, structural characterization, and functional assay of individual proteins in model biological systems. Although such research has been invaluable in advancing functional biochemistry, such approaches are not suitable for interfacing with "global" biological function studies. Over the past several years, a multitude of novel methods have been developed for reducing highly complex protein mixtures derived from, for example, cell lysates and tissue homogenates to (more or less) meaningful datasets of protein identities and relative expression levels. The qualitative characterization methods have in common a general reliance on mass spectrometry as a sensitive and highly versatile analytical tool for protein identification, and its role for quantitative measurements is increasing rapidly. Apart from the methods for characterizing identity and expression levels of protein complements, significant efforts have been made to enable the parallel functional assay of multitudes of individual proteins and protein complexes in a high-throughput fashion.

Several excellent and comprehensive reviews concern various aspects of proteome analysis (1, 18, 36, 46). Our objective is not a comprehensive survey of proteome analysis methods, but instead a review of the current state of the art in the rapidly growing field of proteomics. We first describe recent developments in the separation and analysis of proteins from complex biological samples using electrophoretic, chromatographic, and mass spectrometric techniques, with special regard to methods providing high coverage of known or predicted proteomes and sensitive analysis of low-abundance proteins. Here we discuss the most powerful current methods for assigning individual protein identities in analysis of complex mixtures. We then discuss techniques for relative quantitative analysis of proteomes by isotope labeling and mass spectrometry. Finally, current approaches for the enrichment, identification, and quantitation of proteins that have been modified posttranslationally are described.

QUALITATIVE PROTEOME ANALYSIS

Two-Dimensional Gel Electrophoresis/MALDI-MS

Prior to the mid-1990s, proteins were characterized by amino acid sequencing, using the Edman degradation technique (10). This process, although powerful for individual protein characterization, is laborious and not well suited for qualitative analysis of complex protein mixtures. The first approach to characterization of complex protein mixtures was separation by two-dimensional gel electrophoresis (2DE) followed by in-gel tryptic digestion of visible protein spots and subsequent offline analysis of individual digest by matrix-assisted laser desorption ionization mass spectrometry (MALDI-MS) (26, 54). The fundamental principle is that proteins are separated first by isoelectric point focusing on an immobilized pH gradient strip, then in an orthogonal direction by gel electrophoresis. After digestion with protease (usually trypsin), individual protein spots (now converted to peptides) are removed from the gel matrix and analyzed serially by MALDI-MS. Protein identification is based on the principle of peptide mapping, where observed peptide peak patterns in individual spectra are compared with predicted digest fragments of proteins contained in a database (70). With use of the 2DE/MALDI-MS approach, 502 distinct proteins (of 1742 possible gene products) have been identified in the bacterium *Haemophilus influenzae* (31).

The 2DE/MALDI-MS technique, although robust and relatively straightforward, does suffer from some fundamental limitations for comprehensive proteome characterization. First, there are finite limits to the hydrophobicity, isoelectric point (pI), and molecular weight range of proteins resolvable using conventional two-dimensional gels. Conventional two-dimensional gels can separate on the order of ~1000 components during a single separation; however, incomplete separation of proteins on the gel can lead to overlapping spots and subsequently to problems with protein identification using the MALDI-MS peptide mass mapping approach. In addition, proteins derived from the same gene may migrate to different locations on the gel owing to posttranslational processing. Perhaps the most significant limitation of the 2DE/MALDI-MS approach for qualitative global proteome analysis is the difficulty in observing low copy number gene products such as transcription factors, which may be biologically significant. A general principle of the 2DE approach is that protein spots must be visible after gel staining in order to be subsequently identified by MALDI-MS. 2DE was unable to detect proteins with codon bias values below 0.1 in yeast (generally corresponding to low-abundance proteins, constituting the majority of potential gene products in this organism), up to the protein loading limit of the two-dimensional gel used (21).

Liquid-Phase Separations Coupled to Mass Spectrometry

Given the limitations of the 2DE method outlined above for performing comprehensive qualitative proteome analysis, several techniques have been developed with the aim of enabling more comprehensive, less biased proteome analyses based on liquid separations coupled online with mass spectrometry. This online interfacing has been enabled by the development of the electrospray ionization interface to mass spectrometry in the 1980s (12, 66) and its subsequent refinement for low flow rate capillary interfacing during the 1990s (64). In general, complex mixtures of proteins (or more commonly, peptides) derived from whole cell lysates or tissue homogenates are separated by microcapillary format liquid chromatography or electrophoresis prior to electrospray ionization and subsequent mass analysis.

In most cases, the peptide mass mapping approach described above for 2DE/MALDI-MS is not useful for protein identification using online liquid separations coupled to mass spectrometry for comprehensive proteome analysis because protease digestion of global protein mixtures is typically performed prior to the separation step. A powerful alternative approach is the use of tandem mass spectrometry (MS/MS) to induce fragmentation of individual tryptic peptides after online liquid phase separation. Typically this is performed using triple-quadrupole (TQ), quadrupole ion trap (QIT), Fourier transform ion-cyclotron resonance (FTICR), or quadrupole time-of-flight (QTOF) mass spectrometers. It is now well established that low-energy collision-induced dissociation (CID) fragmentation of peptides provides useful information for identification related to the amino acid sequence. In high-throughput proteome analyses, this information becomes valuable for searching constrained protein databases with defined protease digest and peptide CID fragmentation rules. In the most commonly used approach for global proteome analysis, uninterpreted CID spectra of experimental peptides are matched against theoretical fragmentation patterns of all (typically) tryptic peptides derived from the proteins in the database using either cross-correlation (11, 68) or probability (49) scoring. These methods can provide highly confident protein identifications in complex mixtures, typically requiring only a single to at most several peptide MS/MS spectra to identify a protein in a constrained database (69). Because the MS/MS spectral matching technique does not rely on complete (or near complete) coverage of the tryptic map of all peptides from individual proteins for confident identification, it is much more tolerant of posttranslational modifications and single amino acid substitutions than the peptide mass mapping approach typically used with 2DE/MALDI-MS. Next generation, intelligent database searching tools are currently under development aimed at further extending the utility of the MS/MS spectral matching technique for proteins that have undergone site-specific mutation, posttranslational modification, or nonspecific protease cleavage (58).

While qualitative analyses of tryptic digests can be performed using conventional, low-pressure, single-dimension microcapillary high-performance liquid chromatography (HPLC) coupled with MS and MS/MS, typical applications have so far been limited to simple mixtures of proteins (38). This is in part due to the limited peak capacity typically achieved for simple single-dimension separations. For the analysis of global protein isolates containing potentially thousands of individual proteins, and subsequently tens to hundreds of thousands of tryptic peptides, any truly comprehensive analytical technique must be capable of detecting a great many components over a wide dynamic range.

Two-Dimensional Separations Coupled with Mass Spectrometry

One approach for obtaining high-resolution separation of peptides prior to mass spectrometric detection is an orthogonal combination of cation exchange and reversed-phase chromatography, in a microcapillary format (Figure 1). This approach, termed multidimensional protein identification technology (MudPIT) by its developers (62, 65), consists of repeated ($\sim$15) sequential-step elutions of global tryptic peptides from strong cation exchange (SCX) resin with slow reversed-phase solvent gradients inserted between each salt elution step. On each successive reversed-phase gradient elution profile, the QIT mass spectrometer is operated in data-dependent MS/MS mode, generating data to be searched against a given

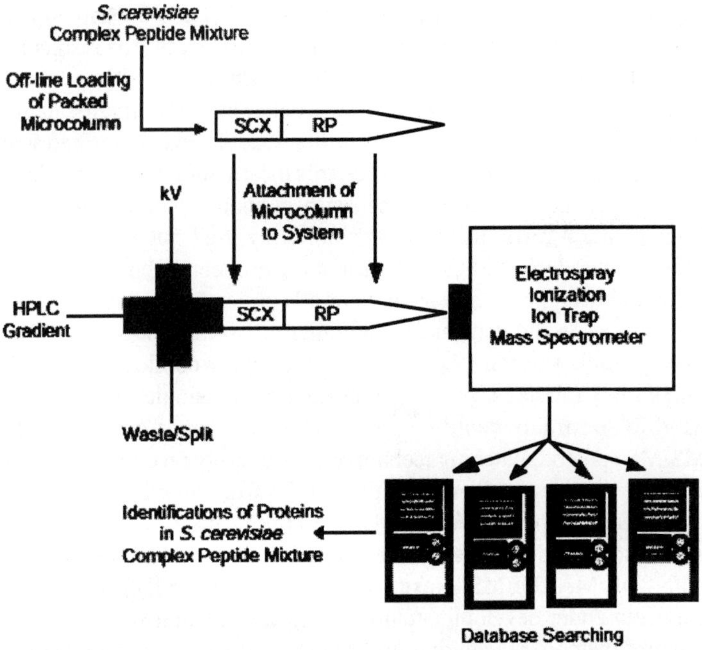

Figure 1 Diagram of the multidimensional protein identification technology (MudPIT) method. This technique involves loading of strong cation-exchange (SCX) beads packed into a microcapillary column with tryptic peptides derived from a global proteome digest (yeast in this case). Peptide fractions are eluted from the SCX beads onto the reversed-phase (RP) packing material in a stepwise manner using a salt gradient. Between salt steps, an organic solvent gradient is used to elute peptides from the RP column into the electrospray MS/MS system. Figure reproduced from Reference 62 by permission of Nature Publishing Group.

database using the SEQUEST algorithm (11). The MudPIT method is effective in managing the complexity of global proteome analyses, with a reported chromatographic peak capacity of ~3000 as currently implemented (65). The approach is generally more robust than an equivalent offline coupling of preparative cation exchange chromatography with capillary reversed-phase HPLC (24), as sample losses due to transfer and manipulation between the separation systems are avoided entirely and the process can be fully automated using readily available analytical equipment.

Using the MudPIT approach, Washburn et al. (62) reported identification of 1484 proteins from yeast (*Saccharomyces cerevisiae*), representing 24% coverage of the predicted ORFs of this organism. Proteins were identified from all known subcellular compartments, with a wide range of functional classifications, and over a dynamic range of $\sim 10^4$ (65). Although some bias toward detection of abundant proteins was noted, a large fraction (53%) of the identified yeast proteins had codon adaptation indices below 0.2, indicating that low-abundance proteins were efficiently detected using MudPIT (62). The sample preparation steps for the MudPIT approach were specifically designed to reduce bias against hydrophobic membrane-bound proteins, and many of these difficult proteins were observed during the analysis (19% of proteins with three or more transmembrane domains) (62).

The MudPIT approach developed by Yates and coworkers does have some limitations. Because it depends on data-dependent MS/MS spectral acquisition for individual tryptic peptides, the technique is ultimately limited by the rate at which the mass spectrometer can switch between MS and MS/MS modes and cycle among potential coeluting precursor ions. Past this limit, any increase in overall peak capacity must be accompanied by a longer overall analysis time. Analysis of a single global tryptic digest already requires nearly 30 h to complete using the MudPIT technique (65); this obviously places a severe strain on high-throughput applications. As currently implemented, protein identification is based solely on SEQUEST score, somewhat limiting confidence in fully automated database searching. Also, posttranslational modifications and site-specific mutations are not routinely accessible in MudPIT analyses.

Several alternative methods for achieving two-dimensional liquid-phase separations of peptides prior to mass spectrometric detection have been proposed for application to proteome analysis. Opiteck et al. (45) interfaced size exclusion chromatography to fast reversed-phase HPLC in order to separate peptides derived from tryptic digests of individual proteins prior to mass spectrometric detection. The chromatographic peak capacity achieved was 495, suitable for analysis of simple protein mixtures when interfaced with quadrupole mass spectrometry (45). The utility of this method could be extended to more complex proteome samples by the substitution of high-performance MS and MS/MS instrumentation (such as QTOF or FTICR-MS) for the quadrupole mass spectrometer.

Step-gradient elution solid-phase microextraction has been coupled to online capillary electrophoresis in order to achieve two-dimensional separation of complex protein digest samples prior to MS/MS analysis (59). The optimized microfluidic design incorporated in this technique enabled highly sensitive (subfemtomole) detection and identification of proteins using the SEQUEST algorithm. The authors reported positive identification of 80%–90% of the proteins associated with the yeast ribosome (75 total proteins) using the solid-phase microextraction/capillary electrophoresis technique. This technique is useful for characterizing proteins from complexes and subcellular fractions. However, the utility of this strategy for global proteome characterization has not yet been demonstrated.

Recently, Lee et al. (31a) have developed a multiplexed capillary HPLC separation method coupled offline to MALDI-QTOF mass spectrometry, designed specifically for the analysis of global proteome samples. This technique incorporates offline preparative cation-exchange chromatography for peptide mixture prefractionation, followed by simultaneous analysis of up to four different fractions by capillary reversed-phase HPLC. This strategy allows a fourfold increase in sample throughput and relies on both MS and MS/MS analysis using the MALDI-QTOF instrument for quantitative and qualitative analysis of proteins. Initial efforts using this method have resulted in identification and relative quantitation of 38 proteins from yeast.

The HPLC-FTICR Accurate Mass and Time Tag Approach to Global Proteome Characterization

Another approach, which has been recently developed to enable high-throughput, comprehensive characterization of global proteomes, is the accurate mass and time tag (AMT tag) strategy. This technology [which has been reviewed in depth recently (55)] consists of a range of individual analytical techniques and experimental protocols designed to rapidly manage the identification of large numbers of proteins in highly complex mixtures. The basic premise is that given the high mass accuracy (low ppm) achievable using FTICR (or perhaps next generation TOF-MS in combination with information from peptide chromatographic elution times), a large fraction of individual tryptic peptides derived from proteins may serve as unique markers of their parent proteins (7). The achievement of routine 1 ppm mass accuracy can provide significant utility for defining unique peptide mass tags in both bacterial and eukaryotic systems (Figure 2). Achievement of higher mass accuracy and the use of selective peptide subfractionation (e.g., cysteine affinity tags) further extend the promise of this approach (7).

In operation, the AMT tag strategy relies on a number of individual technologies for optimum performance when analyzing complex global proteome samples. In general, proteins are isolated from whole-cell lysates and digested using trypsin. Peptides are then separated by a single dimension of high-pressure (10,000 psi), high-efficiency, reversed-phase microcapillary HPLC prior to detection by mass spectrometry (52, 53). This technique provides efficient separation of peptides, over a relatively long (150–200 min) gradient, with reported chromatographic peak capacities between 650 and 1000 (52, 53). Initial validation of peptides for use as AMT tags is a two-stage process (56). First, the separation described above is coupled to a QIT mass spectrometer operated in data-dependent MS/MS mode, and peptides are tentatively identified using the SEQUEST algorithm as described above (Figure 3*A,B*). Unique peptides tentatively identified using this process are designated as potential mass tags (PMTs) pending validation using accurate mass measurements by FTICR. Multiple individual QIT "shotgun" analyses are necessary in order to increase the number of PMTs generated using data-dependent MS/MS (56). These measurements also define the relative HPLC elution times for the PMTs. Identical samples can be analyzed by HPLC-MS/MS with different m/z ranges and "dynamic exclusion" of previously selected MS/MS precursor ions. In addition, cells grown under different environmental conditions and in different growth phases may be subjected to the capillary HPLC-MS/MS analysis described above in order to examine pools of proteins that are differentially expressed in

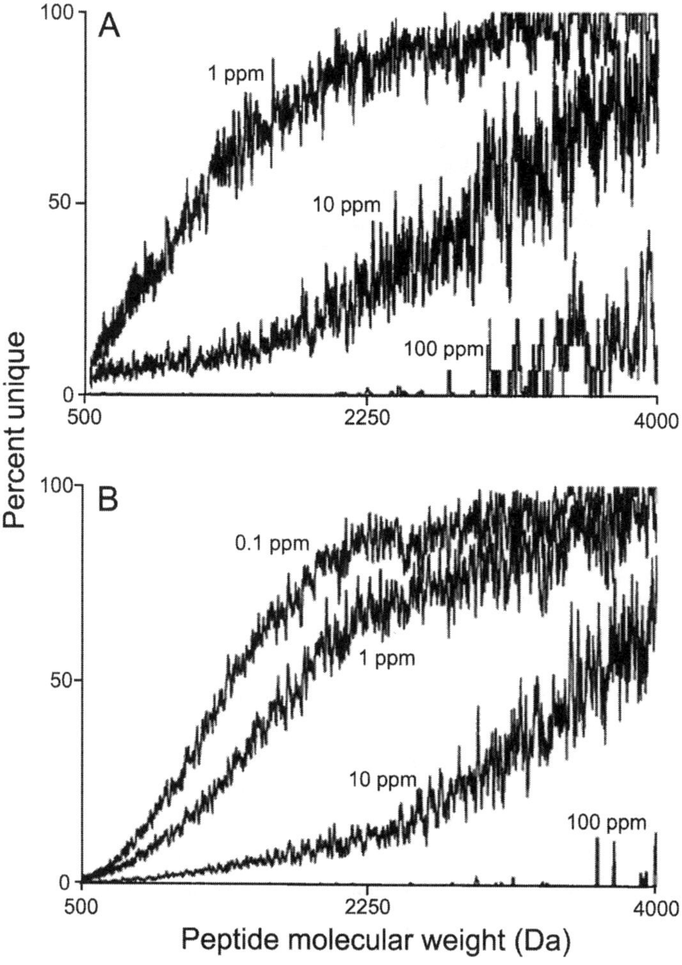

Figure 2 Theoretical percent unique peptides as a function of molecular weight at several levels of mass measurement accuracy for (*A*) the bacterium *D. radiodurans* and (*B*) the yeast *S. cerevisiae*, assuming an ideal tryptic digestion.

cells. Finally, offline prefractionation of peptide mixtures generated from global cell lysate tryptic digests may be performed by cation exchange or size exclusion chromatography prior to HPLC-MS/MS analysis. For the bacterium *Deinococcus radiodurans*, more than 200 individual capillary HPLC-MS/MS analyses were conducted using QIT instrumentation in order to generate a large number of PMTs (56).

PMTs generated using the approach described above are subsequently tested for validation as AMT tags using capillary HPLC as described above coupled to high field (11.4 T) FTICR (Figure 3*C*,*D*). Proteome samples previously analyzed

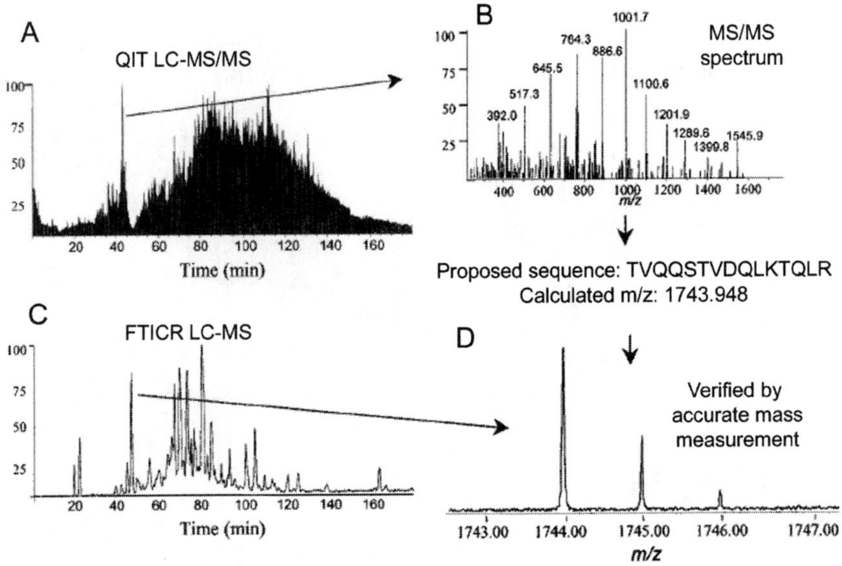

Figure 3 The procedure involved in establishing an accurate mass and time (AMT) tag. (*A*) A tryptically digested proteome sample is analyzed by capillary LC-MS/MS using a QIT instrument. (*B*) Peptides meeting predetermined criteria are automatically selected for MS/MS analysis. The resulting sequence-related information is used to generate a potential mass tag (PMT) via searching of protein or DNA databases using, e.g., the SEQUEST algorithm. (*C*) In order to validate this PMT as an AMT tag, an identical proteome sample is analyzed by capillary LC-MS using a high field FTICR mass spectrometer. (*D*) An AMT tag is generated when a peptide is observed to elute at the same time in both the QIT and FTICR LC-MS analyses, with a measured mass corresponding to the calculated mass (within a set mass accuracy) of the PMT.

by capillary HPLC with QIT MS/MS detection are then run once again using FTICR detection. Unique peptides eluting at the same time and having a measured mass within a set mass accuracy of the calculated value of a PMT generated using the QIT are designated as validated AMT tags, which can subsequently be used to uniquely identify the protein from which it was derived without MS/MS analysis (56). Once a list of validated AMT tags is generated for an organism, the need to perform routine HPLC-MS/MS and SEQUEST searching is obviated, and proteomic investigations may proceed in a high-throughput fashion using single-dimension HPLC-FTICR analyses.

There are obvious advantages to this approach; the only practical limits to the number of coeluting AMT tags that can be identified during a single run are the mass resolution of the mass spectrometer and the dynamic range of a single spectrum. This is in contrast to techniques based on MS/MS, where peptide identification is rate-limited by the speed with which the MS instrumentation can acquire successive tandem mass spectra. The combined peak capacity (number of individual components hypothetically resolvable) of the capillary HPLC-FTICR analysis used in the AMT tag approach has been estimated at $>6 \times 10^7$, providing the highest component resolution capability yet reported in a proteome analysis (52). In addition, the AMT tag approach to global proteome characterization is uniquely suited to analysis of low copy number proteins, as numerous low intensity peptides can be observed in a single complex mass spectrum. This is in contrast to techniques based on MS/MS for identification, where less abundant peptides may be overlooked, as they may not reach sufficient intensity to generate useful product ion spectra, or even to trigger MS/MS acquisition in the first place. As an added benefit, elimination of the requirement to conduct data-dependent MS/MS by use of the AMT tag approach greatly facilitates global quantitative analysis of differential protein expression among proteome samples using stable isotope labeling.

A key enabling technology for obtaining high-sensitivity analysis of low level AMT tags in the FTICR analysis has been the development of efficient ion transfer optics, including the electrodynamic ion funnel (4, 51). Additional technological advances in instrumentation have enabled optimization of the dynamic range of FTICR measurements by techniques such as ion accumulation external to the trapping region (5) and data-dependent, intensity-based selective ion accumulation and ejection prior to mass analysis (3, 25). These techniques combined have the potential to dramatically increase the dynamic range of protein abundance over which identifications can be made using the AMT tag approach, beyond the level of $\sim 10^4$ as currently implemented (52). Another significant technological advance, which reduces the time and effort required to generate the initial set of AMT tags, is the use of capillary HPLC-FTICR with multiplexed, data-dependent MS/MS (33, 40). Essentially, this technique allows simultaneous generation of PMTs and validation of corresponding AMT tags by performing both MS/MS and accurate mass measurement on the same FTICR instrument. The ability of high-resolution FTICR instruments to perform MS/MS experiments on several ions simultaneously provides a substantial throughput and duty-cycle advantage over ion beam instruments (such as TQ and QTOF) and low-resolution ion traps, which can only fragment a single ion at a time.

Current technology development efforts are focused on the use of artificial neural networks for prediction of peptide retention times in the capillary HPLC separation used in the AMT tag approach (30). Such a prediction would provide additional confirmation of peptide identity, beyond that available from MS/MS and accurate mass measurements, and could enable the use of a wider range of AMT tags (e.g., those that are not unique by mass but are chromatographically resolved) and/or the use of mass spectrometers with lower MMA (<5 ppm) such as QTOF instruments. The use of elution time information derived from chromatographic separation of peptides is not limited to identification of proteins by the AMT tag approach. Regardless of the strategy employed (e.g., MS/MS, two-dimensional separation), incorporation of peptide elution time prediction capability would increase the confidence of peptide (and hence, protein) identification from global proteome digests.

The most comprehensive global proteome analysis conducted to date using the AMT tag approach focused on the radiation-resistant bacterium *D. radiodurans* (35). In that work, 1910 gene products were identified, corresponding to 61% of the predicted ORFs (3116) for that organism. This represents the most complete proteome coverage achieved to date for any organism, using any proteome analysis method. Proteins were identified from every major functional category (assigned by TIGR) in *D. radiodurans*. As with the MudPIT approach, the AMT tag approach was slightly biased toward the detection of highly expressed proteins; however, many low-abundance proteins were also efficiently detected, as is evident by the detection and identification of numerous ORFs having codon adaptation indices below 0.2 (Figure 4).

The AMT tag strategy is not presently without disadvantages, however. As previously mentioned, a significant initial time investment is required for the shotgun

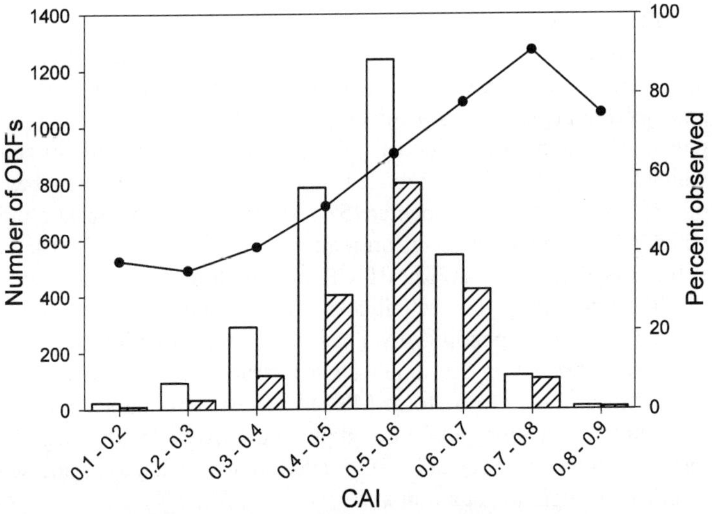

Figure 4 Results of a global proteome analysis of the bacterium *D. radiodurans* using the accurate mass and time (AMT) tag approach. Predicted (*unshaded*) and observed (*shaded*) numbers of proteins corresponding to open reading frames (ORFs) translated from the genome of *D. radiodurans* are shown for various ranges of the codon adaptation index (CAI). Low values of CAI are thought to indicate proteins present in an organism at low abundance. Also plotted is the percent of predicted ORFs observed in the course of the AMT tag proteome analysis, as a function of CAI (•, *solid line*). More than 35% of all predicted low-abundance ORFs (CAI between 0.1 and 0.3) were observed during the course of the analysis. The percent of predicted ORFs observed increased with increasing CAI to a maximum of >90% for ORFs with a CAI between 0.7 and 0.8.

generation of reliable AMT tags for any given organism before the approach can be fully utilized for high-throughput global proteome analysis. The separation methods currently implemented in the AMT tag approach have not been made fully routine as of this time. Indeed, the high-pressure HPLC instrumentation being used is best described as an advanced prototype. This situation will likely be remedied in the near future given recent advances in high-pressure microflow pump and valve technology. Also, because the AMT tag technique depends critically on the ability to perform routine accurate mass measurements, a heavy burden is placed on the mass spectrometry component of the approach. Any degradation in attainable mass accuracy will substantially degrade the performance of the AMT tag strategy. Thus, highly qualified and trained personnel are required for maintenance of the complex FTICR instrumentation, whereas the relatively straightforward QIT instruments require considerably less attention. Finally, the cost of the high-field FTICR instrumentation may prove prohibitive for many potential users (the highest field FTICR instruments now available, equipped with 12 Tesla magnets, sell for ~2 million dollars). This concern will obviously be reduced somewhat if alternative instruments (such as the QTOF) can be made to provide the needed mass accuracy, even if with somewhat reduced dynamic range or sensitivity.

Top-Down Proteomics

The proteome analysis strategies outlined in the sections above depend critically on the conversion of proteins to peptides via enzymatic or chemical digestion prior to mass spectrometric analysis (so called bottom-up approach). It is typically easier to achieve high mass measurement accuracy and to perform routine MS/MS experiments on lower-molecular-weight molecules such as peptides than high-molecular-mass proteins, using conventional mass spectrometers. However, such digestion methods also result in greatly increased sample complexity of global protein extracts. In addition, when using the bottom-up strategy to perform qualitative proteome analysis, complete sequence coverage of proteins is rarely achieved. This limits the ability of the researcher to examine site-specific mutations and posttranslational modifications of individual proteins, which may be important in biological function. Given these limitations, there has been increasing interest in top-down proteome characterization strategies where individual proteins are selected for analysis by mass spectrometry without the need for prior chemical or enzymatic proteolysis.

Capillary isoelectric focusing (CIEF) coupled with high-field (7 Tesla) FTICR mass spectrometry has been used for separation and analysis of global protein isolates of *Escherichia coli* (28). In this work, the CIEF technique served to efficiently separate intact proteins by pI, in a manner analogous to that of the first dimension of 2DE, while simultaneously concentrating individual proteins into narrow bands prior to mass analysis. Subsequent analysis by high resolution, accurate mass FTICR produced a two-dimensional display (pI and molecular weight) similar to that obtained by conventional 2DE, albeit with higher resolution on both axes. Isotope depletion of proteins (37) was utilized to enhance mass measurement accuracy in the FTICR. As noted by the authors, the combination of accurate mass measurement and high-resolution pI separation obtained using the CIEF-FTICR technique provided highly distinctive information regarding the identity of detected proteins. However, for actual global proteome samples, this level of information may not always be sufficient for confident identification of proteins from genome databases, given the possibility of point mutations, posttranslational modifications, and the presence of ORFs having high sequence homology. The need for additional information, such as that obtainable by MS/MS, was suggested for applications where global proteome characterization was the eventual goal.

High-resolution capillary HPLC was coupled to FTICR in order to characterize the proteins associated with the large ribosomal subunit in yeast (32). In this case, the high-efficiency separation afforded by capillary HPLC was used to resolve structurally similar proteins (e.g., highly conserved gene product isoforms and posttranslationally modified analogs). Unlike in the case of global protein isolates, the possible protein complement in the samples was theoretically limited to those gene products associated with the large ribosome subunit and modified versions thereof. This constraint enabled the use of accurate mass measurement to uniquely identify observed proteins. The authors reported positively identifying 42 of 43 proteins associated with the core large ribosomal subunit in yeast having unique masses. An advantage of this approach was the ability to observe modified proteins in their intact state, enabling the identification of modifications that would otherwise be transparent to peptide-level analysis.

Martinovic et al. (39) developed another solution for the global characterization of proteome samples by CIEF-FTICR. In this method, the incorporation of isotopically labeled amino acids into the cellular proteins of unicellular model organisms was used to provide partial amino acid content information for detected proteins. This information, combined with that available using the CIEF-FTICR approach (28), often enabled the identification of proteins from genome databases without the need for MS/MS analysis.

The combination of top-down proteomics approaches such as the CIEF-FTICR approach with bottom-up analysis of tryptic digests by, e.g., LC-MS/MS can provide a powerful method for high-throughput analysis of protein complexes as well as global proteomes. Prefractionation of global proteome samples and simultaneous characterization of the proteins present in individual fractions by LC-MS/MS and CIEF-FTICR can greatly simplify the subsequent routine use of CIEF-FTICR for identification of proteins from fractionated proteome samples. In this way, accurate mass measurements of intact proteins may be used as unique markers for identification from a small database of proteins previously observed in specific fractions using combined LC-MS/MS and CIEF-FTICR, thereby obviating the need to perform routine protein digestion and MS/MS.

Top-down characterization of proteins using FTICR-MS/MS was demonstrated by Kelleher et al. (29) with the 29-kDa protein bovine carbonic anhydrase as a model system. Several limited proteolysis methods (including SORI-CID in the FTICR mass spectrometer) were compared with bottom-up characterization using chymotryptic digestion. Results indicated that the top-down FTICR-MS/MS approach provided 100% sequence coverage of carbonic anhydrase. This approach enabled the efficient localization of posttranslational modifications and/or site-specific mutations in proteins by localizing these modifications to specific fragment ions. This strategy has been further applied for characterization of large proteins (>70 kDa) by incorporating Lys-C proteolysis prior to FTICR-MS analysis (15). This step creates relatively large peptides from high-molecular-weight proteins owing to the relative rarity of potential cleavage sites available to the enzyme Lys-C. For purified proteins, complete sequence coverage was obtained using this technique, enabling identification in genome databases without the need for MS/MS analysis. However, as developed, these techniques were not suitable for characterization of highly complex protein mixtures, such as those present in global proteome samples. Some progress has been made in addressing the top-down analysis of simple protein mixtures by coupling single-dimension separations with FTICR-MS/MS (34, 60). However, additional prefractionation techniques will be required in order to manage the complexity of global proteome samples using a top-down approach.

Recently, Meng et al. (41) have described a top-down proteome characterization method based on two-dimensional separation of intact proteins, followed by infrared multiphoton dissociation (IRMPD)-MS/MS analysis using a tandem quadrupole-FTICR instrument. Complex protein isolates from yeast were first fractionated by continuous-elution gel electrophoresis using a novel acid-labile surfactant. This surfactant was then removed by acid hydrolysis prior to subsequent offline reversed-phase HPLC separation of electrophoresis fractions. As currently implemented, reversed-phase HPLC fractions containing 2 to 5 individual proteins are analyzed offline using either MALDI-TOF or a 9.4 Tesla quadrupole-FTICR hybrid mass spectrometer. Protein identification is performed by probability-based matching of MS/MS fragmentation spectra with proteins present in an ORF database. The authors report a peak capacity of approximately 400 for proteins under 70 kDa. Disadvantages of the method include a lack of automation, limitation of sample throughput, as well as the necessity for rather large initial biological samples (~1 gram of yeast cells were required for a single analysis). In addition, it is unclear what level of ORF coverage has been achieved for yeast using the technique, and no information was provided concerning the ability of the method to reliably detect and identify low-abundance proteins.

A relatively new MS/MS technique, which shows considerable promise for application to top-down proteome characterization studies, is electron capture dissociation (ECD) (73). This technique operates on the principle of charge-neutralization, where charge reduction of a multiply protonated protein ion is accomplished in the gas phase by capture of a single electron, inducing both cleavage of the peptide backbone (generating y and b ions) and the amino acid side chains (generating c and z ions from amine-bond cleavage) (8, 73). This technique provides an additional level of structural information for polypeptides and proteins, complementary to that obtainable by more conventional CID fragmentation alone. The technique has been applied to analysis of peptides derived from enzymatic protein digests (9); however, FTICR ECD-MS/MS shows perhaps the most promise for high-fidelity structural characterization of intact proteins. Ge et al. (17) recently reported elucidation of numerous covalent and noncovalent modifications in proteins derived from *E. coli* using FTICR with ECD-MS/MS analysis. Although ECD-MS/MS produces extremely information-rich spectra of proteins and peptides, the sensitivity of this technique is lower than more "gentle" fragmentation methods such as collision-induced dissociation. In general, the best MS/MS sensitivity is obtained from the method producing the fewest fragment ions.

QUANTITATIVE PROTEOME ANALYSIS

Determination of the relative abundance levels of proteins in organisms or tissues exposed to different physiological or environmental conditions is essential to the study of disease processes and cellular response to stress. Quantitative proteomics has typically been performed by so-called differential display 2DE, where protein spot intensities on paired gels are compared and differences quantified. This approach suffers from the disadvantages noted above for the 2DE qualitative proteome characterization methods in terms of throughput and sensitivity. In addition, spot intensity ratios suffer from limited precision owing to sample handling constraints and sample processing variability. Several alternative methods have been proposed for relative quantitation of protein expression between paired samples. The four most useful methods are outlined in Figure 5.

Metabolic Protein Labeling Using Isotopically Enriched or Depleted Media

Two groups (43, 47) independently developed metabolic labeling of proteins using stable isotopes, followed by mass spectrometric analysis (Figure 5*A*). In one approach first applied for intact proteins, two populations of cells were grown in parallel, one on isotopically depleted media (containing only the most abundant isotopes of N, C, and H) and the other on normal media (47). Cells were harvested from the different media and combined prior to all sample processing and protein analysis by CIEF-FTICR. Comparison of the peak intensities of isotopically depleted and normal proteins allowed determination of relative expression ratios between the two cell populations. This approach was used to study Cd^{2+} stress in the bacterium *E. coli*. Protein abundance ratios between exposed and nonexposed cell populations ranged from <0.1 to >30. In a similar approach applicable to smaller peptide fragments, two parallel cell populations were grown on either isotopically enriched (96% ^{15}N) or normal media (43). Proteins derived from the combined cell populations were then separated by 2DE and analyzed after tryptic digestion by either MALDI or electrospray MS. Abundance ratios were calculated

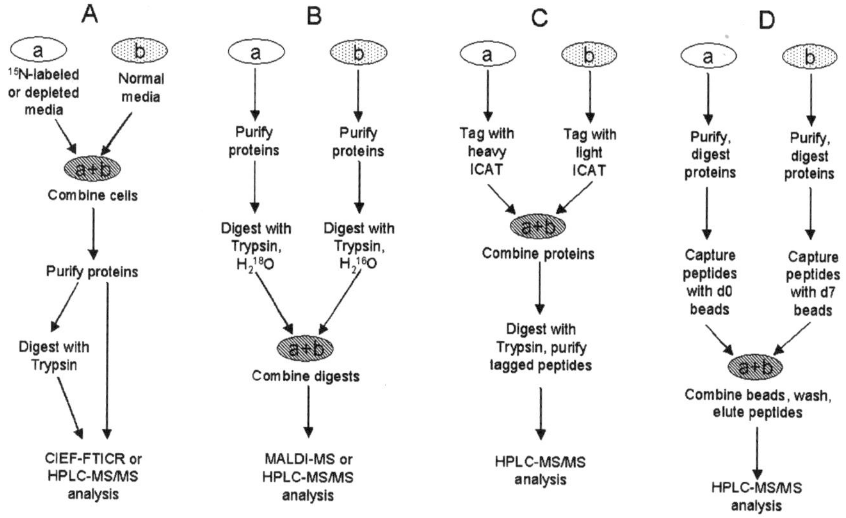

Figure 5 Strategies for quantitation of relative protein abundance using isotope labeling in proteomics. The methods are (*A*) metabolic labeling of proteins by growth of cells on isotopically depleted or enriched media, (*B*) proteolytic labeling of peptides with $^{16}O/^{18}O$, (*C*) the isotope-coded affinity tag approach, and (*D*) the solid-phase peptide capture and isotopic labeling technique. Details of these methods are described in the text and in the associated references.

using the same principle as the isotope depletion method above. Both of these methods provided precise measurements of relative protein abundance between comparable cell populations, since all sample processing and analyses are conducted on both protein versions simultaneously. This approach also has advantages in terms of speed over the differential display 2DE strategy. However, the technique is practically applicable only to cells grown in culture; it is much more difficult to isotopically label an entire living, multicellular organism. This limits the metabolic labeling strategy to in vitro studies, precluding its use, e.g., in many biologically important in vivo assays. Although no large "isotope effects" due to ^{15}N incorporation into protein synthesis were noticed in preliminary studies (47), concerns remain as to whether growth rates and biological function are modified by cell growth on isotopically enriched or depleted media.

Proteolytic Labeling of Peptides Using $H_2{}^{16}O/H_2{}^{18}O$

Another method for comparing relative protein expression levels in paired biological samples is labeling of enzymatically digested proteins with ^{18}O through proteolysis in isotopically labeled water prior to mass spectrometric analysis (42, 57, 67) (Figure 5*B*). In this strategy, proteins are isolated from paired cell or tissue samples and separately digested in either $H_2{}^{16}O$ or $H_2{}^{18}O$. The oxygen atom derived from the aqueous solvent is incorporated into the newly formed C terminus acid functional group in each peptide, thus providing an effective isotope tag for relative quantitation. Advantages of this approach include decoupling of the isotope labeling procedure from the experimental conditions under which the tissues or cells are grown, as well as simplicity of the method and high stability of the isotopically labeled peptides. Also, unlike the metabolic labeling approach, the proteolytic ^{18}O labeling technique works equally well with cultured cells and tissues isolated from living organisms. However, the method requires separate tryptic digestions for paired protein pools, potentially leading to decreased precision owing to increased separate handling of the paired samples. In addition, because the mass difference between ^{16}O and ^{18}O labeled peptides is fixed at 4 Da, the proteolytic labeling technique has limited usefulness for larger peptides, where mass spectral isotope envelopes of labeled and unlabeled peptides separated by only 4 Da begin to overlap.

Isotope-Coded Affinity Tags

The isotope-coded affinity tag (ICAT) approach (22) (Figure 5*C*) is based upon the use of cysteine-specific, isotopically labeled (d0 or d8) biotin affinity tags covalently linked to proteins in paired proteome samples. The "tagged" protein pools are combined and digested enzymatically to yield peptides. Biotinylated peptides (containing cysteine) are then isolated from the complex peptide mixture by avidin affinity chromatography and subsequently analyzed by LC-MS/MS. As with the other isotope labeling techniques noted above, protein abundance ratios are based on relative peak intensities of light and heavy isotope-labeled peptides. The ICAT strategy reduces the peptide mixture complexity by significantly enriching the cysteinyl peptides. This step has the potential to enhance the identification of proteins, as the possible peptides in the mixture are theoretically constrained to those containing one or more cysteine residues. Unfortunately, this is not often achieved, as numerous noncysteine-containing peptides are also isolated during the affinity purification step. Like the proteolytic ^{18}O method, the ICAT technique is amenable to quantitative analysis of protein expression in either cultured cells or tissue isolates. A disadvantage of the technique is that the biotin affinity tag remains linked to the peptides throughout the analysis, leading to shifts in the chromatographic separation, large shifts in m/z, and changes in MS/MS spectra relative to unlabeled peptides. As initially formulated, the ICAT reagent incurred relatively large differences in chromatographic elution times between the deuterated and nondeuterated label versions, resulting in greater uncertainties in peptide relative abundances. This problem can be circumvented by the incorporation of alternative stable isotope labels, such as ^{13}C or ^{18}O into the ICAT reagent.

The ICAT approach has been incorporated into a high-throughput proteome analysis scheme, where identification and characterization of proteins is driven by ICAT-enabled protein abundance ratios between paired tissue and cell treatments (20). This advance represents an effective interface between the proteome characterization methods (designed for qualitative analysis) and bioassay-directed quantitative proteomics. As developed, the coupled technique uses online HPLC-TOFMS (time-of-flight mass spectrometry) to filter ICAT-labeled and purified proteome samples for peptides having expression ratios within defined limits. Expression-dependent fraction collection is performed, with subsequent offline QTOF mass spectrometric characterization of chosen peptides. This technique has been applied to the expression-directed proteomic analysis of cancerous and normal human prostate cells (20). Approximately one third of the 97 peptide pairs selected during the initial analysis exhibited significant differences in expression. Those peptides led to the identification of six proteins that were differentially expressed between the cancerous and nontumorigenic prostate cells. A disadvantage of the method as currently implemented is the offline coupling between the expression ratio determination and the subsequent peptide identification using MS/MS, which leads to sample losses and reduction in sample throughput.

Solid-Phase Peptide Capture and Isotope Labeling

A significant improvement to the ICAT strategy, which reduces the effect of the labeling tag on chromatographic and mass spectrometric performance, is the incorporation of solid-phase capture in the isotope tagging approach (71). In this approach (Figure 5*D*), an isotopically labeled leucine molecule attached to a thiol-specific reactive group and covalently linked to a solid-phase support via a photocleavable linker is used to isolate and label cysteinyl peptides (after proteolytic processing) in global proteome samples. After combining the beads from paired samples, the bound peptides are released by photolysis and analyzed by LC-MS/MS. Unlike the original ICAT approach, the only chemical modification remaining on the peptides at the time of LC-MS/MS analysis is an isotopically labeled (or unlabeled) leucine residue. Further advantages include enhanced sensitivity, reduced sample handling, and facilitation of extensive sample washing protocols prior to peptide elution (71).

PHOSPHOPROTEOME ANALYSIS

In some cases, it may be advantageous to specifically target selected "subproteomes" for analysis. Such is the case with proteins that have been altered by posttranslational modifications. These modifications, especially phosphorylation, can dramatically alter the structure and function of proteins and are important in regulating cellular processes through signaling networks (6, 48). Protein phosphorylation is not observable at the transcriptome level, and consequently there has been considerable interest in developing proteome analysis techniques for selectively enriching and characterizing phosphorylated proteins.

Immobilized Metal Affinity Chromatography

A popular and relatively simple approach to phosphorylated protein enrichment is immobilized metal affinity chromatography (IMAC). In this technique, trivalent metals (typically Fe^{III}) are immobilized on a solid-phase support, and protein (or peptide) mixtures are exposed to the immobilized metals (2). Acidic phosphate residues bind noncovalently with the trivalent metals and can then be selectively eluted and analyzed mass spectrometrically. However, this technique is not completely selective, as relatively acidic, nonphosphorylated proteins may bind through interaction of carboxylic acid functional groups with the immobilized metals. Ficarro et al. (13) developed a modification of the IMAC procedure that largely circumvents this problem. Global proteome samples are first digested tryptically. Residual carboxylic acid functional groups on the resulting peptides are then converted to methyl esters, and phosphorylated peptides are selectively enriched using the IMAC technique described above prior to LC-MS/MS analysis. The presence of phosphoserine or phosphothreonine in isolated peptides may be discovered by observation of phosphoric acid neutral loss during MS/MS analysis. Using this technique, 216 phosphorylated peptides have been observed in global yeast protein digests, corresponding to 171 unique proteins, a significant fraction of which was reportedly present at low abundance based on the codon adaptation index (CAI) (13). Unfortunately, however, low recovery of phosphorylated peptides is generally observed when using the IMAC method, even with incorporation of the methyl-esterification technique.

Chemical Derivatization Methods

An alternative approach for phosphoprotein analysis in global protein isolates is selective derivatization of the phosphate functional group and subsequent analysis using mass spectrometry. An early attempt utilized β-elimination of phosphate followed by Michael addition of deuterated or nondeuterated ethanethiol in tryptic peptides, followed by LC-MS analysis (63). This technique allowed relative quantitation of phosphorylated proteins in comparable samples through examination of stable isotope ratios during MS analysis. However, no enrichment step was incorporated to selectively concentrated phosphorylated proteins or peptides in the mixture. Consequently, although the method was suitable for analysis of pure solutions of phosphorylated proteins, it would likely not be useful for analysis of trace phosphoproteins in complex proteome samples.

Two independent groups (19, 44) recently proposed an operationally similar approach for phosphoproteome analysis. In their work, phosphate β-elimination and Michael addition of a bifunctional linker was performed prior to enzymatic proteolysis. A biotin affinity tag was added to the end of the linker, and the protein mixture was digested tryptically. This step allowed selective isolation of biotinylated, previously phosphorylated peptides from the global protein digest using avidin affinity columns prior to MALDI (44) or electrospray (19, 44) mass spectrometry. Distinct advantages of this technique over the IMAC approach include the inherent ability to perform relative quantitation studies of protein phosphorylation [due to the incorporation of stable isotope tags in the procedure (19)] and the removal of the phosphate functional group prior to mass spectrometry, thus rendering the peptides more amenable to analysis by positive electrospray ionization. Disadvantages include the necessity for increased sample handling and loss of substrate due to incomplete product yield during the reaction chemistry. In addition, the technique as described is applicable only to the enrichment and analysis of phosphoserine- and phosphothreonine-containing proteins.

Zhou et al. (72) have reported selective enrichment and analysis of phosphotyrosine-, phosphoserine-, and phosphothreonine-containing proteins. In this work, a multistep chemical reaction scheme is used to select and covalently link phosphorylated peptides to a solid-phase support after tryptic digestion of protein mixtures. Phosphorylated peptides are then selectively released for analysis by LC-MS/MS. This technique was illustrated to be valid for enrichment of phosphorylated proteins from complex proteome samples; it is the only chemical tag-based method currently applicable to analysis of phosphotyrosine residues. The chemistry can reportedly be modified via incorporation of a stable isotope tag for quantitative analysis. However, the overall reaction yield is only 20%, thereby limiting the method sensitivity, and the reaction chemistry is rather complex.

CONCLUSIONS

The advent of high-throughput, global proteome analysis has enabled rapid, comprehensive studies of biological function. When integrated with complementary techniques such as cDNA microarray analysis and classical biochemical functional assays such as gene knockout/knockdown, proteomics provides a powerful toolkit for probing gene expression in cellular systems. Applications will include high-throughput drug development screening, mechanistic toxicology studies, and studies aimed at elucidating biological signaling pathways related to specific cell functions or disease states. Several current trends are evident in proteome analysis technology, and these trends will likely continue to develop in the near future. Coupling of high-resolution separations and two-dimensional analytical systems with mass spectrometry should continue to expand in use as improvements are made in microfluidics and automation. Much effort is currently being made to increase access by researchers to robust, high mass-accuracy mass spectrometry instrumentation. Further development in mass spectrometer technology will enable more comprehensive global proteome analysis in complex mammalian systems. These advances will include the development of enhanced data station control of mass spectrometers, as well as informatics tools designed for "mining" of large datasets for biologically relevant information. Advancements in chemical-tagging methods and affinity isolation will enable more selective enrichment of modified and mutated proteins from global proteome samples. Increased use of chromatographic elution time information in identifying peptides from LC-MS/MS analyses will lead to more confident qualitative characterization of protein mixtures. Quantitative analysis of proteomes will likely be extended beyond relative expression profiling as techniques for predicting absolute protein concentrations using measured peptide signal intensity in mass spectrometric methods are developed. Finally, development of techniques for rapid, high-throughput functional proteome analysis will be critical. These developments will depend on the interface of sophisticated chemical analysis techniques, such as multidimensional separations combined with tandem mass spectrometry, with robust biochemical methods for isolating intact, high-fidelity protein complexes.

ACKNOWLEDGMENTS

We thank the U.S. Department of Energy Office of Biological and Environmental Research as well as the National Institutes of Health through NCI (CA81654 and CA93306) and NINDS (NS39617) for support of portions of the research described in this review. Pacific Northwest National Laboratory is operated by Battelle Memorial Institute for the U.S. Department of Energy under contract DE-AC06-76RLO 1830. The authors thank Dr. Michael Goshe and Dr. David Camp for helpful comments during the preparation of this manuscript.

The *Annual Review of Biophysics and Biomolecular Structure* is online at http://biophys.annualreviews.org

LITERATURE CITED

1. Aebersold R, Goodlett DR. 2001. Mass spectrometry in proteomics. *Chem. Rev.* 101:269–95
2. Andersson L, Porath J. 1986. Isolation of phosphoproteins by immobilized metal (Fe-3+) affinity-chromatography. *Anal. Biochem.* 154:250–54
3. Belov ME, Anderson GA, Angell NH, Shen YF, Tolic N, et al. 2001. Dynamic range expansion applied to mass spectrometry based on data-dependent selective ion ejection in capillary liquid chromatography Fourier transform ion cyclotron resonance for enhanced proteome characterization. *Anal. Chem.* 73:5052–60
4. Belov ME, Gorshkov MV, Udseth HR, Anderson GA, Tolmachev AV, et al. 2000. Initial implementation of an electrodynamic ion funnel with Fourier transform ion cyclotron resonance mass spectrometry. *J. Am. Soc. Mass Spectrom.* 11:19–23
5. Belov ME, Nikolaev EN, Anderson GA, Auberry KJ, Harkewicz R, Smith RD. 2001. Electrospray ionization-Fourier transform ion cyclotron mass spectrometry using ion preselection and external accumulation for ultrahigh sensitivity. *J. Am. Soc. Mass Spectrom.* 12:38–48
6. Cohen P. 1982. The role of protein-phosphorylation in neural and hormonal control of cellular activity. *Nature* 296:613–20
7. Conrads TP, Anderson GA, Veenstra TD, Pasa-Tolic L, Smith RD. 2000. Utility of accurate mass tags for proteome-wide protein identification. *Anal. Chem.* 72:3349–54
8. Cooper HJ, Hudgins RR, Hakansson K, Marshall AG. 2002. Characterization of amino acid side chain losses in electron capture dissociation. *J. Am. Soc. Mass Spectrom.* 13:241–49
9. Davidson W, Frego L. 2002. Micro-high-performance liquid chromatography/Fourier transform mass spectrometry with electron-capture dissociation for the

analysis of protein enzymatic digests. *Rapid Commun. Mass Spectrom.* 16:993–98
10. Edman P, Begg G. 1967. A protein sequenator. *Eur. J. Biochem.* 1:80–91
11. Eng JK, McCormack AL, Yates JR. 1994. An approach to correlate tandem mass-spectral data of peptides with amino-acid-sequences in a protein database. *J. Am. Soc. Mass Spectrom.* 5:976–89
12. Fenn JB, Mann M, Meng CK, Wong SF, Whitehouse CM. 1989. Electrospray ionization for mass-spectrometry of large biomolecules. *Science* 246:64–71
13. Ficarro SB, McCleland ML, Stukenberg PT, Burke DJ, Ross MM, et al. 2002. Phosphoproteome analysis by mass spectrometry and its application to *Saccharomyces cerevisiae*. *Nat. Biotechnol.* 20:301–5
14. Figeys D. 2002. Proteomics approaches in drug discovery. *Anal. Chem.* 36:413A–19A
15. Forbes AJ, Mazur MT, Patel HM, Walsh CT, Kelleher NL. 2001. Toward efficient analysis of >70 kDa proteins with 100% sequence coverage. *Proteomics* 1:927–33
16. Gavin AC, Bosche M, Krause R, Grandi P, Marzioch M, et al. 2002. Functional organization of the yeast proteome by systematic analysis of protein complexes. *Nature* 415:141–47
17. Ge Y, Lawhorn BG, ElNaggar M, Strauss E, Park JH, et al. 2002. Top down characterization of larger proteins (45 kDa) by electron capture dissociation mass spectrometry. *J. Am. Chem. Soc.* 124:672–78
18. Godovac-Zimmermann J, Brown LR. 2001. Perspectives for mass spectrometry and functional proteomics. *Mass Spectrom. Rev.* 20:1–57
19. Goshe MB, Conrads TP, Panisko EA, Angell NH, Veenstra TD, Smith RD. 2001. Phosphoprotein isotope-coded affinity tag approach for isolating and quantitating phosphopeptides in proteome-wide analyses. *Anal. Chem.* 73:2578–86
20. Griffin TJ, Han DKM, Gygi SP, Rist B, Lee H, et al. 2001. Toward a high-throughput approach to quantitative proteomic analysis: expression-dependent protein identification by mass spectrometry. *J. Am. Soc. Mass Spectrom.* 12:1238–46
21. Gygi SP, Corthals GL, Zhang Y, Rochon Y, Aebersold R. 2000. Evaluation of two-dimensional gel electrophoresis-based proteome analysis technology. *Proc. Natl. Acad. Sci. USA* 97:9390–95
22. Gygi SP, Rist B, Gerber SA, Turecek F, Gelb MH, Aebersold R. 1999. Quantitative analysis of complex protein mixtures using isotope-coded affinity tags. *Nat. Biotechnol.* 17:994–99
23. Gygi SP, Rochon Y, Franza BR, Aebersold R. 1999. Correlation between protein and mRNA abundance in yeast. *Mol. Cell. Biol.* 19:1720–30
24. Han DK, Eng J, Zhou HL, Aebersold R. 2001. Quantitative profiling of differentiation-induced microsomal proteins using isotope-coded affinity tags and mass spectrometry. *Nat. Biotechnol.* 19:946–51
25. Harkewicz R, Belov ME, Anderson GA, Pasa-Tolic L, Masselon CD, et al. 2002. ESI-FTICR mass spectrometry employing data-dependent external ion selection and accumulation. *J. Am. Soc. Mass Spectrom.* 13:144–54
26. Henzel WJ, Billeci TM, Stults JT, Wong SC, Grimley C, Watanabe C. 1993. Identifying proteins from 2-dimensional gels by molecular mass searching of peptide-fragments in protein-sequence databases. *Proc. Natl. Acad. Sci. USA* 90:5011–15
27. Ho Y, Gruhler A, Heilbut A, Bader GD, Moore L, et al. 2002. Systematic identification of protein complexes in *Saccharomyces cerevisiae* by mass spectrometry. *Nature* 415:180–83
28. Jensen PK, Pasa-Tolic L, Anderson GA, Horner JA, Lipton MS, et al. 1999. Probing proteomes using capillary isoelectric focusing-electrospray ionization Fourier transform ion cyclotron resonance mass spectrometry. *Anal. Chem.* 71:2076–84
29. Kelleher NL, Lin HY, Valaskovic GA,

Aaserud DJ, Fridriksson EK, McLafferty FW. 1999. Top down versus bottom up protein characterization by tandem high-resolution mass spectrometry. *J. Am. Chem. Soc.* 121:806–12

30. Konstantinos P, Kangas LJ, Ferguson PL, Anderson GA, Pasa-Tolic' L, et al. 2003. Use of artificial neural networks for the prediction of peptide elution time in liquid chromatography and its application in the proteomic field. *Anal. Chem.* 75:1039–48
31. Langen H, Takacs B, Evers S, Berndt P, Lahm HW, et al. 2000. Two-dimensional map of the proteome of *Haemophilus influenzae*. *Electrophoresis* 21:411–29

31a. Lee H, Griffin TJ, Gygi SP, Rist B, Aebersold R. 2002. Development of a multiplexed microcapillary liquid chromatography system for high-throughput proteome analysis. *Anal. Chem.* 74:4353–60

32. Lee SW, Berger SJ, Martinovic S, Pasa-Tolic L, Anderson GA, et al. 2002. Direct mass spectrometric analysis of intact proteins of the yeast large ribosomal subunit using capillary LC/FTICR. *Proc. Natl. Acad. Sci. USA* 99:5942–47
33. Li LJ, Masselon CD, Anderson GA, Pasa-Tolic L, Lee SW, et al. 2001. High-throughput peptide identification from protein digests using data-dependent multiplexed tandem FTICR mass spectrometry coupled with capillary liquid chromatography. *Anal. Chem.* 73:3312–22
34. Li WQ, Hendrickson CL, Emmett MR, Marshall AG. 1999. Identification of intact proteins in mixtures by alternated capillary liquid chromatography electrospray ionization and LC ESI infrared multiphoton dissociation Fourier transform ion cyclotron resonance mass spectrometry. *Anal. Chem.* 71:4397–402
35. Lipton MS, Pasa-Tolic' L, Anderson GA, Anderson DJ, Auberry DL, et al. 2002. Global analysis of the Deinococcus radiodurans proteome by using accurate mass tags. *Proc. Natl. Acad. Sci. USA* 99: 11049–54
36. Mann M, Hendrickson RC, Pandey A. 2001. Analysis of proteins and proteomes by mass spectrometry. *Annu. Rev. Biochem.* 70:437–73
37. Marshall AG, Senko MW, Li WQ, Li M, Dillon S, et al. 1997. Protein molecular mass to 1 Da by C-13, N-15 double-depletion and FT-ICR mass spectrometry. *J. Am. Chem. Soc.* 119:433–34
38. Martin SE, Shabanowitz J, Hunt DF, Marto JA. 2000. Subfemtomole MS and MS/MS peptide sequence analysis using nano-HPLC micro-ESI Fourier transform ion cyclotron resonance mass spectrometry. *Anal. Chem.* 72:4266–74
39. Martinovic S, Veenstra TD, Anderson GA, Pasa-Tolic L, Smith RD. 2002. Selective incorporation acids for identification proteome-wide level. *J. Mass Spectrom.* 37:99–107
40. Masselson C, Anderson GA, Harkewicz R, Bruce JE, Pasa-Tolic L, Smith RD. 2000. Accurate mass multiplexed tandem mass spectrometry for high-throughput polypeptide identification from mixtures. *Anal. Chem.* 72:1918–24
41. Meng FY, Cargile BJ, Patrie SM, Johnson JR, McLoughlin SM, Kelleher NL. 2002. Processing complex mixtures of intact proteins for direct analysis by mass spectrometry. *Anal. Chem.* 74:2923–29
42. Mirgorodskaya OA, Kozmin YP, Titov MI, Korner R, Sonksen CP, Roepstorff P. 2000. Quantitation of peptides and proteins by matrix-assisted laser desorption/ionization mass spectrometry using O-18-labeled internal standards. *Rapid Commun. Mass Spectrom.* 14:1226–32
43. Oda Y, Huang K, Cross FR, Cowburn D, Chait BT. 1999. Accurate quantitation of protein expression and site-specific phosphorylation. *Proc. Natl. Acad. Sci. USA* 96:6591–96
44. Oda Y, Nagasu T, Chait BT. 2001. Enrichment analysis of phosphorylated

proteins as a tool for probing the phosphoproteome. *Nat. Biotechnol.* 19:379–82
45. Opiteck GJ, Jorgenson JW, Anderegg RJ. 1997. Two-dimensional SEC/RPLC coupled to mass spectrometry for the analysis of peptides. *Anal. Chem.* 69:2283–91
46. Pandey A, Mann M. 2000. Proteomics to study genes and genomes. *Nature* 405: 837–46
47. Pasa-Tolic L, Jensen PK, Anderson GA, Lipton MS, Peden KK, et al. 1999. High throughput proteome-wide precision measurements of protein expression using mass spectrometry. *J. Am. Chem. Soc.* 121:7949–50
48. Pawson T, Scott JD. 1997. Signaling through scaffold, anchoring, and adaptor proteins. *Science* 278:2075–80
49. Perkins DN, Pappin DJC, Creasy DM, Cottrell JS. 1999. Probability-based protein identification by searching sequence databases using mass spectrometry data. *Electrophoresis* 20:3551–67
50. Schena M, Shalon D, Davis RW, Brown PO. 1995. Quantitative monitoring of gene-expression patterns with a complementary-DNA microarray. *Science* 270: 467–70
51. Shaffer SA, Tang KQ, Anderson GA, Prior DC, Udseth HR, Smith RD. 1997. A novel ion funnel for focusing ions at elevated pressure using electrospray ionization mass spectrometry. *Rapid Commun. Mass Spectrom.* 11:1813–17
52. Shen YF, Tolic N, Zhao R, Pasa-Tolic L, Li LJ, et al. 2001. High-throughput proteomics using high efficiency multiple-capillary liquid chromatography with on-line high-performance ESI FTICR mass spectrometry. *Anal. Chem.* 73:3011–21
53. Shen YF, Zhao R, Belov ME, Conrads TP, Anderson GA, et al. 2001. Packed capillary reversed-phase liquid chromatography with high-performance electrospray ionization Fourier transform ion cyclotron resonance mass spectrometry for proteomics. *Anal. Chem.* 73:1766–75
54. Shevchenko A, Jensen ON, Podtelejnikov AV, Sagliocco F, Wilm M, et al. 1996. Linking genome and proteome by mass spectrometry: large-scale identification of yeast proteins from two dimensional gels. *Proc. Natl. Acad. Sci. USA* 93:14440–45
55. Smith RD, Anderson GA, Lipton MS, Masselon CD, Pasa-Tolic L, et al. 2002. The use of accurate mass tags for high-throughput microbial proteomics. *OMICS* 6:61–90
56. Smith RD, Anderson GA, Lipton MS, Pasa-Tolic L, Shen YF, et al. 2002. An accurate mass tag strategy for quantitative and high- throughput proteome measurements. *Proteomics* 2:513–23
57. Stewart, II, Thomson T, Figeys D. 2001. O-18 labeling: a tool for proteomics. *Rapid Commun. Mass Spectrom.* 15: 2456–65
58. Strittmatter EF, Camp DG, Pasa-Tolic L, Anderson GA, Kangas LJ, Smith RD. 2002. *PARALLAX: high throughput software tools for the detection of modified, mutated, and non-specifically cleaved proteins*. Presented at 50th Annu. Conf. Mass Spectrom. Allied Top., Orlando, FL
59. Tong W, Link A, Eng JK, Yates JR. 1999. Identification of proteins in complexes by solid phase microextraction multistep elution capillary electrophoresis tandem mass spectrometry. *Anal. Chem.* 71:2270–78
60. Valaskovic GA, Kelleher NL, McLafferty FW. 1996. Attomole protein characterization by capillary electrophoresis mass spectrometry. *Science* 273:1199–202
61. Vidal M, Endoh H. 1999. Prospects for drug screening using the reverse two-hybrid system. *Trends Biotechnol.* 17: 374–81
62. Washburn MP, Wolters D, Yates JR. 2001. Large-scale analysis of the yeast proteome by multidimensional protein identification technology. *Nat. Biotechnol.* 19:242–47
63. Weckwerth W, Willmitzer L, Fiehn O. 2000. Comparative quantification and

identification of phosphoproteins using stable isotope labeling and liquid chromatography/mass spectrometry. *Rapid Commun. Mass Spectrom.* 14:1677–81
64. Wilm M, Mann M. 1996. Analytical properties of the nanoelectrospray ion source. *Anal. Chem.* 68:1–8
65. Wolters DA, Washburn MP, Yates JR. 2001. An automated multidimensional protein identification technology for shotgun proteomics. *Anal. Chem.* 73:5683–90
66. Yamashita M, Fenn JB. 1984. Electrospray ion-source—another variation on the free-jet theme. *J. Phys. Chem.* 88: 4451–59
67. Yao XD, Freas A, Ramirez J, Demirev PA, Fenselau C. 2001. Proteolytic O-18 labeling for comparative proteomics: model studies with two serotypes of adenovirus. *Anal. Chem.* 73:2836–42
68. Yates JR, Eng JK, McCormack AL, Schieltz D. 1995. Method to correlate tandem mass-spectra of modified peptides to amino-acid-sequences in the protein database. *Anal. Chem.* 67:1426–36
69. Yates JR, McCormack AL, Eng J. 1996. Mining genomes with MS. *Anal. Chem.* 68:A534–A40
70. Yates JR, Speicher S, Griffin PR, Hunkapiller T. 1993. Peptide mass maps—a highly informative approach to protein identification. *Anal. Biochem.* 214:397–408
71. Zhou HL, Ranish JA, Watts JD, Aebersold R. 2002. Quantitative proteome analysis by solid-phase isotope tagging and mass spectrometry. *Nat. Biotechnol.* 20:512–15
72. Zhou HL, Watts JD, Aebersold R. 2001. A systematic approach to the analysis of protein phosphorylation. *Nat. Biotechnol.* 19:375–78
73. Zubarev RA, Kelleher NL, McLafferty FW. 1998. Electron capture dissociation of multiply charged protein cations. A nonergodic process. *J. Am. Chem. Soc.* 120:3265–66

Annu. Rev. Biophys. Biomol. Struct. 2003. 32:425–43
doi: 10.1146/annurev.biophys.32.110601.141807

First published online as a Review in Advance on February 5, 2003

COMPUTER SIMULATIONS OF ENZYME CATALYSIS: Methods, Progress, and Insights

Arieh Warshel
Department of Chemistry, University of Southern California, Los Angeles, California 90089; email: warshel@invitro.usc.edu

Key Words QM/MM, EVB, transition state stabilization, catalysis

■ **Abstract** Understanding the action of enzymes on an atomistic level is one of the important aims of modern biophysics. This review describes the state of the art in addressing this challenge by simulating enzymatic reactions. It considers different modeling methods including the empirical valence bond (EVB) and more standard molecular orbital quantum mechanics/molecular mechanics (QM/MM) methods. The importance of proper configurational averaging of QM/MM energies is emphasized, pointing out that at present such averages are performed most effectively by the EVB method. It is clarified that all properly conducted simulation studies have identified electrostatic preorganization effects as the source of enzyme catalysis. It is argued that the ability to simulate enzymatic reactions also provides the chance to examine the importance of nonelectrostatic contributions and the validity of the corresponding proposals. In fact, simulation studies have indicated that prominent proposals such as desolvation, steric strain, near attack conformation, entropy traps, and coherent dynamics do not account for a major part of the catalytic power of enzymes. Finally, it is pointed out that although some of the issues are likely to remain controversial for some time, computer modeling approaches can provide a powerful tool for understanding enzyme catalysis.

CONTENTS

INTRODUCTION

Enzymes are involved in the catalysis and the control of most life processes. Thus there is a major fundamental and practical interest in finding out what makes enzymes so efficient. Although many crucial pieces of this puzzle were elucidated by biochemical and structural studies, the source of the catalytic power of enzymes is not widely understood. This catalytic power cannot be explained by stating that "the enzyme binds the transition state stronger than the ground state" because the real question is how this differential binding is accomplished.

In order to define our problem, it is convenient to start by describing a typical enzymatic reaction using the generic equation:

$$E + S \leftrightarrow ES \overset{K}{\longleftrightarrow} ES^{\ddagger} \overset{k_{cat}}{\longrightarrow} EP \rightarrow E + P, \qquad 1.$$

where *E*, *S*, and *P* are the enzyme, substrate, and product, respectively, while *ES*, *EP*, and $ES^{\ddagger}$ are the enzyme-substrate complex, enzyme-product complex, and transition state, respectively. Many enzymes were evolved by optimizing k_{cat}/K_M, (59, 69), where $K = k_1/k_{-1}$, $K_M = (k_{-1} + k_{cat})/k_1$ and can be approximated as $K_M \simeq k_{-1}/k_1$.

It is well known (69) that many enzymes evolved by optimizing k_{cat}/K_M. However, this and related findings did not identify the factors responsible to the catalytic effect. As is shown in the next section, the key question is related to the reduction of the activation barrier in the chemical step. Unfortunately, even mutation experiments, which were extremely useful in identifying catalytic factors (35), cannot tell us in a unique way what is the origin of the catalytic effect (60). What is needed is a quantitative tool for structure-function correlation and the ability to decompose the catalytic effect to different energy contributions. It is becoming more and more clear that this requirement can best be accomplished by computer simulation approaches, and this review describes my own insight into the current state of these approaches.

WHAT IS THE PROBLEM?

The discussion of enzyme catalysis is almost pointless without a proper definition of the relevant questions. When talking about catalysis it is essential to clarify what is meant by the term catalysis. Here we must define a reference reaction, and the most obvious reference is the uncatalyzed reaction in water. Because the mechanism in water can be different from that in the enzyme, we should consider the effect of having different mechanisms and having different environments separately. Fortunately, the difference in mechanism can be classified as a "chemical effect" (e.g., the effect of having a general base instead of a water as a base), and such effects are well understood. Therefore, we can focus on the difference in the effect of the environment. In this way we can focus on comparing the rate constant of a reaction that involves the same mechanism and the same chemical groups but is conducted in water. Thus [see also (60)], our question boils down to the difference between the activation barrier in water ($\Delta g_w^{\ddagger}$) and the activation barrier in the protein ($\Delta g_p^{\ddagger}$). However, the enzyme can reduce $\Delta g_p^{\ddagger}$ by binding the substrate equally in the reactant state (RS) and the transition state (TS) (this corresponds to ΔG_{bind}) and by reducing the activation barrier $\Delta g_{cat}^{\ddagger}$ for the chemical step. Because the factors that control binding are well understood, the real puzzle is related to the reduction of the activation barrier of the chemical step ($\Delta g_{cat}^{\ddagger}$). Polanyi (48) and Pauling (47) long ago stated that catalysts and enzymes reduce activation barriers, but this statement does not tell us how the reduction is accomplished.

Even the question of whether the enzyme works by stabilizing the RS or TS is not easily resolved experimentally, although mutational analysis can help in this respect (60). Thus, we should focus on two main questions: (*a*) What contributions are responsible for the difference between ($\Delta g_{cat}^{\ddagger}$) and ($\Delta g_w^{\ddagger}$), and (*b*) how do these contributions operate (i.e., do they destabilize the RS or stabilize the TS).

Possible answers to these two questions have been offered by many proposals [see (56) for a partial list]. However, most of the proposals have not been properly defined or properly examined by their proponents. Moreover, many proposals have not considered a proper thermodynamic cycle. At any rate, as will be shown below, with a clear definition of the problem and with a combination of experimental and computational studies, it is possible to elucidate the source of the catalytic power of enzymes.

In my discussion and analyses of the reference reaction, I refer to $\Delta g_{cage}^{\ddagger}$, which corresponds to the activation barrier in a water cage, where the reactants are already at an interaction distance [see (56) for a rigorous definition].

SIMULATION METHODS

As is clear from the above discussion, we are interested in accurate evaluations of the activation free energies of reactions in enzymes and in solutions. I consider the main options for accomplishing this task and evaluate their effectiveness.

Incomplete Models

One seemingly obvious option is to study the energetics of the reactants in the gas phase. With such a model one can use a relatively high level and rigorous quantum mechanical approaches. Although such studies have been instrumental in providing insight about the reacting system, they are problematic. That is, the entire issue of catalysis is related to the effect of the environment. Thus, omitting the environment from the calculations prevents one from exploring key catalytic effects. Furthermore, the effect of the environment is not a small perturbation but frequently involves enormous energy contributions.

Another option is to consider the reacting system plus a few protein residues. Although this can help in providing some insight, it cannot be considered as a reasonable model of the enzyme-active site. In particular such approaches frequently assume that some residues are ionized (which is true in the complete system), but in the model used the ionized form of these residues can be extremely unstable and would not be ionized. Some specific examples of the problem associated with gas phase models are given below (see Beware of Improper Models, below).

Despite our warning about the risk of using gas phase models, there are clear exceptions. Most notable are the cases of large metal clusters, where the effect of the environment might be relatively unimportant. In such cases one can obtain instructive mechanistic information from ab initio studies that do not include the effect of the environment around the reacting cluster (7).

A seemingly reasonable way to treat a reaction in an enzyme is to use gas phase ab initio calculations to obtain the charges and force field of the substrate (solute). These charges and force field can then be used in free energy calculations in solution and in the enzyme-active site (51). Unfortunately, the solute charges in the enzyme-active site may be different from those in the gas phase. This can lead to major problems in case of charge separation reactions, where the difference between the solvation energies of the correct charges (those obtained in solution) and the solvation of the gas phase charges can be enormous (30). Other problems are considered elsewhere (56).

QM/MM Molecular Orbital Methods

The realization that the effect of the environment of the reacting fragments must be included in studies of enzymatic reactions led to the development of the hybrid quantum mechanics/molecular mechanics (QM/MM) approach (62). This approach divides the simulation system (e.g., the enzyme-substrate complex) into two regions. The inner region (region I) contains the reacting fragments represented quantum mechanically. The surrounding protein (region II) is represented by a molecular mechanics force field. QM calculations require the solution of the Schorodinger equation for a Hamiltonian that represents the sum of the potential energy and kinetic energy of the electrons and nuclei of the system as a mathematical operator (58). In the present case the Hamiltonian of the complete system is given by:

$$H = H_{QM} + H_{QM/MM} + H_{MM}, \qquad 2.$$

where the H_{QM} is the QM Hamiltonian, $H_{QM/MM}$ is the Hamiltonian that couples region I and II, and H_{MM} is the Hamiltonian of region II. H_{QM} is evaluated by a standard QM approach:

$$V_{total} = \langle\Psi\,|H_{QM} + H_{QM/MM} + H_{MM}|\,\Psi\rangle = E_{QM} + \langle\Psi\,|H_{QM/MM}|\,\Psi\rangle + E_{MM}. \quad 3.$$

The QM/MM approach was introduced in the mid-1970s (62). However, the acceptance of this approach took a long time (perhaps because of the difficulty in realizing that medium-range electrostatic effects can be incorporated in quantum treatments using classical concepts). At any rate, the QM/MM method is now widely used and we only mention several applications of it (1, 4, 6, 18, 19, 23, 25, 26, 44, 53, 74).

The connection between the QM and MM regions can be accomplished through the so-called linked-atom treatment (19, 53, 62, 74). Fortunately, as far as enzyme catalysis is concerned, the linked-atom problem is much less serious than commonly assumed. That is, in such studies we compare enzyme and solution reactions and the effect of the linked atoms cancels itself out to a significant extent.

QM/MM approaches that involve several regions of quantum mechanical treatments have also been developed. Typically region I is treated at the most rigorous level, region II is treated by an approximated QM method, and the outside region is represented by a MM force field (an approximated implementation of this idea is presented in Reference 57). It is also possible to represent region II by fixed (frozen) electronic density, while retaining a regular density functional theory (DFT) formulation for region I (see QM Treatment of the Entire Protein, below).

The reliability of QM/MM approaches is far from obvious. In particular it is essential to use accurate QM methods and to perform an extensive configurational sampling of the reacting systems (this is essential for reliable free energy calculations). Unfortunately, regular semiempirical approaches are not sufficiently accurate, although the accuracy of such approaches can be increased by forcing them to reproduce the energetics of the reference solution reaction (14, 56, 71).

The most reliable potential surfaces are obtained by ab initio (ai) QM/MM approaches [referred to here as QM(ai)/MM approaches]. At present, however, the enormous computer time needed for obtaining proper sampling by QM(ai)/MM approaches makes such studies close to impossible. A novel way to reduce this problem is provided by using the empirical valence bond (EVB) potential as a reference for the QM(ai)/MM calculations (6). In this way one evaluates the free energy profile of the EVB surface by free energy perturbation (FEP) calculations and then calculates the free energy ΔG (EVB $\rightarrow$ ai) of moving from the EVB to the ab initio surface. These calculations are usually done by a single-step FEP calculation that involves molecular dynamics (MD) runs on the EVB surface. When the EVB surface is not sufficiently similar to the ab initio surface, it is possible to improve the convergence by using the linear response approximation approach (36). In this way one uses

$$\Delta G(EVB \rightarrow ai) = 0.5\left[\langle E_{ai} - EVB\rangle_{EVB} + \langle E_{ai} - E_{EVB}\rangle_{ai}\right], \quad 4.$$

where $\langle\ \rangle$ designates an MD average of the designated potential (51a).

Zhang et al. (74) used an interesting approach that took into account the effect of the solvent environment in ab initio free energy calculations. This approach, however, constrains the reacting fragments to move along a predefined reaction coordinate obtained from QM(ai)/MM calculations and thus neglects the effect of the solute fluctuations.

In summary, a significant advantage of QM(molecular orbital)/MM [QM(MO)/MM] approaches is that they can be easily integrated with standard QM program packages. However, the problems mentioned above still slow down the progress in realistic studies of enzymatic reactions by such approaches. Nevertheless, advances have been made in studies of different enzymatic reactions (1, 6, 18, 26, 44, 62, 74).

The EVB as a Reliable QM/MM Method

Reliable studies of enzyme catalysis require accurate results for the difference between the activation barriers in enzyme and solution. The early realization of this point led to a search for a method that could be calibrated using experimental and theoretical information about reactions in solution. It also became apparent that in studies of chemical reactions it is more physical to calibrate surfaces that reflect bond properties (i.e., valence-bond-based surfaces) than to calibrate surfaces that reflect atomic properties (e.g., MO-based surfaces). The resulting EVB method (3, 59) is outlined briefly below.

The EVB is a QM/MM method that describes reactions by mixing resonance structures that correspond to classical valence-bond structures. The corresponding Hamiltonian is represented as a matrix of semiempirical integrals (59). The diagonal elements of the EVB Hamiltonian are represented by classical MM force fields that include the interaction between the charges of each state of the reaction region (region I) and the surrounding system (region II). The off-diagonal elements of the Hamiltonian are represented by simple analytical functions that are independent of region II. The ground state energy is obtained by diagonalzing the EVB Hamiltonian.

The EVB treatment provides a natural picture of intersecting electronic states that is useful for exploring environmental effects on chemical reactions in condensed phases (3, 59). The ground state charge distribution of the reacting species (solute) polarizes the surroundings (solvent), and the charges of each resonance structure of the solute then interact with the polarized solvent (59). This coupling enables the EVB model to capture the effect of the solvent on the quantum mechanical mixing of ionic and covalent states of the solute (59).

The EVB provides an extremely convenient and reliable way of obtaining activation free energies. This is done by using a mapping procedure that gradually moves between the different EVB states (3), forcing the system to move gradually from the reactant to the product state. The mapping is performed while taking into account the change in the solute charge distribution and not only changes in the solute structure. In this way the EVB umbrella sampling procedure finds the correct TS in the combined solute-solvent reaction coordinate. This includes the important ability to evaluate nonequilibrium solvation effects. Other advantages of the EVB are discussed elsewhere (56).

The seemingly simple appearance of the EVB method may have led to the initial impression that this is an oversimplified qualitative model rather than a powerful quantitative approach. However, the model has been widely adopted as a general model for studies of reactions within large molecules and in condensed phases [see references in (56)]. Several closely related versions have been put forward with basically the same ingredients as in the EVB method [see discussion in (20)]. Nevertheless, it has been argued (54) that various adaptations of the EVB method are useful because they involve calibration using ab initio surfaces rather than experimental information. However, EVB potential surfaces have been calibrated by ab initio surfaces [e.g., (30)]. The EVB approach has been used extensively in studies of different enzymatic reactions, and some studies are considered in subsequent sections.

QM Treatments of the Entire Protein

A QM treatment of the entire protein/substrate/solvent system is possible, at least in principle. Promising progress in this direction has been offered by the so-called divide and conquer approach (73). This approach, which was originally developed for ab initio DFT studies (73), divides a large system into many subsystems, where the electron density of each subsystem reflects the effect of its surroundings. Unfortunately, the treatment of the entire protein by an ab initio approach is extremely expensive and cannot be used in free energy calculations of enzymatic reactions. Thus, most current efforts in treating the entire protein by QM approaches have been invested in semiempirical treatments with different tricks for accelerating the solution of the large self-consistent field (SCF) problem (16, 38). This approach cannot be used in ab initio studies of enzymes owing to the computational cost of evaluating the relevant integrals.

One of the most promising options for treating larger systems by ab initio approaches is provided by the frozen DFT (FDFT) and constraint DFT (CDFT) approaches (29, 51a, 68). The basic idea behind these approaches is to treat the entire protein solvent system quantum mechanically while freezing (or constraining) the electron density of the groups in region II. In this way the entire system is treated by an ab initio DFT approach, and a formally rigorous nonadditive kinetic energy functional evaluates the coupling between regions II and I.

The FDFT approach presents a general way of coupling two subsystems by means of an orbital-free and first-principle effective potential. This makes it possible to cast the concept of "embedding potential" in DFT terms. The FDFT approach is related in a formal way to the work of Cortona (13), who did not deal, however, with the issue of embedding a subsystem in a larger system, which is described by a more approximate method. Wesolowski & Warshel (68) realized that the coupling term in any hybrid method could be obtained by a partial minimization of the total energy functionals. The CDFT embedding idea has been adopted in related fields such as studies of molecules on metal surfaces [(33); see discussion in (67)].

In concluding this section, it is important to comment on the Car-Parrinello molecular dynamics (CPMD) approach (10), which emerged in recent years as an effective way for studying complex molecular systems. The CPMD approach was applied, for example, in studies of proton transfer in solution (55). However, the use of periodic boundary conditions makes it hard to apply this method to enzymatic reactions. The use of the method in gas phase calculations of the reacting fragments and a few protein residues has led to only confusing results (see Beware of Improper Models, below). Nevertheless, the option of embedding the CP method in an MM surrounding and considering it as a QM/MM method should provide a promising option (15, 70).

REPRODUCING THE OVERALL CATALYTIC EFFECT

A prerequisite for any reliable analysis of enzyme catalysis is the ability to reproduce the observed reduction of the corresponding activation free energies. The progress in addressing these problems is reviewed below.

Beware of Improper Models

The use of standard ab initio computer packages without the protein environment can result in incorrect models. This neglect of solvation effects leads, for example, to incorrect pK_as of key catalytic residues (e.g., lysine residues would tend to be neutral in this model) and completely incorrect energies for proton transfer processes. A glaring example is provided by the study of Futatsugi et al. (24), who studied the mechanism of p21ras. The problem with this work [see discussion in (27)] includes the use of a protonated Lys16 in the RS (without realizing that this residue will not be protonated in the gas phase environment used in the calculations). The unstable lysine thus serves as an artificial proton relay [see discussion in (27)]. More reasonable gas phase treatments have paid more attention to electroneutrality and to the system chosen. However, when such treatments are applied to mechanistic questions they may favor an incorrect mechanism because they ignore the effect of the protein environment [see discussion in (21)]. Other types of problems occur when one studies enzyme catalysis by modeling the reacting system in water. Such approaches preclude any chance to study catalysis because the enzyme and solution reactions are modeled in the same way (see Polar-Preoriented Hydrogen Bonds Versus Low-Barrier Hydrogen Bonds, below).

Finally, an uncritical use of the powerful CPMD model may lead to major problems. An example is given by the work of Cavalli & Carloni (11) who concluded that Gln61 is the general base in the catalytic reaction of the ras/GAP complex [see (27) for a discussion of this reaction]. To determine whether a residue may serve as a base in a nucleophilic reaction, it is essential to evaluate the free energy of proton transfer (PT) between the nucleophile and this residue. It is also useful to examine

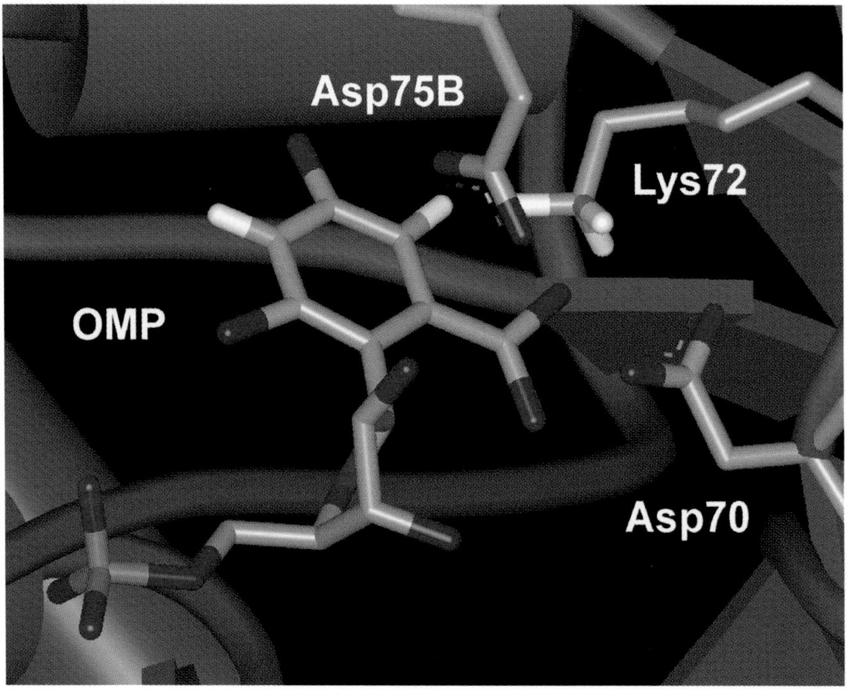

Figure 1 The active-site region of ODCase. The presented structure is based on the crystal structure of ODCase with a TS analog (PDB entry 1DV7) in which the TS analog was converted into orotidine 5′-monophosphate (OMP), and the ODCase-OMP complex was relaxed by a molecular dynamics calculation.

the pKa of the proposed base (in its protein site). Such studies that used the EVB approach [for review see (27)] have shown that Gln61 cannot be the general base in ras and that the γ-phosphate is more suitable for this role. Nevertheless, Cavalli & Carloni studied this system by the CPMD approach without being aware of the previous theoretical study and the need to validate their results by pKa calculations. They started with a subsystem that included the substrate Gln61 and a few other residues and the attacking water molecule (whose orientation was selected based on problematic force field calculations). Next they performed a short MD run while constraining the distance between the oxygen of the attacking water and the γ-phosphate to 1.8 Å. The water proton collapsed to Gln61 during this relaxation, and this led Cavalli & Carloni to conclude that Gln61 must be the general base. Of course, pushing the attacking oxygen to a bonding distance from the phosphate would force the proton to migrate to the closest base. However, this is not an adequate way to examine reaction mechanism or to identify a general base. A more valid study would require (in addition to having a complete system) starting from a relaxed RS and then calculating the free energy profile for different feasible mechanisms. This requires proper equilibration and long simulations in a FEP study.

Systems Studied by Justified Approaches

The ability to obtain reliable free energy (or even energy) profiles for enzymatic reactions was restricted almost exclusively to the EVB method until about 1996. The difficulty of using other approaches can be realized by comparing the reliable EVB profile for the reaction of triose phosphate isomerase (TIM) obtained by Åqvist & Fothergill (2) to the corresponding QM/MM results of Bash et al. (5). The latter results reflected apparently unstable calculations and probably incomplete treatment of long-range electrostatic effects. QM/MM studies with semiemperical QM Hamiltonians started to give stable results following the work of Bash and coworkers (14), who adopted the EVB idea of calibration based on the energetics of solution reactions. The past few years have witnessed a significant progress in the reliability of semiempirical QM/MM calculations with reasonable results and proper sampling, although EVB studies are probably still more reliable. The systems studied by reasonably reliable QM/MM approaches now include a wide range of reactions. Representative studies are reported in References 2, 3, 5, 14, 17, 18, 28, 39, 41, 59, 63, 66, 71, 72. Most of the reported studies involved EVB and QM/MM semiempirical MO approaches, although some studies with QM(ai)/MM methods [e.g., (74)] also gave promising results. It is important to clarify, however, that even reasonable QM(MO)/MM studies of catalysis still treat the reaction in the enzyme and in the solution by different models (e.g., QM/MM model for the enzyme and continuum model for the solution reaction). This makes it hard to assess the validity of the corresponding conclusions.

In all cases that were analyzed correctly, it was found that the catalytic effect was due to electrostatic contributions. The nature of these effects are considered below.

EXAMINING CATALYTIC PROPOSALS

As stated in the introduction, the main open questions regarding the energetics of enzyme catalysis are related to the origin of the difference between $\Delta g^{\ddagger}_{cat}$ and $\Delta g^{\ddagger}_{w}$. Many proposals have been put forward to explain the difference between k_{cat} and k_w [a partial list is given in (56, 60)]. Unfortunately, some of these proposals are so poorly defined that it is essential to reformulate them before examining their validity. It also appears that some proposals violate the law of thermodynamics and thus cannot be examined by using thermodynamic cycles. I contend that the prerequisite to any analysis of a catalytic proposal is the existence of a well-defined logical argument that can actually be tested. With the above comments in mind, I try to provide a critical evaluation of the relative merit of key catalytic proposals.

Electrostatic Preorganization is the Key Catalytic Factor

Simulation studies that consistently compared $\Delta g^{\ddagger}_{cat}$ and $\Delta g^{\ddagger}_{cage}$ have found that $\Delta g^{\ddagger}_{cat}$ is smaller because of electrostatic effects (17, 35, 47, 56, 59). It was also found that enzymes "solvate" their TS more than the corresponding TS in the reference solution reactions (58). However, the origin of this electrostatic stabilization appeared to be quite elusive. That is, the calculated interaction energy between the TS charges of the reacting atoms and the enzyme appeared to be similar to the corresponding interaction energies in solution. This puzzling observation was resolved by the realization that what counts is the entire electrostatic free energy associated with the formation of the TS (58) and not just the electrostatic interaction at the TS. This includes, of course, the penalty for the structural changes (reorganization) upon "charging" the TS. To quantify this point it is useful to express the electrostatic free energy of the TS charges using the linear response approximation with the expression introduced by Warshel and coworkers (36):

$$\Delta G(Q^{\ddagger}) = 0.5(\langle U(Q = Q^{\ddagger}) - U(Q = 0)\rangle_{Q=Q^{\ddagger}} + \langle U(Q = Q^{\ddagger}) - U(Q = 0)\rangle_{Q=0}) = 0.5(\langle \Delta U\rangle_{Q^{\ddagger}} + \langle \Delta U\rangle_0), \quad 5.$$

where U is the solute-solvent interaction potential, Q designates the residual charges of the solute atoms at the TS, and $\langle \Delta U\rangle_Q$ designates an MD average over configurations generated with the given solute charge distribution. The first term in Equation 5 is the abovementioned interaction energy at the TS, where $Q = Q^{\ddagger}$. The second term expresses the effect of the preorganization of the environment. If the environment is randomly oriented toward the TS (as is the case in water) then the second term is zero and we obtain the well-known expression:

$$\Delta G(Q^{\ddagger})_w = \frac{1}{2}\langle \Delta U\rangle_{Q^{\ddagger}}. \quad 6.$$

However, in the preorganized environment the enzyme provides a significant contribution from the second term, and the overall $\Delta G(Q^{\ddagger})$ is more negative than in water. Another way to see this effect is to realize that in water, where the solvent dipoles are randomly oriented around the uncharged form of the TS, it is essential to invest free energy to reorganize these dipoles toward the changed TS. The reaction in the protein costs less reorganization energy because the active-site dipoles (associated with polar groups, charged groups, and water molecules) are already partially preorganized toward the TS charges (59). The above TS stabilization effect is related to the well-known Marcus' reorganization energy (40) for the given step of the reaction, but it is not equal to it. More specifically, the activation energy for the i → j step of a reaction can be related by a modified Marcus equation (59) to the reorganization energy λ_{ij} and the free energy ΔG_{ij}. For example, in the analysis of Reference 17, the enzymes studied catalyze their reactions by reducing both λ_{ij} and ΔG_{ij}. However, both contributions reflect the same preorganization effect. That is, ΔG_{ij} is reduced by the preorganization term of Equation 5 because the enzyme dipoles are already pointing toward the charges of the ith intermediate, while λ_{ij} is reduced because the enzyme dipole tries to minimize the reorientation of its dipoles in the i → j step [e.g., see (72)]. More discussion of the evidence for the preorganization idea is given elsewhere (17, 56, 59). Finally, the reduction of the reorganization energy in an enzyme-active site is not due to a nonpolar environment as was proposed by some (34), but to a polar preoriented active site.

Steric Effects and the NAC Proposal

Several proposals for the origin of enzyme catalysis involve some form of ground state steric strain (22, 52). The original strain hypothesis (22) and related subsequent works (52) invoked the idea of "molding" the substrate toward the TS by strong steric forces. This idea is inconsistent with simulation studies that demonstrated that enzymes are too flexible to apply strong steric forces (59, 62). A related but more reasonable proposal has been put forward by Bruice and coworkers [e.g., (9)], who suggested that the steric confinement of enzyme-active sites brings the substrates to a so-called near attack conformation (NAC), which is closer to the corresponding TS than to the reference reaction in water. Unfortunately, this proposal was not defined in clear terms that would allow its verification. That is, the NAC configuration was taken as a rather arbitrary point along the reaction coordinate, and the location of such a point cannot be related directly to the difference between $\Delta g^{\ddagger}_{w}$ and $\Delta g^{\ddagger}_{cat}$. Furthermore, the NAC idea does not tell us what is the reason for catalysis. The absence of a clear definition motivated Shurki et al. (50) to reformulate the NAC idea by relating the enzyme confinement effect to the corresponding reduction in activation free energy. This was done by adding an external restraint potential, V' to the potential surfaces of the substrate in water, V_w, and forcing the RS probability distribution to satisfy the relationship:

$$\rho(r)^{RS}_{V_w+V'} \cong \rho(r)^{RS}_{V_p}, \qquad 7.$$

where V_p is the potential surface of the substrate in the protein. The reduction of $\Delta g^{\ddagger}_{w}$ upon addition of the restraint potential is the most logical definition of the

NAC effect. The actual effect of V′ can be examined by FEP approaches with several alternative thermodynamic cycles (50).

An actual examination of the effect of the protein restraint was performed in the specific case of haloalkane dehalogenase. The probability distribution in the protein is different from the corresponding distribution in water. However, the free energy needed for changing V_w to V_p by the constraint V′ was found to be rather small. More specifically, V′ reduces to $\Delta g^{\ddagger}_{w}$ by less than 3 kcal/mol, whereas the total catalytic effect is ~7 kcal/mol. More importantly, it appeared that the pure steric component of the NAC contribution is less than 1 kcal/mol. The remaining contribution reflected electrostatic effects, which have little to do with the original idea of steric contributions to enzyme catalysis. The finding of non-negligible electrostatic contributions from the NAC effect might simply reflect the electrostatic stabilization effect of the previous section. That is, when the enzyme stabilizes the TS it pushes down the potential surface along the reaction coordinate. Obviously the average RS configuration, $\langle r \rangle^{RS}$, is shifted toward $\langle r \rangle^{TS}$. Now with this effect $\langle r \rangle^{RS}_{w}$ is displaced farther to the left than $\langle r \rangle^{RS}_{p}$. Thus, if we examine the extreme case where the barrier in the protein is reduced to zero we would conclude that the NAC effect contributes to the entire catalytic effect. This is, however, an artificial consequence of having an ill-defined concept of NAC, since we just considered a model where the catalysis and the reduction of $\Delta g^{\ddagger}$ are due to electrostatic effects.

In summary, it is important to point out that the NAC proposal does not really explain the origin of enzyme catalysis. That is, this proposal tells us that the enzyme reduces $\langle r \rangle^{RS}$ relative to water. It does not tell us, however, how this reduction is being accomplished (steric or electrostatic effect) or whether we have an RS destabilization or a TS stabilization effect. If the NAC proposal implies that we have a pure steric effect (i.e., the catalysis is due to van der Waals forces), then it leads to a small contribution to catalysis (50). The only way for the NAC effect to be significant is for it to reflect a part of the overall TS stabilization by electrostatic effects. In this case the apparent steric effect reflects the response of the solvent to the solute charges. Here, the change in the solute charges is coupled to the changes of both the solvent and the solute components of the reaction coordinate. As demonstrated repeatedly [e.g., (50)], the change in the solvent coordinates leads to the largest contribution to catalysis, and this change is not an NAC effect (it does not depend on the solute coordinate).

Defining and Assessing the Entropic Proposal

Many proposals [e.g., (8, 31, 51)] and many textbooks [see a partial list in (56)] invoke entropic effects as major sources of enzyme catalysis. These proposals assume (correctly) that the large configurational space of the reacting fragments is drastically reduced by the enzyme, but then conclude that this leads to a large increase in k_{cat}. The entropic proposal is frequently presented using appealing but not quantitative terms (e.g., "the entropy trap effect"), and is rarely formulated in a valid way. Recent analysis (55a) demonstrates that the customary formulation leads to a major overestimate because it ignores the entropy of the TS in water (65, 56). A proper computational analysis (55a) demonstrates that the entropic contributions to catalysis are much smaller than previously thought.

Dynamical Proposals

The idea that special "dynamical" effects play a major role in enzyme catalysis has gained significant popularity in recent years [see references in (56)]. Here again, the issue of definition appears to be a major problem. Many proposals overlook the difference between the well-known fact that all reactions involve atomic motions and a requirement from true dynamical contributions to catalysis. This issue and other problems with the dynamical proposal has been analyzed in great length in several recent reviews (56, 65), where it is shown that dynamical effects do not contribute significantly to catalysis. Due to space limitations we refer the readers to the above reviews.

Reactant State Destabilization by Electrostatic Effects: A Lesson from ODCase

The general idea of reactant state destabilization (RSD) has played a major role in various catalytic proposals, including the abovementioned strain and entropy proposals. One of the most popular versions of the RSD concept is the idea that enzymes work by providing a nonpolar environment that destabilizes a highly charged ground state [see references in (66)]. As shown in Reference 59, these proposals involve improper thermodynamic cycles and do not use a proper reference state. This amounts to ignoring the desolvation energy associated with taking the RS from water into a hypothetical nonpolar enzyme site. With a proper reference state, one finds (59) that a polar TS is less stable in nonpolar sites than in water and that the RSD does not help in increasing k_{cat}/k_M. In fact, many desolvation models [e.g., (37)] involve ionized residues in nonpolar environment, but such residues would be un-ionized in nonpolar sites.

The difficulties with the RSD proposal can be illustrated by the case of ODCase [see (42) for a review]. This enzyme has the largest rate acceleration ($k_{cat}/k_w \sim 10^{17}$), and the understanding of this effect is thus a challenging task. Before the evaluation of the structure of ODCase it was proposed (37) that the enzyme provides a nonpolar low-dielectric environment and works by a desolvation mechanism. Warshel & Florian (61) demonstrated that this proposal is based on an incorrect thermodynamic cycle, and proposed that the enzyme must provide a polar environment. The recent elucidation of the structure of the enzyme [see (42) for a review] revealed that the active site contains two aspartic residues whose carbonyl groups are located near the carboxyl group of the substrate (Figure 1, see color insert). Presumably, if both carbonyl groups were negatively charged, their interactions with the orotate carboxyl would be strongly destabilizing. Wu et al. (71) performed QM/MM PMF (potential of mean force) calculations, which reproduce the catalytic effect. They then performed FEP calculations of the binding free energy of the TS and the RS and concluded that the enzyme works by an RSD mechanism. Many workers have almost instantly embraced this idea (49). However, Warshel et al. (66) pointed out that the proposed electrostatic destabilization is questionable because the strong electrostatic interaction of the carboxyl groups would result in protonation of one or the other of these groups. They also pointed out that the binding calculations of Wu et al. (71) did not reproduce the catalytic effect and that the binding calculations that were used to support the RSD proposal are challenging. The FEP calculations of Reference 66, which involved a careful assignment of special boundary condition and a proper treatment of long-range effects, did not reproduce significant RSD for the same reacting system considered by Wu et al. Furthermore, it was pointed out that the reacting system includes the entire pyrimidine ring of the substrate along with a protonated lysine amino group that most probably donates a proton to the ring during the reaction. Stabilizing interactions with the lysine and other residues largely cancels the repulsive interactions with the two asparates. When this larger region is considered as the reacting system, one finds that the dipole moment of the system increases in the TS. The preoriented Asp residues of the enzyme favor this change in dipole moment. These calculations support the view that the catalytic effect of ODCase results mainly from TS stabilization, and additional evidences against the RSD mechanism are available (66).

The RSD proposal was probably put to rest by the recent studies of Miller & Wolfenden (42), who demonstrated that mutations of Asp96 and other residues that were supposed to destabilize the orotate led to weaker rather than stronger binding. As predicted (66), this result is inconsistent with the RSD because destabilization of the RS should result in a reduction of the binding energy.

Polar-Preoriented Hydrogen Bonds Versus Low-Barrier Hydrogen Bonds

It was recognized long ago (35, 58) that hydrogen bonds (HBs) contribute significantly to catalysis. In fact, the preorganization idea includes a major emphasis on TS stabilization by the electrostatic contributions of HBs (58). The experimental findings about TS stabilization by HBs led several research groups [see review in (12)] to come up with an idea that recognized the importance of HBs but tried to attribute the HB contribution to its covalent rather than electrostatic character. As pointed out elsewhere (60), the only new element in the low-barrier (LBHB) idea is the proposal that HBs between the enzyme and charged TS of substrates involve a much larger covalent character ($X^{-\delta} \cdots H^{+\delta} \cdots Y^{-\delta}$) than the corresponding HBs in solution; otherwise we have a regular (X^{-} H-Y) electrostatic HB. The LBHB proposal is inconsistent with energy considerations (64) and is based on somewhat arbitrary interpretations of experiments rather than on direct experimental results (60). Due to space limitations, I comment only on theoretical basis of this proposal. The confusion in the field has been compounded by a series of improper calculations (46) of irrelevant systems (i.e., no protein was studied) that confused ionic HBs with LBHBs. The problems associated with these studies are considered elsewhere (60).

It is also instructive to consider the calculations by Kim et al. (32), which may be taken as a partial support for the LBHB proposal. These calculations considered the reacting fragments in water rather than in the protein. Furthermore, the few protein residues that were immersed in the water model were subjected to an artificial constraint, which led to an artificial preorganization effect. One cannot assess the LBHB contribution without comparing calculations of this effect in water with those in a protein environment.

The LBHB proposal is not supported by calculations that properly considered the effect of the protein environment. These include EVB (17, 63, 64) and QM/MM (26, 43, 45) studies. Although the conclusions are similar, I argue that most of them are tentative. That is, a reliable analysis of the LBHB proposal must involve calculations of the free energy (rather than energy) profile for the transfer of the relevant proton between the donor and acceptor that are postulated to form the given LBHB. It is also essential to compare this profile for the reaction in protein to the corresponding profile in water. This helps to determine whether there is any catalytic advantage to the LBHB charge transfer effect (64) and to validate the reliability of the calculated pKas. At present only EVB studies (17, 64) have explored this issue in a more or less complete way. Nevertheless, in some cases [e.g., (45)] the QM/MM energy differences for the two extreme points on the LBHB proton transfer profile are sufficiently large as to be considered fairly conclusive results. Thus, the LBHB proposal is not supported at present by any consistent simulation study.

CONCLUDING REMARKS

Recent years witnessed accelerated progress in modeling of enzymatic reactions. At present the EVB method still provides the most reliable results and the most careful analyses of enzyme catalysis. However, QM(MO)/MM methods are starting to provide reliable results as well, although these QM/MM approaches do not yet involve calculations of the free energy of the combined solute-solvent systems with ab initio QM surfaces. Unfortunately, there are still many attempts that do not include the enzyme in studies of enzyme catalysis. The fact that such studies are still presented illustrates that the field has not yet matured. Nevertheless, the general direction of the field is promising. There is no doubt that the next few years promise great progress in the use of reliable QM/MM and related approaches. This will involve a wider realization of the importance of proper configurational averaging (51a) in QM(ai)/MM calculations and a better understanding of the importance of clearly defined catalytic proposals. It is also certain that there will be a growing realization that consistent simulation studies provide the ultimate tool for structure-function correlation of enzymes.

ACKNOWLEDGMENTS

This work was supported by NIH grant GM 24492 and NSF grant MCB-0003872.

The *Annual Review of Biophysics and Biomolecular Structure* is online at http://biophys.annualreviews.org

LITERATURE CITED

1. Alhambra C, Gao J, Corchado JC, Villà J, Truhlar DG. 1999. Quantum mechanical dynamical effects in an enzyme-catalyzed proton transfer reaction. *J. Am. Chem. Soc.* 121:2253–58
2. Åqvist J, Fothergill M. 1996. Computer simulation of the triosephosphate isomerase catalyzed reaction. *J. Biol. Chem.* 271:10010–16
3. Åqvist J, Warshel A. 1993. Simulation of enzyme reactions using valence bond force fields and other hybrid quantum/classical approaches. *Chem. Rev.* 93: 2523–44
4. Bakowies D, Thiel W. 1996. Hybrid models for combined quantum mechanical and molecular approaches. *J. Phys. Chem.* 100:10580–94
5. Bash PA, Field MJ, Davenport RC, Petsko GA, Ringe D, Karplus M. 1991. Computer simulation and analysis of the reaction pathway of triosephosphate isomerase. *Biochemistry* 30:5826–32
6. Bentzien J, Muller RP, Florián J, Warshel A. 1998. Hybrid ab initio quantum mechanics/molecular mechanics calculations of free energy surfaces for enzymatic reactions: the nucleophilic attack in subtilisin. *J. Phys. Chem. B* 102:2293–301
7. Bloomberg MRA, Siegbahn PEM. 2001. A quantum mechanical approach to the study of reaction mechanisms of redox active metalloenzymes. *J. Phys. Chem.* 105:9376–86
8. Blow D. 2000. So do we understand how enzymes work? *Structure* 8:R77–R81
9. Bruice TC. 2002. A view at the millennium: the efficiency of enzymatic catalysis. *Acc. Chem. Res.* 35:139–48
10. Car R, Parrinello M. 1985. Unified approach for molecular dynamics and density-functional theory. *Phys. Rev. Lett.* 55: 2471–74
11. Cavalli A, Carloni P. 2002. Enzymatic GTP hydrolysis: insights from an ab initio molecular dynamics study. *J. Am Chem. Soc.* 124:3763–68
12. Cleland WW, Frey PA, Gerlt JA. 1998. The low barrier hydrogen bond in enzymatic catalysis. *J. Biol. Chem.* 273: 22529–32
13. Cortona P. 1991. Self-consistently determined properties of solids without band-structure calculations. *Phys. Rev. B* 44:8454–58
14. Cunningham MA, Ho LL, Nguyen DT, Gillilan RE, Bash PA. 1997. Simulation of the enzyme reaction mechanism of malate dehydrogenase. *Biochemistry* 36:4800–16
15. Eichinger M, Tavan P, Hutter J, Parrinello M. 1999. A hybrid method for solutes in complex solvents: density functional theory combined with empirical force fields. *J. Chem. Phys.* 110:10452–67
16. Ermolaeva MD, van der Vaart A, Merz KM Jr. 1999. Implementation and testing of a frozen density matrix—divide and conquer algorithm. *J. Phys. Chem. A* 103: 1868–75
17. Feierberg I, Åqvist J. 2002. Computational modelling of enzymatic keto-enol isomerization reactions. *Theor. Chem. Acc.* 108:71–84
18. Field M. 2002. Stimulating enzyme reactions: challenges and perspectives. *J. Comp. Chem.* 23:48–58
19. Field MJ, Bash PA, Karplus M. 1990.

A combined quantum mechanical and molecular mechanical potential for molecular dynamics simulations. *J. Comp. Chem.* 11:700–33

20. Florian J. 2002. Comment on molecular mechanics for chemical reactions. *J. Phys. Chem. A* 106:5046–47
21. Florian J, Goodman MF, Warshel A. 2003. Computer simulation studies of the fidelity of DNA polymerases. *Biopolymers.* In press
22. Ford LOJ, Johnson LN, Machin PA, Phillips DC, Tijian RJ. 1974. Crystal structure of lysozyme-tetrasaccharide lactone complex. *J. Mol. Biol.* 88:349–71
23. Friesner R, Beachy MD. 1998. Quantum mechanical calculations on biological systems. *Curr. Opin. Struct. Biol.* 8:257–62
24. Futatsugi N, Hata M, Hoshino T, Tsuda M. 1999. Ab initio study of the role of lysine 16 for the molecular switching mechanism of Ras protein p21. *Biophys. J.* 77:3287–92
25. Gao J. 1996. Hybrid quantum and molecular mechanical simulations: an alternative avenue to solvent effects in organic chemistry. *Acc. Chem. Res.* 29:298–305
26. Garcia-Viloca M, Gonzalez-Lafont A, Lluch JM. 2001. A QM/MM study of the racemization of vinylglycolate catalysis by mandelate racemase enzyme. *J. Am. Chem. Soc.* 123:709–21
27. Glennon TM, Villà J, Warshel A. 2000. How does GAP catalyze the GTPase reaction of Ras? A computer simulation study. *Biochemistry* 39:9641–51
28. Hansson T, Nordlund P, Åqvist J. 1997. Energetics of nucleophile activation in a protein tyrosine phosphatase. *J. Mol. Biol.* 265:118–27
29. Hong G, Štrajbl M, Wesolowski TA, Warshel A. 2000. Constraining the electron densities in DFT method as an effective way for ab initio studies of metal catalyzed reactions. *J. Comp. Chem.* 21: 1554–61
30. Hwang JK, King G, Creighton S, Warshel A. 1988. Simulation of free energy relationships and dynamics of S_N2 reactions in aqueous solution. *J. Am. Chem. Soc.* 110:5297–311
31. Jencks WP. 1986. *Catalysis in Chemistry and Enzymology*. New York: Dover Publ.
32. Kim K, Kim D, Lee JY, Tarakeshwar P, Oh KS. 2002. Catalytic mechanism of enzymes: preorganization, short strong hydrogen bond, and charges buffering. *Biochemistry* 41:5300–6
33. Kluner T, Govind N, Wang YA, Carter EA. 2001. Prediction of electronic excited states of adsorbates on metal surfaces from first principles. *Phys. Rev. Lett.* 86:5954–57
34. Krishtalik LI. 1980. Catalytic acceleration of reactions by enzymes. Effect of screening of a polar medium by a protein globule. *J. Theor. Biol.* 86:757–71
35. Leatherbarrow RJ, Fersht AR, Winter G. 1985. Transition-state stabilization in the mechanism of tyrosyl-trna synthetase revealed by protein engineering. *Proc. Natl. Acad. Sci. USA* 82:7840–44
36. Lee FS, Chu ZT, Bolger MB, Warshel A. 1992. Calculations of antibody-antigen interactions: microscopic and semi-microscopic evaluation of the free energies of binding of phosphorylcholine analogs to mcpc603. *Protein Eng.* 5:215–28
37. Lee JK, Houk KN. 1997. A proficient enzyme revisited: the predicted mechanism for orotidine monophosphate decarboxylase. *Science* 276:942–45
38. Lee TS, Yang W. 1998. Frozen density matrix approach for electronic structure calculations. *Int. J. Quant. Chem.* 69:397–404
39. Lyne PD, Mulholland AJ, Richards WG. 1995. Insights into chorismate mutase catalysis from a combined QM/MM simulation of the enzyme reaction. *J. Am. Chem. Soc.* 117:11345–50
40. Marcus RA. 1956. On the theory of oxidation-reduction reactions involving electron transfer. *J. Chem. Phys.* 24:966–78

41. Marti S, Andres J, Moliner V, Silla E, Tunon I, Bertran J. 2001. Transition structure selectivity in enzyme catalysis: a QM/MM study of chorismate mutase. *Theor. Chem. Acc.* 105:207–12
42. Miller B, Wolfenden R. 2002. Catalytic proficiency: the unusual case of Omp decarboxylase. *Annu. Rev. Biochem.* 71: 847–85
43. Molina PA, Sikorski RS, Jensen JH. 2003. NMR chemical shifts in the low-pH form of a-chymotrypsin. A QM/MM and ONIOM-NMR study. *Theor. Chem. Acc.* In press
44. Monard G, Merz KM. 1999. Combined quantum mechanical/molecular mechanical methodologies applied to biomolecular systems. *Acc. Chem. Res.* 32:904–11
45. Mulholland AJ, Lyne PD, Karplus M. 2000. Ab initio QM/MM study of the citrate synthase mechanism. A low-barrier hydrogen bond is not involved. *J. Am Chem. Soc.* 122:534–35
46. Pan Y, McAllister MA. 1998. Characterization of low-barrier hydrogen bonds. 6. Cavity polarity effects on the formic acid-formate anion model system. An ab initio and DFT investigation. *J. Am. Chem. Soc.* 120:166–69
47. Pauling L. 1946. Molecular architecture and biological reactions. *Chem. Eng. News.* 24:1375–77
48. Polanyi M. 1921. Überadsorptionskatalyse *Z. Elektrochem.* 27:143–52
49. Rouhi AM. 2000. The buzz about a remarkable enzyme. *Chem. Eng. News.* 78:42–46
50. Shurki A, Štrajbl M, Villà J, Warshel A. 2002. How much do enzymes really gain by restraining their reacting fragments? *J. Am. Chem. Soc.* 124:4097–107
51. Stanton RV, Peräkylä M, Bakowies D, Kollman PA. 1998. Combined ab initio and free energy calculations to study reactions in enzymes and solution: amide hydrolysis in trypsin and aqueous solution. *J. Am. Chem. Soc.* 120:3448–57
51a. Štrajbl M, Hong G, Warshel A. 2002. Ab-inition QM/MM simulation with proper sampling: "first principle" calculations of the free energy of the auto-dissociation of water in aqueous solution. *J. Phys. Chem. B.* 106:13333–43
52. Tapia OA, Andrês J, Safront VS. 1994. Enzyme catalysis and transition structures in vacuo. Transition structures for the enolization, carboxylation and oxygenation reactions in ribulose-1,5-bisphosphate carboxylase/oxygenase enzyme (rubisco). *J. Chem. Soc. Faraday Trans.* 90:2365–74
53. Théry V, Rinaldi D, Rivail J-L, Maigret B, Ferenczy GG. 1994. Quantum mechanical computations on very large molecular systems: the local self-consistent field method. *J. Comp. Chem.* 15:269–82
54. Truhlar DG. 2003. Reply to comment on molecular mechanics for chemical reactions. *J. Phys. Chem. A* 106(19):5048–50
55. Tuckerman M, Marx D, Klein ML, Parrinello M. 1997. On the quantum nature of the shared proton in hydrogen bonds. *Science* 275:817–20
55a. Villà J, Štrajbl M, Glennon TM, Sham YY, Chu T, Warshel A. 2000. How important are entropic contributions to enzyme catalysis? *Proc. Natl. Acad. Sci. USA* 97:11899–904
56. Villà J, Warshel A. 2001. Energetics and dynamics of enzymatic reactions. *J. Phys. Chem. B* 33:7887–907
57. Vreven T, Morokuma K. 2000. The oniom (our own n-layered integrated molecular orbital + molecular mechanics) method for the first singlet excited (s1) state photoisomerization path of a retinal protonated Schiff base. *J. Chem. Phys.* 113: 2969–75
58. Warshel A. 1978. Energetics of enzyme catalysis. *Proc. Natl. Acad. Sci. USA* 75: 5250–54
59. Warshel A. 1991. *Computer Modeling of Chemical Reactions in Enzymes and Solutions.* New York: Wiley
60. Warshel A. 1998. Electrostatic origin of

the catalytic power of enzymes and the role of preorganized active sites. *J. Biol. Chem.* 273:27035–38
61. Warshel A, Florian J. 1998. Computer simulations of enzyme catalysis: finding out what has been optimized by evolution. *Proc. Natl. Acad. Sci. USA* 95:5950–55
62. Warshel A, Levitt M. 1976. Theoretical studies of enzymic reactions: dielectric, electrostatic and steric stabilization of the carbonium ion in the reaction of lysozyme. *J. Mol. Biol.* 103:227–49
63. Warshel A, Naray-Szabo G, Sussman F, Hwang J-K. 1989. How do serine proteases really work? *Biochemistry* 28:3629–73
64. Warshel A, Papazyan A. 1996. Energy considerations show that low-barrier hydrogen bonds do not offer a catalytic advantage over ordinary hydrogen bonds. *Proc. Natl. Acad. Sci. USA* 93:13665–70
65. Warshel A, Parson WW. 2001. Dynamics of biochemical and biophysical reactions: insight from computer simluations. *Q. Rev. Biophys.* 34:563–679
66. Warshel A, Villà J, Štrajbl M, Florián J. 2000. Remarkable rate enhancement of orotidine 5′-monophosphate decarboxylase is due to transition state stabilization rather than ground state destabilization. *Biochemistry* 39:14728–38
67. Wesolowski T. 2002. Comment on "prediction of electronic excited states of adsorbates on metal surfaces from first principles." *Phys. Rev. Lett.* 88:209701–1
68. Wesolowski TA, Warshel A. 1993. Frozen density functional approach for ab initio calculations of solvated molecules. *J. Phys. Chem.* 97:8050–53
69. Wolfenden R, Snider MJ. 2001. The depth of chemical time and the power of enzymes as catalysts. *Acc. Chem. Res.* 34:938–45
70. Woo TK, Margl PM, Blochl PE, Ziegler T. 1997. A combined Car-Parrinello QM/MM implementation for ab initio molecular dynamics simulations of extended systems: application to the transition metal catalysis. *J. Phys. Chem. B* 101:7877–80
71. Wu N, Mo Y, Gao J, Pai EF. 2000. Electrostatic stress in catalysis: structure and mechanism of the enzyme orotidine monophosphate decarboxylase. *Proc. Natl. Acad. Sci. USA* 97:2017–22
72. Yadav A, Jackson RM, Holbrook JJ, Warshel A. 1991. Role of solvent reorganization energies in the catalytic activity of enzymes. *J. Am. Chem. Soc.* 113:4800–5
73. Yang W. 1991. Direct calculation of electron density in density-functional theory. *Phys. Rev. Lett.* 66:1438–41
74. Zhang Y, Liu H, Yang W. 2000. Free energy calculation on enzyme reactions with an efficient iterative procedure to determine minimum energy paths on a combined ab initio QM/MM potential energy surface. *J. Chem. Phys.* 112:3483–92

Annu. Rev. Biophys. Biomol. Struct. 2003. 32:445–68
doi: 10.1146/annurev.biophys.32.110601.142433

First published online as a Review in Advance on February 19, 2003

STRUCTURE AND FUNCTION OF THE CALCIUM PUMP

David L. Stokes[1] and N. Michael Green[2]
[1]*Skirball Institute of Biomolecular Medicine, Department of Cell Biology, New York University School of Medicine, New York, New York 10012; email: stokes@saturn.med.nyu.edu*
[2]*National Institute of Medical Research, The Ridgeway, Mill Hill, London NW7 1AA, United Kingdom; email: mgreen@nimr.mrc.ac.uk*

Key Words ion transport, Ca^{2+}-ATPase, reaction cycle, X-ray crystallography, electron microscopy

■ **Abstract** Active transport of cations is achieved by a large family of ATP-dependent ion pumps, known as P-type ATPases. Various members of this family have been targets of structural and functional investigations for over four decades. Recently, atomic structures have been determined for Ca^{2+}-ATPase by X-ray crystallography, which not only reveal the architecture of these molecules but also offer the opportunity to understand the structural mechanisms by which the energy of ATP is coupled to calcium transport across the membrane. This energy coupling is accomplished by large-scale conformational changes. The transmembrane domain undergoes plastic deformations under the influence of calcium binding at the transport site. Cytoplasmic domains undergo dramatic rigid-body movements that deliver substrates to the catalytic site and that establish new domain interfaces. By comparing various structures and correlating functional data, we can now begin to associate the chemical changes constituting the reaction cycle with structural changes in these domains.

CONTENTS

HISTORICAL PERSPECTIVE

The control of ion balance across the cell membrane was originally the province of physiology until almost 50 years ago, when evidence began to emerge that ion gradients were created by ATP-driven cation pumps, opening the way to biochemical analysis. The pumps, or ATPases, were located mainly in the plasma membrane or in the internal membranes of the endoplasmic reticulum, and the resulting gradients were used in a variety of signaling systems mediated by gated ion channels. Although originally discovered in crab sciatic nerve (85), much of the early work on the sodium pump (Na^+/K^+-ATPase) used the resealed ghost of the erythrocyte as a model system (75). This was not a rich source of protein, but it allowed independent control over the ionic environments on either side of the membrane, thus providing a variety of ion exchange parameters (32). From the beginning, work on the calcium pump (Ca^{2+}-ATPasc) (26) utilized the sarcoplasmic reticulum of muscle, a highly enriched source that is well suited to biochemical and structural studies. Parallel studies on Na^+/K^+-ATPase with the renal cortex indicated that the two pumps employed largely similar reaction cycles; this, together with 35% identity in their amino acid sequences, provided the basis for founding the now rather large family of P-type ATPases (28). Extensive studies of kinetics and the effects of chemical modification and site mutation have led to an ever-increasing understanding of pump function [reviewed in (61, 64)]. Nevertheless, fundamental questions related to energy transduction (in this case, the interconversion of chemical and osmotic energy) have only been approachable in the past two years, following the determination of two Ca^{2+}-ATPase structures by X-ray crystallography (93, 94). Together with cryoelectron microscopy, which provides lower-resolution maps that can be modeled at an atomic level with the X-ray structures (51, 104), several conformations have so far been defined. This represents a good start in describing kinetic and chemical properties of the various reaction cycle intermediates in structural terms and thus in better understanding the molecular mechanism.

REACTION CYCLE

From a biochemical perspective, alternating steps of ion binding and phosphate transfer characterize the reaction cycle for these pumps [illustrated in Table 1 and reviewed in (21, 36, 62)]. The central event and the hallmark of this family of P-type

TABLE 1 Reaction intermediates of Ca^{2+}-ATPase and their chemical modifications

Reaction intermediate	**Physiological ligands**[a]	**Stabilizing ligand**	**Chemical modifications**[b]
E_2H[c]		pH6/TG	Two-dimensional crystals scallop/NaK (11, 79) **P-A**: Fe cleavage TGES (71) **P-N**: SH/DTNB ++ (66) Glut. cross-link (80) **A**: T2 cleavage 100 V8/PK cleavage 100 (19, 20) **M**: TG binding (81)
↓	$-nH^+$		
E_1		pH7-8	
↓	$+ 2Ca^{++}$		
E_1Ca_2[c]		Ca	Three-dimensional crystals (91) **P-N**: SH/DTNB +++ (66) Glut. cross-link slow (80) **A**: T2 cleavage 210 V8/PK cleavage 110 (20)
↓	+ ATP		
$E_1MgATPCa_2$		AMPPNP/ AMPPCP	**P**: Mg^{2+} fast exchange (78) **P-N**: SH/DTNB ++ (66) No glut. cross-link (80) **A**: T2 cleavage 60 V8/PK cleavage 10 (20)
↓			
$E_1MgP(Ca_2)ADP$		AlF_4^-/ADP CrATP/Ca^{2+}	**P-N**: SH/DTNB + (66) Glut. cross-link fast (80)[g] **A**: T2 cleavage 70 V8/PK cleavage 0 (20) **M**: TG + CrATP slow slow Ca^{2+} off (81)
↓	−ADP		
E_2MgPCa_2			
↓	$-2Ca^{++} + H^+$		
E_2MgPH[d]		BeF_3/VO_3 CrATP	Two-dimensional crystals $V_{10}O_{29}/VO_4$ (91) **P-N**: no glut. cross-link (80) **P-A**: Fe cleaves TGES (71) **A**: T2 cleavage 0 V8/PK cleavage 0 (19)

(Continued)

TABLE 1 (*Continued*)

Reaction intermediate	Physiological ligands[a]	Stabilizing ligand	Chemical modifications[b]
↓	$+ H_2O$		
E_2MgP_iH[e]		Mg_2F_4	**P-N**: SH/DTNB + (67)
↓	$-Mg^{++}, -PO_4$		
E_2H			

Additional chemical modifications[f]

Unphosphorylated species ($E_2 \ldots E_1Ca_2$): **P**: Mg^{2+} fast exchange (78); **N**: TNPAMP low fluorescence (9)

Phosphorylated species ($E_1MgP(Ca_2)ADP \ldots E_2MgP$): **P**: Mg^{2+} occluded (98); **N**: TNPAMP high fluorescence. (9)

E_1 species: ($E_1 \ldots E_1MgP(Ca_2)ADP$): **P**: NDB-$Cys^{344}$ high fluorescence (97), **P-A**: no Fe cleavage TGES (33, 71), **N**: FITC low fluorescence (74)

E_2 species: ($E_2MgPCa_2 \ldots E_2$): **P**: NBD-Cys^{344} low fluorescence (97), **N**: FITC high fluorescence (74)

[a]Ligands gained or lost at each step of the reaction cycle.

[b]Results for each intermediate, with relevant domains indicated by bold type. The numbers listed for proteolytic cleavages are rate constants relative to unliganded species in EGTA and the pluses listed for DTNB indicate the relative accessibility of cysteines.

[c]Existing X-ray structure.

[d]Existing EM structure.

[e]X-ray structure in progress.

[f]Results not specific for individual intermediates, but for neighboring intermediates in the reaction cycle.

[g]For glutaraldehyde cross-linking, this E_1-P intermediate was generated with acetyl phosphate because nucleotide blocks cross-linking in all states.

TG, thapsigargin; T2, tryptic cleavage site at R^{198}; V8/PK, V8 protease and proteinase K cleavage sites; glut, glutaraldehyde; FITC, fluorescein isothiocyanate; NBD, 4-nitro-2,1,3-benzoxadiazole; DTNB, 5,5′-dithiobis-2-nitrobenzoate.

ATPases is the formation of an acid-stable aspartyl phosphate intermediate. ATP is the preferred substrate for phosphoryl transfer, which is initiated by cooperative binding of two cytoplasmic calcium ions to transport sites. The energy of this phosphoenzyme is postulated to fuel a conformational change that closes the ion gate from the cytoplasm, reduces the affinity of these transport sites for calcium, and opens the ion gate toward the lumenal side of the membrane. After releasing calcium, protons are bound to the transport sites and the aspartyl phosphate is hydrolyzed to complete the cycle. A net exchange of protons for calcium results from the release of these same protons to the cytoplasm prior to binding the next pair of cytoplasmic calcium ions. The coordinated affinity change at the transport sites is the key to active transport as the cytoplasmic calcium ions are bound with μM affinity and released to the lumen with mM affinity.

In broad terms, energy transduction is accomplished by sequential changes in chemical specificity for phosphoryl transfer and in vectorial specificity for ion binding (44, 47). In particular, vectorial specificity refers to whether the pump binds ions from the cytoplasmic or lumenal side of the membrane, and chemical specificity refers to whether the catalytic site reacts with ATP or inorganic phosphate. Early kinetic models for Ca^{2+}-ATPase and Na^+/K^+-ATPase postulated two distinct conformations dubbed E_1 and E_2 (1, 56, 76). The former was accessible to cytoplasmic ions and transferred phosphate to and from ATP, whereas the latter was accessible to extracellular/lumenal ions and reacted directly with inorganic phosphate. More recent models describe the cycle as a series of unique conformations transformed by sequential binding and release of substrates (46). A long-standing goal of structural studies has been to define these conformations and their interactions with the relevant substrates.

ELECTRON MICROSCOPY OF Ca^{2+}-ATPase AND OTHER P-TYPE ATPases

Structural studies of both Ca^{2+}-ATPase and Na^+/K^+-ATPase were initiated in the early 1980s when it was discovered that two-dimensional arrays could be induced within native membranes. In particular, vesicular preparations of porcine kidney Na^+/K^+-ATPase (87) and, later, Ca^{2+}-ATPase from rabbit sarcoplasmic reticulum (23) and H^+/K^+-ATPase from gastric mucosa (77) formed two-dimensional arrays when incubated in vanadate-containing solutions that stabilized an E_2 conformation. Orthovanadate was used because of its characteristic inhibition of many P-type ATPases, presumably as a transition-state analog of phosphate during hydrolysis of E_2-P (22, 73). However, it was later discovered that decavanadate was actually responsible for Ca^{2+}-ATPase crystallization (48, 60, 96) and that the same Na^+/K^+-ATPase crystal form could also be induced by phospholipase A2 in the absence of vanadate (63). A variety of other two-dimensional crystals were studied, including those from scallop adductor muscle in both E_2 (11) and E_2-P (35) conformations and from Na^+/K^+-ATPase stabilized by ATP analogs (86). In addition, both Ca^{2+}-ATPase (25) and H^+-ATPase (17) were crystallized in the E_1 conformation. Conditions for the former required both detergent and lipid and resulted in stacked crystalline layers with Ca^{2+}-ATPase molecules protruding symmetrically from either side thus representing a thin three-dimensional crystal (89). These conditions were optimized to produce larger crystals that were analyzed extensively by electron diffraction (83, 84) and that were ultimately used for the X-ray crystallographic structure of $E_1 \cdot Ca_2$ described below.

These various preparations have resulted in numerous two- and three-dimensional structures by electron microscopy (EM) of Ca^{2+}-ATPase (57, 68, 95), Na^+/K^+-ATPase (37, 59, 79), H^+/K^+-ATPase (103), and H^+-ATPase (7) preserved first in negative stain and later in the frozen, unstained state. Early structures showed that the molecule had a compact, pear-shaped cytoplasmic head that was connected to the membrane by a thinner stalk. Structures of Na^+/K^+-ATPase showed protein domains on both sides of the membrane, consistent with the extracellular contribution of the β subunit. An ongoing controversy over the number of transmembrane helices was finally resolved by structures for Ca^{2+}-ATPase (106) and H^+-ATPase

(7) at 8 Å resolution, which revealed 10 transmembrane helices as originally predicted from the Ca^{2+}-ATPase sequence (55). Comparison of these structures (88), as well as projection structures from the thin, three-dimensional crystals of Ca^{2+}-ATPase (68, 91), revealed large domain movements thought to characterize the conformational change between E_2 and E_1. Currently, the higher-resolution EM structures are proving useful for fitting atomic coordinates determined by X-ray crystallography (51, 79, 104) thus elucidating different conformational states at an atomic level.

MOLECULAR ARCHITECTURE OF Ca^{2+}-ATPase

A major breakthrough for the field came from the X-ray crystal structure of Ca^{2+}-ATPase in the $E_1 \cdot Ca_2$ conformation (93). The resulting atomic model revealed four basic domains (Figure 1) (Figure 2, see color insert). The transmembrane domain is almost entirely helical and includes the short loops on the lumenal and cytoplasmic surfaces; four of the transmembrane helices extend into the cytoplasm to form the stalk seen in earlier EM structures. The three cytoplasmic domains are derived predominantly from two large cytoplasmic loops between transmembrane helices M2/M3 and M4/M5. The latter loop forms the phosphorylation (P) domain, which sits directly on top of M4 and M5, and the nucleotide binding (N) domain, which is an insert within the P domain. These two domains are named for the ligands they carry, namely D^{351}, which forms the phosphoenzyme, and the site near K^{492} that binds ATP. The third cytoplasmic loop, dubbed the transduction or actuator (A) domain, comprises the smaller M2/M3 loop as well as the N terminus.

Transmembrane Domain

The most important landmark of the transmembrane domain is the calcium binding site, which in the E_1 conformation cooperatively binds two calcium ions from the cytoplasm (Figure 3*a*, see color insert). The associated residues correspond remarkably well with those previously identified by site-directed mutagenesis. Initially, these residues were identified by phosphorylation of mutant pumps (15), which require calcium binding when performed in the forward direction with ATP but not when performed in reverse using P_i. Later, the sequential nature of calcium binding was used to distinguish the two sites because binding by the first calcium ion was sufficient to prevent phosphorylation from P_i, whereas binding by both calcium ions was required for phosphorylation from ATP. Thus, individual calcium site mutants displaying P_i phosphorylation that was insensitive to calcium were assigned to the first site, whereas those with normal sensitivity were assigned to the second site (3). This analysis was later corroborated by direct measurement of calcium binding stoichiometries in a more efficient expression system (92).

Specifically, the X-ray structure showed oxygen ligands for calcium provided by residues on M4, M5, M6, and M8. N^{768} and E^{771} on M5, T^{799} and D^{800} on M6, and E^{908} on M8 bound the first calcium ion. The contribution of adjacent residues

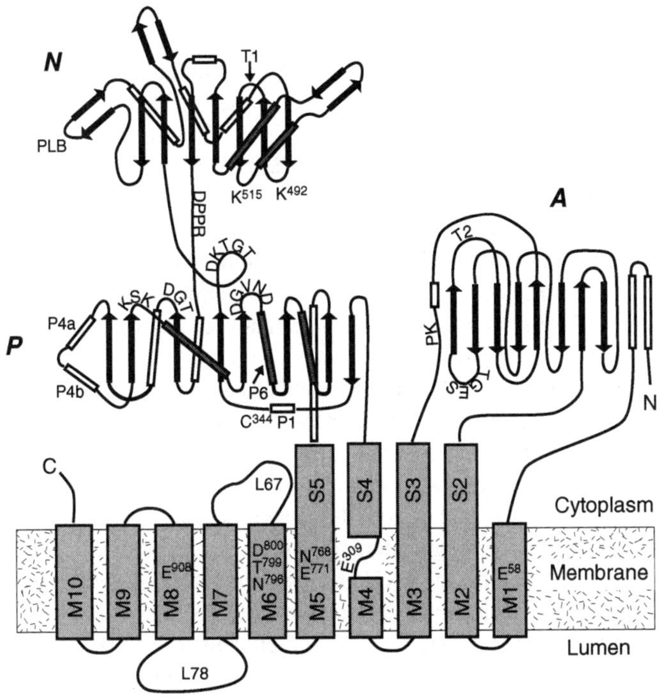

Figure 1 Topology of Ca^{2+}-ATPase indicating the location of key functional sites described in the text. Sequence motifs are indicated by the single-letter code with numbers corresponding to the sequence of rabbit fast-twitch muscle Ca^{2+}-ATPase (SERCA1). Cytoplasmic domains are denoted N, P, and A; P1, P4a,b, and P6 correspond to particular helices within the P domain. Shaded helices in the P and N domains are in front of the central sheet, whereas unshaded helices are behind. Proteolytic sites are indicated as T1 and T2 for trypsin and PK for proteinase K. Minor loops between transmembrane helices are indicated as L67 and L78, and PLB indicates the binding site for phospholamban.

along M6 was aided by flexibility in this helix due to the nonhelical hydrogen bonding of carbonyl oxygens of N^{799} and D^{800}. The second ion binding site was quite different, with extensive contributions from main chain carbonyl oxygens along M4 as well as side chain oxygens from E^{309} on M4 and N^{796} and D^{800} on M6. A highly conserved sequence motif on M4 ($PEGL^{311}$) lies at the heart of this second site and likely represents the key to cooperativity. In particular, binding of the first ion to M5/M6/M8 must somehow induce the favorable configuration of M4 to provide for the cooperative binding of the second ion. This implied structural flexibility of M4, as well as its direct link to the phosphorylation site, places these particular structural elements at the center of the global conformational change that accompanies calcium binding to the enzyme.

Phosphorylation Domain

Phosphorylation occurs on an aspartate ~30 residues beyond the C-terminal end of M4 in a highly conserved region that serves as a signature sequence for P-type ATPases: $DKTGT^{355}$. Initially, this phosphorylation site was identified by chemical means; later, site-directed mutagenesis of the aspartate and its conserved neighbors was shown to interfere specifically with phosphorylation (2). The fold of the P domain had previously been deduced by analogy with bacterial dehalogenases and had been used to define a superfamily of hydrolases that also included small-molecule phosphatases (5). This deduction relied on an alignment of several short, highly conserved sequences that play key roles in the catalytic sites of all these enzymes: $DKTGT^{355}$, KSK^{686}, TGD^{627}, and $DGVND^{707}$ (Figure 1). The resulting structural prediction consisted of a Rossmann fold with an inserted ATP binding domain and was consistent with predictions of secondary structure and with effects of mutagenesis and chemical modification throughout both domains (90). A common catalytic mechanism was also implied by the fact that the nucleophilic aspartates (D^{351}) of both dehalogenases and phosphatases form a covalent intermediate during the reaction cycle (16). These predictions were confirmed by the X-ray structure of Ca^{2+}-ATPase, which revealed not only a Rossmann fold but also a common arrangement of catalytic site residues (Figure 4, see color insert) relative to both dehalogenases (39) and phosphatases (100).

Although the Rossmann fold represents the template for this phosphorylation domain, Ca^{2+}-ATPase has several adaptations relevant to the energy coupling required for calcium transport. The fold is characterized by a central, six-stranded parallel β-sheet flanked by three α-helices on each side (Figures 1 and 4). As predicted from the sequence of Ca^{2+}-ATPase, the α-helices alternate with the β-strands along the peptide chain. Typical of α/β structures, the active site exists at the topological break point dividing the first three strands from the last three strands, and critical residues appear in the loops between each strand and the subsequent helix (10). The N domain interrupts the loop following the phosphorylated aspartate which contains the most highly conserved sequences of the family, namely the signature sequence $DKTGT^{355}$ following the phosphorylation site and $DPPR^{604}$ in the return from the N domain. The P domain is firmly connected to the transmembrane domain by cytoplasmic extensions of M4 and M5, which constitute so-called stalk helices (S4 and S5). In the case of S5, its further cytoplasmic extension represents one of six flanking helices of the Rossmann fold. S4 represents an extra structural element, which is followed by a short, antiparallel β-strand and α-helix leading to the beginning of the Rossmann fold. This preliminary β-strand extends the central β-sheet and probably serves to couple movements of M4 to those of the phosphorylation domain. The preliminary α-helix, dubbed P1, runs underneath the Rossmann fold and interacts with lower parts of S5 as well as with the loop between M6 and M7, thus potentially coupling movements of the membrane components with those of the phosphorylation domain. Finally, there is an insert consisting of a strand and two short helices (P4a and P4b) on the other end of the Rossmann fold. This insert is on the periphery of the structure and is highly variable among P-type ATPases, being considerably larger in Na^+/K^+-ATPase and absent in CadA (90).

Nucleotide Binding Domain

The N domain represents a novel fold comprising a seven-stranded, antiparallel β-sheet sandwiched between two helix bundles (Figure 2*a*). The nucleotide site is tucked under a flap created by one of the α-helices. This site was identified by soaking TNP-AMP into the crystals (93) and is generally consistent with site-directed mutagenesis and chemical modification of nearby residues F^{487}, K^{492}, and K^{515} (2, 8). However, the binding geometry of ATP may be different given that the TNP moiety produces a large increase in affinity and, in the case of TNP-ATP, prevents transfer of γ-phosphate to the catalytic aspartate (101). In any case, the most striking observation is the large distance between this site and the target aspartate (>25 Å), which makes phosphate transfer impossible in this conformation. Given the apparent flexibility of the loops connecting the P and N domains, significant mobility of the N domain has been suggested (104), which is consistent with systematic changes in glutaraldehyde cross-linking of these domains (Table 1) and with direct measurements of the rotational dynamics of the N domain (41). Furthermore, various fluorescence probes and cysteine reactivity indicate lowered solvent exposure of the nucleotide binding site upon phosphorylation (8, 66) (Table 1). A variety of indirect structural evidence also supports the mobility of the N domain. In particular, the three-dimensional crystal packing of Ca^{2+}-ATPase in the $E_1 \cdot Ca_2$ conformation is variable (14), and comparisons of the X-ray structure with a cryo-EM projection map from this same crystal form (68) suggests that this variability is due to different angles between N and P domains. Also, both the N and A domains have variable orientations in a variety of three-dimensional structures of Ca^{2+}-ATPase, Na^+/K^+-ATPase, and H^+-ATPase (Figure 2).

Actuator or Transduction Domain

The third cytoplasmic domain, which was originally called the transduction domain and later the nose, β-strand, and A domain, is composed primarily of β-strands, which form a distorted jelly roll and are tethered to M2 and M3 by two flexible, unstructured loops (Figures 1 and 2). Two α-helices are packed against this jelly roll, which come from the N terminus of the pump, and are connected by another long, unstructured loop to M1. The characteristic $TGES^{184}$ motif is on an exposed loop in the original X-ray crystal structure, but a dramatic reorientation of this domain in other conformations gives this motif a plausible role in catalysis, as implied by mutagenesis studies of these residues (2). Experimental evidence for this reorientation comes primarily from the conformational dependence of proteolytic cleavage. In particular, the tryptic site at R^{198} near the $TGES^{184}$ loop (42, 49) as well as a variety of cleavage sites in the flexible loop of Ca^{2+}-ATPase and Na^+/K^+-ATPase leading to M3 (50, 65) undergo rapid proteolysis in the $E_1 \cdot Ca_2$ conformation relative to the E_2 conformation.

STRUCTURAL CHANGES INDUCED BY CALCIUM BINDING

As stated, a major goal of structural studies has been to elucidate changes that accompany the reaction cycle of the pump. The most recent X-ray structure takes a big step in this direction by revealing the E_2 conformation of Ca^{2+}-ATPase stabilized by thapsigargin (TG) in the calcium-free state, dubbed the $E_2 \cdot TG$ conformation (94). As expected by earlier comparisons with E_2 structures by cryo-EM (91, 93, 104), this conformation has large changes relative to $E_1 \cdot Ca_2$. In particular, the three cytoplasmic domains undergo large, rigid-body movements, namely a 110° rotation of the A domain about an axis normal to the membrane, a 30° rotation of the P domain with respect to the membrane plane, and a further 50° rotation of the N domain relative to the P domain. These large-scale movements were anticipated in a general way by a host of earlier spectroscopic, enzymatic, and biochemical studies (45) and result in a compact cytoplasmic head that contrasts with the markedly open structure of $E_1 \cdot Ca_2$ (Figure 2). Even with their closer association, the interactions between the cytoplasmic domains in $E_2 \cdot TG$ are rather weak and therefore potentially labile during the reaction cycle. Interestingly, the structure within individual cytoplasmic domains is largely unchanged. Although this observation is consistent with the small changes in secondary structure measured by circular dichroism (31), one might have expected some differences at the phosphorylation site to account for its activation after calcium binding.

In contrast to the rigid-body movements of the cytoplasmic domains, the transmembrane domain undergoes extensive deformations along most of its helices. As might be expected, the configuration of side chains surrounding the calcium sites is significantly different in the absence of these ions (Figure 3*b*). In particular, loss of calcium ligands causes M6 to unwind, resulting in a 90° rotation of relevant side chains: N^{796}, T^{799}, and D^{800}. M4 shifts down almost 5 Å and the side chain of E^{309} rotates completely away from the site to face M1. A dramatic bend in M1 pulls E^{58} out of the site altogether (Figure 2*c*). E^{771} and E^{908} in M5 and M8 do not change much, though N^{768} rotates ~30° toward M4. To be sure, the resulting withdrawal of calcium ligands from this site appears to justify the 1000-fold decrease in calcium affinity in the E_2 conformation. Toyoshima & Nomura (94) go further and postulate that this particular arrangement of side chains reflects the evolutionary relationship with Na^+/K^+-ATPase, which would require liganding of two potassium ions at this stage of its reaction cycle. This idea was investigated by identifying plausible Na^+ and K^+ sites in homology models of Na^+/K^+-ATPase based on the $E_1 \cdot Ca_2$ and $E_2 \cdot Tg$ structures (68a). Based on structural studies of the K^+ channel, we know that H_3O^+ and K^+ are virtually interchangeable within its selectivity filter. Thus, the putative K^+ sites in the Na^+/K^+-ATPase model may also correspond to sites of proton countertransport for Ca^{2+}-ATPase, in the form of H_3O^+. Although the resolution of the $E_2 \cdot TG$ structure (3.1 Å) was not sufficient to reveal water molecules, our modeling indicates that O atoms are well accommodated at this site. However, the way in which the the required tetrahedral hydrogen bonding of the oxygen is adapted to the essentially octahedral ligand cage is not clear.

These localized changes at the calcium sites give rise to a larger set of deformations in helices M1–M6, many of which were deduced in fitting the original X-ray structure to a cryo-EM map of the $E_2 \cdot VO_4$ conformation (104). Most of these deformations involve bending or tilting of helices, which depending on the axis of tilt imparts a rocking motion causing several helices to move up or down relative to the bilayer. Perhaps the central movement is the bending of the cytoplasmic end of M5 about a pivot point centered at G^{770}. Because the top of M5 is integrated into the Rossmann fold at the heart of the P domain, the bending of M5 could plausibly induce rotation of the P domain as a whole. Given its rigid link to M4, P domain rotation causes a rocking of M4 about the same pivot point as M5 (G^{770}), producing the observed displacement of M4 normal to the bilayer. M3 undergoes a combination of rocking and bending, such that M3 and M5 end up bowed toward one another in the absence of calcium; both are straight and parallel to one another in $E_1 \cdot Ca_2$. Given the minimal interaction between M3 and the P domain, the changes in M3 are likely induced by van der Waals interactions with M4 and M5 within the bilayer and by interactions between the lumenal loops L34 and L78. Intriguingly, M1 and M2 are connected to the A domain only by flexible loops, yet undergo even larger movements. In the case of M2, there is an inclination about a pivot point at the lumenal end of this long helix, as well as partial unwinding at the cytoplasmic end in $E_2 \cdot TG$. M1 is displaced >10 Å laterally, shifted upward, and bent 90° at the cytoplasmic surface of the bilayer thus pulling E^{58} away from the calcium sites in $E_2 \cdot TG$. Finally, the L67 loop can be considered part of the transmembrane domain and is hydrogen bonded both to the cytoplasmic part of M5 (S5) and to one of the helices in the P domain; thus, a modest movement of this loop is coupled to the rotation of the P domain and the bending of M5. The fact that M7 through M10 remain relatively unchanged is consistent with their absence in the subfamily of P-type ATPases specializing in so-called soft-metal ions such as copper, zinc, and cadmium (54).

STRUCTURAL EFFECTS OF PHOSPHORYLATION

Although there are no X-ray structures for the phosphorylated forms of Ca^{2+}-ATPase, EM has been used to solve a series of structures at intermediate resolution, the latest of which was used to build an atomic model (104). The conformation represented by this structure has been controversial. Initially, the vanadate used for inducing the tubular crystals used for these studies was assumed to stabilize E_2-P (24), but it was later discovered that the decameric form, decavanadate, not the phosphate analog, orthovanadate, was actually the effector for crystallization (60, 96). Also, vanadate-free conditions were sufficient for crystallization of the scallop isoform of Ca^{2+}-ATPase (11), and the ability of TG to promote crystallization was initially ascribed to its trapping of the E_2 conformation (81). Nevertheless, more recent studies of TG document its interaction with E_2-P (82), and proteolysis studies (19) now suggest that the EM structures are indeed representative of E_2-P or $E_2 \cdot PO_4$. Decavanadate appears to occupy two positions in the crystals, one extramolecular site mediating a crystal contact between twofold-related molecules and a second, intramolecular site between the N and A domains (90). This second site was confirmed by crystallizing Ca^{2+}-ATPase labeled by fluorescein isothiocyanate (FITC), which displaced the intramolecular decavanadate and left a corresponding hole in the density map between these domains (104). Although not visible at these resolutions, orthovanadate is undoubtedly also present in these solutions and presumably acts as a transition-state analog at the catalytic site. Finally, a truly phosphorylated form of Ca^{2+}-ATPase can be prepared from the FITC-labeled enzyme (13). After stabilizing this species with TG, tubular crystals can be readily formed by decavanadate (38) and the structure closely resembles that of the unmodified enzyme (D. Stokes, F. Delavoie & J.-J. Lacapere, unpublished results). Taken together, these data suggest that EM structures from the vanadate-induced tubular crystals are indeed representative of E_2-P, providing a nice complement to the X-ray structures of $E_2 \cdot TG$ and $E_1 \cdot Ca_2$ conformations.

The atomic model for $E_2 \cdot VO_4$ was built by fitting atomic coordinates for $E_1 \cdot Ca_2$ to the EM density map at 6.5 Å (104). The three cytoplasmic domains were fitted individually as rigid bodies, whereas the helices composing the membrane domain were bent and displaced to match the corresponding density in the map. The transmembrane and P domains of the resulting structure are rather similar to $E_2 \cdot TG$, despite some spurious displacements of individual transmembrane helices along their axes due to the limited resolution of the EM map. In contrast, the N domain is significantly more vertical, allowing the A domain to more closely approach the P domain in the $E_2 \cdot VO_4$ structure (Figure 5, see color insert). This difference appears to reflect differing interactions between the conserved sequences TGES^{184} and DGVND^{707}, which mediate the interface between A and P domains in $E_2 \cdot TG$. These differences may be due to orthovanadate at the active site, indicating that the position of the A domain is sensitive to formation and hydrolysis of the aspartyl phosphate. Given the documented flexibility of the N domain, it seems likely that its different position in $E_2 \cdot TG$ and $E_2 \cdot VO_4$ structures is governed primarily by the A domain, with which it has the most extensive contacts.

STRUCTURAL EFFECTS OF THAPSIGARGIN

There has been much interest in characterizing the effects of TG, a plant sesquiterpene that binds with exceedingly high affinity and specificity to the SERCA1 isoform of Ca^{2+}-ATPase. Initial enzymatic studies suggested that TG reacts with the calcium-free E_2 state to form a deadend complex, which is then inert with respect to both calcium and P_i (81). Subsequent studies concluded that TG could in fact react with other enzymatic intermediates, namely $E_1 \cdot Ca_2$ (102) and E_2-P (82). In the latter case, measurements of ^{18}O exchange indicated that TG affected the stability of the aspartyl phosphate. Furthermore, TG appears to bind to the

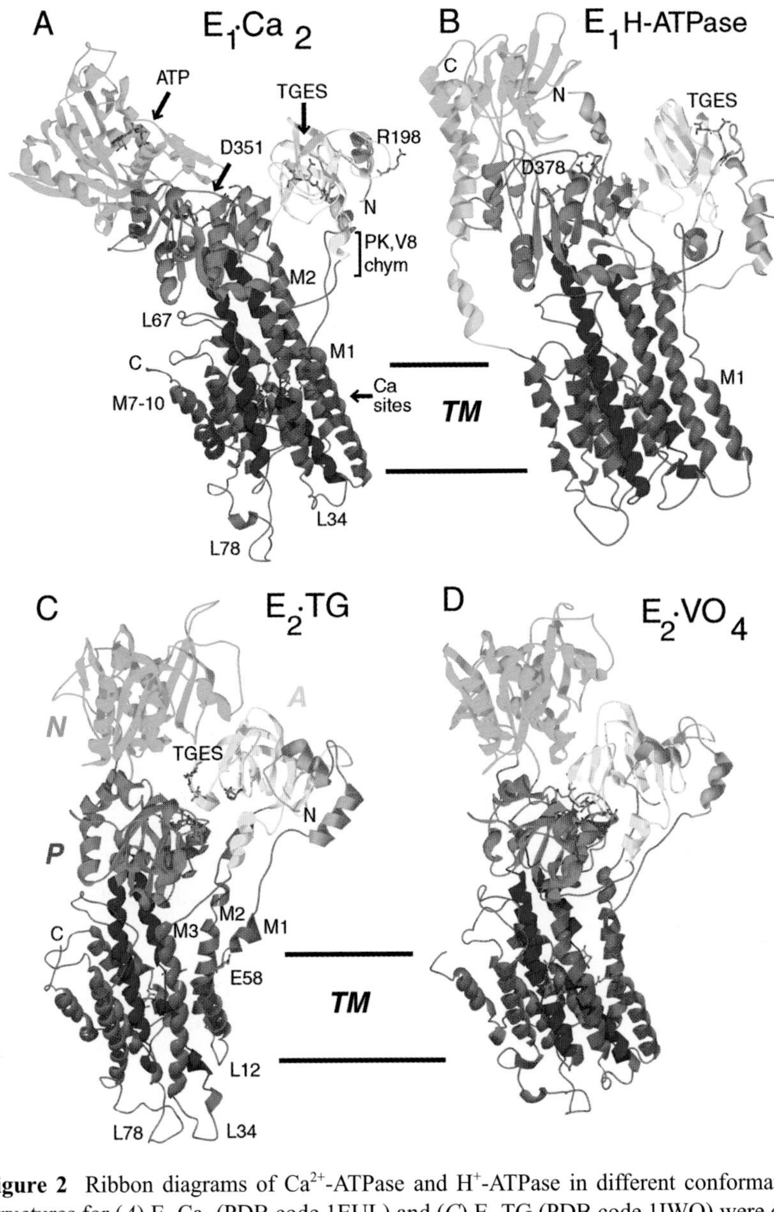

Figure 2 Ribbon diagrams of Ca^{2+}-ATPase and H^+-ATPase in different conformations. Structures for (*A*) $E_1 \cdot Ca_2$ (PDB code 1EUL) and (*C*) $E_2 \cdot TG$ (PDB code 1IWO) were determined by X-ray crystallography, and those for (*B*) H^+-ATPase (PDB code 1MHS) and (*D*) $E_2 \cdot VO_4$ (PDB code 1KJU) were fitted to cryo-EM maps. Several functional sites are indicated (Figure 1) and cytoplasmic domains are color coded green (N), magenta (P), and yellow (A). Transmembrane domain (TM) is gray except for M4 and M5, which are blue. Figure made with SPDBV and rendered with POV-Ray.

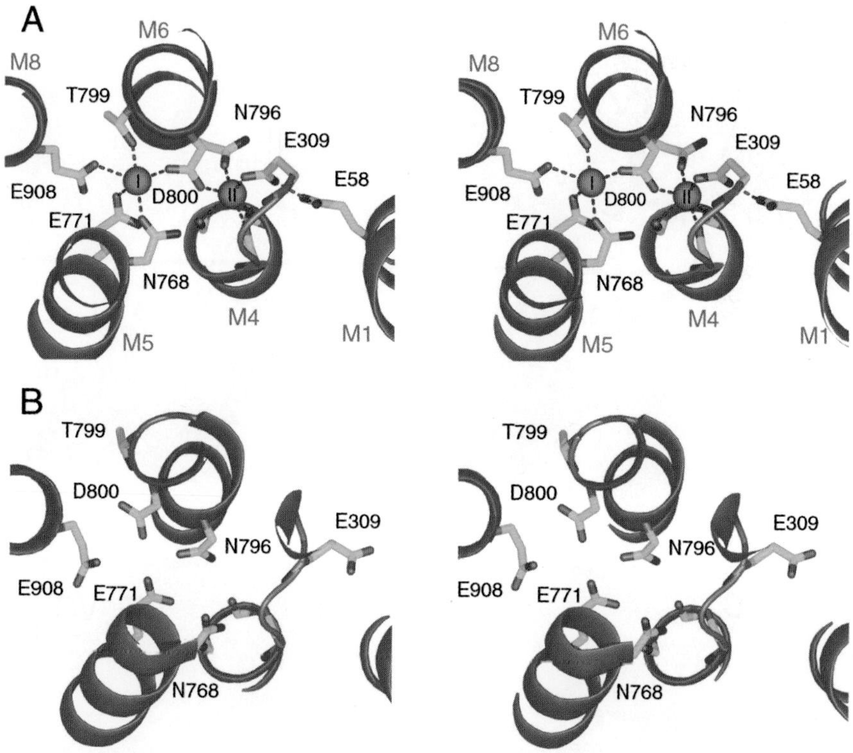

Figure 3 Stereoview of calcium binding residues from Ca^{2+}-ATPase in (*A*) $E_1{\cdot}Ca_2$ and (*B*) $E_2{\cdot}TG$ states. Calcium ions are colored magenta. Dramatic changes in the backbone of M6 and the side chains of E^{309} and E^{58} occur during calcium binding. Figure made with PyMOL.

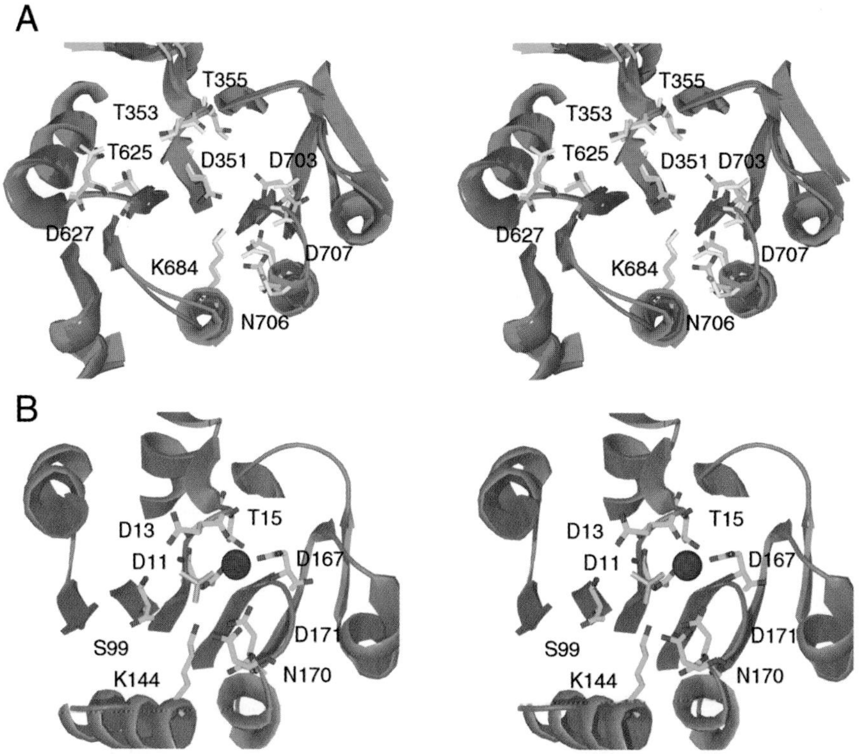

Figure 4 Stereoview of phosphorylation site of (*A*) Ca^{2+}-ATPase and (*B*) phosphoserine phosphatase (PDB code 1J97), the latter with $Mg{\cdot}BeF_3^-$ forming an analog of the aspartyl phosphate. $E_1{\cdot}Ca_2$ and $E_2{\cdot}TG$ structures are overlaid in A with ribbons respectively colored magenta and blue; side chains for $E_2{\cdot}TG$ are colored cyan and, except for small changes in the $DGVND^{707}$ loop, are in virtually identical positions in the two Ca^{2+}-ATPase structures. Compared with phosphoserine phosphatase, this same loop is considerably farther from $Mg{\cdot}BeF_3^-$ (magnesium colored magenta and fluoride in orange), suggesting substantial movements upon phosphorylation of Ca^{2+}-ATPase. The side chain of D^{351} would also be expected to swivel up to match the position of D^{11} in phosphoserine phosphatase. Figure made with SPDBV and POV-Ray.

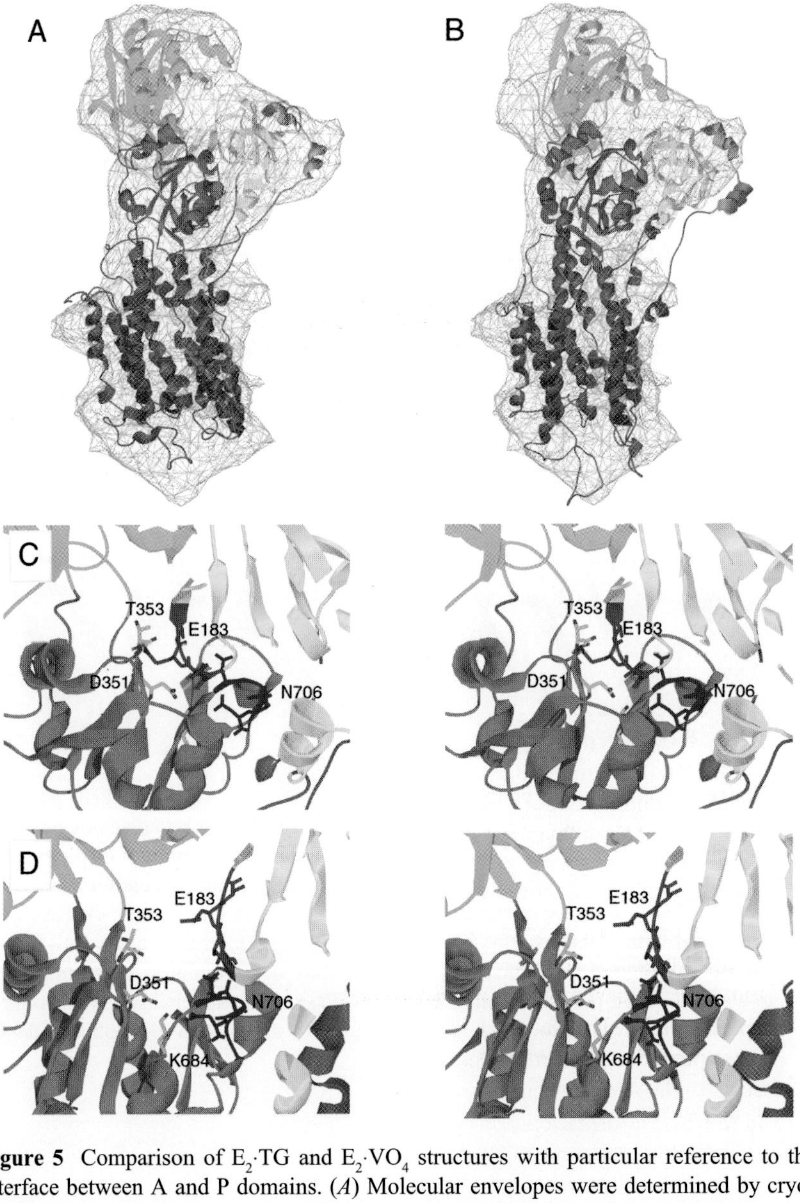

Figure 5 Comparison of $E_2{\cdot}TG$ and $E_2{\cdot}VO_4$ structures with particular reference to the interface between A and P domains. (*A*) Molecular envelopes were determined by cryo-EM of the $E_2{\cdot}VO_4$ state and, after fitting, represent the basis for the $E_2{\cdot}VO_4$ structure; (*B*) due to movements of N and A domains $E_2{\cdot}TG$ coordinates produce a poor fit. These movements are primarily a result of differences in the A-P domain interface, shown in stereo for (*C*) $E_2{\cdot}VO_4$ and (*D*) $E_2{\cdot}TG$. Although the domains are colored according to Figure 2, the $TGES^{184}$ and $DGVND^{707}$ loops are colored red and blue, respectively, to highlight their different interactions in these two states. Figure made with SPDBV and POV-Ray.

stable phosphorylated intermediate produced by FITC-labeled Ca^{2+}-ATPase, making the phosphoenzyme stable for periods up to a week (38). A series of mutagenesis studies implicated a region at the cytoplasmic border of the M3 helix between residues 254 and 262 (107). A contradictory result was obtained by EM of crystals in the presence and absence of TG, which found consistent density differences at the lumenal side of the membrane between M3/M4 and M7/M8 loops, which were assigned to TG (105). The X-ray structure for $E_2 \cdot TG$ supported the mutagenesis data by showing TG bound near the cytoplasmic border in a crevice between M3, M5, and M7. Although still speculative, inhibitory effects of TG are likely to involve a general rigidification of these transmembrane helices thus preventing them from conveying the conformational changes associated with calcium binding. In addition, this X-ray structure showed that the M3/M4 and M7/M8 loops moved substantially closer together relative to the $E_1 \cdot Ca_2$ structure in a way that could explain the difference density observed by EM. Toyoshima & Nomura (94) suggested that these loops controlled the access of transport sites to the lumenal side of the membrane. However, the EM result suggests that TG influences these loops in the $E_2 \cdot VO_4$ crystals, and it would therefore be important to determine their disposition in the uninhibited enzyme and to further study their role in calcium gating.

ACCESS TO THE CALCIUM SITES

Although the structures define the architecture of the calcium sites, they do not show clear entrance and exit paths for Ca^{++}. In the classic E_1/E_2 model, the sites of E_1 would be oriented toward the cytoplasmic side of the membrane, whereas in E_2 they would face the lumen of the sarcoplasmic reticulum. More recent analyses, based primarily on the lack of competition between lumenal and cytoplasmic calcium in formation of $E_1 \sim P$, suggest that calcium sites face the cytoplasm in both E_1 and E_2 and that it is phosphoenzyme formation which serves to reorient these sites (46). The X-ray structure of $E_2 \cdot TG$ supports this latter model by revealing potential access for ions from the cytoplasm, but not from the lumen. In particular, a negatively charged tunnel is visible between M1, M2, and M3, though it is blocked at the bottom by E^{309}. Given the relatively low pH (6) of crystallization, this structure likely represents $E_2 \cdot H_3$ (Table 1), and it is possible that deprotonation of E^{309}, which must precede calcium binding, initiates structural rearrangements to deliver calcium to the site. Although the cryo-EM-based model for *Neurospora* H^+-ATPase in the E_1 conformation also reveals a plausible ion path via conserved polar side chains along M1 and M2 (51), mutagenesis of several glutamate residues along M1 failed to produce any effects on calcium binding by Ca^{2+}-ATPase (27). Also, movements of M1 and M3 block this tunnel in the $E_1 \cdot Ca_2$ structure, creating an inconsistency with the documented exchangeability of calcium by this intermediate. A potential pathway for release of calcium to the lumen was seen at much lower resolution in EM maps of the $E_2 \cdot VO_4$ structure (106). This pathway starts near the calcium sites and widens as it approaches the M3/M4 and M7/M8 loops postulated to act as ion gates toward the lumen. This pathway potentially corresponds to a water-filled tunnel that has been proposed to explain voltage effects on sodium binding to Na^+/K^+-ATPase from the extracellular side of the membrane (30).

BIOCHEMICAL STUDIES OF CONFORMATIONAL CHANGE

Over the past 25 years, a wide variety of biochemical and biophysical techniques have been used to follow conformational changes (Table 1). It is now important to evaluate their results in light of the existing high-resolution structures. On the one hand, these results will help validate the use of artificially stabilized intermediates for crystallographic studies and also fill in information about transient intermediates. On the other hand, comparison will enhance our understanding of these methods and thus facilitate their use in future work.

Changes in the reactivity of sulfydryl groups were studied in several early investigations using the reagent DTNB (5,5′-dithiobis-2-nitrobenzoate) to define subclasses of the 24 cysteine residues. Although it was not possible to assign rate constants to specific residues, three grades of reactivity have been assigned in Table 1, allowing a number of general conclusions. About eight of the cysteines were unreactive in the absence of detergents, most likely corresponding to the six residues in transmembrane helices plus the disulfide linked pair in L78 (18). The remaining sulfydryls were scattered throughout the N and P domains, with only a single cysteine in the A domain, near the N terminus. None were associated with the conserved catalytic loops of the P domain. The changes in reactivity appeared to affect whole classes of cysteines rather than isolated residues, probably reflecting global movements of the N and P domains, and maximal reactivity was obtained with $E_1 \cdot Ca_2$, consistent with its open structure. Over half the reactive cysteines suffered a 70% fall in reactivity when ATP analogs were bound and had even lower reactivity following phosphorylation.

The formation of a cross-link by glutaraldehyde between R^{678} in the P domain and K^{492} near the adenosine binding region of the N domain is also conformation dependent and has given useful information about the proximity of these domains (80). After cross-linking, the affinity for ATP decreased 1000-fold without affecting calcium binding, phosphorylation by acetyl phosphate, or calcium occlusion. The next step, $E_1 \sim P$ to E_2-P, was completely blocked, consistent with the idea that the A domain could not be reoriented after coupling the N and P domains. Cross-linking was most efficient in $E_1 \sim P$, reflecting the close approach of N and P domains during phosphate transfer, and slower in $E_1 \cdot Ca_2$, supporting the mobility of the N domain. The distance between cross-linked residues in the $E_2 \cdot TG$ structure is significantly less than in the $E_2 \cdot VO_4$ structure, consistent with the ability to form a cross-link in E_2 and not in E_2-P. Furthermore, cross-linking blocks phosphorylation of E_2 with P_i, indicating that the observed movements of N and A domains are a necessary adjustment to the presence of the phospho group in the catalytic site.

Oxidative cleavage by iron has been effectively used to investigate the proximity of loops within the catalytic site of Na^+/K^+-ATPase (71). In these experiments, the iron substitutes for magnesium at the phosphorylation site and, under oxidizing conditions, cleaves nearby peptide bonds. Sites of cleavage have been identified in the catalytically important loops of the P domain and, perhaps most importantly, in the TGES loop of the A domain. In particular, cleavage of the TGES loop only occurs in E_2-P and $E_2 \cdot K$, supporting the docking of the A domain next to the P domain in these conformations. Further evidence for this docking comes from numerous proteolytic cleavage studies that show sensitivity of several sites within the A domain in $E_1 \cdot Ca_2$, but not in E_2 conformations (50, 65).

A variety of fluorescent labels have been used to study conformational changes. The best-characterized labels are TNP nucleotides, which show a large increase in fluorescence after phosphorylation, and NBD (4-nitro-2,1,3-benzoxadiazole) and FITC, which have both been used to monitor the E_1/E_2 equilibrium. In addition, advances in infrared difference spectroscopy now allow individual residues or even specific bonds to be studied in the course of the reaction cycle (7a). In future studies, these probes may be useful in monitoring pump dynamics and deducing the structure of intermediates that are not accessible to structural studies.

STRUCTURAL MECHANISM OF TRANSPORT

Given the abundant structural and biochemical evidence for conformational changes, we are now faced with describing how these changes couple the local effects of calcium on the transmembrane sites to phosphorylation of D^{351} some 40 Å away. For some time, the M4/S4 connection between E^{309} and D^{351} has been discussed as most likely to mediate cross-talk between these sites (45). The X-ray structures show us that the links between the P domain and both M4 and M5 are indeed well structured and potentially capable of conveying long-range conformational changes. Nevertheless, the actual sequence of events between binding of calcium and domain movement is still a matter for educated guesswork, guided by the conformational criteria in Table 1.

In the starting conformation, calcium ligands are protonated ($E_2 \cdot H_3$) and a distinct step of deprotonation is required prior to binding cytoplasmic calcium. This step must involve opening the partially occluded proton sites seen in the $E_2 \cdot TG$ structure to the cytoplasm, e.g., by reorientation of E^{309}, and has been followed by kinetic studies of NBD fluorescence (97). This probe labels the P1 helix wedged between the top of M4 and the P domain, and the fluorescence changes indicate that there must be some structural change in this region. However, the T2 tryptic cleavage site in the A domain is still protected at pH 7, even in the presence of nucleotide, indicating that the A domain is still docked with the P domain in this deprotonated state (20, 43). Thus, a high pH, calcium-free conformation, which has typically been called E_1, appears to be intermediate between $E_2 \cdot H_3$ and $E_1 \cdot Ca_2$ with cytoplasmically exposed calcium sites but with the A domain still in the E_2 conformation. Its properties may prove important in defining an access pathway for calcium, which has been problematic from the two existing X-ray structures.

In contrast to NBD fluorescence, increases in tryptophan fluorescence (12) and small changes in circular dichroism (31) occur only after binding of calcium to E_1 and are probably correlated with the cooperative changes required to create the second calcium site, whose occupation activates phosphorylation. From a structural point of view, the X-ray structures reveal an unbending of M5 and displacement of M4, which seem to represent levers for rotating the P domain, but the mechanism is not obvious. Locally, M6 undergoes the largest structural rearrangement with the backbone winding up upon calcium binding (Figure 3). Perhaps this winding puts strain on the L67 loop, which then induces the bending of M5. Binding of calcium by M4 might also induce its vertical movement as the main chain carbonyls in its unwound portion move upward toward D^{800} to provide ligands for cooperative binding of the second calcium ion.

In subsequent steps of the reaction cycle, the P domain is clearly the center of operations and is seen to adopt two distinct orientations in response to calcium binding. In the presence of calcium, the P domain appears to be available for phosphate transfer from ATP bound within the N domain. In the absence of calcium, the P domain is seen to interact with the $TGES^{184}$ loop of the A domain. Thus, it is possible that movement of the P domain represents a kind of switch, under control of calcium binding that selects between the N and the A domains. This simple pivot might represent the mechanism for calcium-induced activation of D^{351} for phosphorylation, which cannot otherwise be explained by rearrangement of residues within the catalytic site (Figure 4*a*). Although we do not yet have a structure of a phosphorylated intermediate, we postulate that consequent changes in the P domain would facilitate its interaction with the A domain and that formation of this A-P domain interface would represent the $E_1{\sim}P$ to E_2-P transition that lowers calcium affinity and induces calcium release to the lumen.

In contrast to the P domain, the N and A domains are covalently tethered by flexible, unstructured loops and their noncovalent interactions with other cytoplasmic domains appear to be transient. Although there is convincing evidence for mobility of N and A domains, there is little indication of what causes their dramatic movements. A likely answer is thermal energy, or Brownian motion, which has been hypothesized as a driving force in the mechanisms of a wide variety of other macromolecular motors such as F_0F_1 (70), myosin (40), kinesin (6), as well as protein translocation into mitochondria (69) and the endoplasmic reticulum (58). Thermal energy would ensure that these weakly bound and flexibly tethered domains would be moving extensively, thus sampling a large range of orientations and potential binding interfaces. In the absence of tethers, these domains could be viewed as separate subunits as seen, for example, in the family of response regulators, although the increased efficiency of a covalently attached domain should be advantageous for the continual turnover of ion pumps.

The specific changes that accompany phosphorylation must be precipitated by local events near D^{351} and, by analogy with G protein switching (34), may well involve changes in the liganding of magnesium. Initially the magnesium is bound by two oxygens from the β and γ phosphates of ATP and by four other ligands provided by the protein (including bound waters). Analogy with the X-ray structures of phosphoserine phosphatase (99) suggests that ligating oxygens of Ca^{2+}-ATPase would come from the phosphate, the D^{351} carboxyl, the D^{703} carboxyl, the T^{353} main chain carbonyl, and two from bound water molecules (Figure 4). Homologous ligands are also found for several members of the CheY response regulator family (53) and appear to be consistent in an unpublished structure for the Mg_2F_4 complex of the Ca^{2+}-ATPase (C. Toyoshima & H. Nomura, unpublished data) and with mutagenesis of the $DGVND^{707}$ loop of Na^+/K^+-ATPase (72). Initially, magnesium bound by Ca^{2+}-ATPase is freely exchangeable (78), but after transfer of phosphate to D^{351} and loss of ADP, the magnesium becomes tightly bound ($k_{off} < 0.5\ s^{-1}$) and is released only after hydrolysis (98); similar results have been obtained with Na^+/K^+-ATPase (29). This represents a substantial difference from the phosphatases and response regulators, which bind magnesium loosely throughout their reaction cycles. We have previously suggested (90) that this occlusion of magnesium by Ca^{2+}-ATPase could imply a change in its ligation, which could initiate further conformational changes required for calcium occlusion and the $E_1 \sim P$ to E_2-P transition. We can now specify that formation of the A-P domain interface may stabilize the magnesium ligand cage and that interactions at this interface might induce further conformational changes within the transmembrane domain.

In the CheY family, a modest conformational change of the α-4, β-4 loop (homologous to TGD^{627} of Ca^{2+}-ATPase) follows formation of a new hydrogen bond to the covalently linked phosphate and induces interactions with responsive subunits (e.g., FliM and CheZ). A similar effect may be occurring in the $DGVND^{707}$ loop in Ca^{2+}-ATPase, which is significantly farther from D^{351} than the analogous loop from response regulators and phosphatases (Figure 4). We hypothesize that the presence of $Mg \cdot PO_4$ will pull this loop closer to provide the requisite ligands, thus producing a binding site for the $TGES^{184}$ loop from the A domain, which would in turn confer tighter magnesium binding. In terms of the reaction cycle, phosphoenzyme formation and movement of the $DGVND^{707}$ loop would produce the ADP-sensitive $E_1 \sim P$ intermediate and the docking of the A domain would initiate the transition to E_2-P.

The elements of this A-P domain interface include the conserved $TGES^{184}$ and $DGVND^{707}$ loops as well as the P6 helix and the $MAATEQ^{244}$ loop connecting the A domain to M3. The P6 helix directly follows $DGVND^{707}$ and contains K^{712}, which makes hydrogen bonds with M^{239}, T^{242}, and Q^{244} in the $E_2 \cdot TG$ structure. Significantly, this $MAATEQ^{244}$ region contains proteolytic cleavage sites that are only accessible in the E_1 conformations (42, 50), consistent with its burial at a domain interface in E_2. Both mutagenesis of $TGES^{184}$ (2) and excision of the $MAATE^{243}$ sequence with proteinase K (65) yield enzymes defective in the $E_1 \sim P$ to E_2-P transition. Results of Fe-catalyzed cleavage of Na^+/K^+-ATPase place $TGES^{184}$ near the magnesium sites in E_2-P (71). Taken together, these results indicate that rotation of the A domain and formation of the A-P domain interface are essential for producing E_2-P. It is tempting to speculate that $TGES^{184}$ contributes ligands to the aspartyl-$PO_4 \cdot Mg$ complex, similar to Q^{147} of CheZ, which ligates magnesium at the active site of CheY and ultimately directs a water molecule to hydrolyze the aspartylphosphate (108). Although the original analyses of $TGES^{184}$ mutations showed no change in phosphorylation levels from either ATP or P_i (4), more thorough studies of magnesium binding might now be possible using larger-scale expression systems.

Both the formation of E_2-P and its subsequent hydrolysis are linked to events at the calcium transport sites. The corresponding 90°–110° rotation of the A domain is likely to place stress on M1, M2, and M3 thereby accounting for their large movements in $E_2 \cdot VO_4$ and $E_2 \cdot TG$ structures. These movements may represent an indirect mechanism for altering the calcium binding properties, by opening the lumenal gate in the M3/M4 loop and by perturbing the ion binding sites between M4, M5, M6, and M8. The dramatic changes in E^{58} probably reflect this perturbation, as this residue is hydrogen bonded to E^{309} in $E_1 \cdot Ca_2$ but is pulled completely out of the site in $E_2 \cdot TG$ owing to the kinking of M1 (Figures 2*c*, 3), perhaps destabilizing E^{309} as a first step in lowering calcium affinity. Release of the calcium ions and/or protonation of the transport sites stimulate hydrolysis of E_2-P, and in analogy with the postulated effects of calcium binding, events at the ion binding sites are likely to be propagated through the M4/M5 helices and the L67 loop, perhaps inducing changes in the A-P domain interface similar to those seen in the structure of $E_2 \cdot TG$ relative to $E_2 \cdot VO_4$ in order to stimulate hydrolysis.

CONCLUDING REMARKS

The early description of E_1 and E_2 conformations of Ca^{2+}-ATPase (56) was of great help in sorting out the complexities of the cycle, but it is now worth considering whether they represent an oversimplification. To start, the switching of calcium sites envisioned by this model now seems to be incorrect, given both kinetic and structural evidence indicating that access by lumenal calcium is blocked in E_2. From the structural evidence, it might be concluded that E_1 and E_2 conformations are distinguished by the location of the A domain, either engaged or disengaged from its docking site with the P domain. However, spectroscopic studies of calcium binding suggest that cytoplasmic domain movements occur upon binding calcium by the E_1 species, not during the deprotonation of $E_2 \cdot H_3$ that produces E_1. Thus, it may be more accurate to describe the reaction cycle as a series of unique intermediates separated by small, reversible steps. This view is certainly consistent with the energetics of the cycle under physiological conditions, which shows most steps to be separated by <10 kJ/mole (52), making them difficult to stabilize for structural studies. The alternative has been to use nonphysiological ligands to trap certain intermediates, but this practice runs the risk of generating conformations that differ from those of the physiological intermediates. Some of these ligands cause the cytoplasmic domains to appear to behave independently from the calcium binding site. For example, glutaraldehyde cross-linking of N and P domains has no effect on calcium binding, and the phosphoenzyme stabilized by

FITC has high-affinity cytoplasmically accessible calcium sites, but is still able to bind TG and form vanadate-induced two-dimensional crystals. This behavior may tell us something about the structural coupling of the molecule, but it also threatens to deceive us about the structural events of the reaction cycle. With this risk in mind, we should now turn our attention to using the several existing structures of Ca^{2+}-ATPase as a guide for formulating more precise experiments to either validate their relevance or test the true nature of the physiological intermediates.

LITERATURE CITED

1. Albers R. 1967. Biochemical aspects of active transport. *Annu. Rev. Biochem.* 36: 727–56
2. Andersen JP. 1995. Dissection of the functional domains of the sarcoplasmic reticulum Ca^{2+}-ATPase by site-directed mutagenesis. *Biosci. Rep.* 15:243–61
3. Andersen JP, Vilsen B. 1994. Amino acids Asn796 and Thr299 of the Ca-ATPase of sarcoplasmic reticulum bind Ca at different sites. *J. Biol. Chem.* 269:15931–36
4. Andersen JP, Vilsen B, Leberer E, MacLennan DH. 1989. Functional consequences of mutations in the beta-strand sector of the Ca^{2+}-ATPase of sarcoplasmic reticulum. *J. Biol. Chem.* 264:21018–23
5. Aravind L, Galperin MY, Koonin EV. 1998. The catalytic domain of the P-type ATPase has the haloacid dehalogenase fold. *Trends Biol. Sci.* 23:127–29
6. Astumian RD, Derenyi I. 1999. A chemically reversible Brownian motor: application to kinesin and Ncd. *Biophys. J.* 77:993–1002
7. Auer M, Scarborough GA, Kühlbrandt W. 1998. Three-dimensional map of the plasma membrane H^+-ATPase in the open conformation. *Nature* 392:840–43
7a. Barth A, Zscherp C. 2000. Substrate binding and enzyme function investigated by infrared spectroscopy. *FEBS Lett.* 477: 151–55
8. Bigelow DJ, Inesi G. 1992. Contributions of chemical derivatization and spectroscopic studies to the characterization of the Ca^{2+} transport ATPase of sarcoplasmic reticulum. *Biochim. Biophys. Acta* 1113:323–38
9. Bishop JE, Nakamoto RK, Inesi G. 1986. Modulation of the binding characteristics of a fluorescent nucleotide derivative to the sarcoplasmic reticulum adenosinetriphosphatase. *Biochemistry* 25:696–703
10. Branden CI. 1980. Relation between structure and function of alpha/beta-proteins. *Q. Rev. Biophys.* 13:317–38
11. Castellani L, Hardwicke PM. 1983. Crystalline structure of sarcoplasmic reticulum from scallop. *J. Cell. Biol.* 97:557–61
12. Champeil P, Henao F, de Foresta B. 1997. Dissociation of Ca^{2+} from sarcoplasmic reticulum Ca^{2+}-ATPase and changes in fluorescence of optically selected Trp residues. Effects of KCl and NaCl and implications for substeps in Ca^{2+} dissociation. *Biochemistry* 36:12383–93
13. Champeil P, Henao F, Lacapere JJ, McIntosh DB. 2001. A remarkably stable phosphorylated form of Ca^{2+}-ATPase prepared from Ca^{2+}-loaded and fluorescein isothiocyanate-labeled sarcoplasmic reticulum Vesicles. *J. Biol. Chem.* 276:5795–803
14. Cheong G-W, Young HS, Ogawa H, Toyoshima C, Stokes DL. 1996. Lamellar stacking in three-dimensional crystals

of Ca^{2+}-ATPase from sarcoplasmic reticulum. *Biophys. J.* 70:1689–99

15. Clarke DM, Loo TW, Inesi G, MacLennan DH. 1989. Location of high affinity Ca^{2+}-binding sites within the predicted transmembrane domain of the sarcoplasmic reticulum Ca^{2+}-ATPase. *Nature* 339:476–78
16. Collet J-F, van Schaftingen E, Stroobant V. 1998. A new family of phosphotransferases related to P-type ATPases. *Trends Biol. Sci.* 23:284
17. Cyrklaff M, Auer M, Kühlbrandt W, Scarborough GA. 1995. 2-D structure of the *Neurospora crassa* plasma membrane ATPase as determined by electron cryomicroscopy. *EMBO J.* 14:1854–57
18. Daiho T, Kanazawa T. 1994. Reduction of disulfide bonds in sarcoplasmic reticulum Ca^{2+}-ATPase by dithiothreitol causes inhibition of phosphoenzyme isomerization in catalytic cycle. This reduction requires binding of both purine nucleotide and Ca^{2+} to enzyme. *J. Biol. Chem.* 269: 11060–64
19. Danko S, Daiho T, Yamasaki K, Kamidochi M, Suzuki H, et al. 2001. ADP-insensitive phosphoenzyme intermediate of sarcoplasmic reticulum Ca^{2+}-ATPase has a compact conformation resistant to proteinase K, V8 protease and trypsin. *FEBS Lett.* 489:277–82
20. Danko S, Yamasaki K, Daiho T, Suzuki H, Toyoshima C. 2001. Organization of cytoplasmic domains of sarcoplasmic reticulum Ca^{2+}-ATPase in E_1P and E_1ATP states: a limited proteolysis study. *FEBS Lett.* 505:129–35
21. deMeis L, Vianna A. 1979. Energy interconversion by the Ca^{+}-dependent ATPase of the sarcoplasmic reticulum. *Annu. Rev. Biochem.* 48:275–92
22. Dupont Y, Bennett N. 1982. Vanadate inhibition of the Ca^{2+}-dependent conformational change of the sarcoplasmic reticulum Ca^{2+}-ATPase. *FEBS Lett.* 139:237–40
23. Dux L, Martonosi A. 1983. Ca^{2+}-ATPase membrane crystals in sarcoplasmic reticulum. The effect of trypsin digestion. *J. Biol. Chem.* 258:10111–15
24. Dux L, Martonosi A. 1983. Two-dimensional arrays of proteins in sarcoplasmic reticulum and purified Ca^{2+}-ATPase vesicles treated with vanadate. *J. Biol. Chem.* 258:2599–603
25. Dux L, Pikula S, Mullner N, Martonosi A. 1987. Crystallization of Ca^{2+}-ATPase in detergent-solubilized sarcoplasmic reticulum. *J. Biol. Chem.* 262:6439–42
26. Ebashi S, Lipmann F. 1962. Adenosine triphosphate-linked concentration of calcium ions in a particulate fraction of rabbit muscle. *J. Cell. Biol.* 14:389–400
27. Einholm AP, Vilsen B, Andersen JP. 2002. *Functional consequences of charge-reversals of acidic residues in M1 of the SR Ca-ATPase*. Presented at the 10th Int. Conf. on Na,K-ATPase and Related Cation Pumps, Elsinore, Denmark
28. Fagan MJ, Saier MH. 1994. P-type ATPases of eukaryotes and bacteria: sequence analysis and construction of phylogenetic trees. *J. Mol. Evol.* 38:57–99
29. Fukushima Y, Post RL. 1978. Binding of divalent cation to phosphoenzyme of sodium- and potassium-transport adenosine triphosphatase. *J. Biol. Chem.* 253: 6853–62
30. Gadsby DC, Rakowski RF, De Weer P. 1993. Extracellular access to the Na, K pump: pathway similar to ion channel. *Science* 260:100–103
31. Girardet JL, Dupont Y. 1992. Ellipticity changes of the sarcoplasmic reticulum Ca^{2+}-ATPase induced by cation binding and phosphorylation. *FEBS Lett.* 296:103–106
32. Glynn I. 1985. The Na^{+},K^{+}-transporting adenosine triphosphatase. In *The Enzymes of Biological Membranes*, ed. A Martonosi, pp. 35–114. New York: Plenum
33. Goldshleger R, Karlish SJ. 1999. The energy transduction mechanism of Na,K-ATPase studied with iron-catalyzed

oxidative cleavage. *J. Biol. Chem.* 274:16213–21
34. Greeves M, Holmes KC. 1999. Structural mechanism of muscle contraction. *Annu. Rev. Biochem.* 687:687–728
35. Hardwicke PMD, Bozzola JJ. 1989. Effect of phosphorylation on scallop sarcoplasmic reticulum. *J. Musc. Res. Cell. Motil.* 10:245–253
36. Hasselbach W, Oetliker H. 1983. Energetics and electrogenecity of the sarcoplasmic reticulum calcium pump. *Annu. Rev. Physiol.* 45:325–29
37. Hebert H, Purhonen P, Vorum H, Thomsen K, Maunsbach AB. 2001. Three-dimensional structure of renal Na,K-ATPase from cryo-electron microscopy of two-dimensional crystals. *J. Mol. Biol.* 314:479–94
38. Henao F, Delavoie F, Lacapere JJ, McIntosh DB, Champeil P. 2001. Phosphorylated Ca^{2+}-ATPase stable enough for structural studies. *J. Biol. Chem.* 276:24284–85
39. Hisano T, Hata Y, Fujii T, Liu J-Q, Kurihara T, et al. 1996. Crystal structure of L-2-haloacid dehalogenase from *Pseudomonas* sp. YL. *J. Biol. Chem.* 271:20322–30
40. Houdusse A, Sweeney HL. 2001. Myosin motors: missing structures and hidden springs. *Curr. Opin. Struct. Biol.* 11:182–94
41. Huang S, Squier TC. 1998. Enhanced rotational dynamics of the phosphorylation domain of the Ca-ATPase upon calcium activation. *Biochemistry* 37:18064–73
42. Imamura Y, Kawakita M. 1989. Purification of limited tryptic fragments of Ca^{2+},Mg^{2+}-adenosine triphosphatase of the sarcoplasmic reticulum and identification of conformation-sensitive cleavage sites. *J. Biochem.* 105:775–81
43. Imamura Y, Saito K, Kawakita M. 1984. Conformational change of Ca^{2+},Mg^{2+}-adenosine triphosphatase of sarcoplasmic reticulum upon binding of Ca^{2+} and adenyl-5′-yl-imidodiphosphate as detected by trypsin sensitivity analysis. *J. Biochem.* 95:1305–13
44. Inesi G. 1985. Mechanism of calcium transport. *Annu. Rev. Physiol.* 47:573–601
45. Inesi G, Lewis D, Nikic D, Hussain A, Kirtley ME. 1992. Long-range intramolecular linked functions in the calcium transport ATPase. *Adv. Enzymol.* 65:185–215
46. Jencks WP. 1989. How does a calcium pump pump calcium? *J. Biol. Chem.* 264:18855–58
47. Jencks WP. 1989. Utilization of binding energy and coupling rules for active transport and other coupled vectorial processes. *Methods Enzymol.* 171:145–64
48. Jorge-Garcia I, Bigelow DJ, Inesi G, Wade JB. 1988. Effect of urea on the partial reactions and crystallization pattern of sarcoplasmic reticulum adenosine triphosphatase. *Arch. Biochem. Biophys.* 265:82–90
49. Jorgensen PL, Andersen JP. 1988. Structural basis for E_1-E_2 conformational transitions in Na,K-pump and Ca-pump proteins. *J. Membr. Biol.* 103:95–120
50. Jorgensen PL, Jorgensen JR, Pedersen PA. 2001. Role of conserved TGDGVND-loop in Mg^{2+} binding, phosphorylation, and energy transfer in Na,K-ATPase. *J. Bioenerg. Biomembr.* 33:367–77
51. Kuehlbrandt W, Zeelen J, Dietrich J. 2002. Structure, mechanism, and regulation of the *neurospora plasma membrane H^+-ATPase. *Science* 297:1692–96
52. Lauger P. 1991. *Electrogenic Ion Pumps.* Sunderland, MA: Sinauer. pp. 240–46
53. Lee SY, Cho HS, Pelton JG, Yan D, Berry EA, et al. 2001. Crystal structure of activated CheY. Comparison with other activated receiver domains. *J. Biol. Chem.* 276:16425–31
54. Lutsenko S, Kaplan JH. 1995. Organization of P-type ATPases: significance of structural diversity. *Biochemistry* 34:15607–13
55. MacLennan DH, Brandl CJ, Korczak B, Green NM. 1985. Amino-acid sequence

of a Ca^{2+} + Mg^{2+}-dependent ATPase from rabbit muscle sarcoplasmic reticulum, deduced from its complementary DNA sequence. *Nature* 316:696–700
56. Makinose M. 1973. Possible functional states of the enzyme of the sarcoplasmic calcium pump. *FEBS Lett.* 37:140–43
57. Martonosi A, Taylor KA, Varga S, Ping Ting-Beall H. 1987. The molecular structure of sarcoplasmic reticulum. In *Electron Microscopy of Proteins: Membranous Structures*, ed. JR Harris, RW Horne, pp. 255–376. London: Academic
58. Matlack KE, Misselwitz B, Plath K, Rapoport TA. 1999. BiP acts as a molecular ratchet during posttranslational transport of prepro-alpha factor across the ER membrane. *Cell* 97:553–64
59. Maunsbach AB, Skriver E, Hebert H. 1991. Two-dimensional crystals and three-dimensional structure of Na,K-ATPase analyzed by electron microscopy. *Soc. Gen. Physiol. Ser.* 46:159–72
60. Maurer A, Fleischer S. 1984. Decavanadate is responsible for vanadate-induced two-dimensional crystals in sarcoplasmic reticulum. *J. Bioenerg. Biomembr.* 16:491–505
61. McIntosh D. 1998. The ATP binding sites of P-type ion transport ATPases. *Adv. Mol. Cell Biol.* 23A:33–99
62. Mintz E, Guillain F. 1997. Ca^{2+} transport by the sarcoplasmic reticulum ATPase. *Biochim. Biophys. Acta* 1318:52–70
63. Mohraz M, Yee M, Smith PR. 1985. Novel crystalline sheets of Na,K-ATPase induced by phospholipase A2. *J. Ultrastruct. Res.* 93:17–26
64. Moller JV, Juul B, le Maire M. 1996. Structural organization, ion transport, and energy transduction of ATPases. *Biochim. Biophys. Acta* 1286:1–51
65. Moller JV, Lenoir G, Marchand C, Montigny C, Le Maire M, et al. 2002. Calcium transport by sarcoplasmic reticulum Ca^{2+}-ATPase: role of the A-domain and its C-terminal link with the transmembrane region. *J. Biol. Chem.* 22: in press
66. Murphy AJ. 1978. Effects of divalent cations and nucleotides on the reactivity of the sulfhydryl groups of sarcoplasmic reticulum membranes. Evidence for structural changes occurring during the calcium transport cycle. *J. Biol. Chem.* 253:385–89
67. Murphy AJ, Coll RJ. 1992. Fluoride binding to the calcium ATPase of sarcoplasmic reticulum converts its transport sites to a low affinity, lumen-facing form. *J. Biol. Chem.* 267:16990–94
68. Ogawa H, Stokes DL, Sasabe H, Toyoshima C. 1998. Structure of the Ca^{2+} pump of sarcoplasmic reticulum: a view along the lipid bilayer at 9-Å resolution. *Biophys. J.* 75:41–52
68a. Ogawa H, Toyoshima C. 2002. Homology modeling of the cation binding sites of $Na^{+}K^{+}$-ATPase. *Proc. Natl. Acad. Sci. USA* 99:15977–82
69. Okamoto K, Brinker A, Paschen SA, Moarefi I, Hayer-Hartl M, et al. 2002. The protein import motor of mitochondria: a targeted molecular ratchet driving unfolding and translocation. *EMBO J.* 21:3659–71
70. Oster G, Wang H. 1999. ATP synthase: two motors, two fuels. *Structure* 7:R67–R72
71. Patchornik G, Goldshleger R, Karlish SJ. 2000. The complex ATP-Fe^{2+} serves as a specific affinity cleavage reagent in ATP-Mg^{2+} sites of Na,K-ATPase: altered ligation of Fe^{2+} (Mg^{2+}) ions accompanies the $E_1P \rightarrow E_2P$ conformational change. *Proc. Natl. Acad. Sci. USA* 97:11954–59
72. Pedersen PA, Jorgensen JR, Jorgensen PL. 2000. Importance of conserved alpha-subunit segment 709GDGVND for Mg^{2+} binding, phosphorylation, and energy transduction in Na,K-ATPase. *J. Biol. Chem.* 275:37588–95
73. Pick U. 1982. The interaction of vanadate ions with the Ca-ATPase from sarcoplasmic reticulum. *J. Biol. Chem.* 257:6111–19
74. Pick U, Karlish SJD. 1980. Indications for

an oligomeric structure and for conformational changes in sarcoplasmic reticulum Ca^{2+}-ATPase labelled selectively with fluorescein. *Biochim. Biophys. Acta* 626:255–61

75. Post R, Merritt C, Kinsolving C, Albright C. 1960. Membrane adenosine triphosphatase as a participant in the active transport of sodium and potassium in the human erythrocyte. *J. Biol. Chem.* 235:1796–802
76. Post RL, Kume S. 1973. Evidence for an aspartyl phosphate residue at the active site of sodium and potassium ion transport adenosine triphosphatase. *J. Biol. Chem.* 248:6993–7000
77. Rabon E, Wilke M, Sachs G, Zampighi G. 1986. Crystallization of the gastric H,K-ATPase. *J. Biol. Chem.* 261:1434–39
78. Reinstein J, Jencks WP. 1993. The binding of ATP and Mg^{2+} to the calcium adenosinetriphosphatase of sarcoplasmic reticulum follows a random mechanism. *Biochemistry* 32:6632–42
79. Rice WJ, Young HS, Martin DW, Sachs JR, Stokes DL. 2001. Structure of Na^+, K^+-ATPase at 11 Å resolution: comparison with Ca^{2+}-ATPase in E_1 and E_2 states. *Biophys. J.* 80:2187–97
80. Ross DC, Davidson GA, McIntosh DB. 1991. Mechanism of inhibition of sarcoplasmic reticulum Ca^{2+}-ATPase by active site cross-linking. *J. Biol. Chem.* 266:4613–21
81. Sagara Y, Wade JB, Inesi G. 1992. A conformational mechanism for formation of a dead-end complex by the sarcoplasmic reticulum ATPase with thapsigargin. *J. Biol. Chem.* 267:1286–92
82. Seekoe T, Peall S, McIntosh DB. 2001. Thapsigargin and dimethyl sulfoxide activate medium Pi $\leftrightarrow$ HOH oxygen exchange catalyzed by sarcoplasmic reticulum Ca^{2+}-ATPase. *J. Biol. Chem.* 276: 46737–44
83. Shi D, Hsiung H-H, Pace RC, Stokes DL. 1995. Preparation and analysis of large, flat crystals of Ca^{2+}-ATPase for electron crystallography. *Biophys. J.* 68:1152–62
84. Shi D, Lewis MR, Young HS, Stokes DL. 1998. Three-dimensional crystals of Ca^{2+}-ATPase from sarcoplasmic reticulum: merging electron diffraction tilt series and imaging the (h, k, 0) projection. *J. Mol. Biol.* 284:1547–64
85. Skou JC. 1957. The influence of some cations on an adenosine triphosphatase from peripheral nerves. *Biochim. Biophys. Acta* 23:394–401
86. Skriver E, Kaveus U, Hebert H, Maunsbach AB. 1992. Three-dimensional structure of Na,K-ATPase determined from membrane crystals induced by cobalt-tetrammine*-ATP. *J. Struct. Biol.* 108: 176–85
87. Skriver E, Maunsbach AB, Jorgensen PL. 1981. Formation of two-dimensional crystals in pure membrane-bound Na^+/K^+-ATPase. *FEBS Lett.* 131:219–22
88. Stokes DL, Auer M, Zhang P, Kuehlbrandt W. 1999. Comparison of H^+-ATPase and Ca^{2+}-ATPase suggests that a large conformational change initiates P-type ion pump reaction cycles. *Curr. Biol.* 9:672–79
89. Stokes DL, Green NM. 1990. Three-dimensional crystals of Ca-ATPase from sarcoplasmic reticulum: symmetry and molecular packing. *Biophys. J.* 57:1–14
90. Stokes DL, Green NM. 2000. Modeling a dehalogenase fold into the 8-Å density map for Ca^{2+}-ATPase defines a new domain structure. *Biophys. J.* 78:1765–76
91. Stokes DL, Lacapere J-J. 1994. Conformation of Ca^{2+}-ATPase in two crystal forms: effects of Ca^{2+}, thapsigargin, AMP-PCP, and Cr-ATP on crystallization. *J. Biol. Chem.* 269:11606–13
92. Strock C, Cavagna M, Peiffer WE, Sumbilla C, Lewis D, Inesi G. 1998. Direct demonstration of Ca^{2+} binding defects in sarco-endoplasmic reticulum Ca^{2+} ATPase mutants overexpressed in COS-1 cells transfected with adenovirus vectors. *J. Biol. Chem.* 273:15104–9
93. Toyoshima C, Nakasako M, Nomura H,

Ogawa H. 2000. Crystal structure of the calcium pump of sarcoplasmic reticulum at 2.6 Å resolution. *Nature* 405:647–55
94. Toyoshima C, Nomura H. 2002. Structural changes in the calcium pump accompanying the dissociation of calcium. *Nature* 418:605–11
95. Toyoshima C, Sasabe H, Stokes DL. 1993. Three-dimensional cryo-electron microscopy of the calcium ion pump in the sarcoplasmic reticulum membrane. *Nature* 362:469–71
96. Varga S, Csermely P, Martonosi A. 1985. The binding of vanadium (V) oligoanions to sarcoplasmic reticulum. *Eur. J. Biochem.* 148:119–26
97. Wakabayashi S, Imagawa T, Shigekawa M. 1990. Does fluorescence of 4-nitrobenzo-2-oxa-1,3-diazole incorporated into sarcoplasmic reticulum ATPase monitor putative E_1-E_2 conformational transition? *J. Biochem.* 107:563–71
98. Wakabayashi S, Shigekawa M. 1984. Role of divalent cation bound to phosphoenzyme intermediate of sarcoplasmic reticulum ATPase. *J. Biol. Chem.* 259:4427–36
99. Wang W, Cho HS, Kim R, Jancarik J, Yokota H, et al. 2002. Structural characterization of the reaction pathway in phosphoserine phosphatase: crystallographic "snapshots" of intermediate states. *J. Mol. Biol.* 319:421–31
100. Wang W, Kim R, Jancarik J, Yokota H, Kim SH. 2001. Crystal structure of phosphoserine phosphatase from *Methanococcus jannaschii*, a hyperthermophile, at 1.8 Å resolution. *Structure* 9:65–71
101. Watanabe T, Inesi G. 1982. The use of 2′,3′-O-(2,4,6-trinitrophenyl) adenosine 5′-triphosphate for studies of nucleotide interaction with sarcoplasmic reticulum vesicles. *J. Biol. Chem.* 257:11510–16
102. Wictome M, Khan YM, East JM, Lee AG. 1995. Binding of sesquiterpene lactone inhibitors to the Ca^{2+}-ATPase. *Biochem. J.* 310:859–68
103. Xian Y, Hebert H. 1997. Three-dimensional structure of the porcine gastric H,K-ATPase from negatively stained crystals. *J. Struct. Biol.* 118:169–77
104. Xu C, Rice WJ, He W, Stokes DL. 2002. A structural model for the catalytic cycle of Ca^{2+}-ATPase. *J. Mol. Biol.* 316:201–11
105. Young H, Xu C, Zhang P, Stokes D. 2001. Locating the thapsigargin binding site on Ca^{2+}-ATPase by cryoelectron microscopy. *J. Mol. Biol.* 308:231–40
106. Zhang P, Toyoshima C, Yonekura K, Green NM, Stokes DL. 1998. Structure of the calcium pump from sarcoplasmic reticulum at 8 Å resolution. *Nature* 392:835–39
107. Zhang Z, Sumbilla C, Lewis D, Inesi G. 1993. High sensitivity to site directed mutagenesis of the peptide segment connecting phosphorylation and Ca binding domains in the Ca transport ATPase. *FEBS Lett.* 335:261–64
108. Zhao R, Collins EJ, Bourret RB, Silversmith RE. 2002. Structure and catalytic mechanism of the *E. coli* chemotaxis phosphatase CheZ. *Nat. Struct. Biol.* 9:570–75

Annu. Rev. Biophys. Biomol. Struct. 2003. 32:469–92
doi: 10.1146/annurev.biophys.32.110601.141704

First published online as a Review in Advance on January 24, 2003

LIQUID-LIQUID IMMISCIBILITY IN MEMBRANES

Harden M. McConnell and Marija Vrljic
Department of Chemistry, Biophysics Program, Stanford University, Stanford, California 94305-5080; email: harden@stanford.edu; marija@stanford.edu

Key Words cholesterol-phospholipid mixtures, phase diagrams, condensed complexes, cell membranes, lipid rafts

■ **Abstract** The observation of liquid-liquid immiscibility in cholesterol-phospholipid mixtures in monolayers and bilayers has opened a broad field of research into their physical chemistry. Some mixtures exhibit multiple immiscibilities. This unusual property has led to a thermodynamic model of "condensed complexes." These complexes are the consequence of an exothermic, reversible reaction between cholesterol and phospholipids. In this quantitative model the complexes are sometimes concentrated in a separate liquid phase. The phase separation into a complex-rich phase depends on membrane composition and intensive variables such as temperature. The properties of defined cholesterol-phospholipid mixtures provide a conceptual foundation for the exploration of a number of aspects of the biophysics and biochemistry of animal cell membranes.

CONTENTS

1056-8700/03/0609-0469$14.00

INTRODUCTION

There is a large literature dealing with the biophysical properties of model membranes containing cholesterol (Chol), particularly membranes with defined lipid compositions. This subject is the focus of one comprehensive book and is also discussed in more recent publications and reviews (15, 19, 42, 44). The present review is rather narrowly focused on liquid-liquid immiscibility in cholesterol-phospholipid mixtures. We do this because studies of this immiscibility provide significant information concerning cholesterol-phospholipid interactions, especially complex formation in model systems. The studies also provide a guide to understanding the physical chemistry of components of cell membranes referred to as lipid rafts (3, 8–10, 30, 69, 70). The review follows one of the editor's recommendations to authors, namely to "track the lineage of ideas and the evolution of thinking, with key results . . . ."

MIXTURES OF CHOLESTEROL AND PHOSPHOLIPIDS ARE NONIDEAL

In 1925 Leathes (38) discovered the nonideality of mixtures of Chol and phospholipids. He found nonadditivity of molecular areas in monolayers at the air-water interface. The area of a mixture was found to be significantly less than the mean area expected from the areas of the two components, all at the same pressure. This was referred to as a condensing effect. This early work has been confirmed in many subsequent studies (49). A related effect has been reported in mixtures of hydrocarbons, when one has a relatively rigid molecular structure and the other a flexible structure (63). In 1972 Hinz & Sturtevant (26) obtained additional evidence of nonideality in studies of bilayer mixtures by scanning calorimetry. The mixtures contained Chol and a phosphatidylcholine, such as dipalmitoylphosphatidylcholine (DPPC), or dimyristoylphosphatidylcholine (DMPC) [see also (43)]. These investigators found a new heat capacity peak not found for either pure component. Feingold (15) reviews other related studies together with theoretical modeling of these mixtures.

MOLECULAR ORIGINS OF NONIDEALITY

There have been many attempts to relate the nonideal physical properties of cholesterol-phospholipid mixtures to molecular structure. There were a number of early proposals for complex formation between Chol and phospholipids

(12, 18, 20, 26, 47). There is extensive evidence that Chol leads to the suppression of gauche conformations of saturated fatty acid chains of phospholipids in bilayers (15). Many models have been proposed concerning how these molecules may be arranged laterally in two dimensions, for example, on a regular hexagonal lattice (28, 72, 76, 83, 85). However, it is not always appreciated that the membranes of interest are liquids, where there is no long-range order. Also, any proposed relation between macroscopic physical chemical properties and molecular order must recognize the possibility of phase separations in the systems studied. The difficulty in obtaining reliable phase diagrams for cholesterol-phospholipid mixtures has been a major obstacle to progress in this field.

THE SEARCH FOR LIQUID-LIQUID IMMISCIBILITY

In early work structure-sensitive spectroscopic techniques were used in an effort to detect immiscibility in cholesterol-phosphatidylcholine bilayers. The methods include the paramagnetic resonance of spin label lipid probes (1) and the deuterium nuclear magnetic resonance of deuterium-labeled phospholipids (84). Variations in spectroscopic signal with membrane composition were used to locate phase boundaries for mixtures of Chol with DPPC (84) and with DMPC (1, 62, 64). A proposed phase diagram for the Chol-DMPC mixture is given in Figure 1 (1). Although the measurements were analyzed in terms of liquid-liquid immiscibility, this interpretation of the data may not be unique. That is, spectroscopic probes and diffusion measurements provide no definitive information on the sizes of the putative domains. Indeed it was pointed out in early work that the spin label data are consistent with very small or "microscopic" domains (64). In this latter case one would question whether thermodynamic phase separation is an accurate description. Moreover, the cooperative formation of a chemically distinct reaction product between Chol and phospholipid molecules (a complex) can lead to a sharp composition-dependent variation of physical properties, especially if complex formation is highly cooperative, involving a large number of molecules. This could easily be mistaken for a phase boundary. To the best of our knowledge, there is no evidence for liquid domains of macroscopic size in DMPC and DPPC bilayer mixtures with Chol. However, as discussed below, there are reasons to suspect that the measurements leading to diagrams such as that in Figure 1 are consistent with the formation of condensed complexes.

DISCOVERY OF IMMISCIBILITY IN MONOLAYER MIXTURES

In 1986 unambiguous liquid-liquid immiscibility was observed in monolayer mixtures of Chol and DMPC at the air-water interface (75). This finding was followed by numerous studies of the physical chemical properties of this and similar

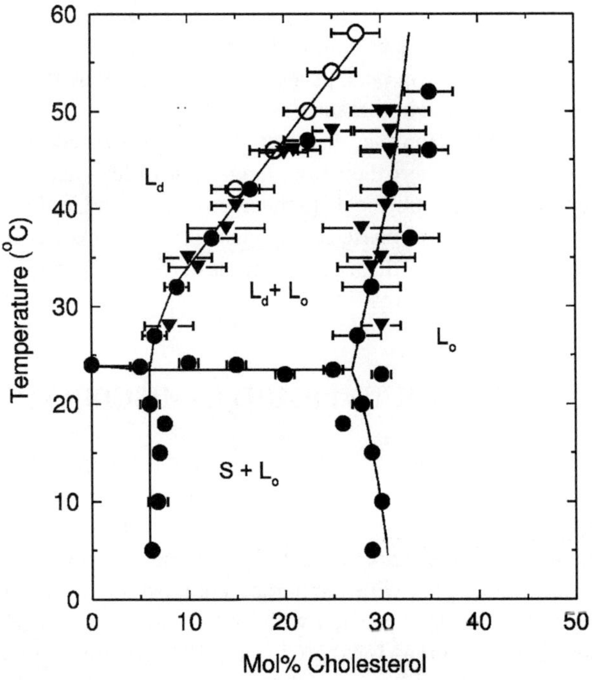

Figure 1 Experimental phase diagram of lipid bilayer consisting of a binary mixture of dimyristoylphosphatidylcholine (DMPC) and cholesterol (Chol). Data points were obtained from spectroscopic probes and diffusion coefficient measurements [reprinted with permission from Almeida et al. (1). Copyright 1992 American Chemical Society]. Macroscopic phase separations were inferred but not observed. The terms L_d, L_o, and S refer to the liquid disordered, liquid ordered, and solid states, respectively. The lines are not a theoretical fit.

mixtures (41). Figure 2 shows two coexisting liquid phases, one rich in Chol and the other rich in DMPC. The monolayer is viewed using an epifluorescence microscope and contains a low concentration of fluorescent lipid probe that partitions unequally between the two liquid phases, providing contrast. The fluorescent probe partitions more favorably into the phospholipid-rich phase, making it bright. The sizes, shapes, electrostatics, and hydrodynamics of such coexisting liquid phases have been investigated extensively (41).

Pressure-composition phase diagrams for these mixtures are obtained by measuring the pressure at which the two phases disappear (appear) on increasing (decreasing) pressure for each membrane composition. A particularly simple, symmetric phase diagram is shown in Figure 3*a*. This describes a binary mixture of Chol and a phospholipid having two unsaturated fatty acid chains (22). The miscibility critical point corresponds to the peak of this curve. This point gives the

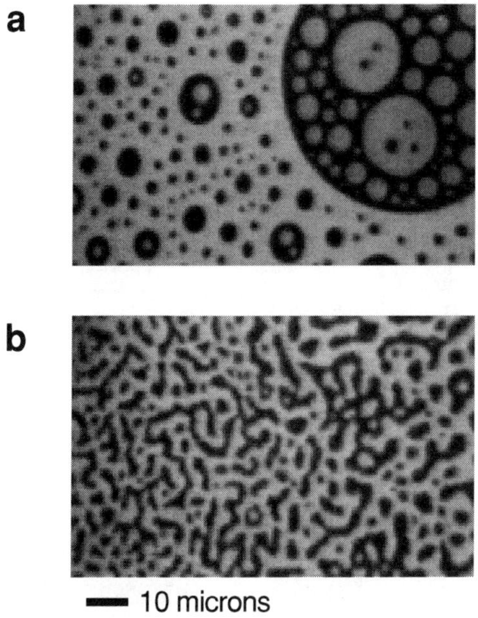

Figure 2 Epifluorescence image of a liquid-liquid immiscibility in a lipid monolayer [for details see (42)]. The monolayer consists of a binary mixture of DMPC (69.8 mole%) and dihydrocholesterol (DChol, 30 mole%) at an air-water interface at 23°C. DChol is frequently used instead of Chol to avoid air- and photo-oxidation. The phase behavior of the two sterols is virtually identical with the exception that DChol yields slightly lower values of the critical pressure (6, 59). The contrast between phases is provided by 0.2 mole% fluorescent lipid analog (Texas Red-dihexanoylphosphatidylethanolamine, TR-DHPE) since the lipid probe preferentially partitions in one of the two phases. (*a*) Image of a monolayer showing two liquid-phase coexistence. The dark phase is a DChol-rich liquid and the brighter phase is a DMPC-rich liquid. The monolayer is at a surface pressure of 2 dyne/cm. The domains are 5–10 μm in diameter and exhibit Brownian motion. (*b*) Image of the monolayer at a surface pressure of 9.4 dyne/cm showing the stripe phase characteristic of proximity to a miscibility critical point. The stripe phase is liquid. The stripes become thinner and disappear at a surface pressure of 9.5 dyne/cm, above which only one homogenous liquid phase can be observed.

critical pressure and critical composition. Figure 3*a* has two insets. The upper inset shows the homogeneous fluorescence at a pressure above the critical pressure, where there is a single phase. The lower inset shows the two coexisting liquid phases when the pressure is below the critical pressure. The lower inset shows that the composition of the monolayer is at the critical composition, in this case near 50 mole% Chol. In this circumstance there is frequent contrast inversion, dark on bright and bright on dark. This is because the two liquid phases are present in

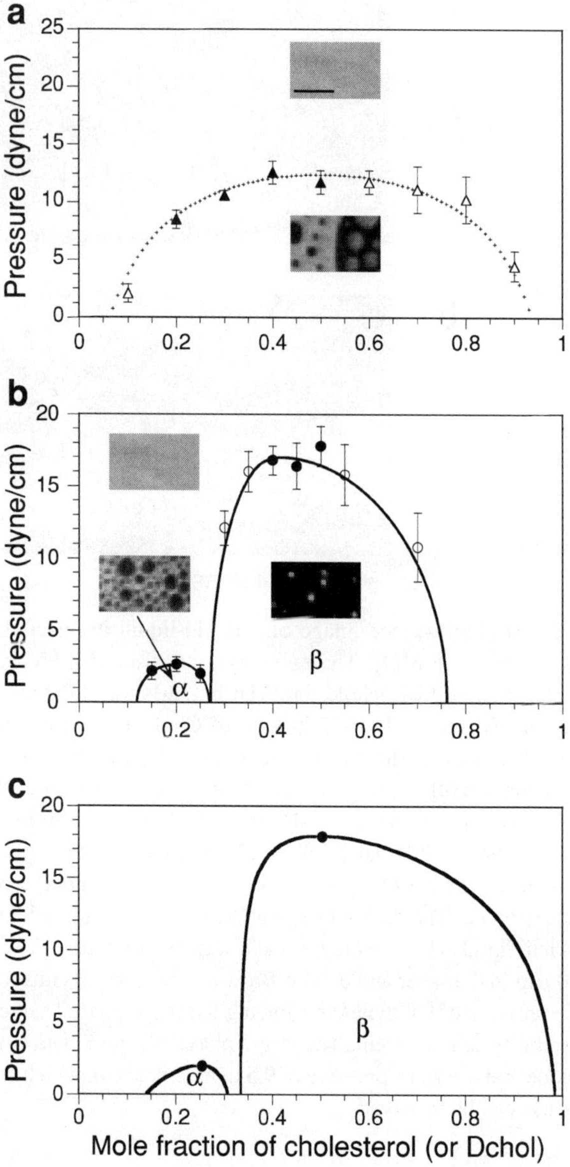

nearly equal amounts. Some phase diagrams for these mixtures are remarkably simple and, as a good first approximation, closely follow regular solution theory. Indeed, the diagram in Figure 3*a* can be thought of as an example of an "ideal immiscibility." The symmetry about the 50:50 composition shows that in these mixtures the Chol and phospholipid play equivalent roles. The thermodynamics

←

Figure 3 Pressure-composition phase diagrams for lipid mixtures in monolayers at the air-water interface. Plotted data points represent transition pressures that mark the disappearance or appearance of two-phase coexistence during monolayer compression. Stripe superstructure phases (Figure 2*b*), which represent proximity to a critical point, were observed at the transition pressures marked by filled symbols (critical pressures) and not at those marked by open symbols. The insets illustrate epifluorescence images of a monolayer at pressures above the critical pressure and below the critical pressure. The insets at pressures below the transition pressures show the coexistence of the two liquid phases, visualized as bright and dark regions of the monolayer. The bright regions are enriched in the fluorescent lipid probe, TR-DHPE, showing that the lipid probe preferentially partitions in one of two phases. The insets at pressures above transition pressures show the one liquid phase visualized by uniform fluorescence intensity. The scale bar represents 10 μm. All measurements were carried out at room temperature (23 +/− 0.5°C) and with a subphase of distilled water (pH 5.4). (*a*) Phase diagram showing liquid-liquid immiscibility for a binary mixture of dipalmitoleoylphosphatidylcholine (DPoPC) and Dchol [modified from (22)]. The shading of the data points is based on measurements of T.M. Okonogi (personal communication). There are two phases below the dotted curve at lower pressures, and one phase above the curve at higher pressures. The one phase is a phospholipid-rich liquid and the other is a cholesterol-rich liquid. The representative epifluorescence images are shown as insets. In the insets the monolayer has a composition close to the critical composition. Note the presence of only one upper miscibility critical point. The dotted curve is a theoretical fit to the data [for details see (22)]. *Recent unpublished work by Okonogi indicates that this phase diagram is incorrect at Dchol concentrations equal to 70 mole% and higher. At these higher concentrations two liquid phases persist to high pressures, and the system shows the same contrast inversion disussed in the text for the β two-phase region.* (*b*) Phase diagram showing liquid-liquid immiscibility for a mixture of Dchol and phospholipids. The phospholipids are comprised of egg-sphingomyelin (SM) and cholesterol (Chol) (59). The two-phase coexistence region at low Dchol mole fractions is labeled α, and the one at high Dchol mole fractions is labeled β. There are two phases below the curves at lower pressures, and one phase above the curves at higher pressures. In the α region one liquid phase is rich in phospholipid (bright), and the other liquid phase is rich in condensed complex (dark). In the β region one liquid phase is rich in condensed complex (bright), and the other liquid phase is rich in cholesterol (dark). Representative epifluorescence images are shown as insets. The curve is not a theoretical fit. The phospholipids have various chain lengths. See text for a discussion of possible errors in the upper boundary of the β two-phase region. (*c*) Calculated pressure-composition phase diagram for a binary mixture showing two upper miscibility critical points (filled circles). The phase diagram was calculated using a regular solution model involving repulsive interactions and components that react to form a condensed complex [for details see (54)]. When a mixture of cholesterol and two or more phospholipids can be modeled as a binary cholesterol-phospholipid mixture, it is referred to as a pseudobinary mixture. The mixture in panel *b* is accordingly interpreted as a pseudobinary mixture.

of immiscibility can be modeled by an average energy of repulsion that is proportional to the product of their concentrations (mole fractions) (22).

MULTIPLE IMMISCIBILITIES AND CONDENSED COMPLEXES

In 1999 unusual phase diagrams were discovered for mixtures of Chol and certain phospholipids (56, 57). These mixtures were reported to have two side-by-side upper miscibility critical points. Although predicted earlier in theoretical studies, phase diagrams of this type had not been observed previously for any liquid mixture in either two or three dimensions (11). The importance of these phase diagrams is that they point to a cooperative reaction between Chol and some phospholipids, particularly sphingomyelins and other phospholipids with saturated fatty acid chains (for details see below). The phase diagram illustrated in Figure 3*b* describes three liquids: One liquid is rich in phospholipid, the other in Chol, and the third liquid, X, has an intermediate composition (about 28 mole% Chol) and is immiscible with the other two liquids at the lower pressures. Liquid X is the condensed complex liquid (57). Obviously, liquid X is significantly different than the other two because it does not mix with them (see the α and β two-phase regions of the phase diagram Figure 3*b*) (59). In general not all three liquids are present simultaneously at the fixed monolayer temperature, pressure, and lipid composition. In the two-phase region labeled α, phospholipid-rich (bright) and condensed complex–rich phase (dark) coexist, while (at lower pressures) in the two-phase region labeled β condensed complex–rich (bright) and Chol-rich (dark) liquids coexist (see below).

The phase diagrams can be calculated using a thermodynamic model in which Chol (C) and phospholipid (P) molecules react reversibly to form a complex $[C_qP_p]_n$ (57, 59) (Figure 3*c*). The fraction of Chol in the complex is $q/(p+q)$, and n is a cooperativity parameter. This fraction is determined by the composition of the cusp between α and β two-phase fields in the phase diagrams. This composition is about 28 mole% dihydrocholesterol (DChol) in Figure 3*b*. For other mixtures the composition has been found to range from 25 to 60 mole% DChol (34, 57). The larger values are associated with phospholipids having fatty acid chains that are less prone to be ordered. These include phospholipids with the shorter fatty acid chains such as those in DMPC and phospholipids containing an unsaturated fatty acid. Values of $q/(p+q)$ observed most often have been around 0.3 to 0.4 when the chains are composed of saturated fatty acids. Values of the cooperativity can be estimated only roughly by fitting the experimental phase diagrams and other data. These values of n are in the range of 2 to 5 for monolayers, but they may be higher in the corresponding bilayers (4).

Some mixtures of Chol together with two or more phospholipids have composition-dependent immiscibilities leading to phase diagrams that mimic those found for binary mixtures. These are referred to as pseudobinary mixtures. For example, the diagram in Figure 3*b* uses data involving multiple components [Dchol and egg-sphingomyelin (SM)]. This phase diagram can be modeled theoretically as a binary mixture, as shown in Figure 3*c*. These and similar data on pseudobinary mixtures indicate that more than one type of phospholipid can participate in forming a single condensed complex.

Studies of the β two-phase regions have presented special problems (42). The domains are often small. The putative critical point properties are difficult to observe and are not observed consistently (42). Recently, an unexpected aspect of the β two-phase region has been uncovered in work with T.M. Okonogi and A. Radhakrishnan (unpublished). The new results can be understood as follows. The boundaries of the β two-phase fields have been defined experimentally by the pressures at which the small fluorescent (white) domains disappear with increasing pressure. These small domains are illustrated in the right-hand micrograph insert in Figure 3b. The recent work indicates that in general such domains do not disappear at these pressures, but the domain fluorescence does. This is contrast inversion, due to a pressure-dependent transfer of fluorescent probe from the small domains to the surrounding liquid. The domains, now dark on a gray background, persist to high pressures, probably as high as 30–40 dynes/cm. Thus the upper boundaries of the β two-phase regions are likely to extend to pressures much higher than indicated by the diagrams in Figures 3b and 5a and in earlier work (42). The enlarged β two-phase fields can be described theoretically by the thermodynamic model of condensed complexes mentioned above, even though it may not be possible to reach the critical points experimentally by pressure changes alone. The complex stoichiometries are not changed by these extensions of the β region.

IMMISCIBILITY IN BILAYER MIXTURES

Clear evidence for liquid-liquid immiscibility in bilayers was first observed in 2001 and 2002 (13, 65, 80). The shapes of the liquid domains and their apparent hydrodynamic properties are remarkably similar to those seen in monolayers. Dietrich et al. (13) have observed phase separation in one of the monolayers of nonsymmetric-supported bilayer composed of multiple lipids. One of the bilayer leaflets was composed of egg-phosphatidylcholine (PC), and the other of 1-palmitoyl-2-oleoyl-sn-glycero-3-phosphocholine (POPC), sphingomyelin (SM), and Chol with 1 mole% of ganglioside GM1. Fluorescent lipid probes were included in the POPC:SM:Chol monolayer. Two liquid phases were observed, the brighter one rich in the lipid probe and the darker one poor in the lipid probe. Depending on the nature of the lipid probe, two liquid phases were visualized as either dark domains on a bright background or bright domains on a dark background. When the lipid probe partitioned equally between the phases, no macroscopic phase separation was observed, as visualized by uniform fluorescence intensity. In laurdan labeled giant unilamellar vesicles (GUVs) composed of dioleoylphosphatidylcholine (DOPC), SM, and Chol with 1 mole% of GM1, the observed phase separation was temperature dependent. The diameters of the observed domains were on the order of a few microns. Samsonov et al. (65) have observed two liquids in bilayers composed of Chol, DOPC, DOPE, and either SM or distearoylphosphatidylcholine (DSPC). Domain formation required the presence of both Chol and SM or DSPC. Smaller domains were 10 to 20 μm in diameter. Veatch & Keller (80) have observed temperature-dependent phase separation in the GUVs and supported bilayers composed of phospholipids, Chol, and SM.

The most striking difference in liquid domain shapes for monolayers and bilayers is that the monolayer phases show pronounced striping near the miscibility critical point, as illustrated in Figure 2*b*. These stripes are present under equilibrium conditions. Stripes with long-range order have not been observed in bilayers. The striping in monolayers is due to the long-range electrostatic forces. These forces are suppressed in bilayers owing to screening by the aqueous phase on both sides of the bilayer. (In principle, transient stripes or other periodic structures could arise from nonequilibrium kinetic effects, such as time-dependent phase separations.)

The free-standing lipid bilayer membranes showing immiscibility that have been studied thus far are symmetric, having the same composition on both sides of the bilayer. The monolayer phases are clearly correlated across the bilayer (13, 65, 80). This might have been expected, since one knows from liquid crystal physics that, in general, neighboring molecules tend to have similar order parameters. Two questions remain open. One is the question of coupling in nonsymmetric bilayers. The second question is whether there is a lower limit on domain size in order for the coupling to persist. This point is illustrated in Figure 4, where individual complexes are not correlated across the membrane in Figure 4*a* but are correlated in 4*b*. Interlayer coupling of complexes might increase the cooperativity parameter, *n*, as suggested by an analysis of the calorimetry of cholesterol-phospholipid mixtures in terms of complexes (4).

CORRESPONDENCE BETWEEN MONOLAYER AND BILAYER PHASES

At present it is not known how the observed immiscibility in bilayers is related to the three liquid phases found in monolayers. Doubtless phase diagrams will ultimately be determined for bilayer mixtures. However, the problem for bilayers is potentially more difficult because there may be composition-dependent shape changes, vesicle budding, and rapid cholesterol flip-flop as well as transmembrane asymmetry.

In the case of bilayers it is difficult to control surface tension. Thus experimental phase diagrams are likely to be temperature-composition diagrams. In comparing these phase diagrams with monolayer pressure-composition phase diagrams, it is convenient to reverse either the pressure or temperature axis. That is, an increase in temperature in bilayers should be compared to a decrease in pressure in monolayers. In this connection, for example, if the bilayer phase diagram in Figure 1 is turned upside down, it will be seen to have the same qualitative appearance as the γ,

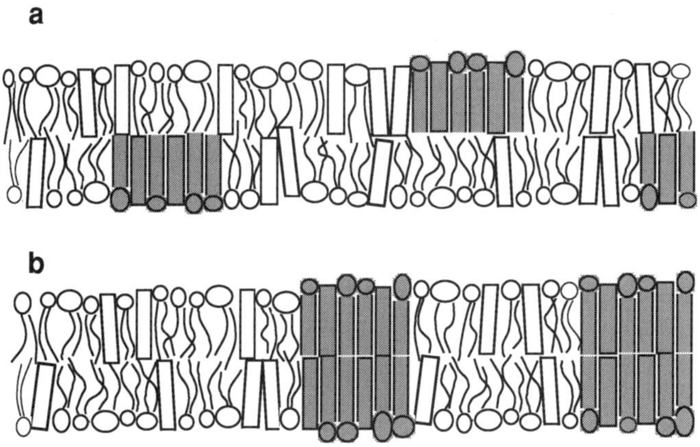

Figure 4 Schematic view of a bilayer region of a plasma membrane. Rectangles represent cholesterol molecules. Lipid molecules shaded in gray are in the condensed complexes and others are not. When phospholipids form condensed complexes with cholesterol, the fatty acid chains of the phospholipids are shown as straight lines. Note that condensed complexes can form between cholesterol and different lipids. Note that the complexes need not form a separate phase. (*a*) Schematic view of a bilayer region of a plasma membrane showing complex formation in both leaflets of the membrane. The condensed complexes are not correlated across the bilayer. (*b*) Schematic view of a bilayer region of a plasma membrane showing complex formation in both leaflets of the membrane. Here the condensed complexes are correlated across the bilayer.

δ, and ε two-phase fields in the monolayer diagram in Figure 5*a*,*b*. We thus suggest that a γ two-phase field of monolayers may be found to correspond most closely to the two-phase liquid-liquid region reported in bilayers (13, 65). In this case the bilayer phase richest in cholesterol would be the condensed complex. The γ two-phase field in monolayers has been suspected for some time (21) but has only been confirmed in recent work (60). When comparing the monolayer and bilayer phase diagrams, note that increasing temperature generally destabilizes condensed complexes because complex formation is strongly exothermic (4, 61). Thus, an upper miscibility critical solution temperature may be affected by complex dissociation in addition to the usual thermal effects tending to increase miscibility.

Theoretical phase diagrams for monolayers and bilayers are shown in Figure 5*b*,*c*. Note that the assumed stoichiometric composition, 33 mole% Chol, provides a sharp demarcation for the various phase boundaries in Figure 5*b*. This theoretical diagram may be compared with the experimental diagram in Figure 5*a*. There are three liquid-liquid two-phase fields in Figure 5*a*,*b* (γ, α, β), and in each case one liquid is rich in condensed complex. Note the strong similarity between γ and ε phase boundaries in Figure 5*a*,*b* and the corresponding boundaries in Figure 1.

The phase diagram in Figure 5*c* is adapted from Ipsen et al. (29). Three of the two-phase fields in Figure 5*c* have been labeled γ, ε, and δ for comparison with the phase diagram in Figure 5*b*. The two coexisting liquid phases in Figure 5*c* were termed liquid-ordered and liquid-disordered in the original work of Ipsen et al. (29). Complex formation was not involved in the model used to create Figure 5*c*.

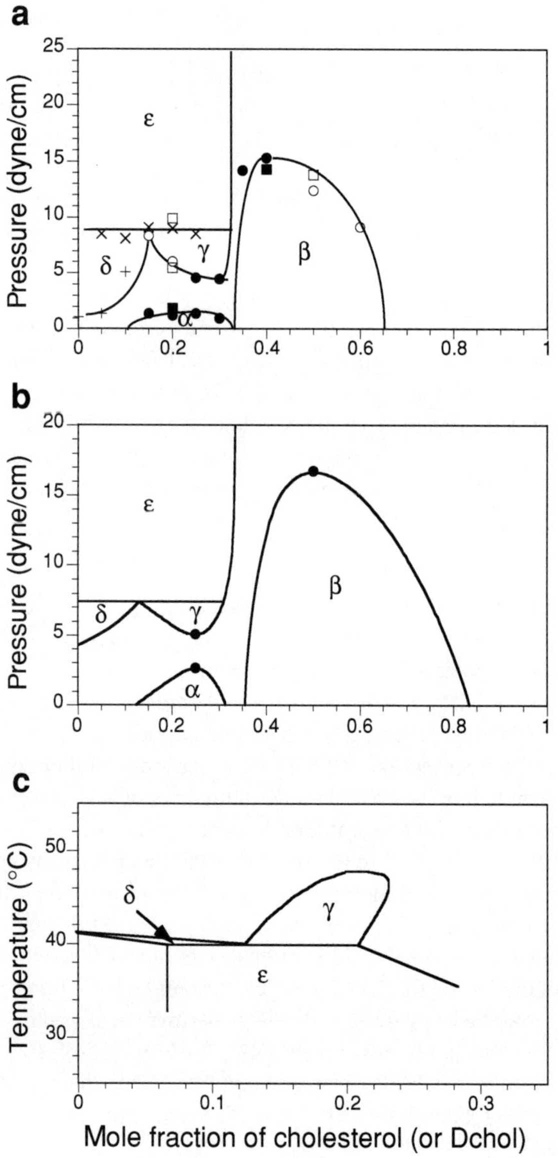

CONDENSED COMPLEXES NEED NOT FORM A SEPARATE PHASE

There is compelling evidence that in monolayers condensed complexes are present even when they do not form a separate thermodynamic phase (42, 54, 57). Their presence has been demonstrated by measurements of molecular area, electrical properties, and chemical activity (42, 59). In fact, as mentioned earlier, some of these properties in the homogeneous liquid phase are so striking they might be due to a "microscopic phase" (64). In this respect some of the early reported phase diagrams (such as Figure 1) may be due to effects of condensed complexes in the homogeneous phase, and not to liquid-liquid phase separations. This point

←

Figure 5 Experimental and theoretical phase diagrams for lipid mixtures in monolayers and bilayers. (*a*) Phase diagram showing liquid-liquid and liquid-solid immiscibility for a lipid monolayer at the air-water interface. The monolayer is a mixture of Dchol, dimyristoylphosphatidylserine (DMPS), and ganglioside GM1. A constant DMPS:GM1 ratio of 7:3 was maintained as the cholesterol concentration was changed [for details see (60)]. Plotted data points represent transition pressures that mark the disappearance or appearance of two-phase coexistence during monolayer compression. The α and β two-phase regions are similar to those in Figure 3*b*. In addition, a new liquid-liquid immiscibility region (γ) and two solid-liquid immiscibility regions (ε and δ) are also observed. In the γ region one liquid phase is rich in condensed complex and the other liquid phase is rich in phospholipid. In the ε region one phase is a liquid rich in condensed complex and the other phase is a solid rich in phospholipid. In the δ region one phase is a solid rich in phospholipid and the other phase is a liquid rich in phospholipid. Stripe superstructure phases, which represent proximity to a critical point, were observed at the transitions marked by filled symbols and not at those marked by open symbols. Squares represent spot checks carried out using Chol instead of Dchol. Note the presence of the two upper miscibility critical points (in α and β regions, respectively) and one lower miscibility critical point (in γ region). *The β two-phase field may extend to significantly higher pressures (see text).* In recent unpublished work by Radhakrishnan the γ two-phase field has been observed in binary Dchol-DMPS mixtures at 37°C in the absence of ganglioside GM1. (*b*) Calculated pressure-composition phase diagram for binary lipid mixtures showing two upper miscibility critical points and one lower miscibility critical point [for details see (60)]. Filled circles represent the critical points. The phase diagram was calculated using a regular solution model involving repulsive interactions and components that react to form a condensed complex. (*c*) Schematic view of calculated phase diagram for a bilayer mixture of DPPC and Chol [reprinted from Ipsen et al. (29); Copyright 1987, with permission from Elsevier Sciences]. The two-phase fields γ, ε, and δ are labeled to show the analogy with the phase diagram in (*a*) and (*b*). In Ipsen et al. (29) the phase fields γ and ε are labeled l_o-l_d and s_o-l_o, respectively.

is made qualitatively in Figure 6. In this top view of a membrane dark circles represent cholesterol, and gray circles represent phospholipid in complexes, having the stoichiometry $q = 1$, $p = 2$, and $n = 2$ or 4. Larger white circles represent phospholipids not in complexes. Figure 6*a* sketches a membrane containing a low concentration of cholesterol, where just a few complexes can be formed. In Figure 6*b* the monolayer is in the α (or γ) two-phase region. Here, one phase is largely phospholipid, and the other phase largely condensed complex. On the other hand, in Figure 6*c* the monolayer is homogeneous single phase, having roughly the same proportion of complexes as in Figure 6*b*. One can easily imagine that a spectroscopic phospholipid-like probe might give two distinct signals in both cases, corresponding to Figure 6*b*,*c*. Thus it could be quite difficult to distinguish the two cases in the absence of observable macroscopic phase separation. In fact, the theoretically calculated area fraction of complexes is sometimes a linear function of composition just as one would expect from application of the phase rule (5). Therefore, in the absence of observed macroscopic phase separations, one must reserve judgment on the significance of reported phase diagrams such as that in Figure 1. From our perspective, the common feature in the data is the evidence for complex formation with or without macroscopic phase separation.

LIQUID-LIQUID IMMISCIBILITY IN ANIMAL CELL MEMBRANES

Membranes of animal cells are formed from lipid bilayers and associated proteins and glycoproteins. There is extensive evidence that the lipid bilayer regions are in a liquid state. In recent years it has been proposed that certain lipids in the plasma membrane promote the formation of specialized regions of the membranes enriched in these lipids and in some proteins and glycoproteins. Chol, SM, and saturated glycerophospholipids and glycosphingolipids are typically associated with these specialized domains (2, 3, 8–10, 69, 70). However, there is considerable uncertainty concerning the size and function of these domains, and even their definition. For example, reports of domain sizes cover a broad range, having radii from 25 to 700 nm (51, 66, 68, 77, 79). These specialized domains are commonly referred to as rafts or detergent-resistant membranes. We enumerate physical chemical aspects of liquid-liquid immiscibility in cholesterol-phospholipid mixtures that may be relevant to studies of putative specialized domains in animal cell membranes.

Immiscibility, Lipid Composition, and Complexes

Keller et al. (33) have shown that the total lipids extracted from the erythrocyte membranes spread as lipid monolayers on the air-water interface show coexistence of two-liquid phases. The lipids from both the inner and outer leaflets show a two-phase coexistence up to pressures of 21 and 29 dyne/cm, respectively. Dietrich et al. (13, 14) have shown that supported lipid monolayers as well as bilayers in

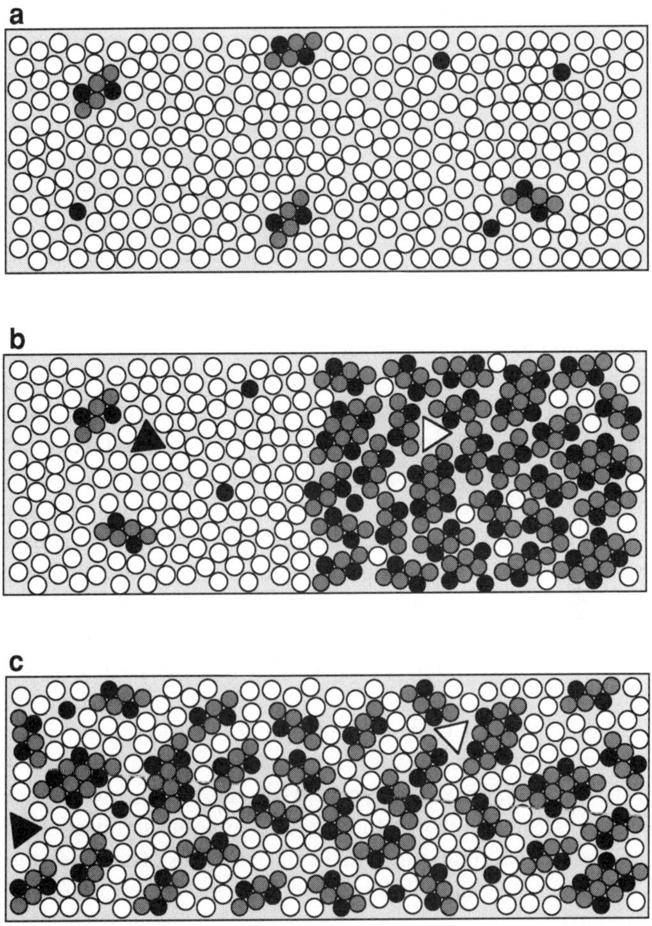

Figure 6 Schematic view of a lipid membrane containing cholesterol and phospholipids. White circles represent phospholipids not in condensed complexes, gray circles represent phospholipids in condensed complexes, and black circles represent cholesterol. The area of the gray circle is smaller than the area of the white circle corresponding to the condensing effect of cholesterol. Clusters composed of gray and black circles represent condensed complexes. The stoichiometry of the complex, $[C_qP_p]_n$, is $q = 1$, $p = 2$, and $n = 2$. (*a*) Schematic view of a lipid monolayer at a low cholesterol concentration showing macroscopic homogenous single phase. Note that most of the cholesterol is in the condensed complex. (*b*) Schematic view of a lipid monolayer showing macroscopic phase separation. The two liquid phases are phospholipid rich (mostly white circles) and condensed complex rich (clusters). The two phases correspond to regions such as α or γ. Some condensed complexes are oligomerized, $n = 2$ and $n = 4$. The black and white triangles represent proteins or other membrane components that have a preference for environments that are largely phospholipids or condensed complexes, respectively. (*c*) Schematic view of a lipid monolayer showing macroscopic homogenous single phase. The number of condensed complexes is the same as in (*b*).

the form of GUVs composed of the lipids extracted from apical membranes from rat kidney cells do show liquid-liquid phase separation. Compared to the basolateral membranes, which are enriched in SM and phosphatidylcholine, the apical membranes are enriched in glycosphingolipids. Concentrations of cholesterol are similar in both membranes (71). The observation of Dietrich et al. (13, 14) was based on the partitioning of TR-DPPE and laurdan fluorescent lipid analogs as well as partitioning of ganglioside GM1 visualized by fluorescein-tagged cholera toxin. Based on the shape of the domains (smooth boundaries), levels of laurdan intensity, and measured diffusion coefficients of TR-DPPE, the observed domains appear to be liquid. Schütz et al. (66) have observed enrichment of saturated fluorescent lipid probes DMPE-Cy5 and DMPE-TMR in certain regions of the HSAM plasma membrane, suggestive of the presence of more than one lipid phase. At the same time, the unsaturated lipid probe DOPE-Cy5, although slightly enriched in the same regions as DMPE probes, showed a more uniform distribution. The monolayer studies show that Chol forms stable complexes with many SMs, phosphatidylcholines, phosphatidylserines, and phosphatidylethanolamines at room temperature (34, 48, 54–57, 59–61). This suggests that condensed complexes could form in both leaflets of cell membranes. Therefore, specialized domains in cell membranes rich in these lipids should contain complexes. Do the complexes form a separate liquid phase? The monolayer studies show that the complexes need not be present as a separate liquid phase in order to be present in high concentration. However, under suitable conditions the complexes can form a separate liquid phase. Thus, in general, there is no need for correlation between the presence of complexes and phase separation. However, if structures related to liquid immiscibility are observed in a membrane containing substantial concentrations of complex formers (e.g., Chol and SM), then the immiscibility is likely to involve complexes with one phase rich in complex.

Cholesterol Extraction

When phase-separated liquid-liquid immiscibility is observed in monolayers, cholesterol extraction with β-cyclodextrin generally changes the relative proportions of the phases, as expected from the phase diagrams. (At a given pressure, or temperature, the observed changes in the relative proportion of the phases are due to the translation along the composition axis in the phase diagram.) Possibly related effects have been reported in cells. As mentioned above, Dietrich et al. (14) have extracted lipids from the apical membranes and spread them at 32–35 dyne/cm (thought to be the bilayer pressure) as supported monolayers. Prior to cholesterol extraction they have observed the coexisting liquid phases, one cholesterol rich and one rich in other lipids. The domains they report resemble those in the α two-phase region shown in the right lower inset in Figure 3*b*. Incubation of the monolayer with the methyl-β-cyclodextrin leads to the disappearance of the dark, cholesterol-rich domains and consequent disappearance of the two-phase region. The monolayer then resembles the one-phase region shown in the top insets in Figure 3*a*,*b*.

Subsequent incubation of the monolayer with the cholesterol-loaded methyl-β-cyclodextrin leads to the reappearance of immiscible domains (much like the domains in the β two-phase region shown in the left bottom inset in Figure 3*b*). A disadvantage of this supported monolayer system is that the pressure is not controlled.

Another example involving cholesterol extraction is the work of Pralle et al. (51). They measured the diffusion coefficients of the proteins localized within and outside lipid rafts, as determined by resistance to detergent solubilization. At normal cholesterol levels, Pralle et al. (51) observed two diffusion coefficients, the faster being nonraft proteins and the slower being raft proteins. After the decrease in cholesterol concentration, all the proteins had the same diffusion coefficient, namely that of nonraft proteins. This led to the conclusion that at normal cholesterol levels, observed diffusion was the diffusion of lipid rafts rather than the diffusion of proteins within the rafts. This could argue that the lipids in the plasma membrane phase separate at normal cholesterol levels. Consequently, there would be no phase separation once cholesterol concentration was lowered.

The studies of Maxfield and coworkers (23) suggest a contrary view for the phase state of the plasma membrane lipids. The authors (23) observed a uniform fluorescence intensity of fluorescent lipid probes at normal cholesterol concentrations indicating a homogenous phase. A nonuniform fluorescence intensity was observed at the decreased cholesterol levels, suggesting the possibility of a two-phase coexistence. These investigators used the lipid probes $DiIC_{16}$ (thought to partition preferentially in liquid-ordered regions) and C_6-NBD-SM and $DiIC_{12}$ (both thought to preferentially partition with liquid-disordered regions). The fluorescence intensity distribution of these probes suggested that plasma membrane lipids become more ordered as cholesterol concentration is decreased. These observations do not support a current model of lipid rafts, in which liquid-ordered domains disappear upon cholesterol depletion, leaving the whole membrane in a disordered state.

Note that in mixtures forming condensed complexes, the most ordered state of the system corresponds to some stoichiometric composition (33 mole% Dchol in Figure 5*a*,*b*). Thus a decrease of cholesterol concentration can lead to less or more order depending on the initial cholesterol concentration. Similar remarks may apply to the appearance or disappearance of immiscible liquid phases.

In addition to the abovementioned work on cells, there are numerous studies that link resistance to detergent solubilization of many raft-resident proteins to plasma membrane cholesterol concentration. Depletion of plasma membrane cholesterol leads to increased detergent solubilization of raft-resident proteins, which in turn has been reported to disrupt functional roles of the above proteins.

The Liquid-Liquid Interface and Domain Boundaries

In monolayers and bilayers there can be a substantial free energy associated with the interface between the liquids. This energy-per-unit length is referred to as line tension. In monolayers this line tension depends on pressure and approaches zero as a critical point is approached. In animal cells a "bare" liquid-liquid interface of this type is unlikely, except near a critical point (33). A large line tension interface would be a trap for impurities. Thus if there is a liquid-liquid interface in animal cell membranes, it is likely to be decorated with specific proteins and/or lipids.

Cell Triggering and Transient Immiscibility

Complex stability (and the associated immiscibility) can be affected by externally applied electric fields. Field strengths comparable to those in nerve cell membranes can lead to the dissociation of complexes and domain formation (58). Immiscibility associated with complex formation is sensitive to ionic strength (distilled water versus phosphate buffer saline) in the subphase under a monolayer containing phosphatidylserine (60). This immiscibility is likely to be particularly sensitive to low concentrations of calcium ion. It is thus plausible that calcium ion fluxes in cells associated with receptor triggering could lead to a change of immiscibility.

There does appear to be a correlation between cell triggering, calcium fluxes, and raft association of membrane components. Ca^{2+} fluxes rapidly follow ligand binding to some receptors. This in turn changes raft-residency of the respective receptors together with other molecules involved in signaling. This correlates with the formation of patches of raft components. In addition, durability of Ca^{2+} fluxes is fine-tuned to the kinetic profile of the ligands. Good examples are signaling complexes of T-cell receptor (TCR) [for details see (7, 36)] and high-affinity IgE receptor (FcεRI) [for details see (35, 37)]. In both cases Ca^{2+} fluxes rapidly follow ligand binding to these receptors and, in the case of T-cells, temporally precede formation of the immunological synapse. Following these early signaling events, the detergent-solubility profile of TCR (46, 86) and FcεRI changes (27, 67). These receptors become resident in rafts and colocalize with other raft components. Abrogation of the Ca^{2+} signal, at least in the case of FcεRI, leads to the dissociation of the protein from detergent-resistant membranes (27, 50). There is some evidence that cytoskeletal events are subsequent to earlier events of Ca^{2+} flux and coalescence of lipid raft-like structures (25, 50, 82).

The possibility of altered lipid immiscibility is suggested by the change in the distribution of lipid probes in association with some cell signaling. Thomas et al. (78) reported that cross-linking of FcεRI as well as cross-linking of gangliosides indirectly induced patching of some [DiI, octadecyl rhodamine B and 5-(N-hexadecanoyl) aminoeosin] but not other (tetramethylammonium diphenylhexatriene and laurdan) fluorescent lipid probes. The lipid probe patches colocalized with the FcεRI or ganglioside patches. A similar observation has been reported following cross-linking of TCR. A colocalization of $DiIC_{16}$ and CD3 was reported (23a).

Transbilayer Correlation in Cell Membranes

There is extensive evidence of transbilayer correlation of proteins, and it is possible that this may include lipids as well (24, 25, 27, 31, 39, 45, 52, 53, 74, 82).

Colocalization of both outer and inner plasma membrane leaflet proteins in the detergent-resistant membranes (16, 17, 52, 81, 86) is consistent with correlation. The detergent-insolubility assay cannot distinguish between the scenarios in Figure 4. As mentioned above, in monolayer studies some of the lipids found in both leaflets of the cell membranes show complex formation. Thus there is the possibility of transbilayer coupling of complexes as well as separate phases.

Lipid-Protein Interaction and Targeting

As far as we know there have been no studies of bilayers with reconstituted proteins where there is also liquid-liquid immiscibility. However, Dietrich et al. (14) have reconstituted the GPI-linked protein Thy-1 in supported lipid monolayers that show liquid-liquid phase separation. The lipids employed were composed of lipids extracted from apical membranes of epithelial cells as well as a mixture of synthetic lipids, DOPC, SM, and Chol. Thy-1 partitions poorly in the dark phase in TR-DPPE. Interestingly, Thy-1 was excluded from the dark phase in the presence of 1 mole% of ganglioside GM1. The partitioning of fluorescent lipid probes between liquid phases appears to be sensitive to the lipid composition, and one anticipates the same will be true for proteins. Extensive spin label studies of reconstituted membranes containing proteins show that proteins can serve to immobilize (order) lipid chains adjacent to the protein (32, 40, 54, 73). Thus, it is plausible that some membrane proteins may preferentially associate with the ordered lipid chains in complexes. This possibility has been discussed by Anderson & Jacobson (3).

The selectivity of lipid-protein interaction is depicted schematically in Figure 6*b*,*c*. In these drawings the dark triangles represent proteins or other components that prefer a cholesterol-poor surrounding. The white triangles depict components that prefer to be surrounded by condensed complexes. As depicted, these selectivities do not require a macroscopic phase separation.

Chemical Activity of Cholesterol

The rate of loss of cholesterol from a monolayer membrane is related to the chemical activity of cholesterol, and this in turn is strongly affected by complex formation (59). In order for two immiscible phases to be in equilibrium, the chemical activity of cholesterol must be the same in both phases. It is plausible that this principle may apply to different membranes in a given animal cell. The plasma membranes normally contain a net concentration of cholesterol that is higher than that in the endoplasmic reticulum. Nonetheless, the activity of cholesterol may be the same in both membranes. In the plasma membrane cholesterol forms complexes with SM and phosphatidylserine, and this complex formation reduces the concentration of free cholesterol and thus lowers its chemical activity. The phospholipids of the endoplasmic reticulum are generally more unsaturated and have less of a tendency to form complexes. In the endoplasmic reticulum its chemical activity is low because the net concentration of cholesterol is low, and thus the concentration of free cholesterol is low.

ACKNOWLEDGMENTS

We are indebted to Arun Radhakrishnan for his help in preparation of this review. We are also indebted to Tom G. Anderson, Tamara M. Okonogi, and Sarah L. Keller for helpful discussions. This work was supported by National Institutes of Health Grant 5R01AI13587-27.

The *Annual Review of Biophysics and Biomolecular Structure* is online at http://biophys.annualreviews.org

LITERATURE CITED

1. Almeida PFF, Vaz WLC, Thompson TE. 1992. Lateral diffusion in the liquid-phases of dimyrostoylphosphatidylcholine cholesterol lipid bilayers: a free volume analysis. *Biochemistry* 31:6739–47
2. Anderson RGW. 1998. The caveolae membrane system. *Annu. Rev. Biochem.* 67:199–225
3. Anderson RGW, Jacobson K. 2002. Cell biology: a role for lipid shells in targeting proteins to caveolae, rafts, and other lipid domains. *Science* 296:1821–25
4. Anderson TG, McConnell HM. 2001. Condensed complexes and the calorimetry of cholesterol-phospholipid bilayers. *Biophys. J.* 81:2774–85
5. Anderson TG, McConnell HM. 2002. A thermodynamic model for extended complexes of cholesterol and phospholipid. *Biophys. J.* 83:2039–52
6. Benvegnu DJ, McConnell HM. 1993. Surface dipole densities in lipid monolayers. *J. Phys. Chem.* 97:6686–91
7. Bromley SK, Burack WR, Johnson KG, Somersalo K, Sims TN, et al. 2001. The immunological synapse. *Annu. Rev. Immunol.* 19:375–96
8. Brown DA, London E. 1998. Functions of lipid rafts in biological membranes. *Annu. Rev. Cell Dev. Biol.* 14:111–36
9. Brown DA, London E. 1998. Structure and origin of lipid domains in biological membranes. *J. Membr. Biol.* 164:103–14
10. Brown DA, London E. 2000. Structure and function of sphingolipid- and cholesterol-rich membrane rafts. *J. Biol. Chem.* 275:17221–24
11. Corrales LR, Wheeler JC. 1989. Chemical reaction driven phase transitions and critical points. *J. Chem. Phys.* 91:7097–112
12. Dervichian DG. 1958. *Surface Science in Chemistry and Biology*, p. 70. Oxford: Pergamon
13. Dietrich C, Bagatolli LA, Volovyk ZN, Thompson NL, Levi M, et al. 2001. Lipid rafts reconstituted in model membranes. *Biophys. J.* 80:1417–28
14. Dietrich C, Volovyk ZN, Levi M, Thompson NL, Jacobson K. 2001. Partitioning of Thy-1, GM1, and cross-linked phospholipid analogs into lipid rafts reconstituted in supported model membrane monolayers. *Proc. Natl. Acad. Sci. USA* 98:10642–47
15. Feingold L. 1993. *Cholesterol in Membrane Models*. Ann Arbor, MI: CRC
16. Field KA, Holowka D, Baird B. 1995. FcεRI-mediated recruitment of p53/56lyn to detergent-resistant membrane domains accompanies cellular-signaling. *Proc. Natl. Acad. Sci. USA* 92:9201–5
17. Field KA, Holowka D, Baird B. 1997. Compartmentalized activation of the high affinity immunoglobulin E receptor within membrane domains. *J. Biol. Chem.* 272:4276–80
18. Finean JB. 1953. Phospholipid-cholesterol complex in the structure of myelin. *Experientia* 9:17–19
19. Fraser DP, Zuckermann MJ, Mouritsen

OG. 1991. Theory and simulations for hard-disk models of binary mixtures of molecules with internal degrees of freedom. *Phys. Rev. A* 43:6642–56
20. Gershfeld NL. 1978. Equilibrium studies of lecithin-cholesterol interactions. I. *Biophys. J.* 22:469–88
21. Hagen JP. 1996. *The physical chemistry of lipid monolayers at the air-water interface*. PhD thesis. Stanford Univ.
22. Hagen JP, McConnell HM. 1997. Liquid-liquid immiscibility in lipid monolayers. *Biochim. Biophys. Acta* 1329:7–11
23. Hao M, Mukherjee S, Maxfield FR. 2001. Cholesterol depletion induces large scale domain segregation in living cell membranes. *Proc. Natl. Acad. Sci. USA* 98:13072–77
23a. Harder T, Kuhn M. 2000. Selective accumulation of raft-associated membrane protein LAT in T-cell receptor signaling assemblies. *J. Cell Biol.* 151:199–207
24. Harder T, Scheiffele P, Verkade P, Simons K. 1998. Lipid domain structure of the plasma membrane revealed by patching of membrane components. *J. Cell Biol.* 141:929–42
25. Harder T, Simons K. 1999. Clusters of glycolipid and glycosylphosphatidylinositol-anchored proteins in lymphoid cells: accumulation of actin regulated by local tyrosine phosphorylation. *Eur. J. Immunol.* 29:556–62
26. Hinz H, Sturtevant JM. 1972. Calorimetric investigation of the influence of cholesterol on the transition properties of bilayers formed from synthetic L-α-lecithins in aqueous suspension. *J. Biol. Chem.* 247:3697–700
27. Holowka D, Sheets ED, Baird B. 2000. Interactions between FcεRI and lipid raft components are regulated by the actin cytoskeleton. *J. Cell Sci.* 113:1009–19
28. Huang J, Feigensen GW. 1999. A microscopic interaction model of maximum solubility of cholesterol in lipid bilayers. *Biophys. J.* 76:2142–57
29. Ipsen JH, Karlstrom G, Mouritsen OG, Wennerstrom H, Zuckermann MJ. 1987. Phase equilibria in the phosphatidylcholine-cholesterol system. *Biochim. Biophys. Acta* 905:162–72
30. Jacobson K, Dietrich C. 1999. Looking at lipid rafts? *Trends Cell Biol.* 9:87–91
31. Janes PW, Ley SC, Magee AI. 1999. Aggregation of lipid rafts accompanies signaling via the T cell antigen receptor. *J. Cell Biol.* 147:447–61
32. Jost PC, Griffith OH. 1978. Lipid-protein interactions: influence of integral membrane proteins on bilayer lipids. In *Biomolecular Structure and Function*, ed. PF Agris, pp. 25–65. New York: Academic
33. Keller SL, Pitcher WH III, Huestis WH, McConnell HM. 1998. Red blood cell lipids form immiscible lipids. *Phys. Rev. Lett.* 81:5019–22
34. Keller SL, Radhakrishnan A, McConnell HM. 2000. Saturated phospholipids with high melting temperature for complexes with cholesterol in monolayers. *J. Phys. Chem. B* 104:7522–27
35. Kinet J. 1999. The high-affinity IgE receptor (FcεRI): from physiology to pathology. *Annu. Rev. Immunol.* 17:931–72
36. Krummel MF, Davis MM. 2002. Dynamics of the immunological synapse: finding, establishing and solidifying a connection. *Curr. Opin. Immunol.* 14:66–74
37. Langlet C, Bernard A, Drevot P, He H. 2000. Membrane rafts and signaling by the multichain immune recognition receptors. *Curr. Opin. Immunol.* 12:250–55
38. Leathes JB. 1925. Condensing effect of cholesterol on monolayers. *Lancet* 208: 853–56
39. Lee K, Holdorf AD, Dustin ML, Chan AC, Allen PM, Shaw AS. 2002. T cell receptor signaling precedes immunological synapse formation. *Science* 295:1539–42
40. Marsh D, Watts A. 1982. *Spin labeling and lipid-protein interactions in membranes*. In *Lipid-Protein Interactions* ed. PC Jost, OH Griffith, 2:53–126. New York: Wiley

41. McConnell HM. 1991. Structures and transitions in lipid monolayers at the air-water interface. *Annu. Rev. Phys. Chem.* 42:171–95
42. McConnell HM, Radhakrishnan A. 2003. Condensed complexes of cholesterol and phospholipids. *Biochim. Biophys. Acta Biomembr.* In press
43. McMullen TPW, Ruthven NAH, Lewis RN, McElhaney RN. 1999. Calorimetric and spectroscopic studies of the effects of cholesterol on the thermotrophic phase behavior and organization of a homologous series of linear saturated phosphatidylethanolamine bilayers. *Biochim. Biophys. Acta* 1416:119–34
44. Miao L, Nielsen M, Thewalt J, Ipsen JH, Bloom M, et al. 2002. From lanosterol to cholesterol: structural evolution and differential effects on lipid bilayers. *Biophys. J.* 82:1429–44
45. Monk CRF, Freiberg BA, Kupfer H, Sciaky N, Kupfer A. 1998. Three-dimensional segregation of supramolecular activation clusters in T cells. *Nature* 395:82–86
46. Montixi C, Langlet C, Bernard A, Thimonier J, Dubois C, et al. 1998. Engagement of T cell receptor triggers its recruitment to low-density detergent-insoluble membrane domains. *EMBO J.* 17:5334–38
47. Needham D, McIntosh TJ, Evans E. 1988. Thermomechanical and transition properties of dimyristoylphosphatidylcholine cholesterol bilayers. *Biochemistry* 27:4468–73
48. Okonogi TM, Radhakrishnan A, McConnell HM. 2002. Two fatty acids can replace one phospholipid in condensed complexes with cholesterol. *Biochim. Biophys. Acta* 1564:1–4
49. Phillips MC. 1972. The physical state of phospholipids and cholesterol in monolayers, bilayers, and membranes. *Prog. Surf. Membr. Sci.* 5:139–221
50. Pierini L, Harris NT, Holowka D, Baird B. 1997. Evidence supporting a role for microfilaments in regulating the coupling between poorly dissociable IgE-FcεRI aggregates and downstream signaling pathways. *Biochemistry* 36:7447–56
51. Pralle A, Keller P, Florin EL, Simons K, Horber JKH. 2000. Sphingolipid-cholesterol rafts diffuse as small entities in the plasma membrane of mammalian cells. *J. Cell Biol.* 148:997–1007
52. Prior IA, Harding A, Yan J, Slumier J, Parton RG, Hancock JF. 2001. GTP-dependent segregation of H-ras from lipid rafts is required for biological activity. *Nat. Cell Biol.* 3:368–75
53. Pyenta PS, Holowka D, Baird B. 2001. Cross-correlation analysis of inner-leaflet-anchored green fluorescent protein co-redistributed with IgE receptors and outer leaflet lipid raft components. *Biophys. J.* 80:2120–32
54. Radhakrishnan A, Anderson TG, McConnell HM. 2000. Condensed complexes, rafts, and the chemical activity of cholesterol in membranes. *Proc. Natl. Acad. Sci. USA* 97:12422–27
55. Radhakrishnan A, Li X-M, Brown RE, McConnell HM. 2001. Stoichiometry of cholesterol-sphingomyelin condensed complexes in monolayers. *Biochim. Biophys. Acta* 1511:1–6
56. Radhakrishnan A, McConnell HM. 1999. Cholesterol-phospholipid complexes in membranes. *J. Am. Chem. Soc.* 121:486–87
57. Radhakrishnan A, McConnell HM. 1999. Condensed complexes of cholesterol and phospholipids. *Biophys. J.* 77:1507–17
58. Radhakrishnan A, McConnell HM. 1999. Electric field effect on cholesterol-phospholipid complexes. *Proc. Natl. Acad. Sci. USA* 97:1073–78
59. Radhakrishnan A, McConnell HM. 2000. Chemical activity of cholesterol in membranes. *Biochemistry* 39:8119–24
60. Radhakrishnan A, McConnell HM. 2002. Critical points in charged membranes containing cholesterol. *Proc. Natl. Acad. Sci. USA* 99:13391–96

61. Radhakrishnan A, McConnell HM. 2002. Thermal dissociation of condensed complexes of cholesterol and phospholipid. *J. Phys. Chem.* 106:4755–62
62. Rechtenwald DJ, McConnell HM. 1981. Phase equilibria in binary mixtures of phosphatidylcholines and cholesterol. 20: 4505–10
63. Rowlinson JS, Swinton FL. 1982. *Liquids and Liquid Mixtures.* London/Boston: Butterworth Sci.
64. Rubenstein JL, Owicki JC, McConnell HM. 1980. Dynamic properties of phosphatidylcholines and cholesterol. *Biochemistry* 19:569–73
65. Samsonov AV, Mihalyov I, Cohen FS. 2001. Characterization of cholesterol-sphingomyelin domains and their dynamics in bilayer membranes. *Biophys. J.* 81:1486–500
66. Schütz GJ, Schindler H, Schmidt T. 1997. Single-molecule microscopy on model membranes reveals anomalous diffusion. *Biophys. J.* 73:1073–80
67. Sheets ED, Holowka D, Baird B. 1999. Critical role for cholesterol in Lyn-mediated tyrosine phosphorylation of FcεRI and their association with detergent-resistant membranes. *J. Cell Biol.* 145:877–87
68. Sheets ED, Lee GM, Simson R, Jacobson K. 1997. Transient confinement of a glycosylphosphatidylinositol-anchored protein in the plasma membrane. *Biochemistry* 36:12449–58
69. Simons K, Ikonen E. 1997. Functional rafts in cell membranes. *Nature* 387:569–72
70. Simons K, Toomre D. 2000. Lipid rafts and signal transduction. *Nat. Mol. Cell Biol.* 1:31–39
71. Simons K, van Meer G. 1988. Lipid sorting in epithelial cells. *Biochemistry* 27:6197–202
72. Somerharju P, Virtanen JA, Cheng KH. 1999. Lateral organization of membrane lipids. The superlattice view. *Biochim. Biophys. Acta* 1440:32–48
73. Sprong H, van der Sluijs P, van Meer G. 2001. How proteins move lipids and lipids move proteins. *Nat. Rev. Mol. Cell Biol.* 21:504–13
74. Stauffer TP, Meyer T. 1997. Compartmentalized IgE receptor-mediated signal transduction in living cells. *J. Cell Biol.* 139:1447–54
75. Subramaniam S, McConnell HM. 1987. Critical mixing in monolayer mixtures of phospholipid and cholesterol. *J. Phys. Chem.* 91:1715–18
76. Sugar IP, Tang D, Chong PL. 1994. Monte Carlo simulation of lateral distribution of molecules in a two-component lipid membrane. Effect of long-range repulsive interactions. *J. Phys. Chem.* 98:7201–10
77. Suzuki K, Sheetz MP. 2001. Binding of cross-linked glycosylphosphatidylinositol-anchored proteins to discrete actin-associated and cholesterol-dependent domains. *Biophys. J.* 81:2181–89
78. Thomas JL, Holowka D, Baird B, Webb WW. 1994. Large-scale co-aggregation of fluorescent lipid probes with cell surface proteins. *J. Cell Biol.* 125:795–802
79. Varma R, Mayor S. 1998. GPI-anchored proteins are organized in submicron domains at the cell surface. *Nature* 394:798–801
80. Veatch SL, Keller SL. 2002. Organization in lipid membranes containing-cholesterol. *Phys. Rev. Lett.* 89:268101
81. Vidalain PO, Azocar O, Servet-Delprat C, Rabourdin-Combe C, Gerlier D, Manie S. 2000. CD40 signaling in human dendritic cells is initiated within membrane rafts. *EMBO J.* 19:3304–13
82. Villaba M, Bi K, Rodrigez F, Tanaka Y, Schoenberger S, Altman A. 2001. Vav1/Rac-dependent actin cytoskeleton reorganization is required for lipid raft clustering in T cells. *J. Cell Biol.* 155:331–38
83. Virtanen JA, Ruonala M, Vauhkonen M, Somerharju P. 1995. Lateral organization of liquid-crystalline cholesterol-dimyristoylphosphatidylcholine bilayers. Evidence for domains with hexagonal and

centered rectangular cholesterol superlattices. *Biochemistry* 34:11568–81
84. Vist MR, Davis JH. 1989. Phase equilibria of cholesterol/dipalmitoylphosphatidylcholine mixtures: D nuclear magnetic resonance and differential scanning calorimetry. *Biochemistry* 29:451–64
85. Wang MM, Sugar IP, Chong PL. 1998. Role of the sterol superlattice in the partitioning of the antifungal drug nystatin into lipid membranes. *Biochemistry* 37:11797–805
86. Xavier R, Brennan T, Li QQ, McCormack C, Seed B. 1998. Membrane compartmentation is required for efficient T-cell activation. *Immunity* 8:723–32

SUBJECT INDEX

CUMULATIVE INDEXES

CONTRIBUTING AUTHORS, VOLUMES 28–32

CHAPTER TITLES, VOLUMES 28–32

Calcium

Carbohydrate Structure

Catalysis

Cations

Cells

Channel

Chromatin

Cre-*loxP*

Cryo-Electron Microscopy

Crystal

Crystal Structure

Crystallography

Cyclooxygenase

DNA

DNA Multiprotein Complexes

DNA Recognition

DNA Repair Mechanisms

DNA Replication

DNA-Transcriptional

Domains

Dynamics

Electrostatics

Enzyme

Protein Dynamics

Protein Folding

Protein Function

Protein and Polypeptide Structure

Protein Structure

Proteins

Ribosome

RNA

RNA Folds

Rotary Enzyme

Signaling

Simulation

Single Molecules